THE FIFTH COMPTON SYMPOSIUM

Related Titles from the AIP Conference Proceedings Subseries on Astronomy and Astrophysics

504 Space Technology and Applications International Forum - 2000: Conference on International Space Station Utilization; Conference on Thermophysics in Microgravity; Conference on Enabling Technology and Required Scientific Developments for Interstellar Missions; Conference on Commercial/Civil Next Generation Space Transportation; 17th Symposium on Space Nuclear Power and Propulsion
Edited by Mohamed S. El-Genk, January 2000
2 vol. print set, hard cover: 1-56396-919-X
CD ROM: 1-56396-920-3

499 Small Missions for Energetic Astrophysics: Ultraviolet to Gamma-Ray
Edited by Steven P. Brumby, December 1999, 1-56396-912-2

471 Solar Wind Nine: Proceedings of the Ninth International Solar Wind Conference
Edited by Shadia Rifai Habbal, Ruth Esser, Joseph V. Hollweg, Philip A. Isenberg, May 1999, 1-56396-865-7

433 Workshop on Observing Giant Cosmic Ray Air Showers from $>10^{20}$ eV Particles from Space
Edited by John F. Krizmanic, Jonathan F. Ormes, and Robert E. Streitmatter, June 1998, 1-56396-788-X

428 Gamma-Ray Bursts: 4th Huntsville Symposium
Edited by Charles A. Meegan, Robert D. Preece, and Thomas M. Koshut, May 1998, 1-56396-766-9

410 The Fourth Compton Symposium
Edited by Charles D. Dermer, Mark S. Strickman, and James D. Kurfess, December 1997, 2 vol. set, 1-56396-659-X

To learn more about these titles, or the AIP Conference Proceedings Series, please visit the webpage
http://www.aip.org/catalog/aboutconf.html

THE FIFTH COMPTON SYMPOSIUM

Portsmouth, NH September 1999

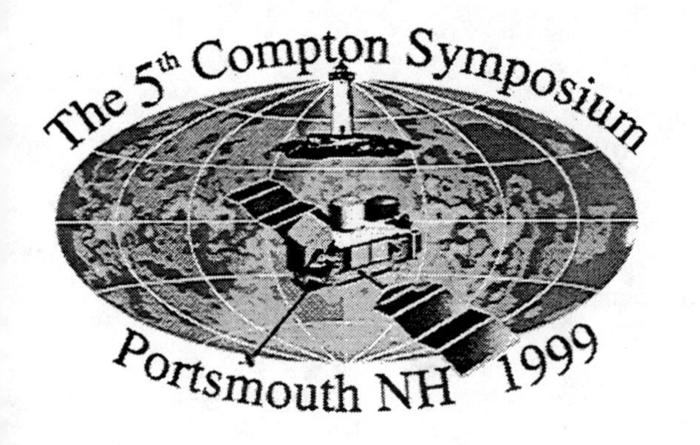

EDITORS
Mark L. McConnell
James M. Ryan
University of New Hampshire

Melville, New York
AIP CONFERENCE PROCEEDINGS ■ 510

Editors:

Mark L. McConnell
James M. Ryan
University of New Hampshire
Space Science Center
Morse Hall
39 College Road
Durham, NH 03824
USA

E-mail: Mark.McConnell@unh.edu
James.Ryan@unh.edu

L.C. Catalog Card No. 00-101590
ISBN 1-56396-932-7
ISSN 0094-243X
Printed in the United States of America

CONTENTS

INTRODUCTION

GAMMA-RAY LINE EMISSION, NOVAE AND SUPERNOVAE

X-RAY BINARIES

PULSARS

DIFFUSE GALACTIC CONTINUUM EMISSION

ACTIVE GALAXIES

CLUSTERS OF GALAXIES

DIFFUSE COSMIC GAMMA-RAY EMISSION

UNIDENTIFIED SOURCES

GAMMA-RAY BURSTS

SOLAR FLARES

CATALOGS AND DATA ANALYSIS

GROUND-BASED GAMMA-RAY ASTRONOMY

FUTURE MISSIONS AND INSTRUMENTATION

ONLINE INFORMATION

PREFACE

Nearly 200 scientists met at the *Fifth Compton Symposium* in Portsmouth, New Hampshire on September 15-17, 1999. The meeting was hosted by the University of New Hampshire and the Compton Gamma-Ray Observatory Science Support Center at the NASA Goddard Space Flight Center.

With the Compton Gamma-Ray Observatory (CGRO) now in its ninth year of operations, the *Fifth Compton Symposium* provided an opportunity for sharing the most recent results not only from CGRO, but also from RXTE, BeppoSAX, ASCA and other high energy experiments. Results from the field of ground-based gamma-ray astronomy were also included in the discussions, as were several related results from optical and radio astronomy. The important topics of gamma-ray bursts and solar flares were treated only briefly, focusing on talks which provided an overview of the current status and future prospects of these fields.

Looking to the future, there were a large number of papers that dealt with planned future missions (such as ASTRO-E, HESSI, INTEGRAL, AGILE and GLAST) as well as new mission concepts and their associated technolgy developments. These new missions will be responsible for continuing the pioneering efforts of the CGRO mission. In addition, there were several papers dealing with instrumentation, both current and future, for ground-based gamma-ray astronomy. With the lower energy thresholds of some of the planned ground-based detectors, it is expected that the spectral gap between space-based and ground-based gamma-ray astronomy will be closing within the next few years, providing coverage of the entire gamma-ray spectrum.

On the day prior to the beginning of the Symposium, the *Compton Workshop on Astronomy Education* provided a forum for local K-12 educators to learn about various activities and resources that might assist them in their educational endeavors. A team of NASA researchers and educators from throughout New Hampshire, aided by D. Duncan of the American Astronomical Society, provided an enlightening experience for roughly 40 local K-12 educators.

The Scientific Organizing Committee for the *Fifth Compton Symposium* consisted of J. Ryan (UNH, Chair), M. Baring (GSFC), M. Cherry (LSU), E. Chupp (UNH), G. Fishman (MSFC), N. Gehrels (GSFC), K. Hurley (UC Berkeley), J. Kurfess (NRL), P. Michelson (Stanford), V. Schönfelder (MPE Garching), H. Völk (MPI Heidelberg), and T. Weekes (SAO). The Local Organizing Committee consisted of M. McConnell (UNH, Co-Chair), C. Shrader (GSFC, Co-Chair), S. Barnes (GRO-SSC & USRA), C. Kustra (UNH), R. Pendexter (UNH, Secretary), and E. Pentecost (USRA). We gratefully acknowledge the support of the UNH Space Science Center, the UNH Institute for the Study of Earth, Oceans and Space, the NASA / Goddard Space Flight Center, and the Universities Space Research Association (USRA). The UNH Research Computing Center and Rocket Science Computing Services (Portsmouth, NH) provided computer services during the symposium. The New Hampshire Space Grant Consortium provided support for the *Compton Workshop on Astronomy Education.* We would also like to acknowledge the assistance of Jane

Fithian (UNH) in the design of the symposium logo.

In assembling the proceedings, we have reordered the sequence of papers from that of the symposium in order that we may follow a more natural thematic sequence. We hope the readers will find this sequence more useful.

Although the scientific advances reported during the symposium were quite stimulating for the participants, perhaps the *Fifth Compton Symposium* will most be remembered for the arrival of Hurricane Floyd on the New Hampshire seacoast, just in time to force an alteration in our planned banquet venue. Although the banquet attendees were not treated to the scenic beauty of the New Hampshire seacoast, as was originally planned, it is hoped that the New England hospitality of the lobster dinner will long be remembered.

Mark L. McConnell
James M. Ryan
December, 1999

INTRODUCTION

Status of the Compton Gamma Ray Observatory

Neil Gehrels and Chris Shrader

Laboratory for High-Energy Astrophysics, NASA Goddard Space Flight Center

Abstract. The Compton Gamma Ray Observatory and three of its four experiment packages continue to function in a nearly flawless manner now well into the eighth year of mission operations. Only the EGRET instrument is operating with reduced capability due mainly to the depleted spark-chamber gas, but it is nonetheless still expected to make significant contributions, notably in the area of Solar flares and AGN variability. We discuss the status of the mission as of mid-1999 as well as the prospects of an extended mission lasting well into the first decade of the next century.

INTRODUCTION

Launched on 5 April 1991, the 15900-kg Compton Gamma Ray Observatory (herein *Compton*), is the second in the current fleet of three NASA Great Observatories. Mission goals include performing broad-band gamma-ray observations with better angular resolution and an order of magnitude better sensitivity than previous missions, performing the first gamma-ray full-sky survey (completed in 1992), and compiling a database of gamma-ray burst measurements unprecedented in size and scope. The scientific theme of the mission is the study of physical processes taking place in the most dynamic sites in the Universe, including supernovae, novae, pulsars, black holes, active galaxies, gamma-ray bursts. Additionally, *Compton* provides a capability to study the high-energy properties of solar flares. It currently serves an international Guest Investigator community with observing time allocated on the basis of peer-reviewed proposals in annual periods or "Cycles". About 100 proposals were received for partiaipation in Cycle-9, which begins in December, 1999. A total mission lifetime of 12-15 years is currently anticipated. The four scientific instrument packages, described briefly below combine to cover the hard x-ray and gamma-ray energy regimes from 15 keV to 30 GeV [1–3].

CP510, *The Fifth Compton Symposium,* edited by M. L. McConnell and J. M. Ryan

THE *COMPTON* INSTRUMENT PACKAGES

The Burst and Transient Source Experiment (BATSE) is optimized to measure brightness variations in gamma–ray bursts and solar flares on time scales down to microseconds over the energy range 30 keV to 1.9 MeV [4,5]. To accomplish this, 8 large–area NaI detectors oriented towards the 8 octants of the sky are used. A smaller spectroscopy detector, optimized for broad energy coverage (15 keV to 110 MeV) and enhanced energy resolution is associated with each of the 8 large–area detectors. BATSE also continuously monitors all transient sources and bright persistent sources in the gamma–ray sky.

The Oriented Scintillation Spectroscopy Experiment (OSSE) is designed to undertake comprehensive spectral observations of astrophysical sources in the 0.05 to 10 MeV range, with capability above 10 MeV for solar gamma–ray and neutron observations [6]. The detectors are NaI/CsI scintillators with a field–of–view determined by a passive tungsten collimator. Each of the 4 detectors has a single axis pointing system which enables a rapid OSSE response to target of opportunities, such as transient X–ray sources, explosive objects, and solar flares.

The Imaging Compton Telescope (COMPTEL) detects gamma rays by the occurrence of two successive interactions: the first one is a Compton scattering collision in a detector of low–Z material (liquid scintillator) followed by a second interaction in a detector with high–Z material (NaI), in which, ideally, the scattered gamma ray is totally absorbed [7]. Source mapping is provided over a field of view of about 1 steradian. COMPTEL has performed the first sky survey in the energy range from 1 to 30 MeV.

The Energetic Gamma Ray Experiment Telescope (EGRET) is the highest energy instrument on *Compton*, and covers the broadest energy range, from 20 MeV to 30 GeV [8,9]. It has a wide field of view (0.5 sr), good angular resolution and very low background. Because it is designed for high–energy studies, the detector is optimized to detect gamma rays when they interact by the dominant high–energy pair–production process which forms an electron and a positron within the EGRET spark chamber.

PROSPECTS FOR AN EXTENDED *COMPTON* MISSION

As of the time of this submission, the prospects for an extended *Compton* mission look very promising. On technical grounds, the spacecraft is functioning in a nearly flawless manner and there is widespread optimism that this will continue to be the case for the foreseeable future. The spacecraft was reboosted to an orbit altitude of 515 km in 1997. This should allow for an extension of the mission to beyond 2005, possibly even to the next (2011) solar maximum!

Recent problems encountered in controlling the high-gain antenna pointing have been solved and diagnosed as problem with software and operational procedures. Earlier problems with the onboard tape recorders have been largely circumvented by the implementation of enhanced real–time telemetry coverage, usage of onboard memory buffers and the low–gain omni antenna. The failure several year ago of one of the spacecraft batteries led to concerns that this problem might be repeated in the remaining 5 batteries – but this now does not seem to be occurring, as the remaining 5 batteries are aging normally. In any case, spacecraft engineers have devised scenarios, involving for example the cycling of spacecraft thermal–control subsystems, so that the mission could continue even in the event of an additional battery failure.

The EGRET instrument uses an expendable neon/argon spark–chamber gas, of which the spacecraft initially carried 5 refills. The gas lifetime is directly related to the spark rate, and thus the number of photons detected. During the current observing cycle, EGRET is generally being configured to an operating mode in which its field of view is reduced by about a factor of 4, leading to a nominal two–fold increase in gas lifetime. In instances where there may be targets within the Z–axis field of view that are of interest for study with COMPTEL, but not with EGRET, the spark chamber is being turned off, again leading to significant gas savings. As of this submission, there is approximately 6 months of gas use remaining with EGRET operated in the mode described above.

GUEST INVESTIGATOR PROGRAM

The scientific productivity of the *Compton* mission results from a rigorous Guest Investigator Program. A total of over 750 scientists from 128 institutions in 23 countries have been involved since launch. All observing time on *Compton* is awarded competitively through scientific peer review held during successive annual cycles. Cycle 9 of the mission begins in December, 1999. GIs are provided extensive assistance by the instrument teams and by the Science Support Center (SSC) at the NASA/Goddard Space Flight Center.

Some recent policy changes will be implemented for Cycle 9, based on recommendations made by the primary advisory body for the mission, the *Compton* GRO User's Committee and the NASA Headquarters Senior Review. Most significantly, support for CGRO science will be moved under the Astrophysics Data Program for Cycle-10 and beyond. The CGRO project will still solicit Guest Observer proposals and conduct a scientifc peer review to determine an optimal viewing plan for each annual observation cycle. The Guest Investigator program is a strong and essential component of the *Compton* mission. For more information refer to the CGRO home page on the World Wide Web: URL http:://cossc.gsfc.nasa.gov.

Compton Public Science Data Archive

All data from *Compton* are available through a public archive managed by the SSC Data are placed in the archive following delivery to the SSC by the instrument teams. In the case of pointed observations from OSSE, COMPTEL or EGRET, delivery follows the expiration of a three-month proprietary period. The data are easily accessible via computer networks through the HEASARC "W3Browse" web-based search and retrival facility. Magnetic tape distribution is also available upon request in the case of extremely large data sets. In addition, a BATSE CD ROM containing the Fourth BATSE Catalog of gamma-ray bursts and a set of the relevant data products and software, is availabe from the SSC. Also available is an EGRET CD ROM containing data from Cycles 1-5 plus refined calibration files and platform-independent "JAVA" based quick-look analysis tools. Refer to the SSC World Wide Web home page (URL given above) for additional information.

Recent Science Highlights

We will not attempt to present a comprehensive summary of the scientific highlights of the mission – that is the purpose of this proceedings volume, and the quality and quantity of its contents attest to the continuing success of the mission. Here we will make brief mention of a few particularly exciting results and refer the reader to more detailed discussions within this volume.

GRB 990123, the bright gamma-ray burst detected by BATSE on 99/01/23 proved to be a landmark breakthrough in the study of gamma-ray bursts (GRBs) [10]. For the very first time, a GRB was observed in the optical band while the burst was still bursting. The University of Michigan's Robotic Optical Telescope Search Experiment (ROTSE) system slewed its CCD cameras to the burst coordinates provided by the GRB Coordinates Network (GCN) and detected the counterpart at 22 seconds after the BATSE signal. In the initial 5 second ROTSE integration, the GRB990123 was detected at 11.8 magnitudes (no filter). A second exposure 25 seconds later recorded a 9th mag object! In gamma-rays, the burst is among the brightest 0.5% of the nearly 3000 bursts recorded by BATSE. In the visual, its "afterglow" emission, distinct from the prompt emission seen by ROTSE, is the brightest seen to date by several magnitudes. A redshift determination places it at z=1.6 or about 4.5 Gpc. If the emission is released isotropically, its total energy release is $\sim 10^{54}$ ergs – far greater than that any other known event, except for the big bang itself!

OSSE has produced revised maps of 511 keV positron annihilation line and positronium continuum radiation from the Galactic center region. The positrons can result from the decay of radioactive elements produced in explosive objects (supernovae and novae), in energetic processes associated neutron stars and black holes. and from interactions of cosmic rays with the interstellar

FIGURE 1. Graphical representation of the Third EGRET Catalog. All 271 sources are plotted, with different symbols indicating different source types as annotated.

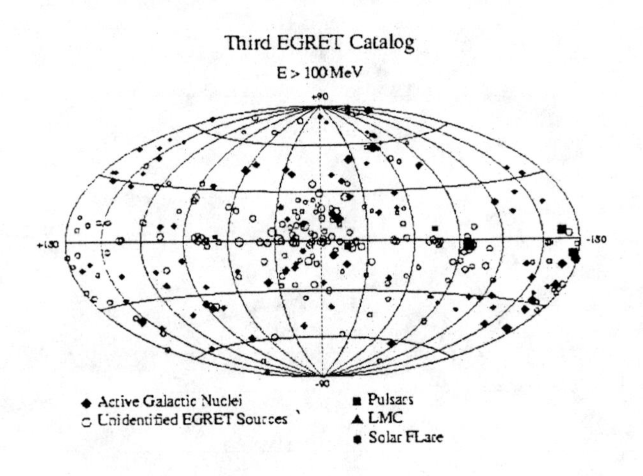

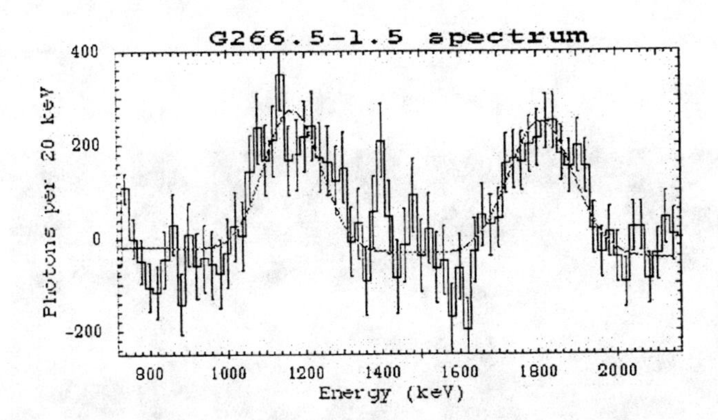

FIGURE 2. COMPTEL line detection from the Vela region. The 1.16-MeV line is due to ^{44}Ti (½-life $\approx 10^2$ years) associated with the newly discovered SNR, which must be younger than ~10^3 years. The 1.8 MeV line is due to ^{26}Al (½-life $\approx 10^6$ years) and it may include contributions from both the Vela SNR and G266.5-1.5.

FIGURE 3. Gamma-ray (BATSE) and optical (ROTSE) light curves for GRB 990123. These observations, made possible by the Global Coordinate Network, resulted the first detection prompt optical emission from a gamma-ray burst.

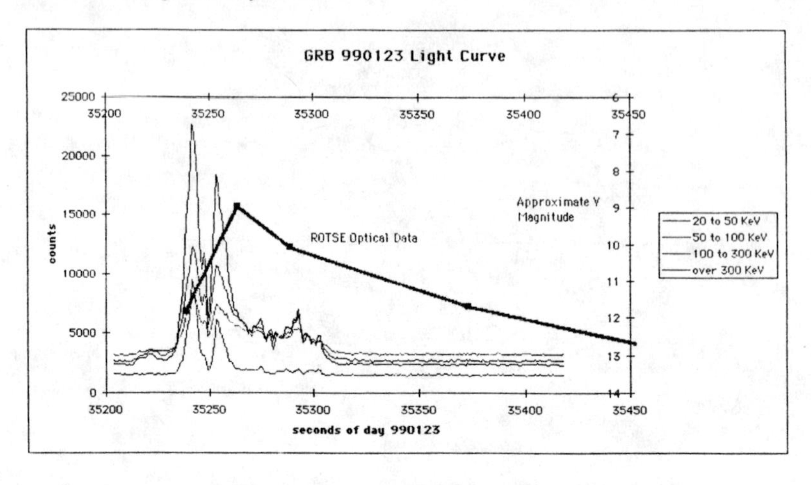

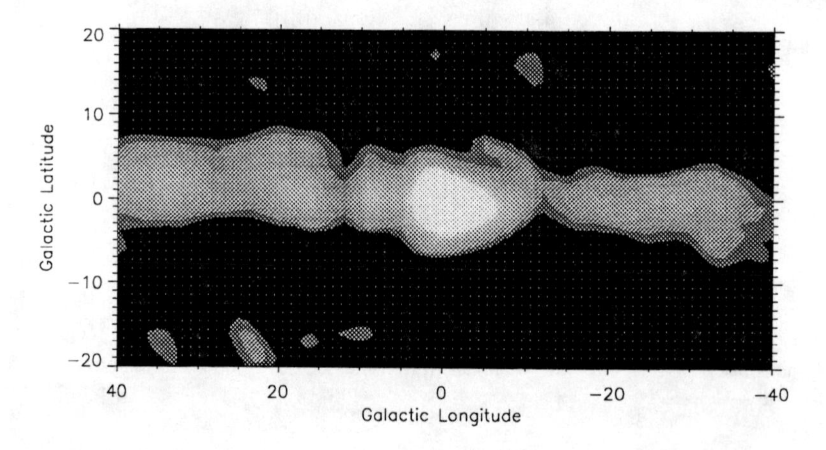

FIGURE 4. Improved spectral fitting has permitted preliminary mapping of the positronium continuum portion of the total annihilation radiation. The bulge and disk components suggested by the 511 keV line maps are also evident in this map.

medium. Differential observations between "source" and "background" measurements using each of the four OSSE detectors are used to map the region. Maps generated using several distinct mapping algorithms, are presented elsewhere in these proceedings [11] and they represent the most comprehensive gamma-ray picture of the Galactic Center region to date. The maps utilize a total of 2×10^7 seconds of observing time. The emission is dominated by an intense, diffuse bulge component, emission along the Galactic plane, and with evidence for excess emission at positive Galactic latitudes. No variability of the 511 keV emission has been observed by OSSE.

Data obtained with EGRET during the first 4 years of the CGRO mission have been reanalyzed in a uniform manner using the most up to date instrumental calibration information to compile a source comprehensive catalog. The resulting Third EGRET Catalog consists of 271 sources: 5 pulsars, 1 solar flare, 66 high-confidence blazar identifications, 27 possible blazar identifications, 1 likely radio galaxy (Cen A), 1 normal galaxy (LMC), and 170 unidentified sources (a sixth EGRET pulsar is also detected, but is seen only in pulsed data, and so is not included in the catalog). The catalog is already proving to be an invaluable resource for a variety of studies, such as AGN and population dynamics and the diffuse gamma-ray background.

Using data from the ROSAT all-sky survey, scientists from the Max Planck Institute found an object slightly offset from the well known Vela Supernova remnant (SNR) which exhibits morphological and spectroscopic characteristics of a young SNR [12]. In lower energy X-rays, it is largely obscured by the Vela SNR. This region of sky has received extensive off-axis exposure with CGRO/COMPTEL since the Vela region has been the subject of intensive study. From the gamma-ray line flux, the X-ray diameter and estimates of the ^{44}Ti yield, estimates of the distance to the SNR and its age can be made. A distance of $\sim$ 200 pc and an age of $\sim$ 680 years are thus derived. At 200 pc, SNR G266.5-1.5 would be associated with the closest supernova event to the Earth in human history. Why then was it not recorded in the historical record? One can only speculate – perhaps it was optically sub-luminous for some reason, or perhaps it occurred at a particularly unfavorable time for viewing.

Future Science

There are a number of exciting future studies that may be accomplished during the remainder of the mission. For example, a deeper and more systematic mapping program for the Galactic center region will be a major objective of OSSE. Coordination of such Galactic center region studies between OSSE, COMPTEL and the forthcoming INTEGRAL mission will also be of great interest.

The next two years of CGRO operations will provide unprecedented high–

energy coverage of a complete solar cycle. All four instruments are expected to contribute – EGRET for the study of high-energy particle acceleration; COMPTEL for study of solar neutron emission and mid-energy gamma-ray coverage; OSSE for nuclear line studies leading to isotopic abundance determinations and BATSE for temporal and monitoring studies.

Despite the recent remarkable breakthroughs in gamma–ray burst science, the basic phenomena underlying the cataclysmic event still remains an enigma. The database that is continuing to be compiled by BATSE represents the most comprehensive set of information on the gamma-ray burst phenomena for the foreseeable future. The possibility of additional prompt-optical counterpart identifications will be enhanced by continuued improvements in the technologies supporting the ground-based facilities, and through the greater sky coverage afforded by additional facilites which are planned.

Most significantly, other totally unanticipated discoveries may await us as well!

REFERENCES

1. Gehrels, N., Fichtel, C.E., Fishman, G.J., Kurfess, J.D., & Schönfelder, V., 1993, Sci.Am. 269, 68.
2. Shrader, C.R., & Gehrels, N., 1995, PASP, 107, 606.
3. Gehrels, N., & Paul, J., 1999, PhysTod, 51, 26.
4. Fishman, G.J., et al, 1992, in proc. "The Compton Observatory Science Workshop", ed. C.R. Shrader, N. Gehrels & B. Dennis, NASA CP-3137.
5. Band, D.L., et al, 1992, Exp.Astr., 2, 307.
6. Johnson, W.N., et al, 1993, ApJSup, 97, 21.
7. Schönfelder, V., et al, 1993, ApJSup, 86, 657.
8. Thompson, D.J., et al, 1993, ApJSup, 86, 269.
9. Esposito, J.A., 1999, ApJSup, 123, 203.
10. Akerlof, C., (these proceedings).
11. Milne, P., et al. (these proceedings).
12. Chen, W., & Gehrels, N., 1999, ApJ, 514, L103.

GAMMA-RAY LINE EMISSION, NOVAE AND SUPERNOVAE

Gamma-Ray Line Astrophysics

Roland Diehl

MPI für extraterrestrische Physik, Postfach 1603, D-85740 Garching, Germany

Abstract. The Compton Observatory instruments have measured γ-ray lines from several individual sources and from the diffuse interstellar medium. At this late phase of CGRO's mission, we review the astrophysical achievements by these observations of solar flare spectra, supernova lines from Co and ^{44}Ti isotopes, 511 keV annihilation radiation from Galactic plane and bulge, and diffuse 1809 keV radioactivity emission in the Galaxy from ^{26}Al. We briefly address other candidate lines from ^{7}Be and ^{22}Na nova radioactivities, and nuclear ^{12}C and ^{16}O de-excitation lines from Orion or the inner Galaxy. An astronomy with γ-ray lines has been established, the derived lessons suggest specific observations with the INTEGRAL observatory and other experiments of the future.

INTRODUCTION

In the last decades, γ-ray line result reports have been spectacular for the astrophysics community: Shortlived radioactivities were seen from supernovae 1987A, SN1991T, and Cas A; ^{26}Al was detected in the interstellar medium and the ^{26}Al sky image showed Galactic-plane emission with hot spots along the Galactic plane; the Orion region appeared to be a site of enormously-enhanced flux density in low-energy cosmic rays, as revealed from ^{12}C and ^{16}O deexcitation lines; the inner Galaxy appeared to feature a highly-variable compact source of positrons, and also produce a fountain in annihilation radiation extending far above the plane; ^{44}Ti emission revealed a supernova remnant in the solar vicinity towards the Vela direction. Although several of those exciting results turned out more modest or even not holding up upon closer looks, much exciting astrophysics remains, even if less-well constrained by γ-ray line data as many of us wish.

Opening a new astronomical window through γ-ray line observations, we expect contributions to these astrophysics issues: Calibration of the nuclear energy sources which power the displays of stars and supernovae, study of nucleosynthesis site details such as temperatures and densities, as they determine efficiencies of specific nucleosynthesis reaction chains, and the study of energetic particle interactions as revealed through nuclear de-excitation lines. NASA's Gamma-Ray Observatory project started in 1978 and held promises for bright science from γ-ray spectroscopy. These were largely reduced, when GRSE, the Ge detector instrument specialized

CP510, *The Fifth Compton Symposium,* edited by M. L. McConnell and J. M. Ryan

for such observations, was eliminated from the Observatory in the initial design phase (1982). Major efforts to close this experimental gap succeeded to establish the INTEGRAL mission, due to be launched in late 2001. We review the γ-ray spectroscopy results from the Compton Gamma-Ray Observatory (CGRO) mission with the COMPTEL and OSSE measurements at their 10^{-5}ph cm^{-2} s^{-1} sensitivity level and $\simeq$10% spectral resolution, and discuss issues for future observations.

SUPERNOVAE

A calibration of the nuclear energy sources through γ-ray lines is crucial for astrophysical models of supernovae, novae, and hydrostatic burning inside stars: The variety of *secondary* processes that are driven by this radioactive energy are difficult to model in suffient detail to consistently constrain the original nuclear burning, while γ-ray line intensity most directly converts into abundance of a specific isotope. In supernovae, the radioactive isotopes of the iron group with ^{56}Ni (decaying rapidly, τ=8.8d, to ^{56}Co with τ=111d) and ^{57}Ni (τ=390d) are generated in large amounts through explosive nucleosynthesis.

SN1987A has been a breakthrough for observations. For the first time γ-ray telescopes were ready to witness a relatively-nearby supernova. In core collapse events a massive stellar envelope is expected to hide the inner radioactivity for months; so the relatively early detection of the radioactivity γ-ray lines with SMM [33] marks undisputable proof of major unexpected mixing processes. OSSE on CGRO detected the γ-ray line at 122 keV from ^{57}Co decay [30]. As the amount of ^{56}Ni is determined from the bolometric light curve to 0.07 $M_\odot$, the 122 keV line could be used to measure the ^{57}Ni/^{56}Ni ratio more precisely than could be inferred from classical spectroscopy at other wavelengths [6]. The resulting 1.5 - 2 times the solar ratio appears more in line with chemical evolution studies than larger ^{57}Ni excesses inferred earlier.

Thermonuclear supernovae (SNIa) produce 10 times or more ^{56}Ni radioactivity than core collapses, therefore should be detectable in γ-ray lines from much larger distances out to the Virgo cluster of galaxies ($\simeq$10 Mpc), in particular in the absence of overlying envelopes. SN 1991T was at the edge of CGRO's range (14-17 Mpc), but unusually bright. The 3σ detection of the ^{56}Co lines by COMPTEL [38,39] was a surprise. But only weak constraints could be derived due to large systematic uncertainty, the inferred Ni mass ranges from 0.6 to (unphysical) 3.3 $M_\odot$. The CGRO user community devoted nearly four months of observing time to the second opportunity of the CGRO mission, SN1998bu in M96. Yet, neither COMPTEL [14] nor OSSE [31] apparently see the expected radioactivity lines. In spite of the distance uncertainty (9-12 Mpc), the γ-ray line limits from the ^{56}Ni decay chain of 2-5 10^{-5}ph cm^{-2} s^{-1} already rule out some of the brighter (because well-mixed) models for thermonuclear supernovae, in particular the 'Helium cap' models with a large amount of ^{56}Ni produced in the outer part of the supernova.

The ^{44}Ti decay chain with γ-ray lines at 1157, 68, and 78 keV has turned out as

a tool to study young supernova remnants. ^{44}Ti decays much less rapid (τ=89y) and causes a supernova to shine brighly in γ-rays long after its optical, X, UV, IR, and radio emission have decayed below visibility, and even until secondary emission from the blast wave interaction with circumstellar and interstellar medium may have brightened already to reveal a young supernova remnant through X-rays and radio emission. Extinct ^{44}Ti was discovered in several interstellar grains, which, from their isotopic anomalies, clearly had formed within a supernova [42]; this establishes independent proof of ^{44}Ti production in supernovae. Cosmic ^{44}Ca is adundant and most likely produced as ^{44}Ti. Theories predict that the dominant ^{44}Ti production is α-rich freeze-out from the inner part of core-collapse supernovae. The COMPTEL 1.157 MeV γ-ray line detection from the Cas A supernova [21] confirms this general picture.
^{44}Ti and co-produced Ni isotope abundances are sensitive probes of entropy and of neutron excess in the inner region of supernovae, addressing current issues in supernova studies: Location of the mass cut separating ejecta from matter accreted onto the compact remnant, and explosion asymmetries suggested e.g. by pulsar runaway or SNR jets; the "collapsar" model for Gamma-Ray Bursts indicates far-reaching implications. Gamma-ray line shape details can be resolved with Ge detectors and will encode kinematics of ejecta from close to the mass cut, thus being a prime task for INTEGRAL in addition to determination of the true ^{44}Ti mass. Such measurements are largely unaffected by processes in the overlying envelope or early supernova, exept for the possible modification of the ^{44}Ti decay time in case of full ionization of ^{44}Ti [37].
The Cas A detection appeared unlikely because it is so old (319 years, hence at 2.7% of the initial ^{44}Ti luminosity) and distant (3.4 kpc). Additionally Cas A was a relatively dim supernova suggesting low ^{56}Ni yields; occultation from circumstellar dust may resolve this paradoxon [17]. If we see Cas A in ^{44}Ti γ-rays, we should see even more ^{44}Ti sources, with a Galactic core-collapse supernova rate of a few events per century. The COMPTEL survey may have shown a second source in the Vela region [23], although too large uncertainty [49] prevents interpretations of possible source parameters from being meaningful. Certainly the 1-3 new source hints in COMPTEL surveys [11,24] will be key targets for deep searches with future experiments. The paucity of ^{44}Ti objects suggests that the ejection of ^{44}Ti is not a common characteristic of supernovae. A simulation study shows that the "^{44}Ti SN rate" appears low, if the inner-Galaxy COMPTEL ^{44}Ti data are compared to the optical record of the last millenium [55]. Rather special conditions may have to conspire to produce this isotope.
The COMPTEL value for the ^{44}Ti mass ejected by the Cas A supernova depends upon the underlying γ-ray continuum emission. Hard non-thermal X-ray emission from SN1006 [29] had stimulated new studies on SNRs as sources of energetic particles and cosmic rays. A by-product of this acceleration process would be continuum γ-ray emission from either synchrotron radiation or Bremsstrahlung [13]. The complex shape of the γ-ray spectrum of Cas A in the MeV regime is now studied in detail with OSSE and COMPTEL [54,51]. The γ-ray flux in the ^{44}Ti line could

reduce down to a level corresponding to a few 10^{-5} $M_\odot$ of ^{44}Ti, in agreement with current models for Cas A, and also with the non-observation of Cas A ^{44}Ti by OSSE [53] and RXTE [48].

NOVAE AND STARS

Novae arise from runaway burning of a slowly accreted layer of material on the surface of a white dwarf. Explosive hydrogen burning not only produces the largely-expanded nova star, but also characteristic nuclear burning products, specifically proton-rich isotopes. Most of these are rather shortlived, β^+-decay produces copious annihilation γ-rays in a flash lasting a few hours; this will happen even before the nova has brightened optically. Radioactive products with longer half lifes produce a hard X-ray continuum for days, driven mainly from ^{13}N and ^{18}F. A few isotopes are expected to be detectable in their decay γ-rays for nearby (few hundred pc) novae (^{7}Be, τ=77d; ^{22}Na, τ=3.75y) [15,18]. The ^{7}Be and 511 keV γ-rays are much brighter, yet will be serendipitous due to their early appearance. Neither the initial annihilation γ-ray flash, nor ^{7}Be, nor ^{22}Na has been detected [22,16]. Large enrichments in intermediate-mass nuclei such as Ne and S (up to 1000 times solar ratios) are revealed through UV spectroscopy and show that nucleosynthesis does occur; yet direct proof is awaited, and nova models are still unsatisfactory [50]. Yields are sensitive to burning temperatures and time profiles and to initial abundances, which again depend on convection and white dwarf masses.

Nuclear burning inside stars remains largely invisible even through penetrating γ-rays. The solar neutrino deficit [26] shows that our understanding of the nuclear burning underlying the main lifetime period of all stars is still patchy. Wolf Rayet stars (and to a lesser extent AGB stars) mark the phases in stellar evolution, where deep convective mixing of the envelope of the star combines with material ejection through stellar winds, and thus reveals products of inner nuclear burning. The spectroscopic identification of Tc (which has no stable isotope) in 1952 was an early direct proof of hydrostatic burning. But γ-ray line isotopic signatures are sparse. Longlived ^{26}Al (τ=1.04 10^6 y) requires nearby sources: For the closest WR star in the γ^2Vel binary system at 260 pc, the non-detection of ^{26}Al γ-rays with COMPTEL is relevant, however: model predictions are significantly above COMPTEL constraints [43] and indicate overly optimistic WR ^{26}Al yields [34].

ENERGETIC PARTICLES

Cosmic rays and radiation processes from AGN and supernova remnants demonstrate the existence of efficient particle accelerators in the universe. Although it is generally believed that Fermi acceleration is the only mechanism capable to provide the observed particle energies, the acceleration process, and the associated injection of suprathermal particles into the acceleration region, is far from being understood. The sun is our closest laboratory for the study of energetic particles, here also directly measured through particle detectors in interplanetary space [7]. Energetic particles produce characteristic γ-rays from nuclear excitation upon their collisions with ambient matter. Detailed measurements of γ-ray flare spectra are available, yet exhibit complex superpositions of narrow solar-atmosphere and broad solar-flare particle de-excitation lines, often burried in an intense electron Bremsstrahlung continuum [40]. The correlated and slower variability of lines originating from neutron interactions, and their difference to line characteristcs from low-energy proton collisions and flare electrons, suggest different acceleration sites: the high-energy component of solar flare particles indeed may result from Fermi acceleration initiated by large-scale processes in the solar magnetosphere, far beyond the loop structures which provide the electron and low-energy particle components, through electrostatic or Fermi accelerators set up in loop reconnection events [20]. – The complexity of solar-flare γ-ray spectra is expected to be greatly simplified in the thin-target configuration which presumably describes the interaction of low-energy cosmic rays with ambient matter in star-forming regions and the general interstellar medium [47]. The COMPTEL discovery of intense nuclear de-excitation γ-rays [2] came as a surprise and great stimulus to studies of low-energy cosmic rays; the experimenters withdrew their discovery however, when they noted that instrumental background may have caused such a signal artifact [3]. Thus, nuclear interaction lines from cosmic ray interactions still are to be discovered, predicted intensities [19] of $\simeq 10^{-6}$ph cm^{-2} s^{-1} leave little hope for CGRO and INTEGRAL.

DIFFUSE RADIOACTIVITIES

After the pioneering detection of the ^{26}Al decay 1.809 MeV γ-ray line with HEAO-C [32], COMPTEL data could map this relatively intense γ-ray line emission over the whole sky [9,44,27,4]. The maps show that ^{26}Al production occurs throughout the Galaxy, roughly as expected from shortlived stars formed from molecular clouds: globally CO data are not too different from the ^{26}Al map. At closer inspection, the dust emission at 240μm as mapped by the COBE DIRBE instrument, but also free-free emission derived from COBE DMR radio data at 53 GHz, provide a better correlation to ^{26}Al emission over the entire range of the Galaxy [28]. Massive stars are expected to be responsible for free-free emission, as their UV luminosity ionizes the surrounding medium; similarly, winds and supernovae from massive stars inject significant turbulence into the interstellar medium, heating dust in those regions in addition to the bright radiation field. It is therefore plausible, that massive stars also are the dominating sources of ^{26}Al, and 1.809 MeV emission traces their spatial distribution throughout the Galaxy. In turn, other candidate sources such as novae and AGB stars apparently are minor contributors. A detection of ^{60}Fe decay γ-rays co-produced with ^{26}Al in core-collapse supernovae is awaited, to strenghten these conclusions. – Maps of ^{26}Al and warm dust differ in detail from the map of molecular gas, and suggest that massive-star activity is not strictly proportional to the gas density on time scales of million years. This may be taken as further evidence that massive stars form in aggregates or clusters, as seen in external galaxies. In our Galaxy, apparently the Cygnus, Carina, and Vela regions are regions of currently-enhanced massive star activity. For the Cygnus region, observations allow to assemble a complete census of massive stars and their associations, as well as inferences of past supernova activity from Cygnus superbubble and Loop. Modeling the expected Cygnus ^{26}Al emission, one finds $\simeq$80% of the observed value; this is a high fraction, considering that the massive stars which terminated their life a million years ago should have left not much observables exept ^{26}Al [45]. In the Vela region the ^{26}Al bright Molecular Ridge (VMR) is marked by signs of recent massive star formation, warm dust and the brightest HII regions of the southern sky. Well-known foreground sources such as the 11000-year-old Vela Supernova Remnant, and the Wolf-Rayet/O-star binary system $\gamma^2 Vel$ are still too far away to be regognizable above this diffuse background, even though recent studies have reduced the quoted distances for these objects in both cases [10,43]. The same is probably true for the newly-discovered X-ray SNR at longitude 266° [1], close to the ^{26}Al emission peak expected from the VMR, although speculations based on the ^{44}Ti signature from this region may make this an even closer (and then relevant) source of ^{26}Al. In the case of the Vela SNR, model predictions are fully consistent with the COMPTEL ^{26}Al data [10].

Observations with Ge detectors have suggested that the 1.809 MeV line from ^{26}Al is broader than shown by the original HEAO-C measurement [41]. Assuming the observed broadening is of cosmic nature, a large fraction of ^{26}Al decays would occur from nuclei with velocities in the 500 km s^{-1} range. It is not easy to imagine environments which can support high velocities over the million-year decay time scale [5]. Yet, aluminium is highly refractory and may preferentially end up on dust grains, which could be a major contributor to cosmic ray nuclei if acceleration in supernova shock fronts of stellar associations is important [12,52]. INTEGRAL with its high-resolution Ge detectors should clearly enlighten this issue.

^{26}Al decays into ^{26}Mg through β^+ decay, in 82% emitting a relativistic e$^+$. There are more radioactivity sources of interstellar e$^+$, most notably the decay chains of ^{60}Co, ^{44}Ti, and nova-produced ^{7}Be, ^{19}F, and ^{22}Na. The disk of the Galaxy has been shown to emit e$^+$ decay radiation, in the form of the 511 keV annihilation line and a 3-photon annihilation spectrum from para-positronium: OSSE scans show that the 511 keV source(s) reported from earlier balloon observations are not restriced to the Galactic Center region [46,25]. Estimations show that the observed disk luminosity of $\simeq$ 10^{-3}ph cm^{-2} s^{-1} in the 511 keV line is about ten times as much as expected from ^{26}Al decay in the ISM. The supernova and nova radioactivity contributions to interstellar e$^+$ are very uncertain, mainly because the e$^+$ transport in young (age $\simeq$years) supernova remnants is unclear [36]. Other sources are expected to contribute to the interstellar e$^+$ budget, e.g. pulsars and accretion sources with plasma jets. OSSE put rather tight limits on variability of such compact sources in the inner Galaxy [46]; variable 511 keV source observed with many earlier balloon experiments must have been inactive during the CGRO mission. The e^+ annihilation emission of the Galactic bulge derives as about 1/3 of the total disk emission. The bulge appears as a bright emission region, extended further out than a standard stellar bulge would suggest. From images derived through deconvolution methods from the set of OSSE pointings towards the inner Galaxy, the existence of a fountain of e^+ in the inner Galaxy had been inferred [46,8]. Recent studies weaken these findings, but still find some north/south asymmetry, specifically in 511 keV emission, while annihilation continuum appears to show a rather symmetric Galactic-disk structure [35].

SUMMARY

The results from the Compton Gamma-Ray Observatory have established γ-ray line spectroscopy as a unique observational window for astronomy. Supernova γ-ray lines have been observed, yet the calibration of the central nuclear power engine needs improvements in absolute detection efficiencies and systematics of the observations. Imaging is a key for diffuse emission studies, and has provided important hints for otherwise untracable processes in the interstellar medium. Observations with high spectral and imaging resolution, such as planned with INTEGRAL, are needed to better constrain arguments based on shapes of γ-ray lines.

REFERENCES

1. Aschenbach B., Nature **369**, 141 (1998)
2. Bloemen H. et al., A&A **281**, L5-8 (1994)
3. Bloemen H. et al., ApJ **521**, L137 (1999)
4. Bloemen H. et al., these proceedings (2000)
5. Chen W. et al., ESA-SP **382**, 105 (1997)
6. Clayton D. D. et al., ApJ **399**, L141 (1992),
7. Cohen C.M.S. et al., Geophysical Research Letters, **26, 17**, 2697 (1999)
8. Dermer C.D., Skibo J.G, ApJ **487**, L57 (1997)
9. Diehl R. et al., A&A **298**, 445 (1995)
10. Diehl R. et al., Astroph. Lett. Commun., in press (1999)
11. Dupraz C. et al., A&A **324**, 683(1997)
12. Ellison D.C. et al., ApJ **487**, 197 (1997)
13. Ellison D.C. et al., ICRC 26, OB2.2.09 (1999)
14. Georgii R. et al., these proceedings (2000)
15. Gomez-Gomar J. et al., MNRAS **296**, 913 (1998)
16. Harris, M. J.et al., ApJ **375**, 216(1991)
17. Hartmann, D. H., et al., Nuc. Physics A **621**, 83(1997)
18. Hernanz M. et al., ApJ **526**, L97 , (1999), and these proceedings (2000)
19. Higdon J.C., Proc. ICRC **20, 1**, 160 (1987)
20. Hudson H. & Ryan J., Ann.Rev.A.A. **33**, 239 (1995)
21. Iyudin A.F. et al., A&A **284**, L1 (1994)
22. Iyudin A.F. et al., A&A **300**, 422(1995)
23. Iyudin A.F. et al., Nature **369**, 142 (1998)

24. Iyudin A.F. et al., Astroph. Lett. Commun., in press (1999)
25. Kinzer R. et al., ApJ **515**, 215 (1999)
26. Kirsten T., Rev Mod. Phys. **71**, 1213 (1999)
27. Knödlseder J. et al., A&A **345**, 813-825 (1999)
28. Knödlseder J. et al., A&A **344**, 68 (1999)
29. Koyama K. et al., Nature **378** 255 (1995)
30. Kurfess J. D. et al., ApJ **399**, L137(1992)
31. Leising M.D. et al., these proceedings (2000)
32. Mahoney W. A. et al., ApJ, **262**, 742 (1982)
33. Matz S. et al., Nature **331**, 416, (1988)
34. Meynet G. et al., A&A **320**, 460 (1997)
35. Milne P.A. et al., these proceedings (2000)
36. Milne P.A. et al., AIP **410**, 1022 (1997)
37. Mochizuki Y. et al., A&A **346**, 831 (1999)
38. Morris, D. J. et al., N.Y. Acad. Sci. **759**, 397 (1995)
39. Morris, D. J., et al., AIP 410, 1084 (1997)
40. Murphy R.J. et al., ApJ **490**, 883, (1997)
41. Naya J. E.et al., Nature **384**, 44(1996)
42. Nittler L. et al., ApJ **462**, L31 (1996)
43. Oberlack U.G. et al., A&A, in press (2000)
44. Oberlack U. et al., A&AS **120**, 311-314 (1996).
45. Plüschke S. et al., these proceedings (2000)
46. Purcell W.R. et al., ApJ **491**, 725 (1997)
47. Ramaty R. et al., ApJ **316**, 801 (1979)
48. Rothschild R.E. et al., AIP **410**, 1089 (1997)
49. Schönfelder V. et al., these proceedings (2000)
50. Starrfield S., Phys. Rep. 311, 371 (1999)
51. Strong A.W. et al., these proceedings (2000)
52. Sturner S. & Naya J.E., ApJ, in press (1999)
53. The L.-S. et al., A&AS **120**, 357 (1996)
54. The L.-S. et al., AIP 410, 1147 (1997)
55. The L.-S. et al., these proceedings (2000)

Investigations of Positron Annihilation Radiation

P.A. Milne[1], J.D. Kurfess[2], R.L. Kinzer[2], M.D. Leising[3] and D.D. Dixon[4]

[1]*NRC/NRL Resident Research Associate, Naval Research Lab, Code 7650, Washington DC 20375*
[2]*Naval Research Lab, Code 7650, Washington DC 20375*
[3]*Clemson University, Clemson, SC 29631*
[4]*Institute of Geophysics and Planetary Physics, Univ. of Cal. Riverside, CA 92521*

Abstract. By combining OSSE, SMM and TGRS observations of the galactic center region, Purcell et al. (1997) and Cheng et al. (1997) produced the first maps of galactic positron annihilation. That data-set has been augmented with additional data, both recent and archival, and re-analyzed to improve the spectral fitting. The improved spectral fitting has enabled the first maps of positronium continuum emission and the most extensive maps of 511 keV line emission. Bulge and disk combinations have been compared with the 511 keV line data, demonstrating that extended bulges are favored over a GC point source for every disk model tested. This result is independent of whether OSSE-only, OSSE/SMM, or OSSE/SMM/TGRS data-sets are used. The estimated bulge to disk ratio (and to a lesser extent the total flux) is shown to be dependent upon the assumption of bulge shape. A positive latitude enhancement is shown to have an effect upon the B/D ratio, but this effect is secondary to the choice of bulge shape.

I INTRODUCTION

One of the primary objectives of the OSSE instrument on NASA's COMPTON Observatory has been to understand the nature of galactic positron annihilation radiation. Through $8\frac{1}{2}$ years of observations, 511 keV line emission has been detected, but the emission has never been unambiguously attributable to a given discrete source. The galactic center (GC) region's 511 keV emission was monitored by the Gamma-Ray Spectrometer on-board the Solar Maximum Mission (SMM) (1980-1988), the Transient Gamma-Ray Spectrometer (TGRS) on-board the WIND mission (1995-1997) and with multiple OSSE observations (1991-present). None of these detections require variable sources in addition to the two component models discussed here to explain the measured fluxes (Share et al. 1990, Harris et al. 1998, Purcell et al. 1997). These results have supported the suggestion that the majority

CP510, *The Fifth Compton Symposium,* edited by M. L. McConnell and J. M. Ryan

of the emission is diffuse. Reported here are preliminary results of the extension of the OSSE analysis into three new areas: the inclusion of observations in regions with no *a priori* expectations of positron annihilation radiation, the extension of the analyzed region to include a larger fraction of the Galaxy, and the mapping of the positronium continuum component (PCONT) of the total annihilation emission. The derived PCONT flux values have a stronger dependence upon fitting the underlying continuum than do the 511 keV line flux values, and thus detailed analyses (model-fitting & 1D cuts) are only performed upon the 511 keV data-set in this preliminary presentation.[1]

Previously published OSSE results have focussed upon the 511 keV line emission emanating from the central radian of the inner Galaxy, mapping emission from $|l| \leq 33°$, $|b| \leq 17°$, and model-fitting on a $|l| \leq 90°$,$|b| \leq 45°$, 1°x1° grid. In a result first reported at the Fourth Compton Symposium and based on 6+ years of OSSE data, Purcell et al. 1997 (hereafter PURC97), showed evidence for three components to the 511 keV line emission, (1) an intense slightly extended emission centered in the direction of the GC, (2) a fainter planar emission, and (3) an unexpected enhancement of emission from positive latitudes (PLE). To generate that data-set, two basic types of data were used; (1) standard & offset pointing data where the scan angle crossed the GC and/or was perpendicular to the galactic plane, and (2) mapping data which searches for emission by observing a large sky region at regularly spaced intervals along a scan path.[2] The live-time from mapping observations is spread over many more pointings (16 or 32 for mapping versus 3-9 for standard), so the sensitivity per source pointing achieved is inferior to the standard or offset observations.

The present work relaxes all selection criteria, initially including all observations whose source and background pointings are within the $|l| \leq 90°$, $|b| \leq 45°$ region. To extract the positronium component from the total emission, three spectral models have been fit to each spectrum (from 60 keV to 700 keV); (1) a single power-law +511 keV line + PCONT (as fit by PURC97, though they fit from 50 keV -4 MeV), (2) a power-law with an exponential fall-off +511 keV +PCONT, (3) a thermal bremsstrahlung model +511 keV +PCONT. If the best-fitting model (of the three) is deemed to provide an acceptable fit, then that result is selected for subsequent mapping analysis. This has been done for the 1153 observations in the data-set. Fewer than 20 observations have been rejected in this preliminary, all-inclusive analysis. The combination of including archival data and the use of new data collected in the 2+ years since the PURC97 paper has increased the total GC exposure from 1.9 x 10^7 det·s to 8.6 x 10^7 det·s, as seen in Figure 1. As a result of this exposure, the fraction of the GC region ($|l| \leq 90°$, $|b| \leq 45°$) mapped above our exposure threshold has increased from 22% to 83%.

The combined OSSE/SMM data-sets are used rather than the OSSE-only data-

---

1) Among the potentially important effects not yet addressed are emission from the diffuse cosmic-ray continuum and a correction for scan-angle dependent background.

2) See PURC97 for details of the OSSE pointing strategies and background techniques.

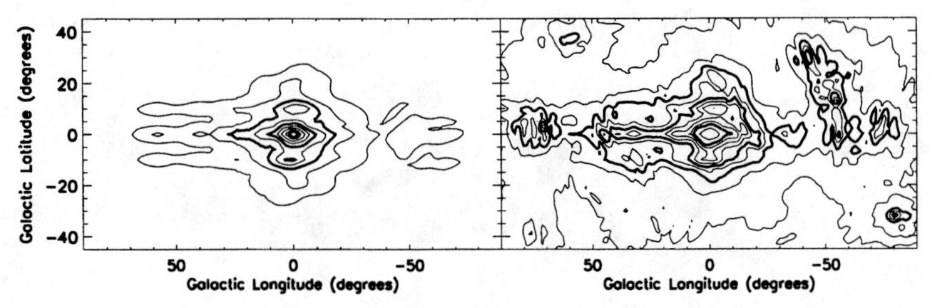

FIGURE 1. The exposure map of the expanded data-set compared with the earlier data-set of Purcell et al. (1997). The maps are in units of 10^9 cm^2 s, with the contour levels 0.05, 0.25, 0.5, 0.75, 1, 1.25, 1.5, 1.75, 2, 2.5.

set because the OSSE background-subtraction technique leads to differential rather than absolute fluxes. OSSE is insensitive to both isotropic emission and modest intensity gradients. The SMM fluxes are not absolute either, being insensitive to isotropic emission. Assuming isotropic emission to be zero, the SMM data contributes an overall normalization (Share et al. 1988). The SMM FoV is wide (~130° FWHM), but it does contribute limited spatial information as the response peak swept through the GC region along the ecliptic. One important difference between the PURC97 data-set and this one is the use of TGRS data. PURC97 used the TGRS data from Teegarden et al. (1996). That data has since been re-analyzed by Harris et al. (1998), with the resultant 511 keV line flux reduced by almost 20%. The Harris et al. (1998) data-set is used for the model-fitting studies shown in Table 1. SVD maps of the combined OSSE/SMM/TGRS data-set were not ready for these proceedings, but a RL map of the OSSE/SMM/TGRS data-sets (not shown) is in general agreement with the OSSE/SMM map.

II MAPPING POSITRON ANNIHILATION EMISSION

Two techniques have been employed to map the OSSE data-set: minimizing χ^2 with an adaptation of the Richardson-Lucy Algorithm (RL), and response matrix inversion using truncated Singular Valued Decomposition (SVD). SVD was described in PURC97, RL will be described in detail in an upcoming work.[3] Differences between the two resulting maps suggest the level of the uncertainties involved.

The upper and middle panels of Figure 2 shows the SVD and RL 511 keV maps of the combined OSSE and SMM data-sets. Both maps show three principle features,

[3] In short, RL adds flux distributed according to the instrument response to the source/background regions to raise/lower individual flux values to match each observation. Through successive iterations, map structure develops and the overall χ^2 lowers.

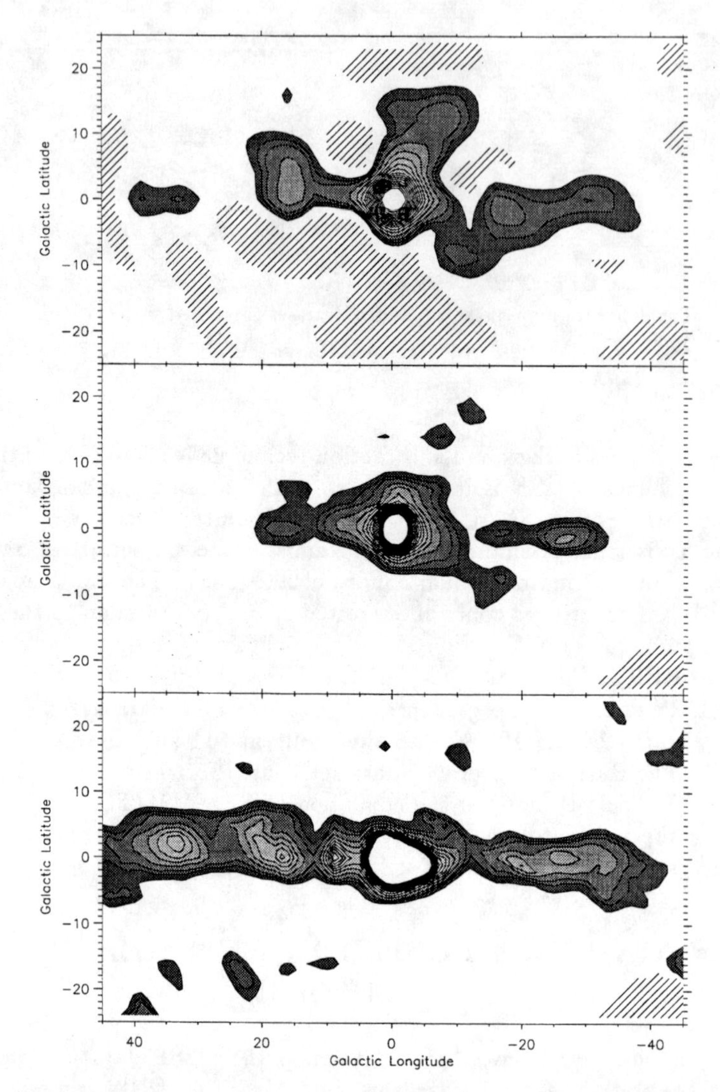

FIGURE 2. The SVD-511 map of the OSSE/SMM data-set is shown in figure 2a, the RL-511 map is shown in figure 2b. The RL-positronium continuum map of the OSSE-only data-set is shown in figure 2c. All contours are in increments of 2.3 x 10^{-3} phot cm^{-2} s^{-1} $ster^{-1}$. For the RL maps, the hatched region is where the OSSE exposure does not meet a minimum threshold, for the SVD map the regions where unphysical negative fluxes occur is also hatched.

intense emission centered near the GC (hereafter called bulge emission), a fainter planar emission (hereafter called disk emission), and emission from the negative longitude/positive latitude region (hereafter called a PLE). The PLE is more pronounced in the SVD map than in the RL map. The hatched region does not meet a minimum exposure threshold (or maps negative intensity). We emphasize the level to which the emission is concentrated in these three components. Although a far larger fraction of the inner radian is mapped than in PURC97, there is no indication of intense emission from any of these newly mapped regions. The apparent dominance of the bulge and disk emissions along with a contribution from the PLE drive the use of two and three component models when model-fitting.

Shown in the lower panel of Figure 2 is a RL map of the fitted positronium continuum fluxes of the OSSE data-set. In most astrophysical environments, more PCONT photons are produced in annihilation events than 511 keV line photons. For a positronium fraction of 0.95, the PCONT:511 ratio is 3.7:1.[4] As a result, the PCONT map is more intense. As explained in the introduction, this map is preliminary as a number of potential biases have not yet been addressed. Nonetheless, the dominance of intense bulge and fainter disk components appears to agree with the 511 keV maps. The principal difference is the lack of evidence of PLE emission, as will be discussed in the next section.

III LONGITUDE AND PLE CUTS

A measure of the planar structure in the 511 keV emission can be seen by taking a cut along the galactic plane of the RL and SVD maps. Cuts for the RL and SVD maps, as well as a model combining an $R^{1/4}$ bulge and the DIRBE 100 disk are shown in the upper panel of Figure 3. Each is the sum of the $|b| \leq 2^{\circ}$ pixels. The SVD data is plotted with 1σ error bars per degree. The SVD and RL 511 maps show rough agreement, except in the $+18^{\circ}$ to $+27^{\circ}$ region, where the peaks and valleys are exaggerated in the SVD map relative to the RL map. The model is more centrally peaked and the centroids of the maps are slightly offset towards negative longitude, but general agreement exists between the model and the maps. A systematic survey of the inner galactic plane scheduled for CGRO Cycle 9 will improve the sensitivity of the longitudinal cut.

The lower panel of Figure 3 shows a cut through the PLE at an angle of 60° relative to the negative-longitude galactic plane (an angle suggested by the SVD map). As seen in the longitudinal cut, the $R^{1/4}$ shape is more centrally peaked. In the anti-PLE direction, all three maps agree, all falling smoothly. Both the SVD-511 and the RL-511 are brighter in the $+3^{\circ} \rightarrow +8^{\circ}$ region than the corresponding negative region. Beyond $+8^{\circ}$, the sensitivity becomes poor, and there is little significance to the differences between the SVD and RL maps. The PCONT map (not shown) does not show a corresponding enhancement, though interpretation of

[4]) OSSE has measured f_{Ps}=0.97±0.03 at the GC (Kinzer et al. 1996). This value is consistent with the TGRS measured value, f_{Ps}=0.94±0.04 (Harris et al. 1998).

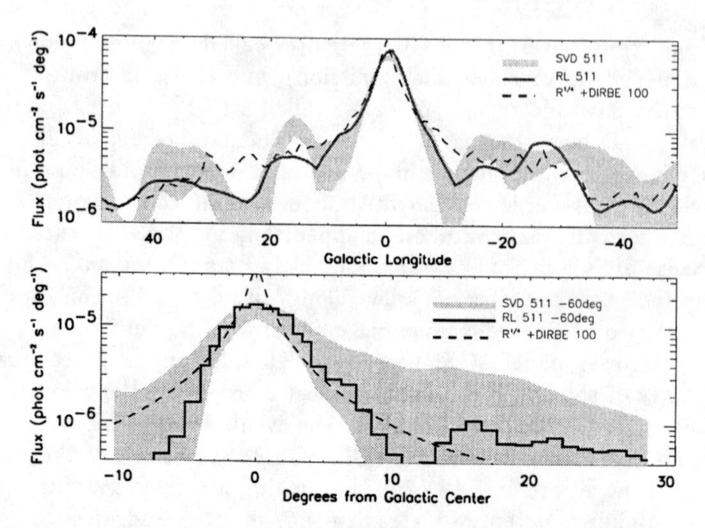

FIGURE 3. Figure 3a shows the integrated intensity from the $|b| \leq 2^\circ$ portion of the galactic plane. The SVD-511 cut is shown with 1σ uncertainties, the RL-511 map and the $R^{1/4}$ bulge and the DIRBE 100 disk model are shown without uncertainties. Figure 3b shows a 1° wide cut through the PLE region, determined to be best characterized as extending at a 60° angle relative to the galactic plane. All maps were generated from the OSSE/SMM data-set.

this result as being due to annihilation physics, or alternatively being due to fitting systematics is not justified in this preliminary analysis.[5]

IV MODEL-FITTING THE GALACTIC CENTER REGION

The maps of 511 keV line emission support the PURC97 representation of the galactic emission as being due to bulge, disk and PLE components. To quantify the contributions by the bulge and disk components, three bulge shapes have been combined with 28 disk shapes and compared with the OSSE, then the OSSE/SMM, then the OSSE/SMM/TGRS 511 keV data-sets. The bulge shapes tested are; a GC point source, a Gaussian, and the projection of a truncated $R^{1/4}$ function. Shown in the left panel of Figure 4 are the chi-squared values for best-fitting bulges of each shape paired with 28 disk models and fit to the OSSE data-set. The disks have

[5] The "annihilation fountain" model (Dermer & Skibo 1997) would have a low positronium fraction due to the high temperature, as direct annihilation with free electrons dominates over radiative recombination above 10^6K (Bussard, Ramaty & Drachman 1979). However, the absence of broad emission in the TGRS spectra provides a constraint to this scenario (Harris et al. 1998).

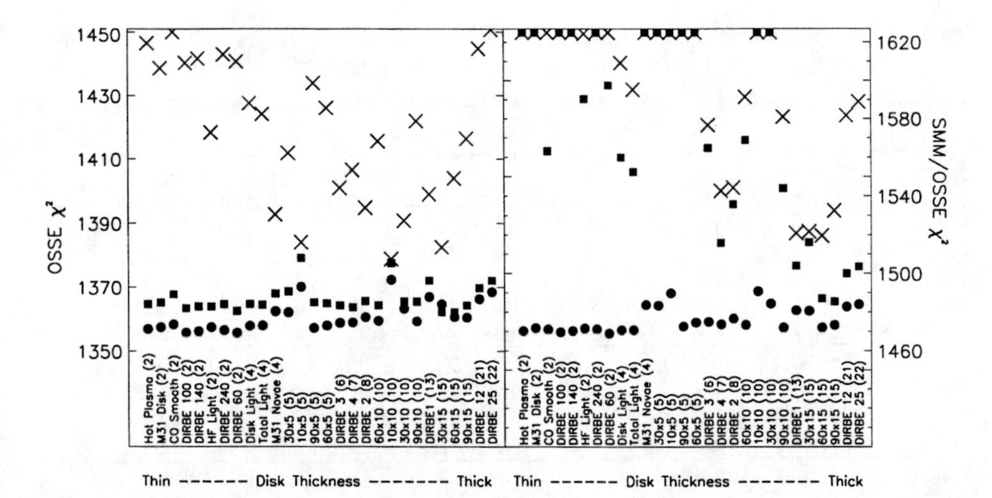

FIGURE 4. Results of bulge-disk fitting to OSSE-only (left panel) and OSSE/SMM (right panel) data-sets. The equivalent latitude FWHM of the disks are shown in parentheses. "90x10", etc. refers to the FWHM of disks with a Gaussian profile in width and thickness. The $R^{1/4}$ bulges (filled circles) and Gaussian bulges (filled squares) fit both data-sets better than does the point source (X). When fitting the OSSE data-set, thin disks are favored for the $R^{1/4}$ shape, but disfavored for the Gaussian shape. The combined OSSE/SMM data-set strengthens these tendencies. Very large χ^2 values for point source & Gaussian fits were truncated to 1622.

been ordered by the equivalent latitude FWHM of a Gaussian profile.[6] In the fits, the FWHM of the Gaussian bulges have been permitted to vary (best-fit FWHM values range from 3.9° to 5.7°). The $R^{1/4}$ radial function has been truncated to a constant value inside of a radius (R_{min}).[7] R_{min} has been constrained to be between 50 pc and 700 pc. The radial distribution has then been projected onto the line of sight assuming the GC to be 8 kpc distant. For all R_{min}, the $R^{1/4}$ has extended wings relative to the Gaussian shape; for small R_{min}, the $R^{1/4}$ is also more centrally peaked. For every disk tested, the extended bulges are strongly favored over the GC point source. The $R^{1/4}$ bulge shows a slight preference for thin disks, while the Gaussian bulge shows a stronger preference for thicker disks. This is interpreted as a hint of emission separated from both the galactic plane and the GC. For the $R^{1/4}$ bulge shape, the extended bulge wings account for this emission. For the Gaussian, a thicker disk is required.

In the right panel of figure 4, the SMM data is combined with the OSSE data. The fits for the $R^{1/4}$ solutions are similar to the OSSE-only plots. However, only

---

[6] The map references are: DIRBE maps (Hauser et al. 1998), the CO map (Dame et al. 1987), the Hot Plasma (Koyama et al. 1989), the $R^{1/4}$, HF Light & Disk Light (Higdon & Fowler 1987), the M31 maps (Ciardullo et al. 1987). "90x10", etc. refer to the FWHM of Gaussian disks.
[7] The $R^{1/4}$ function is $\rho(R \leq 0.24) = A \cdot R^{-6/8} \cdot \{\exp(-B \cdot R^{1/4})\}$, $\rho(R \geq 0.24) = 5/4 \cdot A \cdot R^{-7/8} \{\exp(-B \cdot R^{1/4}) - C \cdot R^{-1/4}]$, where R = radial distance from GC in kpc, and A,B,C are constants.

TABLE 1. Fit parameters for better-fitting models to the OSSE-only, OSSE/SMM & OSSE/SMM/TGRS data-sets (HP=Hot Plasma, DL=Disk Light, D12=DIRBE 12μ). Bulge emission is dominant for $R^{1/4}$ pairs, while disk emission is dominant for Gaussian pairs. The Richardson-Lucy and SVD map parameters are shown for comparison.

Models	OSSE-only				OSSE/SMM				OSSE/SMM/TGRS			
$R^{1/4}$	R_m[a]	B/D	F_T[b]	χ^2/α	R_m	B/D	F_T[b]	χ^2/α	R_m	B/D	F_T[b]	χ^2/α
HP	100	1.6	25.7	1.18	200	1.7	28.3	1.20	200	1.2	27.6	1.20
CO	50	2.6	30.0	1.18	50	1.2	28.9	1.20	50	3.3	31.3	1.20
DL	50	2.7	25.2	1.18	50	1.5	29.0	1.20	50	0.8	28.2	1.20
90x10	50	0.6	27.4	1.18	50	0.9	27.9	1.20	50	0.6	26.8	1.20
Gaus.	W[c]	B/D	F_T[b]	χ^2/α	W[c]	B/D	F_T[b]	χ^2/α	W[c]	B/D	F_T[b]	χ^2/α
60x15	4.4°	0.2	20.9	1.19	4.3°	0.2	26.4	1.21	4.3°	0.2	26.4	1.21
90x15	5.1°	0.3	21.6	1.19	4.8°	0.2	29.6	1.32	5.7°	0.2	29.9	1.21
D12	5.5°	0.1	29.9	1.28	5.7°	0.2	31.3	1.33	5.7°	0.2	31.4	1.22
RL Map	[d]		38.3	1.13		0.63	37.0	1.14				
SVD Map	[d]		31.1	1.15		0.49	25.8	1.18				

[a] The truncation radius in pc.
[b] The total flux in units of 10^{-4} phot cm^{-2} s^{-1}.
[c] W is the Gaussian FWHM.
[d] The degrees of freedom have been set equal to those of the models.

the thicker disk-Gaussian bulge solutions are able to approximate the quality of the $R^{1/4}$ fits. As seen in Table 1, the B/D ratios are larger for $R^{1/4}$ solutions than for Gaussian solutions. The total flux of the $R^{1/4}$ solutions change little when SMM and then TGRS data is added, but for the Gaussian solutions the total flux can change considerably. For many Gaussian bulge-disk combinations, the OSSE data would not permit insertion of the SMM or TGRS-required flux without violating OSSE constraints, leading to poor fits. These solutions do not necessarily span the range of possible bulge-disk combinations, nor are they unique, but they do suggest ranges of plausible B/D ratios and total fluxes. The uncertainties of the parameters are not shown in Table 1 due to concern about them being misinterpreted. The standard approach is to fix all but a single parameter, and calculate the degradation to the fit that results from varying that single parameter. The uncertainties that result do not account for the effect of varying the other parameters, nor does it account for other potential model shapes. In this paper, the uncertainties are evident in Table 1, but are not calculated.

The results shown in Table 1 have ignored the existence of a PLE and its potential influence upon the B/D and F_{Tot} parameters. PURC97 quantified the PLE flux by two methods; (1) by subtracting the mirror region from the outputs of the mapping techniques, and (2) by fitting the data-set with three components (bulge, disk and PLE), all with Gaussian profiles. The two methods yield very different PLE flux results. The PURC97 maps suggest the PLE flux to be (1.3-1.5) x 10^{-4} phot cm^{-2}

s^{-1}. The 3-Gaussian fitting suggests (5.4-8.8) x 10^{-4} phot cm^{-2} s^{-1}. The current maps suggest (1.2±0.5) x 10^{-4} phot cm^{-2} s^{-1}, in agreement with the PURC97 maps. Quantifying the PLE by mirror-region subtraction, the B/D ratios for the RL and SVD maps are 0.63 and 0.49, as shown in Table 1. The B/D values are intermediate to the $R^{1/4}$ and Gaussian solutions, suggesting a general agreement between mapping and two component modeling.

The PURC97 3-Gaussian solution, when inserted with quoted parameters, is not an acceptable solution for the current OSSE/SMM/TGRS data-set ($\chi^2/\alpha = 1.34$), but the same method can be applied to examples of better-fitting models from Table 1. When a Gaussian representing the PLE is added to a 4.3° Gaussian bulge + 60°x15° Gaussian disk, and the B/D re-optimized, the B/D rises from 0.163 to 0.169 ($\Delta\chi^2$=-16.4). The PLE centroid for this Gaussian fit is determined to be (l, b) = -2°, +8°, and the PLE flux is 1.1 x 10^{-4} phot cm^{-2} s^{-1}. When a PLE Gaussian is added to the $R^{1/4}$ +CO Smooth model, the B/D lowers slightly from 3.3 to 2.6 ($\Delta\chi^2$=-12.3), and the parameters become: (l, b) = -2°, +1°, PLE flux = 0.7 x 10^{-4} phot cm^{-2} s^{-1}. These values are given not to suggest a refinement of the PLE, but rather to demonstrate that the B/D depends more on the bulge shape than on the PLE (and vise versa). The current modeling of the PLE lowers the flux values to better agreement with the mapping fluxes, but the interpretation must be that the characteristics of any possible PLE are too poorly constrained and too model dependent to claim flux values and emission centroids (and the corresponding uncertainties).

V DISCUSSION

OSSE investigations of galactic positron annihilation have recently been expanded into three new areas; utilization of archival data near the GC, utilization of archival data away from the GC and mapping the positronium continuum emission. This expansion makes the results more dependent upon the spectral analysis, which is currently in its preliminary stage. The portion of this analysis least dependent upon these new complications are the 511 keV line flux values. Although the spatial coverage of maps from these values has increased, characterization of the emission with three components (bulge, disk and PLE) remains adequate. An extended bulge emission has been shown to be favored over point-like emission from the GC for a collection of 28 disk models. Comparing two approximations of the bulge shape, Gaussian bulges require thick disks to fit the data-sets, while $R^{1/4}$ bulges are less dependent upon the disk thickness, slightly favoring thin disks. The bulge shape must be better resolved with future measurements to improve the constraints upon the B/D ratio (currently ranging from 0.2 -3.3 for the combinations tested). The total flux is better constrained, ranging between (20.9-31.4) x 10^{-4} phot cm^{-2} s^{-1}. The effect of a potential PLE upon these parameters is shown to be less than the bulge-shape uncertainty when the PLE is approximated by a Gaussian.

Positron astronomy is 30 years old but remains in its infancy. Of current generation gamma-ray telescopes, OSSE is the best-suited to investigate the sky distribution of this emission. The time allotted to OSSE annihilation studies has increased in recent cycles of the CGRO mission. This time, combined with the expanded OSSE analysis, and the impending launch of the INTEGRAL telescopes should produce significant improvements of our understanding of the nature of galactic positron annihilation.

REFERENCES

1. Bussard, R.W., Ramaty, R. & Drachman, R.J., *ApJ* **228**, 928 (1979).
2. Cheng, L.-X., et al., *ApJ* **481**, L43 (1997).
3. Ciardullo, R., et al., *ApJ* **318**, 520 (1987).
4. Dame, T.M. et al., *ApJ* **322**, 706 (1987).
5. Dermer, C.D. & Skibo, J.G., *ApJ* **487**, L57 (1997).
6. Harris, M.J., et al., *ApJ* **501**, L55 (1998).
7. Hauser, M.G., et al., *COBE Ref Pub No. 98-A* **Greenbelt, MD:NASA/GSPC**, (1998).
8. Higdon, J.C. & Fowler, W. A., *ApJ* **317**, 710 (1987).
9. Kinzer, R.L., et al., *A &AS* **120**, 317 (1996).
10. Koyama, K. et al., *Nature*, **339**, 603 (1989).
11. Milne, P.A., et al., *Proceedings of the 3rd INTEGRAL Workshop* **in press**, (1998).
12. Purcell, W.R., et al., *ApJ* **413**, L85 (1993).
13. Purcell, W.R., et al., *ApJ* **491**, 725 (1997).
14. Share, G.H., et al., *ApJ* **326**, 717 (1988).
15. Share, G.H., et al., *ApJ* **358**, L45 (1990).
16. Teegarden, B., et al., *ApJ* **463**, L75 (1996).

TGRS MEASUREMENTS OF THE POSITRON ANNIHILATION SPECTRUM FROM THE GALACTIC CENTER

M. J. Harris[1], B. J. Teegarden, T. L. Cline, N. Gehrels, D. M. Palmer[1], R. Ramaty, H. Seifert[1]

Code 661, NASA/Goddard Spaceflight Center, Greenbelt, MD 20771
[1] *Universities Space Research Association*

Abstract.
The TGRS experiment on board the *Wind* spacecraft includes a Ge detector with very high resolution (3–4 keV FWHM) at energies around 511 keV. To take advantage of *WIND*'s fixed pointing at the south ecliptic pole and its 3 s rotation, TGRS also includes a Pb occulter fixed on the spacecraft body subtending an arc of 90° along the ecliptic plane. Spectra of the Galactic center region, identified by this occultation method, have been accumulated since 1994 November. In this paper we present updated results for the Galactic center positron annihilation line from these occulted spectra. From the results prior to fall 1997 we obtained the intensity, width, energy and variability of the line, the positronium (Ps) fraction, and limited information about the spatial distribution of the line, all of which have been published. Since the fall of 1997, degradation of the instrument performance has limited us to improving our results for the spatial distribution. These updated results are in marginal disagreement ($\sim 2.5\sigma$) with a model based on earlier OSSE measurements.

I INTRODUCTION

The occulted spectra from TGRS are received in the form of Fig. 1, which shows the accumulated count rate in the energy range 506–515 keV along the ecliptic (which is divided into 128 "sectors" of 2°.8125 each). The two sources which appear as dips when occulted are the Crab SNR and the Galactic center region. The range of TGRS's energy coverage, 20 keV–1 MeV, is divided into four broad windows, one of which (479–543 keV) is binned at very high resolution (1 keV per channel). Thus a spectrum of this energy range can be obtained by accumulating the occulted count rate as in Fig. 1, but channel-by-channel (6).

Full details of this Galactic spectrum, as measured during 1995–1997, and the inferences for the annihilation source which can be drawn, were given by ref. 1. In

CP510, *The Fifth Compton Symposium,* edited by M. L. McConnell and J. M. Ryan

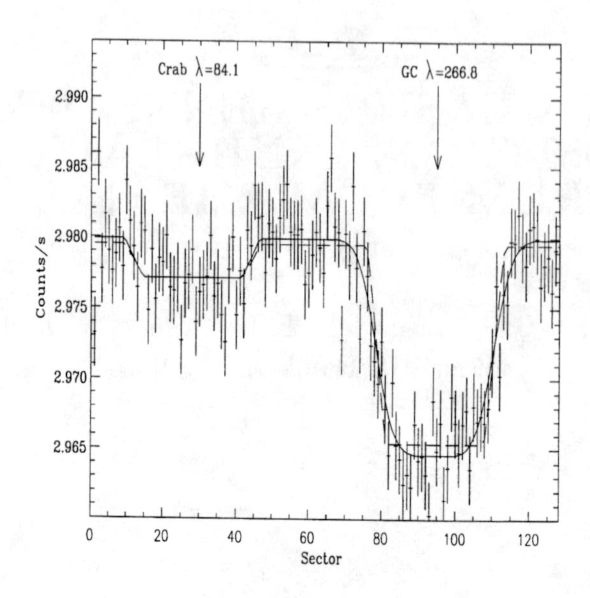

FIGURE 1. Count rate in the energy range 506–516 keV, accumulated by sector bin between 1995 January 1 and 1997 March 13. Dashed line — fitted TGRS response to a point source. Solid line — fitted response to a Gaussian distribution along the ecliptic with FWHM 24°. Fit to a rectangular distribution is almost identical (§II). Arrows — ecliptic longitudes of the Crab and the GC.

addition to the spectrum, some limits can be obtained on the spatial distribution of a diffuse source, such as the Galactic center is supposed to be. In ref. 1 we assumed a Gaussian spatial distribution along the ecliptic, parametrized by its width, intensity, and centroid in ecliptic longitude. This model was fitted to the occultation dip in Fig. 1 after convolution with the occulted instrument response. It differs significantly from the fit to a point source response (dashed line, Fig. 1).

II EFFECT OF DETECTOR DETERIORATION ON THE ANALYSIS

During the period 1995–1997 the spectrometer performed at a level very close to its nominal behavior, in terms of energy resolution, effective area and gain. Since the fall of 1997, however, the detector performance has progressively deteriorated as a result of accumulated cosmic-ray induced damage. Three main effects have occurred: distortion of the response to a narrow line (starting with low-energy tailing, culminating in the formation of a separate low-energy peak); a poorly-quantified loss of efficiency; and an accelerated rate of change of the instrument gain (3).

Of the several quantities measured during the earlier (1995–1997) period, some are worse affected by these problems than others. Those which are derived from high-resolution spectral analysis (the fitted spectrum quantities were in fact the intensity, width and energy of a line at 511 keV, and the intensity of a 3γ Ps annihilation continuum [1]) will be seriously compromised by the loss of efficiency and resolution. Although approximate estimates of these effects are possible (see ref. 2 concerning Nova Velorum 1999), these do not attain sufficient precision to make an improvement upon our earlier measurements. We have not therefore updated our measurements of line intensity, width and energy, and Ps continuum intensity from ref. 1.

The analysis of the position and extension of the source, on the other hand, does not need fine spectral information so long as certain assumptions are made. The position and size of the occultation dip in Fig. 1 (as opposed to its amplitude) depends only on the geometry of the occulter and the source. However, it must be noted that we fit the spectrum by the sum of several components (1), and it cannot be assumed that each component has the same spatial distribution. This problem can be partly overcome by confining the accumulation in angle bins (Fig. 1) to the narrowest possible energy range around the narrowest energy component, i.e. the 511 keV line. In the 1995–1997 data we limited the energy range to 506–516 keV. As the line resolution degraded in the more recent data, a steadily expanding energy range (up to 28° in the worst periods) had to be employed. This increased the possibility of a systematic error due to contamination of the line photons in Fig. 1 by continuum photons. To estimate the size of this effect, we performed an analysis in which the continuum photons were deliberately included; the position and width of the occultation dip were measured for all photons in TGRS's high-resolution energy window.

A different error, which is inevitable in our measurement of the source width, is the impossibility of uniquely inverting the occultation dip profile in Fig. 1. This is a result of the two-dimensional distribution of the diffuse source being covered by a one-dimensional edge, which itself is rather ill-defined ($\sim 7°$ full width at 511 keV; see point source curve in Fig. 1). Therefore a source shape must be assumed *a priori*, and the source "width" has a different meaning depending on the shape chosen. The source shapes currently implemented in the occultation dip fit program are rectangular and Gaussian; we have tried to estimate the error due to the uncertain source shape by comparing the results for the two.

III UPDATED RESULTS AND DISCUSSION

Our updated results for the source position and location are given in Table 1, which also contains the 1995–1997 results (1) for comparison. Although the improvements from the additional two years' data are rather small, it is interesting to compare them with previous knowledge of the spatial distribution of the 511 keV line emission. For this purpose, our measurement of the source centroid is most im-

Quantity	Measured values[a]	Previous values	Comments
Source dimension (FWHM)[b]	$24° \pm 5°\ ^{+3°}_{-4°}$	14° [c]	1995–1997 data (1)
	$23.8° \pm 3.9°\ ^{+5.6°}_{-2.5°}$	14° [c]	Present work
Source centroid[d]	$265.5° \pm 1.2° \pm 1.0°$	264.8° [c]	1995–1997 data (1)
	$267.6° \pm 0.8°\ ^{+1.4°}_{-0.6°}$	264.8° [c]	Present work

[a] Statistical error given first, followed by systematic error.
[b] Assuming a Gaussian distribution along the ecliptic.
[c] Calculated from OSSE model (Purcell et al. 1997).
[d] Ecliptic longitude λ. The GC is at $\lambda = 266.8°$.

TABLE 1. Results from analysis of TGRS measurements 1995–1999.

portant, since it appears to contain fewer systematic errors and since earlier OSSE observations suggested a definite asymmetry with respect to the Galactic Center (5). In the map of ref. 5, this asymmetry arose from a peak of emission, separate from the center, and centered at $l = -2°$, $b = +9°$; relative to the Galactic Center it is at a lower ecliptic longitude λ, so that the centroid of this map is also offset towards smaller λ. We calculated this offset by convolving the TGRS occulted response with a model used by Purcell et al. (5) to describe the map (Table 1).

Our updated result does not show an asymmetry about the Galactic Center; the centroid is consistent with the center at the level 1σ. Moreover, our measured centroid is slightly eastward (i.e. positive λ) from the center. It follows that our measurement of the centroid of the 511 keV line emission is marginally inconsistent ($\simeq 2.5\sigma$ level) with the model describing the map presented in ref. 5. More recent OSSE results (4) suggest that the asymmetric emission arises closer to the center, in an extension of the central peak rather than in a separate feature. The centroid of the map based on these results is therefore closer to the Galactic center, and in better agreement with our value in Table 1.

REFERENCES

1. Harris, M. J., et al. 1998, ApJ, 501, L55
2. Harris, M. J., et. al. 2000, in these Proceedings
3. Kurczynski, P., et al. 1999, Nucl. Instr. Methods. Phys. Research A, 431, 141
4. Milne, P. A., Kurfess, J. D., Leising, M. D., & Dixon, D. D. 2000, in these proceedings
5. Purcell, W. P., et al. 1997, ApJ, 491, 725
6. Teegarden, B. J., et al. 1996, ApJ, 463, L75

COMPTEL 1.8 MeV All Sky Survey: The Cygnus Region

S.Plüschke*, R.Diehl*, V.Schönfelder*, G.Weidenspointner*, H.Bloemen+, W.Hermsen+, M.McConnell†, J.Ryan†, K.Bennett@, U.Oberlack**, J.Knödlseder++

*MPI f. extraterrestrische Physik, Postfach 1603, 85740 Garching, Germany
+SRON-Utrecht, Sorbonnelaan 2, NL-3584 CA Utrecht, The Netherlands
†Universtity of New Hampshire, Durham NH 03824-3525, USA
@Astrophysics Division, ESTEC, NL-2200 AG Noordwijk, The Netherlands
**Astrophysics Laboratories, Columbia University, New York, NY 10027, USA
++CESR, CNRS/UPS, 31028 Toulouse, France

Abstract. We present an updated version of COMPTEL's 1.809 MeV sky survey. Based on eight years of observations we compare results from different imaging techniques using background from adjacent energy bands. We confirm the previously reported characteristics of the galactic 1.809 MeV emission, specifically an extended galactic ridge emission, mainly concentrated towards the inner galaxy, a peculiar emission feature in the Cygnus region, and a low-intensity ridge extending towards Carina and Vela. Because this gamma ray line is due to the decay of radioactive ^{26}Al, predominantly synthesized in massive stars, one anticipates flux enhancements aligned with regions of recent star formation. This is born out by the observations. In particular the Cygnus feature, first presented in 1996 based on three years of COMPTEL data, is confirmed. Based on the stellar population we distinguish three prominent areas in this region, for which we separately derive fluxes, and discuss interpretations.

INTRODUCTION

The imaging gamma-ray telescope COMPTEL [1] aboard the CGRO spacecraft allowed for the first time to survey the entire sky in the MeV regime. One of the mission highlights is the deduction of the first all-sky image in the 1.809 MeV gamma-ray line, first detected by HEAO-C and SMM [2]. This emission line is attributed to the radioactive decay of ^{26}Al with a lifetime of 1.0417 Myr. Based on three years of COMPTEL data Oberlack et al. [3] presented a first 1.809 MeV all-sky image in 1996. The image reconstruction was based on a maximum entropy algorithm (ME) [4] using a background model based on adjacent energy bands for each observation period individually [3,5]. In 1997 Oberlack published a first update of the ^{26}Al all-sky map based on roughly five years of COMPTEL data in his PhD thesis [6]. This map confirmed the non-local character of the detected

CP510, *The Fifth Compton Symposium,* edited by M. L. McConnell and J. M. Ryan

1.809 MeV emission, attributing most of the emission towards young massive stars and active star forming regions.
In addition to this work Knödlseder et al. 1999 [7] introduced a multi-resolution regularized expectation maximization (MREM) algorithm for image reconstruction. The MREM method uses a wavelet based noise reduction. This method suppresses artifacts by allowing only significant structures to appear in the image. Also with this method the prior reported emission characteristics are confirmed, although the MREM map is much less structured than the ME image. Note that the MREM approach seems somewhat conservative with respect to image structures whereas the maximum entropy images tend to show artifacts.
We followed both imaging approaches using all COMPTEL data up to end of mission phase 7. In both cases we applied an improved version of the adjacent energy background model described below. Both methods reveal extended emission structures near Vela/Carina region as well as Cygnus as reported in the previous publications on the ^{26}Al all-sky maps.
In section 2 we give a short description of the data analysis. Section 3 discusses the two imaging approaches and the produced all-sky images, whereas section 4 presents the results of the detail analysis of the Cygnus region. In section 5 we summarize the results and discuss some interpretations.

DATA ANALYSIS & BACKGROUND MODELING

Our present analysis includes all data from beginning of the mission up to the end of mission phase 7. This spans a total observation time from May 1991 to November 1998, split into more than 270 observations periods with typical durations of 10^6s. For the imaging analysis the event data from a 200 keV wide energy band around the line energy is binned in a 3-dimensional dataspace consisting of the scatter direction of the photon (χ,ψ) and the scatter angle $\bar{\varphi}$. For a more detailed description of the dataspace and the event selections see [3].
Due to the very low signal-to-noise ratio of roughly 1%, background modeling is the crucial task in analyzing COMPTEL data. In the case of gamma-ray line analysis the use of adjacent energy bands as basis for a background model seems natural. This background model is applied on the basis of single observation periods as described earlier in [3] and references therein. In contrast to prior applications of this background model an advanced smoothing technique is applied, restoring the original event distribution characteristics. Furthermore this method garantuees an accurate treatment in the galactic polar regions.
A longterm study of count-ratios from the line band relative to a narrow adjacent band, used for background deduction, shows a clear time-dependence in the background normalization (see [6]). This time-variability was attributed to increased instrumental-line background. Details on COMPTEL's instrumental-line background are reported in this proceedings [8] and in [6]. Specifically the contributions from the radioactive decay of ^{22}Na and ^{24}Na contaminate the event distri-

butions in the adjacent bands. To cancel the longterm trend in the background normalization the contributions from various instrumental background components are determined by spectral fits for each observation period separately. From the resulting decay rates a renormalisation coefficient is calculated. These coefficients are used to compute a summed background model for the observation periods under consideration.

THE 1.809 MeV MAPS

We applied two imaging methods to construct all-sky images. First we used a ME algorithm (see [4]) which iteratively extracts sky intensity distributions being compatible with the data. The ME method shows a clear trend to create a lumpy, structured image in late iterations as Oberlack and Knödlseder [6,9] had shown. On the other hand early iterations significantly underestimate the gamma-ray fluxes. For these reasons an intermediate iteration has been chosen as a compromise between flux reproduction and map smoothness (see left panel of figure 1).

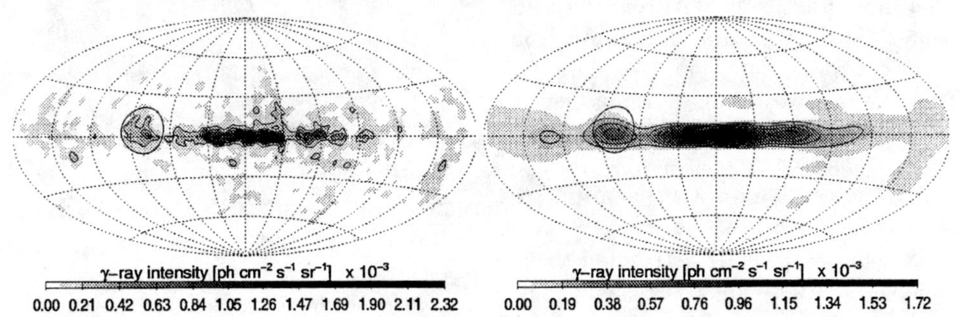

FIGURE 1. 1.8 MeV Allsky Images (Cygnus Region encircled): Maximum Entropy Image (left) and the MREM Image (right)

As second approach we used the convergent MREM method [7] which combines an iterative Expectation Maximisation scheme with a wavelet analysis, filtering low significance structures and artifacts in the reconstruction. This method attempts to produce the smoothest map that is consistent with the data (see [7]). The MREM map is shown in figure 1 (right panel). The main features in both reconstructions are identical: an extended galactic ridge emission, mainly concentrated towards the inner galaxy, a peculiar emission feature towards Cygnus (see encircled region in the maps), and low-intensity ridge extending towards Carina and Vela. However, some low-level emission features towards the galactic anticentre visible in earlier maps disappeared, which might be due to the better statistics in our sample or the improved background treatment. Furthermore, low-level emission features appearing in the ME image between 180° and 215° longitude are suppressed in the MREM map. For the Cygnus analysis we furthermore applied a maximum likelihood (ML) imaging of this region (see figure 2).

1.809 MeV FROM THE CYGNUS REGION

Based on the ME/ML images we analyzed the Cygnus region in detail following the analysis approaches presented in [10,11]. For a detailed description of this analysis see [12].

Figure 2 shows an overlay of our measurements with candidate point sources as well as the Cygnus OB associations. Five of these associations overlap in the Cygnus superbubble region (Cyg X region), which is also our strongest 1.8 MeV source region. We have determined fluxes for three regions ("Cygnus West" [Cyg X], "Cygnus East" [OB 7] and "Cygnus Arc" [arc-like structure north of the equator]), which are listed in table 1. As massive stars are the most promising ^{26}Al source candidates, WR stars and remnants from core-collapse SNe should explain the measured fluxes in the Cygnus region. Del Rio [10] found that WR stars in the regions could reproduce most of the observed flux. Our analysis reveals that up to 80% of the 1.8 MeV flux could be directly attributed to the known isolated point sources [12]. This difference is due to the incorporation of more recent nucleosynthesis models and a different choice of the IMF; we use $\Gamma = 1.35$ instead of $\Gamma = 1.7$.

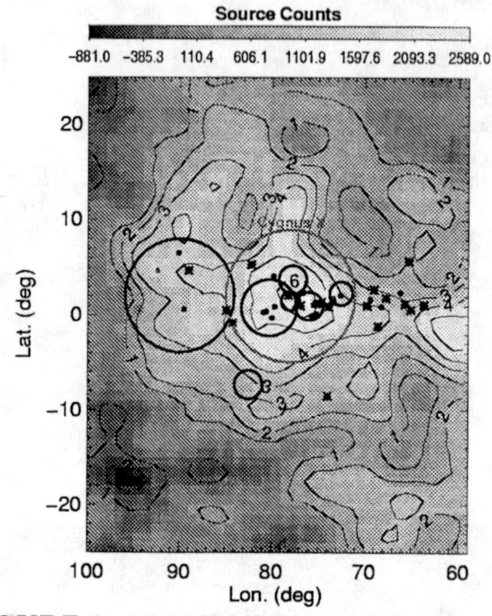

FIGURE 2. 1.8 MeV ML Image of the Cygnus Region: Contours are significances whereas the grey-scale image shows the reconstructed source counts for each bin. The circles mark OB associations in the Cygnus Region whereas WR stars and SNRs are marked with dots and stars, respectively.

TABLE 1. measured 1.809 MeV fluxes for Cygnus subregions; quoted errors are calculated via bootstrapping

Region	Counterpart Region	Longitude deg	Latitude deg	measured flux 10^{-5} cm^{-2} s^{-1}
Cygnus West	Cygnus Superbubble	70 - 86	-7 - 7	3.7 ± 1.1
Cygnus East	Cygnus OB7	86 - 97	-6 - 9	2.0 ± 0.6
Cygnus Arc		76 - 94	9 - 23	2.0 ± 0.6
Cygnus Region		70 - 96	-9 - 25	7.9 ± 2.4

Our model calculations of OB associations in the Cygnus region are still inconclusive due to large uncertainties in the normalisation of the initial mass function and star formation history. However, preliminary analysis indicates, that a large fraction of the observed fluxes can indeed be explained by the cumulative ^{26}Al yield from WR stars and core-collapse SN in these associations [6,12].

SUMMARY AND CONCLUSIONS

In the previous sections we have shown the latest 1.809 MeV maps deduced from COMPTEL data. As was shown in [13] the applied background model may suffer from misdeterminattion of continuum background from galactical sources. A detailed study of this problem is a still on-going task. Our maps confirm the reported structures aligned with areas of recent star formation. These sites are dominated by their massive star content. Specifically in the case of the Cygnus region we have determined fluxes and studied the possible source distribution. In the case of a static point source scenario we found that up to 80% of measured flux could be explained on the basis of recent nucleosynthesis calculations. The dynamic approach using a population synthesis model is somewhat inconclusive up to now due to large uncertainties in details of the population characterisation, but basically this model is also able to explain the measured 1.8 MeV emission from Cygnus.

REFERENCES

1. Schönfelder V., et al., *ApJS* **86**, 657-692 (1993)
2. Mahoney W.A., et al., *ApJ* **262**, 742 (1982)
3. Oberlack U., et al., *A&AS* **120**, 311-314 (1996)
4. Strong A.W., et al., *Data Analysis in Astronomy*, V. Di Gesù et al. (eds.). Plenum Press New York, p. 251 (1992)
5. Knödlseder J., et al., *SPIE Proc. Vol.* **2806**, Ramsey B.D., Parnell Th.A. (eds.), p. 386 (1996)
6. Oberlack U. *PhD Thesis TU München* (1997)
7. Knödlseder J., et al., *A&A* **345**, 813-825 (1999)
8. Weidenspointner G., et al., *in these Proceedings*
9. Knödlseder J., *PhD Thesis U Toulouse* (1997)
10. del Rio E., et al., *A&A* **315**, 237 (1996)
11. Oberlack U., et al., *Proc. of 2nd Oak Ridge Symposium on Atomic and Nuclear Astrophysics, 2-6 Dec. 1997*, p. 179 (1998)
12. Plüschke S., et al., *these Proceedings*
13. Bloemen H., et al. it Proc. Integral Workshop, Taormina (1998)

Gamma-ray line emission from OB associations and young open clusters

Jürgen Knödlseder[*,†], Miguel Cerviño[‡], Daniel Schaerer[‡] and Peter von Ballmoos[†]

[*]*INTEGRAL Science Data Centre, Chemin d'Ecogia 16, 1290 Versoix, Switzerland*
[†]*Centre d'Etude Spatiale des Rayonnements, B.P. 4346, 31028 Toulouse Cedex 4, France*
[‡]*Observatoire Midi-Pyrénées, 14, avenue Edouard Belin, 31400 Toulouse, France*

Abstract. OB associations and young open clusters constitute the most prolific nucleosynthesis sites in our Galaxy. The combined activity of stellar winds and core-collapse supernovae ejects significant amounts of freshly synthesised nuclei into the interstellar medium. Radioactive isotopes, such as ^{26}Al or ^{60}Fe, that have been co-produced in such events may eventually be observed by gamma-ray instruments through their characteristic decay-line signatures. However, due to the sensitivity and angular resolution of current (and even future) γ-ray telescopes, only integrated γ-ray line signatures are expected for massive star associations.

In order to study such signatures and to derive constraints on the involved nucleosynthesis processes, we developed a multi-wavelength evolutionary synthesis model for massive star associations. This model combines latest stellar evolutionary tracks and nucleosynthesis calculations with atmosphere models in order to predict the multi-wavelength luminosity of a given association as function of its age.

We apply this model to associations and clusters in the well-studied Cygnus region for which we re-determined the stellar census based on photometric and spectroscopic data. In particular we study the relation between 1.809 MeV γ-ray line emission and ionising flux, since the latter has turned out to provide an excellent tracer of the global galactic 1.809 MeV emission. We compare our model to COMPTEL 1.8 MeV γ-ray line observations from which we derive limits on the relative contributions from massive stars and core-collapse supernovae to the actual ^{26}Al content in this region. Based on our model we make predictions about the expected ^{26}Al and ^{60}Fe line signatures in the Cygnus region. These predictions make the Cygnus region a prime target for the future INTEGRAL mission.

EVOLUTIONARY SYNTHESIS MODEL

Our evolutionary synthesis model is based on the multi-wavelength code described in [6,1], enhanced by the inclusion of nucleosynthesis yields. In summary, the evolution of each individual star in a stellar population is followed using Geneva evolutionary tracks with enhanced mass-loss rates [7]. In our present implementation, and similar to [6,1], stellar Lyman continuum luminosities are taken from

CP510, *The Fifth Compton Symposium,* edited by M. L. McConnell and J. M. Ryan

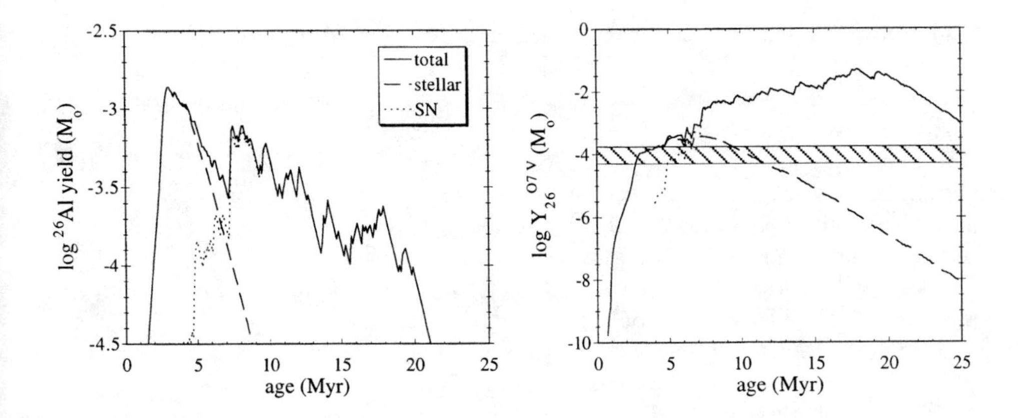

FIGURE 1. ^{26}Al yield (left) and equivalent O7V star ^{26}Al yield (right) as function of association age. The hatched area indicates the COMPTEL measurement of $Y_{26}^{O7V} = (1.0 \pm 0.3)\,10^{-4}\ M_\odot$.

[9,4]. Note, however, that modern atmosphere models including the effects of line blanketing and stellar winds predict enhanced ionising fluxes with respect to these models, hence our predicted ionising luminosities should be considered as preliminary and possibly are somewhat too low[1]. At the end of stellar evolution, stars initially more massive than $M_{WR} = 25\ M_\odot$ are exploded as Type Ib supernovae, while stars of initial mass within $8M_\odot$ and M_{WR} are assumed to explode as Type II SNe. Nucleosynthesis yields have been taken from [8] for the pre-supernova evolution and from [12,13] for Type II and Type Ib supernova explosions, respectively. Note that Type II SN yields have only been published for stars without mass loss and Type Ib yields have only been calculated for pure Helium stars. In order to obtain consistent nucleosynthesis yields for Type II supernovae we followed the suggestion of [5] and linked the explosive nucleosynthesis models of [12] to the Geneva tracks via the core mass at the beginning of Carbon burning. For Type Ib SN we used the core mass at the beginning of He core burning to link evolutionary tracks to nucleosynthesis calculations.

Evolutionary synthesis models were calculated using a stochastic initial mass function where random masses were assigned to individual stars following a Salpeter initial mass spectrum ($d\log\xi/d\log M = -1.35$) until the number of stars in a given mass-interval reproduces the observed population. Typical results for a rich OB association (51 stars within $15 - 40\ M_\odot$) are shown in Fig. 1. ^{26}Al production turns on at about $1 - 2$ Myr when the isotope starts to be expelled by stellar winds into the interstellar medium. Stellar ^{26}Al production reaches its maximum around 3 Myr when the most massive stars enter the Wolf-Rayet phase. Explosive nucleosynthesis sets on around 4.5 Myr for Type Ib and around 7 Myr for Type

[1]) We currently are implementing the CoStar atmosphere models of [11] in our code that consistently treat the stellar structure and atmosphere and include line blanketing and stellar winds.

II supernovae, leading to a second peak in the ^{26}Al yield around 8 Myr. After this peak, a slightly declining ^{26}Al yield is maintained by Type II explosions until the last Type II SN exploded around 20 Myr. Afterwards, the exponential decay quickly removes the remaining ^{26}Al nuclei from the ISM.

We also calculated the time-dependent ionising luminosity ($\lambda < 912$ Å) of the population from which we derived the equivalent O7V star ^{26}Al yield $Y_{26}^{\rm O7V}$. This quantity measures the amount of ^{26}Al ejected per ionising photon normalised on the ionisation power of an O7V star. The analysis of COMPTEL 1.809 MeV data suggests a galaxywide constant value of $Y_{26}^{\rm O7V} = (1.0 \pm 0.3)\,10^{-4}\ M_{\odot}$ [3]. Interestingly, the COMPTEL value is only reproduced for a very young population during a quite short age period (2.5 – 5 Myrs). For younger populations too few ^{26}Al is produced with respect to the ionising luminosity, leading to much lower equivalent yields. For older populations the ionising luminosity drops rapidly, resulting in much higher $Y_{26}^{\rm O7V}$ values. Thus, the equivalent O7V star ^{26}Al yield is a quite sensitive measure of the population age. In particular, the measurement of $Y_{26}^{\rm O7V}$ for individual OB associations or young open clusters provides a powerful tool to identify the dominant ^{26}Al progenitors.

APPLICATION TO THE CYGNUS REGION

We applied our evolutionary synthesis model to the Cygnus region from which prominent 1.809 MeV line emission has been detected by COMPTEL [2]. In [2] the authors modelled ^{26}Al nucleosynthesis in Cygnus by estimating the contribution from individual Wolf-Rayet stars and supernova remnants that are observed in this region. This approach, however, suffers from considerable uncertainties due to the poorly known distances to these objects. In this work we performed a complete census of OB associations and young open clusters in the Cygnus region. Individual association or cluster distances have been estimated by the method of spectroscopic parallaxes, ages have been determined by isochrone fitting. Distance and age uncertainties have also been estimated and were incorporated in the analysis by means of a Bayesian method. The richness was estimated for each association or cluster by building H-R diagrams for member stars and by counting the number of stars within mass-intervals that are probably not affected by incompleteness or evolutionary effects. In total we included 6 OB associations and 19 young open clusters in our analysis which house 94 O type and 13 Wolf-Rayet stars.

For each OB association or cluster, 100 independent evolutionary synthesis models were calculated that differ by the actual stellar population that has been realised by the random sampling procedure. In this way we include the uncertainties about the unknown number of massive stars in the associations that have already disappeared in supernova explosions. From these samples the actual age and distance uncertainties are eliminated by marginalisation, leading to a probability distribution for all quantities of interest. Note that in this approach an age uncertainty is equivalent to an age spread in the cluster formation, hence the possibility of non-instantaneous star formation has been taken into account. The results for all individual associations have been combined by marginalisation to predictions for the entire Cygnus region.

The predicted equivalent O7V star ^{26}Al yield amounts to $(0.3 - 1.2)\,10^{-4}\ M_{\odot}$ and is compatible with the COMPTEL observation, pointing to an extremely young population that is at the origin of ^{26}Al. Indeed, while 90% of the ^{26}Al is produced in our model by stellar nucleosynthesis (during the main sequence and subsequent Wolf-Rayet phase), only 10% may be attributed to explosive nucleosynthesis, mainly in Type Ib SN events. This is also reflected in the low ^{60}Fe yields ($(0 - 7)\,10^{-8}$ ph cm^{-2} s^{-1}) that are predicted by our model since ^{60}Fe is assumed to be only produced in supernovae.

However, in absolute quantities, our model underestimates the free-free intensity in Cygnus by about a factor 3 while the total ^{26}Al flux is low by a factor of 5. This points towards a possible incompleteness of our massive star census which has been based on surveys of OB associations and young open clusters in Cygnus available in the literature. Indeed, while we identify only 95 OB stars in Cyg OB2, [10] estimated 400 OB members in this association, indicating only 25% completeness of our census. Taking 25% as a typical completeness fraction for our OB association census and assuming that the young open cluster census is complete, we obtain a free-free intensity of 0.25 mK and an ^{26}Al flux of $4.3\,10^{-5}$ ph cm^{-2} s^{-1} – values that are in fairly good agreement with the observations (0.26 mK from DMR microwave data and $(7.9 \pm 2.4)\,10^{-5}$ ph cm^{-2} s^{-1} from COMPTEL 1.8 MeV observations; see Plüschke et al., these proceedings). However, we do not predict any noticeable amount of ^{60}Fe for the Cygnus region – a prediction which hopefully will be soon verified by the INTEGRAL observatory. We would like to stress that INTEGRAL has the potential to partially resolve some OB associations and young open clusters in the nearby Cygnus region, and thus may provide important new insights in massive star nucleosynthesis in this area.

REFERENCES

1. Cerviño, M. and Mas-Hesse, J.M., 1994, A&A, 284, 749
2. Del Rio, E., et al. 1996, A&A, 315, 237
3. Knödlseder, J., 1999, ApJ, 510, 915
4. Kurucz, R.L., 1979, ApJS, 40, 1
5. Maeder, A., 1992, A&A, 264, 105
6. Mas-Hesse, J.M. and Kunth, D., 1991, A&AS, 88, 399
7. Meynet, G., et al. 1994, A&AS, 103, 97
8. Meynet, G., et al. 1997, A&A, 320, 460
9. Mihalas, D., 1972, NCAR-TN/STR-76
10. Reddish, V.C., Lawrence, L.C. and Pratt, N.M., PROE, 5, 111
11. Schaerer, D. and de Koter, 1997, A&A, 322, 598
12. Woosley, S. and Weaver, T.A., 1995, ApJS, 101, 181
13. Woosley, S., Langer, N. and Weaver, T.A., 1995, ApJ, 448, 315

On the Massive Star Origin of ^{26}Al in the Cygnus Region

S.Plüschke*, R.Diehl*, U.Wessolowski*,
U.Oberlack†, D.H.Hartmann+

*MPI f. extraterrestrische Physik, Postfach 1603, 85740 Garching, Germany
†Astrophysics Laboratories, Columbia University, New York, NY 10027, USA
+Department of Physics and Astronomy, Clemson University, Clemson, SC 29634, USA

Abstract. The COMPTEL map of the 1.809 MeV line, which is attributed to the radioactive decay of ^{26}Al, shows significant excess emission in the Cygnus region. We investigate counterparts of this emission, based on COMPTEL data accumulated over 8 years. Previous studies suggested that this flux is due to the integrated, but unresolved, nucleosynthesis from young, massive objects like Wolf Rayet (WR) stars and core collapse supernovae. Beside a static scenario of known point sources - WR stars and SNRs - we consider OB associations in a dynamic scenario to model the observed flux in the Cygnus region. We carry out population synthesis studies to determine the time-dependent production of ^{26}Al and also kinetic as well as radiative energy. Furthermore we consider the impact of enhanced ^{26}Al yields probably occuring in a fraction of massive close binary systems. We compare the nucleosynthesis results of our OB association model with recent measurements in the Cygnus region.

INTRODUCTION

The latest COMPTEL 1.809 MeV maps show clear evidence for an emission excess in the Cygnus region (see Fig. 2 in [1]). The measurements reveal fluxes of 7.9 10^{-5} photons cm^{-2} s^{-1} for the Cygnus region and 3.7 10^{-5} photons cm^{-2} s^{-1} for Cygnus Superbubble as well as 2.0 10^{-5} photons cm^{-2} s^{-1} for Cygnus OB7 respectively (see table 1). This gamma-ray line emission is attributed to the radioactive decay of ^{26}Al. Recently, several studies of COMPTEL's all-sky map (e.g. [2]) have revealed convincing evidence for massive, young stars and their final supernovae being the dominant sources of ^{26}Al's interstellar abundance (see recent review [3] for more details).
In the perspective of a massive star origin of ^{26}Al the Cygnus 1.809 MeV excess was previously analyzed under the assumption of a stationary framework [4], taking known point sources as production sites. Furthermore in [5,6] a dynamical prescription of evolving OB associations as production scenario was used. In our

CP510, *The Fifth Compton Symposium,* edited by M. L. McConnell and J. M. Ryan

analysis we use updated point source catalogues [7–9] for modeling the contributions of known Wolf Rayet stars (WR) as well as supernova remnants (SNRs) to the measured ^{26}Al emission. As a second approach we have used enhanced simulations of the evolution of OB associations to calculate the time-dependent emission of newly synthesized ^{26}Al.
In section 2 we summarize our analysis of the ^{26}Al emission from isolated point sources in the Cygnus region. Section 3 gives a short description of our OB association model and the results of its application to the Cygnus associations. Finally we give a summary of the analysis results together with a short discussion of the errors and end up with some conclusions.

CYGNUS POINT SOURCES

The interstellar ^{26}Al abundance is affected by massive stars predominantly during two evolutionary phases - the WR phase [10] and the final core-collapse supernova [11,12]. Because of this, WR stars and young supernova remnants should be considered as source candidates in a point source analysis of the Cygnus emission feature. See Fig. 2 in [1] for the spatial distribution of those point sources in the Cygnus region.

We have taken all WR stars in the Cygnus region from the latest WR catalogues [7,8] and calculated their individual 1.809 MeV flux by applying a mean ^{26}Al yield. This yield value was derived from model calculations [10] by calculating an IMF-weighted average of the model yields $y_{26}(m)$ assuming a Salpeter IMF $\phi(m)$ with $\Gamma = 1.35$ to account for the unknown initial mass of the observed objects (see Eq. 1). The corresponding fluxes are given in table 1.

$$< Y_{26} >= \frac{\int_{M_{WR}^{lo}}^{M_{WR}^{up}} y_{26}(m)\phi(m)dm}{\int_{M^{lo}}^{M^{up}} \phi(m)dm} \tag{1}$$

In the case of the SNR contribution the whole analysis is some what more complicated than in the WR case. In contrast to the analysis presented in [4] we considered all SNR from the recent Green catalogue [9] to contribute. We distinguished between SNRs used in the del Rio analysis and those excluded. For the latter ones no distance and age measurements are available. Therefore we computed an upper limit for their contribution. We assumed all reported SNR to originate in a core-collapse SN, at a rather low age of 10^4 yr and a typical radius of 10.7 pc taken from the Sedov-Taylor solution. Taking the measured angular dimensions we were able to calculate a distance for each of them. The mean ^{26}Al yield for a single SN event was calculated in the same manner as in the WR case using model yields from [11,12]. Based on this yield value and the distances (measurements or approximations, respectively) we finally computed the flux values for each SNR. For significantly lower ages in our approximation sample, say 10^3, the individual contribution is uncertain to a factor of 5. Table 1 shows the fluxes contributed by the SNR component of our Cygnus model. Due to the approximations stated above these values are upper limits of the SNR contribution.

CYGNUS OB ASSOCIATIONS

Beside this static scenario the numerous Cygnus OB associations could be considered as possible sources due to their time-correlated formation of massive stars. Therefore we developed an enhanced model for the evolution of such associations based on an approach previously described in [5,6]. Resting upon a population synthesis model, using a power-law initial mass function (IMF) and a time-variable star formation rate (SFR), we compute integrated light-curves for some types of stellar emission. Beside the emission of nucleosynthesis products such as ^{26}Al and ^{60}Fe the release of kinetic energy as well as the emission of ionizing radiation is modeled. The nucleosynthesis part is modeled under the assumption of singular events in time. This means the WR phase is assumed to be rather short compared to the stellar as well as ^{26}Al lifetime. See Fig. 1 for some details on the nucleosynthesis model. The left panel shows the typical ^{26}Al yield from WR stars and subsequent SNe as function of the initial mass of the star.

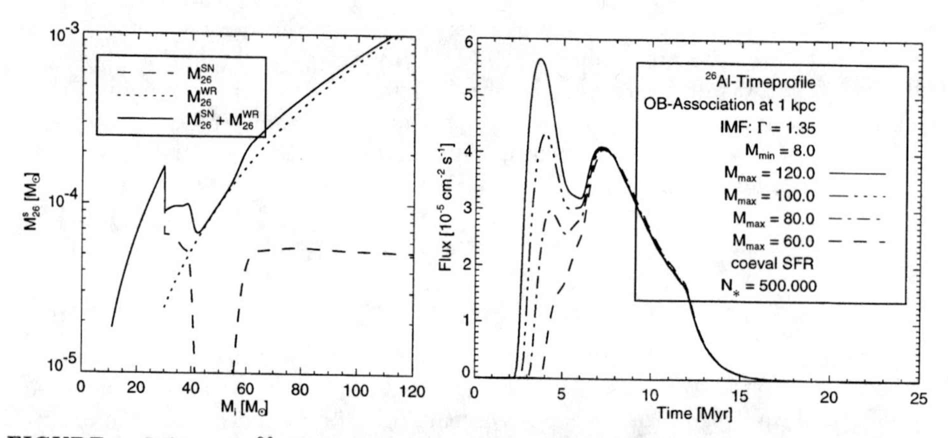

FIGURE 1. Left panel: ^{26}Al yield as function of initial mass; dotted and dashed line show the contributions from WR phase and SN explosion, respectively. In the case of the SN contribution, we coupled the nucleosynthesis results from [11,12] via the He core mass to the stellar evolution models from [10] to account for stellar mass loss during the stellar evolution. Right panel: 1.8 MeV flux light-curves for a OB association in a distance of 1 kpc with a starburst like SF history and 4 different upper mass limits.

The right panel shows the 1.809 MeV light-curve in the case of a starburst scenario. In the case of a starburst the WR- and SN-dominated phases can be clearly distinguished. Furthermore it becomes evident that in this case no 1.8 MeV detection is possible after 15 Myr since end of star formation. In that way measurements

of the ^{26}Al content of an OB association may help to constrain its age.
For analysis of measured radioactivity from star forming regions the problem is somehow the other way around. The star formation history and the age of possible OB associations has to be constrained very accurately, because the IMF normalization and the age spread among association members would be the most severe sources of uncertainty in the flux determination. On the one hand, one could use some other modeled parameters such as the emitted ionizing radiation for a self-consistent determination of the critical parameters. On the other hand, SF-determination and age measurements by other authors might be used as *a priori* knowledge. For our Cygnus analysis we used both approaches. The determination of the IMF normalization is based on [13] as well as on recent membership determinations by HIPPARCOS [14]. All OB associations except Cyg OB7 are found to be consistent with a burst-like or very narrow star formation history. In the case of Cyg OB7 the age spread seems to be more pronounced increasing the uncertainty ones more. Five of the Cygnus OB associations in the Cygnus superbubble region overlay each other partially. So the distinction among them is reduced and the analysis results are somehow degenerated. The cumulative results are displayed in Table 1.
As was pointed out in [15], in some peculiar massive, close binary (MCB) systems a modified evolution of the secondary star may lead to huge enhancement factors of the ^{26}Al yield. Factors as high as 10^3 were discussed. We have included a MCB component in our association model using a MCB frequency of 5% and applying low to intermediate enhancement factors of up to 10^2 for initial masses from 8 to 40 $M_\odot$ of initial masses. This additional component, if existent, will dominate the ^{26}Al light-curve. The required number of stars in an OB association to reproduce the measured flux is highly decreased.

TABLE 1. 1.8 MeV fluxes for Cygnus subregions (measured vs. model results)

Region	measured Flux 10^{-5} cm^{-2} s^{-1}	Flux (WR) 10^{-5} cm^{-2} s^{-1}	Flux (SNR) 10^{-5} cm^{-2} s^{-1}	Flux (OB-Model) 10^{-5} cm^{-2} s^{-1}
Cygnus West	3.7 ± 1.1	≤ 2.9	≤ 0.1	0.3 - 10.2
Cygnus East	2.0 ± 0.6	≈ 0.0	≤ 0.1	0.02 - 4.0
Cygnus Arc	2.0 ± 0.6	?	?	?
Cygnus Region	7.9 ± 2.4	≤ 3.0	≤ 0.7	0.4 - 12.3

SUMMARY & CONCLUSION

Table 1 summarizes the results from the different analysis components together with respective COMPTEL measurements (from [1]) for the distinct regions. As can be seen easily, the contribution from isolated point sources is roughly 80% at maximum and of the order of 50% for the full Cygnus region. Due to the huge errors in the age as well as the stellar content determination in the OB association modeling the summed model fluxes are always consistent with the measurement.

But as stated before, results from the association model are not conclusive up to now. In particular in cases of low-richness associations this attempt is dominated by the statistics. Perhaps the inclusion of some further observables from measurements in other wavelength regimes will help us to get a better constrained model for some of the Cygnus OB associations. In particular, intensities and sizes of image structures in HI or CO maps could help somehow. Also the contribution from MCB systems has to be investigated more extensively.

REFERENCES

1. Plüschke S., et al., *these Proceedings*
2. Knödlseder J., *A&A* **344**, 68-82 (1999)
3. Diehl, R. & Timmes, F.X. *PASP* **110**, 637-659 (1998)
4. del Rio E., et al., *A&A* **315**, 237-242 (1996)
5. Oberlack U., et al., *Proc. of 2nd Oak Ridge Symposium on Atomic and Nuclear Astrophysics, 2-6 Dec. 1997*, p. 179 (1998)
6. Oberlack U. *PhD Thesis TU München* (1997)
7. van der Hucht, K.A., et al., *A&A* **199**, 217-234 (1988)
8. Conti, P.S. & Vacca, W.D., *AJ* **100**, 431-444 (1990)
9. Whiteoak, J.B.Z. & Green, A.J., *A&AS* **118**, 329-380 (1996)
10. Meynet, G., et al., *A&A* **320**, 460-468 (1997)
11. Woosley, S.E. & Weaver, T.A., *ApJS* **101**, 181-235 (1995)
12. Woosley, S.E., Langer., N., Weaver, T.A., *ApJ* **448**, 315-338 (1995)
13. Bochkarev, N.G., Sitnik, T.G., *A&SS* **108**, 237-302 (1985)
14. de Zeeuw, P.T., et al., *AJ* **117**, 354-399 (1999)
15. Langer, N., et al., *Proc. of the 9th workshop on Nuclear Astrophysics* Ed. by W.Hillebrandt & E.Müller (1998)

COMPTEL upper limits for the ^{56}Co γ-rays from SN1998bu

R. Georgii*, S. Plüschke*, R. Diehl*, W. Collmar*, G.G. Lichti*, V. Schönfelder*, H. Bloemen†, J. Knödlseder‡, M. McConnell||, J. Ryan|| and K. Bennett¶

*Max-Planck-Institut für extraterrestrische Physik, Postfach 1603, 85740 Garching, Germany
†SRON, Sorbonnelaan 2, 3584 CA Utrecht, The Netherlands
‡CESR (CNRS/UPS), 9, av. du Colonel-Roche, 31028 Toulouse Cedex, France
||Space Science Center, University of New Hampshire, Durham, NH 03824, USA
¶Astrophysics Division, ESTEC, ESA, 2200 AG Noordwijk, The Netherlands

Abstract. The type Ia supernova SN 1998bu in M96 was observed by COMPTEL for a total of 88 days starting 17 days after the detection of the SN. A special mode improving the low-energy sensitivity was invoked. We obtained images in the 847 keV and 1238 keV lines of ^{56}Co using an improved point-spread function for the low-energies. We do not detect SN1998bu. Sensitive upper limits at both energies constrain the standard supernova model for this event.

INTRODUCTION

On May 9.9 UT SN1998bu was discovered in M96 (NGC 3368) [14]. From wide-band spectrograms it was classified to be of the type Ia [1]. From a prediscovery observation [3] and an estimation for maximum blue light of $t_{Bmax} = 10952.7 \pm 0.5$ TJD (i.e. May 19), P. Meikle [10] estimates the date of the explosion to be May 2.0 $\pm$ 1.0 UT (i.e. TJD 10935 $\pm$ 1). The Cepheid distance to M96 is about 11 Mpc.

Observations of SNe in the optical and neighbouring bands concentrate on information on the light curves from such events. However, due to the creation of the optical photons long after the initial explosion most information of the initial state of a SN-explosion is lost. Therefore a distinction of the various SN-explosion scenarios via their optical light curves alone is very difficult. On the contrary, through the observation of SN in the γ-ray line regime, information from much earlier states in the explosion can be obtained. Considerable differences in the predicted spectra of different SN models, for example the He-Cap or detonation model, exist [6]. They can be used to discriminate between the models and to decide about the extend of mixing in the SN-explosion.

CP510, *The Fifth Compton Symposium,* edited by M. L. McConnell and J. M. Ryan

The sensitivity of existing γ-ray instruments, mainly the instruments on-board CGRO (OSSE and COMPTEL), limits the observations of type Ia SN to such events which occur at distances well below 15 Mpc. Consequently, only one SN of type Ia, SN1991T, was marginally detected with COMPTEL [11]. SN1998bu opens a second opportunity for line searches in SN, since according to some of the models it should be observable with the sensitivities of COMPTEL and OSSE in the Cepheid adopted distance of 11 Mpc.

COMPTEL observations of SN1998bu started on TJD 10952, 17 days after the explosion. Due to the late start of the observations and due to the low sensitivity of COMPTEL for low energies, we missed the decay of the 750 keV and 812 keV lines of ^{56}Ni (τ = 8.8 d). The observations were performed for a total of 88 days with the aim to detect the 847 keV and 1238 keV lines of the daughter nuclei ^{56}Co (τ = 112 d) .

ANALYSIS AND RESULTS

In order to obtain a higher sensitivity at lower energies the observations were performed in the so-called "low mode", where the threshold of the D2 modules were considerably lowered (For a detailed description of COMPTEL see [12]). In the "low mode" most of the thresholds are well below 650 keV, the software threshold in the standard COMPTEL analysis and show a considerable spread. Therefore a homogeneous software threshold at this value for all D2 detectors is no longer a good choice, since we would lose low-energy sensitivity in those modules with low thresholds. On the contrary the hardware thresholds of the D1 modules all lie in a narrow range below 50 keV, thus allowing to use this value as software threshold for all D1 detectors. Therefore the D2 modules were separated in groups with similar thresholds. A new point-spread function (PSF) was calculated, adding up the individual PSFs for each of these groups weighted with the number of the members in each group. Furthermore another PSF was derived from simulations where each module was simulated with its correct hardware threshold. The standard and the two new PSFs were then applied to 3 days worth of Crab low-mode data fortuitously collected during a observation of Geminga. Using the two new PSFs we clearly detect Crab with 5.1 σ of 847 keV, compared to only 3.02 σ with the standard PSF. For the 1238 keV line the values are 1.6 σ and 0.63 σ, respectively. This result reflects that most D2 modules are now sensitive below the standard 650 keV software threshold used for the standard analysis, resulting in a larger sensitivity for the 847 keV line. Since no major difference was found between the added PSF and the simulated one, in the further analysis only the simulated PSF was used, since it reflects the correct treatment of the threshold of each module.

Unfortunately we are not able to make use of the full sensitivity gain in the 847 keV line, due to the low threshold of some D2 modules, we start to see the 511 keV line leading to a very high background. To eliminate this, two approaches are possible, the spectral analysis and the imaging analysis.

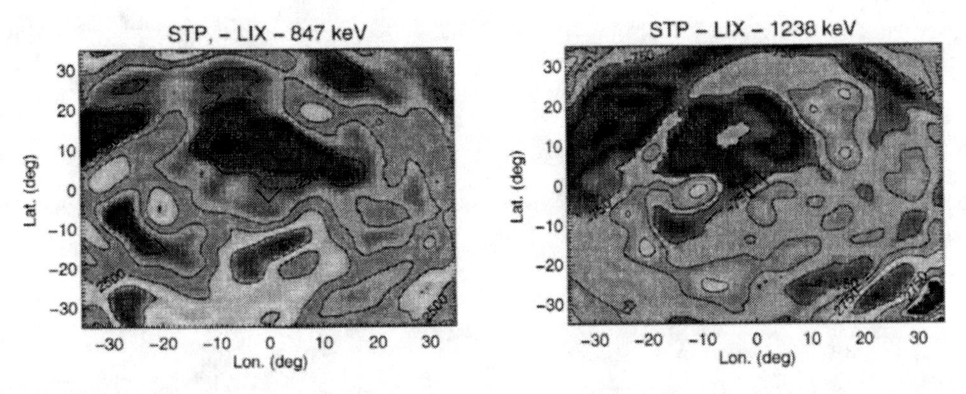

FIGURE 1. Flux map of the SN1998bu region in 847 keV (left) and 1238 keV (right). The SN position is marked by a diamond. The contour levels are in units of 10^{-8} photons/(cm^2 sec). The 1σ contour level corresponds to 2900 for the left and 1600 for the right map, respectively.

In the spectral analysis, a software cut at 600 keV in the D2 modules is used to suppress the 511 keV line background. At the SN-position and at different positions on a grid around the SN position (the positions on the grid allow to obtain off-source spectra) residual spectra, using data from a 3^0 cone in the 3-dimensional data-space as a "source spectrum" and data from a 3^0 - 7^0 cone mantle as "background spectrum" are produced. Subsequently a template Gaussian, with a width corresponding to the instrumental energy resolution, is used to obtain line intensities for both γ-ray lines at every position of the grid. Using this method, no significant difference between the source and the off-source positions can be seen. We derive 2σ upper limits in the following way: histograms for both lines using the fitted intensities from all positions of the grid are derived. The width of the distribution is then interpreted as a measure of the statistical and systematic uncertainty of the method. Together with information on the exposure and the effective area this yields a 2σ upper flux limit of $4.1 \cdot 10^{-5}$ photons/(cm^2 sec) for the 847 keV line and $2.3 \cdot 10^{-5}$ photons/(cm^2 sec) for the 1238 keV line.

In the imaging analysis a $\bar{\varphi}$ cut (see [12]) of 40^0 maximum is applied. Using the simulated PSF and the $\bar{\varphi}$ cut, the two maps for the two γ-ray energies (shown in figure 1) are produced using a maximum-likelihood method. Again it can clearly be seen that there is no signal from the supernova. The 2σ upper limits can be derived as follows: From the histogram of all fluxes in the maps in figure 1 the FWHMs are determined, assuming a Gaussian distribution. Using the Bayesian method described in [4], which also accounts for the systematic and statistical uncertainties, 2σ upper limits of $5.8 \cdot 10^{-5}$ photons/(cm^2 sec) for the 847 keV line and of $3.2 \cdot 10^{-5}$ photons/(cm^2 sec) for the 1238 keV line are found.

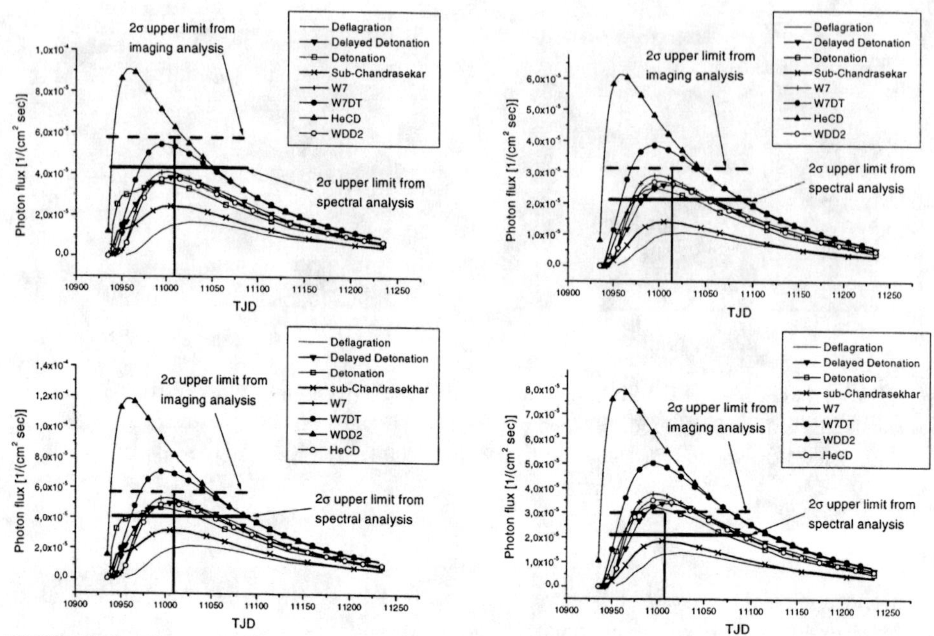

FIGURE 2. The model fluxes for the 847 keV (left) and for the 1238 keV (right) line for different models versus time after the explosion. The upper row is for a distance of 11.3 Mpc, the lower for a distance of 9.9 Mpc.

INTERPRETATION AND CONCLUSIONS

For the comparison of our upper limits, the distance to the SN plays a essential role. Interestingly the host galaxy M96 had already a HST-Cepheid-determinated distance of 11.6 ± 0.9 Mpc [13], later revised to 11.3 ± 0.9 Mpc [5] , leaving SN1998bu as one of seven SN being observed in galaxies with a Cepheid distance. However, a distance determination based on Planetary Nebulae (PN) suggests a distance as close as 9.6 ± 0.6 Mpc [2]. This points to a distance which would be compatible with a Cepheid distance of 9.9 Mpc resulting from recently discussed correction in the distance ladder scale [9].

In figure 2 expected model fluxes for a distance of 11.3 Mpc and 9.9 Mpc to M96 are plotted versus the time after the explosion and are compared to the 2σ upper limits obtained for both lines. The models are taken from Isern et al. [7] and from Kumagai et al. [8] and are scaled to the adopted distances. In figure 3 the expected maximal model fluxes are plotted versus distance in Mpc and are compared to the upper limit of the spectral analysis.

From these figures it can be seen, that we can exclude the HeCD and the W7DT model for SN1998bu: the upper limits are below the peak flux for both lines and distances. For the distance of 9.9 Mpc most other models would be inconsistent

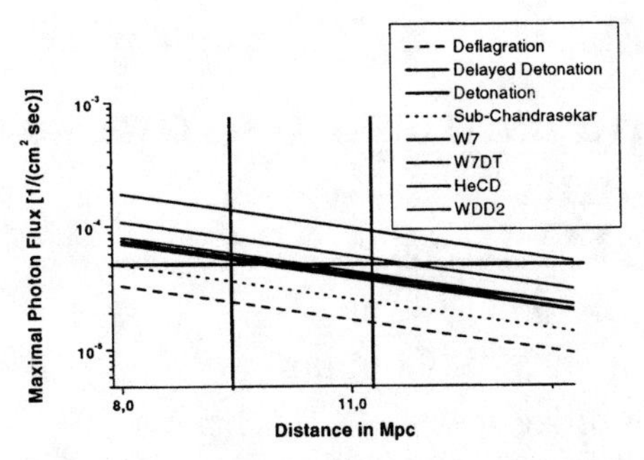

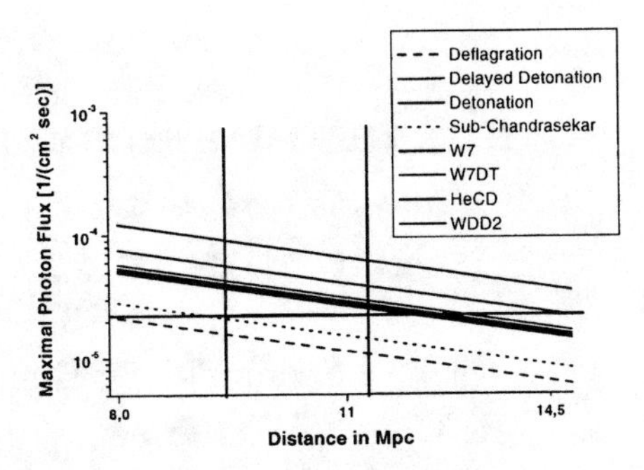

FIGURE 3. The model fluxes for the 847 keV (left) and the 1238 keV line (right) versus distance. The horizontal solid line represents the spectral upper limits. The solid vertical lines show the two different Cepheid distances of 9.9 Mpc and 12.3 Mpc, respectively.

with the 1238 keV measurement only, but be still marginally consistent with the 847 keV flux. The deflagration model and the Sub-Chandrasekar model are consistent with our line measurements for this distance.

In summary we favour the deflagration model and the Sub-Chandrasekar model for SN 1998bu and regard the HeCD model as rather improbable (Otherwise the amount of ^{56}Ni in the outer region of the progenitor CO-white dwarf should be much smaller then predicted by the current model). For detonation models a smaller mixing has to be assumed in order to be compatible with the results of the analysis presented here.

REFERENCES

1. K. Ayani et al. and P. Meikle, IAU Circ. No, **6905**, 1998.
2. Feldmeier, J.J., Ciardullo, R. and Jacoby, G.H., Ap.J. **479**, 231–243, 1997.
3. Fraranda ,C. and Skiff, B.A., IAU Circ. No. **6905**, 1998.
4. Georgii et al., Proc. 2nd INTEGRAL workshop, ESA **SP-382**, 51–54, 1997.
5. Hjorth, J. and Tanvir, N.R., Ap.J. **482**, 68–74, 1997.
6. Höflich, P., Wheeler, J.C. and Khokhlov, A., APJ **492**, 228–245, 1998.
7. priv. communication.
8. priv. communication.
9. Manoz E. et al., astro-ph/9908140, 1999.
10. Meikle, P. and Hernandez, M., astro-ph/990256, 1999.
11. Morris, D.J. et al., Proc. of the 4th Compton Symposium, AIP Conf. proc. **410**, 1084–1088, 1997.
12. Schönfelder et al., Ap.J. Suppl. Ser. **86**, 657–692, 1993.
13. Tanvir, N.R., Shanks, T., Ferguson, H.C. and Robinson, D.R.T., Nature **377**, 27–31, 1995.
14. Villi, M., IAU Circ. No. **6899**, 1998.

^{44}Ti Gamma-Ray Line Emission from Cas A and RXJ0852-4622/GROJ0852-4642

V. Schönfelder*, H. Bloemen†, W. Collmar*, R. Diehl*, W. Hermsen†, J. Knödlseder§, G.G. Lichti*, S. Plüschke*, J. Ryan‡, A. Strong*, C. Winkler‖

**Max-Planck-Institut für extraterrestrische Physik, D-85740 Garching, Germany*
†*SRON-Utrecht, Sorbonnelaan 2, NL-3584 CA Utrecht, The Netherlands*
‡*Space Sience Center, University of New Hampshire, Durham, NH 03824-3525, USA*
‖*Astrophysics Division, ESTEC, NL-2200 AG Noordwijk, The Netherlands*
§*Centre d'Etude Spatiale des Rayonnements (CESR), BP 4336, F-31029 Toulouse Cedex, France*

Abstract. Limitations in COMPTEL ^{44}Ti line searches arise from uncertainties in different background modelling techniques, and in different event selection criteria to suppress a large part of the background. Therefore, the significances of the reported detections of Cas A and RX J0852-4622/GRO J0852-4642 have been reassessed in great detail.

INTRODUCTION

The 1.157 MeV line from the decay-chain of ^{44}Ti $\rightarrow$ ^{44} Sc $\rightarrow$ ^{44}Ca is one of the probes of nucleosynthesis sites in the Galaxy and it is probably the best indicator of young Galactic SNRs. ^{44}Ti is expected to be produced in each of the different types of SNe [1], [2], [3], although with a large variance of abundances per type. The detection of the 1.157 MeV line from young and nearby supernova remnants was one of the prime goals of the Compton mission in the field of gamma-ray line spectroscopy.

COMPTEL [4], indeed, provided the first-ever detection of this line from a supernova remnant, namely from Cas A [5], [6]. This detection confirmed that ^{44}Ti is produced in supernovae, and demonstrated the feasibility of the 1.157 MeV ^{44}Ti line application for uncovering young Galactic supernova remnants.

In a first search for ^{44}Ti line sources along the entire Galactic plane, the COMPTEL team had used the data from the first three years of the mission. No additional source (apart from Cas A) was found [7]. When the survey was repeated with data from the first six years of the mission (from April 1991 to March 1997), a second

CP510, *The Fifth Compton Symposium,* edited by M. L. McConnell and J. M. Ryan

excess of ^{44}Ti line emission appeared in the Vela region at l = 266.5° and b = -1.5° [8]. The source was named GRO J0852-4642. The position of this excess coincides with the position of a previously unknown supernova remnant (RX J0852-4622), which was discovered in ROSAT data above 1.3 keV [9].

After this discovery, the COMPTEL team has performed extensive studies to test the robustness of both ^{44}Ti line detections, if different analysis techniques and/or different event selections are applied. The findings of these studies are subject of this paper.

VERIFICATION TESTS OF THE ^{44}TI LINE EXCESSES

Crucial for the detection of weak celestial sources with signal-to-noise ratios at the one-percent-level are the knowledge of the underlying background and the choice of event-selection parameters.

Two different approaches for background determination below a gamma-ray line source have been developed within the COMPTEL team.

Our first method had been developed and tested mainly for the 1.809 MeV line studies of ^{26}Al [10], [11]. It is based on interpolation of the measurements directly, in adjacent energy bands. This extracts the excess signal above the overall continuum. But it does not distinguish between celestial and instrumental continuum emission, hence will not remove celestial continuum properly as instrumental background dominates [14].

Our second method, derived from our standard method for imaging analysis of continuum point sources, has been significantly advanced recently [13]. The basic concept was already applied in the search for ^{44}Ti sources by Dupraz et al. [7]. In this method, the background is derived by applying a low-pass filter to the data, which smoothes the event distribution and eliminates the celestial signature to first order (e.g. [12]). From results in adjacent energy bands, the line and continuum emission are separated. In contrast to the approach described above, here the celestial emission in narrow adjacent energy bands is now extracted first a.o. by simultaneous fitting of continuum models of celestial emission (see [13] for details). Although some model dependency of this approach cannot be excluded, we consider this method now superior for our ^{44}Ti studies.

The most critical event selection parameters are the following:

- the time of the observation (the instrumental background is different for each viewing period);
- the choice of accepted detector modules (3 D2 detectors (see [4]) of COMPTEL show some degraded performance, because one of the seven photomultipliers is off in these modules);
- the window of accepted events in the time-of-flight forward peak;

- the minimum Earth horizon angle, which determines the contribution of atmospheric gamma rays to the total background;
- the accepted range of the Compton scatter angle $\bar{\varphi}$.

For all these parameters standard values or windows have been determined. We now studied how much are the ^{44}Ti line source detections dependent on these values or windows.

RESULTS OF ROBUSTNESS TESTS

We will see below that the ^{44}Ti signal of Cas A is more robust than that of RX J0852-4622. For the latter, the most significant detection is obtained using the following event selections. These are not optimal for Cas A, but acceptable, as illustrated below:

- data from the three degraded D2 modules were excluded;
- the minimum Earth horizon angle was choosen to be 10°;
- the window of accepted time-of-flight values was chosen to be the standard one;
- data after the second re-boost (after VP 617.1) were excluded from the analysis (after the re-boost the background rate around 1 MeV increased by about a factor of 2).

A set of four likelihood maps, which are based on these event selections, is presented in Fig. 1. For the Cas A region (left) and for the RX J0852-4622 region (right), the two top maps are based on the first of the previously described background methods, in which the background is derived from the interpolation of adjacent energy bands; this method was also used for the reported source discovery [8]. Both sources stand out clearly at about the 6σ level. The two bottom maps were derived recently by using the second method with its recent improvements. Here, both sources are detected at the 4σ level. So, both methods yield similar maps, but the significances are different.

However, if the set of event selection parameters is modified, the picture changes: the significance of the detection of RX J0852-4622 decreases, whereas that of Cas A remains practically stable, as is illustrated in Fig. 2. This figure contains six likelihood maps, which were all derived by the second method with its recent improvements. (The data set is, however, somewhat smaller than that of Fig. 1, only up to VP 522.5.)

Results for three different sets of event selection parameters are compared in Fig. 2. The two bottom maps were derived with the "best" set of parameters that was also used in Fig. 1. In the two middle maps, two selection parameters were changed: the minimum Earth-horizon angle was chosen to be 5° (instead of 10°),

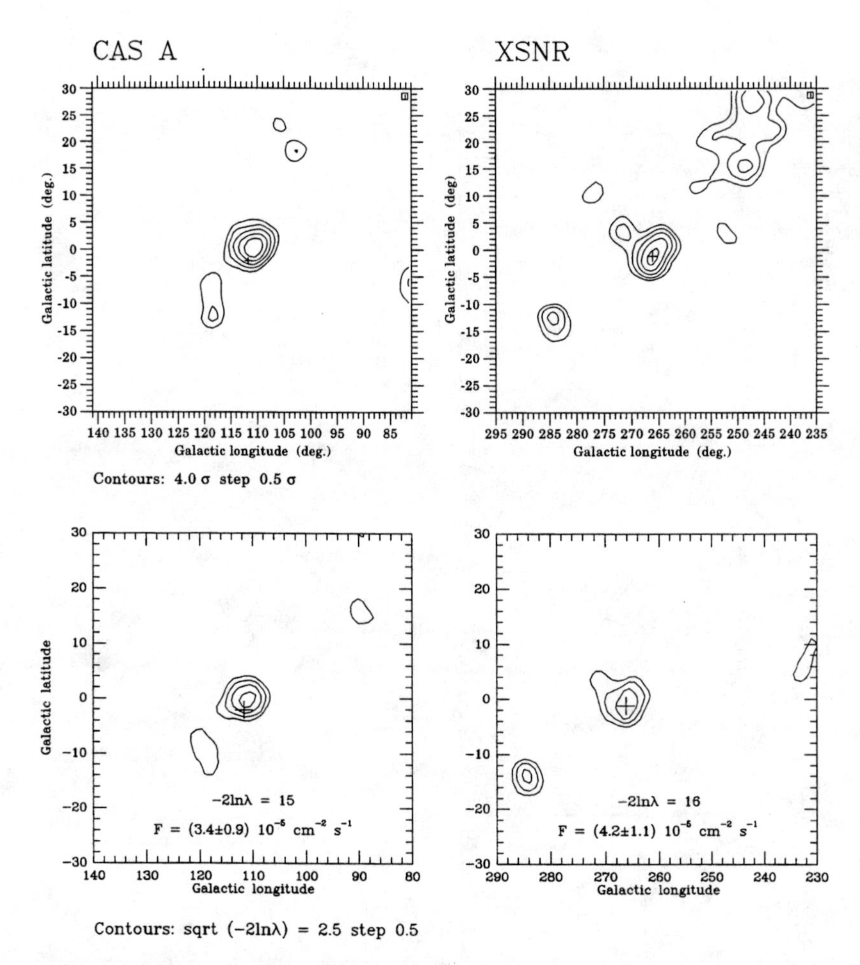

FIGURE 1. Maximum-likelihood maps of the Cas A (left) and RX J0852-4642 region (right) in the ^{44}Ti line interval (1.066 to 1.246 MeV) for the combination of VPs 1.0 to 617.1. Event selections are optimized for the line detection (see text). The two top maps are based on the first background method, and the two bottom maps on the second method with its recent improvements as described in the text.

and the acceptance window of time-of-flight values was somewhat increased. In the two top maps, in addition to the changes in the middle maps, data from the three degraded modules were used as well.

As is evident from Fig. 2, the Cas A detection significance is hardly affected by any of these changes. However, the significance for RX J0852-4622 shows significant variation (from 3.5σ to about 2σ).

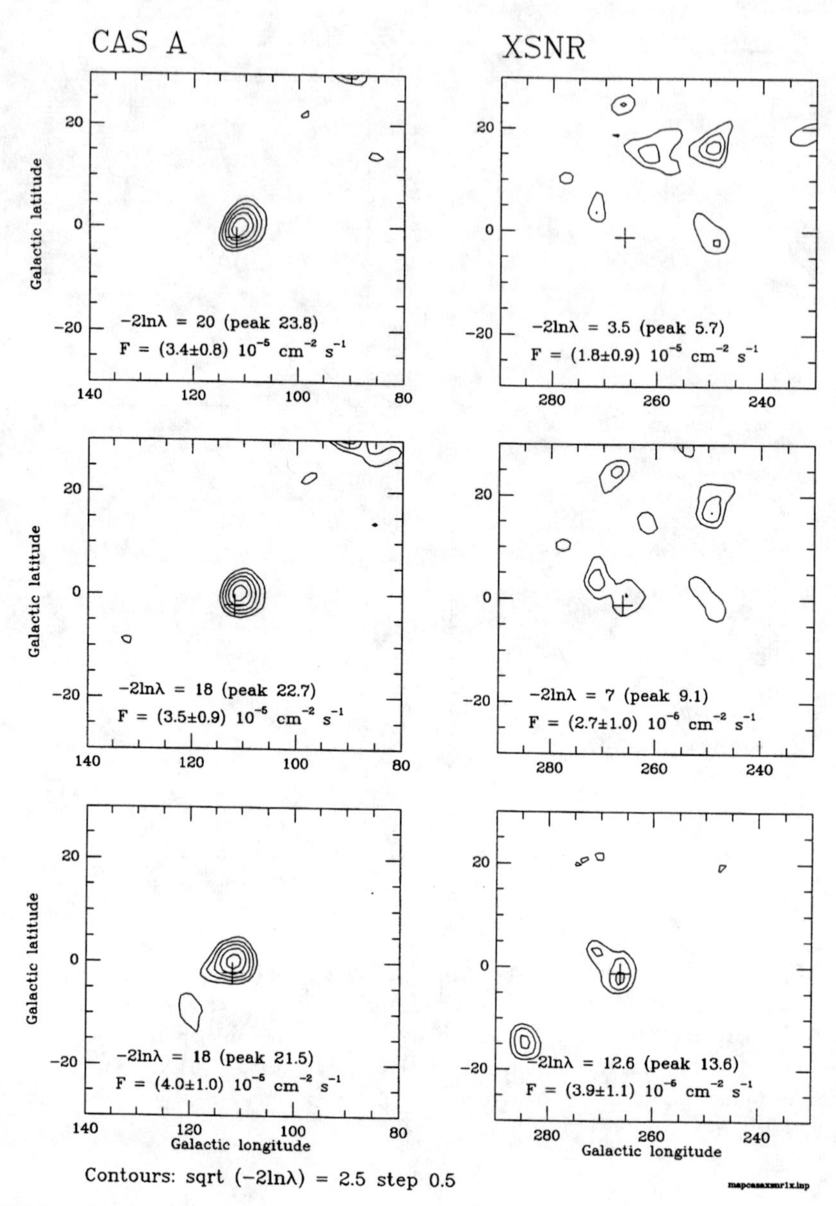

FIGURE 2. Maximum-likelihood maps of the Cas A (left) and RX J0852-4642 region (right) in the ^{44}Ti line interval (1.066 to 1.246 MeV) for the combination of VPs 1.0 to 522.5. All six maps were derived by the second background method with its recent improvements (see text). In the two bottom maps the event selections are identical to those of Fig. 1. The middle and top maps are based on different event selections (see text).

From these tests we conclude:

- the Cas A detection of the 1.157 MeV line is robust against different analysis approaches and data selections;
- promising evidence for 1.157 MeV line emission from RX J0852-4622 remains;
- however, the 1.157 MeV line detection from the RX J0852-4622 is sensitive to the analysis method applied, and to data selections;
- for RX J0852-4622 evidence for a ^{44}Ti signal is found at the 2σ to 4σ significance level, if data selections are modified within the most recently developed analysis method;
- an independent discovery of a ^{44}Ti supernova remnant cannot be claimed, considering the number of trials in an all-sky search and the $(2 \text{ to } 4)\sigma$ result range, derived for a known source from our most recently developed analysis method;
- Cas A and RX J0852-4622 will be important objects to be studied by INTEGRAL, which will not only be able to detect the hard X-ray ^{44}Ti lines, but also the line profile of the 1.157 MeV line.

ACKNOWLEDGEMENT

The COMPTEL project is supported by the German government through DARA grant 50 QV 90968, by NASA under contract NAS5-26645, and by the Netherlands Organization for Scientific Research. AFI acknowledges financial support from the German Bundesministerium für Bildung, Wissenschaft, Forschung und Technologie.

REFERENCES

1. Nomoto, K., Thielemann, F.-K., and Yokoi, K., *ApJ* **286**, 664 (1984).
2. Woosley, S.E. and Weaver, T.A., *ApJS* **101**, 208 (1995).
3. Thielemann, F.-K., Nomoto, K. and Hashimoto, M.A., *ApJ* **460**, 408 (1996).
4. Schönfelder, V. et al., *ApJ* **86**, 657 (1993).
5. Iyudin, A.F., et al., *A&A* **284**, L1 (1994).
6. Iyudin, A.F., et al., *ESA SP-382, Proc. 2nd INTEGRAL Workshop*, p. 37 (1997).
7. Dupraz, C., et al., *A&A* **324**, 683 (1997).
8. Iyudin, A.F., et al., *Nature* **396**, 142 (1998).
9. Aschenbach, B., *Nature* **396**, 141 (1998).
10. Diehl, R. et al., *ApJS* **92**, 429 (1994).
11. Knödlseder, J., *PhD Thesis*, Toulouse Univ. (1997).
12. Bloemen, H. et al., *ApJSS* **92**, 419 (1994).
13. Bloemen, H., et al., *this conference*, (1999).
14. Bloemen, H., et al., *Proc. of 3rd INTEGRAL Workshop*, in press (1999).

Study of MeV continuum from the Cas A SNR with COMPTEL

A. W. Strong*, H. Bloemen†, W. Collmar*, R. Diehl*, W. Hermsen†, A. Iyudin*, V. Schönfelder*, L.-S. The‡

*Max-Planck Institut für extraterrestrische Physik, 85740 Garching, Germany
†SRON-Utrecht, Utrecht, The Netherlands
‡Department of Physics & Astronomy, Clemson University, Clemson, SC 29634-1911, U.S.A.

Abstract. Energetic particles accelerated in supernova remnants provide a variety of continuum emission signatures over the entire electromagnetic spectrum. The measured hard X-ray continuum emission from the Cas A supernova remnant indicates in fact the presence of high-energy electrons and makes it an interesting candidate for study at MeV energies.

We have analysed COMPTEL data with respect to continuum emission in addition to the previously detected ^{44}Ti line. The results provide useful constraints on models of electron shock acceleration in this object.

I INTRODUCTION

Supernova remnants are likely sources of cosmic rays. Energetic particles accelerated in these objects should provide a variety of continuum emission signatures over the entire electromagnetic spectrum. Models predict that electron synchrotron emission may be detected from radio to hard X-ray energies, while at higher energies, in the MeV regime and above, the high-energy electrons are expected to be visible through bremsstrahlung.

The Cas A SNR is the brightest radio source on the sky. It exhibits a power-law spectrum of continuum in X-rays [1–3] suggesting the presence of high-energy (10^{14} eV) electrons emitting synchrotron radiation. OSSE has measured the spectrum in hard X-rays [4], with the most recent results reported at this conference [5]. EGRET has not detected Cas A, upper limits being given in [6]. A detection at TeV energies by the HEGRA instrument has been reported [7].

An extensive updated study of the original Cas A ^{44}Ti line result [8] is given in [9].

The hard X-ray continuum emission from the Cas A supernova remnant indicates in fact the presence of high-energy electrons and makes it an interesting candidate for study at MeV energies; both a detection or upper limits would be of interest in

CP510, *The Fifth Compton Symposium,* edited by M. L. McConnell and J. M. Ryan

constraining models for the emission mechanisms. Detailed models of the emission from this SNR [10] predict fluxes near to COMPTEL sensitivity levels and give the motivation for a study with COMPTEL data, since these can significantly constrain the models. Specifically, a transition of the emission mechanism from synchrotron radiation to bremsstrahlung would be expected to appear as a flattening of the spectral slope towards the MeV region.

II METHOD

The standard COMPTEL source-fitting program SRCFIX was used to fit the data taken from Cycles 1-6. A model for the diffuse Galactic emission was included by full-sky fitting, using total hydrogen column densities (based on HI and CO maps) and an inverse-Compton model. The standard COMPTEL event selections were applied. Fits were made in the broad 'continuum only' energy ranges, 1.3–3.0, 3-10 and 10-30 MeV ranges. The first range excludes the ^{44}Ti line at 1.16 MeV, taking into account the energy resolution of COMPTEL.

In view of the increase of background after the second CGRO reboost, only pre-reboost data (Observation periods 1-617.1) were used here.

III RESULTS

Table 1 summarizes the results of this analysis. Fig 1 shows the contours of log-likelihood ratio (-2ln λ) for 1.3–3.0 MeV. The signature of Cas A remains at the 2σ level after exclusion of the ^{44}Ti line, suggesting a signal in the continuum. However the 1.3–3.0 MeV map shows other features which are actually stronger than that at Cas A, so that an unequivocal detection cannot be claimed based on this analysis. Also a contribution from diffuse line emission, such as 1.8 MeV ^{26}Al, or diffuse continuum emission, cannot be excluded at this stage. For 3–10 and 10-30 MeV only upper limits were obtained.

The derived photon flux spectrum of Cas A is shown in Fig 2, together with RXTE, OSSE and EGRET results.

TABLE 1. Cas A flux fitting results. All fluxes are continuum only.

Energy (MeV)	σ	flux (10^{-5}cm^{-2} s^{-1})
1.3–3.0	2.1	3.0±1.7
3.0–10.0	2σ u.l	<1.4
10.0–30.0	2σ u.l	<0.63

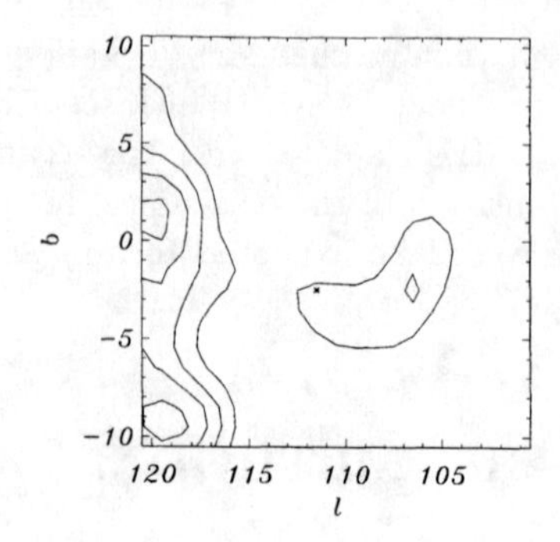

FIGURE 1. Cas A log-likelihood ratio contours (-2ln λ) in the energy range 1.3-3.0 MeV for Observations 1-617.1 . Contour levels: 2.0, 2.5, 3.0, 3.5σ. The asterisk marks the position of Cas A.

IV CONCLUSIONS

The continuum emission from Cas A in the 1.3–3 MeV range appears only at the 2σ level. However there are other features with higher likelihood in the 1.3 - 3 MeV map, which may indicate a poor understanding of the diffuse emission in this region, so the continuum flux should be conservatively regarded as an upper limit. The COMPTEL results are consistent with the OSSE upper limits above 1 MeV given in [5].

Energetic particle shock acceleration models for Cas A [10] are consistent with our results, and suggest that continuum emission, if present, would be bremsstrahlung radiation. Even as upper limits the present results give useful constraints on the parameters of such models.

Even though the continuum level is uncertain, this very uncertainty must reflect on that of the line flux and should always be included explicitly in line studies.

These preliminary results indicate the need for more detailed investigation of the continuum emission of Cas A with COMPTEL data.

REFERENCES

1. Allen, G.E., et al., *ApJ* **487**, L97 (1997)
2. Allen, G.E. *Proc. 26th ICRC* **3**, 480 (1999)
3. Hughes, J.P., et al. *ApJ*, in press, astro-ph/9910474 (1999)
4. The, L.-S.,et al. *A&AS* **120**, 357 (1996)
5. The, L.-S., et al. *this conference*
6. Esposito, J.A., et al. *ApJ* **461**, 820 (1996)
7. Pühlhofer, G. et al. *Proc. 26th ICRC*, **3**, 492 (1999)
8. Iyudin, A.F., et al. *A&A*, **284**, L1 (1994)
9. Schönfelder, V., et al. *this conference*
10. Ellison, D. C., et al. *Proc. 26th ICRC* **3**, 468 (1999) and *this conference*

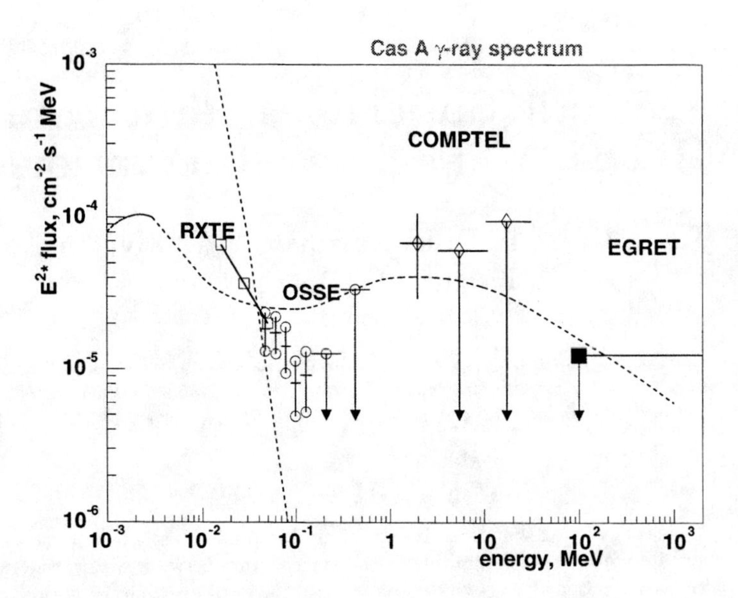

FIGURE 2. Cas A continuum spectrum. Diamonds: COMPTEL points from Obs 1-617.1. Other data: open squares: RXTE (power law fit) [1]; open circles: OSSE [5] ; filled square: EGRET [6]. The dashed curves show the synchrotron and bremsstrahlung emission components in a model for Cas A [10].

The Galactic Supernova Rate from COMPTEL ^{44}Ti γ-line Observations

L.-S The*, R. Diehl‡, D. H. Hartmann*, A.F. Iyudin‡, M. D. Leising*, U. Oberlack$

*Department of Physics and Astronomy, Clemson University, Clemson, SC 29634-0978, U.S.A.
‡Max-Planck Institut für extraterrestrische Physik, Postfach 1603, D-85740 Garching, Germany
$Physics Department, Columbia University, 538 W. 120th St., New York 10027, U.S.A.

Abstract. ^{44}Ti is one of the few radioactive nuclei produced abundantly in supernovae (SNe). Because its life time is comparable to the typical time between supernovae in the Galaxy, γ-ray lines from ^{44}Ti provide a powerful tool for the search for optically obscured supernovae that occured during the last few 100 years. We present our analysis of COMPTEL data (Cycle 1-6), comparing γ-ray line images and flux distributions with Monte Carlo simulations of SNe in the Galaxy. Assuming standard SNe distribution models and theoretical estimates of the ^{44}Ti yields, we constrain the Galactic supernova rate. We compare these constraints with those obtained from the historical record covering the last millenium. Relating the current supernova rate to the observed ^{44}Ti solar abundance through chemical evolution models, agreement can be obtained only if there is a special type of supernova (e.g., helium detonations) that produces large amounts of ^{44}Ti, but with a very low event rate.

INTRODUCTION

The Galactic supernova rate is of great importance for the study of Galactic chemical evolution and other astrophysical questions, so that several methods to estimate this rate have been developed. The SN rate has been estimated from the historical supernova record [4], from SN observations in external galaxies [3], from the statistics of radio supernova remnants, and from the pulsar birth rate. Here, we estimate the Galactic supernova rate using ^{44}Ti gamma-ray measurements.

An estimate of the Galactic supernova rate from ^{44}Ti γ-ray observation was first carried out by Mahoney et al. (1992) [12] using HEAO 3 data with 1σ sensitivity of 8.3×10^{-5} γ cm^{-2} s^{-1}, followed by Leising & Share (1994) [11] using SMM data with 1σ sensitivity of 3.2×10^{-5} γ cm^{-2} s^{-1}. Leising & Share (1994) [11] concluded that a 10^{-4} $M_\odot$ ^{44}Ti production with a rate of $\geq$1 century^{-1} is ruled out at 90% confidence. Recently, COMPTEL's ^{44}Ti γ-line map of the first 6 years of observations has been produced with a 1σ sensitivity of $\simeq0.8\times10^{-5}$ γ cm^{-2} s^{-1} [10],

CP510, *The Fifth Compton Symposium*, edited by M. L. McConnell and J. M. Ryan

~4 times better than that of SMM. Thus COMPTEL can probe the decay of ^{44}Ti farther back in time by $\Delta t \simeq \tau \ln(4)=122$ yr, or two times deeper in distance.

The COMPTEL data set used in this work is the three dimensional data from which the maximum-entropy all-sky map of the Galactic ^{44}Ti presented by Iyudin et al. (1998) [10] was derived. We restrict our analysis to the inner galactic map ($l \leq 90°$, $b \leq 30°$) to reduce possible systematic effects from regions of low exposure.

MONTE CARLO ANALYSIS

For the comparison of the ^{44}Ti γ-ray map with the six historic supernovae of the last millenium, we generate supernova events with Monte Carlo techniques. Supernovae are generated in space according to a distribution described below, and in time with ages distributed uniformly between zero and 1000 y. Hartmann et al. (1993) [6] showed that combining the record of six Galactic supernova in the last millenium and the HEAO 3 upper limits [12] on the ^{44}Ti γ-ray line flux yields better constraints than applying γ-ray line flux constraints alone. In this analysis, we extend the method and improve the constraints with better data.

Supernova events as the source of Galactic ^{44}Ti can either be Type Ia, Ib/c, or II SNe. For this study, we choose a generally accepted (though uncertain) value of the type ratio, Ia : Ib/c : II = 0.1 : 0.15 : 0.75 [4]. Monte Carlo representations of Type Ia SNe are generated using a nova distribution template that traces the blue light distribution in M31 [9]. These populations form an axisymmetric disk and a spherically symmetric bulge. The fraction of SNIa occurring in the spheroid is taken to be ~1/7 of the total SNIa [1].

Supernovae of Type Ib & II are associated with massive stars whose birth places are exponentially distributed in height above the plane with a scale length of ~100 pc. Because the rate of core collapses is larger than the rate of thermonuclear supernovae, we study several Type II distributions to ensure that our results do not depend significantly on this choice. We consider three cases: (1) Model A: Exponential disk with no supernova within 3 kpc of the Galactic center, with radial scale length of 5 kpc [7]. (2) Model B: Exponential disk with radial scale length of 3.5 kpc that produces an acceptable fit to the COMPTEL's ^{26}Al γ-line map [5]. (3) Model C: Gaussian-ring disk at radial distance of 3.7 kpc and radial distance scale length of 1.27 kpc [16].

The peak absolute magnitudes of supernovae in the B-band are roughly distributed as Gaussians. For the mean values and the one standard deviation we adopt for type Ia M_B = -19.4, σ=0.2 [2], for type Ib M_B = -18.2, σ=0.3 [4], and for type II M_B = -17.2, σ=1.2 [17]. Observed magnitudes of simulated SNe are obtained by convolving absolute magnitudes with a Galactic extinction model (we employ the extinction code of Hakkila et al. (1997) [8])

In our models the ^{44}Ti yields of Type Ia SNe are uniformly distributed between 8.7×10^{-6} $M_\odot$ and 2.7×10^{-6} $M_\odot$ so this range covers the ^{44}Ti yield in the deflagration W7 model, the delayed detonation WDD2 model, and the late detonation

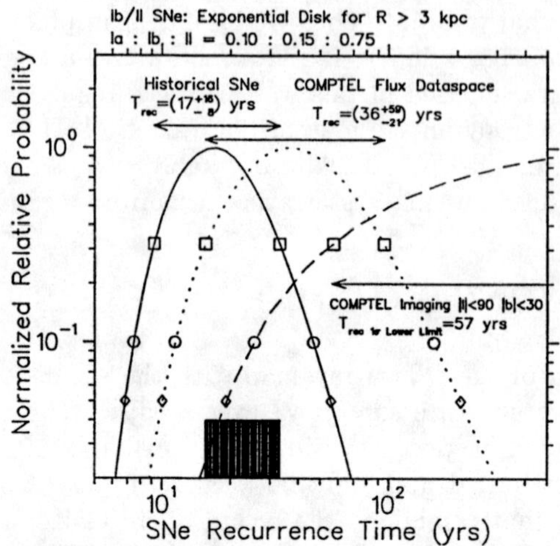

FIGURE 1. *Normalized relative probability of the SN rate, calculated with three different methods, vs. recurrence time for model* **A**. *The solid line is for the model in which the number of SNe with $m_V \leq 0$ is 6 in 1000 years. The dashed-line is inferred by maximum likelihood analysis of COMPTEL's 1.157 MeV γ-line inner Galactic ($l \leq 90°$, $b \leq 30°$) image map. The dotted-line uses the chi-square test that the 1.157 MeV γ-line fluxes in the model are consistent with the observed fluxes. The labels of the curves show the optimal, 1σ ranges, or the 1σ lower limits of the SNe recurrence time. The boxes, circles, and diamonds on the curves show the 1σ, 90%, and 95% relative probabilities. The shaded area shows the range of recurrence time consistent with better than 1σ probability with both the historical record and the COMPTEL flux dataspace.*

W7DT model of Nomoto et al. (1997). The ^{44}Ti yields of Type Ib SNe are uniformly distributed between 3×10^{-5} $M_\odot$ and 9×10^{-5} $M_\odot$ with typical values of 6×10^{-5} $M_\odot$ [15]. Type II supernova yields are generated according to stellar mass, which is drawn from a Salpeter IMF for $M \geq 8$ $M_\odot$, and a ^{44}Ti yield taken from Table 1 of Timmes et al. (1996) [15].

Map/Flux Dataspace and Historical Record Analysis

In map (imaging) analysis, a simulated COMPTEL data set is produced by convolving the directions and the ^{44}Ti γ-line fluxes of the supernovae through the COMPTEL detector response. The probability of the consistency of the simulated data with the measured counts is carried out by calculating the likelihood of the simulated data plus the best-fit background model, and comparing this to the likelihood of the background model only. The change in the likelihood gives the relative probability that the specific realization of the model is consistent with the measured COMPTEL data. This process is repeated for at least 10^4 galaxies to obtain the average relative probability of the model at a particular supernova rate as shown in Fig. 1 as a long-dashed line.

In flux dataspace analysis, in order to include the detected sources, the locations of the ^{44}Ti supernovae are not used, flux information is used exclusively. The flux dataspace consists of the two ^{44}Ti detected fluxes from Cas A SNR and GRO J0852-4642 [10] with other fluxes assumed to be zero (with which they are consistent.) For each Monte Carlo Galaxy, a χ^2 value is calculated by comparing the two strongest fluxes with the fluxes measured from Cas A and GRO J0852-4642 and the other fluxes are compared with null fluxes. The probability of the model to be consistent with the COMPTEL fluxes is determined from the χ^2 and using the number of degrees of freedom as the number of independent $10° \times 10°$ image elements (fields of view) of the supernova positions in the model. This size of independent fields of view is obtained by calibrating the size to produce the results of the analysis of COMPTEL's map of the inner Galaxy. Averaging the probabilities of 10^6 Monte Carlo Galaxies, we obtain the average probability shown in Fig. 1 as a dotted line.

The historical supernova record covering the last millenium shows a total of only six Galactic SNe during that era. Of course, this small number is due to significant losses from extinction and incomplete monitoring of the sky, especially during the early centuries. In fitting this data, we count the fraction of galaxies that have six SNe brighter than magnitude 0. In this approach, we assume that historical SNe were detected if they were brighter than magnitude 0. The average fraction of the model that is consistent with six events brighter than magnitude 0 is shown in Fig. 1 as a solid line.

DISCUSSION & CONCLUSIONS

Our analysis leads to a most probable Galactic supernova recurrence time of ~17, ~16, and ~13 y based on models A, B, and C, respectively. This implies a larger rate than given by previous investigations based on the historical record. For example, Dawson & Johnson (1994) [4] estimated ~3 SNe/century as also obtained by Tammann, Löffler, & Schröder (1994) [14]. The rate is smaller than ours because Dawson & Johnson (1994) [4] considered 7 observed SNe within the last 2000 yrs and assumed that the historical record is 80% complete. A higher SNe rate of 5 SNe/century was obtained by Hatano, Fisher, & Branch (1997) [7] who include a population of "ultradim" SNe in addition to 4 observed Galactic SNe having $V < 0$ and 80% completeness within the last millenium. The two methods (map/flux) of estimating supernova rates based on γ-ray data appear to be significantly different, but they are statistically consistent with each other for a wide range of supernova rates. The imaging analysis gives a smaller rate than the flux analysis, because the data used in the map analysis do not include the Cas A and GRO J0852-4642 γ-line detections. However, the probability estimate from the flux analysis is a somewhat coarser estimate because the size of independent FOVs used in determining the probability is not known exactly. Ignoring the spatial information begs the question as to why the two brightest SN are in the outer Galaxy rather than in the inner Galaxy where we expected them.

Our investigations conclude that:

- Map analysis requires a recurrence time larger than 13 y (99% confidence).
- In the flux dataspace analysis, where we ignore the expected spatial distribution of ^{44}Ti remnants and consider only the measured flux distribution, we find a most probable supernova rate that is more compatible with standard values (such as the rate inferred by the historical record), even for standard ^{44}Ti yields. However, the COMPTEL data indicate a lower SN rate (~1 SN/36 yr) than that suggested by the historical record (~1 SN/17 yr).
- Based on chemical evolution studies, Timmes et al. (1996) [15] estimated that only ~1/3 of the solar ^{44}Ca abundance is accounted for. Models with a SN rate of ~3 SNe/century and standard ^{44}Ti yields fail to produce the solar ^{44}Ca abundance. This rate, when confronted with the gamma-ray data (dotted line in Fig. 1) is too large: the COMPTEL gamma-ray data worsen an already serious problem. Timmes et al. (1996) [15] suggest 3 possibilities: 1) Increase the ^{44}Ti yields by a factor of ~3. 2) Increase the supernova rate by a factor of ~3. 3) There is another source of ^{44}Ca in the Galaxy. Our analysis in Fig. 1 shows that the first and second option are not compatible with COMPTEL's ^{44}Ti γ-line map, which would be brighter by a factor of 3 or exhibit a larger number of ^{44}Ti hot spots than actually observed. We are thus left with the third option, to seriously entertain the idea that there exists some rare type of supernova (i.e., detonation of helium white dwarf), not realized in recent centuries, that produces very large amounts of ^{44}Ti.

REFERENCES

1. Bahcall, J. N., & Soneira, R. M. 1980, ApJS, 44, 73
2. Branch, D. 1998, ARAA, 36, 1
3. Cappellaro, E. et. al. 1997, A&A 322, 431
4. Dawson, P.C. & Johnson, R.G. 1994, J. Roy. Astron. Soc. Can., 88, 369
5. Diehl R. et al. 1995, A&A, 298, 445
6. Hartmann, D. et al. 1993, AAS 97, 219
7. Hatano, K., Fisher, A., & Branch, D. 1997, MNRAS, 290, 360
8. Hakkila, J., Myers, J. M., Stidham, B. J., & Hartmann, D. H. 1997, AJ 114, 2043
9. Higdon, J. C., & Fowler, W. A. 1987, ApJ 317. 710
10. Iyudin, A. et al. 1998, 3rd Integral Workshop, Taormina, Italy.
11. Leising, M. D., & Share, G. H. 1994, ApJ, 424, 200
12. Mahoney, W. A., Ling, J. C., Wheaton, W. A., & Higdon, J. C. 1992, ApJ, 387, 314
13. Nomoto, et al. 1997, Nuclear Physics A621, 467.
14. Tammann, G. A., Lőffler, W., & Schrőder, A. 1994, ApJS, 92, 487
15. Timmes, F. X. et al. 1996, ApJ, 464, 332
16. Taylor, J. H., & Cordes, J. M. 1992, ApJ, 411, 674
17. van den Bergh, S. & Tammann, G.A. 1991, ARA&A 29, 363

A multiple γ-ray source associated with a new supernova remnant

Jorge A. Combi*, Gustavo E. Romero*, Paula Benaglia*, and Justin L. Jonas†

*Instituto Argentino de Radioastronomía, C.C. 5, 1894 Villa Elisa, Argentina
†Department of Physics & Electronics, Rhodes University, Grahamstown 6140, South Africa

Abstract. Only a few of the 170 unidentified EGRET sources cluster within small regions of less than 6° in the sky. Here we present observational evidence supporting the association of such a group of three neighboring γ-ray sources (3EG J1834-2803, 3EG J1847-3219, and 3EG J1850-2652) with a new supernova remnant discovered at radio wavelengths. Two of the sources are coincident with HI clouds, detected through 21-cm line observations, which seem to be overtaken by the expanding shell of the remnant. The discovery of this multiple γ-ray source has profound implications for the origin of the galactic cosmic rays.

I INTRODUCTION

The origin of cosmic rays (CRs) is a long-standing problem in high-energy astrophysics. It has been thought since the 1960s that protons are accelerated up to energies of about 10^{15} eV in the expanding shocks of supernova remnants (SNRs) [9]. However, conclusive observational evidence for the presence of ultra-relativistic hadrons in SNRs has been elusive up till now.

The interaction of shell-type SNRs with molecular or atomic clouds provides a suitable scenario where the existence of a CR nuclear component locally accelerated in the remnants can be revealed by high-energy γ-ray observations [2,3] . The CRs should illuminate the cloud through the decay of neutral pions generated in $p-p$ collisions. The resulting γ-rays, in the EGRET energy range, should provide information on the local enhancement of the CR density, as well as on the energy spectrum of the particles.

Here we report the discovery of a previously undetected SNR in the southern sky (Capricornus region). It is a nonthermal, shell-type source visible at both 408 MHz and 2.3 GHz once the background diffuse emission is removed. Three γ-ray sources are superposed on the remnant, two of them coincident with HI-clouds which were detected through new 21-cm line observations toward the EGRET sources.

CP510, *The Fifth Compton Symposium,* edited by M. L. McConnell and J. M. Ryan

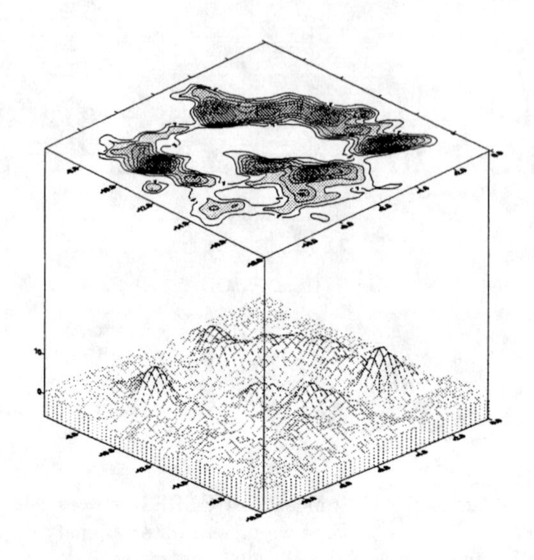

FIGURE 1. Map of the SNR at 408 MHz after background subtraction. Notice the shell-type morphology.

II OBSERVATIONS AND DATA ANALYSIS

We have studied the radio sky toward the best estimate positions of the southern unidentified EGRET sources 3EG J1834-2803, 3EG J1847-3219, and 3EG J1850-2652 using 408- and 2326-MHz continuum data from the surveys by Haslam et al. [10] and Jonas et al. [11]. The diffuse background radiation was eliminated with a well-proven technique [7], which basically consists of an iterative process that operates by successive convolutions and differences on the original map and yields an efficient subtraction of diffuse structure on size scales larger than a fixed filtering beam (see Ref. [6] for details). This procedure accentuates the fine structure of weak and extended sources like SNRs that otherwise would be unobservable.

In addition, we have made new hydrogen line observations toward the EGRET sources with the IAR 30-m single dish telescope at Villa Elisa, Argentina [5]. With the obtained data we made a series of HI brightness temperature maps with a velocity resolution of about 1 km s^{-1} which were examined looking for mass concentrations near the positions of the γ-ray sources.

III RESULTS

The new radio SNR is centered at $(l,\ b) \approx (+6.5^o, -12.0^o)$. It presents a limb brightened shell of size $\sim 8^o \times 8^o$, with an integrated flux of $\sim 180 \pm 20$ Jy at 408

MHz. The average spectral index over the source is $\alpha \sim -0.68 \pm 0.15$, as expected for a typical shell-type SNR.

The distance to the remnant is constrained by the relatively high galactic latitude and the HI observations (according to widely used galactic rotation models) to be $\lesssim 500$ pc. A modified $\Sigma - D$ relation [1] yields a distance estimate of ~ 470 pc. At such a distance the SNR radius is ~ 30 pc, i.e. it should be at the end of its adiabatic phase. The age, assuming a density of $n = 0.05$ cm^{-3} for the intercloud medium, would be of $\sim 16,000$ years.

The estimates for the masses of the HI clouds from the integrated column density map yield values of 1,200 $M_\odot$ (cloud at $l,\ b \approx 5.5^o, -8^0$) and 1,400 $M_\odot$ (cloud at $l,\ b \approx 10^o, -12^o$). More mass could exist in the form of other molecular species.

IV DISCUSSION

Gamma rays are produced in SNRs by electrons in bremsstrahlung and inverse Compton interactions, and by protons through the decay of neutral pions created in proton-nucleus collisions. When a high-density medium is present near the remnant, the latter process can be dominant at energies of hundreds of MeV [3]. The γ-ray flux expected from hadronic interactions in a cloud of mass M_c is $F(E > 100 \text{ MeV}) = q_\gamma M_c / 4\pi\ d^2 m_H$, where q_γ is the γ-ray emissivity in the source, d is its distance, and m_H is the hydrogen atom mass. Using the EGRET flux densities, our estimates for cloud masses, and a distance of 470 pc for the SNR, we determine that there is an enhancement in the CR energy density in the clouds respect to the value observed near the Earth. An enhancement factor of ~ 4.5 is obtained for the larger cloud and of ~ 13 for the smaller one, clearly indicating that protons are accelerated at the remnant shock front. The proton spectrum is $N(E_p) \propto E^{2.6 \pm 0.2}$ at the position of 3EG J1834-2803, and is similar to the electron spectrum determined from the radio data. The expected high-energy cutoff for pro-

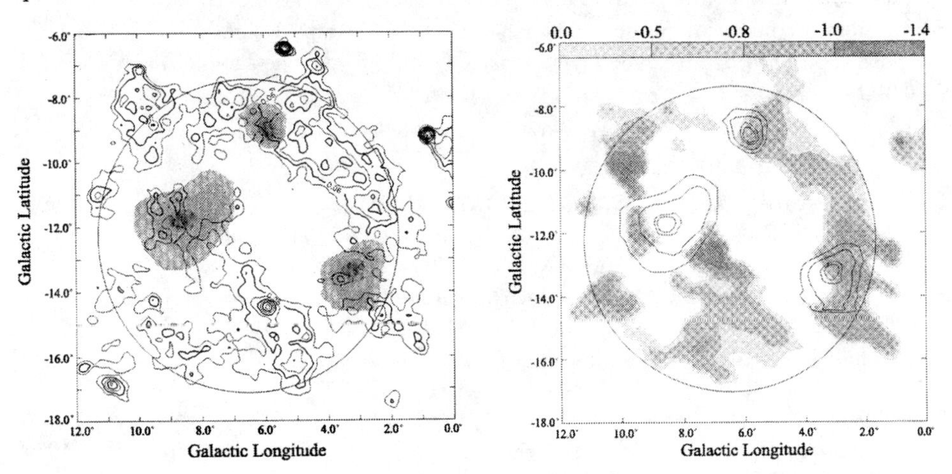

FIGURE 2. (Left) The SNR at 2.3 GHz with the EGRET sources superposed. (Right) Spectral index map.

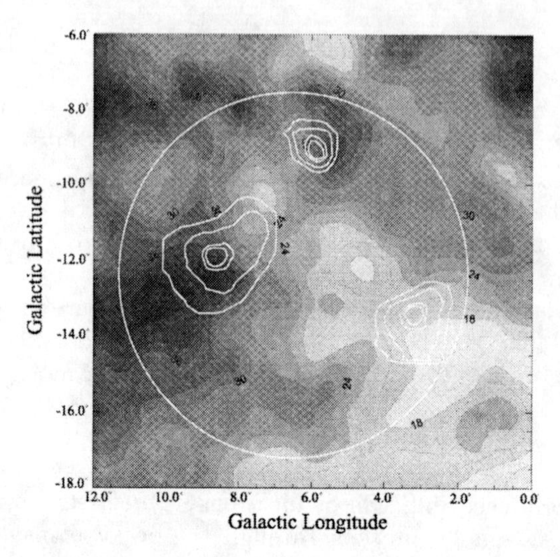

FIGURE 3. Integrated column density HI map. Notice the presence of the clouds near the EGRET sources (superposed in white).

tons is $\sim$ 80 TeV [4]. These observational results are consistent with some recent calculations by Gaisser et al. [8] if the electron-to-proton ratio in the remnant is less than unity.

Future X-ray and TeV observations of this new SNR in the Capricornus region could provide valuable information which would enable the determination of the upper cutoffs in the spectrum of both leptonic and hadronic CR components within a galactic accelerator.

Acknowledgments. J.A.C. and G.E.R. are grateful to the organizers of 5th Compton Symposium for financial support that make possible their participation. This research has been supported by Fundación Antorchas, CONICET, and the agency ANPCT (PICT 03-04881).

REFERENCES

1. Allakhverdiyev A.D., et al., Ap&SS **121**, 21 (1986)
2. Aharonian F.A., Drury L. O'C., Völk H.J., A&A **285**, 645 (1994)
3. Aharonian F.A., Atoyan A.M., A&A **309**, 917 (1996)
4. Biermann P.L., Strom R.G., A&A **275**, 659 (1993)
5. Combi J.A., Romero G.E., Arnal E.M., A&A **333**, 298 (1998)
6. Combi J.A., Romero G.E., Benaglia P., A&A **333**, L91 (1998)
7. Combi J.A., Romero G.E., Benaglia P., ApJ **519**, L177 (1999)
8. Gaisser T.K., Protheroe R.J., Stanev T., ApJ **498**, 219 (1998)
9. Ginzburg V.L., Syrovatskii, The Origin of Cosmic Rays, Pergamon Press, NY (1964)
10. Haslam C.G.T., et al., A&A **100**, 209 (1981)
11. Jonas J.L., Baart E.E., Nicolson G.D., MNRAS **297**, 977 (1998)

Hard X-ray emission from IC 443: the BeppoSAX view

A. Preite-Martinez*, M. Feroci*, R. G. Strom† and T. Mineo‡

*CNR-IAS, Rome, Italy
†NFRA, Dwingeloo, NL
‡CNR-IFCAI, Palermo, Italy

Abstract. Our BeppoSAX observations of IC 443 were primarily directed toward isolating and studying its previously known hard component, thereby to test the hypothesis that this is direct evidence for very energetic ($\simeq$ 100 TeV) cosmic rays. Although our analysis is at an early stage, it was immediately obvious that we have, with the BeppoSAX/PDS, detected a very hard (from 14 to > 30 keV) component which is almost certainly the counterpart of the EGRET γ-ray source 2EG J0618+2234. This object is offset from the prominent soft thermal emission in IC 443. In the same field we found with the BeppoSAX/MECS two hard (5 – 10 keV) hot spots. Both the hot spots and the 2EG source appear to be evidence for energetic cosmic rays produced where the SNR shock is interacting with a dense molecular cloud.

INTRODUCTION

It is generally agreed that cosmic rays (CRs) are accelerated in interstellar shocks by the first-order Fermi process, and there is direct observational evidence that such acceleration occurs in the shock waves of supernova remnants (SNRs). The discovery of a hard, nonthermal component to the X-ray emission from the young remnant associated with the supernova of AD 1006 [8] has been taken as direct evidence for the acceleration of cosmic rays to energies of 100 TeV or more by supernova remnant shocks [9] . The power-law spectrum found dominates the X-rays from, and is morphologically correlated with, the bright radio rims. It has been interpreted as synchrotron emission, although as many as three other mechanisms are potentially responsible.

In their search for EGRET source – SNR associations, Esposito et al. (1996) chose all catalogued radio-strong SNRs ($S_{1\,\mathrm{GHz}}$ > 100 Jy), excluding the Crab and Vela (because of their pulsars) and Sgr A (East) (because of confusion in the galactic center), some 14 objects in all. They found that one-third of the most radio-intense SNRs coincide with EGRET gamma-ray sources. The 5 SNRs with candidate counterparts show some degree of interaction with molecular gas, while several of the others (like Cas A) do not. Moreover, as noted by Claussen et al.

CP510, *The Fifth Compton Symposium,* edited by M. L. McConnell and J. M. Ryan

(1997), those detected are among the least distant in the sample. In particular, they noted that the γ-ray source 2EG J0618+2234 is coincident with IC 443. The EGRET p.s.f. is so large that it cannot be said whether the emission is diffuse or compact, but its position is known to $\pm 12'$ arc, and it does not appear to be centered upon either the brightest soft X-ray component, or the strong nonthermal NE radio radio rim.

We have observed with BeppoSAX one of these objects, IC 443, which is also known to have a hard X-ray component [11]. We detect the power-law spectrum of the EGRET (30 MeV - 6 GeV) source in the hard X-rays, down to nearly 10 keV. We also show that a previously-observed compact feature, and the additional one observed by BeppoSAX, both have a high-energy power-law component which may extend above 10 keV. We argue that the high-energy components in mature remnants are produced when their shocks encounter high-density clouds, that the process is localized and may be transient, and that the emission is not particularly well correlated with the radio synchrotron.

OBSERVATIONS AND DATA ANALYSIS

Our observations were made with the BeppoSAX 0.1 - 300 keV Observatory [1] in October 1997 and 1998, and April 1999. Useful data were obtained with the narrow field imaging and collimated instruments. Imaging and spectral analyses have been carried out using standard software and procedures.

SAX observed four IC 443 fields covering the entire remnant. The PDS detected a hard (14 - > 30 keV) component in all pointings. The MECS (1.5 - 10 keV) detected two high-energy X-ray features on the South (HXR1) and SE (HXR2) boundaries of the remnant, the stronger (HXR1) already discovered by ASCA [7]. In addition, a tenuous bridge connecting the two sources is detected above 5keV.

The HXR1 source at RA = 06 17 05, Dec = +22 22 19 (epoch J2000) has a spectrum consisting of soft ($\simeq 0.9$ keV) and hard (power-law) components. Unlike the ASCA spectrum, we get a significantly poorer fit with a hot thermal component than a power law, in combination with soft thermal emission. The temperature of the soft component, 0.9 keV, is the same as that of the surrounding diffuse emission. ROSAT shows that HXR1 is a bright, compact feature no more than 10 arcsec in size [7].

The HXR2 source is detected for the first time by BeppoSAX (1SAX J0618.1+2227). Also in this case we find a hard power-law component (with index 1.7 ± 0.4) in addition to the soft thermal emission. Note that HXR2 is located within the error circle of the EGRET source.

HXR1 and HXR2 are both associated with IC443. The location of the HXRs coincide with two of the CO clumps detected by Dickman et al. (1992).

It was known that IC 443 has a hard X-ray component from a Ginga spectrum [11], but little could be said about where the emission originates. HXR1 was observed by Keohane et al. (1997) with ASCA (they dubbed it the HXF=hard

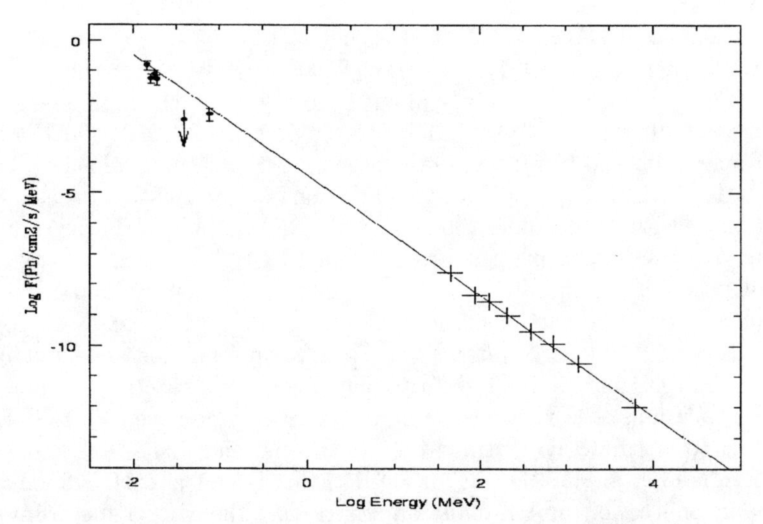

FIGURE 1. The spectrum of the high energy γ source based upon the PDS and EGRET flux densities. (The EGRET errors are not shown, but the size of the crosses roughly indicates their typical magnitude.) The line drawn through the points has a spectral slope of 2.

X-ray feature), in addition to a ridge of hard emission along the southwestern rim. They further noted that the HXF is coincident with a flat spectrum radio component observed by Green (1986; it should however be noted that Green himself suggested that the flat radio spectrum might be emission from a small HII region).

The BeppoSAX high energy PDS detector also observed hard photons from these fields. We assume that this emission is dominated by the EGRET source 2EG J0618+2234, but given the large field of view (1.2deg, FWHM), which includes all of the brighter emission from IC 443, we cannot exclude the presence of other emission in this energy band (the EGRET detector encompasses an even larger area of sky).

We detect emission with the PDS from 14 up to $\simeq$ 50 keV, with a roughly power-law energy dependence. In Fig. 1 we show the resulting spectrum, in combination with the EGRET data on 2EG J0618+2234. The PDS points clearly agree very well with the extrapolation of the EGRET spectrum, and we conclude that all the emission originates in a single, power-law spectral component. We have thus been able to link a gamma-ray component to 14 keV X-ray emission with a single power-law index near 2 over nearly six decades in energy.

We were able to reasonably rule out (long term) variability of the hard X-ray emission. Indeed, if we compute the 2–20 keV flux from a fit of our (still rather poor) PDS spectrum with a power law ($\alpha \simeq 2.1$) we get the same flux (9×10^{-11} ergs/cm^2/s) observed by GINGA seven years earlier, and consistent with the 2–10 keV flux detected by HEAO A-2 19 years earlier.

Recently, Sturner et al. (1997) have modelled the γ-ray (and other) emission from SNR shocks. For remnants of age about 5000 yr or less, the high energy emissions by bremsstrahlung or π^0 decay increase roughly quadratically with ambient density. Synchrotron and Compton emission increase much more slowly (approximately linearly) with density, while all four emission mechanisms are not as strongly dependent on density among older (50 000 yr or more) remnants. These results are for ambient densities which increase from 0.1 to 10 cm^{-3}, so greater γ-ray fluxes might be expected from those shocks interacting with a dense cloud. Indeed in IC443 condensations can reach densities of 10^5 cm^{-3} [3].

Can the two hot spots account for the hard X-ray emission detected by the PDS? Since the PDS and EGRET seem to detect the same spectral component, is in these two locations the origin of the high energy emission seen by EGRET? It was tempting to attribute at least most of the HE emission (> 14 keV) to the hot spots, as Keohane et al. (1997) did for HXR1. But HXR1 and HXR2 cannot be the main/only sources of the HE emission! Extracting the MECS spectrum of the hot spots we find indeed a hard excess over a thermal component, but in order to make this presumably non-thermal excess match the PDS spectrum the hot spots should be a factor >2 brighter than observed.

Our conclusion is that IC 443 is clearly different from (young) SN 1006. EGRET SNRs suggest that the region of shock-molecular cloud interaction produces high-energy emission. The emission is probably coming from the whole interacting region, with an enhancement in denser compact knots as those seen at the location of HXR1 and HXR2.

REFERENCES

1. Boella, G. Butler, R.C., Perola, G.C., et al. *A.& A. Suppl.*, **122**, 299 (1997).
2. Brazier, K.T.S., Kanbach, G., Carraminana, A., Guichard, J. and Merck, M., *M.N.R.A.S.* **281**, 1033 (1996).
3. Claussen, M.J., Frail, D.A., Goss, W.M. and Gaume, R.A., *Ap. J.*, **489**, 143 (1997).
4. Dickman, R.L., Snell, R.L., Ziurys, L.M. and Huang, Y-L., *Ap. J.*, **400**, 203 (1992).
5. Esposito, J.A., Hunter, S.D., Kanbach, G. and Sreekumar, P., *Ap. J.* **461**, 820 (1996).
6. Green, D.A., *M.N.R.A.S.* **221**, 473 (1986).
7. Keohane, J.W., Petre, R., Gotthelf, E.V., Ozaki, M. and Koyama, K., *Ap. J.* **484**, 350 (1997).
8. Koyama, K., Petre, R., Gotthelf, E.V., Hwang, U., Matsuura, M., Ozaki, M. and Holt, S.S., *Nature*, **378**, 255 (1995).
9. Reynolds, S.P., *Ap. J.* **459**, L13 (1996).
10. Sturner, S.J., Skibo, J.G., Dermer, C.D. and Mattox, R., *Ap. J.* **490**, 619 (1997).
11. Wang, Z.R., Asaoka, I., Hayakawa, S. and Koyama, K., *P.A.S.J.* **44**, 303 (1992).

Observation of supernova remnants with the CAT Cherenkov imaging telescope

G. Mohanty*, L.-M. Chounet*, B. Degrange*, P. Fleury*, G. Fontaine*, L. Iacoucci*, F. Piron*, R. Bazer-Bachi†, J.-P. Dezalay†, I. Malet†, A. Musquere†, J.-F. Olive†, G. Debiais‡, B. Fabre‡, E. Nuss‡, X. Moreau||, K. Ragan||, C. Renault||, M. Rivoal||, K. Schahmaneche||, J.-P. Tavernet||, L. Rob¶, A. Djannati-Ataï§, P. Espigat§, M. Punch§, P. Goret**, C. Gouiffes**, I. A. Grenier**, and D. Ellison††

**LPNHE, Ecole Polytechnique, IN2P3/CNRS, France*
†*CESR, Toulouse, INSU/CNRS, France*
‡*GPF, Université de Perpignan, IN2P3/CNRS, France*
||*LPNHE, Paris VI/VII, IN2P3/CNRS, France*
¶*Charles Univ., Prague, Czech Republic*
§*PCC, Collège de France, IN2P3/CNRS, France*
***Centre d'Etudes de Saclay, DSM, DAPNIA/SAp, France*
††*North Carolina State Univ., Raleigh, USA*

Abstract. Supernova remnants (SNR) can be broadly classified into two types: plerionic or plasma-filled SNR, like the Crab Nebula, and shell-type SNR, like Cassiopeia A (Cas A). VHE observations of the Crab Nebula and Cas A made with the CAT atmospheric Cherenkov imaging telescope are used to constrain models for production of gamma rays in SNR. For both plerionic and shell-type SNR, these observations serve primarily to impose a limit on the magnetic field in the region where the particles are accelerated.

THE DETECTOR AND DATA ANALYSIS

The CAT atmospheric Cherenkov imaging telescope [1], located in the French Pyrénées, features a near-isochronous mirror, fast photomultipliers, and high-speed electronics that allow it to reach a low energy threshold of about 250 GeV at the zenith, in spite of a relatively small mirror area of $\sim 18\text{m}^2$. Its high-definition camera allows the use of a powerful image-analysis technique [2], where a semi-analytical model of a gamma-ray shower is used to predict the shower image in the focal plane of the telescope. For each individual image, a χ^2-like minimization

CP510, *The Fifth Compton Symposium,* edited by M. L. McConnell and J. M. Ryan

between the data and the model prediction yields best-fit values for the primary energy, the angular position of the source, and the impact point of the shower core. Excellent background rejection is provided by the standard CAT selection: the χ^2-probability, P $(\chi^2) > 0.35$, the pointing angle, $\alpha < 6°$, along with a cut on the total charge, $Q_{tot} > 30$ pe, that rejects low-light images. The response of the CAT detector has been simulated using a locally-modified version of the KASCADE shower simulation program [3], and the results validated on the strong gamma-ray signal observed from Mk501 in April 1997.

RESULTS

Besides tracking the source, a control OFF-source region covering the same range of right-ascension and declination is also observed; with, typically, more data being taken ON-source than OFF-source. The data quality is constantly monitored for changing weather conditions and electronic problems, and the detector efficiency is regularly calibrated.

The energy spectrum of the Crab Nebula has been derived on the basis of 37.4 hr of ON-source data [4, 5]. For the Cas A data, an upper limit to the emission can be calculated by comparison to a roughly contemporaneous subset of the Crab Nebula data. Both datasets are restricted to zenith angles less than 25°, where the detector behaviour is better understood. Fig. 1 shows the distribution in the pointing angle, α, after selection by the other criteria. The ON-source excess for the Crab Nebula is clearly visible in the peak at small α, whereas there is no evident excess for Cas A. The plots have been normalized in the background region, between 22° and 120° in α. Table 1 tabulates these values, from which a 3-σ upper limit, corrected

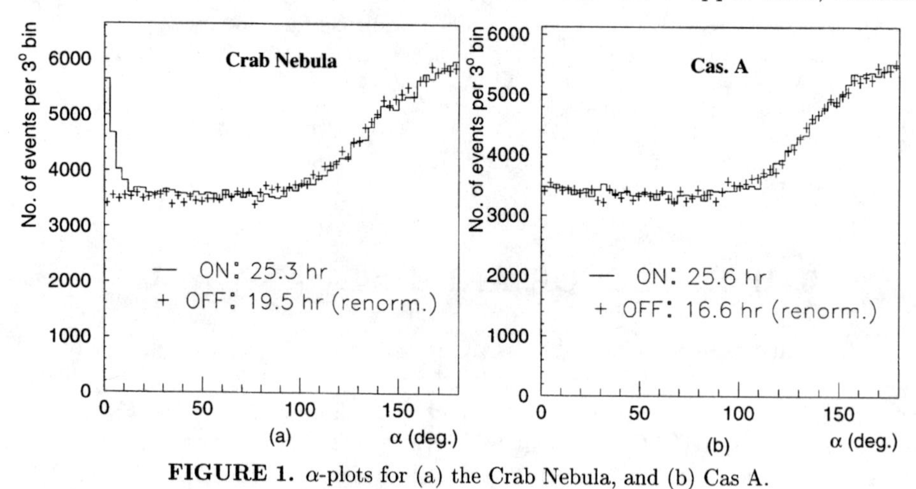

FIGURE 1. α-plots for (a) the Crab Nebula, and (b) Cas A.

for zenith angle effects, of 0.14 times the Crab flux is derived, with the assumption of no emission from Cas A . Using the effective collection area from Monte Carlo simulations, this corresponds to an integral flux of $1.5 \times 10^{-11}\mathrm{cm}^{-2}\mathrm{s}^{-1}$ above a threshold of 250 GeV.

TABLE 1. Analysis results used to derive an upper limit on Cas A emission.

Source	ON dur. (hr)	N_{on}	N_{off}	$N_{exc.}$ ($r_{exc.}$/min.)	$\sigma_{tot.}$	N_{back} ($r_{back.}$/min.)
Crab Nebula (< 25°)	25.3	8 330	5 592	2 738 (1.8)	22.0	74 004 (3.9)
Cassiopeia A (< 25°)	25.6	6 937	6 937	0 (0.0)	0.0	65 393 (4.0)

Results on Cas A were presented earlier by Goret et al [6] at the 26th ICRC, who used slightly different selection criteria to obtain an upper limit of $0.74 \times 10^{-11}\mathrm{cm}^{-2}\mathrm{s}^{-1}$ on the integral flux above a threshold of 400 GeV. This is consistent with our results, given the assumed slope for the Crab spectrum. The HEGRA group [7] has recently reported a preliminary detection of the Cas A SNR at a very low flux level.

MODEL FOR EMISSION FROM THE CRAB NEBULA

The radio to hard X-ray emission spectrum from the Crab Nebula is explained by synchrotron radiation from high-energy electrons in the nebular magnetic field, while the observed GeV to TeV spectra are supposed to derive from Compton up-scattering of ambient infra-red photons by highly relativistic electrons originating at the termination shock of the pulsar wind. Early models [8–12] assumed a constant magnetic field in the source region, while more recent ones [13–16] incorporate a detailed simulation of the hydrodynamic plasma flow, which also carries out the magnetic field lines from the pulsar.

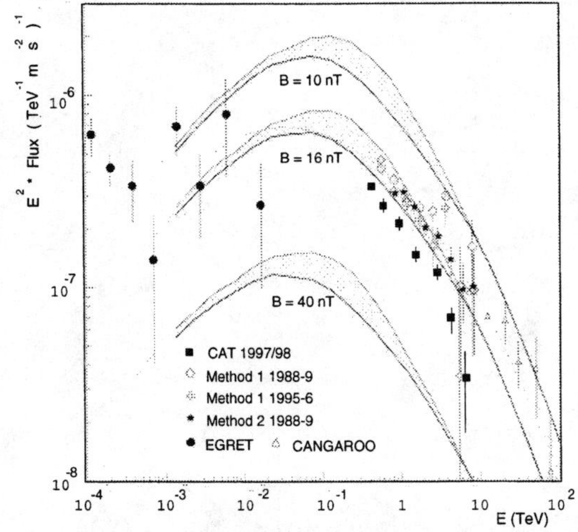

FIGURE 2. Gamma-ray data compared to Hillas et al [17] predicted spectrum for three field strengths Reproduced from Astrophys. J with permission

Here, we use the treatment of Hillas et al [17] that attempts to calculate the magnetic field in the emission region as directly as possible from the observational data over the entire electromagnetic spectrum. It turns out that for a magnetic field of $\sim$ 20 nT needed to explain the observed TeV spectrum, the progenitor electrons must have energies from 2–30 TeV, while the up-scattered photons are from the far-infrared: 0.005–0.3 eV in energy. Thus, the parent electrons should give rise to synchrotron radiation at about 0.4 keV, so that the TeV emission region should be mapped by X-ray observations [18, 19]. From the observed synchrotron flux, the electron spectrum can be deduced. This, along with the infrared photon density allows calculation of TeV spectra for different magnetic field strengths by a full simulation of the exact differential Klein-Nishina cross-section. Fig. 2, from Hillas et al, compares the model results with measured spectra, with a B field of 20 nT (slightly larger than the Hillas et al value) giving a reasonable fit to the CAT data.

MODEL FOR EMISSION FROM CASSIOPEIA A

The emission in Cas A is concentrated in a shell, believed to be the shock front where the supernova expansion impinges on surrounding material. The B field can be amplified in the instabilities at the edge of the expanding shell [20], and particles can be accelerated to high energies in these regions by the Fermi mechanism [21, 22]. There is evidence for 10–100 TeV electrons in SNR [23], the synchrotron emission

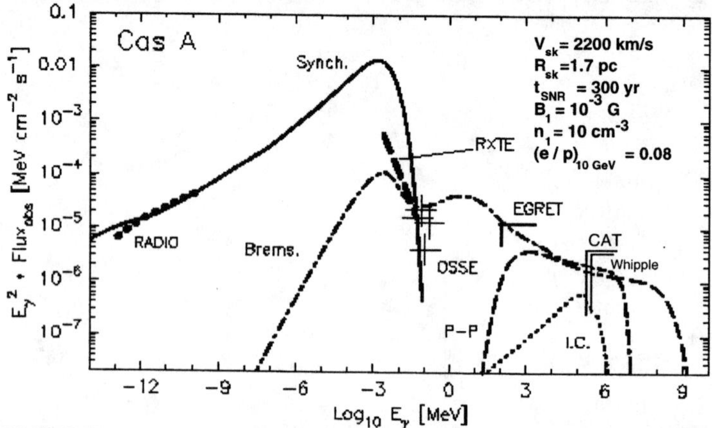

FIGURE 3. Predicted spectrum for Cas A emission [27, 28] compared to data.

from which can adequately explain the radio data. GeV to TeV gamma-rays can be produced by bremsstrahlung and inverse Compton scattering of ambient photons. Cowsik & Sarkar [24] used COS-B upper limits to the gamma-ray flux to set a lower limit of 8×10^{-5} G for the B field in Cas A. EGRET data [25] allows an improved lower limit of 2×10^{-4} G.

The need for a detailed consideration of particle acceleration in the turbulent shocks and the concomitant amplification of the magnetic field led Baring et al [26] to develop a nonlinear diffusive shock acceleration model that predicts the radio to gamma-ray continuum spectrum in SNR, with the particular case of Cas A being addressed by Ellison et al [27,28]. Fig. 3 indicates the predicted contributions of various emission components at different energies, along with relevant data. In the context of this model, the GeV to TeV upper limits imply a lower limit of $\sim 10^{-3}$ G for the B field.

These results may also be used to constrain theories of cosmic-ray origin in SNR [29], as has been done by several authors [30–32]. We are in the process of addressing this issue.

REFERENCES

1. A. Barrau et al., Nucl. Instrum. Meth. A **416**, 278 (1998).
2. S. L. Bohec et al., Nucl. Instrum. Meth. A **416**, 425 (1998).
3. M. P. Kertzman and G. H. Sembroski, Nucl. Instrum. Meth. A **343**, 629 (1994).
4. L. Iacoucci et al, Proc. 16th European Cosmic Ray Conf. , 363 (1998).
5. L. Iacoucci, PhD thesis, Ecole Polytechnique, Palaiseau, France, 1998.
6. P. Goret et al, in *Proc.* 26th *Int. Cosmic Ray Conf.*, volume 3, page 496, 1999.
7. G. Pühlhofer et al, in *Proc.* 26th *Int. Cosmic Ray Conf.*, volume 3, page 492, 1999.
8. R. J. Gould, Phys. Rev. Lett. **15**, 577 (1965).
9. G. H. Rieke and T. C. Weekes, Astrophys. J. **155**, 429 (1969).
10. J. E. Grindlay and J. A. Hoffman, Astrophys. Lett. Commun. **8**, 209 (1971).
11. A. A. Stepanian, Izv. Krym. Astrofiz. Obs. **62**, 79 (1980).
12. A. A. Stepanian, Izv. Krym. Astrofiz. Obs. **84**, 113 (1992).
13. O. C. de Jager and A. K. Harding, Astrophys. J. **396**, 161 (1992).
14. O. C. de Jager et al., Astrophys. J. **457**, 253 (1996).
15. F. A. Aharonian and A. M. Atoyan, Astroparticle Phys. **3**, 275 (1985).
16. A. M. Atoyan and F. A. Aharonian, Mon. Not. R. Astron. Soc. **278**, 525 (1996).
17. A. M. Hillas et al., Astrophys. J. **503**, 744 (1998).
18. F. R. Harnden Jr. and F. D. Stewart, Astrophys. J. **283**, 279 (1984).
19. J. J. Hester et al., Astrophys. J. **448**, 240 (1995).
20. S. F. Gull, Mon. Not. R. Astron. Soc. **161**, 47 (1973).
21. J. S. Scott and R. A. Chevalier, Astrophys. J. **197**, L5 (1975).
22. R. Cowsik and S. Sarkar, Mon. Not. R. Astron. Soc. **207**, 745 (1984), Erratum: [33].
23. G. E. Allen et al, in *Proc.* 26th *Int. Cosmic Ray Conf.*, volume 3, page 480, 1999.
24. R. Cowsik and S. Sarkar, Mon. Not. R. Astron. Soc. **191**, 855 (1980).
25. J. A. Esposito et al, Astrophys. J. **461**, 820 (1996).
26. M. G. Baring et al, Astrophys. J. **511**, 311 (1999).
27. D. C. Ellison et al, in *Proc.* 26th *Int. Cosmic Ray Conf.*, volume 3, page 468, 1999.
28. D. C. Ellison, in *N-01, these proceedings*, 1999.
29. L. O. Drury, F. A. Aharonian, and H. J. Volk, Astron. Astrophys. **287** (1994).
30. J. H. Buckley et al., Astron. Astrophys. **329**, 639 (1998).
31. R. W. Lessard et al, in *Proc.* 26th *Int. Cosmic Ray Conf.*, volume 3, page 488, 1999.
32. T. Tanimori et al., Astrophys. J. **497**, L25 (1998).
33. R. Cowsik and S. Sarkar, Mon. Not. R. Astron. Soc. **209**, 719 (1984).

BATSE observations of classical novae

Margarita Hernanz* David M. Smith†, Jerry Fishman ‡, Alan Harmon‡, Jordi Gómez-Gomar*, Jordi José*, Jordi Isern* and Pierre Jean¶

*Institute for Spatial Studies of Catalonia (IEEC/CSIC/UPC), Gran Capità 2-4 08034 Barcelona (SPAIN).
†Space Sciences Laboratory, University of California, Berkeley, CA 94720
‡NASA/Marshall Space Flight Center, Huntsville, AL 35812
¶Centre d'Etude Spatiale des Rayonnements, 31028 Toulouse Cedex (FRANCE)

Abstract. Detection of gamma-ray emission from classical novae, in the range between 30 and 511 keV, would provide a crucial test of the thermonuclear runaway (TNR) model. This emission results from the annihilation of positrons, emitted by some radioactive nuclei (^{13}N and ^{18}F) synthesized during the TNR; it has a short duration and is produced before the optical maximum. Therefore, it can only be analyzed "a posteriori", once the nova has been discovered optically. The capability to observe all the sky, together with its high sensitivity in the low-energy range, make BATSE an ideal instrument to detect this emission. Data analysis techniques previously applied for BATSE observations of 511 keV transients of short duration have been used. The first results from the systematic search in BATSE background data which is under way are presented, which include nearby novae that have exploded since CGRO launch. Comparison with recent updated theoretical models is made in each case. Although no positive detection has been obtained up to now, upper limits to the emitted flux are presented.

INTRODUCTION

Classical novae are powered by thermonuclear runaways on accreting white dwarfs, which have a main sequence companion in a cataclysmic variable binary system. The explosive burning of hydrogen leads to the synthesis of short-lived β^+-unstable nuclei, such as ^{13}N (τ=862s) and ^{18}F (τ=158 min.). The emitted positrons annihilate either directly, leading to the emission of 511 keV photons, or through positronium formation, which emits photons with a continuum spectrum below 511 keV. In addition, as the nova envelope is still partially opaque to gamma-rays during the decay of ^{13}N and ^{18}F, the emitted photons are comptonized to lower (than 511 keV) energies. Therefore, a strong continuum between 30 and 511 keV and the 511 keV line dominate the early gamma-ray spectrum of classical novae

CP510, *The Fifth Compton Symposium*, edited by M. L. McConnell and J. M. Ryan

TABLE 1. Main properties of the ejecta 1 hour after peak temperature.

Nova type	^{13}N ($M_\odot$)	^{18}F ($M_\odot$)	Mean kinetic energy (erg)
CO 1.15$M_\odot$	2.3 10^{-8}	2.6 10^{-9}	1.1 10^{45}
ONe 1.15$M_\odot$	2.9 10^{-8}	5.9 10^{-9}	1.5 10^{45}
ONe 1.25$M_\odot$	3.8 10^{-8}	4.5 10^{-9}	1.5 10^{45}

[1–5].

Before the launch of the Compton Gamma-Ray Observatory in 1991, Fishman et al. [6] already made a prediction of the detectability of low-energy gamma-rays from novae with the BATSE instrument, based on the models of gamma-ray emission from Leising and Clayton [1]. More recently, we have developed a hydrodynamical code that follows classical nova explosions from the onset of accretion up to the explosion and mass ejection phases [7]. This code has been coupled to a Monte Carlo one, which computes the gamma-ray production and transfer in the expanding nova envelope [4]. Thus, the gamma-ray emission at different epochs is obtained for particular nova models, both of the CO and of the ONe types. The radioactive nuclei ^{13}N and ^{18}F are synthesized in similar quantities in both nova types and the emission related to their decay appears very early (before the maximum in visual light) and has short duration (because of the short lifetimes of those nuclei). The main properties of the ejecta for some nova models are shown in table 1, where the ejected masses of ^{13}N and ^{18}F correspond to 1 hour after peak temperature. The light curves for the relevant energy bands for the BATSE instrument, as well as for the 511 keV line, are shown in figure 1, for a CO nova of 1.15$M_\odot$ at 1 kpc. The times are relative to the maximum temperature one. A comparison between the early-time (6 h after T_{max}) spectrum of a CO nova (1.15$M_\odot$) and an ONe one (1.25$M_\odot$) shows that similar fluxes are expected for both models (see figure 1).

DATA ANALYSIS TECHNIQUES AND TARGETS

Two different types of analyses are being applied, which are mainly sensitive in the ranges (250-511) keV and (40-320) keV. In the (40-320) keV range, the BATSE Earth Occultation technique is applied (see Harmon et al. 1992 [8] for details), with 12 hour sums instead of single occultation steps. Concerning the (250-511) keV range, BATSE LAD data in the HER format are used to produce the spectra. Charged particle information extracted from the DISCLA data format is used for background generation (see Smith et al. 1996 [9,10] for details). Intervals of 12 hours are analyzed at 6-hour spacing, to make sure that the peak of the outburst is not splitted. Background data are taken 24 hours before the period of interest.

The predicted detectability distance of novae with BATSE should be around 2 kpc, but this issue will be better known after the data analysis techniques will be optimized for our particular type of study. Also, novae distances are known with a

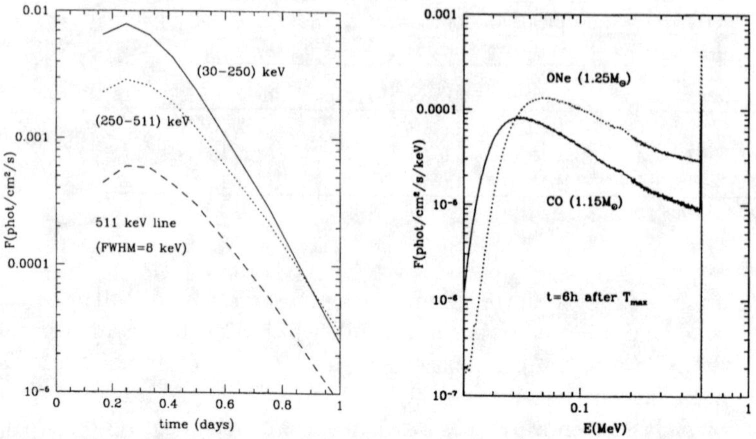

FIGURE 1. (Left) Light curves, in the relevant energy bands for BATSE, of a CO nova ($1.15M_{\odot}$) at a distance of 1 kpc. (Right) Spectra of a CO nova ($1.15M_{\odot}$) and of an ONe nova ($1.25M_{\odot}$), 6 h after peak temperature and for d=1 kpc.

great uncertainty; often they are determined from empirical relationships between the visual magnitude at optical maximum and the rate of decline of their visual light curves, as measured by the time to decline in 2 or 3 magnitudes (t_2 and t_3, respectively). Therefore, we have considered it safe to include all novae that have exploded since CGRO launch which are at distances smaller than 3-4 kpc. The distances adopted are either those from Shafter 1997 [11], when available, or those that we have computed from the data in the IAU circulars and the $m_v(max)$-t_2 relationship. The periods to analyze adopted range from 15 to 20 days before maximum visual luminosity (i.e., roughly before discovery for many cases), since the outburst in γ-rays with $E \leq 511$ keV occurs some hours after peak temperature, which happens well before the maximum in visual luminosity.

RESULTS AND DISCUSSION

The results of our analysis in the (250-511) keV range are shown in figure 2 for two particular cases: Nova Cyg 1992 ($t_2 \sim 16$ days, d$\sim$1.7 kpc) and the recent Nova Vel 1999 ($t_2 \sim 6$ days, d$\approx$2 kpc). Analyses with the Earth Occultation technique for these same novae are still under way. Data are expressed as a fraction of the model flux shown in figure 1 (left). The corresponding fluxes both in the whole range up to 511 keV and in the line ($510 \leq E \leq 520$) keV are displayed on the right y-axis scale of the figure. When data in the whole (250-511) keV range is used to determine the flux in the line (see left panels of figure 2), most of the sensitivity

TABLE 2. 3-σ upper limits to the fluxes for the novae Cyg 1992, Sco 1992 and Vel 1999.

	F(3-σ limit)	F(model)
Nova Cyg 1992 (model: 1.25M$_\odot$ ONe nova at d=1.7 kpc)		
(250-511) keV	5.2 10^{-3}	2.3 10^{-3}
511 keV line[a]	1.0 10^{-3}	4.8 10^{-4}
511 keV line[b]	2.4 10^{-3}	4.8 10^{-4}
Nova Sco 1992 (model: 1.15M$_\odot$ CO nova at d=0.8 kpc)		
(250-511) keV	3.6 10^{-3}	5.3 10^{-3}
511 keV line[a]	7.1 10^{-4}	1.0 10^{-3}
511 keV line[b]	2.3 10^{-3}	1.0 10^{-3}
Nova Vel 1999 (model: 1.25M$_\odot$ ONe nova at d=2 kpc)		
(250-511) keV	5.3 10^{-3}	1.7 10^{-3}
511 keV line[a]	1.0 10^{-3}	3.5 10^{-4}
511 keV line[b]	1.6 10^{-3}	3.5 10^{-4}

[a] using (250-511) keV data with assumed Comptonization
[b] using 511 keV data only

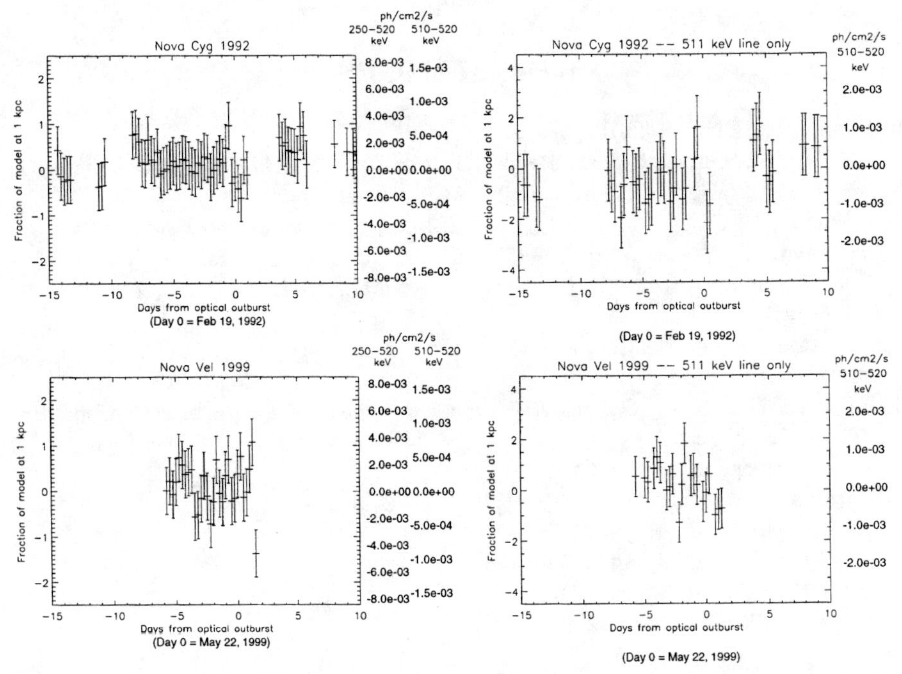

FIGURE 2. Nova Cyg 1992 (top) and Nova Vel 1999 (bottom) observations with BATSE. (Left) Data in the whole range above 250 keV or (right) only in the line (510-520) keV are included (see text for details).

comes from the continuum, where most of the emission is produced according to the theoretical models (figure 1). Another type of analysis has been performed, where only the narrow line is fitted and only the orbits without passage through the South Atlantic Anomaly are used (see right panels of figure 2).

We have evaluated the 3-σ upper limits to the fluxes for the 3 most promising (in principle) novae of the whole sample, i.e. Nova Cyg 1992, Nova Sco 1992 and Nova Vel 1999. For the derivation of the 511 keV line fluxes, the two methods mentioned above have been applied: (a) data in the whole (250-511) keV range with the assumed Comptonization from theoretical models (figure 1) is used to determine the flux in the line, between 510 and 520 keV (see left panels of figure 2); (b) the upper limits are derived from the data in the 510-520 keV only (see right panels of the same figure).

As can be seen in table 2, all upper limits are compatible with theory, except for Nova Sco 1992, which should either have $M < 1.5 M_{\odot}$, d>0.8 kpc or both. The 3-σ sensitivity using the 511 keV data only is similar to that of [12] with WIND/TGRS, but the sensitivity of [12] requires a particular line blueshift, whereas ours is independent of it. The 3-σ sensitivity using the (250-511) keV data with assumed Comptonization is a little more than a factor of 2 better than that of [12].

REFERENCES

1. Leising, M., Clayton, D.D., *ApJ* **323** 157 (1987).
2. Hernanz, M., Gómez-Gomar, J., José, J., Isern, J., *2nd INTEGRAL Workshop, The Transparent Universe*, Noordwijk: ESA SP-382, 1997, p. 47.
3. Hernanz, M., Gómez-Gomar, J., José, J., Isern, J., *Proceedings of the Fourth Compton Symposium*, New York: AIP Conf. Proc. 410, 1997, p. 1125.
4. Gómez-Gomar, J., Hernanz, M., José, J., Isern, J., *Mon. Not. R. Astron. Soc.* **296**, 913 (1998).
5. Hernanz, M., José, J., Coc, A., Gómez-Gomar, J., Isern, J., *ApJ Lett.* **526**, in press (1999).
6. Fishman, G.J. et al., *Gamma-Ray Line Astrophysics*, New York: AIP Conf. Proc. 232, 1991, p. 1125.
7. José, J., Hernanz, M., *ApJ* **494** 680 (1998).
8. Harmon, B.A. et al., *Proceedings of the CGRO Workshop*, 1992, p. 49.
9. Smith, D.M. et al. *ApJ* **458** 576 (1996).
10. Smith, D.M. et al. *ApJ* **471** 783 (1996).
11. Shafter, A. *ApJ* **487** 226 (1997).
12. Harris, M.J. et al. *ApJ* **522** 424 (1999).

TGRS OBSERVATIONS OF POSITRON ANNIHILATION IN CLASSICAL NOVAE

M. J. Harris[1], D. M. Palmer[1], J. E. Naya[1], B. J. Teegarden, T. L. Cline, N. Gehrels, R. Ramaty, H. Seifert[1]

Code 661, NASA/Goddard Spaceflight Center, Greenbelt, MD 20771
[1] *Universities Space Research Association*

Abstract.
The TGRS experiment on board the *Wind* spacecraft has many advantages as a sky monitor — broad field of view ($\sim 2\pi$ centered on the south ecliptic pole), long life (1994–present), and stable low background and continuous coverage due to *Wind*'s high altitude high eccentricity orbit. The Ge detector has sufficient energy resolution (3–4 keV at 511 keV) to resolve a cosmic positron annihilation line from the strong background annihilation line from β-decays induced by cosmic ray impacts on the instrument, if the cosmic line is Doppler-shifted by this amount. Such lines (blueshifted) are predicted from nucleosynthesis in classical novae. We have searched the entire TGRS database for 1995–1997 for this line, with negative results. In principle such a search could yield an unbiased upper limit on the highly-uncertain Galactic nova rate. We carefully examined the times around the known nova events during this period, also with negative results. The upper limit on the nova line flux in a 6-hr interval is typically $< 3.8 \times 10^{-3}$ photon cm^{-2} s^{-1} (4.6σ). We performed the same analysis for times around the outburst of Nova Vel 1999, obtaining a worse limit due to recent degradation of the detector response caused by cosmic ray induced damage.

I INTRODUCTION

Theoretical models of classical novae imply that large quantities of β-unstable proton-rich nuclei are formed during a thermonuclear runaway, with half-lives of the order minutes to hours. The resulting positrons undergo annihilation in the expanding nova envelope, giving rise to a pulse of annihilation γ-rays which is blueshifted and broadened by the nova velocity and lasts for up to ~ 6 hr (2).

All space-borne instruments are subject to a strong background emission line at 511 keV, arising from annihilation of cosmic ray induced β-decay positrons in the instrument, which hinders detection of the same line from cosmic sources. However, a spectrometer with sufficient energy resolution could distinguish between cosmic and background lines because the former is blueshifted, typically by 2–5 keV. The

CP510, *The Fifth Compton Symposium*, edited by M. L. McConnell and J. M. Ryan

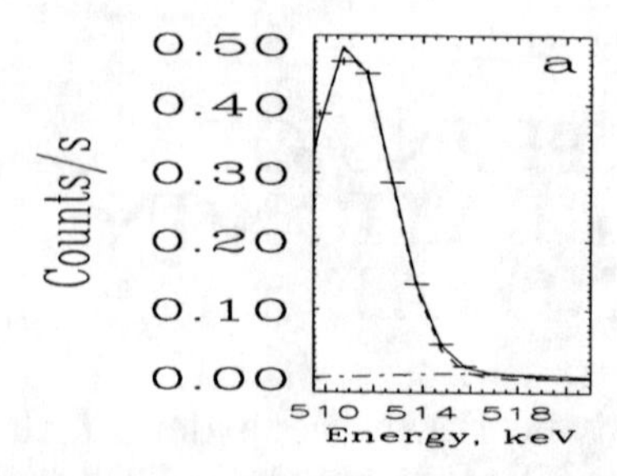

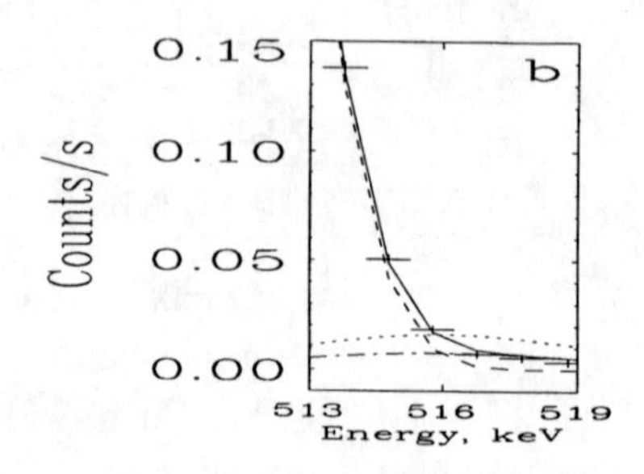

FIGURE 1. Characteristic TGRS background count spectrum at energies around 511 keV, obtained during the interval 3 June 12h–18h UT with a power law continuum subtracted. The components of the fit are a line with the width and position of the nova 6-hr line (2, dot-dashed line) and a Gaussian line fitting the blue wing of the 511 keV background line (dashed line). The total model spectrum is the full line. (b) Expansion of Fig. 2a showing the significance of the fitted nova 6-hr line (dot-dashed line of amplitude $5.2 \pm 0.1 \times 10^{-3}$ photon cm^{-2} s^{-1}; other symbols as in Fig. 2a). Also shown (dotted line) is the theoretically expected level of the nova 6-hr line for a nova at 1 kpc (ref. 6 model HH5).

Ge spectrometer TGRS (FWHM resolution 3 keV at the beginning of its mission) fulfills this requirement. Its other advantages as a detector and monitor of the nova 511 keV line are its broad aperture (nearly 2π sterad) and its high altitude high inclination orbit, in which background count rates are rather low, due to the lack of interaction with Earth's trapped radiation belts, and to the virtual absence of Earth albedo background radiation. The background count rate and spectrum has also been extremely stable throughout the mission.

A disadvantage of Ge detectors for long-term missions is progressive degradation of performance due to cosmic-ray impact damage. For this reason most of the data used here were obtained during 1995–1997, before this effect became serious. In an attempt to avoid the effects of degradation, such as low-energy "tailing", we fitted only energies on the high-energy (blue) side of the background 511 keV line. One such fitted background spectrum is shown in Fig. 1. The dashed line is the strong background 511 keV line. The dot-dashed line is the fit to a theoretical nova line emitted in a 6-hr interval after the event (3). Although the amplitude of this line is significant, we will see shortly that it is characteristic of all our background spectra — in other words, there is a systematic constant positive offset in this measurement. For scale, the dotted line shows how the same line would appear from a model nova at a distance 1 kpc. This line would be detected at a level $\sim 8\sigma$.

II SURVEY OF THE SOUTHERN SKY, 1995–1997

The global rate of classical novae in the Galaxy is highly uncertain, since the two methods used to derive it both suffer from serious systematic problems. One involves correcting the rate of actual discoveries (~ 3 yr^{-1}) for highly uneven sky coverage and interstellar extinction for poorly-known nova distances. The other involves correcting the nova rates measured in external galaxies (according to some parameter such as M_B), which runs into the problem of differences in the stellar populations between galaxies.

The method used here is potentially superior, being free from these biases. The 511 keV γ-rays are not subject to interstellar extinction, and TGRS's sky coverage was almost uniform during the period 1995–1997. Between 1995 January and 1997 October TGRS accumulated about 7.7×10^7 s of background spectra (88% overall). Nor are much data lost due to changing sensitivity across the aperture, which amounts to a factor ~25% between the zenith and the ecliptic plane.

This entire period was divided into 6 hr intervals during which background spectra were accumulated, which were fitted to models of the type shown in Fig. 1. The parameters of the nova line in these fits were fixed at values predicted by theory at epochs $\simeq 6$ hr and $\simeq 12$ hr after the explosion (widths 8 keV FWHM, blueshifts 5 and 2 keV, from ref. 2 and Hernanz 1997, private communication). The fits to the count spectra were normalized by the effective area for the zenith angle of an average nova, which is $\simeq 60°$. The fitted amplitudes for the 6-hr and 12-hr lines were combined in order to maximize the signal.

The fitted nova line amplitudes in the 4005 resulting background spectra are shown in Fig. 2. Since the positive systematic amplitude is constant to a very good approximation, we are justified in subtracting from all measurements. We then see that there is no single 6-hr interval which, after this subtraction, yields a convincing nova signal. We chose a level 4.6σ as the threshold for detection of a candidate signal (Fig. 2, dashed line), which is appropriate for the expected number of chance detections in 4005 events. From Poisson statistics, our measured nova rate is therefore < 1 at the 63% level. The sensitivity of this survey at the 4.6σ level is about 3.8×10^{-3} photon cm^{-2} s^{-1} (4).

If a given nova model predicts a 511 keV line flux ϕ_{pred} at 1 kpc then this survey could have detected it out to a distance $r_{det} = \sqrt{\phi_{pred}(M)/3.8 \times 10^{-3}}$ kpc. Unfortunately the most recent self-consistent hydrodynamic models (5) predict much smaller fluxes ($< 2 \times 10^{-3}$ photon cm^{-2} s^{-1} at 1 kpc) than earlier versions (2), although parametrized models (6) can predict values values as high as 7×10^{-3} photon cm^{-2} s^{-1}. Thus in the most optimistic case $r_{det} = 1.4$ kpc, for an ONeMg nova of model type HH5 (6). Using the Bahcall-Soneira Galactic mass model (1) we can extrapolate our limit < 1 event per 7.7×10^7 s within 1.4 kpc (which includes ~0.8% of the Galaxy) to a Galactic rate < 54 events yr^{-1} of this type.

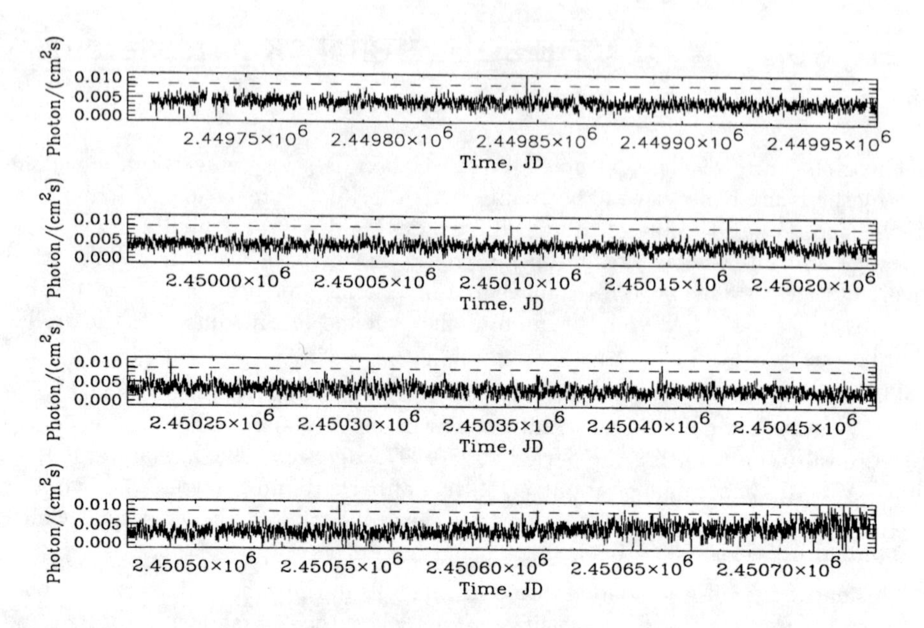

FIGURE 2. Measured fluxes in the nova 6-hr line during the entire interval 1995 January – 1997 October. Dotted line — mean 4.6σ limit, above which candidate line detections would lie.

III NOVA VELORUM 1999

Five novae occurred within TGRS's aperture during the period 1995–1997, before the instrument performance began to deteriorate. Searches for 511 keV line emission around the times of these outbursts were unsuccessful (3), the sensitivities achieved being similar to the values given above for the overall survey. Unfortunately a very bright nova V382 Vel ($\equiv$ N Vel 1999) was discovered in 1999 May, by which time the deterioration of the detector performance had advanced.

Three main effects contribute to this deterioration (7): a distortion of the response to a narrow line (starting with low-energy tailing, culminating in the formation of a separate low-energy peak); a poorly-quantified loss of efficiency; and an accelerated rate of change of the instrument gain. In order to apply our search method to Nova Vel we attempted approximate solutions to these problems. In the count spectrum fitting, the Gaussian line models (Fig. 1) were replaced by more complex shapes reflecting the current detector response (evaluated from the response to the 511 keV and other background lines). Fortunately, during the rather narrow time interval (15–23 May) which we fitted the response did not change significantly, nor did the gain. We estimated the detector effective area by assuming constancy of the true TGRS background line strengths, which we had found to be a very good approximation during the period of stability 1995–1997; a measurement of the 511 keV line intensity, compared with the average from 1995–1997, then enabled an efficiency correction to be made.

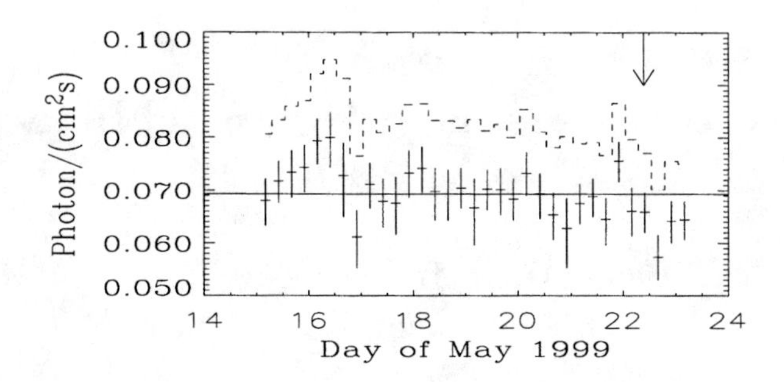

FIGURE 3. Measured fluxes at 6-hr intervals prior to the discovery of V382 Vel (arrow). Full line — assumed constant positive offset. Dashed line — 3σ upper limits on the line flux.

The preliminary results are shown in Fig. 3. Note that, compared to Fig. 2, both the error bars and the systematic positive offset are much worse. The rise to visual maximum probably started at least 1 d before discovery (arrow), and the explosion may have occurred up to 3 d before that. Our fitted background nova line amplitude should show a single 6-hr peak at that epoch, over and above the average background level (solid line). No such event was observed, so that we can only place upper limits on the line flux and a lower limit on the distance to V382 Vel. The average 3σ upper limit above the solid line is $< 1.3 \times 10^{-2}$ photon cm^{-2} s^{-1}; however, the constant positive offset level is much less well-defined than it was earlier in the mission (Fig. 2), so that a systematic error estimate for this must be introduced. We also estimated the systematic error due to the uncertainty in the nova line position arising from the broad and irregular shape of the background 511 keV line relative to which it is measured. We thus obtained an upper limit on the line flux of order $< 1.6 \times 10^{-2}$ photon cm^{-2} s^{-1}. V382 Vel is believed to have occurred on an ONeMg white dwarf (8), in which case we can combine our flux limit with the best-case theoretical flux at 1 kpc of 7×10^{-3} photon cm^{-2} s^{-1} (6) to obtain a lower limit of 0.66 kpc on the distance.

REFERENCES

(1) Bahcall, J. N., & Soneira, R. M. 1984, ApJS, 55, 67
(2) Gómez-Gomar, J., Hernanz, M., José, J. & Isern, J. 1998, MNRAS, 296, 913
(3) Harris, M. J., et al., 1999, ApJ, 522, 424
(4) Harris, M. J., et al., 1999, submitted to ApJ
(5) Hernanz, M., José, J., Coc, A., Gómez-Gomar, J., & Isern, J. 1999, ApJ in press
(6) Kudryashov, A. D., Chugai, N. N., & Tutukov, A. V. 1999, LANL preprint astro-ph/9812100 and these Proceedings
(7) Kurczynski, P., et al. 1999, Nucl. Instr. Methods A, 431, 141
(8) Shore, S. N., Starrfield, S., Gehrz, R. D., & Woodward, C. E. 1999, IAUC 7192

Study of Nova-Produced ^{22}Na with COMPTEL

A. F. Iyudin*, K. Bennett¶, H. Bloemen†, R. Diehl*, W. Hermsen†, J. Knödlseder‖, J. Ryan§, V. Schönfelder*, A. Strong*, C. Winkler¶

*Max Planck Institut für extraterrestrische Physik, Postfach 1603, D-85740 Garching, Germany
† SRON-Utrecht, Sorbonnelaan 2, 3584 CA Utrecht, The Netherlands
‖ CESR (CNRS/UPS), BP 4346, 31028 Toulouse Cedex, France
§ University of New Hampshire, Institute for Studies of Earth, Oceans and Space, Durham NH 03284, USA
¶ Astrophysics Division, ESTEC, 2200 AG Noordwijk, The Netherlands

Abstract.
The COMPTEL telescope on board the Compton Gamma-Ray Observatory (CGRO) is capable of imaging γ-ray line sources, like classical novae, in the MeV regime at a level of sensitivity up to a few 10^{-5} photons cm^{-2}s^{-1}. At this level of sensitivity quite high expectations can be placed on the detection of the predicted ^{22}Na γ-ray line at 1.275 MeV from nearby novae.

We have used COMPTEL data collected up to the 2nd CGRO reboost to update previously published limits for the sodium production by novae. Results of the ^{22}Na line emission studies from the old and the most recent novae are discussed and compared with model predictions.

Introduction

The classical nova outburst has been modelled as a thermonuclear runaway in the accreted hydrogen-rich envelope of the white dwarf companion of a close binary system, e.g. [21, 23]. In general, observations of novae support such models [8, 9].

It is currently believed that neon novae, a distinct subclass of the classical novae associated with an underlying oxygen-neon-magnesium (ONeMg) white dwarf, may be an important source of Galactic ^{22}Na [5]. ^{22}Na decays with a 3.75 yrs life-time to a short lived excited state of ^{22}Ne at 1.275 MeV. An individual nova at 1 kpc from the Sun with a total ejected mass of the order of 10^{-4} M$_\odot$ and a ^{22}Na mass fraction of the order of 10^{-4} could have been seen at a flux value of 4×10^{-5} cm^{-2} s^{-1} [25].

In addition to the ^{22}Na line, both types of novae (CO and O-Ne) are prolific

CP510, *The Fifth Compton Symposium*, edited by M. L. McConnell and J. M. Ryan

producers of 511 keV γ-ray line emission, which accompanies the decay of the β^+-unstable products of the nucleosynthesis in novae, as well as of 478 keV emission originating from the ^{7}Be decay to ^{7}Li. Until now, there have been no positive detections of any of these lines, at 1.275 MeV, 478 keV and 511 keV [12, 13, 15, 16, 18]. Below we will discuss the results of a ^{22}Na line study with COMPTEL.

The latest predictions of the ^{22}Na mass ejected as the result of thermonuclear runaway on the white dwarf in a binary system are as small as $\leq 2.0\times10^{-9}$ $M_\odot$ [11,14, 22]. Unfortunately, these predictions were made using reaction flows where notable uncertainties in cross-sections of several key reactions still exist [6]. **Additionally, none of the above models could reproduce high masses of the ejected nova shell** (see discussion in [22]). The only model that appears capable to reproduce high ejected masses of neon novae is that of Wanajo, Hashimoto and Nomoto (1999). Below we will compare measured ^{22}Na line fluxes (upper limits) from the group of old and recent neon novae with those that were predicted by this model [24].

TABLE 1. Parameters of the Galactic novae discussed in this work.

Nova name	Date, TJD	m_v	t_3, days	E(B-V)	d, kpc	M_{wd}, $M_\odot$	M_{ej} $M_\odot$
V693 CrA	4697	7.0	~14	≥0.5	≤7.2	1.08±0.14	?
V1370 Aql	4995	5.0	~10	0.6	4.2	1.14±0.18	$\geq 2\times10^{-4}$
QU Vul	6073	5.1	28±4	0.60±0.05	1.6±0.4	0.88±0.10	3×10^{-4}
V838 Her	8339	5.3	4	0.5±0.1	~3.0±1.0	0.87±0.12	$1.2\text{-}6.7\times10^{-4}$
V1974 Cyg	8672	4.4	47	0.36±0.04	~1.8	~1.1	$2\text{-}5\times10^{-4}$
V382 Vel	11321	2.60	11.5	0.68	~1.5 ?	1.12±0.13	?

From the large number of Galactic novae observed by COMPTEL we have selected novae with well established Ne over-abundances (Table 1).

Instrument and Data Analysis

COMPTEL, due to its combination of imaging and spectroscopic capabilities [20], provides a unique opportunity to measure line emission from point-like sources or from extended regions (e.g. the Galactic bulge).

Generally, different viewing periods covering the position of the relevant nova (listed in Table 1) were combined to achieve the best possible sensitivity. In calculating the ^{22}Na mass in the ejected shell, the time delays between the nova maximum brightness and the time of the COMPTEL's measurements have been taken into account. Imaging and flux evaluation were done in a ± 2 σ energy window around the 1.275 MeV line, where σ is the instrumental energy resolution for this line. The results are given in Table 2. The background model used in this work is based on the similarity of event distributions in the (χ, ψ) data space in the adjacent energy bands, which is energy independent in first order aproximation, and on the event distribution in the $\bar{\varphi}$ coordinate of the line photons. Some smoothing is then applied in the (χ, ψ) coordinates to reduce statistical fluctuations. The background model derived in this way still contains systematic uncertainties due to the underlying continuum emission and small differences in the event distributions in the (χ, ψ) space.

TABLE 2. ^{22}Na line flux and ejected mass as modelled and measured from novae.

Nova	M_{wd} $M_{\odot}$	M_{ej} msol	^{22}Na line flux, $cm^{-2}s^{-1}$ predicted[a]	measured[b]	XM_{ej}, 10^{-8} $M_{\odot}$ predicted[a]	derived[b]
V693 CrA	1.05	10^{-3}	5.5×10^{-4}	$\leq1.7\times10^{-5}$	2.0×10^{3}	≤170
V1370 Aql	1.1	10^{-3}	7.0×10^{-3}	$\leq3.5\times10^{-5}$	3.1×10^{3}	≤360
QU Vul	1.05–1.1	$10^{-3.5}$–10^{-3}	4.4×10^{-4}	$\leq3.05\times10^{-5}$	89	≤35
V838 Her	1.05	10^{-4}–$10^{-3.5}$	6.3×10^{-6}	$\leq3.3\times10^{-5}$	1.9	≤12
V1974 Cyg	1.1	$10^{-4.5}$	3.0×10^{-5}	$\leq2.1\times10^{-5}$	2.5	≤2.1
V382 Vel	1.1	??	??	$\leq8.1\times10^{-5}$[c]	2.0×10^{3}	≤4.3

[a]) initial flux, or ^{22}Na ejected mass (XM_{ej}) as derived from [24];
[b]) flux value, derived from COMPTEL measurements; XM_{ej} was calculated from the measured flux, time delay and nova parameters of Table 1 ([13], and this work);
[c]) [16].

Results and Conclusions

The derivation of ^{22}Na mass limits from the 1.275 MeV line flux measurements for a single nova is hampered by the fact that in many cases the distance to the nova is unknown. The extinction-based distances are not reliable due to the possible presence of circumstellar dust, or due to a wrong colour determination [26, 1]. Even relatively well studied novae are subject to uncertainties in the distance estimate and in the derived abundances. Examples of such uncertainties were addressed in recent publications [2, 9, 10]. These uncertainties have forced us to restrict our analysis to the cases of old (1981-1984) neon novae and to the most recent bright novae, like Nova Her 1991, Nova Cyg 1992 and Nova Velorum 1999, where overabundance of Ne is firmly established ([17]; [9] and references therein). Table 2 summarises the COMPTEL-limits on the ejected masses from the old novae NCrA 1981, NAql 1982 and NVul 1984, and the more recent neon novae NHer 1991, NCyg 1992 and NVel 1999. The new upper limit of the ^{22}Na yield from Nova Cyg 1992 is based on the improved flux upper limit of COMPTEL and a latest distance estimate of this nova [4]. To derive the distance of d~1.5 kpc to NVel 1999, we have used the standard relation between M_v, m_v, d and A_v=3.3×E(B-V). Here m_v=2.6 (IAUC No. 7177) and E(B-V)=0.2 (IAUC No. 7192) were used. M_v=–8.89±0.02 was evaluated from the relation between t_2 and M_v, as was proposed in [7].

We have to state that the detection of the ^{22}Na gamma-ray line remains an elusive goal for CGRO. In this respect it is useful to reiterate that **until now, no**

firm detection of Na exists in the UV or infrared spectra of novae with the largest Ne over-abundances known [10, 19, 27]. Therefore, an obvious question eventually arises: **Is the ^{22}Ne (neon-E) anomaly [3] related at all to the ^{22}Na production in novae?** All theoretical models of the nova runaway, as well as spectroscopic analysis of the novae abundances are pointing towards a large overproduction of Ne. However, there are no experimental results supporting the over-production of Na in novae!

Conclusion

In order to understand the explosive nucleosynthesis of high-mass white dwarf runaway in a binary system, one has to measure successfully the nova light-curve in the 1.275 MeV γ-ray line, ideally in combination with high-resolution spectroscopy of the sodium coronal IR lines. This hopefully will be achieved with the deployment of the next generation of space-borne γ-ray line and infrared spectrometers like SPI (INTEGRAL) and IRS (SIRTF).

Acknowledgements

The COMPTEL project is supported by the German government through DARA grant 50 QV 90968, by NASA under contract NAS5-266645, and by the Netherlands Organization for Scientific Research. AFI acknowledges financial support from the German Bundesministerium für Bildung, Wissenschaft, Forschung and Technologie.

REFERENCES

1. Andreä, J., Drechsel, H., & Starrfield, S., *A&A* **291**, 869 (1994).
2. Arkhipova, V.P., Esipov, V.F., Sokol, G.V., *Astr. Let.* **23**, No. 6, 819 (1997).
3. Black, D.C., *Geochim. Cosmochim. Acta* **36**, 377 (1972).
4. Chochol, D., et al., *A&A* **318**, 908 (1997).
5. Clayton, D.D., & Hoyle, F., *ApJ* **187**, L101 (1974).
6. Coc, A., et al., *A&A* **299**, 479 (1995).
7. Della Valle, M., & Livio, M., *ApJ* **452**, 704 (1995).
8. Gallagher, J.S. & Starrfield, S., *ARAA* **16**, 171 (1978).
9. Gehrz, R.D., Truran, J.W., Williams, R.E. and Starrfield, S., *PASP* **100**, 3 (1998).
10. Gehrz, R.D., Woodward, C.E., Greenhouse, M.A., et al. *ApJ* **421**, 762 (1994).
11. Gomez-Gomar, J., Hernanz, M., Jose, J. & Isern, J., *MNRAS* **296**, 913 (1998).
12. Harris, M.D., et al., Preprint LHEA, GSFC, NASA (1999).
13. Iyudin, A.F., Bennett, K., Bloemen, H., et al., *A&A* **300**, 422 (1995).
14. Jose, J., Hernanz, M., *ApJ* **494**, 680 (1998).
15. Leising, M.D., Share, G.H., Chupp, E.L. & Kanbach, G., *ApJ* **328**, 755 (1988).
16. Leising, M.D., et al., this conference proceedings (1999).
17. Livio, M., & Truran, J.W. *ApJ* **425**, 797 (1994).
18. Mahoney, W.A., et al., *ApJ* **262**, 742 (1982).
19. Salama, A., Evans, A., Eyres, S.P.S., et al., *A&A* **315**, L209 (1996).
20. Schönfelder, V., Aarts, H., Bennett, K., et al., *ApJS* **86**, 657 (1993).
21. Starrfield, S., Sparks, W.M. & Truran, J.W., *ApJS* **28**, 247 (1974).

22. Starrfield, S., et al., *MNRAS* **296**, 502 (1998).
23. Truran, J.W., in Essays in Nuclear Astrophysics, ed. C.A. Barnes, D.D. Clayton, & D.N. Schramm (Cambridge: Cambridge Univ. Press), 467 (1982).
24. Wanajo, S., Hashimoto, M.-A., & Nomoto, K., astro-ph/9905279 (1999).
25. Weiss, A., & Truran, J.W., *A&A* **238**, 178 (1990).
26. Williams, R.E., *ApJ* **426**, 279 (1992).
27. Williams, R.E., et al. *MNRAS* **212**, 753 (1985).

New theoretical results concerning gamma-ray emission from classical novae

Margarita Hernanz*, Jordi José*, Alain Coc†, Jordi Gómez-Gomar* and Jordi Isern*

**Institute for Spatial Studies of Catalonia (IEEC/CSIC/UPC), Edifici Nexus-201, C/Gran Capità 2-4, 08034 Barcelona (SPAIN).*
†Centre de Spectrométrie Nucléaire et de Spectrométrie de Masse, IN2P3-CNRS, Université Paris Sud, Bât. 104, 91405 Orsay Campus (FRANCE)

Abstract. New results concerning the synthesis of radioactive nuclei in classical nova explosions are presented, together with their influence on the gamma-ray emission from these objects and the prospects for the detectability with present and future instruments. The isotopes involved in gamma-ray emission from classical novae are ^{13}N, ^{18}F, ^{7}Be, ^{22}Na and ^{26}Al. Both ^{13}N and ^{18}F are crucial for the short duration annihilation emission, at 511 keV and below, that is emitted during the first hours after the outburst, before the nova maximum in visual magnitude. New recent results concerning the nuclear reaction rates involved in ^{18}F destruction have been incorporated into our hydrodynamic code, which models the nova explosion with a detailed follow up of the associated nucleosynthesis. As a result of the larger rates of ^{18}F destruction through proton captures, this isotope is produced in smaller quantities (typically by a factor of 10), leading to smaller fluxes in the 30-511 keV range. Concerning the medium and long lived isotopes, the inclusion of the latest reaction rates available in the NeNa-MgAl cycles provides an enhanced ^{22}Na production with respect to previous results, leading to a larger 1275 keV flux emitted through its decay. ^{26}Al is practically unchanged. The implications of our new results for the observability of novae by CGRO and by the future INTEGRAL instruments are addressed.

INTRODUCTION

Classical novae explosions occur in white dwarfs which accrete matter from a companion main sequence star, in a cataclysmic variable system. Hydrogen-rich matter piles up on the top of the white dwarf star, until burning happens in degenerate conditions, leading to a thermonuclear runaway. The explosive burning of H-rich matter (which has been mixed with matter from the underlying CO or ONe core by some unknown mechanism) leads to the synthesis of some radioisotopes, such as ^{13}N, ^{18}F, ^{7}Be, ^{22}Na and ^{26}Al. The decay of ^{13}N (τ=862 s) and ^{18}F (τ=158 minutes) produces no direct gamma-rays but the emission of a positron. Electron-positron annihilation produces a 511 keV line (with some blueshift and

CP510, *The Fifth Compton Symposium,* edited by M. L. McConnell and J. M. Ryan

width related to the expansion of the envelope) and a continuum below it; this continuum comes both from the the positronium continuum and from the Comptonization of photons emitted in the 511 keV line. The decay of ^{7}Be (τ=77 days), ^{22}Na (τ=3.75 years) and ^{26}Al (τ=10^{6} s) produces a photon of 478, 1275 and 1809 keV, respectively, plus positron emission for ^{22}Na and ^{26}Al.

The potential role of novae as gamma-ray emitters has been pointed out some years ago [1–3], but there was a lack of detailed studies of this gamma-ray emission and its relationship with particular models of nova explosions. The use of realistic profiles of densities, velocities and chemical abundances is crucial for the determination of the gamma-ray spectrum at different epochs [4–6].

GAMMA-RAY EMISSION FROM INDIVIDUAL NOVAE

A good knowledge of all the nuclear reaction rates involved in the synthesis of the abovementioned radioactive nuclei is crucial for the computation of their final yields in classical novae explosions. We have adopted the most recent prescriptions for all the rates involved; the main changes obtained with respect to previous works [4–6] affect ^{18}F synthesis (see [7] for details), which is now underproduced by around a factor of 10, when the new rates from [8] for the ^{18}F(p,α) reaction are adopted. However, this rate is still far from being well known at nova temperatures. On the contrary, ^{7}Be and ^{13}N yields are not significantly affected by uncertain or new nuclear reaction rates, while ^{22}Na shows some slight variations. The ejected masses of both ^{22}Na, which is important for the gamma-ray emission of individual novae, and ^{26}Al, which is relevant for their cumulative emission (i.e., to elucidate the possible contribution of novae to the Galactic content of ^{26}Al deduced from CGRO/COMPTEL observations of the diffuse 1809 keV Galactic emission), are displayed in table 1. Values labeled "old rates" correspond to results in [10], whereas the other ones give the range of ejected masses around the nominal value resulting from nuclear physics uncertainties. As can be seen, all new ^{22}Na yields are larger than previous ones, whereas for ^{26}Al similar yields are obtained. The uncertainties (maximum/minimum ratio in table 1) are moderate (from 2 to 7), but large enough to encourage new experiments concerning selected nuclear reactions (see [9] for details). In summary, if we adopt a mass of 1.15 $M_\odot$, CO and ONe novae eject typically the following (in $M_\odot$): ^{7}Be: 10^{-10} (CO) and 10^{-11} (ONe), ^{22}Na: 4 10^{-12} (CO) and 7 10^{-9} (ONe), ^{26}Al: 6 10^{-10} (CO) and 2 10^{-8} (ONe). Finally, the short-lived isotopes ^{13}N and ^{18}F are synthesized by similar amounts in CO and ONe novae: typically, the ejected masses are 10^{-8} $M_\odot$ for ^{13}N and $10^{-9} M_\odot$ for ^{18}F, 1 hour after peak temperature.

Gamma-ray spectra for a CO (1.15$M_\odot$) and an ONe (1.25$M_\odot$) nova at different epochs after the explosion (defined as the time of peak temperature) are shown in figure 1. Two different types of emission appear: lines (511 keV, with very short duration, and 478 and 1275 keV, with moderate and long duration) and a continuum between 30 keV and 511 keV. Concerning the 511 keV line and the continuum (related to ^{13}N and ^{18}F decays), CO and ONe novae display a similar behaviour, although ONe novae emit larger fluxes. For the 478 and 1275 keV lines (^{7}Be and ^{22}Na decays, respectively), the gamma-ray signatures of CO and ONe novae are different, because of the different yields of radioactive elements (see previous paragraph): CO novae will show the ^{7}Be line whereas ONe novae will show the ^{22}Na line.

The time evolution of the different types of emission is better seen in the light curves. For the continuum, 30 keV$\leq$E$\leq$511 keV, and the 511 keV line (figure 2, left panel), a very intense emission (maximum flux around 10^{-2} phot/cm^2/s) emerges, but it disappears very fast, because of the short lifetimes of ^{13}N and ^{18}F. It is worth noticing that the flux in the continuum is larger by more than a factor of 10 than the flux in the line (full width half maximum, FWHM, of the line is 8 keV). The detectability distances for these fluxes should be as large as $\sim$3 kpc with the future SPI instrument onboard INTEGRAL (to be compared with the more promising value of $\sim$10 kpc that we obtained with the old rates for ^{18}F destruction [6]). The main problem for the detectability of this early emission is that it appears well before the maximum in visual magnitude, i.e., well before the visual discovery of the nova. In the future, if some gamma-ray instrument is sensitive enough, it seems clear that Galactic novae could be discovered through their emission in gamma-rays, before their optical discovery. Nova discoveries through gamma-rays would even be made in regions where visual extinction prevents the optical discovery, helping to determine the distribution and the rates of novae in our Galaxy. We are at present analyzing BATSE data for the periods and the locations of nearby recent novae, to see if some signal was detected. Because of its continuous monitoring of the whole sky at the relevant energy range, CGRO/BATSE would be the perfect instrument to detect the early gamma-ray emission of novae, except for its limited sensitivity and energy resolution (see paper by Hernanz, Smith, Fishman et al. in these same proceedings).

The ^{22}Na line at 1275 keV is weaker than the 511 keV line and the continuum, but it has a longer duration, because of the longer lifetime of ^{22}Na (3.75 years). In figure 2 (right panel) we show the light curve for a typical ONe nova (1.25M$_\odot$), with the different possibilities mentioned above for the nuclear reaction rates affecting ^{22}Na synthesis. We obtain a larger ejected mass by a factor of $\sim$6 (for our recommended yield, see table 1), which translates into a larger detectability distance by a factor of $\sqrt{6}$ with respect to our previous results in [6]. Therefore, detectability distances around 1 kpc during months are obtained for INTEGRAL/SPI (where the 20 keV FWHM of the line, which worsens SPI's sensitivity with respect to the narrow line case, has been fully taken into account). It is important to notice that the non detection up to now of the ^{22}Na line with CGRO/COMPTEL [11], from neon novae like Nova Cyg 1992, is compatible with our models, which eject ^{22}Na masses below the threshold for detectability of that instrument (i.e., $\sim 10^{-8}$M$_\odot$, for a distance of $\sim$2 kpc [11]).

For the ^{7}Be line at 478 keV there are no changes with respect to [6]: maximum fluxes of around 10^{-6} phot/cm^2/s are obtained (FWHM 7 keV) for a CO nova of

TABLE 1. Ejected masses (in $M_\odot$) of ^{22}Na and ^{26}Al for ONe novae

M_{wd}	Old rates	Minimum	Nominal	Maximum	Maximum/minimum
			^{22}Na		
1.15	1.0 10^{-9}	3.1 10^{-9}	7.0 10^{-9}	1.4 10^{-8}	4.5
1.25	1.3 10^{-9}	3.4 10^{-9}	6.3 10^{-9}	1.2 10^{-8}	3.5
1.35	2.6 10^{-9}	3.4 10^{-9}	4.4 10^{-9}	6.2 10^{-9}	1.8
			^{26}Al		
1.15	1.8 10^{-8}	8.6 10^{-9}	2.1 10^{-8}	3.1 10^{-8}	3.6
1.25	7.6 10^{-9}	3.6 10^{-9}	1.2 10^{-8}	1.6 10^{-8}	4.4
1.35	3.2 10^{-9}	6.6 10^{-10}	3.2 10^{-9}	4.8 10^{-9}	7.3

1.15$M_\odot$, yielding detectability distances with INTEGRAL/SPI of around 0.5 kpc during a few weeks.

The detection of the gamma-ray emission from classical novae, which is a challenge for the present and future gamma-ray instrumentation, would provide a direct confirmation of the thermonuclear runaway model for nova explosions, with a direct determination of the synthesis of radioisotopes (from the lines) and crucial information about the conditions in the expanding envelope (from the continuum).

REFERENCES

1. Clayton, D.D., Hoyle, F., *ApJ* **187**, L101 (1974).
2. Clayton, D.D., *ApJ* **244**, L97 (1981).
3. Leising, M., Clayton, D.D., *ApJ* **323**, 157 (1987).
4. Hernanz, M., Gómez-Gomar, J., José, J., Isern, J., *2nd INTEGRAL Workshop, The Transparent Universe*, Noordwijk: ESA SP-382, 1997, p. 47.
5. Hernanz, M., Gómez-Gomar, J., José, J., Isern, J., *Proceedings of the Fourth Compton Symposium*, New York: AIP Conference Proceedings 410, 1997, p. 1125.
6. Gómez-Gomar, Hernanz, M., J., José, J., Isern, J., *Mon. Not. R. Astron. Soc.* **296**, 913 (1998).
7. Hernanz, M., José, J., Coc, A., Gómez-Gomar, J., Isern, J., *ApJ Lett.* **526**, in press (1999).
8. Utku, S., et al., *Phys. Rev. C* **57** 2731 (erratum 58, 1354)
9. José, J., Coc, A., Hernanz, M., *ApJ* **520**, 347 (1999).
10. José, J., Hernanz, M., *ApJ* **494**, 680 (1998).
11. Iyudin, A.F. et al., *Astron. & Astrophys.* **300**, 422 (1995).

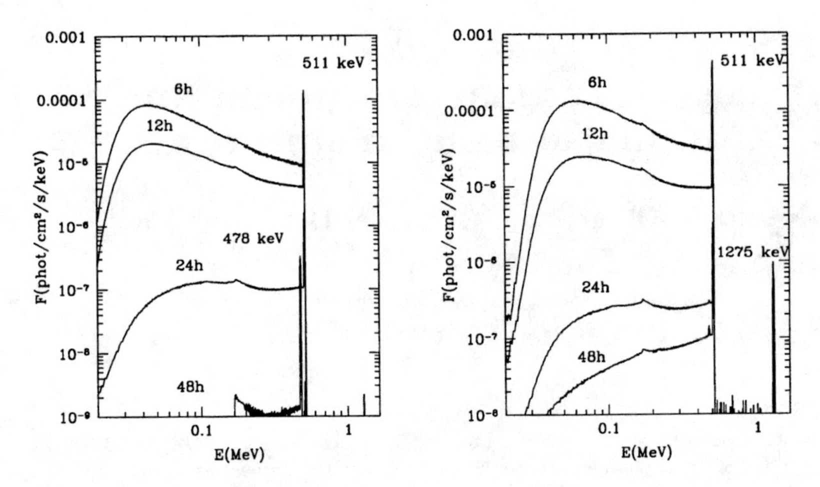

FIGURE 1. (Left) Spectral evolution of a CO nova ($1.15M_{\odot}$) at a distance of 1 kpc. (Right) Same for an ONe nova ($1.25M_{\odot}$)

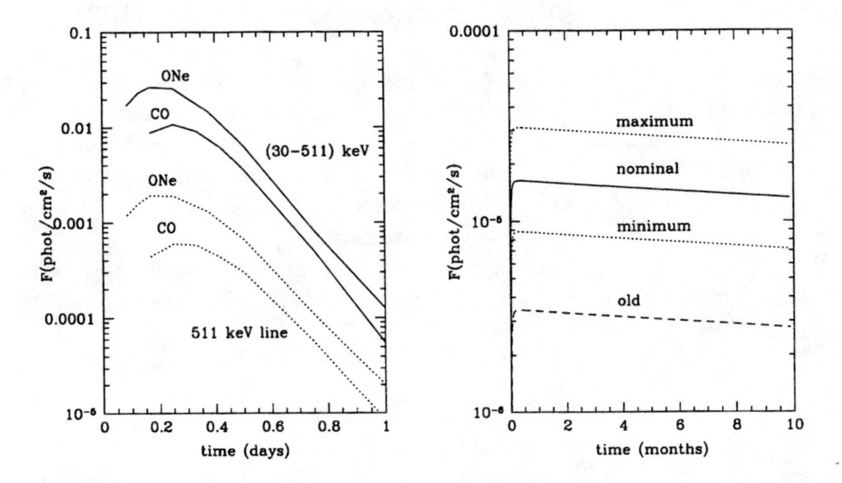

FIGURE 2. (Left) Light curves of the continuum and the 511 keV line (FWHM=8 keV) for a CO nova ($1.15M_{\odot}$) and an ONe nova ($1.25M_{\odot}$) at a distance of 1 kpc. (Right) Light curve of the 1275 keV line (FWHM=20 keV) for an ONe nova ($1.25M_{\odot}$), for different prescriptions of the nuclear reaction rates

Synthesis of radioactive nuclei and the annihilation line from novae

Nikolai N. Chugai* and Alexander D. Kudryashov**

*Institute of Astronomy RAS, Moscow 109017, Russia
** VINITI, Moscow 125219, Russia

Abstract. A possible range of the abundance of radioactive nuclei in CO and ONeMg nova ejecta was found via kinetic computations of the thermonuclear burning in the one-zone hydrogen-rich envelope. We calculate the light curve of the annihilation line for various conditions and found that the principal parameter responsible for the uncertainty of the predicted flux is the mixing degree. The annihilation line observations of ^{18}F and an independent estimate of the total amount of synthesised ^{18}F from ^{18}O may be used to assess the mixing degree of radioactive nuclei in the envelope.

INTRODUCTION

The thermonuclear runaway on accreting white dwarfs (WD) is a conventional model for nova outbursts. High temperature hydrogen burning converts most of initial CNO nuclei to radioactive ^{13}N, ^{14}O, ^{15}O, ^{17}F and ^{18}F (NOF isotopes); their presence in novae may be evidenced by the 511 keV annihilation emission [1–3]. Unfortunately, a theory is not able as yet to describe the very complex nova phenomenon, and therefore the luminosity of annihilation line is poorly predicted. Our main goal here is to investigate a dependence of the annihilation line on various model characteristics, namely, the velocity and density distributions in the ejecta, and a mixing degree.

NUCLEOSYNTHESIS

We compute nucleosynthesis in novae for the one-zone model [4]. Apart from H and He the kinetic network includes nuclei from C to Ca and all the nuclear reactions with charged particles. The temperature and the density are assumed to be constant during the burning, which is terminated at the exhaustion of the hydrogen fraction ΔX=0.1. The assumed temperature is in the range $(1-3)\times 10^8$ K, and the density is 10^4 g cm^{-3}. The composition of CO and ONeMg dwarfs is taken according to [5]. The mixing parameter q (a mass fraction of WD matter in the total envelope mass) is varied in the range 0.1–0.9.

CP510, *The Fifth Compton Symposium,* edited by M. L. McConnell and J. M. Ryan

TABLE 1. Isotope composition for typical ($T_8 = 2$, $q = 0.5$, top) and optimistic ($T_8 = 1$, $q = 0.7$, bottom) cases.

WD	ϵ, 10^{17} erg g^{-1}	^{13}N (862 s)	^{14}O (102 s)	^{15}O (176 s)	^{17}F (95 s)	^{18}F (158 min)
CO	4.5	4.8e-4	2.1e-1	2.5e-1	8.4e-2	2.4e-3
ONeMg	5.6	1.6e-4	1.1e-2	8.1e-2	6.0e-2	1.7e-3
CO	6.1	1.0e-1	2.1e-2	5.8e-2	5.4e-3	7.4e-3
ONeMg	6	2.3e-2	4.2e-3	1.4e-2	1.0e-3	3.9e-2

The obtained amount of NOF isotopes in both types of WD is roughly equal to the total mass of admixed CO matter of WD in the envelope and thus is proportional to the mixing parameter q. The fraction of ^{18}F in most cases is in the range $10^{-3} - 10^{-2}$.

We summarize some results of nucleosynthesis computations in Table 1, where the composition of major radioactive isotopes is given for average values of T and q parameters (typical case) and for parameters, which favour a maximum production of ^{18}F (optimistic case). The computed isotope composition refers to the burning zone, which presumably occupies a fraction ψ of the envelope mass. This fraction is computed from the energy balance (a nuclear energy is a sum of the gravitation and the kinetic energy) for $1M_\odot$ WD. Before the ejection, synthesised isotopes are mixed in the inner fraction $f_{\rm mix}$ of the envelope.

ANNIHILATION LINE LUMINOSITY

To compute the luminosity of escaping 511 keV photons created by a density dependent annihilation we considered a set of models. The kinematics of ejecta is either the homologous expansion ($v = r/t$) or the wind with the constant velocity and the mass-loss rate. The kinetic luminosity of the wind is set at the Eddington limit $L_{\rm k} = (1/2)\dot{M}v^2 = 10^{38}$ erg s^{-1}. Models with different kinematics, density distribution, ejecta mass, velocity, mixing degree, abundance of radioactive nuclei, and dwarf type are presented in Table 2. The table shows also the output parameters, viz., ψ, the maximum flux F of the second luminosity peak ($t \approx 2 \times 10^4$ s) related to ^{18}F, the fluence Φ, and the characteristic duration ($\Delta t = \Phi/F$).

The effects of the variation of the kinematics and the density distribution are shown (Fig. 1a) for models of homologously expanding envelopes (HH1 and HP) and the wind model (W). The light curve consists of the two major maxima related primarily to ^{13}N (first maximum) and ^{18}F (second maximum). The model HH1 with the uniform density produces slightly higher second maximum (^{18}F) compared to the power law density $\rho \propto v^{-7}$ (model HP), while the wind model W is characterized by the lowest luminosity. This is because the mass of the transparent outer layer depends on the density distribution. For the HH1 model the mass of the transparent layer is higher than for HP, while for the wind model W the mass of the transparent outer layer is even lower.

TABLE 2. Parameters of novae and annihilation line flux ($d = 1$ kpc).

Model	WD	M $10^{-5} M_\odot$	V_{max} km s^{-1}	f_{mix}	ψ	F 10^{-4} cm^{-2} s^{-1}	Φ cm^{-2}	Δt 10^4 s
HP	CO	2	2500	1	0.31	19	53	2.8
HH1	CO	2	2500	1	0.34	33	107	3.2
W	CO	2	2500	1	0.28	1	2.6	2.6
HH2	CO	2	2500	0.34	0.34	0.3	1	3.7
HH3	CO	2	2500	0.99	0.34	28	85	3.1
HH4	CO	10	2500	1	0.34	34	116	3.4
HH5	CO	2	3500	1	0.38	70	214	3
HH6	ONeMg	2	2500	1	0.25	17	56	3.2
HH7	CO	2	2500	1	0.34	83	280	3.4
HH8	ONeMg	2	2500	1	0.25	400	1300	3.3

The effects of the variations of mass, velocity, composition, and mixing degree are studied here for homologous homogeneous models HH1–HH6 (Fig. 1b). The variation of mass has little effect on the luminosity since the mass of the outer transparent layer is essentially the same. The velocity is an important parameter since the luminosity is proportional to v^2 (compare HH5 and HH1 models). The composition affects the luminosity in an obvious way: ONeMg dwarf (HH6 model) in the typical case gives a lower luminosity because of the lower abundance of NOF isotopes.

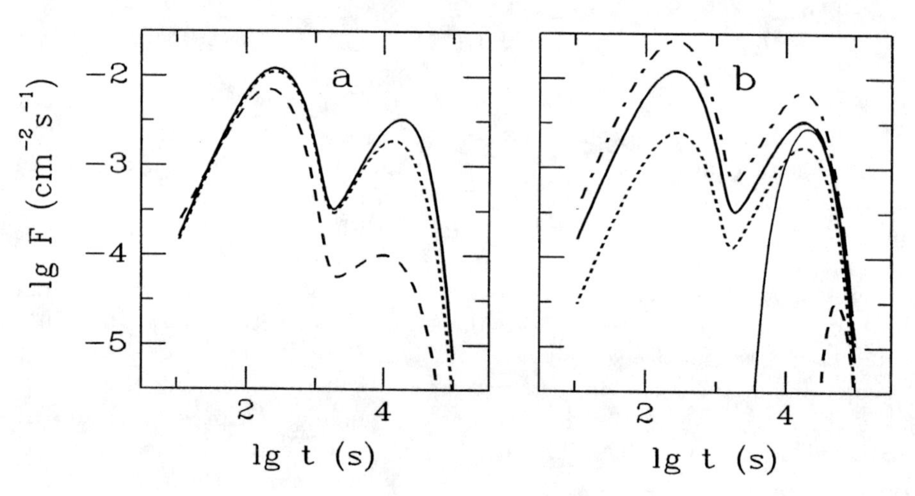

FIGURE 1. Annihilation line flux from model novae. The left panel (a) shows the effect of the kinematics and the density distribution for homologous models HH1 (solid line), HP (dotted), and for the wind outflow model W (dashed). On the right panel (b) the template model HH1 (thick solid line) is compared to the model without mixing (HH2, short dashes), nearly homogeneous mixing (HH3, thin solid line), higher ejecta mass (HH4, long dashes), higher velocity (HH5, dash-dotted), ONeMg WD composition (HH6, dotted).

The most crucial parameter is the mixing degree. The minimal mixing restricted by the burning zone ($f_{\rm mix} = \psi = 0.34$) leads to the severe suppression of both maxima of the light curve. Even nearly complete mixing, $f_{\rm mix} = 0.99$, with only 1% of the unmixed outer shell results in the strong suppression of the gamma-ray luminosity at the initial stage $t < 10^4$ s. Therefore, the incomplete mixing, basically plausible in real novae, is a principal factor to affect the detectability of the annihilation line from novae. The situation with the detection of the annihilation line is somewhat better in the optimistic case of the nuclesynthesis (HH7 and HH8 models). The luminosity of the second maximum is three times higher for the CO dwarf and 20 times higher for the ONeMg dwarf as compared to the typical case (Fig. 2).

The fluxes from novae at 1 kpc presented in Table 2 lie in the range of $10^{-4} - 4 \times 10^{-2}$ cm^{-2} c^{-1}, while the characteristic duration is $(2-4) \times 10^4$ s. The upper limit for the flux corresponds to the optimistic case of nucleosynthesis, the extremely high velocities of novae ejecta and the homogeneous mixing. Note that this upper limit flux and duration are close to those in [2,3]. With the detection limit of BATSE for 0.5 day-long events of 2×10^{-3} cm^{-2} c^{-1} [6], novae with the above maximum luminosity of the annihilation line could be detected at the distances < 4.5 kpc. For future observations of the annihilation line from novae not only the flux evolution but the profile of the line is essential as well. The annihilation line profile for the HP model (see Table 2) with homogeneously mixed isotopes is shown in Fig. 3. The three representative moments correspond to pre-, near-, and post-maximum epochs.

The annihilation line observation from novae generally does not provide the ^{18}F mass, neither in the detection nor in the non-detection case, since the emergent

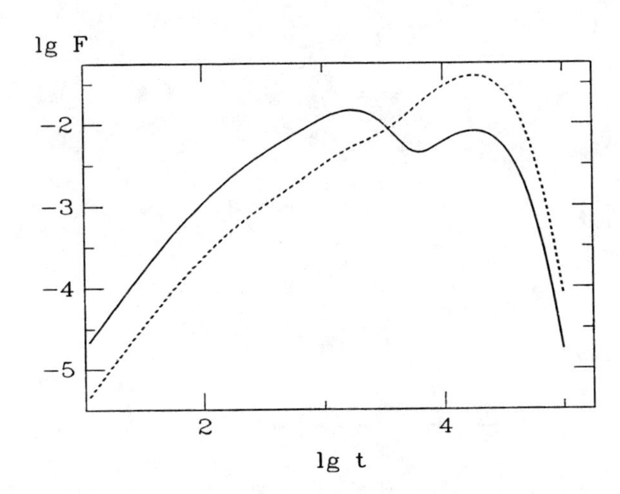

FIGURE 2. Annihilation line flux for CO WD (solid line) and ONeMg WD in the optimistic case.

annihilation luminosity depends on the amount of ^{18}F and also on the mixing degree. The situation with the total mass of ^{18}F and the mixing degree may be clarified from the follow-up determination of ^{18}O amount, the product of ^{18}F decay, using an observation of the CO emission in the fundamental band of 4.6 μm and/or, possibly, first overtone of 2.3 μm. At present only two detections of the CO 4.6 μm emission from novae are known, namely, NQ Vul 1976 [7] and V Cas 1993 [8] at the epoch immediately before the dust formation. One may hope to distinguish the contribution of CO molecules with the different isotope into the emission band. For reference, the isotope shifts ($|\Delta\nu|/\nu$) of molecules $^{12}C^{18}O$, $^{13}C^{16}O$, and $^{13}C^{18}O$ are 0.050, 0.046, and 0.101, i.e., larger than the Doppler shift, which is 0.01 assuming the expansion velocity 3000 km s^{-1}.

REFERENCES

1. Clayton, D.D., and Hoyle, F., *ApJ.* **187**, L101 – L103 (1974).
2. Leising, M.D., and Clayton, D.D., *ApJ.* **323**, 159 – 169 (1987).
3. Gomez-Gomar, J., Hernanz, M., Jose, J., and Isern, J., *MNRAS.* **296**, 913 – 922 (1998).
4. Kudryashov, A.D., and Tutukov, A.V., *Astron. Reports.* **39**, 482 – 488 (1995).
5. Kudryashov, A.D., Chugai, N.N., and Tutukov, A.V., *Astron. Reports.* (in press) (2000).
6. Smith D.M., Leventhal, M., Cavallo, R., et al., *ApJ.* **471**, 783 – 795 (1996).
7. Shenavrin, V.I., and Moroz, V.I., *Pis'ma Astron. Zh.* **2**, 99 – 100 (1976).
8. Evans A., Geballe, T.R., Rawlings, J.M.C., and Scott, A.D., *MNRAS.* **282**, 1049 – 1058 (1996).

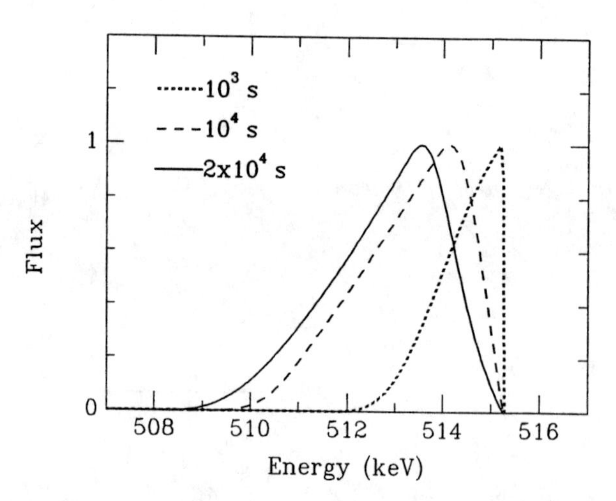

FIGURE 3. Annihilation line profile for three epochs of HP model.

X-RAY BINARIES

Monitoring the Short-Term Variability of Cyg X-1: Spectra and Timing

K. Pottschmidt*, J. Wilms*, R. Staubert*, M.A. Nowak†, J.B. Dove‡,||, W.A. Heindl¶, D.M. Smith§

* *Institut für Astronomie und Astrophysik, Waldhäuser Str. 64, D-72076 Tübingen, Germany,*
† *JILA, University of Colorado, Boulder, CO 80309-440, U.S.A.,*
‡ *CASA, University of Colorado, Boulder, CO 80309-389, U.S.A.,*
|| *Dept. of Physics, Metropolitan State College of Denver, Denver, CO 80217-3362, U.S.A.,*
¶ *CASS, University of California San Diego, La Jolla, CA 92093, U.S.A.,*
§ *SSL, University of California Berkeley, Berkeley, CA 94720, U.S.A.*

Abstract. We present first results from the spectral and temporal analysis of an RXTE monitoring campaign of the black hole candidate Cygnus X-1 in 1999. The timing properties of this hard state black hole show considerable variability, even though the state does not change. This has previously been noted for the power spectral density, but is probably even more pronounced for the time lags. From an analysis of four monitoring observations of Cyg X-1, separated by 2 weeks from each other, we find that a shortening of the time lags is associated with a hardening of the X-ray spectrum, as well as with a longer characteristic "shot time scale". We briefly discuss possible physical/geometrical reasons for this variability of the hard state properties.

INTRODUCTION

For stellar black hole candidates, several distinct states can be identified that differ in their general spectral and temporal properties. Based mainly on spectral arguments these states have been associated with different accretion rates and different geometries of the accretion flow (e.g., Esin et al., 1998; Nowak, 1995). With broad band instruments like the Rossi X-ray Timing Explorer (RXTE) it is possible to study the states with high time resolution over a time base of several years. The focus of this work lies on parameters and functions characterizing the short-term variability (< 1000 s) of the canonical black hole candidate Cyg X-1 and their stability in the hard state.

In 1996 and 1997 observations of Cyg X-1 with the pointed RXTE instruments were not performed regularly and mainly concentrated on the ~3 month long soft state in 1996 (see, e.g., Cui et al., 1998). We initiated a monitoring campaign of the hard state in 1998 (3 ksec exposures), which we expanded to 10 ksec exposures in 1999 in order to allow the calculation of Fourier frequency dependent time lags with

CP510, *The Fifth Compton Symposium,* edited by M. L. McConnell and J. M. Ryan

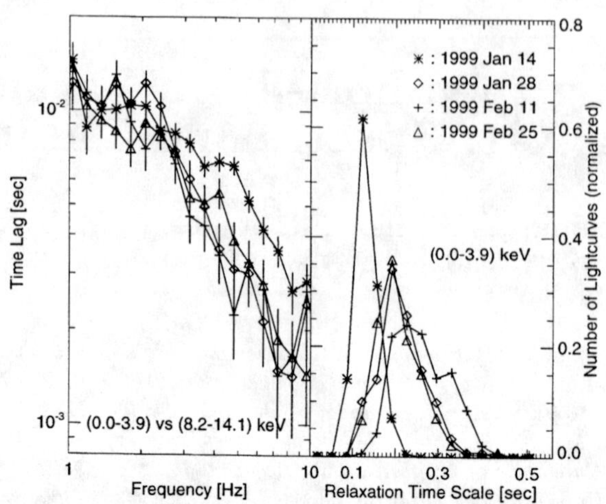

FIGURE 1. Comparison of four consecutive RXTE observations of 1999 January and February, spaced by ~14 d each. **Left:** Time lags as a function of Fourier frequency. The time lag in the 1 to 10 Hz band changes by a factor of three over the course of a month. Note the logarithmic y-axis! **Right:** Distribution of the relaxation time scale τ found from short (32 s long) time segments for these observations. Observations with shorter time lags appear to have larger τ.

sufficient accuracy (Fig. 1). Additionally, the RXTE observations are accompanied by simultaneous radio pointings. The aim of this campaign is to address fundamental questions such as the cause of the long term flux variability in the hard state, namely the 150 d periodic behavior seen in the RXTE All Sky Monitor and in the radio flux (Pooley, Fender & Brocksopp, 1999, Hjellming, priv. comm.). A precessing, interacting disk-jet system has been suggested as one possible explanation for this hard state cycle (Brocksopp et al., 1999).

We have performed spectral and/or temporal analyses on ~ 30% of the available public data measured before 1999. In addition, we have analyzed those of our 2 weekly observations in 1999 that were scheduled before the gain change of the RXTE Proportional Counter Array (PCA) in 1999 March (for data after the gain change the calibration and background models are still uncertain). In this paper we present first results from these monitoring observations. Using the `ftools` 4.2, we extracted PCA spectra and high (2 ms) time resolution PCA lightcurves. We computed periodograms for several energy bands, as well as the time lags, and the coherence function between these energy bands (Nowak et al., 1999a). In addition, we use the linear state space model (LSSM) to model the light curves in the time domain. This method allows one to derive a characteristic time scale, τ, that can explain the dynamics of the lightcurve. τ can be interpreted in terms of a shot noise relaxation time scale, but note that LSSMs only need a single time scale to provide a good fit of the lightcurve (see Pottschmidt et al., 1998, for an application of the LSSM to EXOSAT data from Cyg X-1).

VARIABILITY OF SPECTRAL AND TEMPORAL PROPERTIES

Fourier frequency dependent time lags of up to ~0.1 sec are known to exist between different energy bands in Cyg X-1. While the lags increase with energy, they cannot be explained by the diffusion time scale of photons in a Compton corona alone (Miyamoto & Kitamoto, 1989; Nowak et al., 1999a). Nevertheless, the characteristic time lag "shelves" allow to roughly constrain coronal parameters (Nowak et al., 1999c). We find that over the course of weeks, the typical time lag in the hard state can vary by at least a factor three (Fig. 1, left panel). The first three observations show a gradual decrease in the time lags, while the fourth observation has intermediate values. This systematic development is mirrored by the shot relaxation time scale τ, which gets larger for observations with smaller time lag (Fig. 1, right panel).

At the same time, the X-ray spectrum also changes systematically (Fig. 2). Spectral fitting of black hole candidate spectra with the PCA is severely affected by the uncertainty of the PCA response matrix. Although the data exhibit a clear and varying hardening above $\sim$ 10 keV, it is difficult to associate these changes with physically interpretable spectral parameters. For example, both, a broken power-law and a power law reflected from cold matter result in acceptable fits. This behavior is similar to GX 339−4 (Wilms et al., 1999). In order to characterize the spectral variability of Cyg X-1 independently of any spectral model, therefore, we directly compared the data in detector space. Fig. 2 displays the relative deviation of the four observations with respect to the observation of 1999 Jan 28. Cyg X-1 is clearly spectrally variable on a time basis of 14 d (note that part of the variation could be due to orbital modulation).

Comparison of Figs. 1 and 2 shows that a spectral hardening of the source correlates with a decrease of the time lags and with an increase of the relaxation time τ. Recently, Gilfanov, Churazov & Revnivtsev (1999) also analyzed several of the public RXTE observations of Cyg X-1. They found a variability of the spectral hardness of the same order as presented here and an increase of the PSD break frequency with the reflection fraction. They also confirmed for Cyg X-1 a correlation between the intrinsic spectral slope and the reflection fraction (Zdziarski, Lubiski & Smith, 1999), as well as a relationship between two temporal parameters, namely the PSD "break frequency" and the PSD "hump frequency" (Wijnands & van der Klis, 1999; Psaltis, Belloni & van der Klis, 1999).

Due to the long time basis of the available RXTE data it is also possible to compare observations that are widely spaced in time, e.g., the 1999 monitoring observations with an observation made more than two years earlier, in 1996 Oct 23. The latter has previously been published in a series of papers (Dove et al., 1998; Nowak et al., 1999a,c). It was performed shortly after the soft state of 1996, and we cautioned, therefore, that the observation might still have been "contaminated" by soft state peculiarities. But, the comparison with the observation of 1999 Feb

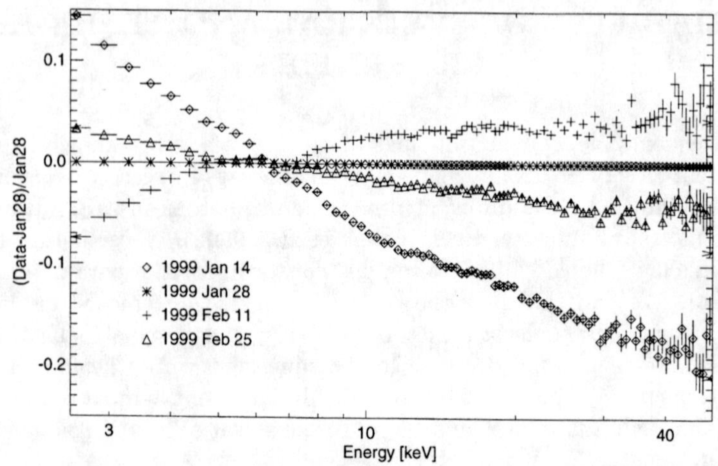

FIGURE 2. Relative deviation of the shape of the RXTE PCA count rate spectra from the observation of 1999 Jan 28.

25 shows almost identical PSDs and time lags. So, we see that the source really was in its hard state and that the hard state timing properties can be reproduced with great accuracy on the time scale of years.

DISCUSSION

We have presented first results from our systematic analysis of RXTE data of Cyg X-1 in the hard state. Apparently, during the canonical hard state this source can vary by up to a factor of ~ 2 in 2–50 keV flux and by up to a factor of three in the associated time lags within a few weeks. On the other hand, we were also able to identify data with almost identical spectral and temporal behavior spaced by more than two years.

As we noted in the previous section, there is possible evidence for a correlation of the changes in the spectral and temporal behavior of the source. Harder spectra appear to be associated with shorter time lags, similar to the hard state of GX 339−4 (Nowak et al., 1999b). A possible interpretation would be that the accretion disk penetrates to smaller disk radii at times of harder flux, thereby increasing the reflection fraction of the Comptonized radiation (see also Gilfanov, Churazov & Revnivtsev, 1999), i.e., hardening the spectrum, and shortening the time-delay of the harder photons (with the smaller system geometry corresponding to shorter lags). Alternatively, the harder spectra might be due to changes in the coronal parameters: our results might indicate that coronae with larger optical depth and/or temperature are physically smaller. This is also consistent with the development of the shot time scale in the sense that more scattering events lead to longer relaxation times.

Acknowledgments

We thank all participants in the 1999 broad band campaign for their continued effort to obtain simultaneous radio through X-ray observations of Cygnus X-1. This work has been partially financed by DFG grant Sta 173/22 and a travel grant by the Deutsche Forschungsgemeinschaft to JW.

REFERENCES

Brocksopp, C., Fender, R. P., Larionov, V., Lyuty, V. M., Tarasov, A. E., Pooley, G. G., Paciesas, W. S., & Roche, P., 1999, MNRAS, in press (astro-ph/9906365)

Cui, W., Ebisawa, K., Dotani, T., & Kubota, A., 1998, ApJ, 493, L75

Dove, J. B., Wilms, J., Nowak, M. A., Vaughan, B., & Begelman, M. C., 1998, MNRAS, 289, 729

Esin, A. A., Narayan, R., Cui, W., Grove, J. E., & Zhang, S.-N., 1998, ApJ, 505, 854

Gilfanov, M., Churazov, E., & Revnivtsev, M., 1999, A&A, in press (astro-ph/9910084)

Miyamoto, S., & Kitamoto, S., 1989, Nature, 342, 773

Nowak, M. A., 1995, PASP, 107, 1207

Nowak, M. A., Vaughan, B. A., Wilms, J., Dove, J. B., & Begelman, M. C., 1999a, ApJ, 510, 874

Nowak, M. A., Wilms, J., & Dove, J. B., 1999b, ApJ, 517, 355

Nowak, M. A., Wilms, J., Vaughan, B. A., Dove, J. B., & Begelman, M. C., 1999c, ApJ, 515, 726

Pooley, G. G., Fender, R. P., & Brocksopp, C., 1999, MNRAS, 302, L1

Pottschmidt, K., König, M., Wilms, J., & Staubert, R., 1998, A&A, 334, 201

Psaltis, D., Belloni, T., & van der Klis, M., 1999, ApJ, 520, 262

Wijnands, R., & van der Klis, M., 1999, ApJ, 514, 939

Wilms, J., Nowak, M. A., Dove, J. B., Fender, R. P., & di Matteo, T., 1999, ApJ, 522, 460

Zdziarski, A. A., Lubiski, P., & Smith, D. A., 1999, MNRAS, 303, L11

The Spectral Variability of Cygnus X-1 at MeV Energies

M.L. McConnell*, K. Bennett**, H. Bloemen†, W. Collmar††, W. Hermsen†, L. Kuiper†, B. Phlips‡, J.M. Ryan*, V. Schönfelder††, H. Steinle††, A.W. Strong††

*Space Science Center, University of New Hampshire, Durham, NH 03824, USA
**Space Science Department, ESTEC, Noordwijk, The Netherlands
†Space Research Organization of the Netherlands (SRON), Utrecht, The Netherlands
††Max Planck Institute for Extraterrestrial Physics (MPE), Garching, Germany
‡George Mason University, Fairfax, VA 22030, USA

Abstract. In previous work, we have used data from the first three years of the CGRO mission to assemble a broad-band γ-ray spectrum of the galactic black hole candidate Cygnus X-1. Contemporaneous data from the COMPTEL, OSSE and BATSE experiments on CGRO were selected on the basis of the hard X-ray flux (45–140 keV) as measured by BATSE. This provided a spectrum of Cygnus X-1 in its canonical low X-ray state (as measured at energies below 10 keV), covering the energy range from 50 keV to 5 MeV. Here we report on a comparison of this spectrum to a COMPTEL-OSSE spectrum collected during a high X-ray state of Cygnus X-1 (May, 1996). These data provide evidence for significant spectral variability at energies above 1 MeV. In particular, whereas the hard X-ray flux *decreases* during the high X-ray state, the flux at energies above 1 MeV *increases*, resulting in a significantly harder high energy spectrum. This behavior is consistent with the general picture of galactic black hole candidates having two distinct spectral forms at soft γ-ray energies. These data extend this picture, for the first time, to energies above 1 MeV.

INTRODUCTION

Observations by the instruments on CGRO, coupled with observations by other high-energy experiments (e.g., SIGMA, ASCA and RXTE) have provided a wealth of new information regarding the emission properties of galactic black hole candidates. An important aspect of these high energy radiations is spectral variability, observations of which can provide constraints on models which seek to describe the global emission processes. Based on observations by OSSE of seven transient galactic black hole candidates at soft γ-ray energies (i.e., below 1 MeV), two γ-ray spectral shapes have been identified that appear to be well-correlated with the soft

CP510, *The Fifth Compton Symposium,* edited by M. L. McConnell and J. M. Ryan

X-ray state [1,2]. In particular, these observations define a *breaking* γ-ray spectrum that corresponds to the low X-ray state and a *power-law* γ-ray spectrum that corresponds to the high X-ray state. (Here we emphasize that the 'state' is that measured at soft X-ray energies, below 10 keV.)

At X-ray energies, the measured flux from Cyg X-1 is known to be variable over a wide range of time scales, ranging from msec to months. It spends most of its time in a low X-ray state, exhibiting a breaking spectrum at γ-ray energies that is often characterized as a Comptonization spectrum. In May of 1996, a transition of Cyg X-1 into a high X-ray state was observed by RXTE, beginning on May 10 [3]. The 2–12 keV flux reached a level of 2 Crab on May 19, four times higher than its normal value. Meanwhile, at hard X-ray energies (20-200 keV), BATSE measured a significant *decrease* in flux [4]. Motivated by these dramatic changes, a target-of-opportunity (ToO) for CGRO, with observations by OSSE, COMPTEL and EGRET, began on June 14 (CGRO viewing period 522.5). Here we report on the results from an analysis of the COMPTEL data from this ToO observation.

OBSERVATIONS AND DATA ANALYSIS

COMPTEL has obtained numerous observations of the Cygnus region since its launch in 1991, providing the best available source of data for studies of Cyg X-1 at energies above 1 MeV. Figure 1 shows a plot of hard X-ray flux, as obtained from BATSE occultation monitoring, for each day in which Cyg X-1 was within 40° of the COMPTEL pointing direction.

In previous work, we have compiled a broad-band spectrum of Cyg X-1 using contemporaneous data from BATSE, OSSE and COMPTEL [5,6]. The observations were chosen, in part, based on the level of hard X-ray flux measured by BATSE, the goal being to ensure a spectral measurement that corresponded to a common spectral state. In Figure 1, the data points from the selected observations are indicated by open diamonds. The resulting spectrum, corresponding to a low X-ray state, showed evidence for emission out to 5 MeV. The spectral shape, although consistent with the so-called breaking spectral state [1,2], was clearly not consistent with standard Comptonization models. The COMPTEL data provided evidence for a hard tail at energies above ~1 MeV that extended to perhaps 5 MeV.

During the high X-ray state observations in May of 1996 (VP 522.5), COMPTEL collected 11 days of data at a favorable aspect angle of 5.3°. The hard X-ray flux for these days is denoted by open triangles in Figure 1. An analysis of COMPTEL data from this observation revealed some unusual characteristics. The 1–3 MeV image (Figure 2) showed an unusually strong signal from Cyg X-1 when compared with other observations of similar exposure. The flux level was significantly higher than the average flux seen from earlier observations [5,6]. In the 1–3 MeV energy band, the flux had increased by a factor of 2.5, from $8.6(\pm 2.7) \times 10^{-5}$ cm^{-2} s^{-1} MeV^{-1} to $2.2(\pm 0.4) \times 10^{-4}$ cm^{-2} s^{-1} MeV^{-1}. The observed change in flux is significant at a level of 2.6σ. In addition, unlike in previous measurements, there

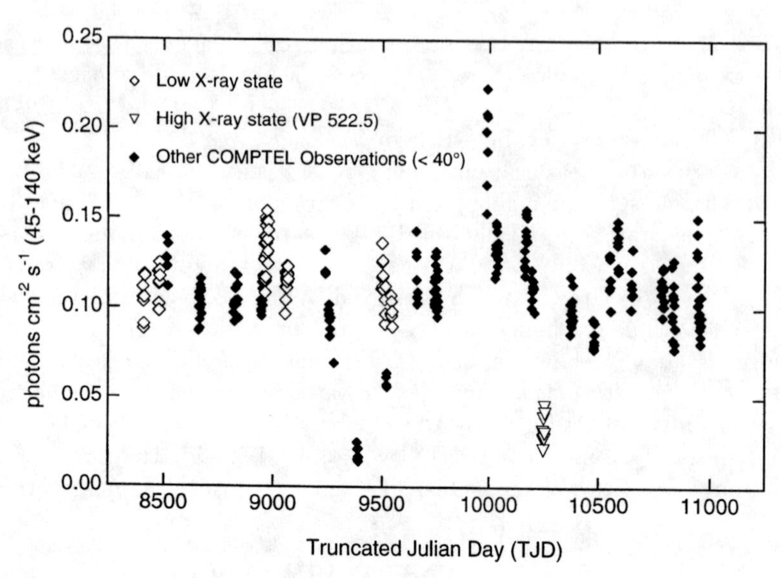

FIGURE 1. Hard X-ray time history (from 45–140 keV BATSE occultation data) for COMPTEL observations of Cyg X-1. Open diamonds indicate those data used to generate the low-state γ-ray spectrum. Open triangles correspond to CGRO viewing period 522.5.

was no evidence for any emission at energies *below* 1 MeV. This fact is explained, in part, by a slowly degrading sensitivity of COMPTEL at energies below 1 MeV due to increasing energy thresholds in the lower (D2) detection plane. Part of the explanation, however, appears to be a much harder source spectrum.

A more complete picture of the MeV spectrum is obtained by combining the COMPTEL results with results from OSSE, extending the measured spectrum down to ~50 keV. Unfortunately, a comparison of the COMPTEL and OSSE spectra for VP 522.5 shows indications for an offset between the two spectra by about a factor of two, with the OSSE flux points being lower than those of COMPTEL in the overlapping energy region near 1 MeV. A similar offset between OSSE and COMPTEL-BATSE is also evident in the contemporaneous low soft X-ray state spectrum [5,6]. The origin of this offset is not clear. Here we shall assume that there exists some uncertainty in the instrument calibrations and that this uncertainty manifests itself in a global normalization offset. We have subsequently increased the flux for each OSSE data point by a factor of two. This provides a good match between COMPTEL and OSSE at 1 MeV for both the low-state and high-state spectra, but we are left with an uncertainty (by a factor of two) in the absolute normalization of the spectra.

We compare the resulting COMPTEL-OSSE spectra in Figure 3 (with the data points in both OSSE spectra increased by a factor two). The low-state spectrum shows the breaking type spectrum that is typical of most high energy observations

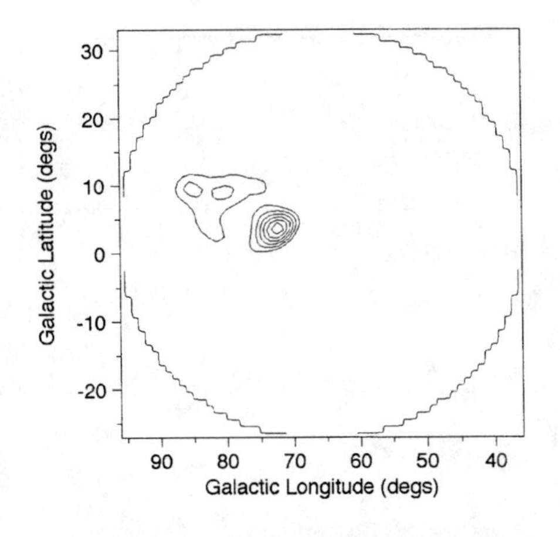

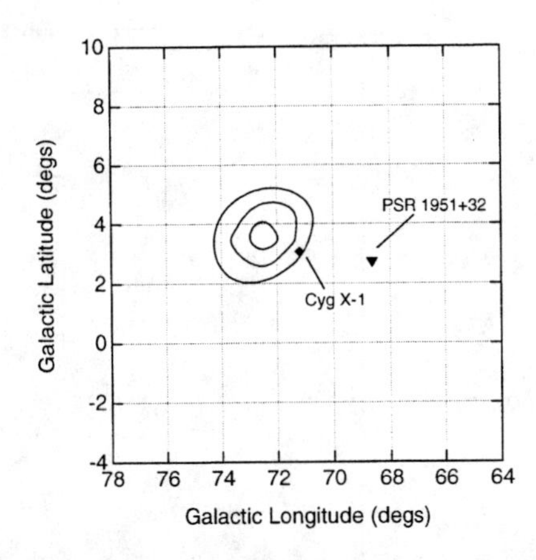

FIGURE 2. COMPTEL imaging of the Cygnus region as derived from 1–3 MeV data collected during high X-ray state of VP 522.5. The left-hand figure shows the maximum likelihood map. The right-hand figure shows the 1, 2 and 3-σ location contours. The emission is consistent with a point source at the location of Cyg X-1, with no significant contribution from PSR 1951+32.

of Cyg X-1. The high-state spectrum, on the other hand, shows the power-law type spectrum that is characteristic of black hole candidates in their high X-ray state. This spectral behavior had already been reported for this time period based on observations with both BATSE [7] and OSSE [8]. The inclusion of the COMPTEL data provides evidence, for the first time, of a continuous power-law (with a photon spectral index of -2.6) extending beyond 1 MeV, up to ~10 MeV.

A power-law spectrum had also been observed by both OSSE and BATSE during February of 1994 [9,10], corresponding to the low level of hard X-ray flux near TJD 9400 in Figure 1. In this case, however, the amplitude of the power-law was too low for it to be detected by COMPTEL.

DISCUSSION

We can use the COMPTEL data alone to draw some important conclusions regarding the MeV variability of Cyg X-1. Most importantly, the flux measured by COMPTEL at energies above 1 MeV was observed to be higher (by a factor of 2.5) during the high X-ray state (in May of 1996) than it was during the low X-ray state. The lack of any detectable emission below 1 MeV further suggests a relatively hard spectrum.

Inclusion of the OSSE spectra clearly show an evolution from a breaking type spectrum in the low X-ray state to a power-law spectrum in the high X-ray state. The COMPTEL data are consistent with a pivot point near 1 MeV. The power-law appears to extend to ~10 MeV with no clear indication of a cut-off.

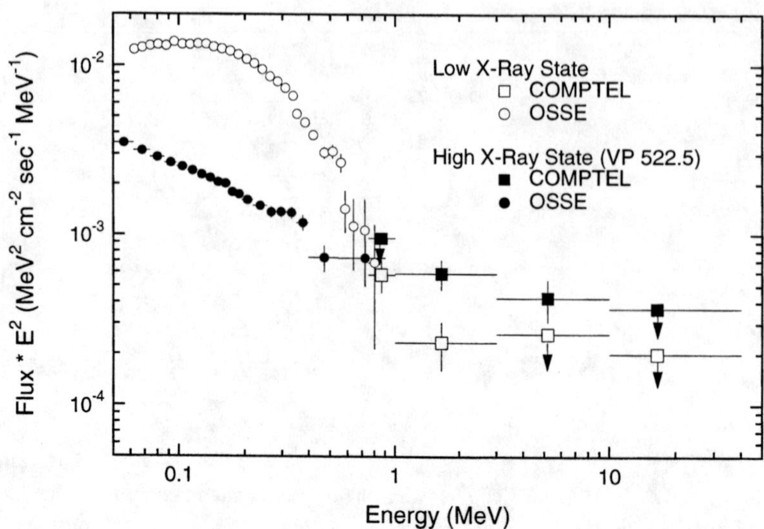

FIGURE 3. Spectra of Cyg X-1, shown as E^2 times the photon flux. OSSE flux levels have been increased by a factor of two and OSSE upper limits have been removed for the sake of clarity.

ACKNOWLEDGEMENTS

The COMPTEL project is supported by NASA under contract NAS5-26645, by the Deutsche Agentur für Raumfahrtgelenheiten (DARA) under grant 50 QV90968 and by the Netherlands Organization for Scientific Research NWO. This work was also supported by NASA grant NAG5-7745.

REFERENCES

1. Grove, J.E. et al., *Proceedings of the Fourth Compton Symposium*, ed. C.D. Dermer, M.S. Strickman, and J.D. Kurfess (New York, AIP), 1997, p. 122.
2. Grove, J.E. et al., *Ap.J.*, **500**, 899 (1998).
3. Cui, W. et al., *Ap.J.*, **474**, L57 (1997).
4. Zhang, S.N. et al., *Ap.J.*, **477**, L95 (1997).
5. McConnell, M.L. et al., *Proceedings of the 26th International Cosmic Ray Conference*, **4**, 119 (1999).
6. McConnell, M.L. et al., *Ap.J.*, submitted (2000).
7. Zhang, S.N. et al., *Proceedings of the Fourth Compton Symposium*, ed. C.D. Dermer, M.S. Strickman, and J.D. Kurfess (New York, AIP), 1997, p. 839.
8. Gierlinski, M. et al., *Proceedings of the Fourth Compton Symposium*, ed. C.D. Dermer, M.S. Strickman, and J.D. Kurfess (New York, AIP), 1997, p. 844.
9. Phlips, B. et al., *Ap.J.*, **465**, 907 (1996).
10. Ling, J.C. et al., *Ap.J.*, **484**, 375 (1997).

RXTE Monitoring of LMC X-3: Recurrent Hard States

J. Wilms*, M.A. Nowak†, K. Pottschmidt*, W.A. Heindl‡, J.B. Dove||,¶, M.C. Begelman†,§, R. Staubert*

* *Institut für Astronomie und Astrophysik, Waldhäuser Str. 64, D-72076 Tübingen, Germany*
† *JILA, University of Colorado, Boulder, CO 80309-440, U.S.A.*
‡ *CASS, University of California, San Diego, La Jolla, CA 92093, U.S.A.*
|| *CASA, University of Colorado, Boulder, CO 80309-389, U.S.A.*
¶ *Dept. of Physics, Metropolitan State College of Denver, Denver, CO 80217-3362, U.S.A.*
§ *APS, University of Colorado, Boulder 80309, U.S.A.*

Abstract. The black hole candidate LMC X-3 varies by a factor of four on a timescale of either 200 or 100 days (Cowley et al., 1991). We have monitored LMC X-3 with RXTE in three to four week intervals starting in December 1996, obtaining a large observational database that sheds light on the nature of the long term X-ray variability in this source. In this paper we present the results from this monitoring campaign, focusing on evidence of recurring hard states in this canonical soft state black hole candidate.

INTRODUCTION

Long term variability on timescales of months to years is seen in many galactic black hole candidates. In analogy to the 35 d cycle of Her X-1, this variability has been identified with the precession of a warped accretion disk in some objects. Possible driving mechanisms include radiation pressure (Maloney, Begelman & Nowak, 1998; Wijers & Pringle, 1999) or accretion disk winds (Schandl, 1996). In this poster we present a spectral and temporal analysis of the long term variability of the canonical soft state black hole candidate LMC X-3. Together with LMC X-1, this source is the only persistent black hole candidate which has only been observed in the soft state. While LMC X-1 does not exhibit any long term variability, LMC X-3 was known to be variable on a ~100 d timescale (Cowley et al., 1991, 1994). Detailed results of our campaign are presented elsewhere (Wilms et al., 1999b).

CP510, *The Fifth Compton Symposium,* edited by M. L. McConnell and J. M. Ryan

LONG TERM VARIABILITY

Our analysis of the long-term RXTE All Sky Monitor light curve (Fig. 1, top) indicates a complex long term behavior. Analysis with the Lomb (1976)-Scargle (1982) Periodogram indicates that the variation is dominated by epochs of low luminosity, which are recurring on the $\sim$100 d timescale found previously (Cowley et al., 1994). In addition, a long term periodicity is apparent in the data. Contrary to the 100 d timescale, the long term periodicity is not stable: Depending on what time interval of the ASM light curve is studied, the long term period varies between 200 and 300 d. This periodicity is caused by the times of average to high luminosity seen in the light curve and causes a broadening of the $\sim$250 d peak in the Lomb Scargle PSD.

SPECTRAL VARIABILITY

We analyzed the RXTE data using the newest RXTE `ftools`, as well as XSPEC, Version 10.00ab. The spectral model used for the data analysis was the standard multi-temperature disk blackbody (Mitsuda et al., 1984; Makishima et al., 1986), plus a power-law component. Adding a Gaussian iron line resulted in upper limits for the line equivalent width only. Typical reduced χ^2 values were $\chi^2_{\rm red} < 2.5$, with the residuals fully consistent with the uncertainty of the detector calibration (Wilms et al., 1999a,b).

In Fig. 1 we present the variation of the spectral parameters found during the analysis as a function of time. During episodes of high ASM flux, the source behaves like any other source in the classical soft state: The accretion disk temperature, $kT_{\rm in}$ varies freely to accommodate the variable luminosity of the source, while the the normalization of the multi-temperature disk black body is constant. At the same time, the photon index Γ varies independently of $kT_{\rm in}$. See, e.g., Tanaka & Lewin (1995) for similar examples in other soft state black hole candidates.

On the other hand, for times of low ASM count rate, the disk temperature decreases to $kT_{\rm in} \ll 1$ keV from its usual value of ~ 1 keV, while at the same time the photon index hardens dramatically from ~ 4 to ~ 1.7. We interpret these changes as evidence for transitions to the hard state in LMC X-3.

HARD STATES IN LMC X-3

Fig. 2 displays the spectral evolution of LMC X-3 from 1998 June through August. In Obs28 the source had the lowest flux of all monitoring observations. No evidence for a soft spectral component is present in the data, the spectrum is consistent with a pure power-law spectrum with photon-index 1.8. After Obs28, the soft component slowly emerged until the standard soft state spectrum was reached.

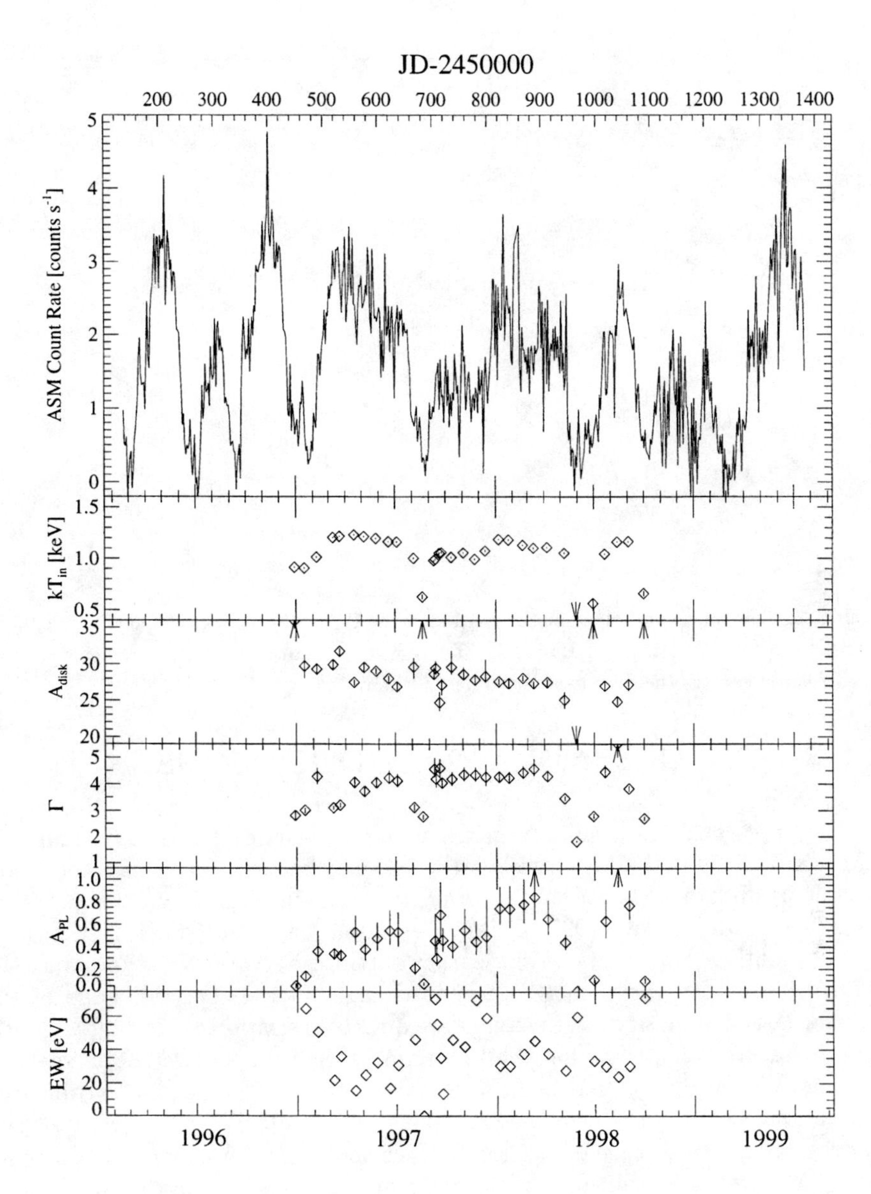

FIGURE 1. Temporal variability of the spectral parameters of LMC X-3. Note that the power law is harder and the inner disk temperature is smaller during the recurring episodes of low source luminosity.

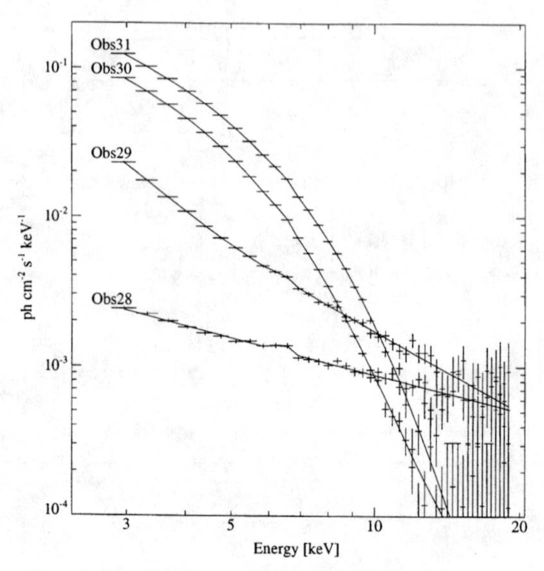

FIGURE 2. Spectral evolution from the hard state of 1998 May 29 (JD=2450962; Obs28) to the normal soft state spectrum in 1998 July and August (Obs30 and Obs31;). Shown are the unfolded photon spectra the lines denote the best fit model.

DISCUSSION AND CONCLUSIONS

We have presented the results from the first two years of our RXTE campaign on LMC X-3. This is the first campaign where a systematic study of a soft state black hole candidate with monthly coverage was possible (earlier campaigns, such as those of Cowley et al., 1991; Ebisawa et al., 1993, were constrained by the inflexible pointing constraints of the earlier satellites). We have found that the long-term luminosity variations are due to changes in the spectral shape of the source, for large luminosity these changes are due to a variation of the characteristic disk temperature, $kT_{\rm in}$, while for small luminosity the source undergoes a spectral hardening. We have presented the first clear case for a soft to hard state transition in this canonical soft state black hole candidate.

Our results are a challenge to models in which the long term variability of sources such as LMC X-3 is explained in the context of warped accretion disk models. In these models, no clear spectral evolution with source intensity is expected, with the exception of possible changes in $N_{\rm H}$ due to covering effects. In black hole candidates such as Cyg X-1, the hard- to soft-state transitions are attributed to changes in the accretion disk geometry, e.g., the (non-) existence of a hot and Comptonizing electron cloud in the center of the source. These changes are typically attributed to a varying mass accretion rate, $\dot{M}$. Our result makes such a geometry also probable for LMC X-3. A possible cause for the quasi-periodicity of the soft to hard transitions, therefore, might be periodic changes in $\dot{M}$.

Acknowledgements

We thank the RXTE schedulers for their patience in scheduling a total of up to now 40 observations on LMC X-3 and 32 observations on LMC X-1 during the course of the past three years. We also thank those who are responsible for the almost non-existent observing constraints of RXTE for providing us with the ability to perform campaigns such as this one. The attendance of JW at the Compton symposium was made possible by a travel grant from the Deutsche Forschungsgemeinschaft.

REFERENCES

Cowley, A. P., et al., 1991, ApJ, 381, 526

Cowley, A. P., Schmidtke, P. C., Hutchings, J. B., & Crampton, D., 1994, ApJ, 429, 826

Ebisawa, K., Makino, F., Mitsuda, K., Belloni, T., Cowley, A. P., Schmidtke, P. C., & Treves, A., 1993, ApJ, 403, 684

Ebisawa, K., Mitsuda, K., & Hanawa, T., 1991, ApJ, 367, 213

Lomb, N. R., 1976, Ap&SS, 39, 447

Makishima, K., Maejima, Y., Mitsuda, K., Bradt, H. V., Remillard, R. A., Tuohy, I. R., Hoshi, R., & Nakagawa, M., 1986, ApJ, 308, 635

Maloney, P. R., Begelman, M. C., & Nowak, M. A., 1998, ApJ, 504, 77

Mitsuda, K., et al., 1984, PASJ, 36, 741

Scargle, J. D., 1982, ApJ, 263, 835

Schandl, S., 1996, A&A, 307, 95

Shimura, T., & Takahara, F., 1995, ApJ, 445, 780

Tanaka, Y., & Lewin, W. H. G., 1995, in X-Ray Binaries, ed. W. H. G. Lewin, J. van Paradijs, E. P. J. van den Heuvel, Cambridge Astrophysics Series 26, (Cambridge: Cambridge Univ. Press), Chapt. 3, 126

Wijers, R. A. M. J., & Pringle, J. E., 1999, MNRAS, 308, 207

Wilms, J., Nowak, M. A., Dove, J. B., Fender, R. P., & di Matteo, T., 1999a, ApJ, 522, 460

Wilms, J., Nowak, M. A., Pottschmidt, K., Heindl, W. A., Dove, J. B., & Begelman, M. C., 1999b, ApJ, submitted

GRS 1915+105: The X-Ray Spectrum Following a Radio Flare

D.C. Hannikainen*, O. Vilhu*, L. Alha*, R.W. Hunstead† and D. Campbell-Wilson†

* *Observatory, PO Box 14, FIN-00014 University of Helsinki, Finland*
† *School of Physics, University of Sydney, NSW 2006, Australia*

Abstract. We present simultaneous RXTE (PCA and HEXTE) and CGRO (OSSE) spectra of GRS 1915+105 following a radio flaring episode. The outcome of the spectral fits, using both a thermal sombrero and a hybrid thermal/nonthermal model, imply a Kerr hole with a large population of relativistic electrons. The radio-infrared excess (over the disk model) may arise from this population, or it could be a relic from major plasmoid ejections or due to frequent "baby" blobs.

INTRODUCTION

The Galactic superluminal X-ray transient GRS 1915+105 was discovered by the GRANAT satellite in 1992 [1], and has been extensively monitored at all wavelengths ever since. In 1994, Mirabel & Rodríguez [2] observed the multiple ejection of radio-emitting plasmoids at apparent superluminal velocities. Another series of ejections from GRS 1915+105 was reported by Fender et al. [3] in 1997 which were coincident with a double-peaked radio flare (observed with the Green Bank Interferometer). The Molonglo Observatory Synthesis Telescope (MOST) recorded a similar double-peaked radio flare in 1996; however, no radio imaging was conducted at that time and hence it is not known whether ejections occurred. Due to the similarity in the RXTE/All-Sky Monitor (ASM) light curves during both radio flaring events (eg. [4]), it is tempting to suggest that blob ejections occurred also at the time of the MOST flare. We examine the combined Proportional Counter Array (PCA), High Energy X-ray Timing Experiment (HEXTE) and Oriented Scintillation Spectrometer (OSSE) spectrum of GRS 1915+105 obtained shortly after the MOST flaring event within the framework of a thermal sombrero and a hybrid thermal/nonthermal model.

CP510, *The Fifth Compton Symposium,* edited by M. L. McConnell and J. M. Ryan

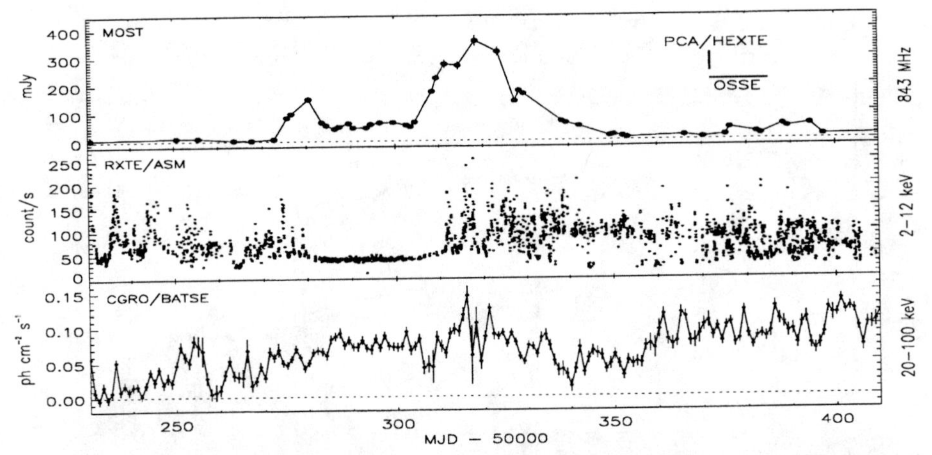

FIGURE 1. Time history of the observations. The epochs of the PCA and OSSE observations are marked in the top panel.

OBSERVATIONS

The data were obtained with the PCA and HEXTE on board RXTE, and with OSSE on board CGRO. The energy ranges covered by the three instruments are 2–60 keV (PCA), 15–250 keV (HEXTE) and 50 keV–10 MeV (OSSE). Figure 1 shows the MOST 843 MHz radio light curve, the 2–12 keV RXTE/All-Sky Monitor and the CGRO/BATSE 20–100 keV light curves. The duration of the OSSE observations is shown in the top panel, as is the epoch of the PCA/HEXTE observation. One must bear in mind that GRS 1915+105 is a higly variable object at all wavelengths, and as the OSSE observations are accumulated over $\sim$ two weeks, the spectra shown here actually represent an *average*.

SPECTRAL MODELS

Analyses of broad band data of Galactic black hole candidates have shown that the spectra are well described by a disk blackbody and successive Compton upscatterings of soft photons in a hot cloud surrounding the central source (see eg. [5] and references therein). Here we fit the broad band $\sim$ 2–400 keV spectrum of GRS 1915+105 with two models: a thermal sombrero model and a hybrid thermal/nonthermal model, as was done by Vilhu et al. [6] for a later observation. The fitting was carried out using codes implemented into the XSPEC package.

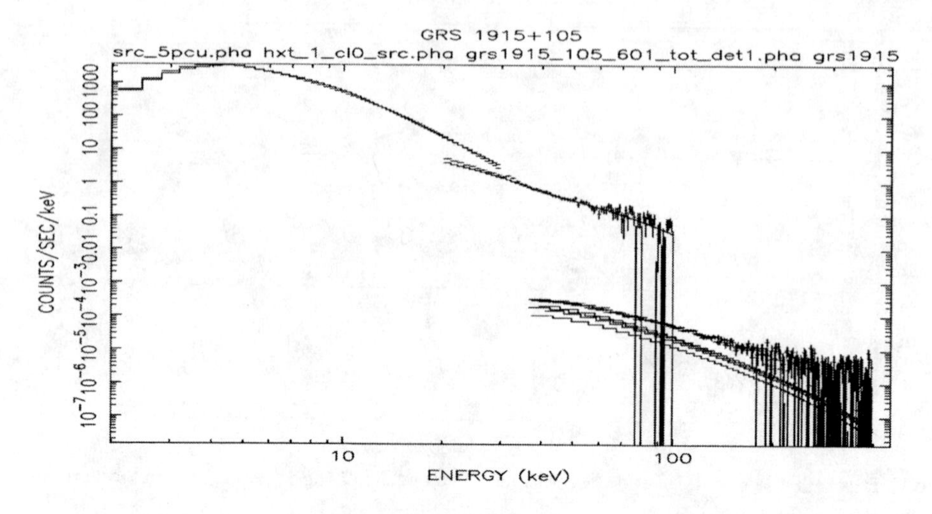

FIGURE 2. The thermal sombrero model fit to the PCA, HEXTE and OSSE data.

The thermal sombrero model

In the thermal sombrero model the optically thick cold disk penetrates into the central region of an optically thin rarefied hot cloud of electrons, or corona [7]. Soft photons from the disk are successively inverse Compton scattered in the corona to produce the observed hard X-ray emission. Figure 2 shows the result of applying the thermal sombrero model to the combined PCA, HEXTE and OSSE spectra. The best fit to the spectra yielded an inner radius value of $R_{in} = 30$ km (with $N_H = 2.2 \times 10^{22}\,\mathrm{cm}^{-2}$, $T_e \approx 70$ keV, $T_{in} \approx 1.9$ keV and $\tau \approx 0.5$) which corresponds to 1 R_g ($R_g = 2GM/c^2$) for a 10 $M_\odot$ black hole (and 3 R_g for 3 $M_\odot$, 3 R_g being the innermost stable orbit of a nonrotating black hole). Previous mass estimates have shown GRS 1915+105 to be harboring a ~ 33 $M_\odot$ black hole (eg. [8]). Hence, either the mass of the compact object in GRS 1915+105 is much less than that thought before, or it is indeed a Kerr black hole (as proposed eg. by [9]), where in the extreme case the innermost stable orbit can be as small as 0.5 R_g. As Figure 2 shows, the higher energies (> 100 keV) are not well accommodated for by the thermal sombrero model. This is reflected in the rather high value of $\tilde{\chi}^2 \approx 6$.

The EQPAIR hybrid thermal/nonthermal model

Thermal blackbody models have been successful in describing the spectra of black holes, especially at the lower energies However, the high energy tail present in many of the spectra probably indicates that the electron energy distribution deviates from

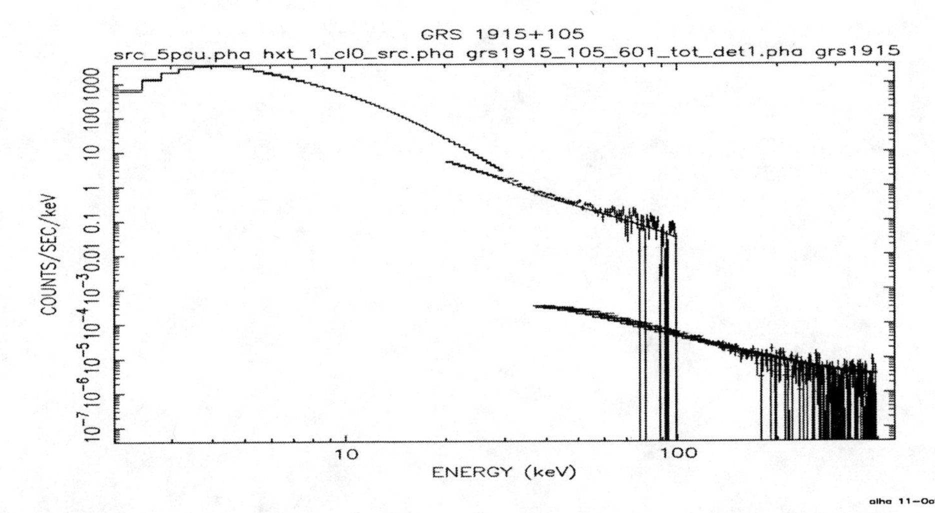

FIGURE 3. The EQPAIR hybrid thermal/nonthermal fit to the PCA, HEXTE and OSSE data.

a purely Maxwellian distribution. A model has been recently introduced to explain the spectra in terms of a hybrid plasma involving thermal and nonthermal processes (see eg. [10] and references therein). In this model, a fraction of the energy input can be injected into the source in the form of relativistic electrons and/or positrons. The outcome of applying the EQPAIR fitting routine to the PCA, HEXTE and OSSE spectra yielded a nonthermal efficiency of $\sim$ 33% ($l_{nth}/l_{th} \approx 0.33$), and a power law injection index of $\sim$ 2.1 (ie. $N_e \propto \gamma^{-2.1}$, where γ is the Lorentz factor). Figure 3 shows the resulting best-fit of the EQPAIR model to the spectra, where one can see that the fit now takes into account the higher energies. The resulting $\tilde{\chi}^2$ of the fit ($\tilde{\chi}^2 = 2.3$) is still rather high, though better than that for the ISMBB model. [1]

THE RADIO TO X-RAY SPECTRUM

Figure 4 shows the radio to X-ray spectrum of GRS 1915+105. The radio data are the minima and maxima from the epochs of the OSSE observations (the Ryle 15 GHz data courtesy of R. Fender and G. Pooley), and the infrared (IR) observation (MJD 50379) is taken from Fender et al. [11]. The dashed line portrays the disk blackbody model. As Figure 4 clearly shows, there is an excess in the radio-IR portion of the spectrum for a pure disk blackbody fit, especially during the radio

[1] We would like to point out that at the time of the fitting we were unaware of the new the PCA responses. A paper is in preparation where the fitting is being undertaken using the updated versions of the PCA responses.

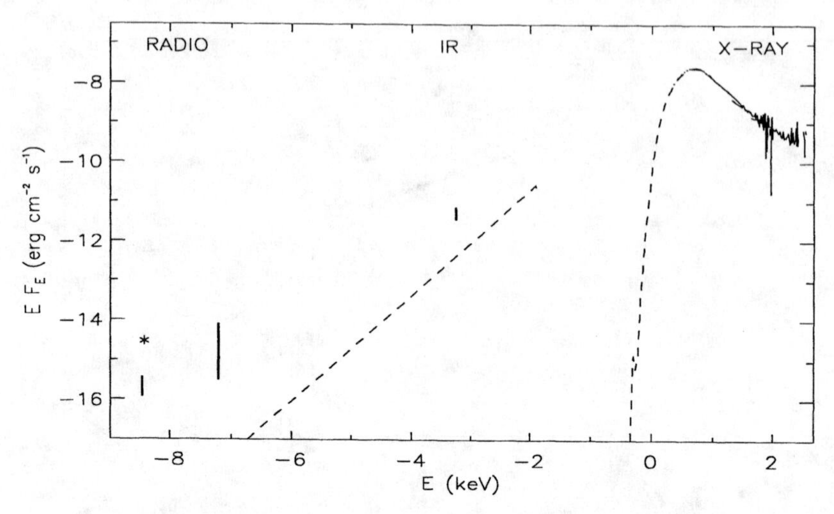

FIGURE 4. The MOST 843 MHz — OSSE 400 keV spectrum. The dashed line represents the disk blackbody model with interstellar absorption. The star marks the maximum flux during the MOST radio flare (~ MJD 50370).

flare (the star). As radio emission is often attributed to nonthermal synchrotron radiation, we propose three alternative scenarios here. The first is one in which the electrons that give rise to the hard power law tail and to the radio emission arise from the same population. A second scenario is that we speculate that this excess may be attributed to relic synchrotron electrons arising from possible plasmoid ejections which occurred in conjunction with the MOST flaring episode. A third possibility is that "baby" blobs are continuously ejected from the system. A paper is in preparation with further modelling of the data.

REFERENCES

1. Castro-Tirado, A., Brand, S. & Lund, N., *IAUC* **5590**, (1992).
2. Mirabel, I.F. & Rodríguez, L.F., *Nature* **371**, 46, (1994).
3. Fender, R. et al., *MNRAS* **304**, 865, (1999).
4. Hannikainen, D.C. et al., *PASP Conf. Ser.*, **161**, 88, (1999).
5. Liang, E., *Physics Reports* **302**, Nos. 2–3, (1998).
6. Vilhu, O. et al., *Proc. 3rd INTEGRAL Workshop*, in press.
7. Poutanen, J., *Theory of Black Hole Accretion Disks*, CUP, (1998).
8. Morgan, E.H., Remillard, R.A. & Greiner, J., *ApJ* **482**, 993, (1997).
9. Zhang, S.N., Cui, W. & Chen, W., *ApJL* **482**, L155, (1997).
10. Coppi, P.S., *PASP Conf. Ser.*, **161**, 375, (1999).
11. Fender, R.P., Pooley, G.G., Brocksopp, C. & Newell, S.J., *MNRAS* **290**, L65 (1997).

Accretion-Ejection Instability and a "Magnetic Flood" scenario for GRS 1915+105

Michel Tagger*

*Service d'Astrophysique (CNRS URA 2052)
C.E.A. Saclay, 91191 Gif sur Yvette (France)
e-mail tagger@cea.fr

Abstract. We present an instability, occurring in the inner region of magnetized accretion disks, which seems to be a good candidate to explain the low-frequency QPO observed in many X-ray binaries. We then briefly show how, in the remarkable case of the microquasar GRS 1915+105, identifying this QPO with our instability leads to a scenario for the $\sim$ 30 mn cycles of this source. In this scenario the cycles are controlled by the build-up of magnetic flux in the disk.

INTRODUCTION

This contribution comes from two different lines of work: the first one is purely theoretical, and has led us to present recently [1] an instability which may occur in the inner region of magnetized disks. We have called it Accretion-Ejection Instability (AEI), because one of its main characteristics is to extract energy and angular momentum from the disk, and to emit them *vertically* as Alfven waves propagating along magnetic field lines threading the disk. These Alfven waves then may deposit the energy and angular momentum in the corona above the disk, providing an efficient way to energize winds or jets from the accretion energy.

The second approach has consisted in the comparison of this instability with the observed properties of the low-frequency (.5 - 10 Hz) Quasi-Periodic Oscillation (QPO) observed in the low and hard state of the micro-quasar GRS 1915+105. The very large and fast growing number of observational results on this source gives access to many aspects of the physics of the disk. They allow this comparison to rely on basic properties of the instability, and on more detailed ones, such as the correlation between the QPO and the evolution of the disk and coronal emissions (identified respectively as multicolor black body and comptonized power-law tail in the X-ray spectrum).

CP510, *The Fifth Compton Symposium,* edited by M. L. McConnell and J. M. Ryan

This comparison encourages us to consider that the AEI may indeed be the source of the QPO. Thus we proceed by considering the ~ 30 mn cycles of GRS 1915+105. These cycles are the most spectacular in the gallery of behaviors and spectral states of this source, in particular because multi-wavelength observations have shown IR and radio outbursts coinciding with them, consistent with the synchrotron emission from an expanding cloud ejected at relativistic speeds. These cycles have been analyzed in great details, and the QPO shows a very characteristic and reproducible behavior. We have thus built a scenario, starting from the identification of the QPO with the AEI, and considering how this could explain the evolution of the source during the cycle. We refer to it as a *magnetic flood* scenario, because we are led to believe that the cycle is controlled by the build-up of the vertical magnetic flux stored within the disk. The scenario is compatible with the available information on this type of cycle, explains a number of results in existing data, and leads to intriguing considerations on the behavior of GRS 1915+105.

ACCRETION-EJECTION INSTABILITY

We will present the instability here only in general terms, and refer to our recent publication [1] for detailed derivation and results. It appears in disks threaded by a vertical magnetic field, of the order of equipartition with the gas pressure ($\beta = 8\pi p/B^2 \lesssim 1$). The instability appears essentially as a spiral density wave in the disk, very similar to the galactic ones, but driven by the long-range action of magnetic stresses rather than self-gravity. The main difference lies in the amplification mechanism: instability results from the coupling of the spiral density wave with a *Rossby wave*. Rossby waves, associated with a gradient of vorticity, are best known in planetary atmospheres, including the other GRS – the Great Red Spot of Jupiter. In the present case differential rotation allows the spiral wave to grow by extracting energy and angular momentum from the disk, and transferring them to a Rossby vortex at its corotation radius. This radius is constrained by the physics of the spiral to be a few times ($\sim 2-5$ for the azimuthal wavenumber $m = 1$, *i.e.* a one-armed spiral) the inner radius of the disk.

A third type of wave completes the description of the instability: it is an Alfven wave, emitted along the magnetic field lines towards a low-density corona above the disk. The mechanism here is simply that the Rossby vortex twists the footpoints of the field lines in the disk. This twist will then propagate upward, carrying to the corona the energy and angular momentum extracted from the disk by the spiral.

The mechanism is thus quite complex; this comes essentially from differential rotation, which allows a mixing of waves which would otherwise evolve independently. It results in an instability, growing on a time scale of the order of r/h times the rotation time (where h is the disk thickness and r its radius). We will present here its main characteristics, which will be essential in what follows:

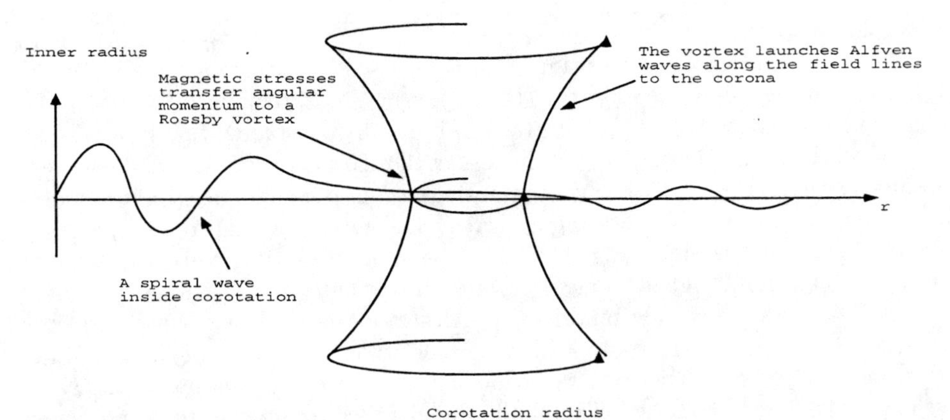

FIGURE 1. The propagation of the wave in the disk and its corona. A spiral wave grows by extracting energy and angular momentum from the disk, and transferring them to a Rossby vortex. The latter in turn transfers them to an Alfven wave toward the corona.

• It occurs when the vertical magnetic field B_0 is near equipartition ($\beta \lesssim 1$) and presents a moderate or strong radial gradient.

• The efficiency of the coupling to the Rossby wave selects modes with low azimuthal wavenumbers (the number of arms of the spiral), $m = 2$ or $m = 1$ usually, depending on a number of parameters (density and temperature profiles, field strength, etc.)

• For a given m, the mode frequency is close to $\omega \approx (m-1)\Omega_{int}$, the rotation frequency at the inner disk radius. In the special case of the $m = 1$ mode, the frequency is usually of the order of $\sim .2 - .5$ times Ω_{int}.

• By analogy with galactic spirals, we can expect that these properties result in the formation of a large scale, quasi-stationary spiral structure rather than in a turbulent cascade to small scales.

• This should strongly affect the structure of the disk. Indeed, underlying the usual model of turbulent viscous transport in a disk (leading to Shakura and Sunyaev's model of α disks) is the assumption of small scale turbulence. This leads to a *local* deposition of the accretion energy, efficiently heating the disk. Here on the other hand, the accretion energy is transported away by *waves*: extracted from the disk by the spiral wave, it is first transferred to the Rossby vortex, then to Alfven waves. Thus, here as in galaxies, the connection between gas accretion and disk heating is not as straightforward as in α-disks.

MAGNETIC FLOOD SCENARIO

The low-frequency QPO in GRS 1915+105 has been the object of many recent studies [2,3]. During the $\sim$ 30 mn cycles of this source, the QPO appears only during the low state, and its frequency varies in a repetitive manner during that phase. Let us convert its frequency ν_{QPO} to a keplerian radius r_{QPO}, and compare it to the color radius r_{color} resulting from a multi-color black body model of the disk emission: observations show that the ratio r_{QPO}/r_{color} remains of the order of 5 while both radii vary during the low state. It is usually considered that r_{color} gives a measure of the internal radius r_{int} of the disk, although the ratio r_{color}/r_{int} is subject to some uncertainties. It is thus very tempting to consider that the QPO originates from a pattern in the disk, rotating at a frequency corresponding to a radius r_{QPO} of the order of a few times r_{int}. This may be supported by a correlation, found between ν_{QPO} and a higher frequency feature in various binary systems, including neutron star and black hole binaries [4]. Although the evidence is fragile in the case of GRS 1915+105, it would lead to consider that the ratio r_{QPO}/r_{int} is of the order of 5, in agreement with the previous result, and corresponding to the value we expect for the $m = 1$ AEI. This, and more detailed arguments to be presented elsewhere, leads us to tentatively identify the AEI as the source of the QPO, and to consider how this could fit with the 30 mn cycles of this source.

We start from the conditions responsible for the onset of the instability, *i.e.* a change in B_0 or its radial gradient. We find better agreement with the former, and in this case the sudden transition from the "high and soft" state to the "low and hard" one would find a natural explanation: one has to remember that the best candidate to explain accretion in a magnetized disk is the magneto-rotational instability (MRI) [6]. It appears in disks with low magnetization ($\beta > 1$), and results in small-scale turbulence which causes viscous accretion, in agreement with a standard α disk.

Let us consider that in the high state the disk extends down to its last stable orbit at r_{LSO}, as suggested by the consistent minimal value of r_{color}, and that accretion is caused by the MRI, following an α prescription. The MRI might be responsible for the "band-limited noise" observed in power density spectra (below the QPO frequency, *i.e.* farther in the disk, when the QPO is present). Although numerical simulations of the MRI give estimates of the resulting α, *i.e.* turbulent viscosity, they are not able at this stage to give the associated turbulent magnetic diffusivity, so that the evolution of the magnetic flux in the disk cannot be prescribed. Our main assumption is that in these conditions vertical magnetic flux builds up in the disk: either because it is dragged in with gas flowing from the companion, or from a dynamo effect [5]. This is actually the configuration observed near the center of the Galaxy.

Then the field must grow in the disk, so that β decreases until it reaches $\beta \simeq 1$, at which point the MRI stops and our instability sets in, appearing as the low-frequency QPO. The most important consequence is that turbulent disk heating stops, so that the disk temperature should drop, further reducing β. The abrupt transition from the high to the low state thus finds a natural explanation, as a sharp transition between a low magnetization, turbulently heated state to a high magnetization one, where disk heating stops and accretion energy is redirected toward the corona (although estimating what fraction of this energy is actually deposited in the corona depends on the physics of Alfven wave damping).

The content of the space between the disk (when it does not extend down to r_{LSO}) and the black hole is not known. It might be an ADAF, or a large-scale, force-free magnetic configuration holding the magnetic flux frozen in the black hole (following the Blandford-Znajek mechanism). In both cases the condition which determines the inner disk radius r_{int} must be complex, but it is reasonable to assume that a drop in the disk pressure could explain the increase of r_{color} at the onset of the low state. Continuing accretion from the outer disk region would then move the disk back toward the last stable orbit, as seen during the low state.

The light curves show an "intermediate spike", halfway through the low state. At this time r_{color} is back to its minimal value, the QPO stops, and the coronal emission decreases sharply. This is also when the infra-red synchrotron emission, presumably from a "blob" ejected at relativistic speed, begins [7]. It is then natural to consider that at this time a large-scale magnetic event, possibly reconnection with the magnetic flux surrounding the black hole, causes ejection of the coronal plasma. This allows the disk to return to a lower magnetization state, so that once it has fully recovered it can start a new cycle in the high and soft state.

CONCLUSION

The properties of the low-frequency QPO in GRS 1915+105 have led us to tentatively identify it with the Accretion-Ejection Instability. This has allowed us to build up a scenario for the 30 mn cycles of this source. In contrast with global descriptions, such as α disks, this does not allow us to predict specific spectral signatures: in the same manner, knowledge of Rossby waves would hardly allow one to predict the existence and appearance of the Great Red Spot on Jupiter. On the other hand, the scenario is qualitatively compatible with all the information we have about these cycles. It explains why and how the QPO appears, how its frequency varies with the color radius, and why the transition from the high to the low state has to be a sharp one. Future work will be devoted to the QPO behavior at other times in GRS 1915+105, and then to other sources (black hole or neutron star binaries) where the identification of the QPO might give access to additional physics.

REFERENCES

1. Tagger, M. and Pellat, R., A&A **349**, 1003
2. Markwardt, C.B., Swank, J., and Taam, R.E., *ApJ* **73**, 47 (1999)
3. Muno, M.P., Morgan, E.H., and Remillard, R.A., submitted to *ApJ*, 1999
4. Psaltis, D., Belloni, T., and Van der Klis, M., *ApJ* **520**, 262 (1999)
5. Brandenburg, A., Nordlund, A., Stein, R.F., and Torkellson, U., *ApJ* **458**, L45 (1996)
6. Balbus, S.A., and Hawley, J.F., *ApJ* **376**, 214 (1991)
7. Mirabel, I.F., Dhawan, V., Chaty, S., Rodriguez, L.F., Marti, J., Robinson, C.R., Swank, J., and Geballe, T., *A&A* **330**, 9 (1998)

Broadband Observations of the Black Hole Transients XTE J1550–564 and GRS 1915+105

J. E. Grove[1], W. Cui[2], and Y. Ueda[3]

[1]*Naval Research Lab*
[2]*Massachusetts Institute of Technology*
[3]*ISAS*

Abstract. Previous work with OSSE has shown that galactic black hole candidates (BHCs) in the X-ray high, soft state have power-law emission above 50 keV that extends at least to energies comparable to the electron rest mass. We report on recent contemporaneous observations of the transients XTE J1550–564 and GRS 1915+105 with ASCA, RXTE, and OSSE. During the OSSE observations of both objects, the emission above 50 keV was slowly declining, covering a range of a factor of three in intensity for the former and a factor of ten for the latter. This is the first time that high-sensitivity observations above 50 keV have been made of these objects over such a wide range in intensity.

Coordinated X-ray and γ-ray spectral and timing observations with ASCA, RXTE, and OSSE were performed on the galactic black hole transients XTE J1550–564 in 1999 March and GRS 1915+105 in 1999 April. We report here on an initial, broadband spectral analysis of these data. Note that all of the spectral parameters given below are from fits to the individual instruments, and that the flux normalizations between instruments were free parameters in all of the figures. Simultaneous broadband fits will be reported elsewhere.

XTE J1550–564

The black hole transient XTE J1550-564 was discovered in 1998 Sep during an intense X-ray flare (peak: 6.8 Crab in 2-10 keV). It has a low-mass companion, and lies at a distance R ~ 2.5 kpc [1]. X-ray spectral evolution during the discovery outburst is given by Sobczak et al. [2].

Fig. 1a shows lightcurves of the X-ray and γ-ray emission during the epoch studied here. The 2-10 keV flux measured by the RXTE/ASM (shown in the upper panel as daily averages) was generally declining from a ~3-month plateau of ~2.5 Crab flux units. The 50-150 keV flux measured by OSSE (middle panel, 94-minute averages) dropped from 1/3 Crab to 1/10 Crab within 24 hours. The fractional drop in the 150-300 keV flux (bottom panel, daily averages) was not as large, indicating spectral hardening above 50 keV.

CP510, *The Fifth Compton Symposium,* edited by M. L. McConnell and J. M. Ryan

ASCA Observations

ASCA observed for 21 ksec on 1999 Mar 16-17, during which time the 2-10 keV flux was 2.9 x 10^{-8} erg s^{-1} cm^{-2}, or about 1.3 Crab, and exhibited little variability. The 1-10 keV spectrum was well described by a disk-blackbody model plus a power law and an emission line at 6.5 keV, modified with interstellar absorption (Fig. 1b). The power-law component dominated above 4 keV. The disk-blackbody apparent temperature and radius of the inner edge, T_{in} = 1 keV and R_{in} = 10 km (D/2.5 kpc) (1/sqrt(cos *i*)), are typical of BHCs. The strong ultra-soft component and weak variability are indicative of the X-ray high, soft state.

RXTE Observations

RXTE observed for 3 ksec on 1999 Mar 17, coincident with ASCA. The flux and spectral shape observed by the PCA were consistent with ASCA, and the photon number index below 30 keV was 2.48 (Fig. 1b).

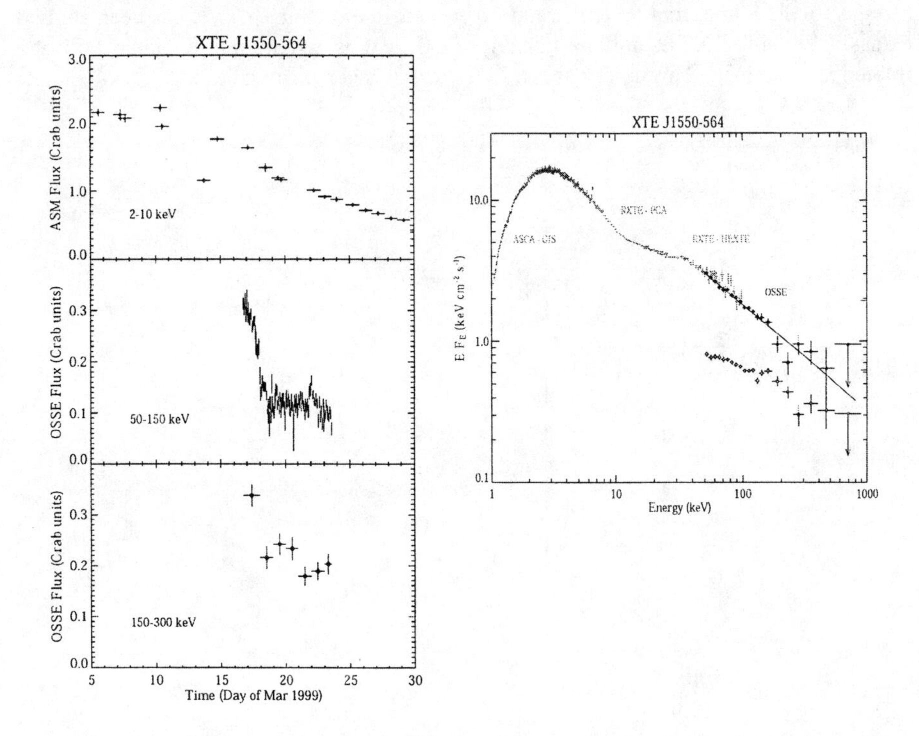

FIGURE 1. (a) At left, lightcurves from the RXTE/ASM (2-10 keV) and OSSE (50-150 keV and 150-300 keV) for XTE J155-564. (b) At right, broadband spectrum from contemporaneous observations by ASCA, RXTE, and OSSE. Two epochs are shown for OSSE. Normalization between instruments was a free parameter.

OSSE Observations

OSSE observed during a 1999 Mar 16-23 Target of Opportunity, as the hard X-ray and γ-ray emission rapidly declined. The spectrum above 50 keV was at all times consistent with a single power law, and the spectral hardness was anti-correlated with the flux at 50 keV. High-flux days, Mar 16-17, have photon number –2.75 ± 0.0, while low flux days, Mar 19-23, have photon number –2.38 ± 0.05 . There is excellent agreement in spectral shape between HEXTE and OSSE. There is no evidence for a break between 50 keV and at least 400 keV. We note that a longer observation would have provided a strong test for emission above the electron rest-mass energy.

GRS 1915+105

One of only two galactic objects clearly showing superluminal radio jets, the transient GRS 1915+105 was first discovered in X-rays in 1992 [3]. It is located at a distance of 12.5 kpc, and its jets are at 70° to our line of sight [4]. It has been studied extensively with RXTE, and its behavior <30 keV is remarkably variable ([5] and references therein). Gamma-ray spectra were discussed in Grove et al. [6].

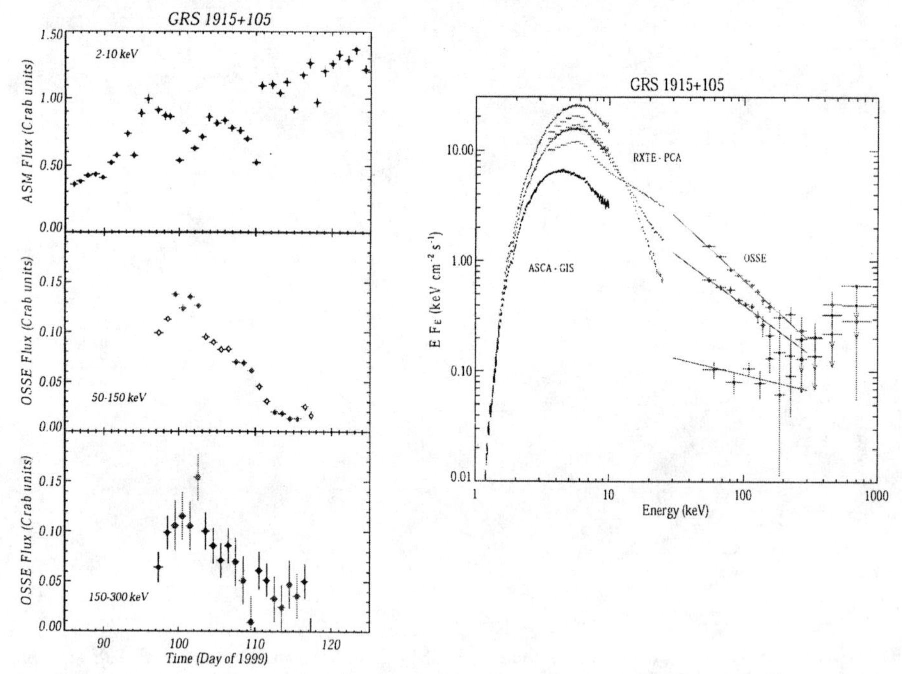

FIGURE 2. (a) At left, lightcurves from the RXTE/ASM (2-10 keV) and OSSE (50-150 keV and 150-300 keV) for GRS 1915+105. (b) At right, broadband spectrum from contemporaneous observations by ASCA, RXTE, and OSSE at several epochs. Normalization between instruments was a free parameter.

Fig. 2a shows lightcurves of the X-ray and γ-ray emission during the epoch studied here. The 2-10 keV flux from the RXTE/ASM is slowly rising (upper panel, daily averages), although it is highly variable from sample to sample and day to day. Since the launch of RXTE, this source has varied between ~0.3 Crab and ~2.5 Crab in daily averages. The 50-150 keV flux rose by ~40% and then fell by 90% (middle panel, daily averages). The 150-300 keV lightcurve is similar, although its peak-to-peak variation is smaller, again suggesting spectral hardening in γ rays. Shades indicate epochs of OSSE spectral analysis.

ASCA Observations

ASCA observed for 29 ksec on 1999 Apr 15-16, during which time the 2-10 keV flux was 2.2 x 10^{-8} erg s^{-1} cm^{-2}, or ~1.0 Crab. It exhibited strong quasi-periodic variability on a time scale of ~60 sec. Similar behavior and hardness correlation (i.e. the X-ray spectrum is hard when the flux is high) have been reported from RXTE (e.g. [7], [5]).

The 1-10 keV spectrum in the high-flux state was well described by a disk-blackbody, modified with interstellar absorption (Fig. 2b, uppermost ASCA data). No power law was required. The disk-blackbody apparent temperature and radius of the inner edge, T_{in} = 2.1 keV and R_{in} = 28 km (D/12.5 kpc) (sqrt(cos 70 / cos *I*)). In the low-flux state, the temperature decreased to T_{in} = 1.2 keV, the radius increased to R_{in} = 35 km, and a power-law tail was required (lowermost ASCA data). The average spectrum is shown (Fig 4, middle dataset) for comparison with RXTE and OSSE. We have not yet examined the RXTE and OSSE data for bimodality. While the ASCA and OSSE observations spanned the same days, by an unfortunate twist of fate the spacecraft orbits gave entirely complementary viewing times, with no overlapping time.

RXTE Observations

We analyzed RXTE observations on 1999 Apr 10-11 (day 100-101), 1999 Apr 16 (day 106), and 1999 Apr 23 (day 113). Only PCA data have been analyzed. In all cases, the spectra were fit with a disk-blackbody and power-law tail. The spectrum pivoted near 13 keV, with the flux below 10 keV increasing with time, while the flux above 10 decreased with time (Fig. 2b). The power law softens with time, from photon number index 2.7 to 3.2. This behavior is well known.

OSSE Observations

GRS 1915+105 has been observed repeatedly by OSSE, and its emission >50 keV has always been consistent with a power law of varying index [6]. The OSSE data here have been grouped around the epochs of RXTE observations and are color-coded in Fig. 5 to match. The photon number index above 50 keV is only weakly varying with flux, *hardening* with time from 3.1 ± 0.1, to 2.9 ± 0.1, to 2.3 ± 0.3 (95% conf. intervals). In each epoch, a spectral break is suggested between PCA and OSSE. We note that the detection in the weakest γ-ray epoch is highly statistically significant, and that the spectrum is harder than that of the diffuse galactic plane emission and most accreting neutron stars.

CONCLUSIONS

We have presented broadband spectra for two galactic BHCs in the X-ray high, soft state. While simultaneous, multi-instrument fitting is still in progress, a number of features are already apparent.

For XTE J1550-564, the presence of an iron line and the broad bump near 40 keV are strongly suggestive of Compton reflection in a cool disk. Ebisawa, Titarchuk, & Chakrabarti [8] have proposed that in the high, soft state, the high-energy emission arises from bulk-motion Comptonization in the convergent accretion flow from the inner edge of the disk. The Comptonization takes places within the advecting, Thomson-thick medium above the disk. We note, however, because the medium is optically thick, the reflected component would be trapped in the advecting flow and would not escape to the observer. Thus confirmation of reflection in the high, soft state would cast doubt on the validity of bulk Comptonization in this source.

ACKNOWLEDGMENTS

J.E.G. acknowledges support under NASA contract S-10987-C.

REFERENCES

1. Sanchez-Fernandez et al. 1999, A&A, 348, L9.
2. Sobczak et al. 1999, ApJ, 517, L121.
3. Castro-Tirado et al. 1992, IAUC 5590.
4. Mirabel & Rodriguez, 1994, Nature, 371, 46.
5. Belloni et al. 1997, ApJ, 488, L109.
6. Grove et al. 1998, ApJ, 500, 899.
7. Greiner et al. 1996, ApJ, 473, L107.
8. Ebisawa, Titarchuk, & Chakrabarti, 1996, PASJ, 48, 59.

ASCA Observations of the Superluminal Jet Source GRO J1655–40

K. Yamaoka*, Y. Ueda*, H. Inoue*, F. Nagase* Y. Tanaka†, K. Ebisawa‡, T. Kotani‡, C.R. Robinson||, and S.N. Zhang||

*The Institute of Space and Astronautical Science, †Max Planck Institut fur extraterresteische Physik, ‡ NASA/Goddard Space Flight Center, || NASA/Marshall Space Flight Center

Abstract.

We report on the results of *ASCA* observation of the Galactic jet source GRO J1655–40 performed in 1997 Feb. Previous *ASCA* observations of GRO J1655–40 in 1994 and 1995 revealed the presence of K_α resonance absorption lines from highly ionized iron ions in the X-ray spectra [1]. The center energy of the absorption line varied in association with the change of X-ray intensity: it was that of K_α line of H-like irons and of He-like irons when the X-ray intensity was high (2.2 Crab) and low (0.27–0.57 Crab), respectively. These results suggest that the high temperature plasma producing the absorption lines is located in a non-spherical configuration and is photo-ionized by X-ray radiations. In order to investigate the geometry of the plasma in detail, we performed a long observation of this source from 1997 Feb. 26 to Feb. 28 to cover its full orbital period (2.6 days). The average X-ray intensity was 0.9 Crab and a blend of two absorption lines from H-like and He-like irons was detected. This is consistent with the photo-ionization scenario in relation to the previous observations. We found that the absorption line features exist over the full orbital phase. This fact indicates that the plasma responsible for line absorptions distributes symmetrically around the rotation axis of the accretion disk.

INTRODUCTION

GRO J1655–40 (Nova Scorpii 1994) is one of the Galactic superluminal jet sources. Since its discovery by the Burst and Transient Source Experiment (BATSE) onboard *the Compton Gamma Ray Observatory (CGRO)* on 27 July 1994 [2], hard X-ray outbursts have been repeated for several times. Correlated with these outbursts, plasma ejection episodes at superluminal speed are detected [3] [4]. The intrinsic velocity of the relativistic jets was found to be 92% of the light speed [4]. The large mass function derived from optical observations indicates that the compact object is a black hole [5]. Such objects, called "microquasar" [6], are the best targets to understand the relation between jet formation and a black hole.

The Advanced Satellite for Cosmology and Astrophysics (ASCA) [7], carrying two Gas Imaging Spectrometers (GIS) and two Solid-State Imaging Spectrometers (SIS), made five pointed observations of GRO J1655–40 up to present. Ueda et

CP510, *The Fifth Compton Symposium,* edited by M. L. McConnell and J. M. Ryan

TABLE 1. Summary of *ASCA* observations of GRO J1655–40

Observation Epoch	Intensity (2–10 keV)[a]	Iron K_α Absorption Lines
Aug. Sep. 1994[b]	4.5 – 12	He–like
Aug. 1995	47	H–like
Mar. 1996	0.0002	—
Feb. 1997	24	H–like + He–like

[a] Mean intensity (10^{-9} erg/cm^2/s)
[b] Dip-like events whose mean intensity is 1.7 $\times 10^{-9}$ erg/cm^2/s in 2–10 keV are excluded.

al.(1998) [1] detected K-absorption lines from highly ionized irons, which is the first detection among accretion powered sources. They suggest the presence of highly ionized plasma in a non-spherical configuration with an estimated size of $\sim 10^{10}$cm, which is photo-ionized by X-ray irradiation. The similar features were also detected from another microquasar GRS 1915+105 [8].

OBSERVATION AND ANALYSIS

We summarize all the *ASCA* observations of GRO J1655–40 in Table 1. We performed a long observation of GRO J1655–40 for a net exposure of 100 ksec over a full orbital period (2.6 days [9]) from 1997 Feb. 26 to Feb. 28. Main purpose of this observation is to make a clear picture about the geometry of the plasma that produces the absorption line by investigating correlation of the X-ray spectrum with orbital phase. In the 1997 observations, the average X-ray intensity was 2.4 $\times 10^{-8}$ ergs/cm^2/s in 2–10 keV (0.9 Crab), which is intermmedate between that of the 1994 and 1995 observations. We used SIS data for study of absorption line structures because of its good energy resolution, but used GIS data for continuum spectra to avoid "pile-up" effects in the SIS data (see [1]).

Light Curve and Continuum Spectra

The left panel in Fig. 1 shows GIS light curve with hardness ratio in the 1997 observation. As noticed, the intensity was almost constant, with slight variations at 3.1% rms level, and any dip-like event as observed in 1994 August [1] was not detected. The history of the hardness ratio suggests that the spectral shape did not change significantly during the observation. Hence, we made spectrum by integrating the data over the whole observation. We found that the spectrum in the 1997 obsevations can be well explained by the MCD model (MultiColor Disk [10]) with the innermost temperature of 1.18$\pm$0.01 keV. An additional power law component is not necessary unlike the case of the 1995 observation [11]. The right panel in Fig 1 shows the GIS unfolded spectra in 1.0–10 keV when fit with the above continuum model. We also plot the spectra in the previous observations for comparison.

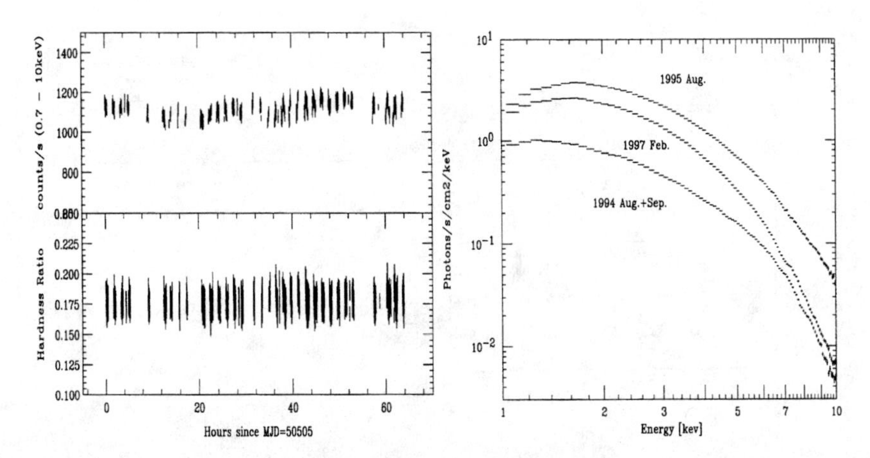

FIGURE 1. Left panel: The GIS light curve of GRO J1655–40 in 0.7–10.0 keV (the upper panel) and the hardness ratio (the lower panel) between 0.7–4.0 keV and 4.0–10 keV. Right panel: The GIS unfolded Spectra in 1.0–10 keV in comparison with previous observation.

TABLE 2. The best fit parameter of iron K_α absorption lines.

Observation Epoch	Enegy[keV][a]	1 σ Line Width[eV]	E.W.[eV]	Line Identification
Aug. 1995	6.95±0.11	<150	25^{+25}_{-13}	H-like iron K_α
Feb. 1997	6.86±0.08	160±37	62±11	H+He-like iron K_α
Aug.+Sep. 1994	6.63±0.07	<75	61^{+15}_{-13}	He-like iron K_α

NOTE.— Errors are 90% confidence limits for a single parameter.
[a] Includes 1% systematic and statistical error at 90% confidence level.

Iron Absorption Lines

The left panels of Fig. 2 show the SIS folded spectra and the ratio of the data to the continuum model for the three different observations. The best fit parameters for the iron-K_α absorption lines are summarized in Table 2. When the X-ray intensity was high and low, we detected H-like Fe K_α and He-like K_α absorption lines, respectively [1]. On the other hand, when it was medium, we detected apparently broad absorption line feature centered at 6.86±0.08 keV with a 1σ width of 160±37 eV. If this line width is attributed to thermal motion, then the corresponding iron-ion temperature becomes about 14 MeV, which is too high to keep the ions not being fully ionized. Hence, by comparison with the results of 1994 and 1995 observations, we interpret this feature as a blend of two K_α absorption lines from H-like and He-like irons. The change of ionization state of iron ions with X-ray intensity confirms the scenario that the highly ionized plasma that produces the absorption lines is photo-ionized by X-ray irradiations.

To investigate orbital-phase dependence of the absorption lines, we next divided

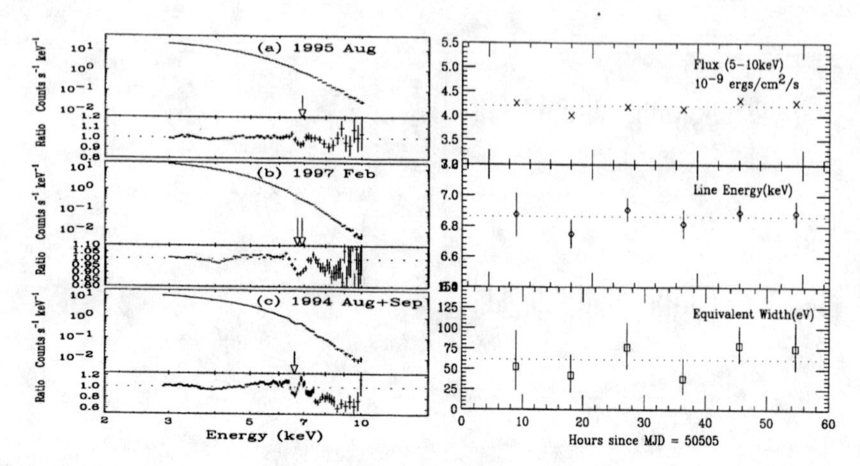

FIGURE 2. Left panel: the SIS raw spectra in 3.0–10 keV (the upper panel) and the data to model (cutoff power law model) ratio (the lower panel) for each observations. Fe K_α absorption lines are denoted by arrows. Right panel: Dependence of the equivalent width and the center energy of iron K_α absorption lines on orbital phase. The dotted line shows the best fit parameter derived from the whole observation.

the whole observation into six periods (0.45 days for each) and fit each spectrum with a cutoff power law model plus a negative gaussian in the 5.0–10 keV range. The right panel in Fig. 2 shows the results. As noticed from the figure, we *always* detected iron K_α absorption line and found no significant changes of its parameters, such as line center energy and equivalent width, in association with the orbital phase. This means that the highly ionized plasma has an axial symmetric (but spherically asymmetric) configuration around the black hole. This suggests that the plasma is likely to be geometrically thin accretion-disk corona extending above the entire accretion disk.

REFERENCES

1. Ueda, Y., et al. 1998, ApJ, 492, 782
2. Zhang, S.N., et al. 1994, IAU Circ., No 6046
3. Harmon, B.A. et al. 1995, Nature, 374, 703
4. Hjellming, R.M., & Rupen, M.P., 1995, Nature, 375, 464
5. Orosz, J.A., & Bailyn, C.D., 1997, ApJ, 477, 876
6. Mirabel, I.F., & Rodriguez, L.F., 1994, Nature, 371, 46
7. Tanaka, Y., et al. 1994, PASJ, 46, L37
8. Kotani, T., et al. 1999, submitted to ApJ
9. Bailyn, C.D., et al. 1995, Nature, 378, 157
10. Mitsuda, K., et al. 1984, PASJ, 36, 741
11. Zhang, S.N., et al. 1995, ApJ, 479, 381

Observations of some X-ray transients with RXTE

Konstantin N. Borozdin*, William C. Priedhorsky*, Mikhail G. Revnivtsev**, Artem N. Emelyanov**, Sergey P. Trudolyubov**, Lev G. Titarchuk†, and Alexey A. Vikhlinin††

*NIS-2, Los Alamos National Laboratory, Los Alamos, NM 87545
**Space Research Institute, 117810 Moscow, Russia
† Goddard Space Flight Center, Greenbelt, MD
†† Harvard-Smithsonian Center for Astrophysics

Abstract. We present results of observations of several X-ray transients with RXTE in 1996-1998, namely, GRS 1739-278, XTE J1748-288, GS 1354-64, 2S1803-245 and XTE J0421+560 (CI Cam). We studied light curves and spectra of their outbursts and compared them with observations of other X-ray transients. We discuss fits of high state spectra with BMC model, and similarities and differences between black holes and neutron stars in their low state. Special attention is paid to CI Cam as possible legate for new class of X-ray transients.

INTRODUCTION

Many X-ray binaries with low mass companions (typically K or M dwarf) are characterized by episodic irregular X-ray outbursts. The recurrence time for most of these systems is estimated to be 10-100 years. The typical duration of an outburst is several weeks, but some outbursts last just few days, while some other sources remain bright for years after an initial outburst. Because the time of the outburst cannot yet be predicted, data from X-ray all-sky monitors (ASM) are used to trigger multi-wavelength studies. In recent years most outbursts of X-ray transients were initially detected by ASM on-board Rossi X-ray Timing Explorer (RXTE) satellite [1]. Several examples of the outbursts observed with ASM/RXTE are shown in Fig. 1. LMXB transients are of special interest because most reliably identified black hole binaries belong to this class [2]. During an outburst, the flux from such source varies by orders of magnitude, typically accompanied by dramatic changes in spectral shape. These sources are natural laboratories for studying of mass accretion onto the compact object in very different regimes. Here we present some results obtained from RXTE observations of several X-ray outbursts.

CP510, *The Fifth Compton Symposium,* edited by M. L. McConnell and J. M. Ryan

UNUSUAL VERY HIGH STATE IN BLACK HOLE X-RAY TRANSIENTS

Many black hole X-ray transients demonstrate common evolution pattern during their outbursts with several typical states and transitions between them. We studied this pattern recently for XTE 1748–288 [3] and before - for KS 1730-312 and GRS 1739-278 [4–6]. Light curves of GRS 1739-278 and XTE 1748–288 (Fig. 1 - upper left and bottom right panels) were of FRED type [7]. XTE 1748–288 was observed by RXTE in *very high*, *high* and *low* spectral states, typical for black hole systems (Fig.2). During several observations corresponding to the peak X-ray flux, the spectrum of the source was in an unusual VHS, with a very bright hard component. Similar spectra were detected at the beginning of the outbursts of X-ray black hole Novae GS/GRS 1124–68 [8,9], KS/GRS 1730–312 [4,5] and GRS 1739–278 [6]. One may consider this spectrum to be representative of a new state or sub-state in black hole X-ray transients.

We studied the relationship of the spectral and temporal properties of XTE J1748–288 in its VHS [3]. The general correlation between the main parameters describing the power density spectrum of the source was observed. Most striking is the established close relation between the evolution of the spectral and timing parameters of the source. In particular, there is a clear trend of increasing the QPO centroid frequency with rise of the soft component flux. This type of correlation holds on the wide range of time scales from seconds to several days.

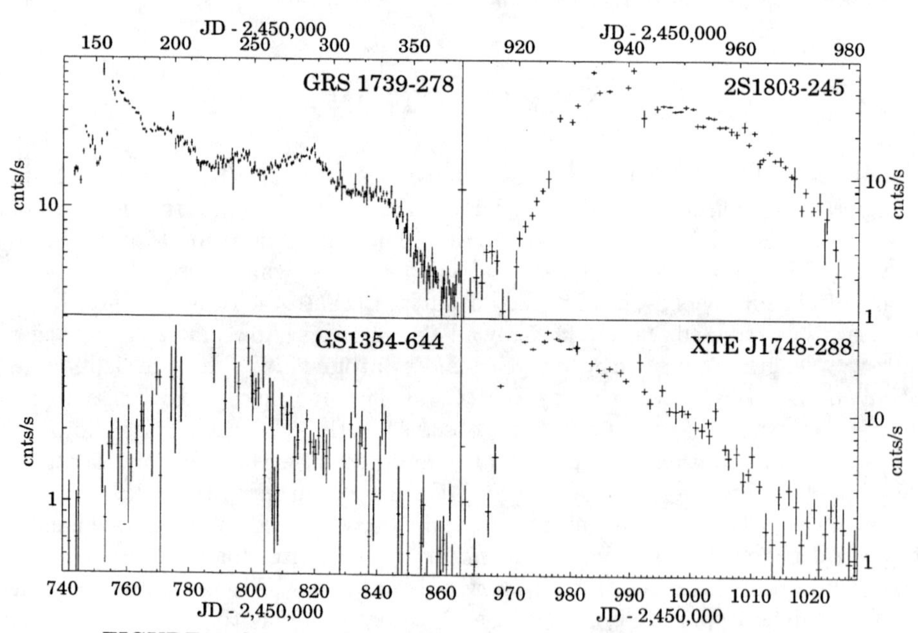

FIGURE 1. Some outbursts of X-ray Novae detected with ASM/RXTE.

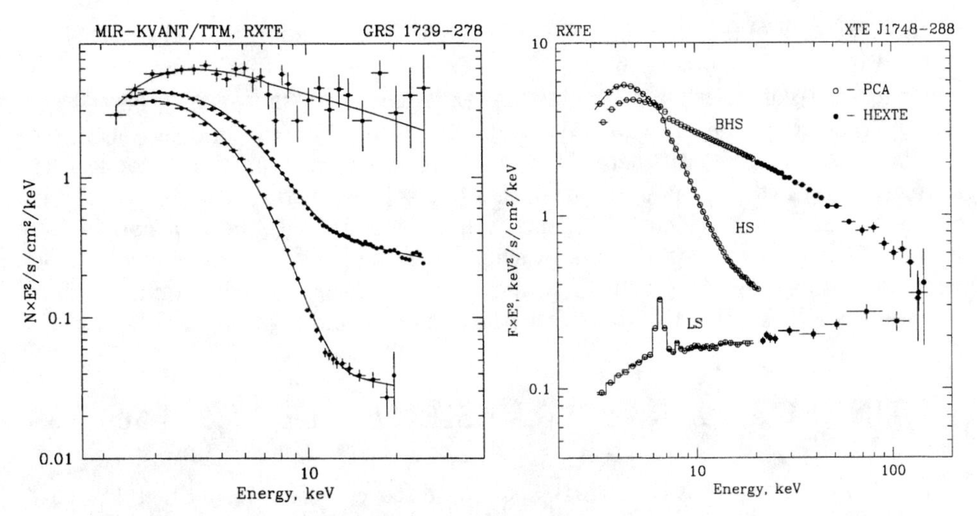

FIGURE 2. Typical broad-band energy spectra of GRS 1739–278 [6] and XTE J1748–288 [3] during the different stages of their outbursts: BHS – bright hard state (VHS), HS – high state and LS – low state. Upper spectrum on left panel taken with Mir-Kvant, two lower - with RXTE. Hollow and filled circles on right panel represent the data of PCA and HEXTE instruments respectively.

HIGH STATE

In *high* state black hole (BH) and neutron star (NS) binaries exhibit qualitatively different spectra. Neutron star binaries spectra have a soft thermal component characterized by color temperatures $\simeq 1 - 2$ keV, which is interpreted as radiation emanating from the neutron star surface and/or boundary layer. In the BH case we observe a thermal component from accretion disk and an additional hard power law component dominating at energies higher than 10–20 keV. One of the possible explanations for the origin of this hard emission is offered by *Bulk Motion Comptonization* model [10,11]. In this model hard X-ray emission is generated by up-scattering of soft X-ray photons on relativistic electrons rushing into black hole. This model is explaining the difference in spectra of BH and NS binaries in *high* and *very high* state. The application of BMC model to spectra of several black hole transients observed by the RXTE showed that the model is able to fit the spectra [12].

LOW STATE

Low state spectra for BH and NS binaries are very similar. We observed *low* state with RXTE from BH transients XTE J1748-288, GS 1354–64 [3,17], and from NS transient 2S1803–245 [13]. The overall power-law-with-high-energy-cutoff spectral shape may be approximated in each case by Comptonization models based on up-scattering of soft photons on energetic electrons in a hot plasma cloud.

The rapid time variability of the X-ray flux was in general very similar to the low/hard state of other Galactic black hole systems, such as Cyg X-1, Nova Persei 1992, and GX 339-4 [14–16]. We found that amplitude of *rms* variability for 2S1803–245 increased with energy. Our analysis of RXTE archival data for several sources, confirmed that an anti-correlation of fractional variability with energy is typical for Galactic black holes in their low spectral state, but a positive correlation is typical for neutron star systems in their low state [17]. This difference can be very useful for segregating neutron star binaries from black hole systems in their *low* state, which is otherwise difficult with X-ray data only.

UNUSUAL X-RAY TRANSIENT XTE J0421+560

The X-ray transient XTE J0421+560 was discovered on Mar. 31, 1998 with ASM/RXTE. The light curve of the source in 1.3-12 keV was characterized by extremely fast rise [18] with subsequent, also unusually fast decay of the flux. The observations by ASCA, BeppoSAX, and RXTE satellites showed an unusual spectrum with the strong emission line around 6.7 keV [19–21]. The optical counterpart of the X-ray source was identified as relatively close to the Sun ($\sim$1 kpc) symbiotic star CI Cam [22]. Following the detection of X-ray emission many researches suggested that compact object of the system is neutron star or black hole. The outburst however was distinctly different from any other X-ray transients. First of all it was much shorter than typical and the source returned to quiescence in few days instead of weeks or months. X-ray spectrum of the source can be described as the emission of two-temperature optically thin plasma, with prominent iron emission line around 6.5–6.7 keV and weaker, but detectable line around 8 keV [21]. Strong absorption was observed with RXTE in first two days of the outburst, but became undetectable later. The effective temperature and the center of iron line evolved with time significantly. Altogether the evolution of the X-ray spectrum can be interpreted as an emission of non-equilibrium plasma, which started inside larger cold cloud and later expanded. Radio observations started during the outburst and continued during at least half of a year after it showed the evidence of expanding quasi-spherical shell [23]. All these data confirmed that it was the first X-ray transient of such kind. The outburst somewhat resembles down scaled supernova explosion. Hopefully future observations will reveal more objects of this class and allow to understand their nature.

ACKNOWLEDGMENTS

For this work we used results provided by the ASM/RXTE teams at MIT and at the RXTE SOF and GOF at NASA's GSFC.

REFERENCES

1. Levine A.M., Bradt H., Cui W., Jernigan J.G., Morgan E.H., Remillard R., Shirey R.E., and Smith D.A., *Ap. J.*, **469**, L33 (1996).
2. Tanaka Y., and Lewin W.H.G., *X-ray Binaries* eds. W.H.G.Lewin, J.van Paradijs, & E.P.J. van den Heuvel, Cambridge Univ.Press, 126 (1995).
3. Revnivtsev M. G., Trudolyubov S. P., and Borozdin K. N., *M.N.R.A.S.*, accepted, preprint astro-ph/9903306 (1999).
4. Borozdin K. N., Aleksandrovich N. L., Arefiev V. A., Sunyaev R. A., and Skinner G. K., *Astr.Lett.*, **21**, 212 (1995).
5. Trudolyubov S. P., Gilfanov M. R., Churazov E. M., Borozdin K. N., Aleksandrovich N. L., Sunyaev R. A., Khavenson N. G., Novikov B. S., Vargas M., Goldwurm A., Paul J., Denis M., Borrel V., Bouchet L., Jourdain E., and Roques J.-P., *Astr.Lett.*, **22**, 664 (1996).
6. Borozdin K. N., Revnivtsev M. G., Trudolyubov S. P., Aleksandrovich N. L., Sunyaev R. A., Skinner G. K., *Astron.Lett.*, **24**, 435 (1998).
7. Chen W., Shrader C. R., and Livio M., *Ap. J.*, **491**, 312 (1997).
8. Ebisawa K., Ogawa M., Aoki T., Dotani T., Takizawa M., Tanaka Y., Yoshida K., Miyamoto S., Iga S., Hayashida K., Kitamoto S., and Terada K., *PASJ*, **46**, 375 (1994).
9. Miyamoto S., Kitamoto S., Iga S., Hayashida K., Terada K., *Ap. J.*, **435**, 398 (1994).
10. Chakrabarti S. K., and Titarchuk L. G., *Ap. J.*, **455**, 623 (1995).
11. Titarchuk L. G., Mastichiadis A., and Kylafis N. D. *Ap. J.*, **487**, 834 (1997).
12. Borozdin K., Revnivtsev M., Trudolyubov S., Shrader C., and Titarchuk L. G., *Ap. J.*, **517**, 367 (1999).
13. Revnivtsev M. G., Borozdin K. N., and Emelyanov A. N., *Astron. Astrophys.*, **344**, L25 (1999).
14. Nowak M. A., Vaughan B. A., Wilms J., Dove J. B., and Begelman M. C., *Ap. J.*, **510**, 874 (1999).
15. Vikhlinin A., Churazov E., Gilfanov M., Sunyaev R., Finoguenov A., Dyachkov A., Kremnev R., Sukhanov K., Ballet J., Goldwurm A., Cordier B., Claret A., Denis M., Olive J.F., Roques J.P., and Mandrou P., *Ap. J.*, **441**, 779 (1995).
16. Nowak M. A., Wilms J., and Dove J. B., *Ap. J.*, **517**, 355 (1999).
17. Revnivtsev M. G., Borozdin K. N., Priedhorsky W. C., and Vikhlinin A. A., *Ap. J.*, accepted, preprint **astro-ph/9905380** (2000).
18. Smith D., Remillard R., Swank J., Takeshima T., and Smith E., *IAU Circ.*, **6855** (1998).
19. Orr A., Parmar A.N., Orlandini M., Frontera F., Dal Flume D., Segreto A., Santangelo A., and Tavani M., *Astron. Astrophys.*, **340**, L19 (1998).
20. Ueda Y., Ishida M., Inoue H., Dotani T., Greiner J., and Lewin W.H.G., *Ap. J.*, **508**, L167 (1998).
21. Revnivtsev M.G., Emel'yanov A.N., and Borozdin K.N., *Astron. Lett.*, **25**, 294 (1999).
22. Chkhikvadze Ja., *Astrofizika*, **6**, 65 (1970).
23. Mioduszewski A., *LAAstro Seminar*, April 28 (1999).

Physical Parameter Estimation in Black Hole X-Ray Binaries

C.R. Shrader & L.G. Titarchuk

Laboratory for High-Energy Astrophysics, NASA Goddard Space Flight Center

Abstract. We describe a method for extracting physical information on black-hole binaries systems from their X- and gamma-ray spectral properties. The high-energy continuum is interpreted as thermal emission from an accretion disk and Comptonized emission from a relativistic bulk inflow. Application of this methodology to recent X- and gamma- ray observations are presented, with emphasis on the recently discovered X-ray nova XTE J1550-564.

INTRODUCTION: BULK-MOTION COMPTONIZATION

Accreting stellar-mass black holes in Galactic binaries have long been known to exhibit a "bi-modal" spectral behavior - namely the so called high-soft and low-hard spectral states [1]. One of the major contributions of CGRO has been to extend this picture into the high-energy domain. An important point is that the high-soft state spectra are apparently a *unique* black hole signature, whereas the low-hard state spectral form has been seen in neutron star binaries under certain conditions – typically in low luminosity states of bursters [2], although note [3]. It thus seems reasonable to speculate that this unique spectral signature is directly tied to the black hole event horizon. This is the primary motivation for the Bulk Motion Comptonization Model (BMC) introduced in several previous papers, and recently applied successfully to observational data. We will not describe the model in detail here, but refer the interested reader to the literature and briefly state its essential components: Compton scattering of thermal X-rays from bulk-motion infall in close proximity to the black-hole event horizon, and a hot Compton cloud obscuring the inner disk during periods of low-mass-accretion rate [4–9].

It has been further pointed out, that from the spectral shape and normalization, one can calculate an effective disk radius and the mass-to-distance ratio m/d [8,9]. Other quantities such as a black hole mass and mass-accretion rate can be determined if the distance or mass are known independently.

CP510, *The Fifth Compton Symposium,* edited by M. L. McConnell and J. M. Ryan

DESCRIPTION OF PARAMETER ESTIMATION METHOD

From the inferred color temperature and absolute normalization, the effective area of the emission region can be obtained in terms of the distance. The observable surface area can be related to the black hole mass using, a particular disk model [10] which we have modified to treat electron scattering and free-free absorption and emission [11]. The emergent spectrum is then represented by the integral of a diluted blackbody function characterized by a hardening factor T_h^{-4}. As shown previously [8,9], by following this basic prescription and applying the mean-value theorem, a simple expression for the emergent flux is derived:

$$g_\nu(E) = A_N \frac{E^3(1.6 \cdot 10^{-9})}{\exp[E/T_{col}] - 1}, \quad (1)$$

where $A_N = 0.91(m/d)^2 cos(i) T_h^{-4} r_{eff}^2$. Using the value of T_h derived by [8], $T_h \simeq 2.6$, we can then calculate the mass-to-distance ratio

$$\left(\frac{d}{m}\right)^2 = \frac{0.91 T_h^{-4} r_{eff}^2}{A_N} \cos i \quad (2)$$

APPLICATION TO XTE J1550-564

We have applied this formalism to recent observational data for X-ray nova XTE J1550-564 [12]. The analysis presented here derives from the CGRO OSSE and BATSE instruments, as well as the RXTE ASM, PCA and HEXTE instruments during the 1999 March time frame. The composite hard-soft X-ray light curve (Figure 1) illustrates roughly the evolution of that particular event – its rather gradual linear decay from about 2.5-to-2 Crab, followed by an exponential decay in the 2-10 keV band, with more erratic flaring behavior in the BATSE 20-100 keV band. The time of the OSSE integration, Viewing Period 808.5, is indicated by the dotted vertical lines.

Application of the BMC model to the composite RXTE/OSSE spectrum was attempted, and a successful fit is illustrated in Figure 2. It must be noted however, that the OSSE spectrum presented here represents an integration over the 1-week viewing period, whereas the RXTE soft X-ray spectra are typically $\sim 10^3 - s$ "snapshots". In particular, we found that in a number of cases that although the basic high-soft state spectral form was present, the hard power law showed clear indications of significant steepening in the 25-30 keV range.

Applying our parameter estimation methods analysis for XTE J1550-56, we obtain $d/m = 0.0402 \pm 0.0040$, where m is in solar units and d is in units of

FIGURE 1. Composite soft- and hard-X-ray light curve for the March 1999 time frame. The dotted vertical lines indicate the OSSE observation interval.

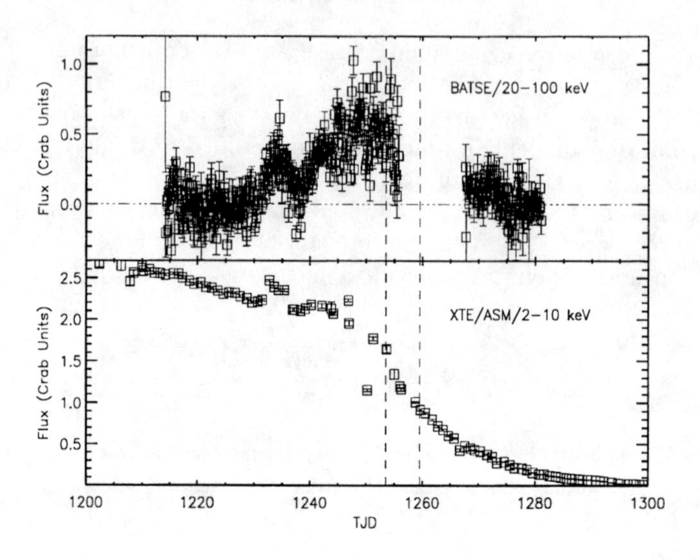

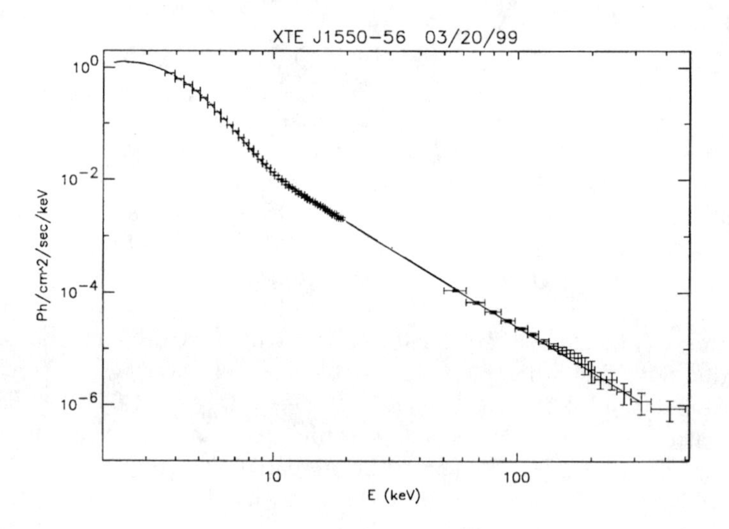

FIGURE 2. Combined OSSE and RXTE/PCA spectrum. One must note that the while the OSSE spectrum as presented here represents an integration of about 1 week duration, while the RXTE data are a 10^3-second "snapshot. In this case the paramters derived were $(\alpha,T,f)=(1.7,0.86,0.67)$.

FIGURE 3. Substantial spectral evolution is seen over the course of the February-March 1999time frame. Initially, the source is very bright and very soft, as evidenced in panel (a). The hard spectral index, although poorly constrained is quite steep, and the fraction of upscattered photons is small (<10%). Later, panel (b), the hard flux has increased substantially, as has the fraction of photons upscattered.

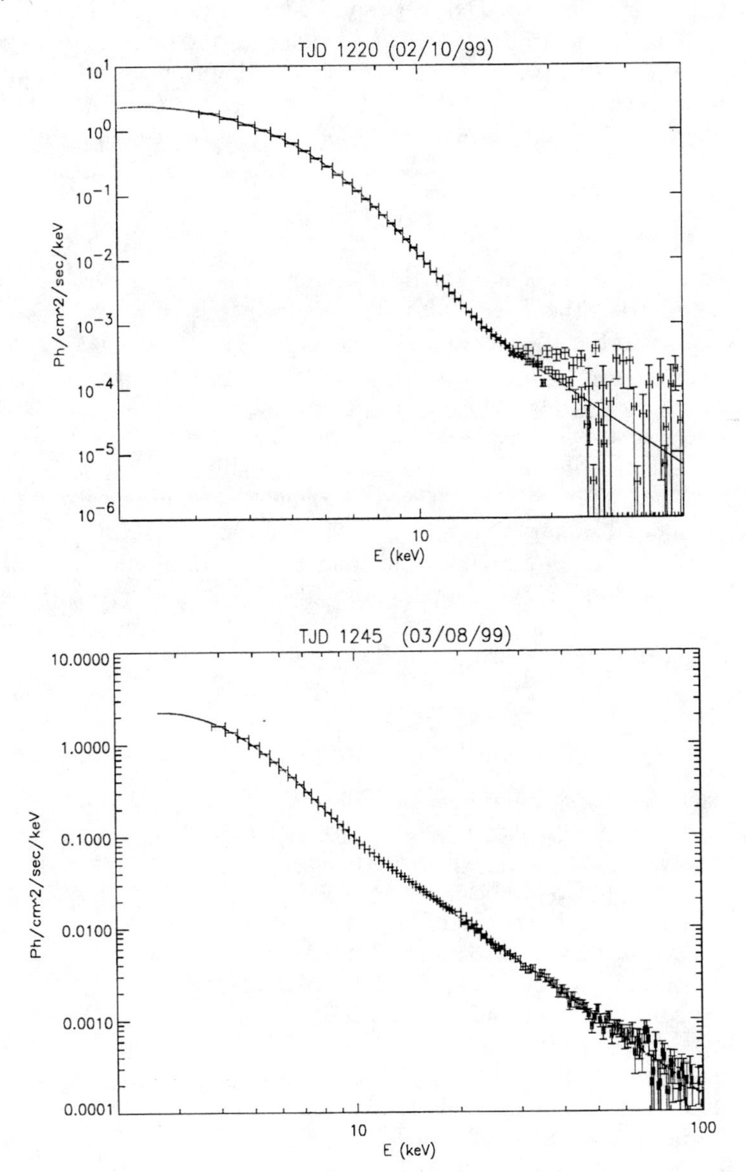

10 kpc. A $cos(i) = 0.5$ inclination term has been assumed in this calculation. This suggests a large black-hole mass, perhaps $m \simeq 10 - 15\ M\odot$ if the source is in the 5 kpc distance range. The distance is unknown, but given the large column density $log(N_H) \simeq 2$ it is probably more than a few kpc. We further note that for $d = 5$ kpc and our corresponding mass estimate, the source would have been radiating in the $5 - 7\%\ L_{edd}$ range during early March 1999 when the $2 - 10$ keV flux was in the $2 - 2.5$ Crab. However, given the apparently large mass accretion rate $10 - 15\%$ Ledd is more plausible and the distance may be in the $7 - 8$ kpc range calling for a larger black-hole mass.

Spectral Evolution

From Figure 1, it is evident that the source is very bright, yet in a very soft state during the TJD 1200-1230 time frame, as the BATSE 20-100 keV flux remains at a level consistent with zero. This indicates that direct thermal emission is predominant relative to the scattered component - or in the context of our model $f \sim 0$ [4]. Referring to Figure 3a, which illustrates the spectrum of TJD 1219, this indeed seems to be the case. Here we find that $f < 0.1$ and the spectral index, although only marginally constrained is very steep ($\alpha \simeq 3.5$). In Figure 3b on the other hand (spectrum of TJD 1245), the spectral index has hardened to $\alpha = 1.77 \pm 0.03$, while f has increased to 0.6.

This particular X-ray nova event, and another dramatic event discovered at about the time of this submission, XTE J1859+226, given the extensive coverage available and the large flux dynamic range offer the possibility of further clarifying this picture.

REFERENCES

1. Liang, E.P. 1998, Phys. Rep., 302, 66.
2. Barret, D., & Grindlay, J.E. 1995, ApJ, 440, 841.
3. Strickman, M.S., et al., (these proceedings).
4. Titarchuk, L. G., Mastichiadis, A., Kylafis, N. D. 1997, ApJ, 487, 834
5. Titarchuk, L., & Zannias. T. 1998, ApJ, 493, 863
6. Shrader, C.R., & Titarchuk,L.G. 1998, ApJ, 499, L31
7. Laurent, Ph., & Titarchuk, L.G. 1999, ApJ, 511, 289
8. Borozdin, K., Revnivtsev, M., Trudolyubov, S., Shrader, C.R., & Titarchuk, L.G. 1999, ApJ, 517, tbd.
9. Shrader, C.R., & Titarchuk, L.G., 1999, ApJL, 521, L121.
10. Shakura, N.I., & Sunyaev, R.A. 1973, A&A, 24, 337
11. Titarchuk, L.G. 1994, 429, 340
12. Sobczak, G.J., et al., 1999, ApJ, 517, L121.

Her X-1: Correlation between the histories of the 35 d cycle and the 1.24 sec pulse period

R. Staubert, S. Schandl, J. Wilms

Institut für Astronomie und Astrophysik, Waldhäuser Str. 64, D-72076 Tübingen, Germany

Abstract. In Her X-1 the history of the 35 day turn-ons is strongly correlated with the histories of the 1.24 s pulsations and the X-ray flux. We suggest a common origin for these correlated variations, namely a variable mass transfer rate.

INTRODUCTION

We study the long-term (1971–1999) behavior of the 1.24 sec pulse period and the 35 day precession period of Her X-1 and show that both periods vary in a highly correlated way (see also Schandl et al. 1997): When the spin-up rate decreases, the 35 day turn-on period gets shorter. This correlation is most evident on long time scales (~2000 days), e.g., around two extended spin-down episodes, but also on shorter timescales (a few 100 days) on which quasi-periodic variations are apparent. We argue that the likely common cause is variations of the mass accretion rate onto the neutron star. This is supported by a corresponding correlation with X-ray flux over part of the historical time considered. The data since 1991 allow a continuous sampling (by BATSE and RXTE/ASM) and indicate a lag between the turn-on behavior and the spin behavior, in the sense that changes are first seen in the spin, about one cycle later in the turn-on. We suggest a physical explanation within the framework of the coronal wind model (Schandl and Meyer 1994). A variation in the mass transfer rate from the companion of a few percent is sufficient to alter the torque on the NS and the shape of the warped disk leading to a different precessional frequency.

THE CORONAL WIND MODEL

Heating of the accretion disk surface by X-ray illumination yields a hot corona. This is hydrostatically layered above inner disk regions and escapes from the system at outer radii where its inner energy gets larger than the gravitational potential.

CP510, *The Fifth Compton Symposium,* edited by M. L. McConnell and J. M. Ryan

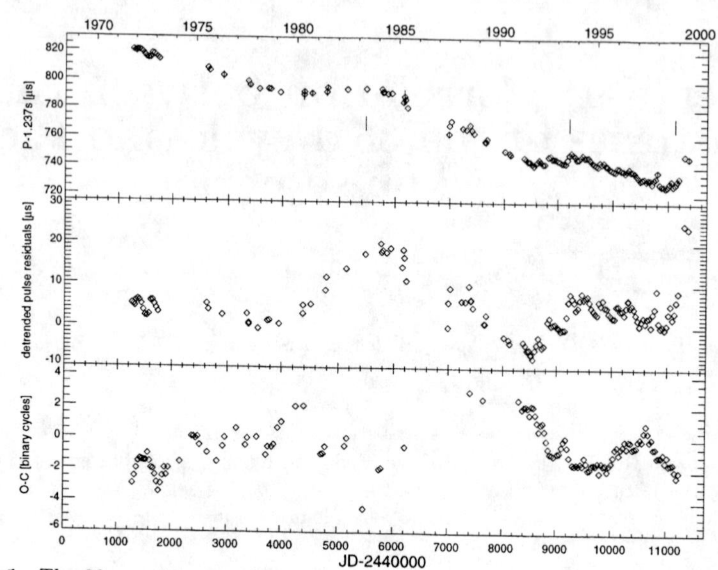

FIGURE 1. The 30 year history of the Her X-1 pulse period (upper panel), the pulse period derivatives with the linear spin-up trend removed (middle panel) and the turn-on data (lower panel), where O−C is the difference between observed and calculated turn-on times (see text). – The dashes mark the start times of anomalous low periods (Parmar et al. 1985, Vrtilek et al. 1994, Levine et al. 1999.

Above the surfaces of a warped disk such a coronal wind is asymmetrical. The repulsive forces of the leaving matter yield therefore a torque which does not vanish. This torque acts on the inclination of each disk ring as well as on the precessional period.

The stationary torque equation includes viscosity, tidal forces and coronal wind torque. For the boundary condition torques from outside are excluded. The resulting solution for the disk shape is discussed in Schandl and Meyer 1994 and Schandl 1996.

THE SPIN PERIOD DATA

The historical data for the 1.24 s X-ray period (the spin of the NS) are compiled from the original literature. A complete list will appear in Staubert, Schandl and Wilms (1999). Previous compilations (which contain several errors) are by Nagase (1989), Sunyaev et al. (1988) and Kunz (1996). Since 1991 pulse periods were regularly measured by BATSE onboard of GRO. We have made use of the publically distributed pulsar data as well as lists kindly provided by R. Wilson, and data from pointed RXTE observations (e.g., Stelzer et al. 1997). The pulse period development from its discovery until today is shown in Fig. 1 (top). The average spin-up trend dP/dt amounts to $\sim$9 ns/day, but with clear deviations, including episodes of spin-down. This is emphasized in Fig. 1 (middle) which shows the pulse period derivatives (with a linear spin-up trend removed). The pulse period during the current anomalous low (1.2377485(3) s at JD2451295.0) was determined from RXTE observations (Coburn et al. 1999a and 1999b) and from BeppoSAX observations (1.237747(2) s at JD 2451368.8, Parmar et al. 1999).

THE TURN-ON DATA

Her X-1 shows a 35 day flux modulation supposedly due to the occultation of the NS by a precessing warped accretion disk. Turn-ons, the rise of the X-ray signal towards the Main-On, are observed since the discovery of Her X-1 in 1972 (Tananbaum et al. 1972). Our historical data set is based on Staubert et al. (1983) and Kunz (1996). Additionally, we determined turn-ons from the occultation and pulsed flux data of BATSE and from the RXTE All Sky Monitor. These data were taken from the HEASARC archive at NASA/GSFC. Details of the determination of the turn-on times and a complete list of turn-ons will be given in Staubert, Schandl and Wilms (1999). The fluctuations of the turn-on times can be expressed by the "(O−C)"-diagram (Fig. 1, lower panel) which shows the difference between the observed turn-on time and the calculated turn-on time. The calculated time follows the epoch of Staubert et al. (1983), the 31st turn-on being on T_{31} = JD 2442410.349 d with a 35 day model period of 20.5 P_{orbit} (P_{orbit} = 1.700167788 d, Deeter et al. 1981). Positive values of O-C correspond to turn-ons which are observed later than the calculated one. A clear anti-correlation is seen between the evolution of the pulse period and that of (O−C).

OBSERVATIONAL RESULTS

The mean general spin-up of the neutron star ($\sim$9 nsec/d) is modified by significant structure: Most apparent are two periods of extended spin-down followed by an increased spin-up. These events happen over a time scale of about 2000 days and occur at times of the two previously observed anomalous low periods (Parmar et al. 1985, Vrtilek et al. 1994), as marked by the dashes in Fig. 1. The appearance of the current, still ongoing anomalous low was repoted by the RXTE/ASM team on 1999 April 7 (Levine et al. 1999). In addition, there are deviations on shorter timescales (a few hundred days) which are of quasi-periodic nature (on time scales of 400–600 days, see Fig. 2).

The (O−C) diagram is a representation of the history of the 35 day period. It appears from Fig. 1 that on the longest time scale the average period is consistent with 20.5 P_{orbit}, a prediction made 16 years ago by Staubert et al. (1983) when only the first increasing leg in the (O−C) diagram was known. Individual 35 day cycles (from one turn-on to the next) are either, 20, 20.5, or 21 times P_{orbit}.

It is quite apparent that the large features in the development of the pulse period and in the (O−C) diagram are highly correlated: Whenever the NS enters a period of (relative) spin-down, (O−C) goes into a minimum (that is the turn-on period is short). This is also true for short-term features as is shown in Fig. 2. We note that this kind of correlation has been seen earlier by Bochkarev et al. (1981) and by Ögelman et al. (1985) on the basis of smaller sets of data. Furthermore, there is a clear correlation between pulse period residuals and X-ray flux (e.g., occultation data and pulsed flux data from BATSE) in the sense that (relative) spin-down correlates with lower flux. This fact has also been noted by Wilson et al. (1994), but debated by Wilson et al. (1997).

We find evidence for a lag between the pulse period evolution and the (O−C) evolution. Using the BATSE pulse data (an average pulse period for each Main-On), we cross-correlated the pulse period residuals and the (C-O) data (see Fig. 2). Our overall conclusion is that structures are seen first in the pulse period and then in (C-O). The lag is hard to measure accurately, but appears to be of the order of one 35 day period (which is equal to the resolution of the correlated time series).

DISCUSSION

We interpret the observed correlations in the following way: The driver for the apparent variability is the optical companion which provides more or less material at the inner Lagrangian point. The average mass accretion rate of the NS is such that the system operates close to the equilibrium period with a slight bias towards spin-up. When the mass accretion rate drops also the spin-up rate drops and may even turn to spin-down, in accordance with standard accretion theory (e.g., Ghosh and Lamb, 1997). As expected, the X-ray flux is reduced at such times. The consequences for the observable turn-on times can be understood within the framework of the coronal wind model (Schandl and Meyer 1994, Schandl 1996): The reduced X-ray irradiation of the outer parts of the disk reduces the coronal wind and its torque on the accretion disk leading to a generally less inclined disk (causing the blocking of our line of sight to the NS and hence the X-ray low). The flatter disk geometry adjusts itself to keep the balance between mass loss rate in the wind and matter accreted by the neutron star nearly constant. As a result the ratio of the disk inclination at outer radii (the wind region) to the inclination of the inner disk increases. This causes a faster disk precession as observed. For details see Staubert, Schandl and Wilms (1999). When the accretion rate increases again the inverse process is initiated.

It is noted that Vrtilek and Cheng (1996) in discussing results of optical, UV- and X-ray observations of Her X-1 around the second anomalous low had speculated that the apparent irregularities in the 35 day clock may be related to a variable mass accretion rate around an average value. The above analysis and physical interpretation shows that this is indeed the case.

When the current low will have ended we predict to find a pulse period of less

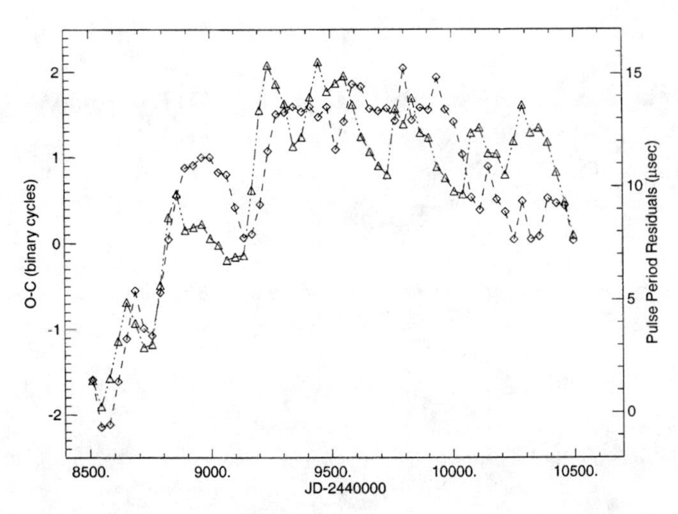

FIGURE 2. Pulse period residuals (triangles, right y-axis) and (C-O) data (diamonds, left y-axis) for JD-2440000 from 8500 to 10500.

than 1.237750 s and a positive value for (O−C).

This work was supported by DLR under grant No. 50OR9205.

REFERENCES

Bochkarev N.G., et al., 1981, Sov.Astron.Let. 14(6), 421
Coburn W., et al., 1999a, this volume
Coburn W., Heindl W.A., Wilms J., et al., 1999b, submitted to ApJ
Deeter J.E., Boynton P.E., Pravdo S.H., 1981, ApJ 247, 1003
Gosh P., Lamb F.K., 1979, ApJ 234, 296
Kunz M., 1996, PhD Thesis, University of Tübingen, Germany
Levine A.M., Corbet R., et al. 1999, IAU Cir. No. 7139
Nagase F., 1989, PASJ 41, 1
Ögelmann H., et al., 1985, Sp.Sc.Rev. 40, 347
Parmar A.N., Pietsch W., McKechnie S., et al., 1985, Nat 313, 119
Parmar A.N., Oosterbroeck T., DalFiume D., et al., 1999, A&A, 350, L5
Schandl S., 1996, A&A, 307, 95
Schandl S., Meyer F., 1994, A&A 289, 149
Schandl S., Staubert R., König 1997, Proc. 4th COMPTON Symp., CP410, 763
Staubert R., Bezler M., Kendziorra E., 1983, A&A 117, 215
Staubert R., Schandl S., Wilms J., 1999, to be submitted
Sunyaev R., Gilfanov M., Churazov E., Loznikov V., Efremov V., 1988, SvAL 14, 418
Tananbaum H., Gursky H., Kellog E.M., et al., 1972, ApJ 174, L143
Vrtilek S.D., Mihara T., Primini F.A., et al., 1994, ApJ 436, L9
Wilson R.B., et al., 1994, Proc. The Evolution of X-ray Binaries, AIP CP308, 475
Wilson R.B., 1997, Proc. 4th COMPTON Symp., AIP CP410, 739

The 1999 Her X-1 Anomalous Low State

W. Coburn, D. E. Gruber, W. A. Heindl, M. R. Pelling, R. E. Rothschild

Center for Astrophysics and Space Sciences, University of California at San Diego, La Jolla, CA, 92093-0424, USA

R. Staubert, I. Kreykenbohm, P. Risse, J. Wilms

Institut für Astronomie und Astrophysik - Astronomie, University of Tübingen, Waldhäuser Strasse 64, D-72076, Tübingen, Deutschland

Abstract. Recently an Anomalous Low State in the 35d cycle of Her X-1, which is thought to be caused by the tilted accretion disk, was observed with the RXTE. This has been seen only twice before; in 1983 and again 1993. We present timing and spectral results of this latest anomalous low obtained 1999 April 26. Pulsations were observed in the 2-20 keV band but with a pulsed fraction down by a factor of 10 from the main-on. Spectral analysis indicates 70% absorbed flux ($N_H = 7 \times 10^{23}$) with the remainder unabsorbed. This is consistent with continuous screening of the X-ray source by the accretion disk causing the anomalous low.

I INTRODUCTION

Her X-1 is a well known accreting X-ray pulsar in a binary orbit with the A/F star HZ Her. It is characterized by a 1.24s pulse period, 1.70d binary orbital period, and a 35d intensity cycle. The 35d cycle normally has a ~10d main-on and a ~5d short-on state separated by two ~10d low states. This is thought to be due to the precession of a tilted, warped accretion disk viewed nearly edge on which periodically obscures the line of sight to the pulsar.

On 1999 March 23, the 35d cycle vanished and Her X-1 entered an Anomalous Low State(ALS) [3,5] (Fig 1). The ALS has been seen only twice before; in 1983 [6] and again in 1993 [9,10] and is characterized by a substantial drop in X-ray flux with little or no change in UV and optical fluxes [9]. In these earlier events the neutron star was believed to be steadily obscured due to a temporary change in the accretion disk [6]. We discuss spectral and temporal analyses of the 1999 April 4 *RXTE* observation which support the idea that the X-ray source is being continuously screened.

CP510, *The Fifth Compton Symposium,* edited by M. L. McConnell and J. M. Ryan

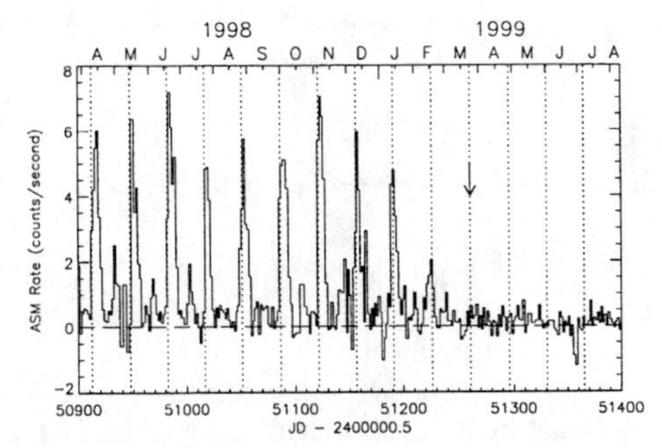

FIGURE 1. The *RXTE*/ASM 0.2-12 keV light curve of Her X-1 for the last 10 35 d cycles before the onset of the ALS. The dotted lines indicate predicted main-on turn-ons using the ephemeris of Scott & Leahy (1999). The bins are the average count rate in each 1.7d orbit of the system. The start of the ALS can be clearly seen at the arrow. Note that the last few high states show evidence for a gradual decrease of flux.

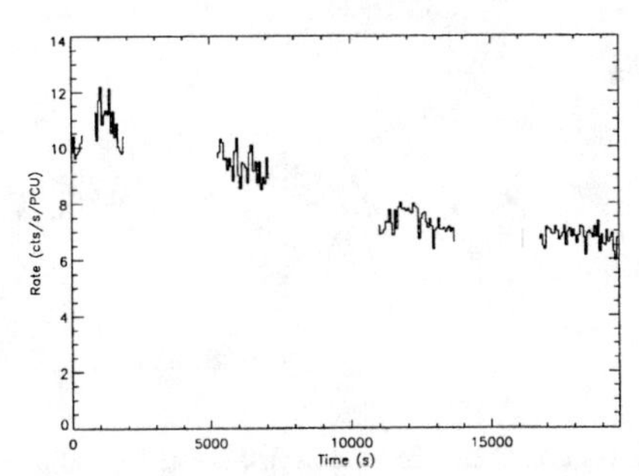

FIGURE 2. The *RXTE*/PCA 2.5-20 keV light curve of our observation, with 64s time resolution, beginning on MJD 51294.40 (1999 April 26, 9:40 UT). Although the flux decreased during the observation, the spectra and pulsed fractions of each *RXTE* orbit remain unchanged. The gaps between the data are due to earth occultations.

II OBSERVATIONS

Our Anomalous Low State observation was made using the Proportional Counter Array (PCA) [2] and the High Energy X-ray Timing Experiment (HEXTE) [7] on board the *RXTE*. The source was observed on 4 consecutive *RXTE* orbits spanning an elapsed time of 20 ks centered at MJD 51294.52 and during phases 0.38-0.52 of the Her X-1 binary orbit. The PCA acquired 4.6 ks of on-source data with 4 Xenon proportional counters (PCUs) while HEXTE's on-source livetime was 2.8 ks. We compared this observation to archival *RXTE* data taken during a main-on and normal low state. The main-on observation was on 1997 September 14 with 15.1 ks of PCA data (all five PCUs) and 9.2 ks in the HEXTE. The low state was observed on 1997 September 12 for 3.3 ks in the PCA (all five PCUs) and 2.0 ks of livetime in the HEXTE.

We barycenter corrected the event times, searched for pulsations, and derived a period of 1.237748(1) seconds on MJD 51294.5 using a z^2 search [1]. This is consistent with the 1.237747(2)s period found by Parmar et al. (1999) with the *Beppo*SAX satellite during a 1999 July 8-10 observation. Pulsed fractions were calculated by dividing the number of counts in the peak of the folded light curve (above a DC level defined by the lowest bin) by the total number of counts. The ALS, main-on, and low state pulsed fractions are 0.047 ± 0.002, 0.5180 ± 0.0002, and 0.089 ± 0.003 respectively.

Initially we fit the ALS spectrum with a standard, high energy cutoff continuum (HECUT), photoelectric absorption, and FeK line of the form

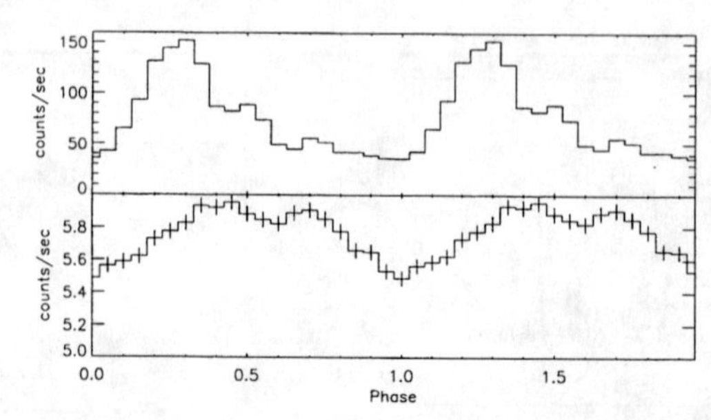

FIGURE 3. Top: The main-on 2.5-20 keV PCA folded light curve. Bottom: The ALS folded light curve with arbitrary phase alignment . In addition to being much shallower (note offset in y axis) the peak has been broadened.

$$F(E) = e^{-\sigma(E)N_H}(FeK + HECUT) \tag{1}$$

where $\sigma(E)$ is the photoabsorption cross section [4], N_H is the column density, the FeK line is a simple Gaussian, and the HECUT is

$$HECUT(E) = A\ E^{-\Gamma} \begin{cases} 1 & (E < E_c) \\ e^{\frac{-(E-E_c)}{E_f}} & (E > E_c) \end{cases} \tag{2}$$

Γ is the photon powerlaw index and E_c and E_f are the cutoff and folding energies respectively. This fit gave $\Gamma = 0.37 \pm 0.05$, $E_c = 16.7 \pm 0.4$, $E_f = 11.3 \pm 1.0$, and no photoelectric absorption. This spectral index is quite different than what is seen during the main-on (see Table 1).

To test if the intrinsic, main-on spectrum of the source was unchanged, we fit the data with a partially absorbing column. We fit the ALS spectrum to a model of the form

$$F_{pa}(E) = (1 + f \times e^{-\sigma(E)N_H})(FeK + HECUT) \tag{3}$$

We define a "covering factor" $\mathcal{F}$=f/(1+f), which is the fraction of observed flux coming to us through the absorbing medium. The best fit parameters for the HECUT in Eq 3 are consistent with those from the main-on (see Table 1). The small improvement in the fit is not statistically significant, but the model makes more physical sense. When the HECUT was constrained to be the same shape as the main-on we also got a reasonable fit (see Fig 4).

The low state fit with Eq. 3 and the main-on HECUT continuum shape gives something similar; $N_H = (50.3 \pm 4.2) \times 10^{22}$ cm^{-2} and $\mathcal{F} = 60. \pm 3.\%$. The equivalent width of the Iron line in the low state is 0.5 keV, compared to 0.7 keV in the ALS. The ALS and normal low state 20-40 keV fluxes are the same. The ALS spectrum observed by *Beppo*SAX was found to be indistinguishable from that observed during the normal low state [5]. So, the ALS spectra are consistent with the same mechanism causing both the normal low and anomalous low states.

TABLE 1. Fit parameters for the Her X-1 main-on, anomalous low, and low states

Parameter	Main – On[a]	ALS[a]	ALS[b]	ALS[c]	Low[c]
N_H (10^{22} cm^{-2})	1.8 ± 0.2	0.0	$45. \pm 6.$	$56. \pm 3.$	50.3 ± 4.2
Γ	1.07 ± 0.01	0.37 ± 0.04	0.90 ± 0.07	1.07^d	1.07^d
E_c (keV)	21.0 ± 0.1	16.8 ± 0.4	20.1 ± 0.7	21.0^d	21.0^d
E_f (keV)	10.2 ± 0.1	$11. \pm 1.$	$10. \pm 1.$	10.2^d	10.2^d
FeK E (keV)	6.48 ± 0.09	6.43 ± 0.08	6.44 ± 0.08	6.48^d	6.48^d
FeK EW (keV)	0.3 ± 0.1	0.7 ± 0.1	0.7 ± 0.1	0.7 ± 0.1	0.5 ± 0.1
Flux[e]	24.4	1.4	1.4	1.4	1.4
$\mathcal{F}$	N/A	N/A	0.56 ± 0.03	0.70 ± 0.03	0.60 ± 0.03
χ^2_{red}/DOF	4.71/221	1.12/114	1.08/113	1.09/117	1.18/112

[a] Fit with Eq 1
[b] Fit with Eq 3 and the HECUT shape free
[c] Fit with Eq 3 and the main-on HECUT shape
[d] Not allowed to vary
[e] 20 – 40 keV flux in units of 10^{-10} ergs/cm^2/s

III DISCUSSION/CONCLUSIONS

The 2.5-50 keV *RXTE* spectrum of the ALS can be fit with a standard pulsar continuum model; however, the derived parameters are significantly different from the main-on state. These fits can be marginally improved by the addition of a partially covering screen of absorbing matter (Eq 3), which is consistent with a simple and reasonable physical picture. The HECUT continuum in these fits is almost identical in shape to the high state spectrum, but requires a drop in flux similar to what is seen in the normal low state. The same partial absorbing model can also be used to describe the low state and gives similar results (Table 1). Thus both the anomalous and low state spectra are highly absorbed and heavily scattered versions of the main-on spectrum.

We compared the spectra from the four individual *RXTE* orbits of the ALS and find that the spectral shape remains the same. So, there is no evidence for a change in amount of cold absorber, which would indicate an inhomogeneous screen of material moving across the line of sight. Because of this, we interpret the unabsorbed spectral component as scattering from a hot corona rather than transmission through a patchy, partially covering disk. This can also explain the change in the pulsations. The pulsed fraction in the ALS is reduced by a factor of ~10 from the main-on state and the pulse is broader (Fig 3). A corona of moderate optical depth around the pulsar that is smaller than ~1.2 light seconds (the pulse period) would reflect some of the pulsed emission into the line of sight without completely washing out pulsations.

Because the ratio of ALS pulsed fraction to that of the normal low state is ~0.5, there is either an intrinsic pulse shape variation at the X-ray source, or the ALS corona is either slightly larger and/or denser than in our normal low

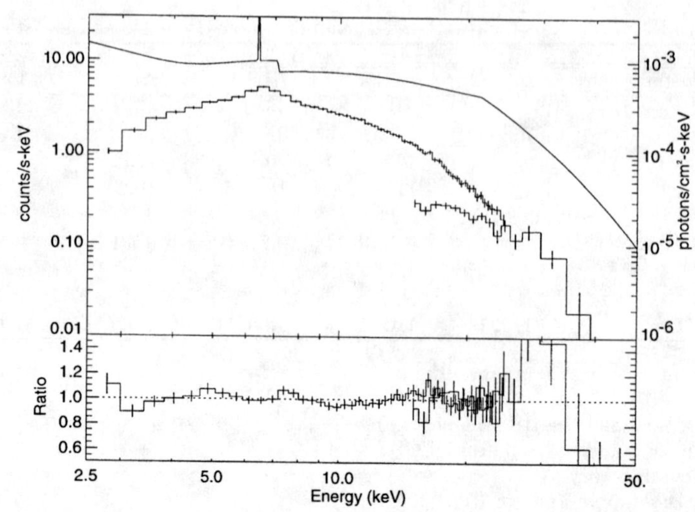

FIGURE 4. Top: The joint *RXTE* PCA/HEXTE spectrum of Her X-1 in the 1999 ALS (crosses). The fit with Eq 3 with the shape of the HECUT constrained to that of the main-on is shown as histograms, along with the model photon spectrum (smooth curve). Bottom: The ratio of the data to the fit model.

state observation. If the accretion disk changed slightly such that it allowed more illumination of the surface of the disk by X-rays from the neutron star, then there would be a corresponding increase in the size and/or density of the corona. The increased scattering would then account for the decrease in pulsed fraction seen from the normal low state to the ALS.

REFERENCES

1. Buccheri, R., et al. 1983, *A&A*, **128**, 245
2. Jahoda, K., Swank, J. H., Giles, A. B., Stark, M. J., Strohmayer, T., & Zhang, W. 1996, *SPIE*, **2808**, 59
3. Levine, A. M. 1999, *IAUC*, No. 7139
4. Morrison, R., & McCammon, D. 1983, *ApJ*, **270**, 119
5. Parmar, A. N., Oosterbroek, T., Dal Fiume, D., Orlandini, M., Santangelo, A., Segreto, A., & Del Sordo, S. 1999, astro-ph/9909039, *A&A*accepted
6. Parmar, A. N., Pietsch, W., McKechnie, S., White, N. E., Trümper, J., Voges, W., & Barr, P. 1985, *Nature*, **313**, 119
7. Rothschild, R., et al. 1998, *ApJ*, **496**, 538
8. Scott, D. M., & Leahy, D. A. 1999, *ApJ*, **510**, 974
9. Vrtilek, S. D., & Cheng, F. H. 1996, *ApJ*, **465**, 915
10. Vrtilek, S. D., et al. 1994, *ApJL*, **436**, L9

Disappearing Pulses in Vela X-1

P. Kretschmar[1,2], I. Kreykenbohm[2], J. Wilms[2], R. Staubert[2],
W. A. Heindl[3], D. E. Gruber[3], R. E. Rothschild[3]

[1] *INTEGRAL Science Data Centre, Ch. d'Ecogia 16, 1290 Versoix, Switzerland*
[2] *Institut für Astronomie und Astrophysik – Astronomie, Waldhäuser Str. 64, D-72076 Tübingen, Germany*
[3] *CASS, University of California at San Diego, La Jolla, CA 92093, U.S.A.*

Abstract. We present results from a 20 h RXTE observation of Vela X-1, including a peculiar low state of a few hours duration, during which the pulsation of the X-ray emission ceased, while significant non-pulsed emission remained. This "quiescent state" was preceded by a "normal state" without any unusual signs and followed by a "high state" of several hours of increased activity with strong, flaring pulsations. While there is clear spectral evolution from the normal state to the low state, the spectra of the following high state are surprisingly similar to those of the low state.

INTRODUCTION

Vela X-1 (4U 0900–40) is an eclipsing high mass X-ray binary consisting of the 23 $M_\odot$ B0.5Ib supergiant HD 77581 and a neutron star with an orbital period of 8.964 d and a spin period of about 283 s (van Kerkwijk et al., 1995, and references therein). The persistent X-ray flux from the neutron star is known to be very variable exhibiting strong flares and low states. Inoue et al. (1984) and Kreykenbohm et al. (1999) have observed low states of near quiescence where no pulsations were seen for a short amount of time. Before or after these low states normal pulsations were observed. During an observation of Vela X-1 for 12 consecutive orbits in January 1998 by the Rossi X-ray Timing Explorer (*RXTE*), we have by chance observed such a quiescent state for the first time from the beginning to the end, preceded and followed by the usual pulsations.

LIGHTCURVES AND PULSE PROFILES

As Fig. 1 demonstrates, the source flux suddenly decreased between orbits 2 and 3, reaching its minimum during orbit 4. At the same time *the source pulsations decreased strongly, while significant non-pulsed source flux remained.* This is shown in detail in Fig. 2. The pulsed fraction decreased from 30%–50%, depending on the energy band, to 7%–9%. Note that even at the lowest state, the overall source flux was >5 times the predicted background level in the energy range used.

CP510, *The Fifth Compton Symposium,* edited by M. L. McConnell and J. M. Ryan

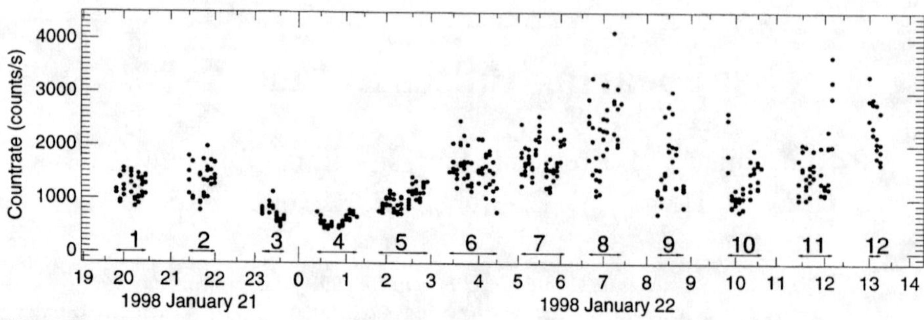

FIGURE 1. Light curve of the complete observation. The individual *RXTE* orbits are indicated below the light curve.

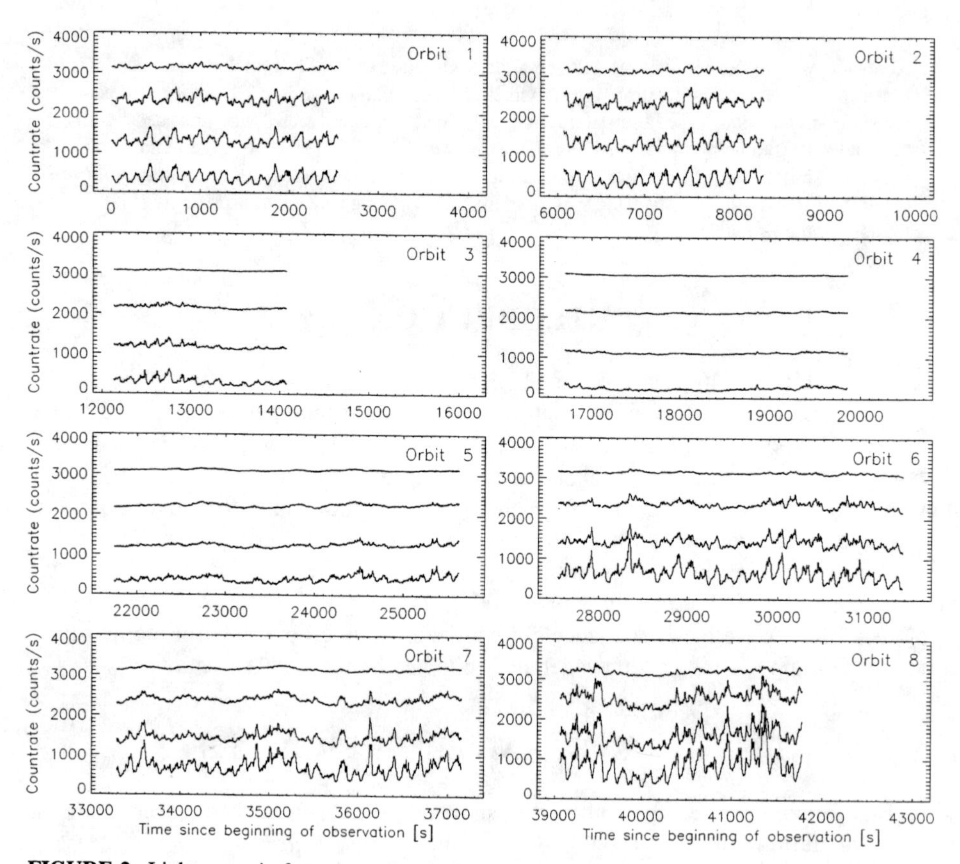

FIGURE 2. Lightcurves in four energy bands for the individual orbits 1 to 8. The energy bands are the same as for the pulse profiles (see Fig. 3). The lightcurve for the highest energy band is plotted at the bottom with its actual count rate, the others have been shifted upwards by 1000, 2000 and 3000 counts/s respectively.

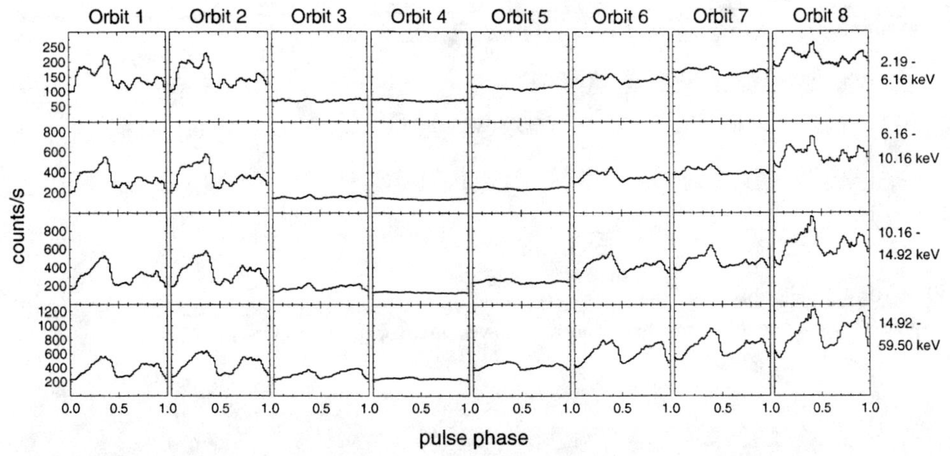

FIGURE 3. Evolution of the pulse profile for the individual orbits 1 to 8. The energy bands are the same used in Fig. 2. The lightcurves were folded with a fixed period (283.3 s) and a common zero time.

Fig. 3 presents the pulse profiles obtained from the individual orbits 1 to 8. The profiles of the first two orbits correspond to the well-known, complex shape usually obtained when integrating over many pulse periods, with a clear transition from a five-peaked profile at low energies to a double-peak structure at high energies (Raubenheimer, 1990). In contrast, orbits 3 to 5 show much less pronounced profiles, with the profile of orbit 4 being essentially flat. The pulse profile of the source during "recovery" (orbits 6 & 7) is similar to those observed before the low state at higher energies but much less pronounced at energies <10 keV.

HARDNESS RATIOS AND SPECTRA

Fig. 4 shows the evolution of the spectral hardness, both at energies up to 10 keV (energy band 2 vs. band 1) and at energies beyond 10 keV (energy band 4 vs. band 3). The hardness ratios were calculated using $(H - S)/(H + S)$ where H and S are the fluxes in the hard and soft band respectively. There are three apparent properties of this plot:

1. *The two hardness ratios are very clearly anticorrelated.*
2. *The disappearence of the pulses in orbit 3 goes hand in hand with an abrupt spectral change.*
3. The reemergence of pulsations is accompanied by a "normalization" of the hardness ratios, *but during the high state the spectrum stays significantly harder than in the normal state at the beginning.*

The strong spectral changes at the onset of the low state are also apparent in the quotient spectra shown in Fig. 5. There are clear indications for strong absorption and at the same time increased flux in the iron line and a soft excess at the lowest energy range.

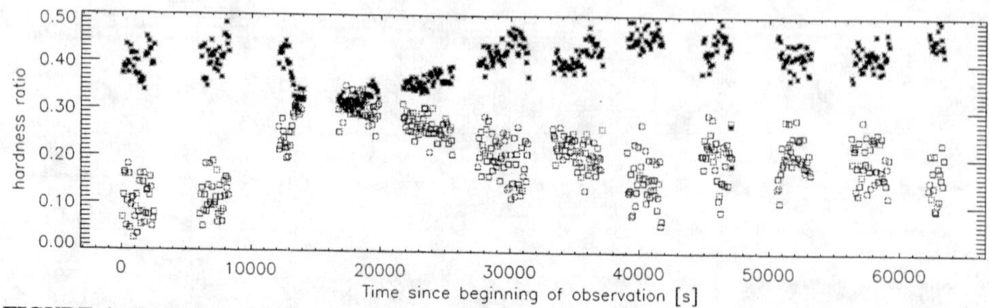

FIGURE 4. Evolution of the spectral hardness over the complete observation. The upper curve (stars) displays the ratio of energy band 2 to band 1, the lower curve (open squares) the ratio of band 4 to band 3.

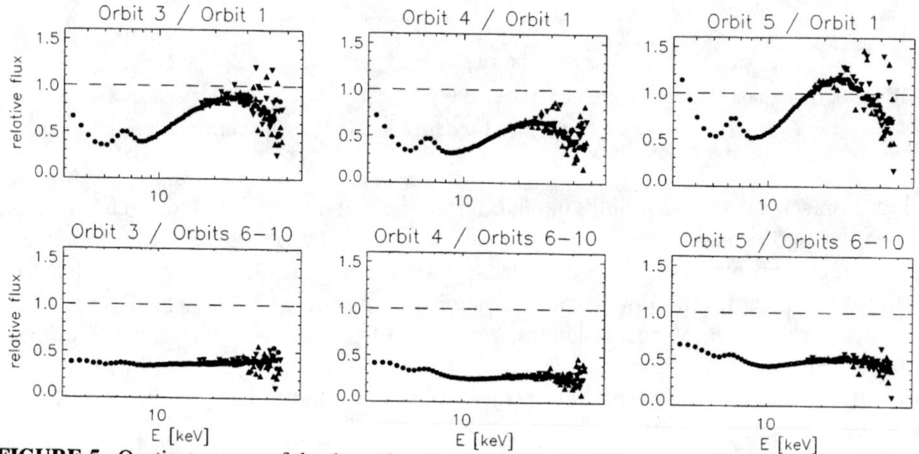

FIGURE 5. Quotient spectra of the three "low-state" orbits, with *PCA* data plotted as filled circles and *HEXTE* data as triangles: 1. upper panels: ratio to the spectrum of the normal state before the low state, 2. lower panels: ratio to the average spectrum of the active high state following the low state.

In contrast there is little change in the global spectral shape as the source begins to pulsate again. Except for a slight soft excess, the spectrum during the low state is rather well described by simply scaling the spectrum of the following flaring state.

Attempts to fit the spectra turned out to be quite difficult, even allowing for the known complexity of the Vela X-1 continuum. Detailed results have been presented on two posters at the X-ray'999 conference in Bologna, the following paragraphs summarize the results.

We used a partial covering model with two additive components, using the same continuum, one heavily absorbed and one scattered unabsorbed into the direction of the observer. The continuum used was the NPEX (Negative Positive EXponential Mihara 1995). This model had to be further modified by an additive iron line and two coupled cyclotron lines. For the high state a cyclotron line at $\sim$55 keV is required by the data, a second feature at $\sim$21 keV may be due to uncertainties in the *PCA* response matrix.

Modeling the spectra of the individual orbits in the first part of our observation, we found the N_H value of the absorbed component varying between $40\times10^{22}\,cm^{-2}$ and $230\times10^{22}\,cm^{-2}$ with a clear maximum during orbit 4. The relative importance of the scattered component appears also to be maximal during the low state. There is no clear correlation of the other continuum parameters with source flux, but their values are quite different before and after the low state.

DISCUSSION

The results of spectral fitting are somewhat in contrast with the finding above that the global spectral shape remains more or less constant after orbit 3. Within the framework of our spectral model this similarity is obtained by parallel changes in the column depth and in the spectral continuum. Further analysis will have to show if this is an artefact or reality.

A possible scenario to explain the disappearing pulses is that a very thick blob in the surrounding stellar wind – which is known to be clumpy (e.g., Nagase et al., 1986) – temporarily obscures the pulsar. Taking our fit results from above as basis ($N_{H,max}\approx2\times10^{24}\,cm^{-2}$), the optical depth for Thomson scattering of such a blob would be $\sim$1.6, reducing the direct component to $\sim$20%. The scattered radiation would need to come from a relatively large region ($\sim10^{13}$ cm) to destroy coherence. The large fraction of scattered radiation would also explain the relatively increased Fe-line emission and soft excess.

After the quiescent state, when pulsations begin again, the emission in this scenario would be a combination of heavily absorbed direct radiation – the accretion being fueled by some part of the blob – and scattered radiation from a wide region. This would also explain the reduced pulse fraction, due mainly to a higher "pedestal" during the high state as compared to the normal state at the beginning of the observation (see Fig. 3).

References

Inoue H., Ogawara Y., Ohashi T., et al., 1984, PASJ 36, 709
Kreykenbohm I., Kretschmar P., Wilms J., et al., 1999, A&A 341, 141
Mihara T., 1995, *Ph.D. thesis*, University of Tokyo
Nagase F., Hayakawa S., Sato N., 1986, PASJ 38, 547
Raubenheimer B.C., 1990, A&A 234, 172
van Kerkwijk M.H., van Paradijs J., Zuiderwijk E.J., et al., 1995, A&A 303, 483

Very High Energy Gamma Rays from X-ray Binaries

P. M. Chadwick, K. Lyons, T. J. L. McComb, K. J. Orford, J. L. Osborne, S. M. Rayner, S. E. Shaw, and K. E. Turver

Department of Physics, Rochester Building, Science Laboratories, University of Durham, Durham, DH1 3LE, U.K.

Abstract. We have reported the discovery of VHE gamma ray emission from the X-ray binary Cen X-3. We report further VHE observations of this object taken in 1998 and 1999. While these new data also show VHE gamma ray emission, no evidence of modulation at either the orbital or pulsar periods is found.

INTRODUCTION

Results of observations of the accreting X-ray binary Cen X-3 using ground based gamma ray telescopes have been reported which have included evidence for sporadic outbursts of strong pulsed emission in the > 1 TeV band [1,2] and a constant but weaker unpulsed emission at > 400 GeV [3]. These results, together with the *CGRO* EGRET measurement of an outburst of pulsed GeV emission [4], indicate that Cen X-3, an accurately measured system containing a 4.8 s pulsar in a 2.1 day orbit around an O-type supergiant, is a sporadic source of high energy gamma rays.

We report the results of analysis of data taken during 1998 March and April and 1999 February. We present the results of a search for a possible correlation between > 400 GeV gamma rays recorded by the University of Durham Mark 6 telescope and X-ray emission according to measurements made with the *RXTE* and *CGRO*/BATSE experiments. We also present the results of searches for variation of the emission at both the orbital and spin periods.

RECENT OBSERVATIONS OF VHE GAMMA RAYS

We report observations made with the University of Durham Mark 6 imaging gamma ray telescope operating at Narrabri NSW, Australia. The telescope has been described in [5] and the results of initial observations of Cen X-3 have been reported [3]. Our Cen X-3 dataset now comprises data from 31 hrs of observation during 23 exposures in 1997 March and June (JD 2450508 – JD 2450606), 1998

CP510, *The Fifth Compton Symposium,* edited by M. L. McConnell and J. M. Ryan

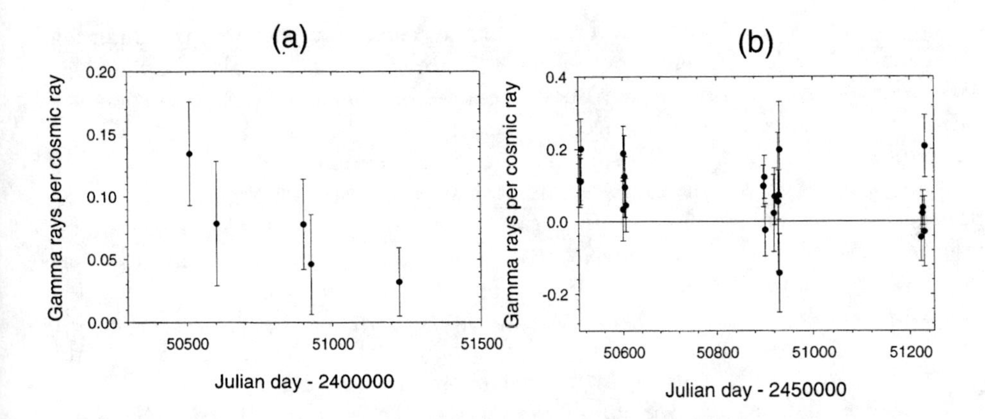

FIGURE 1. (a) The VHE gamma ray flux from Cen X-3 averaged over observing periods. (b) The VHE gamma ray flux from Cen X-3 plotted on a day-by-day basis.

March and April (JD 2450899 – JD 2450932) and 1999 February (JD 2451220 – JD 2451230).

Our earlier report [3] was based on data recorded in 1997 March and June (JD 2450508 – JD 2450606) only. Assuming a collection area of 10^9 cm^2 and that our selection procedure retained $\sim$ 50% of the original gamma ray events, the time averaged flux was estimated to be $(2.0 \pm 0.3) \times 10^{-11}$ cm^{-2} s^{-1} for $>$ 400 GeV. We concluded that the measurements in these two months were consistent with a constant flux. Ongoing simulations suggest that our current selection procedure retains 20% of the gamma rays. On this basis, the flux for the 1997 March and June (JD2450508 – JD2450606) data would be $(5.0 \pm 0.9) \times 10^{-11}$ cm^{-2} s^{-1}. The additional data taken in 1998 and 1999 provide fewer gamma ray candidates suggesting weaker TeV emission when analysed using the same selection procedure as previously [3]. An analysis of the total data yields a time averaged flux of $(2.8 \pm 1.4_{sys} \pm 0.6_{stat}) \times 10^{-11}$ cm^{-2} s^{-1}; the significance of the detection based on the total dataset is 4.7σ.

RESULTS

Our recent work on PKS 2155–304 [6] has demonstrated a method of assessing the signal strength of gamma rays recorded by Cerenkov telescopes. It is suited to measurements made at different epochs and at different zenith angles when the telescope may have different sensitivities and consequently a varying background cosmic ray detection rate. We have estimated the signal strength of TeV gamma ray emission by expressing it as a fraction of the cosmic ray background remaining after image shape and orientation selection [7]. In so doing we make allowance for first order variations in sensitivity due to changes in efficiency of the telescope and variations in telescope performance with zenith angle. It is also assumed that the slopes of the gamma ray and cosmic ray spectra are similar.

In the present study, the average gamma ray signal strength from Cen X-3, expressed as a percentage of the cosmic ray background remaining after shape and orientation selection is $(7.0\pm1.5)\%$. The most straightforward, but not most powerful, test for constancy of emission is to repeat this process for the data recorded in each of the 5 dark periods as shown in Figure 1(a). On the basis of this test we find no internal evidence for monthly variability of the VHE signal; the data treated this way are consistent with a constant signal strength ($\chi^2 = 4.5$, 4 df).

Cen X-3 is a strong but variable X-ray emitter. For example, the average daily rates for X-rays detected with the *RXTE*/ASM during 1997 and 1998 range from 0 to 32 counts s^{-1}; the data are variable on a time scale of days. The daily average for the *RXTE*/ASM count rates are available for 22 of the 23 days when TeV gamma ray observations were made[1].

The strength of pulsed X-ray emission was also available as a daily average from the BATSE archive[2] for 1997. During the 1998 and some of the 1999 VHE observations, the X-ray flux was low and less than the threshold for BATSE detection. The BATSE data provide a series of independent X-ray measurements, including a measurement on the single day of the TeV gamma ray observations for which there is no corresponding *RXTE*/ASM measurement.

The VHE gamma ray signal plotted on a day by day basis is shown in Figure 1(b). There is no evidence for outbursts of TeV gamma ray emission on a timescale of days and the data are consistent with a constant TeV gamma ray flux ($\chi^2 = 22.1$, 22 df).

In Figure 2(a) we show the relation between the count rate of the *RXTE*/ASM data and our gamma ray signals. In Figure 2(b) we show a similar plot between the individual BATSE pulsed X-ray fluxes and our gamma ray signals. There is no significant evidence for a correlation, although it is interesting to note that the day of highest detected gamma ray flux coincides with the day of most X-ray activity in the dataset (1997 Mar 4).

We have looked for modulation of the gamma ray signal at the orbital period of the binary system. The orbital phase of each of our observations has been calculated using the ephemeris of Kelley et al. [8]. The results are shown in Figure 3. From this evidence we conclude that there is no modulation of the VHE gamma ray emission at the orbital period.

The data have been subjected to a Rayleigh test for periodicity at a small range of periods around the BATSE period. Phase coherence between observations was not assumed. No significant periodicity was detected, leading to a 3σ upper limit

---

1) Available on the web at http://space.mit.edu/XTE/asmlc/srcs/cenx3.html
2) Original data obtained from the web at http://www.batse.msfc.nasa.gov/data/pulsar

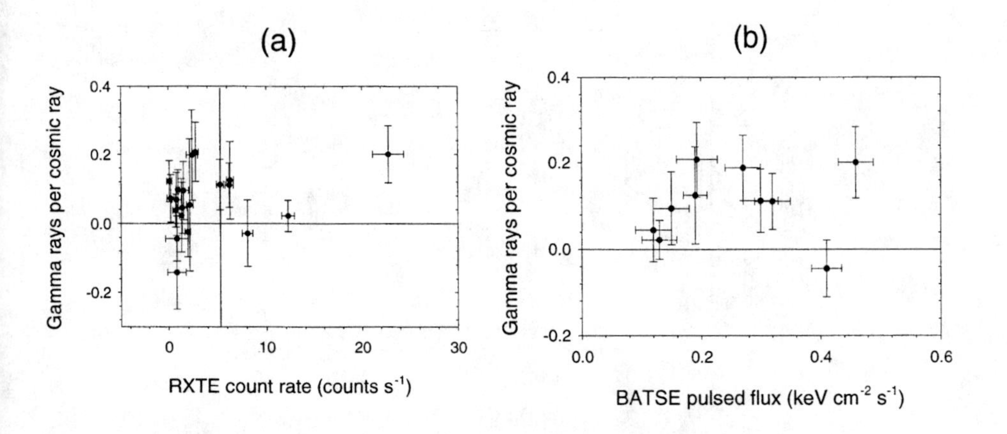

FIGURE 2. The relation between the daily VHE gamma ray flux from Cen X-3 and (a) the X-ray flux detected by ASM/*RXTE* and (b) the X-ray pulsed flux detected by BATSE.

to the pulsed flux of 2.0×10^{-12} cm^{-2} s^{-1} in the total dataset.

DISCUSSION

We have detected VHE gamma ray emission from Cen X-3 during each dark moon period that we have observed this object. The data are consistent with a weak but persistent emission, both when the VHE data are averaged over dark moon periods or when considered observation by observation. Although the observation that yields the strongest gamma ray flux occurs on the day when the daily averaged *RXTE*/ASM X-ray flux was the highest of any day on which we observed Cen X-3, there is no evidence for a formal correlation between the VHE gamma ray and X-ray fluxes.

We have also tested for modulation of the VHE gamma ray flux at the orbital period of the binary system and at the pulsar period. We have no evidence for modulation of the VHE gamma ray emission at either period.

We are grateful to the UK Particle Physics and Astronomy Research Council for support of the project. The Mark 6 telescope was designed and constructed with the assistance of the staff of the Physics Department, University of Durham. This paper uses quick look results provided by the ASM/*RXTE* and BATSE teams. Dr. Mark Finger is thanked for his kind help with BATSE data.

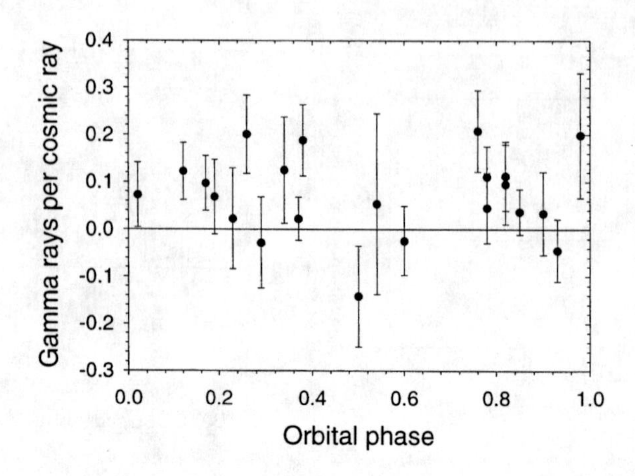

FIGURE 3. The measured VHE gamma ray rate during each observation of Cen X-3 plotted as a function of orbital phase.

REFERENCES

1. Carraminana, A., et al., *Timing Neutron Stars* ed. H. Ögelman & E. P. J. van den Heuvel (Dordrecht: Kluwer Academic Press), p 369 (1989).
2. Raubenheimer, B. C., et al., *Ap. J.*, **336**, 349 (1989).
3. Chadwick, P. M., et al., *Ap. J.*, **503**, 391 (1998).
4. Vestrand, W. T., Sreekumar, P. & Mori, M., *Ap. J.*, **483**, L49 (1997).
5. Armstrong, P. et al., *Experimental Astron.* **9**, 51 (1999).
6. Chadwick, P. M., et al., *Ap. J.*, **513**, 161 (1999)
7. Fegan, D. J., *J. Phys. G. Nucl. Part. Phys.*, **23**, 1013 (1997).
8. Kelley, R. L., et al., *Ap. J.*, **268**, 790 (1983).

Multiple Cyclotron Lines in the Spectrum of 4U 0115+63

W.A. Heindl*, W. Coburn*, D.E. Gruber*, M. Pelling*, R.E. Rothschild* J. Wilms†, K. Pottschmidt†, and R. Staubert†

**Center for Astrophysics and Space Sciences, University of California San Diego, La Jolla, CA, 92093, USA*
†*Institut für Astronomie und Astrophysik – Astronomie, Waldhäuser Str. 64, D-72076 Tübingen, Germany*

Abstract. We report phase resolved spectroscopy of the transient accreting pulsar, 4U 0115+63. For the first time, more than two cylotron resonance scattering features are detected in the spectrum of an X–ray pulsar. The shape of the fundamental line appears to be complex, and this is in agreement with predictions of Monte-Carlo models. As in other pulsars, the line energies and optical depths are strong functions of pulse phase. One possible model for this is an offset of the dipole of the neutron star magnetic field.

I INTRODUCTION

4U 0115+63 is a transient accreting X–ray pulsar in an eccentric 24 day orbit [1] with an O9e star [9]. A cyclotron resonance scattering feature (CRSF) was first noted near 20 keV by Wheaton, *et al.* (1979) with the UCSD/MIT hard X–ray (10-100 keV) experiment aboard *HEAO-1.* White, Swank & Holt (1983) analyzed concurrent data from the lower energy (2-50 keV) *HEAO-1*/A2 experiment and found an additional feature at ~12 keV, making 4U 0115+63 the first pulsar with two cyclotron line harmonics.

We discuss here phase-resolved spectra derived from an observation of the 1999 March outburst [12,2] obtained with the *Rossi X-Ray Timing Explorer* (*RXTE*). First results of this work are detailed in Heindl *et al.* (1999). *Beppo*SAX has also made detailed observations of this outburst [7,8].

II OBSERVATIONS AND ANALYSIS

Observations were made with the Proportional Counter Array (PCA) [4] and High Energy X-ray Timing Experiment (HEXTE) [6] on board *RXTE.* The PCA is a set of 5 Xenon proportional counters sensitive in the energy range 2–60 keV

CP510, *The Fifth Compton Symposium,* edited by M. L. McConnell and J. M. Ryan

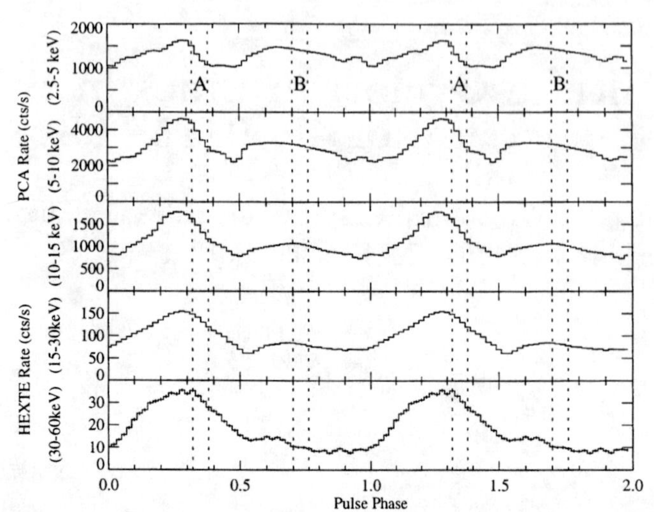

FIGURE 1. The folded light curve (or "pulse profile") of 4U 0115+63 in five energy bands. We report detailed spectral analyses in phase ranges A and B which are indicated by the dashed lines.

with a total effective area of $\sim$7000 cm^2. HEXTE consists of two arrays of 4 NaI(Tl)/CsI(Na) phoswich scintillation counters (15-250 keV) totaling $\sim$1600 cm^2. The HEXTE alternates between source and background fields in order to measure the background. The PCA and HEXTE fields of view are collimated to the same 1° full width half maximum (FWHM) region.

We performed four long pointings (duration $\sim$15-35 ks to search for CRSFs. The second observation, on 1999 March 11.87-12.32, spanned periastron passage at March 11.95 ([1]) and preceded the outburst maximum by about 2 days. The results presented here are from this observation.

The spectrum of 4U 0115+63 varies significantly with neutron star rotation phase, making fits to the average spectrum difficult to interpret. To avoid this problem, we accumulated spectra as a function of pulse phase. Figure 1 shows the folded light curve in 5 energy bands. The pulse is double peaked at low energies, but the second peak nearly disappears at high energy. Two phase bands, "A" and "B", which we selected for detailed analysis, are indicated. Our spectral fitting process consists of fitting the joint PCA/HEXTE spectra to various heuristic models (see Kreykenbohm, *et al.* 1998) which have been successful in fitting pulsars with no cyclotron lines. When none of these models can adequately describe the spectrum, and line-like residuals are present, we add Gaussian absorption lines to the spectrum as required. For a detailed description of our analysis technique, see Heindl, *et al.* (1999).

Figure 2 shows the resulting best fit spectra for phases A and B. Both fits have a Fermi-Dirac cutoff powerlaw continuum, a low energy excess in the form of a black body with kT$\sim$0.4 keV, *no* FeK line, and multiple cyclotron lines. In phase A, the falling edge of the main pulse, five cyclotron lines are required, while in phase B,

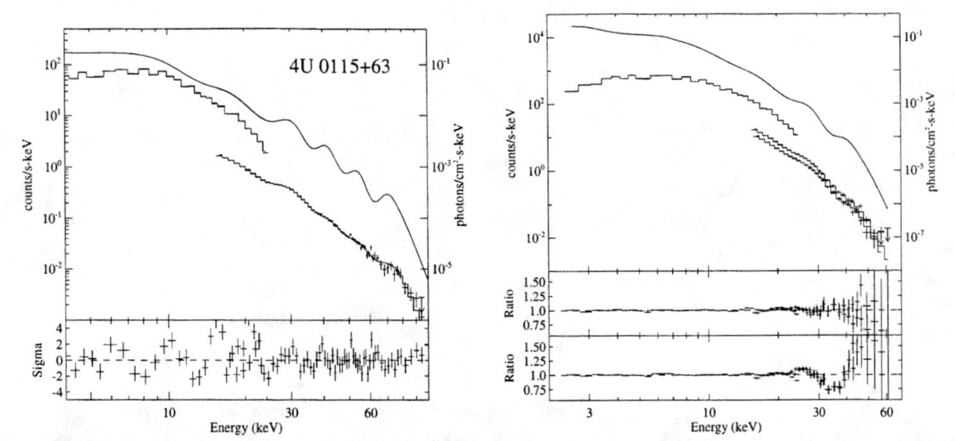

FIGURE 2. Best fit spectra to (left) phase A and (right) phase B. The top panel shows the inferred incident spectrum (smooth curve), the measured count spectra (data points), and the model count spectra (histograms). The bottom panels show residuals. For phase B, residuals fits with three (middle) and two (bottom) cyclotron lines.

the fall of the second pulse, only three lines are necessary. Figure 3 illustrates how we determine how many lines are present. The five panels show the residuals to phase A fits with increasing numbers of lines. With fewer than five lines, significant residuals are present. These residuals follow the pattern of a dip at the energy of the first missing line and a gross underprediction of the continuum at high energies. The fit continuum tends to conform to the low side of the missing line, because the statistics are rapidly decreasing with energy. The too steeply falling continuum can then never recover the high energy data. In phase A, adding five harmonics significantly improves the fit, while a sixth is not statistically required.

III RESULTS AND DISCUSSION

Table 1 lists the best fit cyclotron line parameters. Two interesting phenomena are apparent from these results. First, within the individual phases, the lines are not harmonically spaced. And, second, the line energies vary significantly with pulse phase.

A harmonic relation (with small modifications due to relativistic effects) between line energies is expected in simple Landau theory. However, models of cyclotron line formation (e.g. Isenberg, Lamb, & Wang 1998) predict that the fundamental line shape can be quite complex, even having wings resembling *emission* features. On the other hand, the higher harmonics should have relatively simple line shapes. Thus, it may be that our simple Gaussian absorption does not give an accurate measure of the fundamental line energy. Evidence for this appears in the residuals near 10–15 keV (see Fig. 3), which are the most significant remaining deviations in

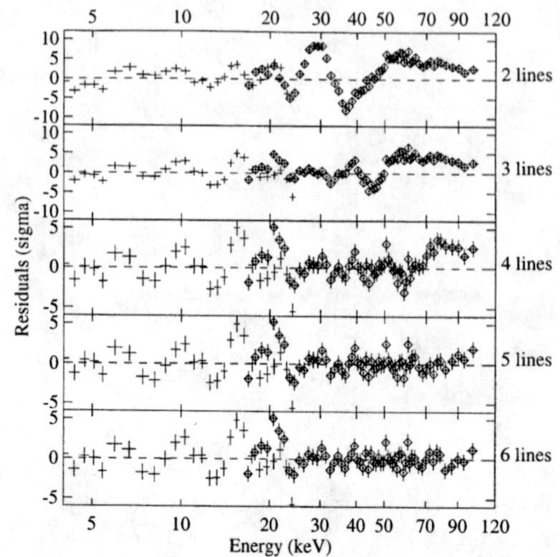

FIGURE 3. Residuals to fits to the phase A spectrum with different numbers of cyclotron line harmonics. Five lines are required to adequately fit the data.

our best fit. Furthermore the third and higher harmonics' energies are consistent with integer multiples of *half* of the second harmonic energy. Thus it seems likely that *half* of the second harmonic energy gives a more accurate measure of the magnetic field. In the case of phase A, this is $B = 1.3 \times 10^{12}$ G, assuming a gravitational redshift of 0.3 to the neutron star surface.

Figure 4 shows the HEXTE flux and the ~20 keV line energy and optical depth as a function of pulse phase. These parameters were determined from the HEXTE data alone in 20 phase bins. The second harmonic line was not required in the HEXTE data *alone* at phases greater than 0.7. Both the line energy and optical depth are maximal not at the pulse peak, but on the falling edge of the main pulse. The line energy varies by ~20%. This behavior is also seen in Cen X-3 [3]. In

TABLE 1. Best fit cyclotron line parameters from the two phase bins shown in Fig. 1.

Harmonic	Phase A Energy (keV)	Phase A Width[a] (keV)	Phase A Optical Depth	Phase B Energy (keV)	Phase B Width[a] (keV)	Phase B Optical Depth
1[b]	$13.35^{+0.08}_{-0.06}$	$3.29^{+0.13}_{-0.07}$	$1.17^{+0.06}_{-0.04}$	$12.40^{+0.65}_{-0.35}$	$3.3^{+0.19}_{-0.4}$	$0.72^{+0.10}_{-0.17}$
2	23.7±0.1	$5.43^{+0.18}_{-0.27}$	$2.68^{+0.05}_{-0.07}$	$21.45^{+0.25}_{-0.38}$	$4.5^{+0.7}_{-0.9}$	$1.24^{+0.04}_{-0.06}$
3	$36.4^{+0.4}_{-0.5}$	$4.3^{+0.4}_{-0.6}$	$2.41^{+0.11}_{-0.13}$	$33.56^{+0.70}_{-0.90}$	$3.8^{+1.5}_{-0.9}$	$1.01^{+0.13}_{-0.14}$
4	$47.8^{+0.4}_{-0.7}$	$5^{+\infty}_{-1.2}$	2.3±0.2	–	–	–
5	61.7±1.1	5*fixed*	1.8±0.3	–	–	–

[a] One standard deviation of the Gaussian optical depth profile
[b] also called the "fundamental".

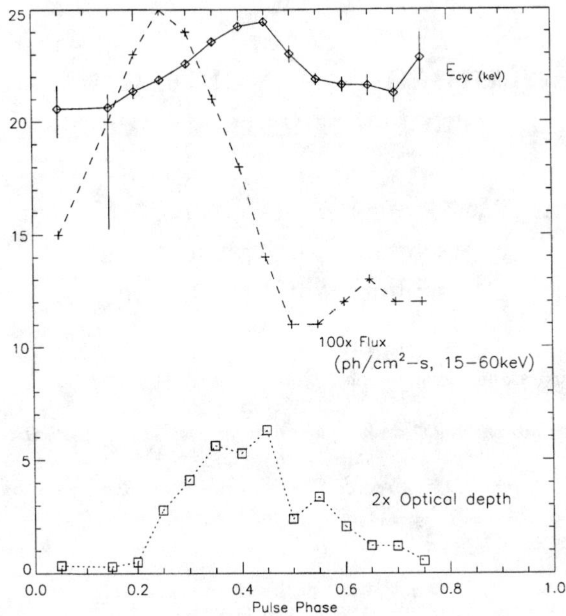

FIGURE 4. As a function of pulse phase, the energy (top) and depth (bottom) of the ~20 keV cyclotron line. Also plotted is the flux in the HEXTE band.

Cen X-3, Burderi et al. (1999) have modeled the variation of the line energy as due to an offset of the magnetic dipole moment from the center of the neutron star, and this may also be the case in 4U 0115+63.

REFERENCES

1. L. Bildsten et al., ApJS **113** (1997), 367.
2. W. A. Heindl et al., ApJ **521** (1999), L49.
3. W.A. Heindl and D. Chakrabarty, 1999, to appear in MPE Report: "Highlights in X-ray Astronomy in Honour of Joachim Trümper's 65th Birthday".
4. K. Jahoda et al., SPIE **2808** (1996), 59.
5. I. Kreykenbohm et al., A&A **341** (1998), 141.
6. R.E. Rothschild et al., ApJ **496** (1998), 538.
7. A. Santangelo et al., 1999, in this proceedings.
8. ______, ApJ **523** (1999), L85.
9. S.J. Unger, P. Roche, I. Negueruela, F.A. Ringwald, C. Lloyd, and M.J. Coe, A&A **336** (1998), 960.
10. W. A. Wheaton et al., Nature **282** (1979), 240.
11. N.E. White, J.H. Swank, and S.S. Holt, ApJ **270** (1983), 711.
12. R.B. Wilson, B.A. Harmon, and M.H. Finger, IAU Circ. (1999), No. 7116.

RXTE Studies of Cyclotron Lines in Accreting Pulsars

W.A. Heindl*, W. Coburn*, D.E. Gruber*, M. Pelling*, R.E. Rothschild* P. Kretschmar†,‡, I. Kreykenbohm†, J. Wilms†, K. Pottschmidt†, and R. Staubert†

*Center for Astrophysics and Space Sciences, University of California San Diego, La Jolla, CA, 92093, USA

†Institut für Astronomie und Astrophysik – Astronomie, Waldhäuser Str. 64, D-72076 Tübingen, Germany

‡INTEGRAL Science Data Centre, Ch. d'Ecogia 16, 1290 Versoix, Switzerland

I INTRODUCTION

Cyclotron lines in accreting X-ray pulsar spectra result from the resonant scattering of X-rays by electrons in Landau orbits on the intense ($\sim 10^{12}$ G) magnetic fields near the neutron star poles. For this reason they are known as cyclotron resonance scattering features (CRSFs). Because Landau transition energies are proportional to field strength ($E_{cyc} = 12$ keV approximately corresponds to $B = 10^{12}$ G), CRSF energies give us our most direct measures of neutron star magnetic fields. Other line properties, such as depths, widths, and presence of multiple harmonics, are strongly dependent on the details of the geometry and environment at the base of the accretion column. CRSFs therefore give us a sensitive (if difficult to interpret) diagnostic of the accretion region.

In this paper, we summarize the *RXTE* measurements of CRSFs in 8 accreting pulsars. The wide bandpass and modest resolution of the *RXTE* instruments make them ideal for measuring these generally broad features. In particular, the high energy response of HEXTE provides a window for discovery of new lines not detectable with proportional counters such as *Ginga* or the PCA alone.

Some highlights of this work are:

- New lines in the well known pulsars Cen X-3 and 4U 1626-67 [4,11,9].
- The discovery of more than two cyclotron line harmonics in 4U 0115+63 [3].
- The *RXTE* picture of the correlation between cyclotron line energy and width.

CP510, *The Fifth Compton Symposium,* edited by M. L. McConnell and J. M. Ryan

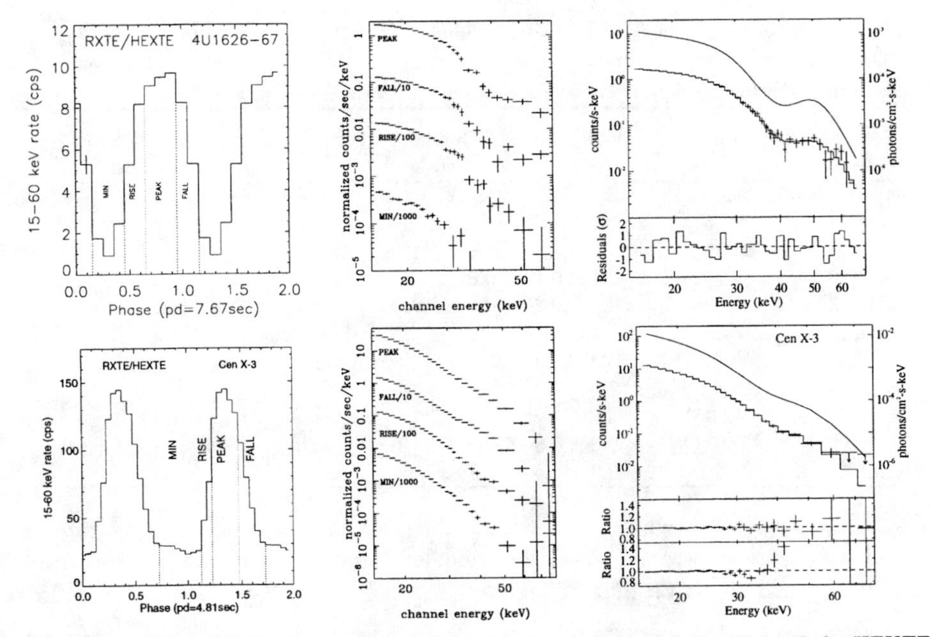

FIGURE 1. *RXTE* observations of (Top) 4U 1626-67 and (Bottom) Cen X-3 . Left: HEXTE light curves folded on the pulse period. Middle: The counts spectra from the four phase intervals. The CRSFs are apparent as inflections in the spectra near 30–40 keV. Right: the inferred incident spectrum from the peak phase, HEXTE count spectrum with model, and ratios to models with (and for Cen X-3 without) a cyclotron line.

II OBSERVATIONS AND ANALYSIS

The results discussed here are based on observations made with the Proportional Counter Array (PCA) [6] and High Energy X-ray Timing Experiment (HEXTE) [10] on board *RXTE*. The PCA is a set of 5 Xenon proportional counters, while HEXTE consists of two arrays of 4 NaI(Tl)/CsI(Na) phoswich scintillation counters. The PCA and HEXTE share a 1° field of view.

Figures 1 and 2 show the measured count spectra together with model fits and inferred incident spectra for eight accreting pulsars. To emphasize the presence of cyclotron lines, residuals to fits made without lines are also shown. In general, these residuals show a dip-like structure at the line energy and then a gross underprediction of the continuum above the line. This is caused by the better statistics on the low side of the feature forcing the model to fit the falling edge of the line. Because no adequate prediction of accreting pulsar continua exists, we employ empirical continuum models. These models (high energy cut-off power law (HECUT), Fermi-Dirac cut-off times a power law (FDCO), and Negative and Positive power law Exponential (NPEX); see [7]) have been successful in fitting pulsars with no cyclotron lines. Only when none of these continuum models provided an acceptable

TABLE 1. Summary of *RXTE* Cyclotron Line Measurements. 4U 0115+63 and Vela X-1 require multiple CRSFs in their spectra. The surface B-fields assume a gravitational redshift of 0.3 for the emitting region.

Source	Energy (keV)	Sigma (keV)	Optical Depth	Pulse Phase	Surface B Field (10^{12} G)
4U 0115+63	$12.40^{+0.65}_{-0.35}$	$3.3^{+0.1.9}_{-0.4}$	$0.72^{+0.10}_{-0.17}$	2nd Fall	1.4
	$21.45^{+0.25}_{-0.38}$	$4.5^{+0.7}_{-0.9}$	$1.24^{+0.04}_{-0.06}$	2nd Fall	
	$33.56^{+0.70}_{-0.90}$	$3.8^{+1.5}_{-0.9}$	$1.01^{+0.13}_{-0.14}$	2nd Fall	
4U 1907+09	19.7 ± 0.1	2.6 ± 0.1	0.87 ± 0.05	Average	2.2
4U 1538-52	21.3 ± 0.3	3.05 ± 0.20	0.86 ± 0.07	Average	2.3
Vela X-1	$23.7^{+0.4}_{-0.3}$	5 (fixed)	$0.29^{+0.03}_{-0.04}$	Main Pulse	2.3
	59.7 ± 3.7	12.6 ± 0.8	$1.41^{+0.67}_{-0.61}$	Main Pulse	
Cen X-3	31.8 ± 0.3	7.5 ± 0.9	$0.77^{+0.16}_{-0.11}$	Peak	3.5
GX 301-2	$42.9^{+0.9}_{-2.6}$	$10.0^{+1.9}_{-2.3}$	$0.8^{+0.7}_{-0.3}$	Average	4.8
4U 1626-67	$39.3^{+1.9}_{-1.1}$	6.4 ± 0.8	$2.3^{+0.6}_{-0.4}$	Peak	4.4
Her X-1	41.0 ± 1.0	9.8 ± 0.5	1.84 ± 0.05	Average	4.6

fit, did we allow cyclotron line(s). We modeled the cyclotron lines with a simple Gaussian optical depth profile.

III RESULTS AND DISCUSSION

Table 1 summarizes the cyclotron line parameters in eight accreting pulsars. In general, pulsar spectra (including line parameters) vary with pulse phase. For this reason, cyclotron lines are often best measured in spectra from limited pulse phase ranges. This is called "pulse phase spectroscopy". Table 1 indicates the phase relative to the pulse profile for which the given parameters apply.

New Lines in Cen X-3 and 4U 1626-67 Cen X-3 and 4U 1626-67 are two of the earliest known accreting pulsars; however, it is only recently that CRSFs were discovered in their spectra [4,2]. Figure 1 shows folded light curves, phase resolved count spectra, and fits to the peak phase spectra. In Cen X-3, the CRSF energy moves by about 20% with phase. This variation has been modeled as resulting from an offset of the magnetic dipole moment from the center of the neutron star [1]. In both sources, the CRSF appears strongest in the "peak" and "fall" spectra. This may also be related to an offset dipole.

4U 0115+63: Multiple Cyclotron Line Harmonics 4U 0115+63 is a transient accreting pulsar in an eccentric orbit with a massive star, a so-called "Be X-ray binary". It was also the first pulsar to show two cyclotron line harmonics [13]. 4U 0115+63 underwent a two month long outburst in 1999 February–April. Observations made with *RXTE* and *Beppo*SAX reveal for the first time more than two harmonics in a single pulsar [5,12] (see Fig. 2).

Correlation between Line Energy and Width The width of a CRSF depends on the temperature of the emitting region (kT), the line energy (ω_B) and the

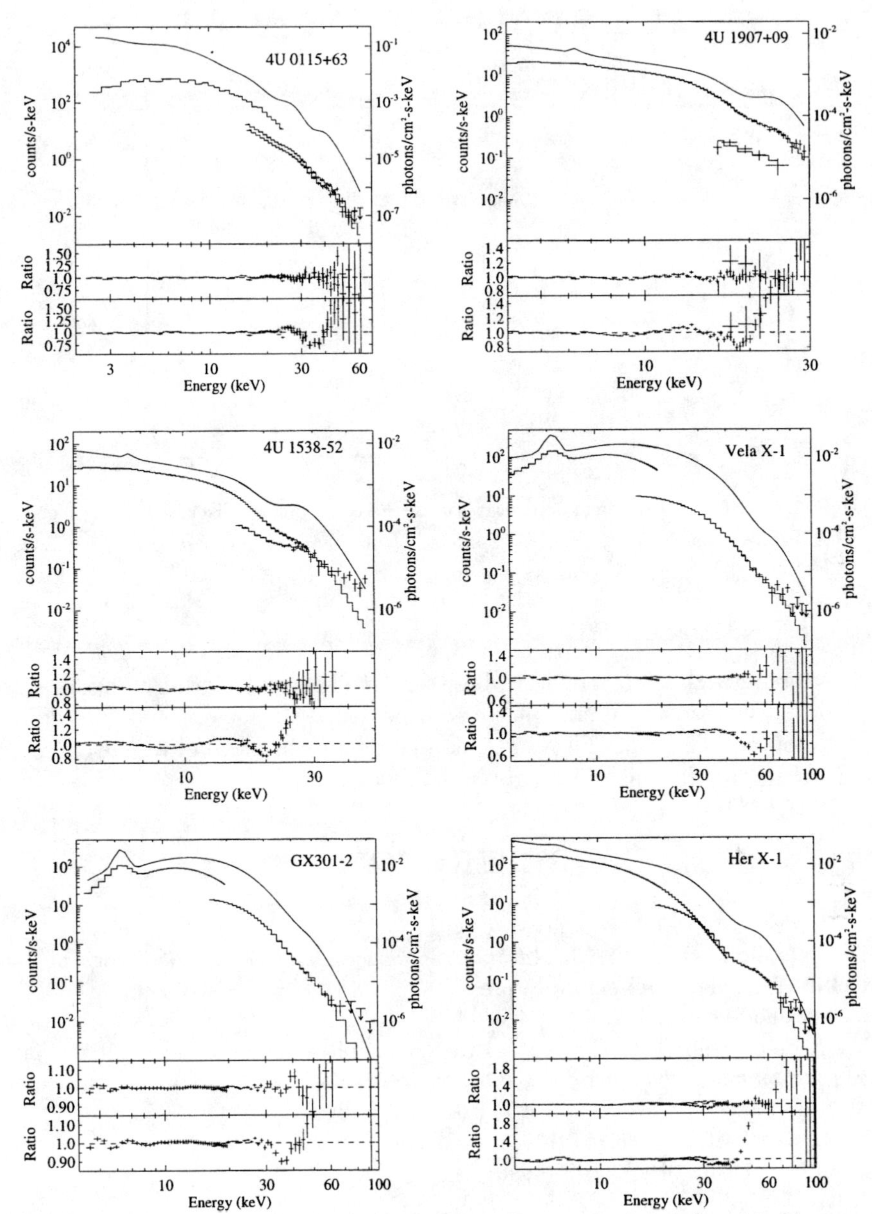

FIGURE 2. *RXTE* spectra of six accreting pulsars. Top panels show inferred incident spectra (solid lines) and PCA and HEXTE counts spectra. Middle panels show the ratio of the best fit model, which includes a CRSF, to the data. Bottom panels show this ratio for a model which has no CRSF. In the case of 4U 0115+63, which has three lines in this spectrum, we show residuals to models with three and two lines included. For line parameters, see Table 1.

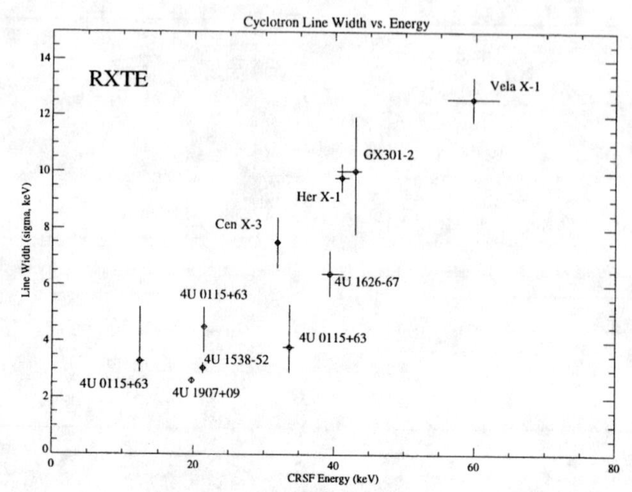

FIGURE 3. CRSF width versus line energy for all the eight pulsars discussed here. A positive correlation between energy and with is apparent. Note that for the three cyclotron lines in 4U 0115+63, the correlation does not appear to apply.

viewing angle (θ) with respect to the magnetic field as: $\Delta\omega \propto \omega_B(kT/mc^2)^{1/2}cos\theta$ [8]. This predicts that a correlation between line energy and width should exist, provided that viewing angles and plasma temperatures do not vary too greatly from source to source. These *RXTE* results, as well as measurements by *Beppo*SAX [2], show a strong correlation, which supports this picture.

REFERENCES

1. L. Burderi et al., 1999, submitted to ApJ.
2. D. Dal Fiume et al., 1999, *Advances in Space Research* - Proceedings of 32nd COSPAR Scientific Assembly, in press.
3. W. A. Heindl et al., ApJ **521** (1999), L49.
4. W.A. Heindl and D. Chakrabarty, Proceedings of the Symposium *Highlights in X-ray Astronomy in honour of Joachim Trümper's 65th birthday* (B. Aschenbach and M.J. Freyberg, eds.), 1999, MPE Report 272, p. 25.
5. W.A. Heindl et al., 1999, in this proceedings.
6. K. Jahoda et al., SPIE **2808** (1996), 59.
7. I. Kreykenbohm et al., A&A **341** (1999), 141.
8. P. Mészáros and W. Nagel, ApJ **298** (1985), 147.
9. M. Orlandini et al., ApJ **500** (1998), L165.
10. R.E. Rothschild et al., ApJ **496** (1998), 538.
11. A. Santangelo et al., A&A **340** (1998), L55.
12. ———, ApJ **523** (1999).
13. N.E. White, J.H. Swank, and S.S. Holt, ApJ **270** (1983), 711.

Cyclotron lines in X-ray pulsars as a probe of relativistic plasmas in superstrong magnetic fields

D. Dal Fiume[1], Filippo Frontera[2], Nicola Masetti[1], Mauro Orlandini[1], Eliana Palazzi[1], Stefano Del Sordo[3], Andrea Santangelo[3], Alberto Segreto[3], Tim Oosterbroek[4], Arvind N. Parmar[4]

[1]*Istituto TESRE/CNR, via Gobetti 101, 40129 Bologna, Italy*
[2]*Istituto TeSRE and Dipartimento di Fisica, Universitá di Ferrara, via Paradiso 1, 44100 Ferrara, Italy*
[3]*IFCAI/CNR, via U. La Malfa 153, 90146 Palermo, Italy*
[4]*Space Science Department, ESA, ESTEC, Noordwjik, The Netherlands*

Abstract. The systematic search for the presence of cyclotron lines in the spectra of accreting X-ray pulsars is being carried on with the BeppoSAX satellite since the beginning of the mission. These highly successful observations allowed the detection of cyclotron lines in many of the accreting X-ray pulsars observed. Some correlations between the different measured parameters were found. We present these correlations and discuss them in the framework of the current theoretical scenario for the X–ray emission from these sources.

INTRODUCTION

Accreting magnetized neutron stars are an ideal cosmic laboratory for high energy relativistic physics. Cyclotron resonance features are the signature of the presence of a superstrong magnetic field, following the first discovery in Her X-1 (Trümper et al. [1]). These features are due to the discrete Landau energy levels for motion of free electrons perpendicular to the field in presence of a locally uniform superstrong magnetic field. A slight deviation from a pure harmonic relationship in the spacing of the different levels is expected due to relativistic effects ($\frac{\omega_n}{m_e} = ((1 + 2n\frac{B}{B_{\rm crit}}\sin^2\theta)^{\frac{1}{2}} - 1)/\sin^2\theta$, e.g. Araya and Harding [2]). Therefore the detection of these features in the emitted X–ray spectra is in principle a direct measure of the field intensity.
As the number of sensitive measurements in the hard X–ray interval (above $\sim$ 10 keV) is continuously growing, a sample is available to search for possible correlations between the observed parameters. A detailed modeling is difficult and a parametrized shape of the continuum still is not available from theoretical models, but substantial advances in our understanding of the radiation transport in strongly magnetized atmospheres were done in the last decade (e.g. Alexander et al. [3], Alexander and Mészáros [4], Araya and

CP510, *The Fifth Compton Symposium*, edited by M. L. McConnell and J. M. Ryan

Harding [2], Isenberg et al. [5,6], Nelson et al. [7]). Some of these new results focused on the properties of the cyclotron resonance features observed in the spectra of accreting X–ray pulsars. In this report we discuss the current status of the measurements of cyclotron lines, with emphasis on the possible correlations between observable parameters.

THE DATA

The BeppoSAX satellite has observed all the bright persistent and three bright transient (recurrent) accreting X–ray pulsars. Apart from the case of X Persei (Di Salvo et al. [8]), a source with a luminosity substantially lower than the other sources in the sample, the spectra observed by BeppoSAX can be empirically described using the classical power–law–plus–cutoff spectral function by White et al. [9]. The sensitive broad band BeppoSAX observations also allowed the detailed characterization of low energy components below a few keV (like in Her X–1, Dal Fiume et al. [10], and in 4U1626–67, Orlandini et al. [11]) and the detection of absorption features in the hard X–ray range of the spectra, interpreted as cyclotron resonance features.

A summary of the properties of the broad band spectra and of the cyclotron lines as measured with BeppoSAX is reported in Dal Fiume et al. [12]. From these measurements we obtained evidence of a correlation between the centroid energy of the feature and its width. This correlation is presented and discussed elsewhere (Dal Fiume et al. [12,13]).

Transparency in the line

A straightforward parameter to be obtained from observations is the transparency in the line, defined as the ratio between the transmitted observed flux and the integrated flux from the continuum without the absorption feature. This ratio likely depends on the harmonic number of the feature we are observing (e.g. Wang et al. [14]) and on the physical parameters of the specific accretion column. From an observational point of view, this ratio is strongly affected by the modelization of the "continuum" shape, that is by the spectral shape used to describe the differential broad band photon number spectrum. In Figure 1 we report the observed transparencies obtained dividing the observed by the expected flux, both integrated in a $\pm 2\sigma$ interval around the line centroid (here σ is the Gaussian width of the measured cyclotron feature). To further emphasize the uncertainty in this estimate, we added a 10% error to the data. The purely statistical uncertainties are substantially smaller. This measured transparency is related to a simple physical parameter, the opacity to photons with energy approximately equal to the cyclotron resonance energy. However the emerging integral flux and the shape of the line itself are non trivially related to the radiation transport in this energy interval, a rather difficult problem to be solved.

From the phenomenological point of view, one can observe that the measured transparencies cluster around 0.5–0.6, with the notable exception of Cen X–3.

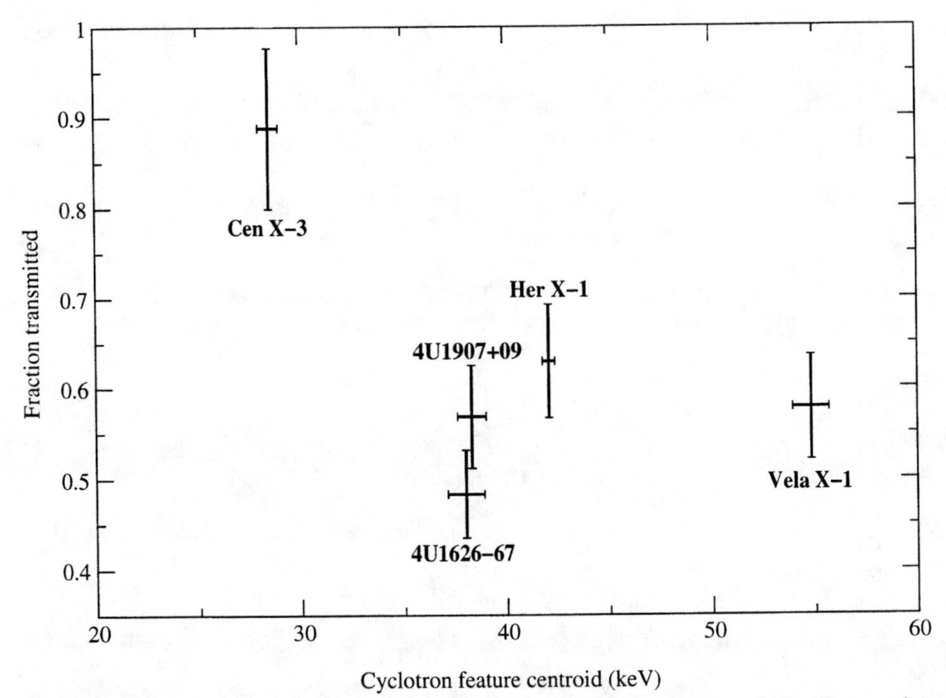

FIGURE 1. Transparency in the observed cyclotron resonance features with BeppoSAX . The error bars are *NOT* statistical, but rather indicate the uncertainty in the determination of the shape and intensity of the expected continuum flux (with no line absorption).

Magnetic field intensity and spectral hardness

The influence of the magnetic field intensity on the broad band spectral shape is debated. Early attempts to estimate a possible dependence of electron temperature, and therefore of broad band spectral shape, on the magnetic field intensity were done by Harding et al [15]. Actually they conclude that *"the equilibrium atmospheres have temperatures and optical depths that are very sensitive to the strength of the surface magnetic field"*. If this is the case and if the broad band spectral hardness is related, as one could naively assume, to the temperature of the atmosphere, some correlation between this hardness and the cyclotron line energy should appear in data.

This seems to be the case shown in Figure 2. Here we report the ratio between photon fluxes in two "hard" bands (the flux in 20–100 keV divided by the flux in 7–15 keV) versus the cyclotron line centroid. The ratio between the two fluxes is affected by the choice of the continuum, as in Figure 1. We therefore also in this case added 10% error bars that indicate this uncertainty. The statistical errors are substantially smaller.

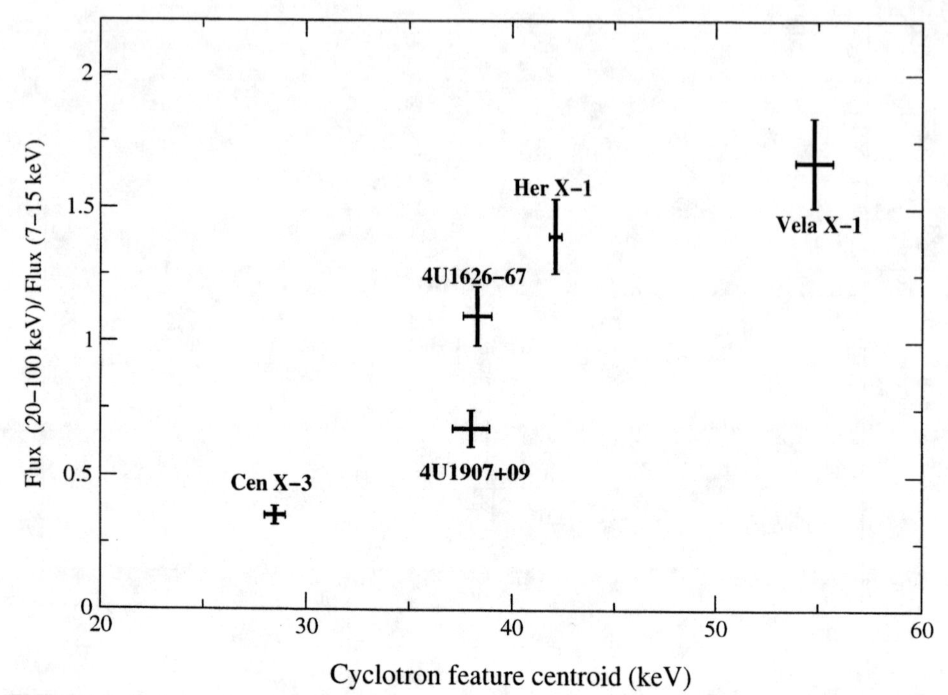

FIGURE 2. Ratio between the measured photon flux in two energy band versus the cyclotron line energy with BeppoSAX . The error bars are *NOT* statistical, but rather indicate the uncertainty in the determination of the shape and intensity of the measured photon flux.

The number of sources in this plot is still very limited and therefore one cannot exclude that this apparent correlation is merely due to the limited size of the sample. Nevertheless the apparent correlation is in the right direction, i.e. harder spectra are observed for higher field intensities.
We parenthetically add that no cyclotron resonance feature was observed in the pulse–phase averaged spectra of the two hardest sources of this class observed with BeppoSAX (GX1+4 Israel et al. [16] and GS1843+00 Piraino et al. [17]). If this correlation proves to be correct, this may suggest that cyclotron resonance features in these two sources should be searched at the upper limit of the BeppoSAX energy band or beyond.

Conclusions

In conclusion, even if no complete, parametric theoretical approach to model the observed spectra of accreting X–ray pulsars is still available, some quantitative measures of parameters of hot plasmas in superstrong magnetic fields are possible.
Modeling the transparency in the cyclotron resonance feature is a complex problem. Fur-

ther information will be extracted from maps of this transparency as a function of pulse phase.
The correlation between spectral hardness and field intensity is in agreement with theoretical models. This correlation, if confirmed, can be used as a rough estimate of the magnetic field intensity from the measured spectral hardness.

ACKNOWLEDGMENTS

This research is supported by Agenzia Spaziale Italiana (ASI) and Consiglio Nazionale delle Ricerche (CNR) of Italy. BeppoSAX is a joint program of ASI and of the Netherlands Agency for Aerospace Programs (NIVR).

REFERENCES

1. Trümper, J. et al. 1978 *Ap. J. Letters*, **219**, L105
2. Araya, R. A. and Harding, A. K. 1999 *Ap. J*, **517**, 334
3. Alexander, S. G. et al. 1996 *Ap. J.*, **459**, 666
4. Alexander, S. G. and Mészáros, P. 1991 *Ap. J.*, **373**, 565
5. Isenberg, M. et al. 1998a, *Ap. J.*, **493**, 154
6. Isenberg, M. et al. 1998b, *Ap. J.*, **505**, 688
7. Nelson, R. W. et al. 1995 *Ap. J. Letters*, **438**, L99
8. Di Salvo, T. 1998 *Ap. J.*, **509**, 897
9. White, N. E. et al. 1983 *Ap. J.*, **270**, 711
10. Dal Fiume, D. et al. 1998 *Astron. Astrophys*, **329**, L41
11. Orlandini, M. et al. 1998 *Ap. J. Letters*, **500**, L165
12. Dal Fiume, D. et al. 1999 *Adv. Sp. Res.*, in press (*astro-ph/9906086*)
13. Dal Fiume, D. et al. 1998 *Nucl. Physics B*, **69**, 145
14. Wang, J. C. L. et al. 1993 *Ap. J.*. **414**, 815
15. Harding, A. K. et al. 1984 *Ap. J.*, **278**, 369
16. Israel, G. L. et al. 1998 *Nucl. Phys. B*, **69**, 141
17. Piraino, S. et al. 2000 *Astron. Astrophys.*, submitted

Evidence for a 304-day Orbital Period for GX 1+4

João Braga, Marildo G. Pereira and Francisco J. Jablonski

Divisão de Astrofísica, Instituto Nacional de Pesquisas Espaciais, CP 515, 12201–970, São José dos Campos, Brazil

Abstract. In this paper we report strong evidence for a $\sim$ 304-day periodicity in the spin history of the accretion-powered pulsar GX 1+4 that is very likely to be a signature of the orbital period of the system. Using BATSE public-domain data, we show a highly-significant periodic modulation of the pulsar frequency from 1991 to date which is in excellent agreement with the ephemeris proposed by Cutler, Dennis & Dolan in 1986 [1], which were based on a few events of enhanced spin-up that occurred during the pulsar's spin-up era in the 1970s. Our results indicate that the orbital period of GX 1+4 is 303.8±1.1 days, making it by far the widest low-mass X-ray binary system known. A likely scenario for this system is an elliptical orbit in which the neutron star decreases its spin-down rate (or even exhibits a momentary spin-up behavior) at periastron passages due to the higher torque exerted by the accretion disk onto the magnetosphere of the neutron star.

INTRODUCTION

GX 1+4 is a unique accretion-powered pulsar in a low-mass x-ray binary system (LMXB). In the 1970s the pulsar exhibited a spin-up behavior with a rate of $\dot{P} \sim -2\,\mathrm{s/year}$, the hightest among all persistent X-ray pulsars, and was one of the brightest and hardest X-ray sources in the sky. After an extended low-intensity state in the early 1980s, GX 1+4 re-emerged in a spin-down state [2] and has produced occasional short-term variations of $\dot{P}$ ever since. The optical counterpart is a M5 III giant star, V2116 Oph, in a rare type of symbiotic system [3–5]. The identification was made secure by a ROSAT accurate position [6] and by the discovery of optical pulsations consistent with the spin period of the neutron star [7,8]. In 1991, BATSE initiated a continuous and nearly uniform monitoring of GX 1+4, confirming the spin-down trend with occasional dramatic spin-up/down torque reversal events [9,10]. GX 1+4 has a much longer (factor of $\sim$ 100) spin period than the other four known LMXB accretion-powered pulsars and its orbital period has been known to be at least one order of magnitude longer than the periods of the other systems [5]. Attempts to find the orbital period by Doppler shifts of the pulsar pulse timing [9] or optical lines [4,11,12] have both been inconclusive

CP510, *The Fifth Compton Symposium*, edited by M. L. McConnell and J. M. Ryan

so far. Using a small number of X-ray measurements carried out during the spin-up phase of GX 1+4 in the 1970s, Cutler, Dennis & Dollan [1] produced an ephemeris for predicting periodical enhancements in the spin-up rate of the neutron star and claimed that this could be due to an elliptical orbit with a 304-day period. Here we report the discovery of a 304-day modulation in the BATSE frequency data and discuss its implications to the models for this source.

DATA ANALYSIS AND RESULTS

The frequency and the pulsed flux data between Julian Day (JD) 2448376.5 and 2451138.5 (i.e., 1991 April 29 to 1998 October 20) used in this work were obtained from Chakrabarty [9] and from the BATSE public domain data. The 20–50 keV pulsed signals are extracted from DISCLA 1.024s channel 1 data. 15-day mean values for the fluxes and pulse frequencies of GX 1+4 were calculated for the entire dataset.

A dataset of GX 1+4 residual pulsation frequencies was obtained from the frequency history by subtraction of a standard cubic spline function to remove low frequency variations in the spin-down trend. The fitting points are mean frequency values calculated over suitably chosen time intervals. The results of the spline fitting are fairly insensitive to intervals greater than $\sim$ 200 days between fitting points (we have used Δt = 215 days). The pulsed X-ray flux, frequency history and residual frequencies are shown in Figure 1 as functions of time.

We have carried out a power spectrum analysis to search for periodicities of less than 1000 days in both the residual frequency and the pulsed flux data. A Lomb-Scargle periodogram [13], suitable for time series with gaps, shows a significant periodic signal at 302.0 days (Fig. 2) in the residual frequency time series. The power spectrum shows a red noise with an approximate power-law index index of -2. In order to estimate the statistical significance of the detection, a series of numerical simulations of the frequency time series with 1-sigma gaussian deviations were performed [14]. The simulations show that the use of the 215-d spline, besides providing an effective filter for frequencies below $\sim 2 \times 10^{-3} \mathrm{d}^{-1}$, does not produce power in any specific frequency in the range of interest. By comparing the amplitude of our 302-day peak with the local value obtained by the mean of the numerical simulations, we obtain a statistical significance of 99.98% for the detection. Epoch folding the data using the 302-day period yields a 1-σ uncertainty of 1.7 days.

By analyzing the variation of the period of GX 1+4 during the spin-up phase in the 1970s, Cutler, Dennis & Dolan [1] proposed a 304-day orbital period and an ephemeris to predict the events of enhanced spin-up: $T = \mathrm{JD}\ 2,444,574.5 \pm 304\, n$, where n is an integer. This ephemeris is based on four events discussed by the authors, whose existence was inferred from ad-hoc assumptions and extrapolations of the observations. The projected enhanced spin-up events derived from that ephemeris for the epochs contained in the BATSE dataset, represented as solid vertical lines in the lower panel of Fig. 1, are in excellent agreement with

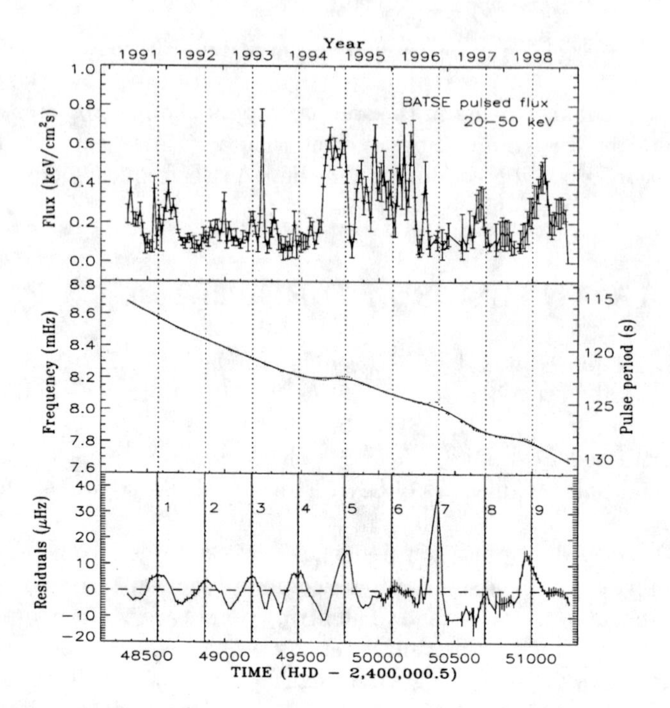

FIGURE 1. *Upper panel*: Light curve of the 20-50 keV pulsed flux of GX 1+4 as measured by BATSE from 1991 to 1998; *middle panel*: GX 1+4 frequency measurements by BATSE over the same period. The error bars are in general smaller than the size of the dots. The solid curve is a cubic spline fit to the data; *lower panel*: frequency residuals. The dotted vertical lines mark the times times predicted by the ephemeris calculated in this work, whereas the solid vertical lines show the predictions of Cutler, Dennis & Dolan (1996). The events of positive residual frequency modulation are labeled for reference in the text.

the BATSE reduced spin-down and spin-up events. The BATSE dataset is obviously significantly more reliable than the one given by Cutler, Dennis & Dolan [1] since it is based on 9 well-covered events measured with the same instrument as opposed to the 4 events discussed by those authors. The striking agreement of their ephemeris with the BATSE observations is very conspicuous and give a very strong support to the claim that the orbital period of the system is indeed $\sim$ 304 days. Taking integer cycle numbers, with the $T0$ epoch of Cutler, Dennis & Dolan [1] as cycle -23, and performing a linear least-squares fit to the frequency residuals seen in the lower panel of Fig. 1, we find that the following ephemeris can represent the time of occurrence T of the maxima in the frequency residuals: $T =$ JD $2,448,571.3(\pm 3.2) \pm 303.8(\pm 1.1)\ n$, where n is any integer. The events predicted by the above ephemeris are shown as vertical dotted lines in the three panels of Fig. 1. The value of 303.8 ± 1.1 days for the orbital period is consistent

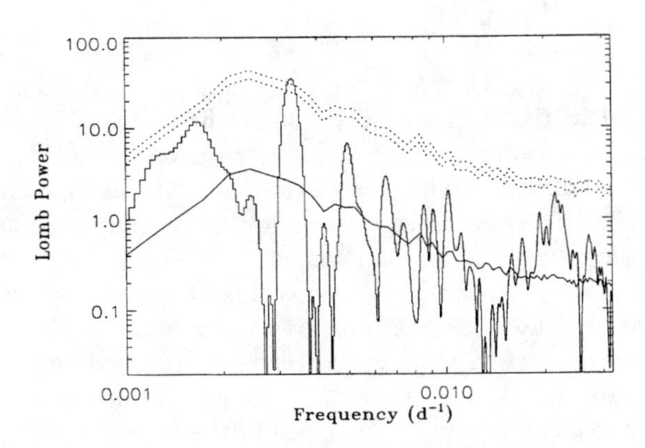

FIGURE 2. Lomb-Scargle periodogram of the frequency residuals of GX 1+4 from 1991 to 1998, represented by the histogram-type solid line. The standard solid line is the mean of 1500 numerical simulations carried out in order to calculate the significance level of the detection. The upper dotted line indicates a significance level of 0.001, whereas the lower dotted line indicates a significance level of 0.01.

with the one obtained through power spectrum analysis performed on the BATSE data, which gives further support for the period determination.

DISCUSSION

In the 1970s, when the measurements used by Cutler, Dennis & Dolan [1] were carried out, the source was in a spin-up extended state. They proposed that the periodic occurrence of enhanced spin-up events was due to the fact that the system was in a elliptical orbit and the periastron passages would occur when $\dot{P}$ is maximum, as expected in standard accretion from a spherically expanding stellar wind. However, it is widely accepted today that the system has an accretion disk. Since the neutron star is currently spinning-down, the radius at which the magnetosphere boundary would corotate with the disk is probably smaller than the magnetosphere radius. Since the pulse period is $\sim$ 120 s and the luminosity is typically $\lesssim 10^{37}$ erg/s, it can be shown [14] that the period is probably close to the equilibrium value, for which the two radii are equal. This allows spin-down to occur even though accretion continues, the centrifugal barrier not being sufficiently effective [15]. Assuming that the elliptical orbit is the correct interpretation for the presence of the modulation, the mass accretion rate (and hence the luminosity) should increase as the neutron star approaches periastron. The spin-down torque then gets smaller and the neutron star decelerates at a slower rate [14]. Occasionally, due to the highly variable mass loss rate of the red giant, the neutron star will *spin-up* for a brief period of time during periastron, as observed in the BATSE frequency curve in events 5, 7 and 9. According to this picture, one would expect an increase in X-ray luminosity at periastron. Although this is only marginally indicated in the BATSE pulsed flux light curve, it should be pointed out that total flux data from the ASM/RXTE for the epoch MJD 50088 to 51044 does not correlate significantly with the BATSE pulsed flux, indicating that the pulsed flux may not be a good tracer of the accretion luminosity in this system. Furthermore, the periodic $\sim 5\mu$Hz excursions in the residual frequency would lead to very low-significance variations in the X-ray flux measured by the ASM [14].

An alternative interpretation for the observed modulation would be the presence of oscillation modes in the red giant star. However, the stability of the infrared magnitudes of V2116 Oph [5] preclude it from being a long-period variable, since these stars undergo regular $\gtrsim 1$ mag variations in the infrared [16].

We conclude by pointing out that, given the 304-day orbital period and the spectral and luminosity characteristics of V2116 Oph, it can be shown that the companion in this system is probably not filling its Roche lobe and the accretion disk forms from the slow, dense stellar wind of the red giant [14]. A more thorough covering of the X-ray luminosity of the system, with high sensitivity and spanning several cycles, will be very important to test the elliptical model for GX 1+4.

We thank Dr. Bob Wilson from NASA Marshall Space Flight Center for gently providing us BATSE frequency and flux data on GX 1+4. M. P. is supported by a FAPESP Postdoctoral fellowship at INPE under grant 98/16529-9. J. B. thanks CNPq for support under grant 300689/92-6. F. J. acknowledges support by PRONEX/FINEP under grant 41.96.0908.00.

REFERENCES

1. Cutler, E.P., Dennis, B.R., and Dolan, J.F., *Astrophys. J.*, **300**, 551 (1986).
2. Makishima, K., et al., *Nature*, **333**, 746 (1988).
3. Glass, I.S., and Feast, M.W., *Nature Phys. Sci.*, **245**, 39 (1973).
4. Davidsen, A., Malina, R., and Bowyer, S., *Astrophys. J.*, **211**, 866 (1977).
5. Chakrabarty, D., and Roche, P., *Astrophys. J.*, **489**, 254 (1997).
6. Predehl, P., Friedrich, S. and Staubert, R., *Astr. Astrophys.*, **294**, L33 (1995).
7. Jablonski, F.J., et al., *Astrophys. J.*, **482**, L171 (1997).
8. Pereira, M., et al., *IAU Circ.*, **6794** (1997).
9. Chakrabarty, D., Ph.D. thesis, California Institute of Technology (1996).
10. Chakrabarty, D., et al., *Astrophys. J. Lett.*, **481**, L101 (1997).
11. Doty, J.P., Hoffman, J.A., and Lewin, W.H.G., *Astrophys. J.*, **243**, 257 (1981).
12. Sood, R.K., et al., *Adv. Space Res.*, **16(3)**, 131 (1995).
13. Press, W.H., et al., *Numerical Recipes in FORTRAN* (2nd ed.), Cambridge: Cambridge Univ. Press, 1992.
14. Pereira, M.G., Braga, J., and Jablonksi, F.J., *Astrophys. J. Lett.*, in press.
15. White, N.E., *Nature*, **333**, 708 (1988).
16. Whitelock, P.A., *Pub. Astr. Soc. Pacific*, **99**, 573 (1987).

SAX J0635+0533: Detection of 33.8 ms X-ray pulsations

Giancarlo Cusumano*, Maria C. Maccarone*, Luciano Nicastro*, Bruno Sacco* and Phil Kaaret†

*Istituto di Fisica Cosmica con Applicazioni all'Informatica, CNR, Palermo 90146, Italy
†Harvard-Smithsonian Center for Astrophysics, Cambridge, MA 02138, USA

Abstract. We have detected 33.8 ms pulsations from SAX J0635+0533, a newly discovered Be/X-ray binary suggested as a possible counterpart to the gamma-ray source 2EG J0635+0521. We interpret the periodicity as the spin period of a neutron star in a binary system with a Be companion.

INTRODUCTION

The X-ray source SAX J0635+0533 was discovered by Kaaret et al. (1999) thanks to a *Beppo*SAX observation within the error box of the unidentified Galactic gamma-ray source 2EG J0635+0521 [1], a candidate gamma-ray pulsar as suggested by its hard gamma-ray spectrum [2]. The X-ray source is characterized by quite hard X-ray emission detected up to 40 keV [3]. Its energy spectrum is consistent with a power-law model with a photon index of 1.5, an absorption column density of $2.0 \times 10^{22}\,\mathrm{cm}^{-2}$, and a flux of $1.2 \times 10^{-11}\,\mathrm{erg\,cm^{-2}\,s^{-1}}$ in the $2 - 10$ keV energy band. A search for pulsed emission over a period range from 0.030 s to 1000 s did not detect any pulsed signal. Due to the large error box of the gamma-ray source, the identification of SAX J0635+0533 with 2EG J0635+0521 is not definitive: such an identification could only be made through pulsed detection in both X-ray and gamma-ray emission or a much improved gamma-ray position. Follow up optical observations [3] suggest as a counterpart of SAX J0635+0533 a Be star with a V-magnitude of 12.8, located within the $1'$ X-ray source error box. The estimated distance is in the range $2.5 - 5$ kpc.

We revisited the *Beppo*SAX observation of SAX J0635+0533, available from the public archive (obs.code #30326001). In our analysis, we use only data coming from the two NFI imaging instruments, namely the Low Energy Concentrator Spectrometer (LECS) operating in the energy range 0.1–10 keV [4] and the Medium Energy Concentrator Spectrometer (MECS) operating in the energy range 1.3–10 keV [5]. In this letter we present new timing results. Our analysis has revealed a 33.8 ms pulsation of the X-ray source.

CP510, *The Fifth Compton Symposium,* edited by M. L. McConnell and J. M. Ryan

TIMING ANALYSIS

The SAX J0635+0533 light–curve (1000 s bin size) for the MECS is shown in Fig. 1 (gaps are present due to non–observing time intervals during South Atlantic Anomaly and Earth occultation). Fig. 1 shows that the emission of SAX J0635+0533 is variable up to a factor of 10.

In order to search for periodicity, the arrival times of all selected events have been converted to the Solar System Baricentric Frame. The Z_1^2 test [6] on the fundamental harmonics with the maximum resolution ($\delta f = 1/72\,714$ Hz) applied to the MECS baricentered arrival times does not reveal significant deviations from a statistically flat distribution up to 50 Hz.

If SAX J0635+0533 is a binary pulsar of rotational spin period P_s and orbital period P_o, the observed P_s is modulated by the orbital motion. Thus, a direct search for a coherent oscillation at P_s can be successful only if the modulation amplitude is small over the time interval ΔT in which the search is performed. This condition is satisfied if $\Delta T \ll P_o$. To reduce the effect of a possible orbital motion in the periodicity search, we divide the whole data span into M subintervals, calculating the Z_1^2 statistics for each trial period in each subinterval, and then adding together the M statistics for each trial period. This procedure results in a less noisy spectrum. We selected time slices corresponding to intervals of continuous observation taken between two Earth occultation periods. The total number of these slices is M = 13, each one lasting $\sim$ 3300 s. We adopted a frequency step

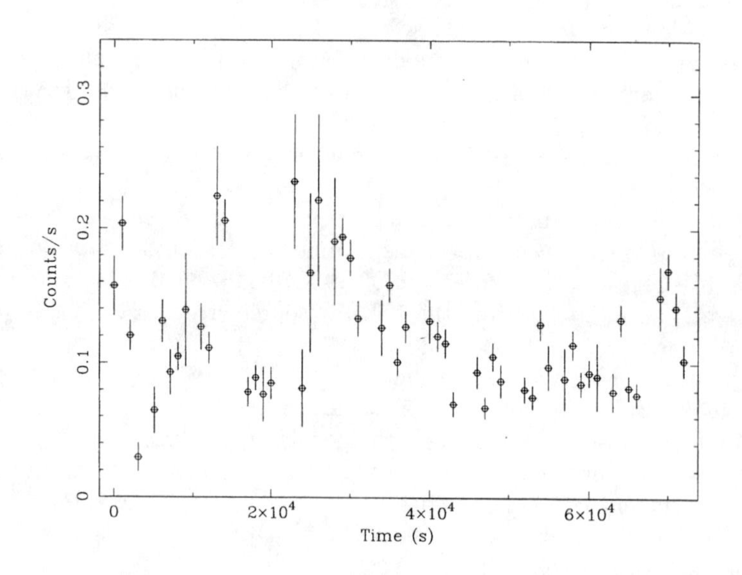

FIGURE 1. The MECS light curve of SAX J0635+0533.

$\delta f = 2 \times 10^{-4}$ Hz, spanning 50 Hz of search range with 250 000 trial frequencies.

Fig. 2 shows the power spectrum obtained with the MECS and LECS data in the $Z_1^2(\nu = 26)$ statistics as a function of frequency, where an evident excess appears at $f_0 = 29.5364 \pm 0.0001$ Hz. The value of $Z_1^2(\nu = 26)$ at this frequency is equal to 99.6. Because the $Z_1^2(\nu = 26)$ follows the χ^2 statistics with 26 degrees of freedom, the single trial chance occurrence probability to have an excess greater than 99 is 2×10^{-10}. Taking into account the number of trial frequencies used, the probability is 5×10^{-5}, corresponding to 4 standard deviations of the Gauss statistics.

From the $Z_1^2(\nu = 26)$ value we can estimate the pulsed fraction. For N_p pulsed counts over N_t total counts, $Z_1^2(\nu = 26) = 2\alpha N_p^2/N_t + \nu$, where α is a shape constant. In our case, $\alpha = 0.25$ (sinusoidal shape), and the pulsed fraction is then about 0.2.

DISCUSSION

The coherence of the detected periodicity is high, $Q = f/\delta f \sim 10^5$. The high coherence induces us to interpret this periodicity as a neutron star spin period. Timing analysis results indicate that the neutron star could orbit around a companion star.

The X-ray emission may be powered either by accretion or by spin-down of the neutron star. We consider these possibilities in turn.

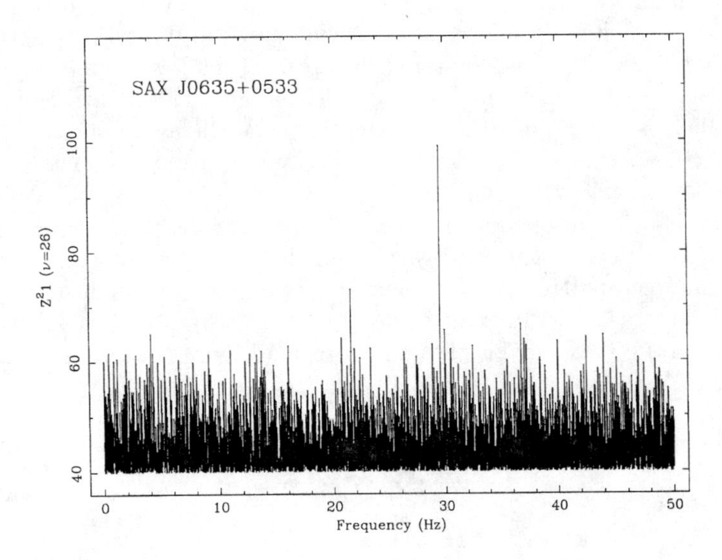

FIGURE 2. The $Z_1^2(\nu = 26)$ statistic as a function of frequency for MECS+LECS events.

The SAX J0635+0533 system may consist of a rotation-powered pulsar orbiting the Be star. In this case, the X-ray emission could be magnetospheric emission similar to the power-law component in the X-ray emission of known X-ray/gamma-ray pulsars. The pulsation frequency we have found and the X-ray luminosity of the source are similar to those of the known X-ray/gamma-ray pulsars, but high X-ray variability of SAX J0635+0533 on time scales of 1000 s is unlike the steady X-ray emission seen from isolated pulsars. However, it could be produced by variable X-ray absorption caused by matter in the binary system or a wind from the Be star.

An alternative, but still rotation-powered, scenario is that SAX J0635+0533 is similar to the Be radio pulsar PSR J1259–63. These two sources have similar spin frequencies and similar X-ray spectra [7]. In this case, the X-ray emission of SAX J0635+0533 should arise from a shock interaction of the energetic particles from the pulsar with the wind from the Be star. However, the upper bound, 8%, on the X-ray pulsed fraction from PSR J1259–63 in the 2–10 keV band (Kaspi et al. 1995) is well below the value estimate for SAX J0635+0533 in the same energy band.

The X-ray emission from SAX J0635+0533 may be powered by accretion. Strong X-ray variability would naturally occur in such a system. Following this interpretation we can infer the magnetic field strength of the neutron star. For accretion to proceed, the centrifugal force on the accreting matter co-rotating in the magnetosphere must be less than the local gravitational force [9]. Assuming a bolometric luminosity of $1.2 \times 10^{35}\,\mathrm{erg\,s^{-1}}$ (0.1–40 keV) estimated from spectra results given in Kaaret et al. (1999) for a 5 kpc distance, and a neutron star mass of $1.4\,M_{\odot}$ and radius of 10 km, we can set an upper limit on the magnetic field strength of 2×10^9 G. This is a factor 10^3 lower than those measured in typical accreting X-ray pulsars, but similar to the fields inferred for the 2.49 ms low-mass X-ray binary SAX J1808.4−3658 [10] and for millisecond radio pulsars. The X-ray luminosity of SAX J0635+0533 is a factor of 10 below that of most Be/X-ray binaries or the peak luminosity of SAX J1808.4−3658, but may simply indicate a low mass accretion rate.

A definitive association of SAX J0635+0533 with the EGRET source requires detection of a periodicity in gamma-rays at the pulsar spin period. Due to the long integration time required to obtain a detectable gamma-ray signal, only a priori knowledge of the binary parameters would permit a sensitive search for periodicity in gamma-rays. This can be obtained with additional X-ray observations of SAX J0635+0533.

REFERENCES

1. Thomson, D. J., et al., *ApJS* **101**, 259 (1995).
2. Merck, M., Bertsch., D. L., Dingus, B. L., et al., *A&AS* **120**, 465 (1996).
3. Kaaret, P., Piraino, S., Halpern, J., & Eracleous, M., *ApJ* in press.

4. Parmar A.N., Martin D.D.E., Bavdaz M., et al., *A&AS* **122**, 309 (1997).
5. Boella, G., Chiappetti, L., Conti, G., et al., *A&AS* **122**, 327 (1997).
6. Buccheri, R., Bennett, K., Bignami, G. F., et al., *A&A* **128**, 245 (1983).
7. Nicastro, L., Dal Fiume, D., Orlandini, M., et al., *Nucl.Phys.B* **69/1-3**, 257 (1998).
8. Kaspi, V. M., Tavani, M., Nagase, F., et al.,*ApJ* **453**, 424 (1995).
9. Stella, L., White, N. E., & Rosner, R., *ApJ* **308**, 669 (1986).
10. Wijnnands, R., & van der Klis, M., *Nature* **394**, 344 (1998).

Iron line and soft excess properties of GX301-2 in different orbital phases

S. Del Sordo[(1)], M. Orlandini[(2)], D. Dal Fiume[(2)], A. Parmar[(3)], T. Oosterbroek[(3)], A. Santangelo[(1)], and A. Segreto[(1)]

(1) IFCAI, Consiglio Nazionale delle Ricerche, via Ugo La Malfa 153, Palermo, Italy
(2) ITESRE, Consiglio Nazionale delle Ricerche, via Gobetti 101, Bologna, Italy
(3) Space Science Department ESA, ESTEC The Netherlands

Abstract. The high mass X-ray binary pulsar GX301-2 was the target of a campaign of observations with the Italian-Dutch satellite BeppoSAX. The source was observed at six different orbital phases in order to monitor its spectral and timing behavior along the 41.5 days orbit. The pulse-averaged spectrum is very complex and rich in features. Here we mainly discuss the low energy part (0.4 - 10 keV) of the spectrum and, in particular, the large variability observed in the iron line and in the soft energy excess, along the orbital period. The detection of pulsations at low energy (below 3 keV) seems to be in contrast with the scattering model successfully used until now to explain the soft excess. We also made a comparison with the results of an ASCA observation of this source.

INTRODUCTION

GX301-2 was discovered as a hard and variable X-ray source by Lewin et al. [1] and McClintock et al. [2]. It was identified as an X-ray binary system comprising a B1.5 Ia supergiant primary [3] and a neutron star. The system with an orbital period of 41.5 days [4] has the most eccentric orbit (e=0.47) among the binary X-ray pulsar with well established orbital parameters. X-ray pulsations with a period of 11.6 minutes were discovered by White et al. [5]. The system shows no eclipses.
GX301–2 is a typical slow X–ray pulsar that accretes matter from the strong wind of its massive companion. It shows a variable X–ray luminosity in the range $(2–400)\times10^{35}$ ergs/s. This variability probably depends on the fraction of the stellar wind captured by the neutron star. Due to these characteristics GX301-2 is well suited to study different regimes of the accretion on the neutron star. The X–ray spectra of GX301–2 show variable photoelectric absorption with column density up to 2×10^{24} H cm^{-2}. The absorption column seems to be highest around periastron passage and then decline around phase 0.7.
Recently an ASCA observation of this source [6] performed at orbital phase between 0.27 and 0.31 shows, for the first time, the presence of a strong low energy excess below 4 keV and a high photoelectric absorption. The ASCA spectrum also reveal a strong iron K–line and an absorption K–edge.

CP510, *The Fifth Compton Symposium,* edited by M. L. McConnell and J. M. Ryan

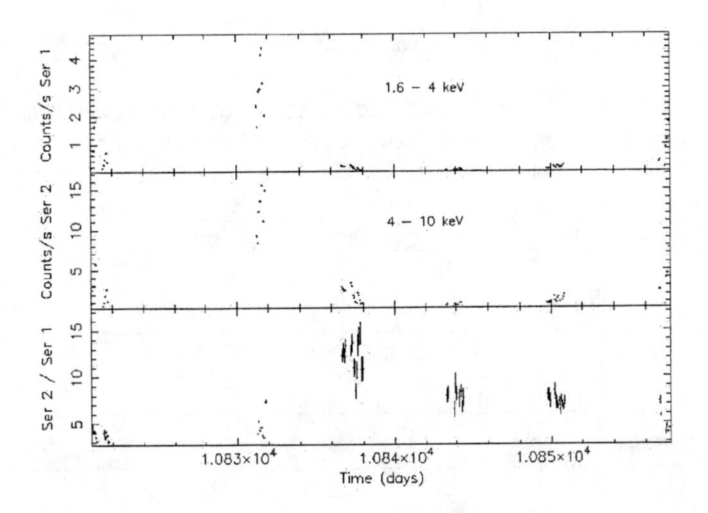

FIGURE 1. Total light curve in two different energy band and hardness ratio. It is interesting to note the rapid increase in the hardness ratio near the periastron.

DATA ANALYSIS AND RESULTS

BeppoSAX satellite observed GX301–2 in 6 different orbital phases. The average duration of each pointing is about 20 ks. All four Narrow Field Instruments (NFIs) aboard BeppoSAX worked nominally during all the observations, i.e., LECS (0.1 - 10 keV), MECS (1.6 - 10 keV), HPGSPC (5 - 120 keV) and PDS (15 - 200 keV).
The LECS and MECS light curves and spectra have been extracted following the standard procedure and the background subtraction was performed using the standard BeppoSAX background files.
In Tab.1 are shown the different observations together with their count rate in the 2–10 keV energy band and the corresponding orbital phase. In Fig.1 are shown the total light curve in two different energy band and the hardness ratio.
To have a first look to the variation of the spectral properties of this source along the orbital cycle we performed a spectral fitting in the 1.6-10 keV energy band with a simple model composed by a absorbed power law plus an iron gaussian line. In Tab.1 are shown the fitting parameters for the different observations with the relevant values of the reduced χ^2. From an inspection of the table is evident that this simple model is not adequate to fit the spectra even in this restricted energy band. In particular the fit of spectra observed around the periastron (orbital phase 0.0) are unacceptable. In Fig. 2 we report, as an example, the spectral fitting at orbital phase 0.6 with the above model. Two things are particularly evident : the strong soft excess below 4 keV and the

structured iron K-line. The following explanations have been put forward to explain the soft excess :

1) X–ray scattered by the wind around a dense gas stream trailing the neutron star
2) A partial covering by a non uniform clumpy circumstellar medium
3) Reduced opacities due to photoionization of the wind by the X–ray source

OP	Cts/s	α	$n_H * 10^{22}$	χ^2	Orb. Phase
3275	3.6	1.2	19.8	1.7	0.6
3373	15.6	0.7	17.4	8.1	0.85
3428	2.8	-0.6	24.6	8.6	0.95
3429	1.3	-0.5	31.4	1.7	0
3503	0.7	0.66	35.8	1	0.12
3514	0.7	0.37	30.7	1.9	0.15
3588	1.3	1.0	35.1	2.3	0.3
3650	7.1	1.1	22.4	1.2	0.45

TAB. 1. Summary of the different observations and spectral fitting parameters for the absorbed power law plus gaussian line model.

In the scattering model a gas stream originating from the primary trails the pulsar. If the line of sight from the observer to the compact object passes through the enhanced density region it results in an increased column density. The soft excess seen in the spectrum can be explained by the X–ray scattered in the stellar wind of the massive star. The scattered component is absorbed by a lower column density and causes the low energy excess. This model was successfully used to fit the spectrum of the ASCA observation (orbital phase about 0.3) in the 2–10 keV energy band. Due to the scattered origin of the low energy component this model foresee no pulsation at low energy. In the BeppoSAX observations the pulse profiles shows large variations along the orbital phase. It is high structured at orbital phase 0.85 and it is more smoothed at orbital phase 0.3. This difference in the pulse shape may be linked to the different luminosity in these two orbital phases (8.5×10^{35} and 0.8×10^{35} ergs/s).
In Fig. 3 are shown the pulse profiles in different energy bands for the orbital phase 0.85. The presence of pulsations well below 3 keV is evident. It is interesting to observe that in the orbital phase corresponding to the ASCA observation no (or very little) pulsations are observed at low energy in agreement with the prediction of the scattering model. The other suggested spectral model is the partial covering. This model assumes that the continuum source is partially obscured by cold clumpy matter with a uniform column density which covers a fraction of the continuum source and the soft excess arises from the continuum that leaks through this partial cover. This model justifies the presence of low energy pulsations. The partial covering model gives good fit in the 2–10 keV energy band in all the orbital phases where we detect pulsations at low energy. The spectrum in the 0.5 - 10 keV energy band clearly shows that another component is needed. There exists an ultrasoft component below 2 keV that can be modeled quite well by emission from a thermal plasma. This emission probably derives from the outer part of the atmosphere and has a large emission area.

In Fig. 4 is shown the spectral fitting in the range 5 - 8 keV for the observation at orbital phase 0.85. The fit performed with a single iron line is clearly inadequate. Adding a second gaussian line centered at 7.05 keV greatly improves the fit (the reduced χ^2 goes from 5.7 to 1.24) and gives physically acceptable parameters.

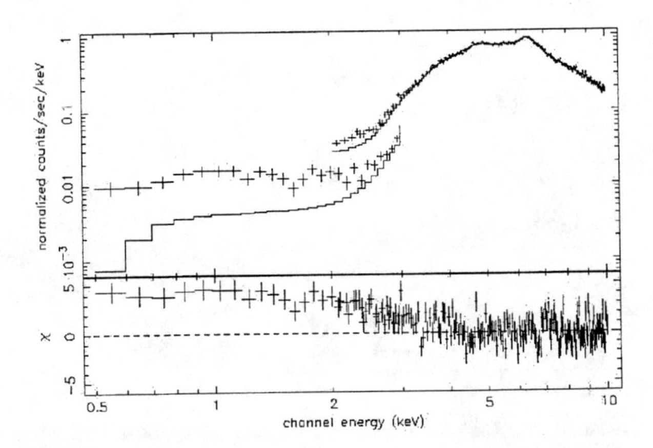

FIGURE 2. Spectrum at orbital phase 0.6 in the 0.5 - 10 keV energy range fitted with an absorbed power law plus a gaussian line.

The EQWI is higher in the more absorbed phases in agreement with the fluorescence origin of the lines. The presence of a mixture of two lines is particularly evident at orbital phase 0.85 and near the periastron where the neutron star is quite close to the primary and a strong stellar wind is expected.

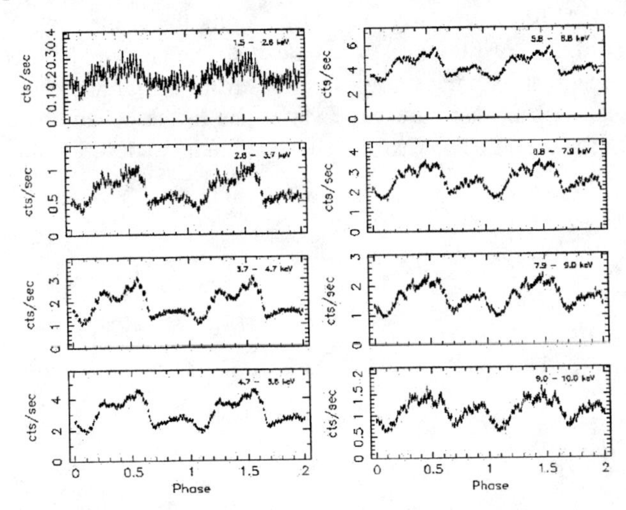

FIGURE 3. Pulse profiles in different energy bands for the observation at orbital phase 0.85. The presence of pulsations at low energy (below 3 keV) is evident.

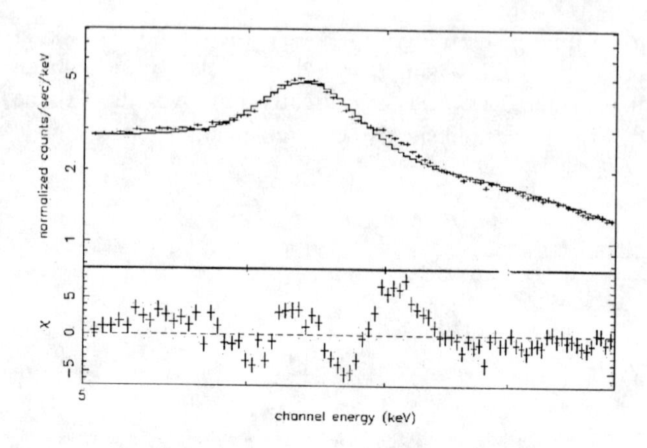

FIGURE 4. Fitting of the iron line at orbital phase 0.8 with a single gaussian. The structure of the line is evident.

CONCLUSIONS

BeppoSAX observation performed at orbital phase 0.3 is in good agreement with the ASCA observation in the same orbital phase. The scattering model plus an ultrasoft component and a single iron line gives a good fit to the data. Low energy pulsations are very small. In various orbital phases (0.45, 0.6, 0.85) BeppoSAX observations detect, for the first time, evident pulsations in the 2-3 keV energy band. Also in these observations the scattering model gives good fit of the spectra but it does not explain pulsations at low energy. Partial covering model plus a blackbody gives good fit in the 0.5 - 10 keV energy range and has the advantage to explain quite well the presence of pulsations at low energy. The iron line shows large variability along the orbital phase and, in some cases, is clearly a mixture of K_α and K_β components. The energies of the two components are in good agreement with the energies of lines from cold iron.
The above results seems to indicate that there are large variations in the behavior of this source as a function of the orbital phase and that the physical model to describe the emission mechanism is not unique.

References

[1] Lewin, W.H.G., et al., 1971, ApJ, 166, L69
[2] McClintock, J.E., et al., 1971, ApJ, 166, L73
[3] Bradt, H.V., et al., 1977, Nature, 269, 21
[4] Sato, N., et al., 1986, ApJ, 304, 241
[5] White, N., et al., 1976, ApJ, 209, L119
[6] Saraswat, P., et al., 1996, ApJ, 463, 726

Clocked Bursts From GS 1826-238

M. Cocchi[1], A. Bazzano[1], L. Natalucci[1], P. Ubertini[1]
J. Heise[2], J. J. M. in't Zand[2], J. M. Muller[2,3], M. J. S. Smith[2,3]

(1) Istituto di Astrofisica Spaziale, C.N.R., Via del Fosso del Cavaliere 00133 Roma- Italy
(2) Space Research Organisation of the Netherlands (SRON), Utrecht, the Netherlands
(3) also BeppoSAX Science Data Centre, Nuova Telespazio, Roma, Italy

ABSTRACT. We report on the long term monitoring of the GINGA transient source GS 1826-238 performed with the BeppoSAX Wide Field Camera (WFC) instrument, during four different observing campaigns covering October 1996 - September 1999.
WFC detection of type-I bursting activity from the source ruled out its proposed Black Hole candidacy and clearly suggested the compact object related to GS 1826-238 to be a weakly magnetized neutron star. The analysis of the arrival times of the observed 78 bursts lead to the discovery of a recurrence of ~5.75 hours with a spread of 38 minutes (FWHM) along more than 3 years monitoring data [15].
We performed a more detailed analysis of the whole available data, and evidence of shortening of the recurrence time, together with a drastic narrowing in the spread (down to a few minutes) was observed on a one year time scale. Possible relation with the source X-ray persistent emission is discussed.

INTRODUCTION

GS 1826-238 was serendipitously discovered by GINGA on September 8, 1988 with an average X-ray intensity of about 26 mCrab in the 1-40 keV range [9]. During September 9-16, 1988, rapid fluctuations (flickering) were observed on time scales down to 2 ms with an rms variation of 30% [12]. In the GINGA All Sky Monitor the source was below 50 mCrab all along the period from August through October 1988 [13]. Since no detection was available from previous satellite observations, GS 1826-238 was tentatively reported as transient. Similarity with Cyg X-1 and GX 339-4 in the low (hard) state suggested to consider the source as a black hole candidate [12]. The GINGA spectrum was in fact well fitted by a single power law with photon index ~1.7.
In 1989 the source was detected by TTM on March 17 [6] at a flux of about 32 mCrab (2-28 keV).
Later on the source was observed by the ROSAT PSPC (October 1990, June and October 1992, see [1]) and no X-ray bursts were detected during 8 hours of net exposure time. The spectrum was well fitted by a power law with a photon index in the range 1.5-1.8 and an absorption $N_H = 5\times10^{21}$ H cm^{-2}. Follow-up optical studies led to the identification of a V=19.3 optical counterpart.

CP510, *The Fifth Compton Symposium,* edited by M. L. McConnell and J. M. Ryan

The ROSAT source was inside the GINGA error box and a larger one of an unidentified X-ray burster [16] observed with OSO-8 [4], OSO 7 Catalogue [10], and containing the source 4U1831-23 [5], which is also present in the ARIEL V catalogue [17].
During November 1994 the source was detected with OSSE at 7.5 standard deviations in the 60-200 keV range [11].
The BeppoSAX WFC discovery of type-I bursting activity from GS 1826-238 since August 1996 [3],[15]) solved the question about the source nature, clearly pointing to a neutron star Low Mass Binary (LMXB) scenario. Moreover its very stable X-ray emission over more than 3 years suggested the authors to classify GS 1826-238 as a weak persistent source rather than a transient. The hypothesis of a neutron star LMXB system was previously discussed by Strickman et al. (1996) and Barret et al. (1996).

OBSERVATION AND DATA ANALYSIS

The Wide Field Cameras (Jager et al., 1997) on board BeppoSAX are designed for performing spatially resolved simultaneous monitoring of X-ray sources in crowded fields enabling studies of spectral variability at high time resolution. The mCrab sensitivity in 2-28 keV over a large (40°×40° deg.) field of view (FOV) and the near-to-continuos operation over a period of years offers for the first time the unique opportunity to measure continuum emission as well as bursting behavior from new as well as already known (weak) transients. For this reason the Galactic Bulge has been monitored over 1 to 2 months during each of the visibility periods since the beginning of the BeppoSAX operational life in July 1996. During those observations, amounting to a total of ~2.1 Ms live time [14], at least 45 sources and about 700 bursts have been detected.

The sky region containing GS1826-238 was monitored since August 20th 1996. Whenever observed with the WFCs the source shows an average 2-28 keV persistent emission of 31 mCrab, varying from 22 to 39 mCrab. This corresponds to an average flux of 1.1×10^{-9} erg cm^{-2} s^{-1} in the 2-28 keV range. The long term 2-10 keV monitoring performed with the ASM experiment on board RXTE confirms the WFC results.

An extensive search for X-ray bursts from GS 1826-238 was performed on the whole available data set, thus leading to the identification of 78 events in the period August 1996 - April 1998. A preliminary time analysis of the bursts occurrence time lead to the discovery that the wait time between the events is almost constant, being ~5.75 hours with a spread of only 38 minute FWHM [15]. Regular type-I bursting was already found in other sources, e.g. 1658-298, 1820-303 and 1323-619 [8]. Nevertheless, the stability of the clock for over 3 years is a unique feature of GS1826-238.

The burst recurrence was investigated in more detail by analyzing the 1996, 1997 and 1998 data separately, to search for possible differences in the clock on one year time scale. The results, which are plotted in Figure 1, show some interesting features.

First of all, the spread of the wait times in 1997 and 1998 is much less (only ~5 minutes FWHM) than the one obtained by Ubertini et al. 1999 on the total 3 years data set (~38 minutes FWHM, see Table 1). Besides that, the wait times decreased significantly in 1998 with respect to 1997 and 1996. Despite the narrow distributions in 1997 and 1998, the wait times distribution in 1996 is very broad and, together with the period decrease in 1998, accounts for the much larger FWHM observed on the 3 years time span.

Such regular bursting behavior on a wide time interval imposes tight constraints on the parameters of the burst engine, and so the standard model of type-I bursts needs to be checked to explain thc features of this almost unique binary system.

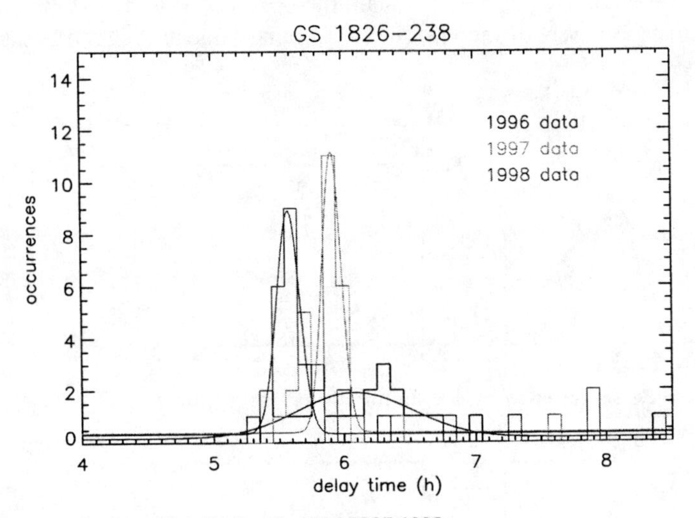

FIGURE 1. Burst wait times distribution in 1996,1997,1998.

Table 1. Burst wait times and persistent emission distributions in 1996,1997,1998

	1996	1997	1998
Recurrence time (h)	6.08 ± 0.14	5.92 ± 0.01	5.58 ± 0.01
Dispersion (h, 1 sigma)	0.44 ± 0.15	0.07 ± 0.01	0.09 ± 0.01
WFC 2-28 keV average intensity (mCrab)	30.3 ± 2.2	31.4 ± 2.3	31.9 ± 2.0
Dispersion (1 sigma)	5.5 mCrab	4.5 mCrab	2.9 mCrab
ASM 2-10 keV average intensity (mCrab)	24.6 ± 0.2	28.1 ± 0.3	27.8 ± 0.5
Dispersion (1 sigma)	9.2 mCrab	12.0 mCrab	11.9 mCrab

DISCUSSION

Despite its X-ray variability, GS 1826-238 is the first example of clocked burst activity for more than 3 years. This, in terms of standard burst model, points to a very stable matter accretion for a long time.

It is a well established statement that the X-ray luminosity is correlated to the accretion rate on the compact object in the binary system. So, under the hypothesis that the burst energetics did not vary, we investigated if slight changes in the source persistent emission influenced the variation of the recurrence period or the spread of the burst wait times distribution. In particular, one could expect increased accretion, and so higher average X-ray intensity, in 1998 when the burst period decreased from 5.92 h to 5.58 h. Moreover, less stable accretion, and so higher X-ray variability, could explain the observed 1996 wait time distribution. A preliminary analysis of the GS 1826-238 bursts fluences supports the assumption made.

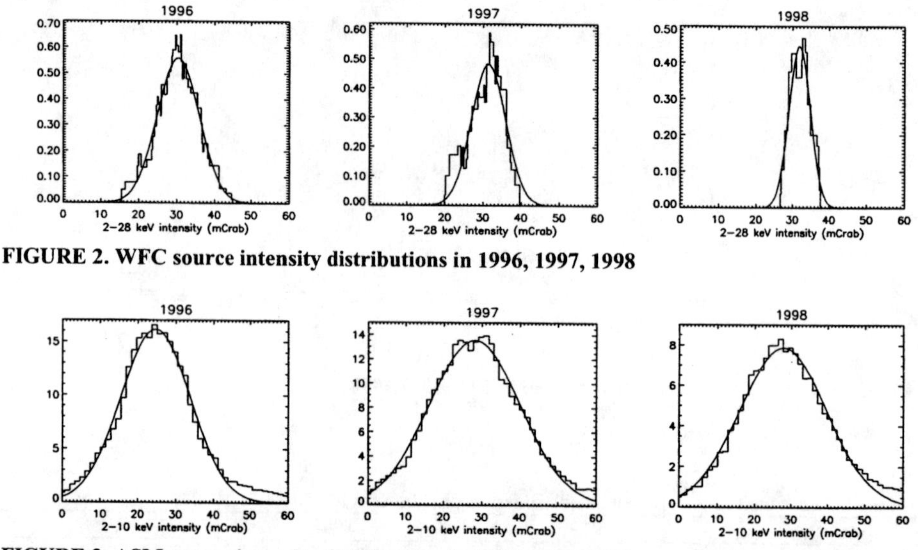

FIGURE 2. WFC source intensity distributions in 1996, 1997, 1998

FIGURE 3. ASM source intensity distributions in 1996, 1997, 1998

We plotted (Figure 2 and 3) the distribution of the source X-ray intensities measured by BeppoSAX WFC in 2-28 keV and by RXTE-ASM (2-10 keV) in 1996, 1997 and 1998 respectively. We observe no evident correlation between the burst wait times, which varied by ~8% in 1 year, and the source persistent intensity. There is also no evident correlation between the "dispersions" of the wait times and the X-ray intensity, but perhaps only marginal evidence in the 1996

WFC data, possibly implying an energy dependence effect which needs further investigation.

If the wait times are strictly related with the accretion rate on the neutron star, M_{dot} variation is expected in 1997-1998. This is not apparent from the measured persistent X-ray intensity, which is almost constant. One could deduce that the X-ray source intensity is only a rough indicator of the mass transfer rate.

Further analysis is needed to better understand the peculiar features of GS 1826-238. For example, all WFC 1999 data (25% of the total available) need still to be included in the burst wait times analysis. Moreover, more accurate spectral data on a wide energy band of both the bursts and the persistent emission could better explain the source behavior. To this end, a new wide band BeppoSAX NFI observation campaign is planned within October 1999.

ACKNOWLEDGEMENTS

We also thank Team Members of the BeppoSAX Science Operation Center and Science Data Center for their continuos support and timely actions for quasi "real time" detection of new transient and bursting sources and the follow-up TOO observations.

REFERENCES

[1] Barret D., Motch C., and Pietch W., 1995, *A&A*, **303**, 526
[2] Barret D., Mc Clintock J.E., and Grindlay J.E., 1996, *ApJ*, **473**, 963
[3] Bazzano A., et al. 1997, AIP Conf. Proc. N. **410**, 729
[4] Becker R.H., et al. 1976, IAU circ. 2953
[5] Forman W. et al. 1978, *ApJ Suppl.*, **38**, 357
[6] in 't Zand J.J.M., et al. 1989, 23rd ESLAB Symposium, 693
[7] Jager R., et al. 1997, *A&A*, **125**, 557
[8] Lewin W.H.G., et al. 1995, in "*X-ray Binaries*" ed. W.H.G. Lewin, J. van Paradijs & E.P.J. van den Heuvel, Cambridge University Press, 175
[9] Makino F., et al. 1998, IAU circ. 4653
[10] Markert T. H., et al. 1977, *ApJ*, **218**, 801
[11] Strickman M., et al. 1996, *A&A Suppl. Ser.*, **120**, 217
[12] Tanaka Y., 1989, proc. 23rd ESLAB Symp, 3
[13] Tanaka Y. & Lewin, W.G.A., 1995, in "*X-ray Binaries*" ed. W.H.G. Lewin, J. van Paradijs & E.P.J. van den Heuvel, Cambridge University Press, 126
[14] Ubertini P., et al. 1998, proc. 3rd Integral Workshop, Taormina, in press
[15] Ubertini P., et al. 1999, *ApJ*, **514**, L27
[16] van Paradijs J., 1995, in "*X-ray Binaries*" ed. W.H.G. Lewin, J. van Paradijs & E.P.J. van den Heuvel, Cambridge University Press, 536
[17] Warwick R.S., et al. 1981, *MNRAS*, **197**, 865

GRO J2058+42 Observations with BATSE and RXTE

Colleen A. Wilson, Mark H. Finger[†] and D. Matthew Scott[†]

SD 50 Space Science Department
NASA Marshall Space Flight Center
Huntsville, AL 35812
[†] *Universities Space Research Association*

Abstract. GRO J2058+42 is a 196-second pulsar discovered with the Burst and Transient Source Experiment (BATSE) on the Compton Gamma-Ray Observatory (CGRO) during a giant outburst in 1995. It underwent a series of 9 weaker outbursts from 1995 to 1997 that alternated in peak pulsed intensity, with a 110-day cycle in the 20-50 keV band [4]. These outbursts did not show the same intensity variations in the 2-10 keV observations with the Rossi X-ray Timing Explorer's (RXTE) All-Sky Monitor (ASM) [2]. Additional outbursts after this series were observed with BATSE and with the RXTE Proportional Counter Array (PCA) and ASM [3]. These outbursts do not appear to continue the alternating peak pulsed intensity pattern seen with BATSE in the first 9 outbursts. Histories of pulse frequency, pulsed flux, and total flux are presented. Pulse profiles and spectra from PCA observations are also presented.

INTRODUCTION

In the last 30 years, about 70 accretion-powered X-ray pulsars have been detected. More than half of these are transient and are thought to have a Be star companion. Neutron stars with Be companions are believed to accrete material from the slow, dense, stellar outflow confined to the equatorial plane of the Be star. These systems exhibit two types of outbursts: "giant" outbursts accompanied by high luminosities and high spin-up rates and "normal" outbursts which are often modulated at the orbital period and have lower luminosities and spin-up rates. Recent long-term studies with BATSE revealed that giant outbursts are often interspersed with normal outbursts and that an accretion disk is often present in giant outbursts. While the peak of the giant outburst is often delayed in orbital phase with respect to the normal outburst peaks, the onsets of outbursts, regardless of type, are limited to a narrow range in orbital phase [1]. GRO J2058+42 underwent a giant outburst in 1995 where the pulse period changed from 198 seconds to 196 seconds in 46 days. This giant outburst was followed by a series of

CP510, *The Fifth Compton Symposium,* edited by M. L. McConnell and J. M. Ryan

normal outbursts that initially alternated in peak 20-50 keV pulsed intensity with a 110-day cycle [4].

OBSERVATIONS

GRO J2058+42 pulse frequencies were determined from a grid search in frequency and frequency derivative using 4-day intervals of BATSE DISCLA 20-50 keV data. Each 4-day interval was split into segments of 1200 seconds. The data in each segment were fit with an empirical background model and a 6 harmonic Fourier expansion in the pulse phase model. The background model was a quadratic spline fit with a continuous slope and value across segment boundaries, but not across gaps. The pulse phase model initially used a constant frequency and was iteratively improved as frequencies were measured. Excess frequency dependent aperiodic noise, primarily from Cygnus X-1, caused the variances on the Fourier coefficients to be larger than for Poisson statistics. Corrections to Poisson variances were estimated by multiplying the variances on the mean Fourier coefficients for each 4-day interval by the reduced χ^2 of a fit of the mean coefficients for each harmonic to those from the 1200-s segments within each 4-day interval. Then a modified Z_n^2 statistic was computed using the corrected variances. For the BATSE frequency measurements in Figure 1, a grid of 161 frequency offsets (range: ± 2 cycles day^{-1}) from the pulse phase model by 65 frequency derivatives (range: ± 0.32 cycles day^{-2}) was searched using a modified Z_2^2 test. Frequency measurements where $Z_2^2 > 33$ are plotted in Figure 2. Root-mean-squared pulsed fluxes are measured from the best fit pulse profile (with errors corrected for aperiodic noise), assuming an exponential spectrum with an e-folding energy of $kT = 20$ keV.

GRO J2058+42 was observed in two series of 7-8 observations with the RXTE PCA in 1998 January and 1998 March. Unfortunately, the ephemeris we used to predict these observations was not good enough, so both outburst peaks were missed. The background count rate was modeled with *pcabackest*, using the faint source models, and was then subtracted from Standard 2 data for each observation. The times were barycentered and the count rates were normalized to the count rate in 5 proportional counter units (PCU), allowing intervals with different numbers of PCUs to be combined. For each observation, a grid of pulse frequency offsets was searched. The data in each observation were fit with a constant plus a 3 harmonic Fourier expansion in pulse phase model. The pulse phase model used a constant frequency. To estimate the effects of frequency dependent aperiodic noise, Leahy normalized background (pulse subtracted) power spectra were estimated from 2000-s contiguous segments from each observation. If multiple segments were available within an observation, their power spectra were averaged. The average background power, $\bar{P}_n$ for the region $n\nu_0/2$ to $3n\nu_0/2$, where ν_0 is the model frequency and $n = 1, 2, ...$ is the harmonic number, was computed for each harmonic. The variance on the Fourier coefficient for harmonic n was multiplied by $\bar{P}_n/2$ (2 is the expected power for Poisson statistics) to correct for excess frequency dependent aperiodic noise. For the RXTE PCA frequency measurements shown in Figure 1, a grid of 161 frequency offsets (±16 cycles day^{-1}) was searched using a modified Z_3^2 test. Root mean squared pulsed fluxes were computed from the best fit pulse profile for each observation.

A period search of the RXTE ASM[1] data revealed an outburst period of 54.9 days, consistent with [2]. A fit to the BATSE outburst onsets yielded the outburst ephemeris T = MJD 510411.3(5)±55.03(6)N where T is the onset time of outburst N in MJD.

PHASE RESOLVED SPECTRA

Phase resolved spectra were generated using the *fasebin* software with Good Xenon data. Given Good Xenon event files (after *make_se* has been run), a pulsed phase model, RXTE orbit files, and a good time interval file (GTI) specifying when the source was not occulted and 5 PCUs were present, *fasebin* barycentered the data and generated a spectrum for each of 12 phase bins. An average background spectrum and an average Good Xenon spectrum were generated using the same GTI file. The statistical errors in the background file were zeroed and a systematic error of 1% was included to correctly treat the background model errors. The average Good Xenon spectrum was used to generate a response file. Three models were tried in XSPEC, a power law, a thermal Bremsstrahlung, and a power law with a high energy cutoff. The power law with a high energy cutoff was the best fit model. The phase dependence of spectral parameters on 1998 February 5 is shown in Figure 2.

SUMMARY AND DISCUSSION

The odd-even pattern in the BATSE peak pulsed flux, which was present for the first 11 outbursts of GRO J2058+42, [4] has stopped or possibly reversed. All of the outbursts show frequency increase which if attributed mainly to orbital effects, require an orbital period of 55 days rather than 110 days. However, the presence of the odd-even pattern for 11 outbursts is difficult to explain in a 55-day orbit.

Pulse shape differences were found between an odd and an even outburst observed with the RXTE PCA, but insufficient data were available to determine if these differences were intensity dependent or not. The spectrum of GRO J2058+42 appears to vary with pulse phase, with a much softer spectrum near the pulse minimum than near the pulse peak.

REFERENCES

1. Bildsten, L. et al., *ApJS*, **113**, 367 (1997).

1) Results provided by the ASM/RXTE teams at MIT and at the SOF and GOF at GSFC.

2. Corbet, R., Peele, A., and Remillard, R., *IAU Circ*, **6556** (1997).
3. Wilson, C. A., Ph.D. Thesis, University of Alabama in Huntsville (1999).
4. Wilson, C. A. et al., *ApJ*, **499**, 820 (1998).

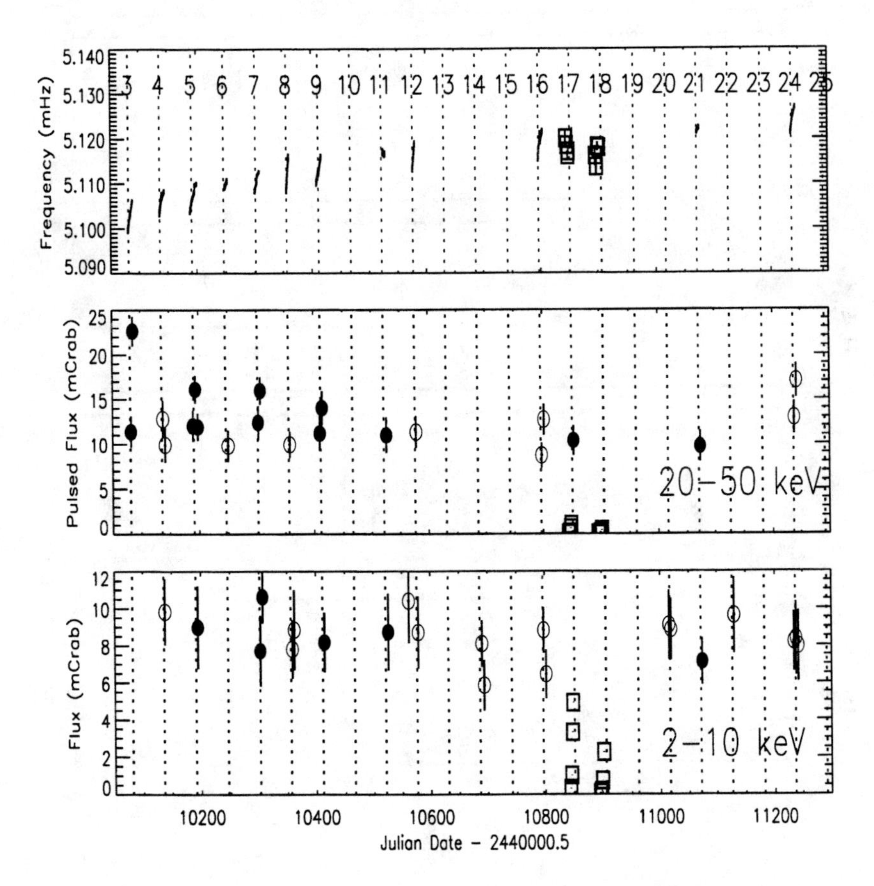

FIGURE 1. The normal outbursts of GRO J2058+42. Outbursts are numbered starting with the giant outburst (not shown) as 1. (*Top*): Pulse frequencies were determined using BATSE data at 4-day intervals. PCA frequency measurements are denoted by squares. (*Center*): Pulsed flux measurements with BATSE in the 20-50 keV band (1 mCrab $\simeq 10^{-11}$ erg cm^{-2} s^{-1}) are denoted odd numbered (filled circles) or even numbered (open circles). PCA pulsed flux measurements in the 2-30 keV band (1mCrab $\simeq$ 14 cts s^{-1}) are denoted by squares. (*Bottom*): Total flux measurements from averaging 4-days of 90-second dwell observations with the RXTE ASM in the 2-10 keV band (1mCrab $\simeq$ 0.075 cts s^{-1}) are denoted odd (filled circles) or even (open circles). Only $\geq 4\sigma$ measurements are shown. PCA average fluxes are also shown as squares.

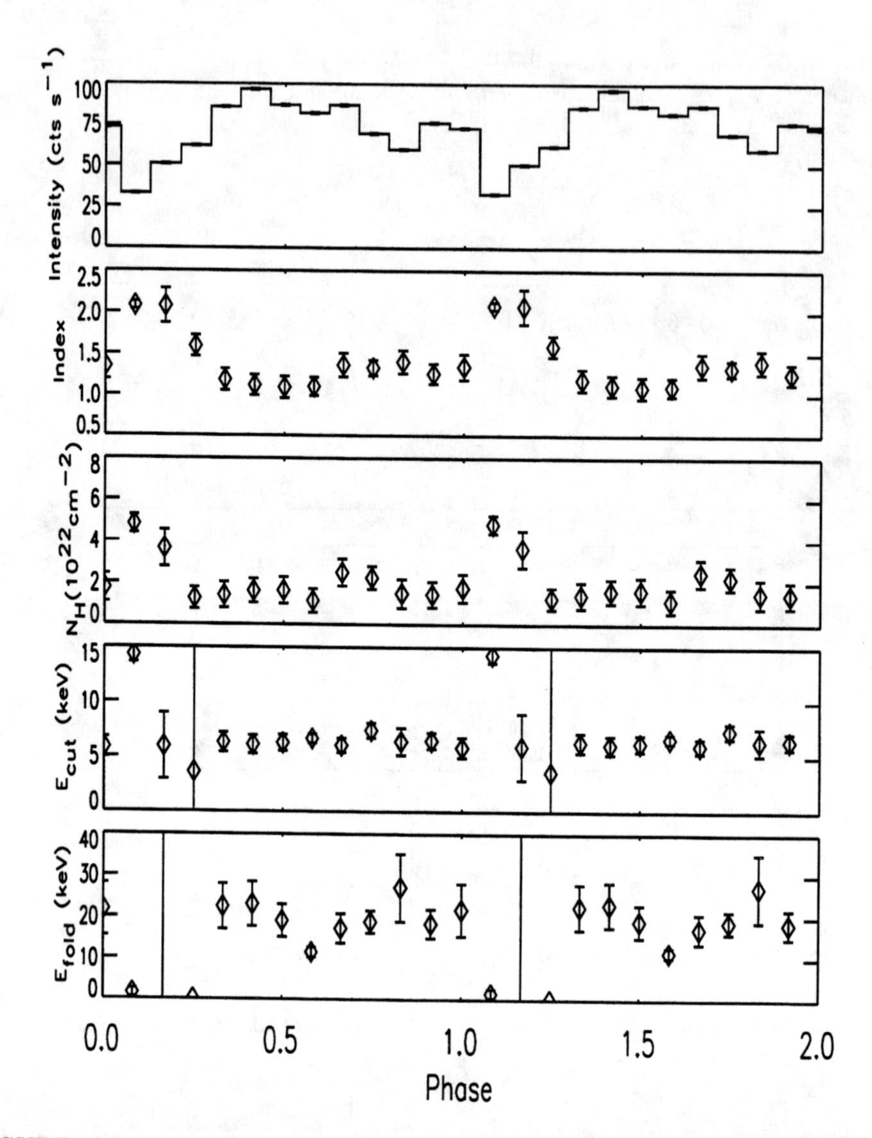

FIGURE 2. The pulse phase dependence of the February 5, 1998, PCA spectrum. The photon spectra in each of 12 phase bins were best fitted with a power law with low energy absorption and a high energy cutoff. The top panel shows the 2-30 keV intensity in each of the 12 bins. The photon index (2nd panel) and the absorption (3rd panel) showed significant pulse phase dependence. The cutoff (4th panel) and folding energies (5th panel) were not as well determined, so their phase dependence is less clear.

BeppoSAX Observation of the X–ray Pulsar 4U 1538–52

Natale R. Robba[1], Tiziana Di Salvo[1], Luciano Burderi[2], Antonino La Barbera[1] and Giancarlo Cusumano[3]

[1]*Dipartimento di Scienze Fisiche ed Astronomiche, Via Archirafi 36, 90123 Palermo, Italy.*
[2]*Osservatorio di Monteporzio, Roma, Italy.*
[3]*Istituto di Fisica Cosmica con Applicazioni all'Informatica, CNR, Palermo 90146, Italy*

Abstract. We report preliminary results of the temporal and spectral analysis performed on the X–ray pulsar 4U1538–52 out of eclipse observed by BeppoSAX. We obtain a new estimate of the pulse period of the neutron star $P = 528.24 \pm 0.01$ s (corrected for the orbital motion of the X–ray source): the source is still in the spin-up state, as since 1988. The broad band (0.12–100 keV) spectral analysys shows the presence of an absorption feature at $\sim$ 21 keV is present, interpreted as due to cyclotron resonant scattering. Another absorption feature at $\sim$ 51 keV seems also to be present (at 99% confidence level). If confirmed this might be interpreted as a cyclotron line, but its energy is not compatible with being double than the energy of the first line.

INTRODUCTION

4U 1538–52 is a wind–fed X–ray binary system formed by a massive B0 star and a neutron star spinning with a period of about 529 s ([1]; [2]). The X-ray luminosity has been estimated $\sim 2\ 10^{36}$ erg/sec for a distance of ≈ 6.4 kpc ([2]). The orbit, in an almost edge-on plane, is characterized by a period of ≈ 3.75 $days$, and a well defined X-ray eclipse lasting ≈ 0.6 $days$ ([1]; [2]). The pulse period measurements show that the neutron star was in a spin-down state before 1988 with a $|\dot{P}/P| \sim 10^{-11}$ s^{-1} and in a spin-up state with the same $|\dot{P}/P|$ after 1988 ([3]).

Before *Ginga* observations the X-ray spectrum has been well modeled by a power-law modified by a high-energy exponential cutoff and a weak iron line emission at 6.7 *keV* (*EW* $\sim$ 100 *eV*). A phase-dependent absorption feature, in the X-ray spectrum, around 20 *keV* observed by *Ginga* ([4], [5]) is explained as a cyclotron resonance absorption. A cut-off in the spectrum above about 30 *keV* was interpreted as the second harmonic but the *Ginga* energy range precludes any definitive conclusion.

CP510, *The Fifth Compton Symposium,* edited by M. L. McConnell and J. M. Ryan

OBSERVATIONS AND ANALYSIS

BeppoSAX observed 4U 1538–52 with its Narrow Field Instruments (NFI) on 1998 from 29th July to 1st August. The source was in its eclipse state during the first $\sim$ 52 ks and out of eclipse in the last $\sim$ 177 ks. In this paper we concentrate our analysis on data out of eclipse. Adopting a distance of 6.4 kpc ([2]), the source luminosity out of eclipse is $\sim 4.7 \times 10^{36}$ ergs/s in the whole range 0.1 – 100 keV.

We used MECS data, which have a better statistics, to perform a temporal analysis. The arrival times of all the events were reported to the solar system barycentre. Moreover we corrected the arrival times for the orbital motion of the source. We performed a folding search for the best pulse period on these corrected arrival times. The best period obtained was 528.24 ± 0.01 s, demonstrating that the source is still in a spin-up state. Figure 1 shows the pulse profiles during the post-egress phase in different energy bands, namely 0.1–1.8 keV (upper panel), 1.8–10.5 keV (middle panel), 15–100 keV (lower panel). A double peaked pulse profile is present in the 1.8–10 keV band, progressively disappearing at lower energies. Indeed the secondary peak is almost absent in the soft range 0.1–1.8 keV. In the PDS range (15–100 keV) the emission is pulsed without significant variations in shape, although the secondary peak is less pronounced.

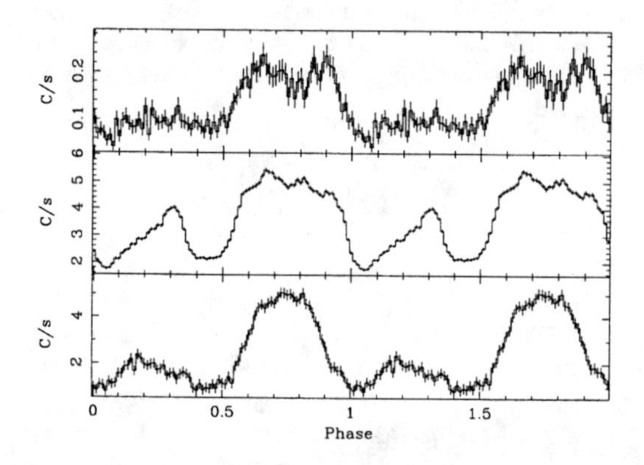

FIGURE 1. Pulse profiles during the post-egress phase in different energy bands, namely 0.1–1.8 keV (LECS, upper panel), 1.8–10.5 keV (MECS, middle panel), 15–100 keV (PDS, lower panel).

We performed spectral analysis on the post-egress energy spectrum of 4U 1538–52 in the energy range 0.12–100 keV. The high energy part of the spectrum (above $\sim$ 10 keV) is not well fitted by a power law modified only by two absorption features as proposed by [4] in the analysis of *Ginga* data. The broader band of the BeppoSAX instruments unambiguously demonstrates that *both* an absorption feature and an exponential cutoff are needed to adequately model the spectrum.

We fit the spectrum with an absorbed power–law continuum modified by a high energy rollover and by an absorption cyclotron line of gaussian shape plus a gaussian iron emission line at ~ 6.4 keV. This model gave a good fit with a $\chi^2/d.o.f. = 733/700$. The best fit parameters are shown in Table 1 (model 1); the observed spectrum and the residuals (in units of σ) are shown in Figure 2, upper and lower panel respectively. With respect to this model a residual is still visible in the PDS range, between 40 and 60 keV. Since this is the range of energy in which we expect to find harmonics of the cyclotron line we added another cyclotron line to the model above. In this way we obtain a reduction of the χ^2 to 721/697. An F–test gives a probability of 99% that the improvement is real. The results with this model are also show in Table 1 (model 2).

TABLE 1. Results of the fit of the pulse averaged spectrum in the energy range 0.1-100 keV. Uncertainties are at 90% confidence level for a single parameter.

Parameter	Model 1	Model 2
$N_H \times 10^{22}\ cm^{-2}$	1.632 ± 0.040	1.636 ± 0.040
Photon index	1.125 ± 0.013	1.127 ± 0.012
E_{cut} (keV)	16.41 ± 0.72	$14.1^{+2.4}_{-0.56}$
E_{fold} (keV)	10.03 ± 0.47	$11.59^{+0.48}_{-0.80}$
$A_{cyc,1}$	0.491 ± 0.038	$0.412^{+0.085}_{-0.027}$
$E_{cyc,1}$ (keV)	21.09 ± 0.21	$21.49^{+0.29}_{-0.44}$
$\sigma_{cyc,1}$ (keV)	3.45 ± 0.27	$2.82^{+0.85}_{-0.42}$
$A_{cyc,2}$	-	$0.86^{+0.13}_{-0.36}$
$E_{cyc,2}$ (keV)	-	$51.87^{+4.2}_{-3.0}$
$\sigma_{cyc,2}$ (keV)	-	$4.4^{+3.5}_{-2.3}$
E_{Fe} (keV)	6.374 ± 0.057	6.374 ± 0.057
σ_{Fe} (keV)	< 0.2	< 0.2
Fe Equiv. Width (eV)	58	58
χ^2/d.o.f.	733/ 700	721/697

DISCUSSION AND CONCLUSIONS

We performed temporal and broad band (0.1–100 keV) spectral analysis on the out-of-eclipse data of 4U 1538–52 observed by BeppoSAX NFIs. We obtained a new measurement of the spin period $P_{spin} = 528.24 \pm 0.01$ s, which indicates that the neutron star is still spinning–up.

The broad band energy spectrum is well fitted by a power–law with high energy cutoff continuum (the typical continuum of the HMXB systems), with low energy absorption by cold matter, an emission line due to fluorescence from iron in low ionization states, and an absorption feature around 20 keV that we interpreted as due to cyclotron scattering (in agreement with [4]). Another absorption feature seems to be present around 50 keV (at 99% confidence level). If confirmed, this feature might be interpreted as the second harmonic of the 20 keV cyclotron line.

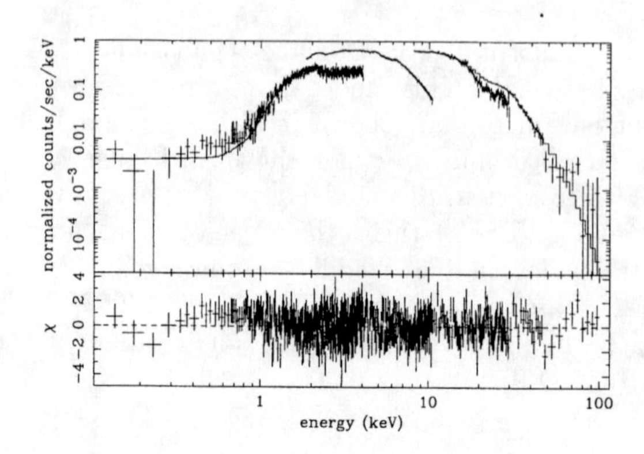

FIGURE 2. Data and model with one absorption line (upper panel) and residuals in units of σ with respect to this model (lower panel).

However the energy of this second feature ($51.9^{+4.2}_{-3.0}$ keV) is not compatible with being double than that of the first line ($21.49^{+0.29}_{-0.44}$ keV). Considering that the optical depth of the fundamental and second harmonic are different we expect that the two lines should form at different heights. In particular the fundamental, with a larger cross section, will form higher in the atmosphere where the magnetic field is weaker. We calculate that a difference in the heights of ~ 0.07 neutron star radii are sufficient to explain the difference in the energy. Another possibility is that we see two different lines coming from the two magnetic poles. A displacement of the magnetic dipole momentum by about 0.15 neutron star radii from the center is sufficient to explain the difference of energy of the two lines produced at the two magnetic poles. In this case we expect a phase dependence of the strength of these two lines. The phase resolved spectral analysis that we performed (whose results we will present elsewhere) are not conclusive as regard the second line (because of the low statistics). Further observations are needed to address the question of the presence of this second high energy absorption feature.

REFERENCES

1. Davison, P.J.N., *MNRAS*, **179**, 35 (1997)
2. Becker, R.H., Swank, J.H., Boldt, E.A., Holt, S.S., Pravdo, S.H., Saba, J.R., and Serlemitsos, P.J., *ApJ*, **216**, L11 (1977)
3. Rubin, B.C., Finger, M.H., Scott, D.M., Wilson, R.B., *ApJ*, **488**, 413 (1997)
4. Clark, G. W., Woo, J. W., Nagase, F., Makishima, K., Sakao, T., *ApJ*, **353**, 274 (1990)
5. Mihara, T., *Ph. D. thesis*, Tokyo University (1995)
6. Brainerd, J. J., Meszaros, P., *ApJ*, **369**, 179 (1991)

Bright X-ray bursts from 1E1724-3045 in Terzan 2

M. Cocchi[(1)], A. Bazzano[(1)], L. Natalucci[(1)] and P. Ubertini[(1)]
J. Heise[(1)], E. Kuulkers[(1)] and J. J. M. in 't Zand[(1)]

[(1)]*Istituto di Astrofisica Spaziale), via Fosso del Cavaliere, 00133 Roma, Italy*
[(2)]*Space Research Organisation Netherlands (SRON), Sorbonnelaan 2, 3584 CA, Utrecht, the Netherlands*

ABSTRACT. During about 3 years wide field monitoring of the Galactic Centre region by the WFC telescopes on board the *BeppoSAX* satellite, a total of 14 type-I X-ray bursts were detected from the burster 1E 1724-3045 located in the globular cluster Terzan 2. All the observed events showed evidence of photospheric radius expansion due to Eddington-limit burst luminosity, thus leading to an estimate of the source distance (~7.2 kpc). Preliminary results of the analysis of the bursts are presented.

INTRODUCTION

Since its discovery [14], the globular cluster source 1E 1724-3045, has been repeatedly studied by several satellite experiments both in the "classical" (~1-20 keV) X-ray energy band (e.g. *EINSTEIN, EXOSAT, TTM, ROSAT, BeppoSAX, RXTE, ASCA*) and in the hard X-rays (*GRANAT, BeppoSAX, RXTE*).

The source is persistently bright, though variable, and is one of the few X-ray bursters showing persistent emission up to the soft Gamma-rays [1],[17],[16].

Type-I X-ray bursts from 1E 1724-3045 were observed since the very first observations performed with OSO-8 [14]. In particular, *EINSTEIN* detected a bright burst showing photospheric radius expansion [7],[15], allowing an estimate of the source distance (~7 kpc) which is consistent with what is obtained taking into account the measured reddening of the globular cluster Terzan 2 [13].

Bursting activity identifies the source as a weakly magnetized neutron star in a low mass binary (*LMXB*) system.

Soft X-ray observations have shown that the persistent spectrum is best-fitted by a power law of photon index ~2.0-2.4 [9],[12],[18]. A higher spectral index (~3.0) is obtained in the hard X-rays by *SIGMA* [5],[6]. More recently, simultaneous *SAX* and *RXTE* observations demonstrated that the wide band (1-200 keV) spectrum of 1E 1724-3045 is actually power law below ~30 keV, attenuated at high energies by an exponential cutoff at ~70 keV [8],[2]. This is interpreted as the result of the Comptonization of soft photons in a spherical scattering region of electron temperature ~30 keV and optical depth ~3. Besides that, the *RXTE* and *SAX*

CP510, *The Fifth Compton Symposium,* edited by M. L. McConnell and J. M. Ryan

observations suggested the presence of an additional soft component, most likely a multicolor disk blackbody, as confirmed by recent *ASCA* results [3].

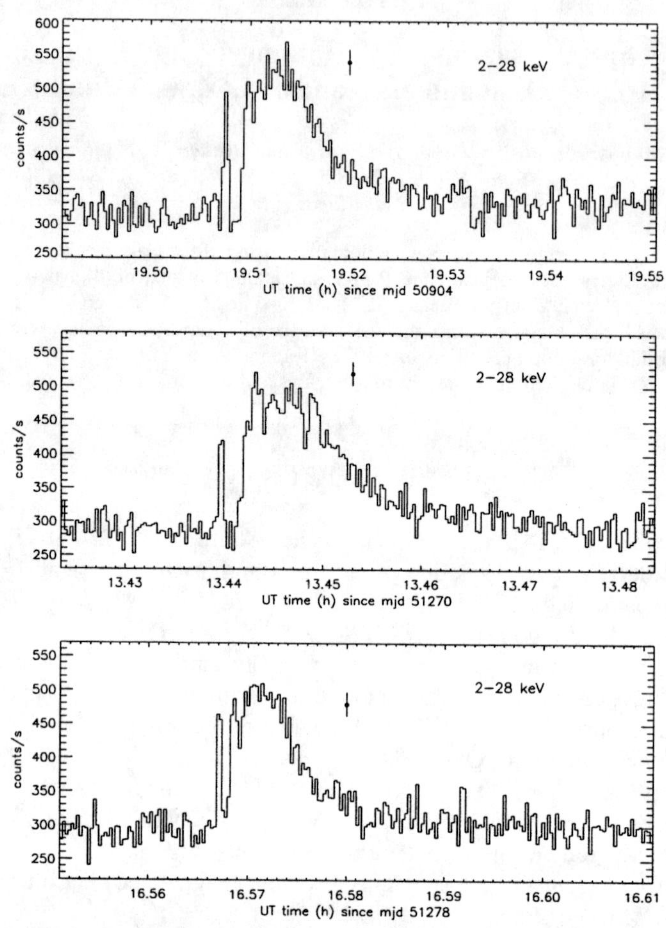

FIGURE 1. 2-28 keV time history of three different bursts of 1E 1724-3045.

OBSERVATION AND DATA ANALYSIS

The *Wide Field Cameras (WFC)* on board *BeppoSAX* satellite consist of two identical coded mask telescopes [10]. The two cameras point at opposite directions each covering 40°×40° field of view. With their source location accuracy in the range 1'-3', a time resolution of 0.244 ms, and an energy resolution of 18% at 6 keV, the *WFC*s are very effective in studying hard X-ray (2-28 keV) transient phenomena.

The imaging capability, combined with the good instrument sensitivity (5-10 mCrab in 10^4s), allows accurate monitoring of complex sky regions like the Galactic Bulge.

The data of the two cameras is systematically searched for bursts or flares by analyzing the time profiles of the detector in the 2-11 keV energy range with a 1s time resolution. Reconstructed sky images are generated in coincidence with any statistically meaningful enhancement, to identify possible bursters. The accuracy of the reconstructed position, which of course depends on the intensity of the burst, is typically better than 5'. This analysis procedure allowed for the identification of ~700 X-ray bursts in a total of about 2 Ms WFC net observing time (see e.g. [4]).

Among the detected bursts, a total of 14 was observed from a sky position consistent with that of Terzan 2.

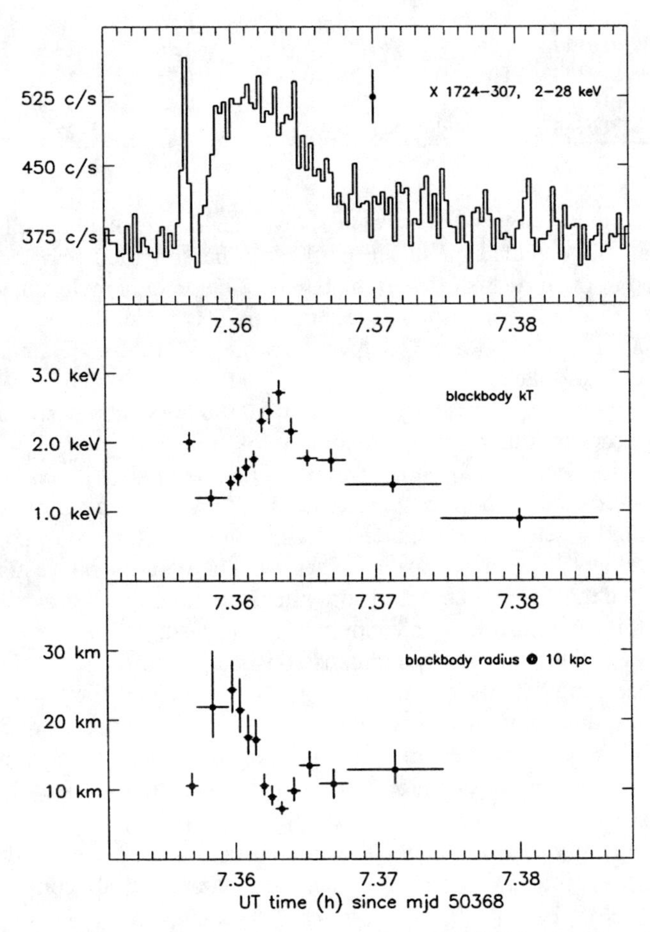

FIGURE 2. Results of time resolved spectroscopy on the mjd 50368 burst. Upper panel: 2-28 keV time history of the burst; Middle panel: time history of the measured blackbody temperature; Lower panel: time history of the blackbody radius, assuming a 10 kpc distance.

TABLE 1. Summary of the results of the time resolved spectral analysis of the mjd 50368 burst (assumed $N_H = 10^{22}$ cm^{-2}).

Data set	Time range (T0= 7.35648 UT)	blackbody kT (keV)	χ2R (27 d.o.f.)	R (km) @10 kpc
Whole burst	T0 – T0+68 s	1.80 ± 0.04	1.14	11.2 ± 0.5
A	T0 – T0+3 s	2.01 ± 0.14	0.65	10.6 +1.8 / -1.4
B	T0+3 s – T0+11 s	1.20 ± 0.12	1.20	21.9 +8.1 / -4.4
C	T0+11 s – T0+13 s	1.42 ± 0.10	0.73	24.3 +4.2 / - 3.3
D	T0+13 s – T0+15 s	1.51 ± 0.13	1.11	21.4 +4.5 /- 3.2
E	T0+15 s – T0+17 s	1.64 ± 0.12	1.08	17.5 +3.2 / -2.4
F	T0 +17s – T0+19 s	1.76 ± 0.12	1.68	17.2 +2.9 / -2.2
G	T0+19 s – T0+21 s	2.31 ± 0.16	1.01	10.6 +1.7 / -1.3
H	T0+21 s – T0+23 s	2.45 ± 0.19	0.94	9.0 +1.6 / -1.2
I	T0+23 s – T0+26 s	2.72 ± 0.17	0.96	7.3 +1.0 / -0.8
J	T0+26 s – T0+29 s	2.16 ± 0.18	0.91	9.8 +1.9 / -1.4
K	T0+29 s – T0+34 s	1.77 ± 0.11	1.36	13.5 +2.0 / -1.6
L	T0+34 s – T0+41 s	1.74 ± 0.16	0.75	10.9 +2.1 / -2.1
M	T0+41 s – T0+65 s	1.39 ± 0.12	0.65	12.9 +2.8 / -2.0
N	T0+65 s – T0+95 s	0.90 ± 0.14	1.09	22.6 +8.1 / -8.1

A preliminary analysis of the burst data showed evidence of photospheric radius expansion on all the detected bursts. In particular, the so-called *precursor event* is always observed in the time histories of the bursts. The precursors last in average 1-2 s, and are followed a few seconds later (~5 s in average) by the main burst. In Figure 1 and Figure 2 a sample of 4 bursts is shown; the time profiles are obtained in the full WFC energy band (2-28 keV).

Detailed time resolved spectroscopy of the mjd 50368 burst was performed (Figure 2, Table 1), in order to study the time evolution of relevant burst parameters. To better constrain the fits, the N_H parameter was kept fixed to 10^{22} cm^{-2}. The burst spectra are all consistent with an absorbed blackbody model, typical of type-I X-ray bursts, originating by helium thermonuclear flashes onto the surface of a neutron star (see [11] for a review). The time history of the blackbody temperature also points to type-I bursts, since the observed spectral softening is commonly regarded as the result of the cooling of the neutron star photosphere after the flash.

The time history of the blackbody radius is consistent with an adiabatic expansion of the photosphere, most likely due to Eddington luminosity of the burst; the data is consistent with a radius expansion of a factor ~2.5 just after the precursor event. After the subsequent contraction of the emitting region, the burst behaves as a more typical type-I, showing the color softening and an almost constant blackbody radius. The burst decay is exponential with a characteristic time of 24.1 ± 1.7 s.

Eddington-luminosity X-ray bursts allow for an estimate of the source distance. The 2-28 keV peak intensities of the 4 analyzed bursts are all consistent with a weighted average of 1.01 ± 0.03 Crab, which extrapolates to a bolometric intensity of $(3.23 \pm 0.10) \times 10^{-8}$ erg cm^{-2} s^{-1}. The consistency of the peak intensities of all the analyzed bursts with a constant value supports the adoption of the peak bolometric luminosities of super-Eddington bursts as a standard candle. Assuming a 2×10^{38} erg s^{-1} Eddington luminosity for a 1.4 M_0 neutron star and using standar burst parameters, we obtain for 1E 1724-307 a distance value d = 7.2 ± 0.2 kpc in agreement with previous results [15].

From the average blackbody radius of the cooling track and for the estimated distance of 7.2 kpc we derive an average radius of ~7 km for the blackbody emitting region during the burst. On the other hand, if we assume the average luminosity of Eddington-limit bursts proposed by Lewin, van Paradijs and Taam [11], i.e. (3.0 ± 0.6) $\times10^{38}$ erg s^{-1}, we obtain a slightly larger distance, d = 8.8 ± 1.0 kpc.

REFERENCES

[1] Barret, D., et al. 1991, *ApJ* **379**, L21
[2] Barret, D., et al. 1999, in press
[3] Barret , D., Grindlay, J.E., Harrus, I.M., and Olive, J.F. 1999, *A&A* **341**, 789
[4] Cocchi, M., et al. 1998, *Nucl.Phys. B* **69**/1-3, 232
[5] Goldwurm, A., et al. 1993, proc. 2nd Compton symp., *AIP* **304**, 421
[6] Goldwurm, A., et al. 1994, *Nature* **371**, 589
[7] Grindlay, J.E., et al. 1980, *ApJ* **240**, L121
[8] Guainazzi et al., 1998, *A&A* **339**, 802
[9] In 't Zand, J.J.M., 1992, PhD Thesis, University of Utrecht
[10] Jager R., et al. 1997, *A&A* **125**, 557
[11] Lewin, W.H.G., van Paradijs, J., and Taam, R.E. 1995, in *"X-ray Binaries"*, ed. W. Lewin, J. van Paradijs & E. van den Heuvel, Cambridge University Press, Cambridge, p. 175
[12] Mereghetti, S., et al. 1995, *A&A* **302**, 713
[13] Ortolani, S., Bica, E., and Barbuy, B., 1997, *A&A* **326**, 620
[14] Swank, J., et al. 1977 *ApJ* **212**, L73
[15] Tanaka, Y., 1981, IAU Symp. No 125, 161
[16] Tavani, M., and Barret, D., 1997, proc. 4th Compton symp., *AIP* **410**, 75
[17] Vargas, M., et al. 1996, proc. 2nd INTEGRAL Workshop, ESA SP-**382**, 129
[18] Verbunt, F., et al. 1995, *A&A* **300**, 732

Detection of Multiple Hard X-ray Flares from Sco X-1 with OSSE

Mark Strickman

Naval Research Laboratory, Washington DC, USA

Didier Barret

CESR, Toulouse, FRANCE

Abstract. CGRO/OSSE observations of the bright LMXB Sco X-1 have resulted in the discovery of transient episodes of hard X-ray emission extending to energies above 200 keV. During these periods, the emission spectrum above 80 keV deviates significantly from the extrapolation of the soft thermal spectrum observed at lower X-ray energies. The intervals of hard emission last ~4-days with a duty cycle ~30%. Using RXTE/ASM data from the same intervals to establish the state of Sco X-1 on the color-color diagram "Z" is in progress, but preliminary results indicate that the hard X-ray episodes occur over a limited range of mass accretion rates.

OBSERVATIONS

The luminous low mass X-ray binary (LMXB) Sco X-1 is the brightest celestial X-ray source yet observed. However, it is generally not considered a source of hard X-rays. While hard X-ray tails have been observed from the lower-luminosity LMXB bursters (1), and early, mostly unconfirmed reports have been made of hard emission from a few bright LMXB, e.g. (2-4), conventional wisdom has it that hardness is inversely correlated with luminosity across the range of LMXB (1,5).

In order to further test this correlation we have observed Sco X-1 on two occasions, with a third observation pending, using the OSSE instrument on board the Compton Gamma-Ray Observatory. The existing observations took place during 20 September – 27 September 1995 (CGRO viewing period 429) and 5 January – 2 February 1999 (CGRO viewing periods 807.5, 806.5 and 806.7). At this time, an observation is pending for 26 October – 16 November 1999 (CGRO viewing period 832). In each case, the OSSE background positions have been placed such that no significant confusion with know sources or diffuse emission regions occurs. During the 1995 observation, contemporaneous BATSE LAD occultation data have been used to extend the spectra down to 20 keV. During the 1999 observation, in addition to the BATSE data, we have data from RXTE/ASM and the Green Bank Interferometer, each sampling ~2-6 times per day.

CP510, *The Fifth Compton Symposium,* edited by M. L. McConnell and J. M. Ryan

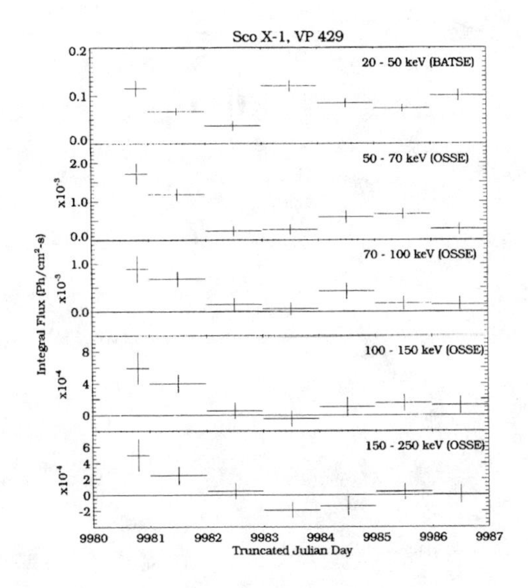

FIGURE 1. OSSE and BATSE broad-band light curves for 1995 observation (VP 429).

RESULTS

For all observations, we integrated the OSSE and BATSE data into 1-day intervals for analysis. Shorter intervals were tried but did not supply any new information due to poorer counting statistics. The results of the 1995 observation, in the form of broad band lightcurves, are shown in Figure 1. We detected the tail of the thermal emission in the 20-50 keV BATSE and 50-70 keV OSSE bands. However, during the first two days of the observation, we detected significant emission above 70 keV. The photon spectrum, well-described by an $E^{-2.5}$ power-law model above 70 keV and not consistent with an extension of the lower energy thermal emission, was observed to ~250 keV.

Although exciting, the 1995 result yields no information on the frequency or duration of these events. The much longer observation in Jan 1999 has addressed these concerns, as shown by the lightcurves in Figure 2. OSSE observes two complete hard X-ray flaring episodes, each about four days long, with a 12 day separation. Based on this observation interval, the flaring "duty cycle" is ~30%. These flares were detected to approximately 250 keV, with limited evidence of emission to 500 keV.

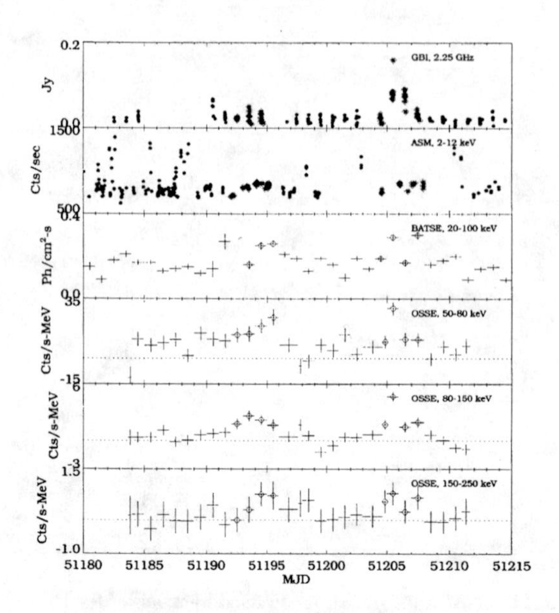

FIGURE 2. Multiwavelength lightcurves for the Jan 1999 observation (VP 807.5/806.5/806.7). Diamond symbols delineate the hard X-ray flaring episodes.

During day 51205 (the second day of the second episode), a period of especially high emission occurs across all wavebands with data for that day. In particular, GBI observes a flare at 2.25 GHz in a single sample. GBI cannot detect Sco X-1 in its normal, non-flaring state, so the detections before and after the flare (during 51205 and 51206) are also significant. No similar event was observed during the first hard X-ray episode, although, not knowing the duration of the radio event during the second interval, it is possible that the radio sampling times missed the flare. Similarly, the lack of detection of a narrow flare by ASM is not significant due to a lack samples on 51205. We examined OSSE data on a 6-hour timescale and saw no evidence of variability during 51205. On a broader timescale, we see no evidence of correlation between hard X-ray events and features in the ASM lightcurve. However, the hard X-ray episodes do both appear to take place during times of quiescent X-ray emission.

In order to study spectral properties of the hard X-ray emission from Sco X-1, we fit BATSE and OSSE spectra together, using a thin thermal Bremsstrahlung model ("thermal component") for the non-flare days, and the same model plus a power law ("hard component") for the flare days. Note that the latter is a convenient representation of the spectral shape, but is not unique. In particular, the hard component may be thermal as well. A representative flare-day spectrum is shown in Figure 3.

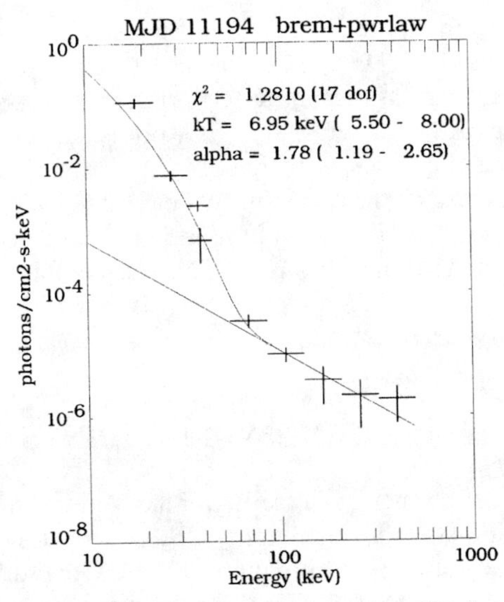

FIGURE 3. Sample OSSE (above 50 keV) and BATSE (below 50 keV) spectrum during a hard X-ray flare episode.

The shape of the thermal component does not appear to correlate with the presence of absence of the hard emission. In particular, the mean value of kT for the flaring and non-flaring days differs by less than 2 standard deviations. Overall, though, kT does exhibit significant variability. All flare days are consistent with a constant power law index of –2.6. These parameters are shown as a function of time in Figure 4.

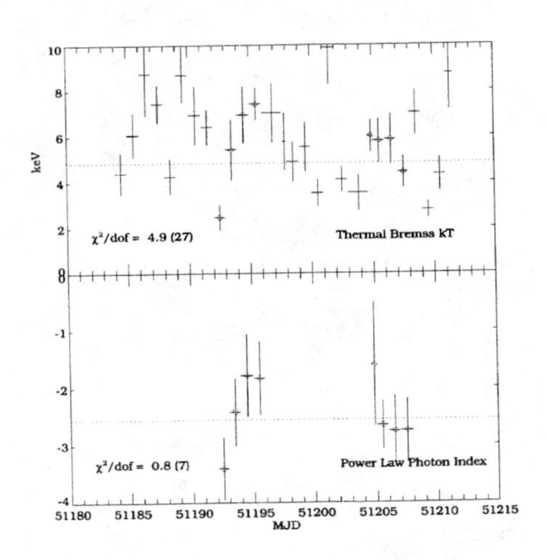

FIGURE 4. Best fit thermal bremsstrahlung temperatures and power law indices vs. time. Diamonds indicate hard X-ray flare days.

In order to determine if the X-ray state of the source was correlated with the hard X-ray flaring, we used RXTE/ASM color data to form a color-color diagram. This procedure has significant systematic effects, not the least of which is disagreement between the various ASM cameras, and the work is still very preliminary. However, we do appear to see clustering of the hard X-ray flaring times in a limited part of the characteristic "Z" traced on the diagram. We cannot tell whether these times are in the normal or flaring branch of the "Z", but it seems likely that they are near the normal-flaring branch junction. If the clustering of hard X-ray flaring on the color-color chart holds up with further analysis, the implication is that hard X-ray flares occur over a limited range of mass accretion rate.

ACKNOWLEDGMENTS

The Green Bank Interferometer is a facility of the National Science Foundation operated by the National Radio Astronomy Observatory. It is operated in support of NASA High Energy Astrophysics programs. ASM/RXTE data come from both quick look data provided by the ASM/RXTE team and data from the NASA/HEASARC archive.

REFERENCES

1. Barret, D and Vedrenne, G., ApJ S 92, 505-510 (1994).
2. Greenhill, J.G. et al., MNRAS 189, 563-570 (1979).
3. Duldig, M.L. et al., Ap and Sp Sci 95, 137-144 (1983).
4. Matt, G. et al., ApJ 355, 468-472 (1990).
5. Van Paradijs, J. and van der Klis, M., A&A 281, L17-L20 (1994).

Correlated Optical and X-ray Emission from Sco X-1

Bernard McNamara, Thomas Harrison, Robert Zavala, Omar Mirales, Diana Olivaras, Javier Galvan, Edward Galvan

New Mexico State University, Las Cruces, New Mexico 88003

Abstract. This paper presents preliminary results of the correlated x-ray and optical behavior of Sco X-1 based on two years of data. The optical data was obtained during 20 nights of observing time using the Cerro Tololo InterAmerican Observatory 1 meter Yale telescope. These measurements were obtained simultaneously with data collected by a BATSE Spectroscopy Detector (SD) on the CGRO. The gain of this SD was adjusted to provide measurements in the energy bandpass 8-16 keV. The intent of our study is to quantify the correlated x-ray and optical behavior of Sco X-1 over a variety of time frames. Here we discuss the data sets to be employed in that study. We also discuss the method used to reduce the BATSE SD x-ray data. Some preliminary simultaneous x-ray and optical light curves are presented and their correlated behavior over time frames from tens to a few hundred seconds is briefly discussed.

INTRODUCTION

Much of what we know about x-ray binaries comes from a study of their correlated high and low energy emission. Differences in the locations of the high and low energy emission sites can be obtained by measuring the time delay between these two signals. The efficiency at which x-rays are reprocessed into optical photons can also be estimated by comparing the energy content in correlated emission events. Long term variations in the emission from x-ray binaries have been used to determine their orbital periods, the geometry of their accretion disks, homogeneity of the mass flow within their disks, and disk precession.

THE OPTICAL AND X-RAY DATA

All of the optical data used in this study was collected using the CTIO Yale 1 meter telescope and the People's Photometer. Differential Johnson B and V measurements were obtained relative to two similar brightness reference stars located within a few arcminutes of Sco X-1. One of these reference stars served as a check star and was monitored on an occassional basis. The primary reference star

CP510, *The Fifth Compton Symposium,* edited by M. L. McConnell and J. M. Ryan

was measured approximately once every 20 minutes. Each optical measurement consisted of a two second integration and had a typical S/N of 50.

DATA REDUCTION

The optical data was reduced using existing software at New Mexico State University. Since the primary comparison star was measured frequently during the course of a night, extinction related corrections were rarely larger than 1 percent. No color corrections were applied since the reference star and Sco X-1 have similar colors. The BATSE X-ray data was reduced using the orbital subtraction technique developed by McNamara et al. (1995). In that method a temporally nearby orbit, called a reference orbit, is identified in the SD data. During these orbits the flux from Sco X-1 is stable as determined from the similar size of the rise and set step fluxes and a visual inspection of light curve. A target Sco X-1 orbit is then subtracted from this reference orbit to remove common background effects. The fluxes obtained in this fashion are then normalized to the rise and set step fluxes reported in the long term Sco X-1 light curve published by McNamara et al. (1998).

EXAMPLES OF THE CORRELATED SD AND OPTICAL SCO X-1 EMISSION

Figures 1 and 2 provide examples of the correlated x-ray and optical emission from Sco X-1. Each data point represents a two second integrated flux measurement. The first figure shows a time period when Sco X-1 was especially active. These observations started about 24,400 seconds after the beginning of TDJ 9512.

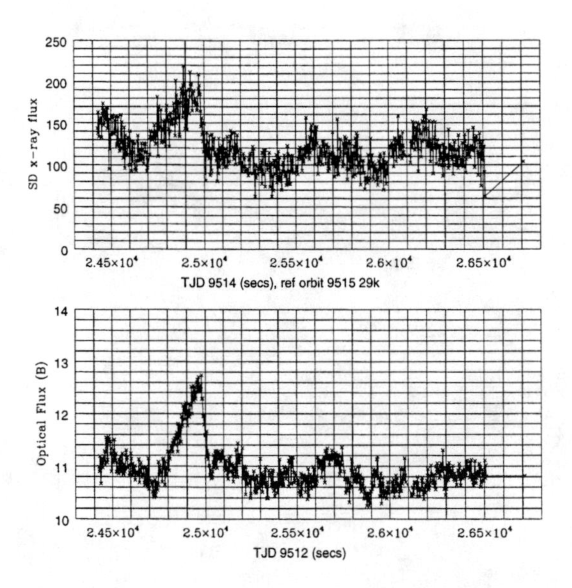

Figure 1: Emission from Sco X-1 during an active state. The top panel is the background subtracted 8-16 keV SD data. The bottom panel shows the simultaneous optical light curve in the Johnson B.

The top panel shows the reduced SD data and the bottom panel shows the simultaneous optical light curve. Note that a 30 second delay exists between the strong x-ray and optical feature centered near t= 24,900 seconds. A second, weaker x-ray event occurs at t= 25,700 seconds. This feature is also seen in the optical light curve although its morphology is different. A third x-ray feature occurred at t=26,200 seconds, but it did not produce an optical counterpart. We suspect that the presence or absence of an optical feature may be related to the x-ray hardness of the event. We intend to examine this hypothesis using the BATSE LAD data.

The second figure shows a comparison between the x-ray and optical emission from Sco X-1 during an inactive period. Unlike the prior active period where the optical emission followed the x-ray emission, the continuum variations are not well correlated. In this figure the x-ray flux is initially constant and then rises in a linear fashion. The optical flux is also initially constant but then decreases. It is clear that the x-ray/optical behavior of Sco X-1 exhibits a variety of behaviors. The simple hypothesis that the optical flux follows the x-ray flux is not always true. The additional simultaneous optical and SD/LAD x-ray data sets we have already collected will be used to examine these correlations more closely.

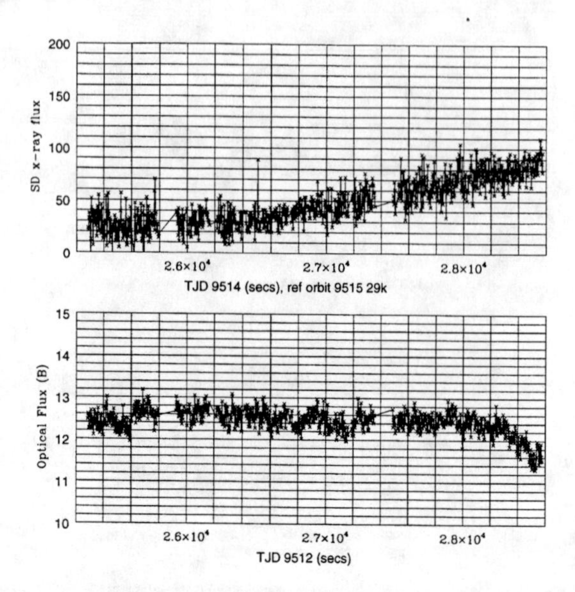

Figure 2: Emission from Sco X-1 in an inactive state. The top panel presents the background subtracted BATSE SD data. The bottom panel shows the simultaneous optical Johnson B light curve.

REFERENCES

1. McNamara, B.J., Harmon, B.A., Harrison, T.E. *Astron and Astrophy.* **111**, 587 (1995).
2. McNamara, B.J. et al. *Ap.J. Suppl* **116**, 287 (1998).

SSS: More photometric observations of RX J0019.8+2156

C. Bartolini*, A. Guarnieri*, G. Iannone*, A. Piccioni* and L. Solmi*

Dipartimento di Atronomia
Università degli Studi di Bologna
40126 Bologna, Italy

Abstract. We present new photometric data of the galactic supersoft X-ray source (SSS) RX J0019.8+2156 taken in Johnson's V band at Loiano Observatory with a two head photoelectric photometer between December 1994 and October 1997. Our observations confirm those already reported by Greiner & Wenzel (1995), Matsumoto (1996), Deufel *et al.* (1998) and Will & Barwig (1996), as far as orbital period, depth of the light curve, and presence of a secondary minimum are concerned. Considering the years 1955-1997 we found no indication for a period change.

The H burning white dwarf model (van den Heuvel *et al.*, 1992) seems to be most capable of explaining the observed phenomena, although the results by Meyer-Hofmeister *et al.* (1998) are of further interest. Also we propose here that the system is seen at a high orbital inclination, high enough to produce partial eclipses of the accretion disk.

INTRODUCTION

Supersoft X-ray sources (SSS) have been first discovered by the Einstein satellite, but most of them were detected by ROSAT in its surveys of the Magellanic Clouds. The spectra of these objects are well described by black-body emission of very low temperature (kT $\approx$ 25-40 eV) and a luminosity close to the Eddington limit (Trümper *et al.*, 1991). Several SSS exhibit X-ray variability. Most of these objects could be binaries in which a white dwarf accretes matter at a sufficiently high rate to allow a more or less stable burning of the accreted material (van den Heuvel *et al.*, 1992; Iben and Tutukov, 1993), but this scenario doesn't seem to be applicable to all SSS, thus suggesting that they do not represent a homogeneous class of objects.

RX J0019.8+2156 was discovered by the ROSAT satellite during its search for galactic SSS in the All-Sky survey and later optically identified with a $m_V = 12.2$ object (Beuermann *et al.*, 1995). This star is a member of our galaxy and its coordinates are characterized by a high galactic latitude (b=-40). Its optical spectrum resembles that of Cal 83 (Cowley *et al.*, 1994), although its X-ray luminosity seems to be lower. RX J0019.8+2156 displays periodical quasi-sinusoidal modulations of

CP510, *The Fifth Compton Symposium,* edited by M. L. McConnell and J. M. Ryan

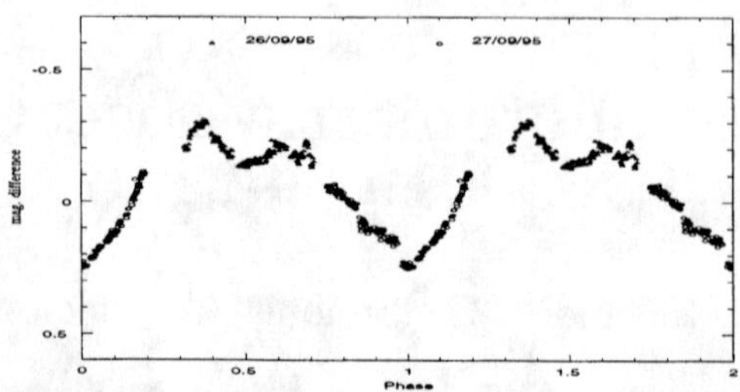

FIGURE 1. The complete V band light curve of RX J0019.8+2156, obtained on 2 nights in September 1995. Note the secondary minimum at $\phi = 0.5$, the irregular variations near top brightness and the asymmetry of the curve about the deep primary minimum.

its optical, UV and X-ray light curves, with a period of $15^h.85$ and $\Delta m \simeq 0.5$ peak-to-peak amplitude (in the optical waveband; amplitudes are smaller at shorter wavelengths). A quasi-periodic pulsation with a period of roughly 2 hours and an amplitude smaller than 0.1 mag was also discovered (Beuermann *et al.*, 1995). Greiner & Wenzel (1995) have analyzed the long term behavior of the object (considering a time span of over 100 years) and conclude that this period has been stable at least over the last 100 years.

I OBSERVATIONS AND RESULTS

All observations were carried out at Loiano Observatory, using the 152 cm telescope. To this Ritchey-Chrétien reflector we applied a two head photometer, described in Piccioni *et al.* (1979). The data obtained are given in Table 1; as can be seen the most reliable and/or useful measurements are those taken during the September/October 1995 series and the October 1997 ones.

The photometer was used only with Johnson's V filter, with a sampling time of 1 second; one head was pointed at the X-ray source, while the other observed a comparison star of $m_v \sim 13$: its year 2000.0 coordinates are $\alpha = 0^h22^m21^s.13$, $\delta = 21°56'57.3''$, putting it some 38' distant from RX J0019.8+2156. The data were reduced using the Stumpff (1980) algorithm. As output of the reduction program we obtained the instrumental magnitudes of each of the two stars, the standard deviation of such magnitude values, the magnitude difference between the 2 stars, and the standard deviation of this difference.

One first comment must be made on the observations of Jan 7-9, 1995. On each of the 3 nights the moon was very close to the observational field (from a minimum distance of 15 to a maximum distance of 27 degrees) and near first quarter; these two factors strongly affected the counts from both stars so we believe the measurements obtained by Bartolini *et al.* (1995) are not reliable.

Using the period, accurately derived by Greiner & Wenzel (1995) from over 100 years of photographic observations of RX J0019.8+2156, in Fig. 1 we have plotted the observations of September 26 and 27. As is apparent, these measurements

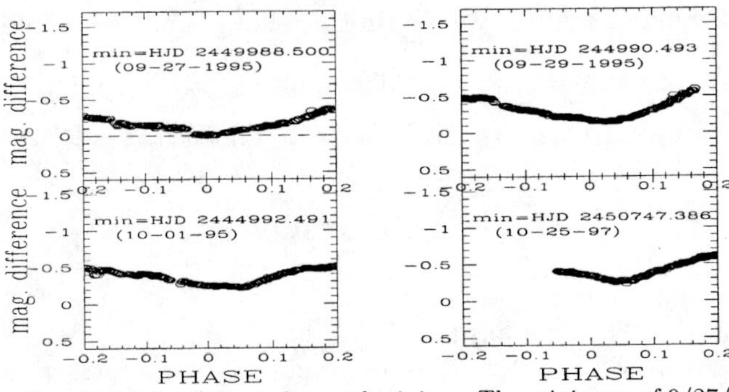

FIGURE 2. Complete runs of the 4 observed minima. The minimum of 9/27/95 was arbitrarily set to $mag = 0$ to emphasize the long term variations of the system.

cover almost an entire orbital cycle of the binary and they show the characteristic quasi-sinusoidal variations of this system, with a peak-to-peak amplitude of ≈ 0.5 mag (same as that given by Greiner & Wenzel). Also visible are the secondary variations of small amplitude ($\lesssim 0.1$ mag) and of a few hours duration. Fairly evident is what seems to be some sort of secondary minimum, also observed by Will & Barwig (1996), Matsumoto (1996) and Deufel *et al.* (1998). In agreement with these last authors we also observed that the maximum following the secondary minimum, at $\phi = 0.65$, is fainter by 0.1 mag than the preceding one (again, see Fig. 1). The times of the 4 minima we observed (see Fig. 2) are listed in Table 1. These timings are in good agreement with those predicted by Deufel *et al.* (1998) (refer to the O-C values in column 7 of the same table).

Using the 1955 photometric minimum and ephemeris reported by Greiner &

TABLE 1. Log of the observations

Date	phase coverage	σ (mag)	Comments	T_{min} (d)	σ_{min} (d)	O-C (d)
Dec 26, 1994	0.32-0.46	0.059	Moon absent	-		
Dec 27, 1994	0.76-0.90	0.057	Moon absent	-		
Jan 7-9, 1995	*	*	See text	-		
Sep 26, 1995	0.31-0.71	0.068	Moon absent	-		
Sep 27, 1995	0.75-0.19	0.070	Moon absent	2449988.500	0.022	-0.011
Sep 29, 1995	0.78-0.17	0.068	Moon absent	2449990.493	0.026	0.000
Oct 01, 1995	0.79-0.24	0.065	Moon near 1st qrt	2449992.491	0.030	0.017
Oct 18, 1995	0.67-0.68	0.054	Moon absent	-		
Oct 31, 1996	0.43-0.62	0.100	Moon 2^{d} before last qrt	-		
Nov 02, 1996	0.45-0.51	0.065	Moon absent	-		
Oct 25, 1997	0.95-0.20	0.062	Moon absent	2450747.386	0.023	0.009

Wenzel (1995) with our 4 minima we obtain the following orbital period:

$$\text{HJD} = 2435799.247 + 0.6604575(\pm 11 \cdot 10^{-7}) \times E$$

This value is similar to the more accurate one determined by Deufel *et al.* (1998).

II DISCUSSION

To date the most complete study on this object is surely the one by Deufel *et al.* (1998). In their work these authors derive for the inclination i of the orbital plane a value between 50° and 90°. Here we would like to go further and propose an even more stringent condition on i. Our basic theory is that the main light variations are due to partial eclipses of the accretion disk, leaving the innermost part of the disk (the one, presumably, responsible for the X-ray flux) visible to us. This idea is based on the fact that we observe a deep primary minimum, a secondary minimimum and also on the observation that the X-ray light curve (Beuermann *et al.* 1995, Bartolini *et al.*, in preparation) is very shallow. In his graduation Thesis, Iannone (1996) used the tables from Plavec & Kratochvíl (1964) to determine the values i may have if we're to see partial eclipses. It was found that the inclination is roughly indepedent from the mass ratio $q = M_{sec}/M_{WD}$, keeping between $\sim 70°$ and $\sim 77°$ (see Table 2). A high orbital inclination would imply a low mass for the donor star; we can see this in Fig. 3, which is simply a plot of the donor's mass function for given WD masses; the secondary star would have a mass between $0.3M_{\odot}$ (for a white dwarf mass of $0.6M_{\odot}$) and $0.5M_{\odot}$, for a white dwarf mass of $1.2M_{\odot}$. This conclusion is the same as that obtained by Deufel *et al.* (1998).

In closing we mention that our results for i are consistent with those of Tomov *et al.* (1998). Meyer-Hofmeister *et al.* (1998) used the model by Schandl *et al.* (1997) on CAL 87 to give a good fit to the optical light curve of RX J0019.8+2156. The only problem we see is with their assumption of $M_2 = 1.5\,M_{\odot}$ for the donor star: at 2.3 kpc (the distance given in their paper) such a star would have $m_V \simeq 11$ but this is obviously not the case, since we see no signature whatsoever (in either the spectra or in the light curve) of the secondary star's presence. The donor must be a low mass star.

In summary we believe that RX J0019.8+2156 is a white dwarf binary system seen at high inclinations, with the compact star accreting mass from a low mass donor. The inclination is the major responsible of the photometric behaviour of the star while the secondary variations could be due to the variable height of the accretion disk rim, or the varying position of the hot spot on the disk rim.

TABLE 2. Orbital inclination values necessary to produce partial eclipses of the accretion disk, as a function of the mass ratio $q = M_{sec}/M_{WD}$; from Plavec & Kratochvíl (1964).

q	0.20	0.30	0.40	0.50	0.60	0.70	0.80	0.90
i	76.9	75.4	74.2	73.3	72.6	71.9	71.3	70.2

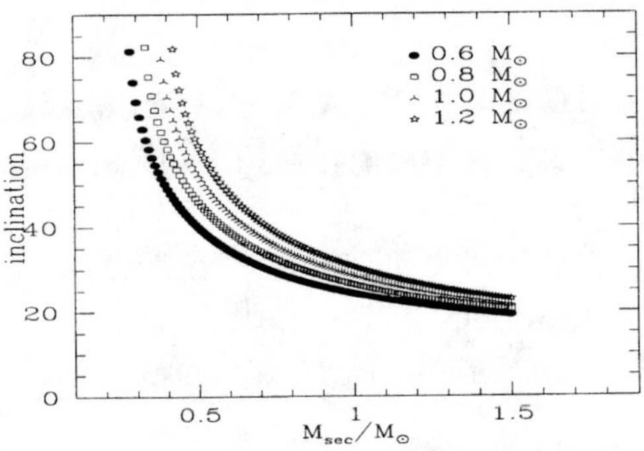

FIGURE 3. Relation between i and M_{sec} for several WD masses. From Iannone (1996), but using the Deufel *et al.* (1998) revised mass function $f_{opt}(M_{sec}) = 0.0274M_{\odot}$.

ACKNOWLEDGEMENTS

The authors thank the University of Bologna for the selected topics funds and Dr. Roberto Silvotti of the Observatory of Napoli, for his improvements on the data reduction software.

REFERENCES

1. Bartolini C., Guarnieri A., Piccioni A., Solmi L., in: "Cataclysmic Variables and Related Objects", IAU Coll. 158, 427, 1995.
2. Bartolini C., Dal Fiume D., Guarnieri A., Iannone G., Piccioni A., Solmi L., in preparation.
3. Beuermann K., Reinsch K., Barwig H. et al., *A&A* **294**, L1, 1995.
4. Cowley A. P., ASP Conf. Ser. Vol 56 p. 160, 1994.
5. Deufel B., Barwig H., Šimić D., Wolf S., Drory N., *A&A* **343**, 455, 1999.
6. Greiner J., Wenzel W., *A&A* **294**, L5, 1995.
7. Iannone G., Graduation Thesis, Bologna University, Dept. of Astronomy, 1996.
8. Iben I. & Tutukov A. V., *Ap.J.* **418**, 343, 1993.
9. Matsumoto K., *PASJ* **48**, 827, 1996.
10. Meyer-Hofmeister E., Schandl S., Deufel B., and Meyer F., *A&A* **331**, 612, 1998.
11. Piccioni A., Bartolini C., Guarnieri A., Giovannelli F., *Acta Astronomica*, **29**, 423, No.3, 1979.
12. Schandl S., Meyer-Hofmeister E., and Meyer F., *A&A* **318**, 73, 1997.
13. Stumpff P., *A&A Suppl.* **41**, 1, 1980.
14. Tomov T., Munari U., Kolev D., Tomasella L., Rejkuba M., *A&A* **333**, L67, 1998.
15. Trümper J., Hasinger G., Aschenbach B., et al., *Nature* **349**, 579, 1991.
16. van den Heuvel, E.P.J., Bhattacharya D., Nomoto K., Rappaport S.A., *A&A* **262**, 97, 1992.
17. Will T., Barwig H., Workshop on Supersoft X-ray Sources, Garching, ed. J. Greiner, Lecture Notes in Physics No. 472, Springer Verlag, p. 99, 1996.

ASCA and BeppoSAX observations of the peculiar X–ray source 4U1700+24/HD154791

D. Dal Fiume[1], N. Masetti[1], C. Bartolini[2], S. Del Sordo[3], F. Frontera[4], A. Guarnieri[2], M. Orlandini[1], E. Palazzi[1], A. Parmar[5], A. Piccioni[2], A. Santangelo[3], A. Segreto[3]

[1]*Istituto TESRE/CNR, via Gobetti 101, 40129 Bologna, Italy*
[2]*Dipartimento di Astronomia, Universitá di Bologna, via Ranzani 1, 40127 Bologna, Italy*
[3]*IFCAI/CNR, via U. La Malfa 153, 90146 Palermo, Italy*
[4]*Istituto TeSRE and Dipartimento di Fisica, Universitá di Ferrara, via Paradiso 1, 44100 Ferrara, Italy*
[5]*Space Science Department, ESA, ESTEC, Noordwjik, The Netherlands*

Abstract. The X-ray source 4U1700+24/HD154791 is one of the few galactic sources whose counterpart is an evolved M star [1–3]. In X-rays the source shows extreme erratic variability and a complex and variable spectrum. While this strongly suggests accretion onto a compact object, no clear diagnosis of binarity was done up to now. We report on ASCA and BeppoSAX X–ray broad band observations of this source and on ground optical observations from the Loiano 1.5 m telescope.

I INTRODUCTION

In optical astronomy the identification of a binary system comes in most cases from the observation of photometric and/or radial velocity variations. As not all X-ray binaries have known optical counterparts, a further effective criterium in galactic X–ray astronomy for the identification of a binary system with an accreting compact object was often based on the observed X–ray luminosity. For X–ray binaries harbouring a neutron star or possibly a black hole, luminosities L_X of the order of $10^{34} - 10^{35}$ erg s^{-1} are easily reached. The diagnosis of the presence of a neutron star in most cases is directly confirmed by the observation of pulsations or thermonuclear bursts, apart from bright persistent Low Mass X–Ray Binaries (LMXRBs). X–ray binaries harbouring white dwarfs also show some distinctive features. As an example in polars and intermediate polars optical and UV observations often reveal the distinctive signatures of the presence of a white dwarf in the system. Orbital periods and light curves also add unambiguous and reliable evidence of the presence of white dwarfs in this class of X–ray binaries.
For a number of X–ray sources the identification of a class or even the diagnosis of binarity is rather difficult, especially when the observed X–ray luminosity is $\leq 10^{33}$ erg s^{-1} . 4U1700+24/HD154791 belongs to this class. The optical counterpart was identified by

CP510, *The Fifth Compton Symposium,* edited by M. L. McConnell and J. M. Ryan

Garcia et al. [2] as a late type giant on the basis of the positional coincidence with a HEAO1–A3 error box. The optical spectrum of this giant looks quite normal [1,2], even if Gaudenzi and Polcaro [3] find some interesting and variable features in its spectrum. Variable UV line emission was detected [1,2] in different IUE pointings, showing at last some unusual features in the emission from this otherwise normal giant. These high excitation lines are likely linked to the same mechanism that produces the observed X–ray emission. In spite of various attempts, no evidence of a binary orbit was obtained from radial velocity analysis of optical spectra.
The X–ray source shows extreme erratic variability, but no pulsations were detected. The rapid (10–1000 s) time variability is strongly suggestive of turbulent accretion, often observed in X–ray binaries. The X–ray spectrum is rather energetic and was measured up to 10 keV. The X–ray luminosity $L_X \sim 10^{33}$ erg s^{-1} at an assumed distance of 730 pc [2] may be marginally consistent with coronal emission, even if an evolved giant is not expected to be a strong X-ray emitter. Therefore the picture emerging from observations gives only hints in favour of a binary system, given that no "classical" feature to be associated to the presence of a compact object was found.
We have observed this source for ~15 years both with X–ray satellites (EXOSAT, ASCA and BeppoSAX) and with ground optical observations from the Loiano 1.5 m and 0.6 m telescopes of the Bologna Astronomical Observatory. Here we report on the ASCA and BeppoSAX observations, performed respectively on March 8, 1995 and on March 27, 1998. We also report on photometric optical UBVRI monitoring.

II OBSERVATIONS

In Figure 1 we show the observed 1.5–9 keV count rate from the GIS2 and GIS3 instruments on board ASCA and the 1.5–10 keV observed count rate from the MECS2 and MECS3 instruments on board BeppoSAX . A clear increase of the 4U1700+24 count rate was detected in November 1997 by RXTE/ASM (http://space.mit.edu/XTE/ASM_lc.html). The BeppoSAX observation was performed approximately five months after this event, when the source had already recovered its quiescent flux. The substantial erratic variability already detected with EXOSAT [1] is clearly present also in both observations. The source flux in the BeppoSAX observation is significantly lower than that in the ASCA observation. The erratic source variability is clearly visible in the Power Spectral Density (PSD) shown in Figure 2, calculated on the time series of GIS2 and GIS3 count rate binned on 0.1 s. The spectra were calculated for runs with typical duration of 3000s. The PSD shown in Figure 2 is obtained by averaging the spectra of different runs and by summing adjacent frequencies with a logarithmic rebinning. The observed X–ray source luminosity (2–10 keV) was $L_X = 1.7 \times 10^{33}$ erg s^{-1} in the ASCA observation and $L_X = 6 \times 10^{32}$ erg s^{-1} in the BeppoSAX observation assuming a distance of ~700 pc [2].
The X–ray energy spectrum cannot be fitted by simple single component models. The high energy (>2 keV) spectrum can be fitted by an absorbed thermal continuum, but the extrapolation of such a model at lower energies lies significantly below the measured

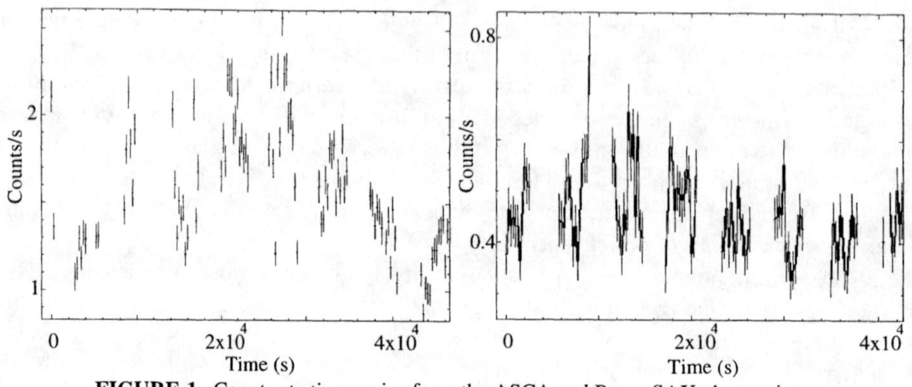

FIGURE 1. Count rate time series from the ASCA and BeppoSAX observations

spectrum, both in ASCA and in BeppoSAX observations.
For a thermal model, similar to that used in the low luminosity source γ Cas (a suspected Be/white dwarf binary [4,5]), the addition of a complex absorber (e.g. a partial absorber) is needed to model the low energy part of the spectrum. The lack of Fe emission line however requires a very low Fe abundance.
As an example the count rate spectra from the ASCA and BeppoSAX observations fitted with an optically thin thermal bremsstrahlung continuum with partial absorber (*"bremss"* and *"pcfabs"* models in XSPEC) are shown in Figure 3. The BeppoSAX spectrum is softer than that observed with ASCA , and very similar to that observed with EXOSAT [1] at almost exactly the same flux level of the BeppoSAX observation. Optical observations were performed at the Loiano 1.5 m telescope of the Bologna Astronomical Observatory during the last 15 years. HD154791, the optical counterpart of the X–ray source, is a M2–M3 giant [2,3] with a rather normal optical spectrum. A simple comparison with M1–M3 III templates shows a close match with the M2 template of HD104216. In Figure 4 we report the long term UBVRI photometry of HD154791. No clear long term trend is visible. Some variability is present, in particular in the U measurements, that may be intrinsic to the source. The long-term spectral/photometric monitoring of the source is continuing.

III DISCUSSION

The observations we report still cannot be used to perform a "classical" diagnosis of binarity. We nevertheless note some interesting similarities with other low luminosity X–ray sources. In particular some interesting similarities can be found with the X–ray emission from γ Cas. The power spectrum is strikingly similar and the energy spectrum shows a similar shape, even if no iron line is detected in 4U1700+24 .

However this close resemblance of the properties of the X–ray emission does not help to determine the presence of a compact object in a binary system, as for γ Cas itself the diagnosis of binarity is not completely assessed. In fact Owens et al. [5] favour the

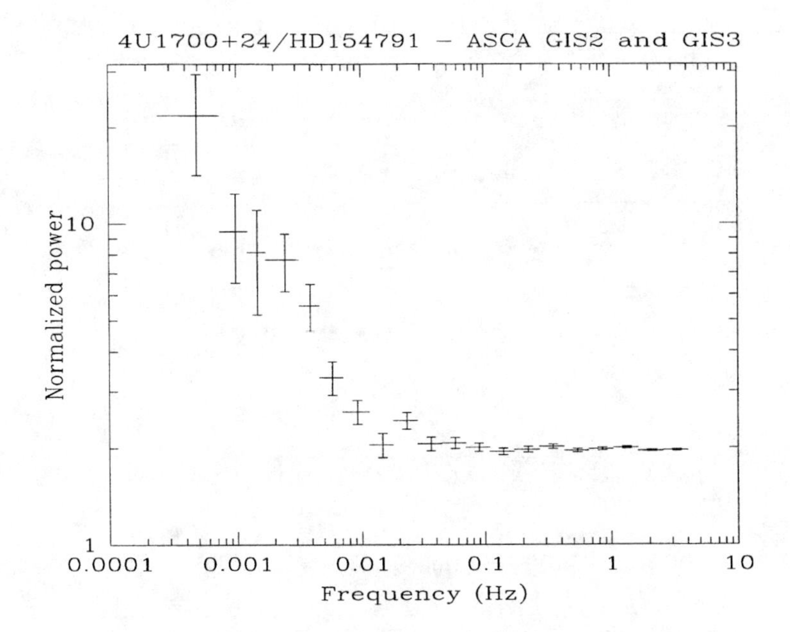

FIGURE 2. Power Spectral Density from ASCA observation

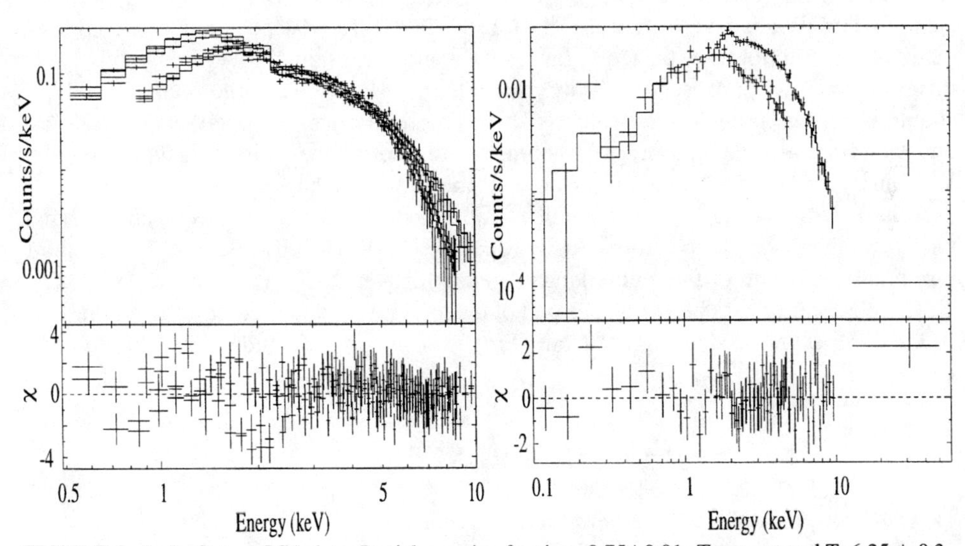

FIGURE 3. Left: fit to ASCA data. Partial covering fraction: 0.75±0.01. Temperature kT=6.25 ± 0.2. Reduced χ^2_{dof}: 1.55 Right: fit to BeppoSAX data. Partial covering fraction: 0.72±0.04. Temperature kT=3.6 ± 0.3. Reduced χ^2_{dof}: 0.9

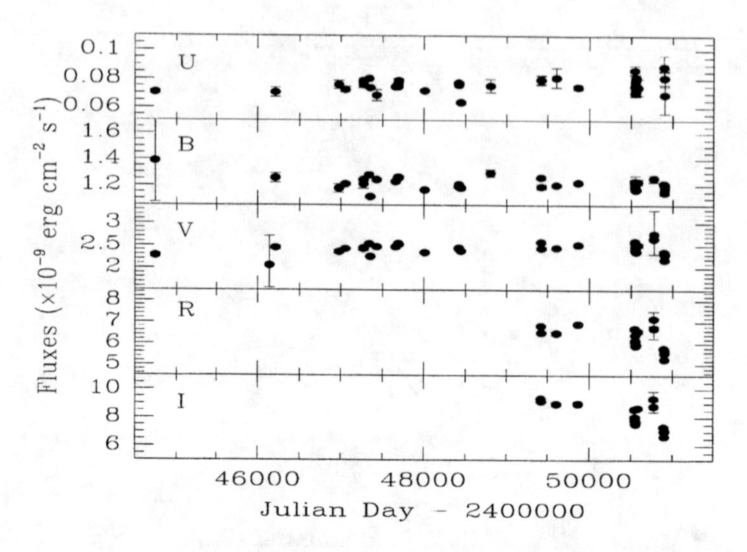

FIGURE 4. UBVRI long term variability of HD154791

hypothesis of a WD binary, but a completely different point of view is based on recent UV/X–ray observations of γ Cas [6]. Smith et al. support the hypothesis that the X–ray emission of γ Cas comes from continuous flaring from the Be star. This hypothesis cannot be easily adapted to the case of 4U1700+24/HD154791 , as the much colder M giant star should not be expected to have strong and persistent X–ray flaring activity. If this is the case, i.e. the observed X–ray emission from 4U1700+24 is coronal, HD154791 should be an exception in its own class. If the similarity of the properties of the X–ray emission from 4U1700+24 and γ Cas comes from a common origin, we suggest that the WD binary hypothesis is much more comfortable and more easily met.

Acknowledgements. This research is supported by the Agenzia Spaziale Italiana (ASI) and the Consiglio Nazionale delle Ricerche (CNR) of Italy. BeppoSAX is a joint program of ASI and of the Netherlands Agency for Aerospace Programs (NIVR). The ASCA observation was performed as part of the joint ESA/Japan scientific program. CB, AG and AP acknowledge a grant from "Progetti di ricerca ex-quota 60%" of Bologna University.

REFERENCES

1. Dal Fiume, D. et al. 1990 *Il Nuovo Cimento C*, **13**, 481
2. Garcia, M. et al. 1983, *ApJ*, **267**, 291
3. Gaudenzi, S. F., Polcaro, V. F. 1999 *Astron. Astrophys.*, **347**, 473
4. Kubo, S. et al. 1998 *PASJ*, **50**, 417
5. Owens, A. et al. 1999 *Astron. Astrophys*, **348**, 170
6. Smith, M. A. et al. 1998 *ApJ*, **503**, 877.

On the Rapid Spin-down of AE Aquarii

Chul-Sung Choi* and Insu Yi†

*Korea Astronomy Observatory, 36-1 Hwaam, Yusong, Taejon 305-348, Korea
†School of Physics, Korea Institute for Advanced Study, 207-43 Cheongryangri, Dongdaemun, Seoul 130-012, Korea

Abstract. The exact nature of the large spin-down power of AE Aqr has not been well explained mainly due to the fact that the observed luminosities in various energy ranges are much lower than the spin-down power. We consider an unconventional picture of AE Aqr in which an accreting white dwarf, modeled as a magnetic dipole whose axis is misaligned with the spin axis, is rapidly spun-down via gravitational radiation emission and therefore the spin-down power is not directly connected to any observable electromagnetic emission. We propose that the rapid spin-down is caused by the non-axisymmetric polar mounds of accreted material slowly spreading away from the magnetic poles over the surface of the star.

INTRODUCTION

AE Aqr, which consists of a magnetic white dwarf and a late-type companion star with a spectral type of K3 – K5, is a nova-like object classified as a DQ Her type magnetic cataclysmic variable or an intermediate polar. It is an unusual close binary system (P_{orb} = 9.88 hr) with a very short white dwarf spin period (P_s = 33.08 s), a high spin-down rate ($\dot{P}_s = 5.64 \times 10^{-14}$ s/s), a relatively low quiescent X-ray luminosity ($L_x \approx 7 \times 10^{30}$ erg/s), and clear pulse signals [10]. The spin-down power, $L_{sd} = I\Omega_* \dot{\Omega}_* \approx 2 \times 10^{34}$ erg/s, exceeds the bolometric luminosity of the source by an order of magnitude. A long standing question for AE Aqr is what is the exact nature of the large spin-down power.

Wynn et al. [11] proposed magnetic propeller model in which most of the accreted matter is expelled from the binary system at the expense of the spin-down power of the white dwarf. Although the propeller model offers a good explanation of the observational features of AE Aqr, there is no direct evidence of the high-velocity gas stream escaping from the system. In addition, the propeller model requires a high mass-transfer rate, $\dot{M} \gtrsim 7 \times 10^{18}$ g/s, in order to account for the observed spin-down rate [4]. This mass-transfer rate is greater by an order of magnitude or more than the rate expected from the empirical $\dot{M}$–period relation [8]. Alternatively, several studies have invoked the pulsar-like spin-down mechanism, where the spin-down power can be used for the generation of electromagnetic dipole radiation and

CP510, *The Fifth Compton Symposium,* edited by M. L. McConnell and J. M. Ryan

particle acceleration [3,4]. The rapid spin-down rate of AE Aqr can be explained by this mechanism if the white dwarf has a strong surface magnetic field of ~50 MG. This field strength exceeds the upper limit of ~ 5 MG derived by Stockman et al. [9]. There has been some sporadic attempts to connect the observed spin-down power to the TeV γ−ray emission of $\sim 10^{32}$ erg/s [1,7]. However, according to a recent observation by Lang et al. [6], there is no evidence for any steady, pulsed or episodic TeV γ−ray emission. The exact nature of the spin-down power in AE Aqr therefore remains yet to be clarified.

SPIN-UP AND SPIN-DOWN IN AE AQR

AE Aqr is an unusual white dwarf as its spin period is quite close to the theoretically maximum break-up spin period. Such a short spin period could be achieved through accretion only if the accreted mass ΔM over time Δt is at least as high as

$$\Delta M \gtrsim I_* \Omega_* / (G M_* R_*)^{1/2} \sim 0.1 \ \mathrm{M}_\odot, \tag{1}$$

where this estimate has to be taken as a lower bound since we have assumed that the accreted material has the specific angular momentum $(GM_* R_*)^{1/2}$ which is realized only when the Keplerian accretion disk extends all the way down to the stellar surface. For an accretion rate $\dot{M} = 10^{16} \dot{M}_{16}$ g/s, the accretion of angular momentum has to occur for the duration of

$$\Delta t \gtrsim 6 \times 10^7 (\Delta M / 0.1 M_\odot) \dot{M}_{16}^{-1} \ \mathrm{yr}. \tag{2}$$

There have been discussions on possible spin-down mechanisms without a clear favorite. The main problem for various spin-down mechanisms arises due to the fact that the inferred large spin-down power is not observed in any detectable forms such as high velocity gas or high luminosities. The relatively low quiescent luminosities also pose a serious problem for any mechanisms involving accretion. If the white dwarf's dipole-type magnetic field is strong enough, the electromagnetic power due to dipole radiation could account for the rapid spin-down as discussed by Ikhsanov [4]. The electromagnetic power is estimated as

$$L_{em} = 2\mu_*^2 \sin^2\theta \Omega_*^4 / 3c^3 \sim 2.5 \times 10^{30} \sin^2\theta B_{*,6}^2 R_{*,9}^6 \Omega_{*,-1}^4 \ \mathrm{erg/s}, \tag{3}$$

where $\mu_* = B_* R_*^3$ is the magnetic moment of the dipole stellar field, $\Omega_{*,-1} = \Omega_*/0.1$ s^{-1}, $B_{*,6}$ is the stellar polar surface field strength in units of 10^6 G, and θ is the misalignment angle between the rotation axis and the magnetic axis. For AE Aqr with $\Omega_* = 0.19$ s^{-1}, we expect $L_{em} \sim 4 \times 10^{30} B_{*,6}^2$ erg/s or for the observed upper limit $B_{*,6} \sim 5$, $L_{em} \lesssim 1 \times 10^{32}$ erg/s which is at least two orders of magnitude lower than the observed spin-down power.

The rapid spin-down has been widely attributed to the propeller action in which the inflowing material is flung out at a radius R_x. This radius is likely to be beyond the corotation radius and is conventionally understood as the magnetic truncation radius. The spin-down power due to the propeller action could be estimated as

$$L_{prop} \sim \dot{M} R_x^2 (GM_*/R_x^3) \propto R_x^{-1}, \tag{4}$$

where R_x is the radius at which the accretion flow is expelled. If the propeller action occurs at the magnetospheric radius $R_{mag} \sim 1.4 \times 10^{10} \dot{M}_{16}^{-2/7} M_{*,1}^{-1/7} B_{*,6}^{4/7}$, L_{sd} is accounted for by the propeller action if $L_{sd} \sim GM\dot{M}R_o^{-1}$ or

$$\dot{M} \sim 8 \times 10^{17} B_{*,6}^{4/9}. \tag{5}$$

In short, the mass accretion rate required for the propeller action to account for the spin-down power is likely to be considerably higher than $\sim 10^{17}$ g/s. The observed luminosities indicate that the mass accretion rate is considerably lower than $\sim 10^{17}$ g/s. We therefore conclude that the observed luminosities and the spin-down power are incompatible if the spin-down power results in the emission of the observable radiation.

SPIN-DOWN DUE TO GRAVITATIONAL RADIATION

We consider the gravitational radiation emission as an alternative spin-down mechanism. This mechanism could be an attractive one since the resulting spin-down power does not need to go into the observable electromagnetic radiation, which effectively avoids the long standing question of non-detection of the observed large spin-down power.

We have argued that a significant mass accretion must have occurred in order to account for the observed unusual spin period, as shown in Eqs. (1) and (2). One of the plausible possibilities is that the accreted material mostly lands on a small fraction of the total surface area near the magnetic poles. If this is the case as expected in a strongly magnetized accretion case, the accreted material would provide a source of non-zero quadrupole moment if the magnetic axis is misaligned with the rotation axis. That is, the magnetically channeled material would spread from the magnetic poles while its spread is partially hindered by the strong stellar magnetic field. Conceivably, in a steady state achieved in the high mass accretion rate episode while the rapid spin-up occurred, prior to the present spin-down, the accreted material could form accretion mounds at the magnetic poles [5].

If we assume that the accreted material is present at the magnetic poles in the form of the spatially limited blobs or mounds while the rotation axis and the magnetic axis are misaligned by an angle θ, the time averaged rate of gravitational radiation power could be estimated as

$$L_{gr} = \frac{336G}{5c^5} \delta m^2 R_*^4 \sin^4 \theta \Omega_*^6, \tag{6}$$

where δm is the amount of mass accumulated on one magnetic pole. We have assumed for simplicity that the accumulated material exists at the magnetic poles without any significant spatial spread and it remains unperturbed during each stellar rotation. We note that the large coefficient in the formula effectively amounts to a large eccentricity despite the fact that the bulk of the stellar mass is not perturbed and distributed in such a way that the contribution to the quadrupole moment is negligible.

This power becomes rapidly negligible as the star slows down due to the sensitive dependence of the gravitational radiation power on the spin frequency. For the AE Aqr parameters, $L_{gr} > L_{em}$ occurs if

$$\delta m > 2.1 \times 10^{-5} \sin^{-1}\theta B_{*,6} \, \mathrm{M}_\odot, \tag{7}$$

or for $\delta m \sim 10^{-3}$ $\mathrm{M}_\odot$

$$P_* < 0.4 \sin\theta B_{*,6}^{-1} \text{ hr}. \tag{8}$$

AE Aqr could well have been continuously spun-down after reaching a high spin frequency resulting from the high accretion phase. The above estimate indicates that the AE Aqr's current rapid spin-down could continue to periods much longer than the present short spin period.

CONCLUSION

By removing the spin-down power from the observable luminosities, we have constructed a self-consistent model which in essence incorporates all the existing ingredients. We propose that the gravitational radiation emission mainly drives the spin-down power while accretion, ejection, and electromagnetic radiation from the spinning white dwarf are responsible for the observed luminosities in various energy bands. The details are presented in a separate paper [2].

REFERENCES

1. Bowden, C. C. G., et al., *Ap. Phys.* **1**, 47 (1992).
2. Choi, C. S., and Yi, I., *ApJ*, submitted (1999).
3. de Jager, O. C., *ApJS* **90**, 775 (1994).
4. Ikhsanov, N. R., *A&A* **338**, 521 (1998).
5. Inogamov, N., and Sunyaev, R., *Astron. Lett.*, submitted (1999).
6. Lang, M. J., et al., *Ap. Phys.* **9**, 203 (1998).
7. Meintjes, P. J., et al., *ApJ* **434**, 292 (1994).
8. Patterson, J., *ApJS* **54**, 443 (1984).
9. Stockman, H. S., et al., *ApJ* **401**, 628 (1992).
10. Welsh, W. F., Horne, K., and Gomer, R., *MNRAS* **298**, 285 (1998).
11. Wynn, G. A., King, A. R., and Horne, K., *MNRAS* **286**, 436 (1997).

Monte Carlo Simulations of Radiation from Compact Objects

Edison Liang, Markus Boettcher, Dechun Lin and Ian Smith

Rice University, Houston, TX 77005-1892

Abstract. We review the space-and-time-dependent Monte Carlo code we have developed to simulate the relativistic radiation output from compact astrophysical objects. We also highlight some major recent results obtained with this code, which are most relevant to the observations of CGRO and other high energy astrophysics space missions.

1. INTRODUCTION

The high energy radiation from compact astrophysical objects is emitted by relativistic or semi-relativistic thermal and nonthermal leptons (electrons and pairs) via synchrotron, bremsstrahlung, and Compton processes, plus bound-bound and bound-free transitions of high-Z elements. Since Compton scattering is a dominant radiation mechanism in this regime, the most efficient and accurate method to model the transport of high energy radiation is the Monte Carlo (MC) technique. During the past decade we have developed a versatile state-of-the-art space-and-time-dependent MC code to model the radiative output of compact objects. More recently we have added the self-consistent evolution of the leptons coupled to the photon transport. Here we highlight some sample results of the radiative output using our MC code.

2. PHYSICS OF PHOTON AND LEPTON EVOLUTION

We use the Monte Carlo technique (Podznyakov et al 1984, Canfield et al 1987, Liang 1993, Hua et al 1997) to simulate relativistic photon transport. We include the full (energy-and-angle-dependent) Klein-Nishina cross section for Comptonization, relativistic bremsstrahlung from lepton-ion and lepton-lepton scattering (Dermer 1984), and cyclo-synchrotron processes (Brainerd 1984) in magnetic fields. Pair production and annihilation processes are handled off-line but can be iterated with the photon distribution from the MC code to compute the steady-state output of e.g. pair-balanced plasmas (Liang and Dermer 1988). Similarly, the MC code can be coupled to the XSTAR code of Kallman and Krolik (1998) to compute spectral line transport. However, for the purpose of Compton scattering line photons are treated the same way as continuum photons. The MC photon transport is fully space-and-time-dependent. Photons are born with a certain "weight" which is diminished by absorption and escape, until the weight drops below a user-specified limit, at which point the photon is "killed". Surviving photons are sampled at boundaries to provide time-and-frequency-dependent spectral output. In addition to self-emitted photons from the plasma, soft photons can be injected at zone boundaries and inside volume elements.

Currently the code can handle 1-D spherical, cylindrical or slab geometries with an arbitrary number of spatial zones. However since the photon ray tracing is done with full angle informations, the generalization to 2-and-3-D transport is straight forward. The maximum number of photon frequency bins is 128 and the default number of cosine polar angle bins is 10. For more details of this code see Canfield et al (1987). A typical MC run with a million particles at a Thomson depth of a few

CP510, *The Fifth Compton Symposium*, edited by M. L. McConnell and J. M. Ryan

takes 10s of minutes on a DEC alpha server. Since the CPU time usage scales as the square of the Thomson depth (~number of scatterings), Thomson thick runs can be quite time consuming. We are currently adopting a random walk approximation for Wien photons trapped in Thomson thick zones, which can save large amounts of CPU time without introducing too much error.

The lepton population can be computed locally in each spatial zone using the Fokker-Planck approximation (Dermer et al 1996, Li et al 1996), taking into account coulomb and Moller scattering, stochastic acceleration by Alfven and whistler wave turbulence, and radiative cooling (plus pair processes if necessary). In general the lepton population consists of a low energy thermal population plus a quasi-power law tail truncated at high energy by radiative cooling. The photon and lepton evolutions are coupled to each other via implicit time schemes. Since the lepton distribution typically evolves much faster than the photon distribution, each photon cycle contains many lepton cycles. The user, however, can always turn off the nonthermal lepton acceleration and assume a strict thermal population whose temperature can be computed self-consistently from energy balance alone.

In the following we review some major recent results obtained with this MC code.

3. SPECTRAL OUTPUT OF STRATIFIED THERMAL CORONA

Boettcher et al (1998) computed the steady state output of a slab stratified thermal corona using the MC code, allowing for a self-consistent transition zone and thermal balance between the corona and the blackbody disk (Fig.1). Such results have been applied to model the broadband spectrum of GX339-4 (Fig.2). Here we see that the corona + disk system can produce a significant Fe fluorescence line without producing a strong reflection hump in the hard x-rays. However, in reality, the Fe line would be smeared out by Doppler broadening and appear much less prominent than is depicted in Fig.2. Such simulations can be used to provide diagnostic models for future broadband data from OSSE, RXTE, Chandra and XMM.

4. ENERGY-DEPENDENT VARIABILITY OF 2-PHASE CORONA

Boettcher and Liang (1999) proposed a model for the energy dependent variability of Galactic Black Hole Candidates (GBHC) in which the time-dependent soft photon source is provided by embedded cool blobs drifting inward in a hot optically thin but inhomogeneous corona (Krolik 1998). Fig.3 gives the sample output of one such model. Here the hard-lag spectrum and PDS break frequency is related to the drift time scale, not the light crossing time of the corona as in the model of Hua et al (1997) and Kazanas et al (1997). Thus it can reproduce the results observed for Cyg X-1 and many other GBHCs with a much smaller compact corona. Note that the dependence of time lag with photon energy, the linear dependence of the hard lag on fourier peroid, and the Lorentzian shape of the PDS all roughly agree with those observed in many GBHCs. These results represent the energy-dependent variability caused by a single embedded drifting cool blob. In reality there will likely be many drifting blobs at any given time, and the very low frequency behavior of the PDS and coherence will depend on the number and randomness of the blobs.

5. SPECTRA OF HYBRID THERMAL-NONTHERMAL MAGNETOSPHERIC SHELL

We next present the results of ~keV stellar blackbody soft photons reprocessed by a spherical shell of magnetized hybrid thermal-nonthermal plasma located near the

Alfven radius of a 10^{10}G neutron star. We specify the central blackbody temperature T_{bb}, the shell ion temperature T_i and density ρ_i, Thomson depth τ_T, (nondirectional) magnetic field B, wave turbulence level $(\delta B/B)^2$ and wave power index q. Fig.4 shows the photon output spectrum without wave turbulence ($\delta B/B=0$), but with increasing B, τ_T, n_i, T_i and T_{bb} corresponding to increasing accretion rate onto the neutron star. Note that the spectrum hardens significantly with increasing accreting rate, despite the increasing soft photon flux. Figs.5-7 show output of similar cases but with increasing wave turbulence level. The hard tails above 100 keV is primarily emitted by the nonthermal leptons accelerated by wave turbulence.

6. SUMMARY

The versatility of our MC radiation code allows us to model not only the steady-state spectral output of hybrid thermal-nonthermal plasmas but also their energy-dependent time variability. By performing systematic study of the input parameter space we can begin to understand how the spectral variability depends on the interplay between the different physical parameters. This will provide us with a powerful tool to diagnose the emission environment in GBHC and neutron star binary systems, using the increasingly high quality data from the current and pending NASA high energy astrophysics missions.

ACKNOWLEDGEMENT

This research was partially supported by NASA grant NAG 5-7980.

REFERENCES

1. Boettcher, M. et al. 1998, A&A 339, 87.
2. Boettcher, M. and Liang, E. 1999, ApJ 511, L37.
3. Brainerd, J. 1985, Harvard U. Ph.D. Thesis.
4. Canfield, E. et al. 1987, ApJ 33, 565.
5. Dermer, C. 1984, U.C.S.D. Ph.D. Thesis.
6. Dermer, C. et al. 1996, ApJ 456, 106.
7. Hua, X. et al. 1997, ApJ 428, L57.
8. Kallman, T. and Krolik, J. 1998, The XSTAR User's Guide.
9. Kazanas, D. et al. 1997, ApJ 480, 735.
10. Krolik, J. 1998, ApJ 498, L13.
11. Li, H. et al. 1996, ApJ 460, L29.
12. Liang, E. and Dermer, C. 1988, ApJ 325, L39.
13. Liang, E 1993, AIP Conf. Proc. No. 280, p.418, ed. Friedlander, M. et al (AIP, NY).
14. Pozdnyakov, L. 1983, Sov. Sci. Rev. 2, 189.

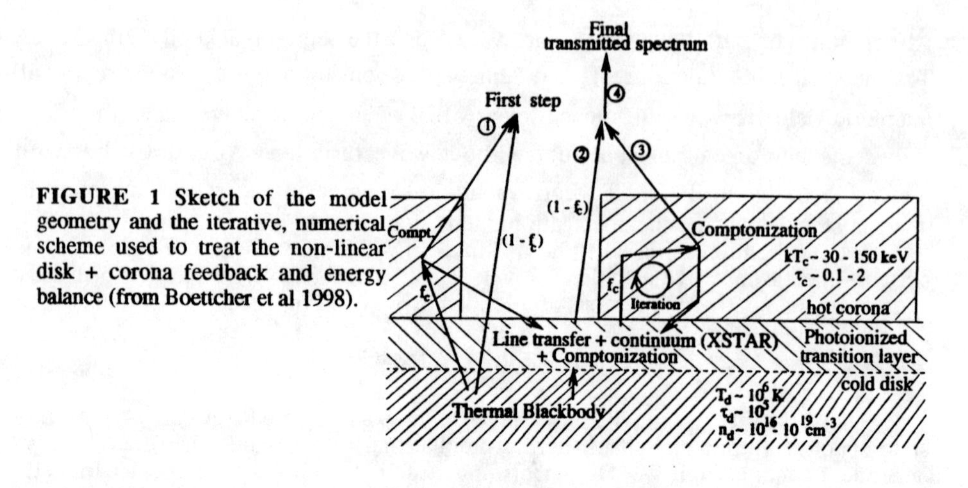

FIGURE 1 Sketch of the model geometry and the iterative, numerical scheme used to treat the non-linear disk + corona feedback and energy balance (from Boettcher et al 1998).

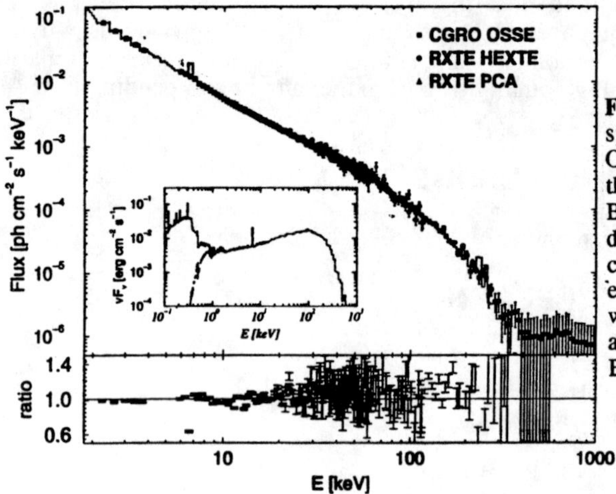

FIGURE 2 Joint RXTE-OSSE spectrum for the 1996 July Observation of GX339-4 fitted with the disk + corona model of Fig.1. Bottom panel shows the ratio: data/model. Inset: the total disk + corona model spectrum including low energy emission with (dashed) and without (solid) ISM absorption assuming $N_H = 5.10^{21}$ cm^{-2} (from Boettcher et al 1998).

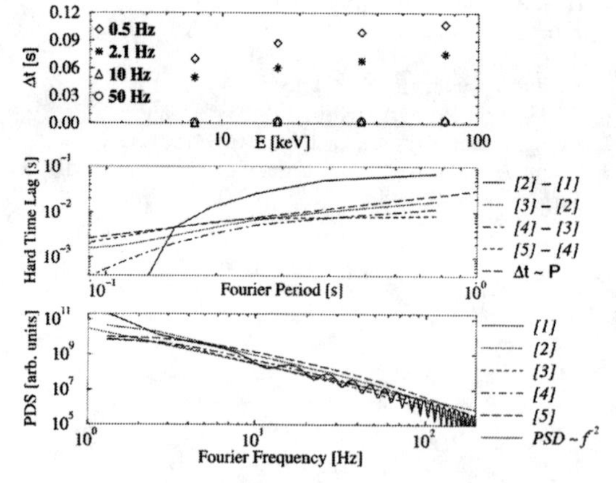

FIGURE 3 The time-lag and power density spectra of different energy photons versus fourier frequency (lower panels) and time-lag versus energy (top panel) based on the time-dependent Comptonization simulation of an embedded cool blackbody soft photon source drifting inward through an inhomogeneous corona (from Boettcher and Liang 1999).

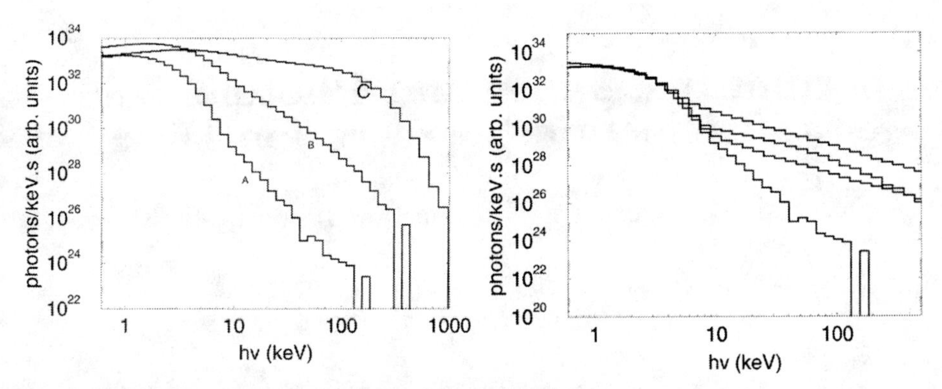

FIGURE 4 Spectra of ~keV blackbody stellar photons reprocessed by a hybrid thermal-nonthermal magnetospheric shell at the Alfven radius of a 10^{10} G neutron star without wave turbulence. Left to right: L/L_{edd} = 0.01 (A), 0.1 (B) and 1 (C).

FIGURE 5 Same as case (A) of Fig.4 but with $(\delta B/B)^2$ = 0.01, 0.001, 0.0001 and 0 (top to bottom).

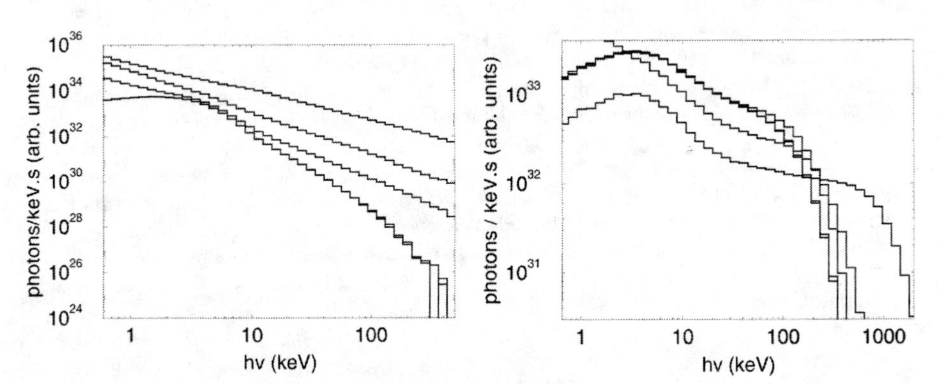

FIGURE 6 Same as case (B) of Fig. 4 but with $(\delta B/B)^2$ = 0.1, 0.01, 0.005, 0.001 and 0 (top to bottom).

FIGURE 7 Same as case (C) of Fig.4 but with $(\delta B/B)^2$ = 1, 0.3, 0.1, 0.01, and 0 (hard to soft).

Thermal Instability and Photoionized X-ray Reflection in Accretion Disks

Sergei Nayakshin, Demosthenes Kazanas and Timothy R. Kallman

Laboratory for High Energy Astrophysics, NASA Goddard Space Flight Center.

Abstract. We study the illumination of accretion disks in the vicinity of compact objects by an overlying X-ray source. Our approach differs from previous works of the subject in that we relax the simplifying assumption of constant gas density used in these studies; instead we determine the density from hydrostatic balance which is solved simultaneously with the ionization balance and the radiative transfer in a plane-parallel geometry.

We find that the self-consistent density determination makes evident the presence of a thermal ionization instability. The main effect of this instability is to prevent the illuminated gas from attaining temperatures at which the gas is unstable to thermal perturbations. In sharp contrast to the constant density calculations that predict a continuous and rather smooth variation of the gas temperature in the illuminated material, we find that the temperature profile consists of several well defined thermally stable layers. In particular, the uppermost layers of the X-ray illuminated gas are found to be almost completely ionized and at the local Compton temperature ($\sim 10^7 - 10^8$ K); at larger depths, the gas temperature drops abruptly to form a thin layer with $T \sim 10^6$ K, while at yet larger depths it decreases sharply to the disk effective temperature. The results of our self-consistent calculations are both quantitatively and qualitatively different from those obtained using the constant density assumption. We believe that usage of the latter can be completely misleading in attempts to understand the accretion disk structure from observations of iron lines and the reflection component.

Introduction

Basko, Sunyaev & Titarchuk (1974) discussed the reprocessing of X-rays from an accreting neutron star incident on the surface of the adjacent cooler companion, while Lightman & White (1988) studied similar reprocessing on the surface of a geometrically thin, optically thick accretion disk. Soon thereafter, a number of authors included the effects of ionization on the structure of the iron lines and the reflected continuum (e.g., George & Fabian 1991; Ross and Fabian 1993, Matt et al. 1993; Życki et al. 1994). Magdziarz & Zdziarski (1995) computed angle-dependent neutral reflection.

CP510, *The Fifth Compton Symposium,* edited by M. L. McConnell and J. M. Ryan

Except for Basko et al. (1974), all these studies made use of a simplifying assumption that the density in the illuminated gas be constant and equal to the disk mid-plane value. The justification for this assumption was the fact that in the simplest version of radiation-dominated Shakura-Sunyaev disks, the gas density is roughly constant in the vertical direction. However, the approximation of the constant gas density was made by Shakura & Sunyaev (1973; hereafter SS73) for disks heated by viscous dissipation only and for large optical depths. Further, a self-consistent density determination is especially important here because of the existence of a thermal instability (see Krolik, McKee & Tarter 1981 – hereafter KMT; Raymond 1993; Ko & Kallman 1994; Różańska & Czerny 1996, and Różańska 1999). In this paper, we present an X-ray illumination calculation that relaxes the assumption of the constant gas density. We adopt a plane-parallel geometry and gravity law appropriate for the standard geometrically thin SS73 disk, and solve for the gas density via hydrostatic pressure equilibrium.

Thermal Ionization Instability and X-ray Reflection

A full account of the analytical and numerical methods with which we solve the equations for the ionization, energy and hydrostatic balance for the illuminated gas is given in Nayakshin, Kazanas & Kallman (1999; hereafter paper I). The radiation transfer is done by using variable Eddington factors that allows us to find the angle-dependent reflected spectra. The hydrostatic balance equation can be parameterized through the "gravity parameter" A as

$$\frac{\partial \mathcal{P}}{\partial \tau_H} = \left[A\,\frac{z}{H} - \frac{\Delta\sigma}{\sigma_t} - \frac{F_d}{F_{\rm x}}\right], \tag{1}$$

where $d\tau_H \equiv -n_H \sigma_T dz = (n_H/n_e) d\tau_t$, τ_t is Thomson depth measured from the top of the atmosphere, and $\Delta\sigma$ is the difference in the upward and down-ward full cross sections (which is typically very small for the ionized skin). By varying A, one can cover all the parameter space to study weak and strong illumination limits.

Figure 1 shows several ionization equilibrium curves. Parts of the curve with a negative slope are thermally unstable (Field 1965, Krolik et al. 1981). Figure 2 shows the temperature profiles and corresponding spectra for both the self-consistent (upper panels) and constant density (lower panels) calculations. We now summarize the important points that one can learn from our results (see paper I for details).

1. The thermal ionization instability plays a crucial role in determining the density and temperature structure of the illuminated gas. Its effects can be taken into account by a proper treatment of the hydrostatic balance or dynamics of winds if the latter are important. Because of a (nearly) discontinuous behavior of the gas density and temperature caused by the thermal instability, one may not assume the illuminated gas to have either a constant or a Gaussian density profile.

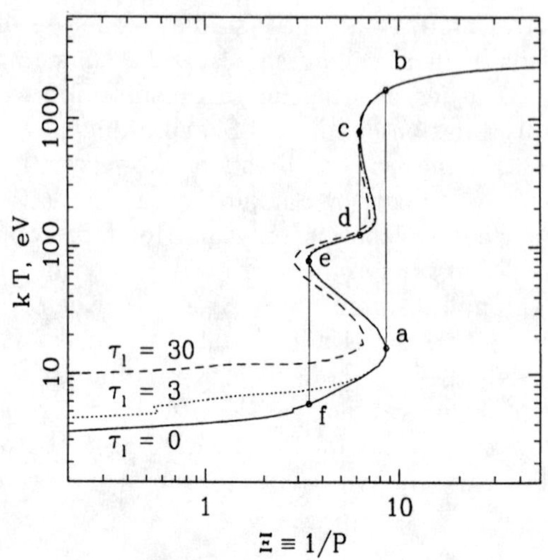

FIGURE 1. Several illustrative ionization equilibrium curves ("S-curves") showing dependence of the gas temperature, T, on the ionization parameter Ξ for three different values of the optical depth in resonance lines.

For **hard X-ray spectra** with $\Gamma \gtrsim 2$, and large X-ray flux (i.e., $F_x/F_d \gg 1$), the thermal instability "forbids" the illuminated layers from attaining temperatures at which the gas in unstable. The main contributor to the reprocessing features (lines, edges and the characteristic reflection bump itself) is the cool layer with temperature close to the effective one. The overall structure of the illuminated layer is then approximately two-phase one: (i) the Compton-heated "skin" on the top of the disk, and (ii) cold, dense, and hence weakly ionized layers below the hot skin.

The iron is completely ionized in the Compton-heated material (for hard spectra), and the only important radiative processes there are Compton scattering and bremsstrahlung emission. As a result, the existence of the hot material always reduces the strength of the reprocessing features, because the incident X-rays may scatter back (and out of the disk atmosphere) before they reach the cold layers where atomic processes could imprint the characteristic reprocessing features. We can distinguish the following characteristic limits:

• Low illumination, i.e., high values of A ($A \gg 1$), results in $\tau_h \ll 1$ and hence the iron lines, edges and continuum reflection characteristic of neutral material.

• Moderate illumination, which corresponds to values of A in the range $0.1 \lesssim A \lesssim 10$, produces moderately thick Thomson-heated layers. The reprocessed spectrum represents a combination of that for the high and low illuminated cases discussed above. We believe that because the Fe atomic features are created in the cold nearly "neutral" material, and yet their appearance is reduced by Compton scatterings in the skin, some narrow band X-ray telescopes may confuse the mildly illuminated reflector covering a full 2π solid angle with a *non-ionized* reflec-

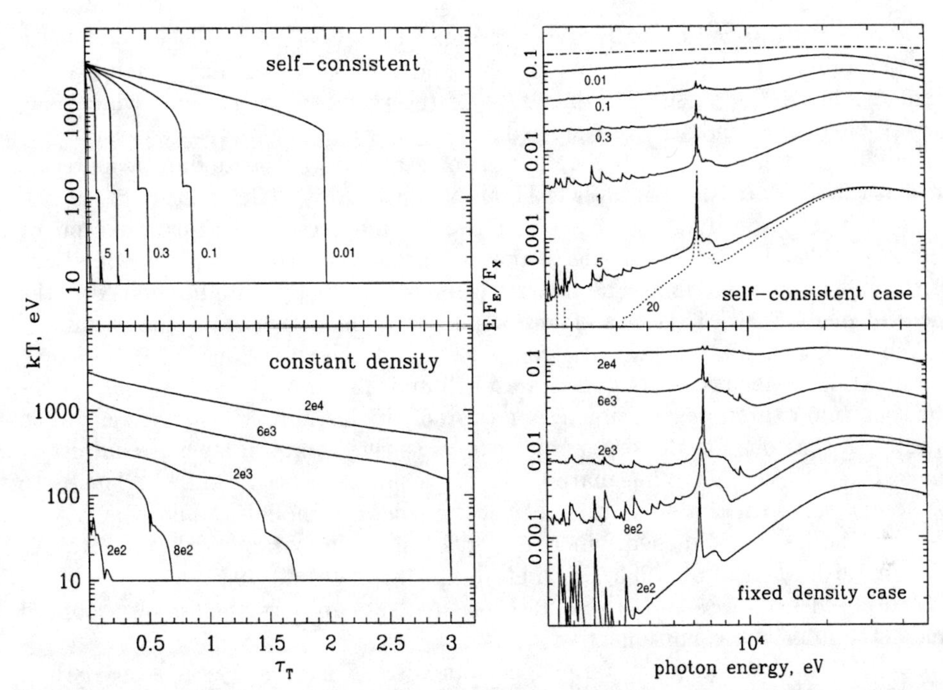

FIGURE 2. (LEFT) *Lower:* Temperature of the illuminated gas atmosphere as a function of the Thomson optical depth computed with the constant density assumption. The parameter values are $F_x = 10^{16}$ erg cm^{-2} s^{-1} and $\Gamma = 1.9$ for all the tests. The value of the density ionization parameter ξ is shown next to the corresponding curve. *Upper Panel:* Temperature of the illuminated gas atmosphere computed with our self-consistent approach. The curves differ by their respective values of the gravity parameter A, whose value is shown to the right of the corresponding curve. The self-consistent solution "avoids" the unstable regions of the S-curve, whereas the constant density solutions unphysically cover the whole temperature range. (RIGHT) Reflected spectra for the tests shown on the left. Note that evolution of the temperature and the reprocessing features in the constant density cases is distinctly different from that for the self-consistent calculations.

tor covering a fraction of 2π. Further, because Compton scattering will influence line photons more than the continuum, the solid angle of the cold material deduced from the line's EW is smaller than that of the reflected continuum, which may actually explain pequliarity of the hard state spectra of Cyg X-1.

• High illumination ($A < 0.1$). X-radiation ionizes a substantial amount of material, so that the Compton-heated layer is Thomson-thick. Very little of either of the 6.4, 6.7 or 6.9 keV iron lines are created. Thus, the reprocessing features, except for the roll-over at few tens keV due to the Compton down-scattering are wiped out of the reflection spectrum. The latter is a power-law with a similar index as the incident one for $E \lesssim 30$ keV, and can be undetectable in the lower energy data.

• The evolution of the reflection component and the iron lines from the weak illumination limit to the strong illumination limit is monotonic and in no point does the spectrum exhibit observational signatures of "highly ionized" matter (as do the spectra of the constant density studies), because the Compton layer is completely ionized and the line-creating material is very cold, effectively neutral. This is to be contrasted with the predictions of constant density models, where the EW of the line and its centroid energy increases with ionization parameter ξ (e.g., Ross & Fabian 1993; Matt et al. 1993; Życki et al. 1994; Ross et al. 1999).

2. Ionized iron lines and strong absorption edges can nevertheless be produced under the following conditions:

• If $A \ll 1$, and the incident X-ray spectrum is steep, as in soft states of GBHCs, i.e., $\Gamma \gtrsim 2$, then the Compton temperature even on the top of the reflecting layer can be lower than ~ 1 keV, which can then lead to the appearance of 6.7 and 6.9 keV iron lines.

• If $A \ll 1$, and the incident X-ray spectrum is hard, $\Gamma \lesssim 2$, but the disk intrinsic flux exceeds the X-ray illuminating flux, i.e., $F_d \gg F_x$.

REFERENCES

Basko, M.M., Sunyaev, R.A., & Titarchuk, L.G. 1974, A&A, 31, 249

Field, G.B. 1965, *ApJ*, 142, 531.

George, I.M., & Fabian, A.C. 1991, *MNRAS*, 249, 352.

Ko, Y-K, & Kallman, T.R. 1994, *ApJ*, 431, 273.

Krolik, J.H., McKee, C.F., & Tarter, C.B. 1981, *ApJ*, 249, 422.

Lightman, A.P., & White, T.R. 1988, *ApJ*, 335, 57.

Nayakshin, S., Kazanas, D., & Kallman, T. 1999 (paper I), submitted to ApJ (astro-ph/9909359).

Raymond, J.C. 1993, *ApJ*, 412, 267.

Ross, R.R., & Fabian, A.C. 1993, *MNRAS*, 261, 74.

Ross, R.R., Fabian, A.C., & Brandt, W.N. 1996, *MNRAS*, 278, 1082.

Ross, R.R., Fabian, A.C., & Young, A.J. 1999, *MNRAS*, 306, 461.

Różańska , A., & Czerny, P.T. 1996, Acta Astron., 46, 233

Różańska , A., 1999, submitted to MNRAS, astro-h/9807227

Życki , P.T., Krolik, J.H., Zdziarski, A.A., & Kallman, T.R. 1994, *ApJ*, 437, 597.

PULSARS

Observation of the millisecond pulsar PSR J0218+4232 by EGRET

W. Hermsen[1], L. Kuiper[1], F. Verbunt[2], A. Lyne[3], I. Stairs[3], D.J. Thompson[4], G. Cusumano[5]

[1]*SRON-Utrecht, Sorbonnelaan 2, 3584 CA, Utrecht, The Netherlands*
[2]*Astronomical Institute Utrecht, 3508 TA, Utrecht, The Netherlands*
[3]*University of Manchester, Jodrell Bank, Macclesfield SK11 9DL, UK*
[4]*Code 661, LHEA, NASA GSFC, Greenbelt, MD 20771, USA*
[5]*IFCAI CNR, Via U.La Malfa 153, I-90146, Palermo, Italy*

Abstract. We report on the likely detection of pulsed high-energy γ-rays from the binary millisecond pulsar PSR J0218+4232 in 100-1000 MeV data from CGRO EGRET. Imaging analysis demonstrates that the highly significant γ-ray source 2EG J0220+4228 ($\sim 10\sigma$) is for energies > 100 MeV positionally consistent with both PSR J0218+4232 and the BL Lac 3C66A . However, above 1 GeV 3C66A is the evident counterpart, whereas between 100 and 300 MeV PSR J0218+4232 is the most likely one. Timing analysis using one ephemeris valid for all EGRET observations yields in the 100-1000 MeV range a double-pulse profile at a $\sim 3.5\sigma$ significance level. The phase separation is similar to the component separation of ~ 0.47 observed at X-rays with ROSAT HRI (0.1-2.4 keV) and BSAX MECS (1.6-10 keV). A comparison in absolute phase of the γ-ray profile with the 610 MHz radio profile shows that the two γ-ray pulses coincide with two of the three emission features in the complex radio profile. The luminosity in high-energy γ-rays appears to amount $\sim$10% of the total pulsar spin-down luminosity.

INTRODUCTION

PSR J0218+4232 is a 2.3 ms radio-pulsar in a two-day orbit around a low mass ($\sim 0.2\ M_\odot$) white dwarf companion [5]. A striking feature was that the radio profile appeared complex and very broad.

The pulsar was first detected on positional arguments as a soft X-ray source between 0.1-2.4 keV using ROSAT HRI data, with only indications for a pulsed signal [8]. This stimulated a long targeted observation with the ROSAT HRI instrument establishing also the pulsed nature in the soft X-ray window: a double peaked lightcurve with a main emission feature phase separated by ~ 0.47 from a second less prominent pulse [2].

CP510, *The Fifth Compton Symposium,* edited by M. L. McConnell and J. M. Ryan

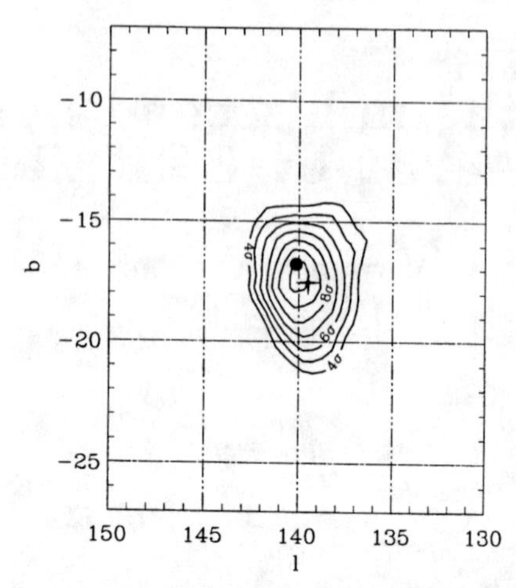

FIGURE 1. MLR image in galactic coordinates for energies in excess of 100 MeV of the sky region containing 2EG J0220+4228, combining data from 5 separate observations. A detection significance of $\geq 10\sigma$ is reached. The contours start at 4σ in steps of 1σ for 1 degree of freedom. PSR J0218+4232 is indicated by a star symbol and 3C66A by a bullet.

In a recent observation at harder X-rays (1.6-10 keV) with the BSAX MECS instruments the double peaked nature with phase separation ~ 0.47 of the X-ray profile was confirmed [4]. Spectral analysis showed that the pulsed emission has a very hard spectrum with a power-law photon-index of ~ -0.6, the hardest pulsar X-ray spectrum reported sofar.

At high-energy (100 MeV - 10 GeV) γ-rays the positional coincidence of the CGRO EGRET source 2EG J0220+4228 [7] with PSR J0218+4232 was noticed by Verbunt et al. [8]. These authors found also indications for pulsed emission at energies above 100 MeV. Since then, PSR J0218+4232 was again twice in the field-of-view of EGRET. In this work all available EGRET observations of PSR J0218+4232 between April 1991 and November 1998 with off-axis angles < 30 deg have been used to obtain maximum statistics. In addition, for the timing analysis we used one single very accurate ephemeris (rms error $85\mu s$) with a validity interval of about 5 years, allowing direct phase folding of all selected events.

IMAGING ANALYSIS

We have combined data from CGRO viewing periods 15, 211, 325, 427 and 728.7/9 and binned the measured γ-ray arrival directions in a galactic $0.^\circ5 \times 0.^\circ5$ grid after applying "standard" EGRET event selections. The measured distribution is compared with an expected model distribution, composed of galactic and extra-galactic diffuse model components and established high-energy γ-ray sources within

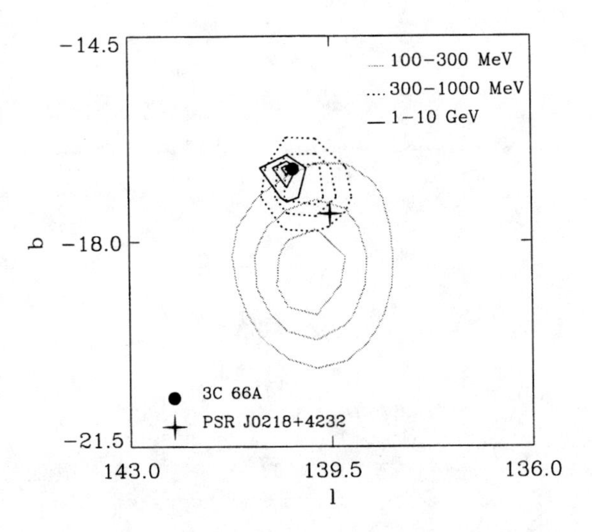

FIGURE 2. 1, 2 and 3σ location confidence contours of γ-ray source 2EG J0220+4228 in three different broad energy intervals. Between 100-300 MeV 3C66A is located outside the 3σ contour, whereas between 1-10 GeV this is the case for PSR J0218+4232 .

a 30°.0 radius around PSR J0218+4232 , by applying a Maximum Likelihood Ratio (MLR) test for the presence of a source at each grid position [3].

The MLR-map for energies > 100 MeV is shown in Figure 1 with superimposed the positions of 2 candidate counterparts, PSR J0218+4232 and 3C66A. The detection significance of the γ-ray source reaches a $\geq 10\sigma$ level. The number of counts (> 100 MeV) assigned to this excess is $\sim$ 230. We also produced MLR-maps in the broad "standard" EGRET differential energy windows: 100-300 MeV, 300-1000 MeV and 1-10 GeV. The resulting location confidence contours of the γ-ray source are shown in Figure 2 for all 3 energy windows.

This figure shows that 3C66A is the evident counterpart for the 1-10 GeV window (consistent with the third EGRET catalogue result [1]), whereas PSR J0218+4232 is the most likely counterpart for the 100-300 MeV window. Between 300 and 1000 MeV both sources contribute to the excess.

TIMING ANALYSIS

For the timing analysis we have selected events in a circular aperture around the PSR J0218+4232 position with an energy dependent extraction radius. This radius has been determined a priori from a signal-to-noise optimization study taking into account the energy dependent point source distribution and the best fit total sky background model as derived in the imaging analysis.

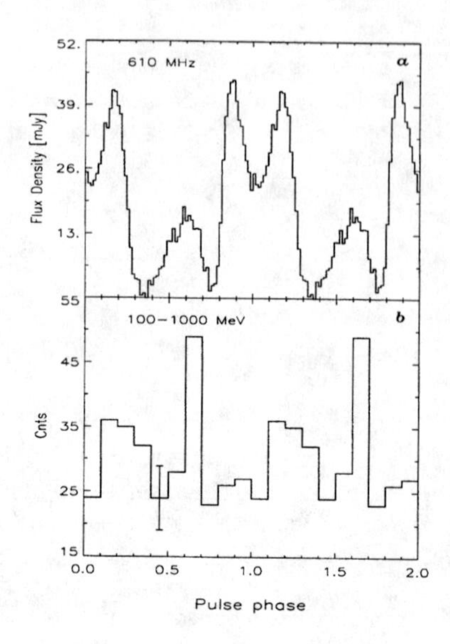

FIGURE 3. Comparison in absolute time of the radio 610 MHz profile (*a* [6]) and the 100-1000 MeV EGRET lightcurve (*b*) of PSR J0218+4232 . The EGRET lightcurve deviates from a flat distribution at the $\sim 3.5\sigma$ significance level. A typical error is indicated in the lower panel. Notice the (near) alignment of the 2 high-energy pulses with 2 of the 3 radio-pulses.

We folded the barycentric arrival times of 100-1000 MeV events with the pulsar timing parameters from one single ephemeris taking into account the binary nature of the system. We obtained a 3.5σ signal in a Z_4^2-test and the lightcurve showed one prominent pulse between phases 0.6 and 0.7 following a broader less prominent pulse between phases 0.1 and 0.4 (see Figure 3b). Phase folding of events with energies above 1 GeV gives a statistically flat light curve. Moreover, a pulse phase resolved imaging analysis [3] shows that the 100-300 MeV spatial signal is concentrated in the phase intervals of the 2 pulses. This reinforces the conclusion drawn in the imaging analysis that 3C66A is the dominant counterpart of the high-energy EGRET source above 1 GeV, and PSR J0218+4232 below 300 MeV.

A comparison with the X-ray BSAX MECS [4] and ROSAT HRI [2] lightcurves shows that the phase separation of the pulses in the γ-ray lightcurve is similar to the separation of ~ 0.47 found at X-rays. The large uncertainties in the absolute timing of the X-ray profiles does sofar not allow conclusions on the absolute phases of the X-ray and γ-ray pulses.

However, we can compare in absolute phase the 100-1000 MeV lightcurve with the 610 MHz radio profile (Figure 3a) and find that the 2 γ-ray pulses coincide with 2 of the 3 radio-pulses within the absolute timing uncertainty of the radio Jodrell Bank observations and of CGRO (the latter $\leq 100\mu s$).

SUMMARY

This study shows that we obtained good circumstantial evidence for the first detection of high-energy γ-rays from a millisecond pulsar, PSR J0218+4232 :

- A double-peaked lightcurve in the 100-1000 MeV energy interval with a $\sim 3.5\sigma$ modulation significance.
- The phase separation between the 2 γ-ray pulses is similar to that at hard X-rays ($\sim$0.47); a comparison in absolute time with the 610 MHz radio-profile shows alignment of the γ-ray pulses with 2 of the 3 radio pulses.
- Between 100 and 300 MeV the EGRET source position is consistent with PSR J0218+4232 with the signal concentrated in the phase intervals of the 2 pulses. Above 1 GeV the BL Lac 3C66A is in the sky map the evident counterpart, and, consistently, no pulsed signal is found in the timing analysis for energies above 1 GeV.

Finally, we confirm the earlier indications [8], that the energy loss in γ-rays corresponds to $\sim$10% of the total pulsar spin-down luminosity.

The full analysis and implications of our findings will be presented in detail in a forthcoming paper [3].

REFERENCES

1. Hartman R.C., Bertsch D.L., Bloom S.D., et al., *ApJS* **123**, 79 (1999).
2. Kuiper L., Hermsen W., Verbunt F., et al., *A&A* **336**, 545 (1998).
3. Kuiper L., Hermsen W., Verbunt F., et al., *A&A*, in preparation (2000).
4. Mineo T., Cusumano G., Kuiper L., et al., *A&A*, submitted (2000)
5. Navarro J., de Bruyn A.G., Frail D.A., et al., *ApJ* **455**, L55 (1995).
6. Stairs I.H., Thorsett S.E., Camilo F., *ApJS* **123**, 627 (1999).
7. Thompson D.J., Bertsch D.L., Dingus B.L., et al., *ApJS* **101**, 259 (1995).
8. Verbunt F., Kuiper L., Belloni T., et al., *A&A* **311**, L9 (1996).

A Search for VHE Gamma Rays from Young Pulsars and Supernova Remnants in the Southern Hemisphere

P. M. Chadwick, K. Lyons, T. J. L. McComb, K. J. Orford, J. L. Osborne, S. M. Rayner, S. E. Shaw, and K. E. Turver

Department of Physics, Rochester Building, Science Laboratories, University of Durham, Durham, DH1 3LE, U.K.

Abstract. Observations have been made with the University of Durham Mark 6 telescope of a number of supernova remnants and young pulsars (PSR B1706–44, Vela pulsar, PSR B1055–52, PSR J1105–6107, PSR J0537–6910 and PSR B0540–69). Although steady emission has been seen from PSR B1706–44, no VHE gamma ray emission, either steady or pulsed, has been detected from the other objects. Implications of these results for theories of high energy gamma ray production in plerions and young pulsars are discussed.

INTRODUCTION

Emission from young pulsars is a well established feature of the gamma ray sky. The *Compton Gamma Ray Observatory* (*CGRO*) telescopes have detected pulsed gamma radiation from 9 pulsars: the Crab, Vela, Geminga, PSR B1509–58, PSR B1706–44, PSR B1951+32, PSR B1055–52, PSR B0656+14 and PSR B1046–58. Of these medium and high energy gamma ray pulsars, the Crab [1] and PSR B1706–44 [2,3] are confirmed very high energy (VHE) gamma ray emitters, and the Vela remnant has been detected by the CANGAROO group [4]. Although the gamma ray emission from each of the pulsars detected with the EGRET telescope has a pulsed component, thus far no imaging VHE gamma ray telescope has detected pulsed radiation at TeV energies from any of the EGRET pulsars.

Pulsed gamma ray emission from young pulsars at the neutron star rotation period is probably produced inside the magnetosphere by particles accelerated to high energy. There are two competing models for the acceleration process — the polar cap models (e.g. [5,6]) and the outer gap models (e.g. [7,8]). One powerful discriminant between these two classes of model is the energy at which the pulsed gamma ray spectrum is cut off. Polar cap models predict a cutoff at comparatively low energies, typically between a few GeV and a few 10's of GeV, while outer

CP510, *The Fifth Compton Symposium,* edited by M. L. McConnell and J. M. Ryan

gap models predict pulsed emisson extending up to TeV energies. Thus VHE observations can constrain models for pulsed emission from young pulsars.

We have previously reported limits on pulsed VHE gamma ray emission from a number of Southern hemisphere pulsars using the University of Durham Mark 3 non-imaging telescope [9–11]. We present here the results of VHE gamma ray observations of five plerions using the Mark 6 imaging telescope; two EGRET sources (Vela and PSR B1055–52) and three X-ray emitting pulsars, PSR J1105–6107, PSR J0537–6910 and PSR B0540–69. We have searched for both steady and pulsed emission from these objects.

OBSERVATIONS

The Durham University Mark 6 telescope is described in detail elsewhere [13]. Data from all objects except the PSR J0537–6910/PSR B0540–69 field were taken in 15-minute segments. Off-source control observations were taken by alternately observing regions of sky which differ by ± 15 minutes in RA from the position of the object to ensure that on- and off-source segments have identical zenith and azimuth profiles and cosmic ray background response. The choice of alternate off-source segments which preceed and follow the on-source segment allow for any small residual secular effects. Data were accepted for analysis only if the sky was clear and stable and the gross counting rates in each on-off segment were consistent at the 2.5 σ level. In the case of PSR J0537–6910 and PSR B0540–69, the field containing the two objects was tracked and kept in the field of view at all times during the observations. The mean source position was placed 0.25° off-center, alternately left and right of the center at 3 minute intervals. The image rejection parameters were calculated with respect to the true source position and its mirror image about the camera center.

A total of 36.5 hours of on-source observations under clear skies of the 5 objects was completed.

DATA ANALYSIS

Data reduction and analysis followed our standard procedure, which has been described in detail previously [14]; this is a set of criteria developed from our observations of PKS 2155–304, and allows for the variation of image parameters with image size.

The threshold energy for the observations has been estimated on the basis of preliminary simulations [15], and is in the range 300 to 400 GeV for these objects, depending on the elevation at which observations were made. The collecting areas which have been assumed are 5.5×10^8 cm^2 at an energy threshold of 300 GeV and 1×10^9 cm^2 at 400 GeV. These are subject to systematic errors estimated to be $\sim 50\%$. We have assumed that our current selection procedures retain $\sim 20\%$

TABLE 1. Flux limits for observations of pulsars made with the University of Durham Mark 6 Telescope.

Object	Estimated Threshold (GeV)	Flux Limit (DC) ($\times 10^{-10}$ cm^{-2} s^{-1})	Flux Limit (pulsed) ($\times 10^{-11}$ cm^{-2} s^{-1})	Ephemeris Reference
Vela pulsar	300	0.50	1.3	[16]
PSR B1055–52	300	1.3	6.8	[17]
PSR J1105–6107	400	0.22	0.53	[18]
PSR J0537–6910	400	0.61	1.0	[19]
PSR B0540–69	400	0.61	1.1	[20]

of the γ-ray signal. All steady flux limits are 3 σ limits, based on the maximum likelihood ratio test.

To check for the presence of a pulsed signal, the phase of each event was evaluated using the ephemeris nearest the observation date from the Princeton database [16] or other published sources. For data from the Vela pulsar, PSR B1055–52 and PSR J1105–6107 the events were then binned in 20 phase bins. Rayleigh and χ^2 tests were performed on the binned data. The pulsed flux limits for these objects are based on the pulsed flux that would be required to yield a 3 σ excess in a single bin of a 20 bin lightcurve.

The data from PSR J0537–6910 and PSR B0540–69, for which no sufficiently accurate ephemerides were available, were subjected to a Rayleigh test over a small range of periods about the most likely period. Pulsed flux limits for PSR J0537–6910 and PSR B0540–69 are based on the percentage pulsed flux which would be required to produce a 3 σ pulsed detection using the Rayleigh test.

RESULTS

The Vela pulsar dataset has been tested for the presence of a steady gamma ray signal as described above. No source is detected. In addition, as the Vela SNR source reported by the CANGAROO group is $\sim 0.13°$ from the pulsar position, a false source analysis has been performed for these data. Again, there is no evidence for DC emssion from this offset position or, indeed, from anywhere within the SNR. The data have been folded at the contemporary radio pulsar period [16]. We find no evidence for pulsed emission of VHE gamma rays from this source. The flux limits obtained are shown in Table 1.

The PSR B1055–52 dataset has been tested for the presence of a steady VHE gamma ray signal. No signal has been found and the DC flux limit is given in Table 1. An accurate ephemeris [17] allows us to form a light curve by epoch folding our data. The resulting light curve again shows no evidence for pulsed VHE emission. The flux limit for pulsed emission is given in Table 1.

Data from PSR J1105–6107 show no evidence for a steady VHE gamma ray signal when tested in the manner described above. The radio ephemeris of Kaspi et al. [18] has been used to epoch fold the VHE data. Flux limits for steady and pulsed emission are given in Table 1.

The available ephemerides for PSR J0537–6910 are not accurate enough to enable data taken at our observing epoch to be epoch-folded and so we have searched for periodicity using the Rayleigh test around the predicted period [19]. No significant evidence for periodicity was found and the result is given in Table 1, along with the limit for steady emission.

The ephemeris of Deeter et al. [20] for PSR B0540–69 is of limited accuracy when extrapolated to our observing epoch and we have again had to search a range of periods when testing for pulsed emission from this object. No significant Rayleigh power is found and the resulting limit is given in Table 1, along with the limit for constant emission.

DISCUSSION

The only object considered in this study which has been detected at TeV energies is the Vela pulsar/nebula. An extrapolation of the flux detected with the CANGAROO telescope at 2.5 ± 1.5 TeV to our threshold energy of about 300 GeV suggests that we might expect to have detected the offset source described by Yoshikoshi et al. [4]. However, taking into account the errors on our flux and threshold energy estimates, CANGAROO's flux and energy threshold estimates and the errors on the measured spectral index, it is possible these results may be compatible.

Harding et al. [21] have predicted the unpulsed TeV flux that might be expected from a number of pulsars (including Vela and PSR B1055–52) on the basis of the model of de Jager et al. [22]. Our measured flux limit is not in conflict with these predictions for PSR B1055–302. Taking into account the errors in the flux and energy threshold measurements, our upper limit for steady emission from Vela is also not in conflict with the prediction.

Both PSR J0537–6910 and PSR B0540–69 are good candidates for steady TeV emission on the basis of their radio and X-ray characteristics. However, their distance from the earth (they are both situated in the LMC) means that an extended exposure will be necessary to detect a significant flux.

Early versions of the outer gap model [7] predicted large fluxes of pulsed TeV gamma rays from Vela-type pulsars, produced via the inverse Compton scattering of infra-red photons by primary electrons. The observations reported here emphasise that this class of model is unable to reproduce the TeV observations.

Detailed predictions of the expected spectrum of pulsed high-energy photons from several pulsars have been made for a number of models. Daugherty and Harding [6], using the polar cap model, have predicted a very sharp cut-off in the pulsed high energy gamma ray spectrum of the Vela pulsar, with no emission occurring above 10 GeV. The polar cap model of Sturner et al. [23], where the high energy gamma rays are produced via inverse-Compton scattering rather than curvature radiation, also predicts that no pulsed TeV emission should be seen from the Vela pulsar.

Modern versions of the outer-gap model [8] predict a cut-off in the pulsed specrum of the Vela pulsar at around 10 GeV, due to the cut-off in the curvature radiation spectrum. However, this model predicts another component in the pulsed high energy spectrum due to inverse Compton scattering of the primary electrons on soft photons from the pulsar gap. This will result in this additional component peaking at an energy of a few TeV. The results reported here present no support for such an additional component; however, it has been pointed out [12] that the absence of such a peak does not rule out this model since the appearance of the TeV peak depends on the density of local soft photons, which may not be correctly estimated.

We are grateful to the UK Particle Physics and Astronomy Research Council for support of the project. Pulsar ephemerides were extracted from the Princeton GRO/Radio Timing Database.

REFERENCES

1. Weekes, T. C., et al., *Ap. J.*, **342**, 379 (1989).
2. Kifune, T., et al., *Ap. J.*, **438**, L91 (1995).
3. Chadwick, P. M., et al., *Astropart. Phys.*, **9**, 131 (1998).
4. Yoshikoshi, T., et al., *Ap. J.*, **487**, L65 (1997).
5. Daugherty, J. K., & Harding, A. K., *Ap. J.*, **252**, 337 (1982).
6. Daugherty, J. K., & Harding, A. K., *Ap. J.*, **458**, 278 (1996).
7. Cheng, K. S., Ho, C., & Ruderman, M. A., *Ap. J.*, **300**, 500 (1986).
8. Romani, R. W., *Ap. J.*, **470**, 469 (1996).
9. Brazier, K. T. S., et al., *Proc. 21st ICRC (Adelaide)*, **2**, 304 (1990).
10. Bowden, C. C. G., et al., *Proc. 22nd ICRC (Dublin)*, **1**, 424 (1991).
11. Bowden, C. C. G., et al., *Proc. 23rd ICRC (Calgary)*, **1**, 294 (1993).
12. Burdett, A. M., et al., *astro-ph/9906318* (1999).
13. Armstrong, P., et al., *Exp. Astron.* **9**, 51 (1999).
14. Chadwick, P. M., et al., *Ap. J.*, **513**, 161 (1999).
15. Chadwick, P. M., et al., *Proc. 26th ICRC (Salt Lake City)*, paper OG 4.3.10 (1999).
16. Arzoumanian, Z. et al., *GRO/Radio Timing Database* Princeton: Princeton Univ., (1992).
17. Kaspi, V. M., et al., unpublished (1996).
18. Kaspi, V. M., et al., *Ap. J.*, **485**, 820 (1997).
19. Wang, Q. D. & Gotthelf, E. V., *Ap. J.*, **509**, L109 (1999).
20. Deeter, J. E., et al., *Ap. J.*, **512**, 300 (1999).
21. Harding, A. K., & de Jager, O. C., *Towards a Major Atmospheric Čerenkov Detector – V*, Potchefstroom: Potschefstroom University, ed. O. C. de Jager, p. 64 (1997).
22. de Jager, O. C. et al., *Proc. 24th ICRC (Rome)*, **2**, 528 (1995).
23. Sturner, S. J., et al., *Ap. J.*, **445**, 736 (1995).

Gamma Ray Pulsar Luminosities

Maura A. McLaughlin and James M. Cordes

Department of Astronomy, Cornell University

Abstract. We apply a likelihood analysis to pulsar detections, pulsar upper limits, and diffuse background measurements from OSSE and EGRET to constrain the γ-ray pulsar luminosity law. We find a steeper dependence on period and magnetic field at OSSE than at EGRET energies. We also find that pulsars may be an important component of the OSSE diffuse flux, but are most likely not important at EGRET energies. We estimate that as many as half of the 170 unidentified EGRET sources may be γ-ray pulsars. Furthermore, we predict that GLAST will detect roughly 1000 γ-ray pulsars, only 100 of which are currently known radio pulsars.

INTRODUCTION

Because pulsed γ-rays have been detected from only 8 spin-driven pulsars, many questions in pulsar γ-ray astronomy remain unanswered. The γ-ray pulsar (GRP) emission mechanism is not well-understood. The number of unidentified EGRET sources which are actually GRPs and the GRP contribution to the diffuse background are uncertain. Furthermore, it is important and timely to predict the pulsar population that next generation γ-ray telescopes will see.

We therefore have developed an analysis which makes use of all the available data about GRPs to constrain their luminosity law and population parameters. We describe the data used, the model for pulsar luminosity and population evolution, the method of likelihood analysis, and our results.

DATA

We include 3 OSSE pulsar detections (B0531+21, B0833−45, B1509-58), 27 upper limits to the pulsed OSSE flux for known pulsars ([5] [6]), and 3 measurements of the diffuse OSSE flux in the Galactic plane at longitudes 0, 25, and 95 degrees.

We include 6 EGRET pulsar detections (B0531+21, J0633+1746, B0833−45, B1055−52, B1706−44, B1951+32), 354 upper limits ([4] [2] [1]), and 3 measurements of the diffuse background in the plane at longitudes 20, 40, and 60 degrees.

CP510, *The Fifth Compton Symposium,* edited by M. L. McConnell and J. M. Ryan

LUMINOSITY AND POPULATION MODEL

We model a pulsar's γ-ray luminosity L as

$$L = \gamma P^{-\alpha} {B_{12}}^{\beta} \tag{1}$$

where P is the spin period in seconds and B_{12} is the surface dipole magnetic field in units of 10^{12} Gauss. Given this luminosity law and assuming a spindown law $\dot{\Omega} \propto \Omega^n$, we may calculate a population-averaged γ-ray luminosity

$$\langle L_\gamma \rangle = \frac{10^{15} \gamma {B_{12}}^{\beta-2} P_0^{2-\alpha}}{T_g(\alpha - 2)} \left[1 - \left(1 + \frac{T_g}{\tau_0} \right)^{\left(\frac{2-\alpha}{n-1}\right)} \right], \tag{2}$$

which depends upon initial spin period P_0, magnetic field B_{12}, Galactic age T_g, initial spindown time τ_0, braking index n, and the luminosity law parameters. Assuming a constant pulsar birthrate and a pulsar spatial distribution, we may use Eq. 2 to calculate the total diffuse flux due to GRPs in any direction of the Galaxy.

LIKELIHOOD ANALYSIS

We calculate the total likelihood for one set of model parameters as

$$\mathcal{L}_{\rm tot} = \mathcal{L}_{\rm det} \mathcal{L}_{\rm up} \mathcal{L}_{\rm dif}. \tag{3}$$

The total likelihoods for the detections, upper limits, and diffuse measurements are $\mathcal{L}_{\rm det} = \prod_{i=1}^{N_{det}} \mathcal{L}_{i,det}$, $\mathcal{L}_{\rm up} = \prod_{i=1}^{N_{up}} \mathcal{L}_{i,up}$, and $\mathcal{L}_{\rm dif} = \prod_{i=1}^{N_{dif}} \mathcal{L}_{i,dif}$, where N_{det}, N_{up}, and N_{dif} are the number of included detections, upper limits, and diffuse pointings, respectively. Figure 1 shows the form of these functions. We do a grid search over a range of parameters to find the combination of parameters which maximizes the total likelihood. For each parameter, we calculate marginalized PDFs and confidence intervals by integrating over all other parameters.

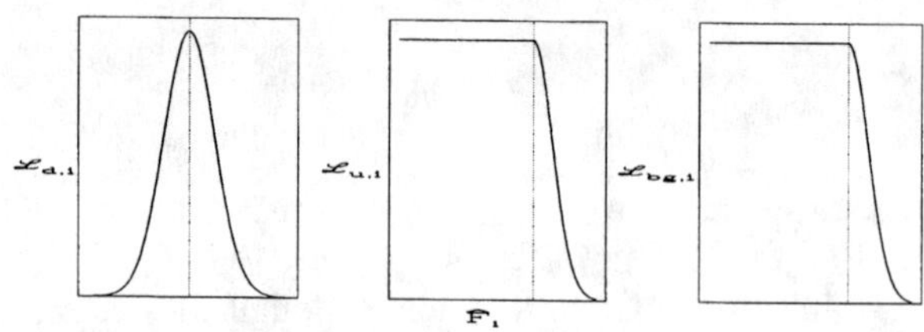

FIGURE 1. Likelihood functions for detections, upper limits, and diffuse measurements are plotted against predicted model flux, $\hat{F}_i$. Dotted lines mark where predicted flux $\hat{F}_i$ equals the measured detection, upper limit, or maximum allowed fraction ϵ of the total diffuse measurement.

RESULTS

We assume that pulsars are distributed in a disk with Gaussian radial scale 10 kpc and exponential vertical scale 0.1 kpc. We include a molecular ring at radius 4 kpc with width 1.5 kpc. We take the age of the Galaxy to be 10^{10} years and set the maximum diffuse flux attributable to pulsars, ϵ, equal to 0.5. The 6 parameters α, β, γ, P_0, B_{12}, and n are varied to find best-fit values.

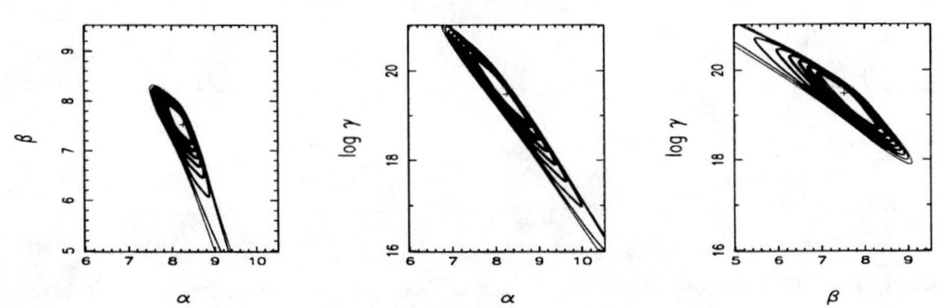

FIGURE 2. Contours of equal log likelihood for OSSE data. Crosses mark the maximum.

OSSE: We calculate the likelihood across a range of parameter values and find a well-defined likelihood maximum. Contours of equal log likelihood are shown in Figure 2. Best parameter values for α, β, and log γ, with 95% confidence intervals, are $8.3^{+0.7}_{-0.5}$, $7.6^{+0.8}_{-0.4}$, and $19.4^{+0.9}_{-1.5}$, respectively, leading to a best luminosity law of

$$L = 10^{19.4} P^{-8.3} {B_{12}}^{7.6} \text{ ergs/s}. \tag{4}$$

Figure 3 illustrates that undetected pulsars may be quite important in contributing to the OSSE diffuse background. Figure 4 shows the predicted histogram of pulsar OSSE fluxes. We note that the pulsar with the fourth highest predicted flux is B0540−69, the Crab-like pulsar in the LMC, illustrating the capability of a more sensitive low-energy γ-ray instrument for detecting young, distant pulsars.

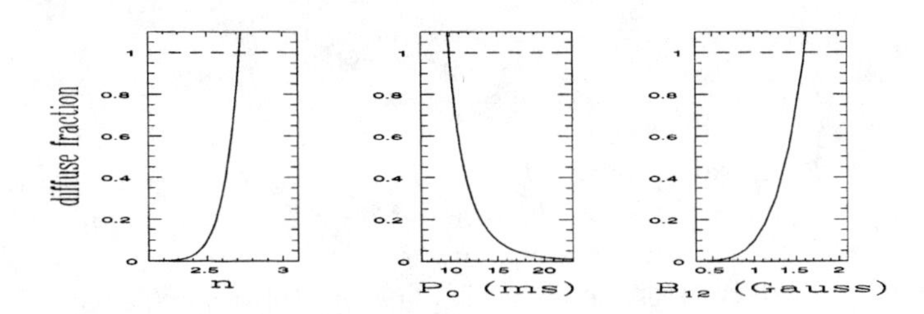

FIGURE 3. The fraction of OSSE diffuse flux attributable to undetected pulsars as a function of braking index, initial spin period, and surface magnetic field. As one parameter is varied, the other parameters are kept constant at nominal values of $n = 2.5$, $P_0 = 15$ ms, and $B_{12} = 1.0$.

EGRET: Figure 5 shows contours of log likelihood for the EGRET analysis. Best values for α, β, and log γ are 1.9 ± 0.1, 1.5 ± 0.2, and 32.0 ± 0.1, leading to

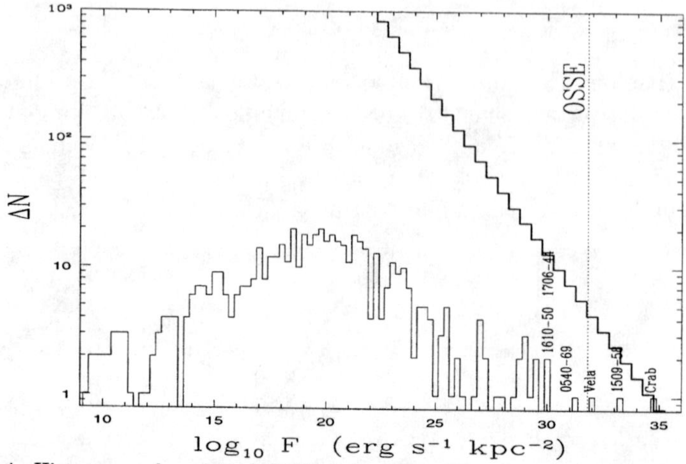

FIGURE 4. Histogram of predicted OSSE pulsar flux. The thin solid line shows the predicted fluxes for known pulsars. Pulsars with highest predicted fluxes are labelled. The thick solid line shows the predicted pulsar flux distribution of our model. The dotted line shows OSSE's sensitivity.

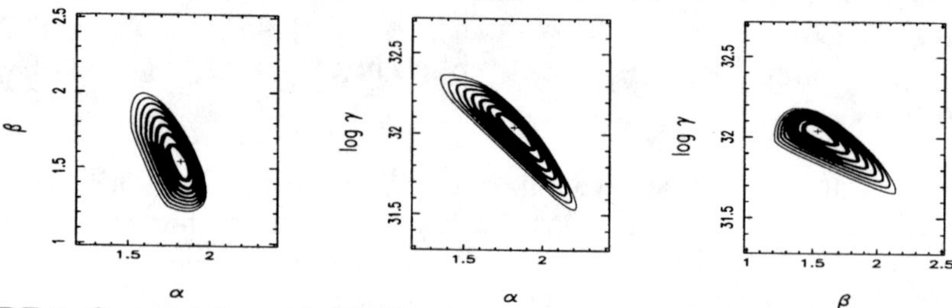

FIGURE 5. Contours of equal log likelihood vs. pairs of parameters for EGRET data. Crosses mark the maximum.

$$L = 10^{32.0} P^{-1.9} {B_{12}}^{1.5} \text{ ergs/s}. \tag{5}$$

Figure 6 shows that undetected pulsars are likely not an important contributer to the EGRET diffuse background. Figure 7, the predicted histogram of pulsar EGRET fluxes, shows that EGRET should have detected roughly 80 pulsars as steady sources. This number is tantalizing, as there are 74 unidentified EGRET sources within 10 degrees of the Galactic plane. We therefore suggest that as many as half of the unidentified EGRET sources may be GRPs. As searches for pulsars of unknown period are not possible with the sparse EGRET data, detecting these unidentified sources as pulsars will likely have to wait for GLAST. The projected sensitivity of GLAST to point sources in the Galactic plane for a 2-year survey is shown in Figure 7 ([3]). According to our model, GLAST should detect roughly 1000 GRPs, allowing us to form a complete picture of the GRP population. About

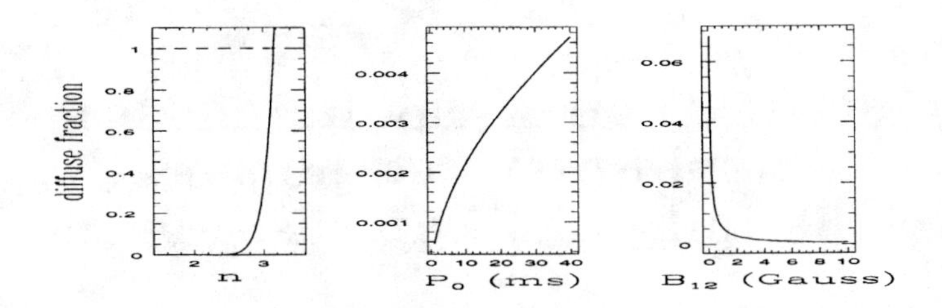

FIGURE 6. The fraction of EGRET diffuse flux attributable to undetected pulsars as a function of braking index, initial spin period, and surface magnetic field. As one parameter is varied, other parameters are kept constant at values of $n = 2.5$, $P_0 = 15$ ms, and $B_{12} = 1.0$.

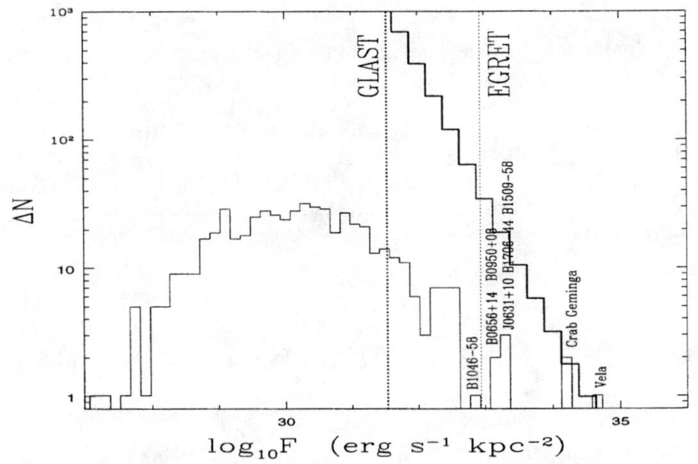

FIGURE 7. Histogram of predicted EGRET pulsar flux. The thin solid line shows the predicted fluxes for known pulsars. The thick solid line shows the predicted pulsar flux distribution of our model. EGRET's sensitivity and the predicted sensitivity of GLAST are shown.

100 of these pulsars are currently known radio pulsars. Comparing their radio and γ-ray pulse profiles will enable us to determine the GRP emission mechanism and the relationship between radio and high-energy pulsar beams.

REFERENCES

1. Arzoumanian et al., in preparation.
2. Fierro, J. M., et al., 1995, ApJ, 447, 807.
3. Gehrels, N., & Michelson, P., 1999, APh, 11, 277.
4. Nel, H. I., et al., 1996, ApJ, 465, 898.
5. Schroeder et al., 1995, ApJ, 450, 784.
6. Ulmer et al., 1999, in preparation.

Gamma-ray and X-ray luminosities from spin-powered pulsars in the full polar cap cascade model

Bing Zhang and Alice K. Harding

Laboratory of High Energy Astrophysics, NASA Goddard Space Flight Center.

Abstract. We modify the conventional curvature radiation (inverse Compton scattering) + synchrotron radiation polar cap cascade model by including the inverse Compton scattering of the higher generation pairs. Within the framework of the space-charge-limited-flow acceleration model with frame-dragging proposed by Harding & Muslimov (1998), such a full polar cap cascade scenario can well reproduce the $L_\gamma \propto (L_{\rm sd})^{1/2}$ and the $L_x \sim 10^{-3} L_{\rm sd}$ dependences observed from the known spin-powered pulsars. According to this model, the "pulsed" soft ROSAT-band X-rays from most of the millisecond pulsars might be of thermal origin, if there are no strong multipole magnetic components near their surfaces.

INTRODUCTION

Eight and 35 spin-powered pulsars have been also detected in γ-ray and X-ray bands, respectively. Despite their great diversity of emission features, the luminosities of these pulsars seem to obey the empirical laws $L_\gamma \propto (L_{\rm sd})^{1/2} \propto B/P^2$ (Thompson et al. 1997), and $L_x \sim 10^{-3} L_{\rm sd} \propto B^2/P^4$ (Becker & Trümper 1997), where $L_{\rm sd}$ is the spin-down luminosity of the pulsar. The spectra of the γ-ray emission and the X-ray emission from most of the pulsars are non-thermal, while full surface thermal emission components are identified from Vela, Geminga, PSR 1055-52 and PSR 0656+14, and possible hot polar cap thermal emission components are reported from PSR 1929+10 and PSR J0437-4715.

Two competing models for pulsar high energy emission, i.e., the polar cap models (Daugherty & Harding 1996; Sturner, Dermer & Michel 1995) and the outer gap models (Cheng, Ho & Ruderman 1986; Zhang & Cheng 1997) were proposed. Canonical polar cap cascade models involve the curvature radiation (CR) or inverse Compton scattering (ICS) of the primary particles and the synchrotron radiation (SR) of higher generation pairs. Here we modify such a cascade picture by including the ICS of the higher generation pairs, which is important since it usually occurs in the resonant regime. A more detailed presentation of this study is shown in Zhang & Harding (1999).

CP510, *The Fifth Compton Symposium*, edited by M. L. McConnell and J. M. Ryan

THE MODEL

The "full-cascade" picture

The "full-cascade" scenario is: primary particles accelerated from the inner gap emit primary γ-rays via CR or ICS, these γ-rays will pair produce in strong magnetic fields. The secondary pairs have non-zero pitch angle with respect to the field lines. The perpendicular energy of the pairs will go to high energy radiation via SR, and the parallel energy of the pairs will also convert to radiation via ICS with the soft thermal photons from either the full neutron star surface or the hot polar cap. Under the condition of $\gamma \geq \gamma_{\rm res} = 48B_{12}{\rm Max}[(1-\beta\mu_{\rm s,max})T_{s,6},(1-\beta\mu_{\rm h,max})T_{h,6}]^{-1}$ (i.e. the "resonant scattering condition", Dermer 1990), the efficiency of converting particles' kinetic energy to radiation by ICS is almost 100%, so that the total high energy emission luminosity is approximately the polar cap particle luminosity. Here T is the temperature of the soft photons, and $\mu_{\rm max}$ is the cosine of the maximum scattering angle. Note that two components, i.e., a soft full surface thermal emission (denoted by 's') and a hard hot polar cap thermal emission (denoted by 'h'), are adopted.

The basic ingredients in constructing an analytic description of such a full cascade scenario are (1) the energy distribution in SR (perpendicular) and ICS (parallel) branches, and (2) the recursion relations between different generations. For the former, $\eta_\parallel = \gamma_{i,\parallel}/\gamma_i = [1+(\gamma_{i+1}^2-1)\sin^2\theta_{\rm kB}]^{-1/2}$ and $\eta_\perp = 1-\eta_\parallel$ are the energy portions for the parallel (ICS) and the perpendicular (SR) branches, respectively, where γ_i and $\gamma_{i,\parallel}$ are the total and parallel Lorentz factors of the i-th generation pairs, and $\theta_{\rm kB}$ is the impact angle between the photon and the magnetic field line. With this, one can get the reduction factor of the typical photon energies for the adjacent generations, e.g., $\kappa_{\rm SR} = \epsilon_{i+1,\rm SR}/\epsilon_i = (3/4)\chi$, and $\kappa_{\rm ICS} = \epsilon_{i+1,\rm ICS}/\epsilon_i = \eta_\parallel B_e'$, where $B_e' = B_e/B_{cri}$, and χ is the parameter to describe the $\gamma - B$ pair production process. The photon escaping energy is $E_{\gamma,\rm esc}({\rm nPSR}) \simeq 2.0{\rm GeV}\,B_{e,12}^{-1}P^{1/2}r_{e,6}^{-1/2}\chi_{1/16}$ for normal pulsars, and $E_{\gamma,\rm esc}({\rm msPSR}) \simeq 73{\rm GeV}\,B_{e,9}^{-1}P_{-3}^{1/2}r_{e,6}^{-1/2}\chi_{1/12}$ for millisecond pulsars, where $\chi_{1/16} = \chi/(1/16)$, and $\chi_{1/12} = \chi/(1/12)$, and $r_{e,6}$ is the emission height in units of 10^6cm. Given the typical primary photon energy E_0 (which is model-dependent), we can then get some non-integer generation order parameters (Lu et al. 1994; Wei et al. 1997), e.g. $\zeta_{\rm SR} = \frac{\log(E_{\rm esc}/E_0)}{\log(\kappa_{\rm SR})}+1$ (for pure SR generations), and $\zeta_{\rm ICS} = \frac{\log(E_{\rm esc}/E_0)}{\log(\kappa_{\rm ICS})}+1$ (for pure ICS generations), which can describe the complex cascade process analytically.

Harding & Muslimov acceleration model

Harding & Muslimov (1998) has improved the space-charge-limited flow acceleration model (Arons & Scharlemann 1979) by incorporating the frame-dragging $E_\parallel$, upper and lower pair formation front and both the CR and ICS of the primary electrons. It was found that a stable accelerator is located at an effective "radius" of $R_{\rm E} \sim (1.5-2)R$ for normal pulsars when ICS energy loss is less than that of CR, since the ICS of the upward versus downward particles with the soft thermal photons are anisotropic due to different geometries. The typical length of the CR-controlled gap is $S_c \simeq 4.8 \times 10^4 {\rm cm} B_{p,12}^{-4/7} P^{4/7} R_{E,6}^{16/7} (\cos\alpha)^{-3/7}$, and the typical Lorentz factor of the primary particles is $\gamma_0 = 4.7 \times 10^7 B_{p,12}^{-1/7} P^{1/7} r_{e,6}^{4/7} (\cos\alpha)^{1/7}$, where α denotes the inclination angle of the neutron star. Thus the typical energy of the primary photons is $E_0 \simeq 33.2({\rm GeV}) B_{p,12}^{-3/7} P^{-1/14} r_{e,6}^{17/14} (\cos\alpha)^{3/7}$, with which one can get explicit expressions for the generation parameters, and complete the analytic description of the full-cascade model.

LUMINOSITY PREDICTIONS

Gamma-ray luminosity

An interesting feature of the Harding & Muslimov model is that γ_0 is insensitive to pulsar parameters, so that the polar cap luminosity $L_{\rm pc} = \gamma_0 mc^2 \dot{N}_p$ is roughly proportional to $(L_{\rm sd})^{1/2}$. One advantage of the full cascade model is that the ICS branches can convert the "lost" parallel kinetic energies of the particles also to radiation, so that the total high energy luminosity (mainly γ-ray luminosity) is also roughly proportional to $(L_{\rm sd})^{1/2}$. More specifically, the model predicts

$$L_\gamma({\rm full}) \simeq L_{\rm pc} = 5.4 \times 10^{31} {\rm erg \cdot s^{-1}} B_{p,12}^{6/7} P^{-13/7} r_{e,6}^{4/7} (\cos\alpha)^{8/7} \quad (1a)$$

$$= 1.7 \times 10^{16} B_{p,12}^{-1/7} P^{1/7} r_{e,6}^{4/7} (\cos\alpha)^{8/7} (L_{\rm sd})^{1/2}. \quad (1b)$$

This nearly reproduces the observed $L_\gamma \propto (L_{\rm sd})^{1/2}$ feature.

Thermal X-ray luminosity

There are two thermal components in a pulsar's X-ray spectrum (though they might be buried under the non-thermal component). For the full surface thermal component, we have adopted a simple rough "standard" cooling model, which is not inconsistent with the observations.

For the hot polar cap thermal component, we treated it with the self-consistent polar cap heating in the Harding & Muslimov model. Since the flow is space-charge-limited so that the deviation of local charge density (ρ) from the Goldreich-Julian density ($\rho_{\rm GJ}$) is small, the backflow particle luminosity should be only a small portion (a factor of $\mid \rho - \rho_{\rm GJ} \mid / 2\rho_{\rm GJ}$) of the polar cap luminosity, which reads

$$L_{x,th,{\rm max}} \simeq 5.9 \times 10^{29} {\rm erg \cdot s^{-1}} B_{p,12}^{2/7} P^{-9/7} r_{e,6}^{-22/7} (\cos\alpha)^{5/7}, \quad (2)$$

and the maximum polar cap temperature (assuming an area of πr_p^2, where $r_p = \theta_{\rm pc} R = 1.45 \times 10^4 P^{-1/2} {\rm cm}$) is

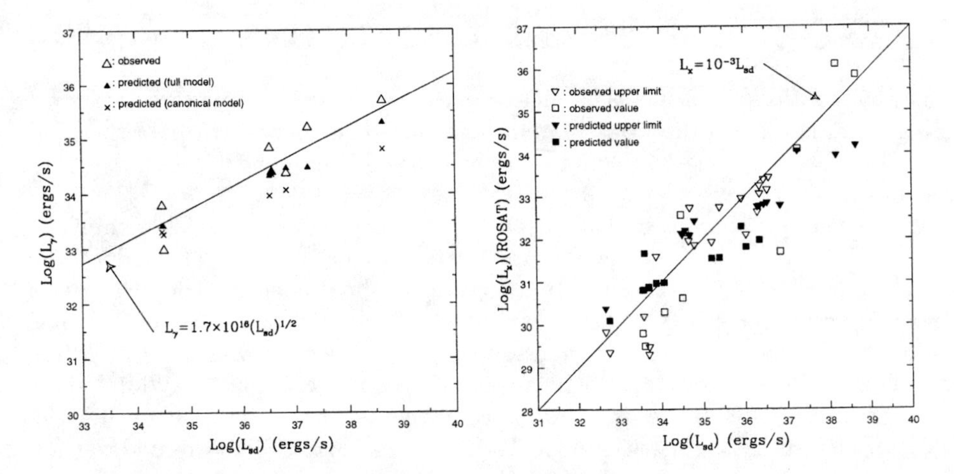

FIGURE 1. Observation versus theory: (a) γ-ray luminosities (b) X-ray luminosities.

$$T_{\rm pc,max} = 2.0 \times 10^6 {\rm K} B_{p,12}^{1/14} P^{-1/14} r_{e,6}^{-11/14} (\cos\alpha)^{5/28}. \tag{3}$$

Non-thermal X-ray luminosity

The SR spectra of all SR branches can not get down to X-ray energies as those observed by ROSAT and ASCA, since there is a low energy cut-off at the blueshifted local resonant frequency (corresponding to the transition between the ground and the first excited Landau levels). Another advantage of the full cascade scenario is that, the ICS branches, which were neglected in the canonical cascade model, can naturally give a non-thermal component extend to the X-ray band. For normal pulsars, the maximum non-thermal luminosity below a certain energy E_c could be estimated as

$$L_{x,nth}(E_c) \le L_{\rm pc} \sum_{k=1}^{\rm int(\zeta_{SR})} \left[\eta_\perp^{k-1} \left(\sum_{j=1}^{\rm int(\zeta_{ICS,k-1})} \eta_\parallel^j \eta_{c,k,j} \right) \right], \tag{4}$$

where $\zeta_{\rm ICS,k} = \frac{\log(E_{esc}/E_k)}{\log(\kappa_{\rm ICS})} + 1$ is the number of pure ICS generations for the typical energy of the k-th SR generation, $E_k = E_0 \kappa_{\rm SR}^k$, to reduce to the escaping energy $E_{\rm esc}$, $\eta_{c,k,j} = \gamma_c/\gamma_{k,j}$ (but $= 1$ when $\gamma_c \ge \gamma_{k,j}$), $\gamma_c = E_c/[(1-\beta\mu)2.8kT]$, and $\gamma_{k,j} = (\epsilon_0/2)\kappa_{\rm SR}^{k-1}\kappa_{\rm ICS}^{j-1}\eta_\parallel$. For millisecond pulsars, a slightly different formula is adopted (see Zhang & Harding 1999). When calculating the X-ray luminosity within a certain band, $E_{c1} < E < E_{c2}$, we then have $L_{\Delta E} = L_{x,nth}(E_{c2}) - L_{x,nth}(E_{c1})$.

RESULTS AND DISCUSSIONS

The observation versus model prediction of the γ-ray and X-ray pulsars are shown in Fig.1 and Fig.2. For the X-ray luminosities, three components (the non-thermal, the full surface thermal and the hot polar cap thermal components) are taken into account.

An obvious conclusion is that the full polar cap cascade model within the framework of Harding & Muslimov acceleration model can both reproduce the $L_\gamma \propto (L_{\rm sd})^{1/2}$ and $L_x(ROSAT) \sim 10^{-3} L_{\rm sd}$ feature simultaneusly. The former was not done by the outer gap model which interprets non-thermal X-ray emission as the SR of the downward cascade from the outer gap (Cheng & Zhang 1999; Zhang & Cheng 1997). In our model, we also compare the non-thermal X-ray luminosity with the luminosities of the two thermal components. It is found that for middle-aged pulsars such as Vela, Geminga, PSR 1055-52, and PSR 0656+14, the full surface thermal luminosity is comparable to the non-thermal one, so that such a component should be detectable from these pulsars. Such a feature is actually observed. For the hot polar cap thermal component, our model shows that it is detectable in relative old pulsars such as PSR 0950+08 and PSR 1929+10, although the non-thermal component is also detectable. The outer gap model actually only predicts pure thermal components in these pulsars, since the thick outer gap does not exist in their magnetospheres (Cheng & Zhang 1999). For the millisecond pulsars, our model predicts that the thermal emission from polar cap heating is the dominant component in the "pulsed" ROSAT-band spectra, while the outer gap model predicts non-thermal emission, though they assume strong multipole magnetic fields near the surfaces of the millisecond pulsars. Future observations and spectral analyses can distinguish the riveling models.

REFERENCES

Arons, J., & Scharlemann, E.T. 1979, ApJ, 231, 854

Becker, W., & Trümper, J. 1997, A&A, 326, 682

Cheng, K.S., Ho, C. & Ruderman, M.A. 1986, ApJ, 300, 500

Cheng, K.S., & Zhang, L. 1999, ApJ, 515, 337

Daugherty, J.K., & Harding, A.K. 1996, ApJ, 458, 278

Dermer, C.D. 1990, ApJ, 360, 214

Harding, A.K., & Muslimov, A.G. 1998, ApJ, 508, 328

Lu, T., Wei, D.M., & Song, L.M. 1994, A&A, 290, 815

Sturner, S.J., Dermer, C.D., & Michel, F.C. 1995, ApJ, 445, 736

Thompson, D.J., Harding, A.K., Hermsen, W., & Ulmer, M.P. 1997, in C.D. Dermer, M.S. Strickman, & J.D. Kurfess (eds.), Proc. Fourth Compton Symposium, AIP Conf. Proc. 410, 39

Wei, D.M., Song, L.M., & Lu, T. 1997, A&A, 323, 98

Zhang, B., & Harding, A.K. 2000, ApJ, 532, in press (astrp-ph/9911028)

Zhang, L. & Cheng, K.S. 1997, ApJ, 487, 370

Pulsar radiation, quantum gravity, and testing fundamental physics with astrophysics

Philip Kaaret

Harvard-Smithsonian Center for Astrophysics
60 Garden St., Cambridge, MA 02138, USA

Abstract.
I describe recent efforts towards using high-energy astrophysical observations to measure energy dependent variations in the speed of light predicted in some forms of quantum gravity.

INTRODUCTION

Testing fundamental physics with high-energy astrophysics observations is of great current interest. Successful tests of fundamental physics, particularly the discovery of genuinely new phenomena, will require a productive exchange between theoretical physicists and observational astrophysicists. For each suggested fundamental phenomenon, it will be important to search for many different astrophysical observational tests. Multiple astrophysical tests of a single fundamental phenomenon will be critical in distinguishing between astrophysical and fundamental effects.

SPEED OF LIGHT IN QUANTUM GRAVITY

Recently, it was suggested that quantum gravity may induce an observable energy dependence of the speed of light [1]. The dispersion relation for an electromagnetic wave can be written as

$$c^2p^2 = E^2\left[1 + f\left(\frac{E}{E_{QG}}\right)\right] \tag{1}$$

where c is the speed of light for very low (approaching zero) energy photons, p is the photon momentum, E is the photon energy, E_{QG} is the energy scale for quantum gravity, and the function f describes corrections to the dispersion relation produced by quantum gravity.

CP510, *The Fifth Compton Symposium,* edited by M. L. McConnell and J. M. Ryan
2000 American Institute of Physics 1-56396-932-7

As the expected value for E_{QG} is of the order of 10^{19} GeV and is, thus, much larger than currently observable photon energies, one can expand f as a power series,

$$f = a_1 \left(\frac{E}{E_{QG}} \right)^1 + a_2 \left(\frac{E}{E_{QG}} \right)^2 + ... \tag{2}$$

with coefficients, $a_1, a_2, ...$ Because the expansion parameter, E/E_{QG}, is small, typically less than 10^{-14}, the first non-zero coefficient in the expansion will dominate the behavior of the quantum gravity corrections to the speed of light.

Amelino-Camelia et al. [1] suggested that the first order term might be non-zero. In this case, there would be an observable time delay,

$$\Delta t = \frac{L}{c} \left(\frac{\Delta E}{E_{QG}} \right) \tag{3}$$

between photons with energy difference ΔE which have traveled a distance L. This suggestion that the speed of light may have a first order dependence on the photon energy is controversial. However, the current status of theories of quantum gravity cannot definitively exclude the possibility that the dependence is first order. Thus, measurements to check the energy dependence of the speed of light are of interest.

MEASURING OF THE SPEED OF LIGHT

Amelino-Camelia et al. [1] proposed observations of fine time structure in gamma-ray bursts as a potentially sensitive probe of the energy dependence of the speed of light. An analysis of bursts having fine time structure and those detected at high energies was carried out by Schaefer [4]. The best lower limit on the energy scale for quantum gravity, 8×10^{16} GeV, comes from a burst from which GeV photons were detected. However, a redshift measurements is not available for this burst and the distance was inferred by an indirect means.

Biller et al. [2] noted that TeV gamma-ray flares from active galaxies could also be used to place useful limits on the energy scale of quantum gravity. TeV emission has been observed from AGN at distances up to a few 100 Mpc and rapid variability has been observed in the TeV band on time scales as fast as 100 s. Biller et al. (1999) used observations of Markarian 421 obtained with the Whipple telescope to derive a lower limit on the energy scale for quantum gravity of 6×10^{16} GeV.

Kaaret [3] suggested that gamma-ray pulsars might afford an additional test of the energy dependence of the speed of light. Comparison of pulse profiles for the Crab pulsar derived from EGRET data in various energy bands above 70 MeV, with the highest band containing only photons with energies above 2 GeV, led to a limit on the energy scale for quantum gravity of 2×10^{15} GeV. This limit may be improved if observations of the Crab pulsar can be extended to higher energies as may be possible with new air Cherenkov telescopes, such as STACEE or CELESTE, or with new space-based gamma ray detectors with increased sensitivity at high energies, such as AGİLE or GLAST.

CONCLUSION

All of the astrophysical measurements of the energy dependence of the speed of light have reported only upper limits on any energy dependent time delay. The current limits imply lower bounds on the energy scale for quantum gravity below the theoretically expected scale, so detections of an energy dependence were not anticipated. However, in the future it should be possible to improve gamma-ray bursts measurements in the GeV band, refine TeV AGN flare measurements to shorter time scales, and extend the pulsar measurements to higher energies. These new astrophysical observations should bring measurements of the energy dependence of the speed of light which are sensitive to quantum gravity corrections at the expected energy scale (if the suggestion [1] that the correction is first order in photon energy is correct).

If the detection of an energy dependence of the speed of light is reported from an astrophysical observation, then it will be critical to be able to determine whether that energy dependence is, indeed, evidence for a new fundamental physics or is simply an energy dependence in the emission time or location in the astrophysical phenomenon. Given any one astrophysical observation, it is difficult to distinguish between these two possibilities. Multiple observations are required. While multiple observations of one particular class of astrophysical phenomena, i.e. many of observations of different gamma-ray bursts, may suffice to disentangle the astrophysical and fundamental energy dependence of the photon time delays, the most robust demonstration of a fundamental energy dependence in the speed of light would come from a combination of several different astrophysical phenomena. If gamma-ray burst, AGN, and pulsar observations all demonstrated an energy dependence in the speed of light which imply the same quantum gravity energy scale, then the combination of the observations could be robust evidence for new fundamental physics.

REFERENCES

1. Amelino-Camelia G., Ellis J., Mavromatos N.E., Nanopoulos D.V., and Sarkar S., *Nature* **393**, 763 (1998).
2. Biller, S.D. *et al.*, *Phys. Rev. Lett.*, **83**, 2108-2111 (1999).
3. Kaaret, P., *Astron. Astrophys.*, **345**, L32-L34 (1999).
4. Schaefer, B.E., *Phys. Rev. Lett.*, **82**, 4964-4966 (1999).

DIFFUSE GALACTIC CONTINUUM EMISSION

Diffuse Galactic Continuum Gamma Rays

Andrew W. Strong*, Igor V. Moskalenko*†‡, and Olaf Reimer*

*Max-Planck Institut für extraterrestrische Physik, 85740 Garching, Germany
†Institute for Nuclear Physics, M.V. Lomonosov Moscow State University, Moscow, Russia
‡LHEA NASA/GSFC Code 660, Greenbelt, MD 20771, USA

Abstract. Galactic diffuse continuum γ-ray emission is intricately related to cosmic-ray physics and radio astronomy. We describe recent results from an approach which endeavours to take advantage of this. Information from cosmic-ray composition constrains the propagation of cosmic rays; this in turn can be used as input for γ-ray models. The GeV γ-ray excess cannot be explained as π^o-decay resulting from a hard nucleon spectrum without violating antiproton and positron data; the best explanation at present appears to be inverse-Compton emission from a hard interstellar electron spectrum. One consequence is an increased importance of Galactic inverse Compton for estimates of the extragalactic background. At low energies, an additional point-source component of γ-rays seems to be necessary.

I INTRODUCTION

This paper discusses recent studies of the diffuse continuum emission and their connection with cosmic-ray physics. The basic question concerns the origin of the intense continuum emission along the Galactic plane observed by EGRET, COMPTEL and OSSE. The answer is surprisingly uncertain. A comprehensive review can be found in [1]. The present work uses observational results given in [2–4]; new imaging and spectral results from COMPTEL are reported in [5]. Most of the analysis reported here is based on the modelling approach described in [6,7]. First we present some results from cosmic-ray isotopic composition which bear directly on the γ-ray models. We then discuss the problems which arise when trying to fit the γ-ray spectrum, and present possible solutions, both at high and low energies. The low energy (1–30 MeV) situation is addressed in more detail in [8], and additional references can be found at [9].

Our basic approach is to construct a unified model which is as far as possible realistic, using information on the gas and radiation fields in the Galaxy, and current ideas on cosmic-ray propagation, including possible reacceleration; we use these to predict many different types of observations: direct measurements in the heliosphere of cosmic ray nuclear isotopes, antiprotons, positrons, electrons; and

CP510, *The Fifth Compton Symposium,* edited by M. L. McConnell and J. M. Ryan

astronomical measurements of γ-rays and synchrotron radiation. Any given model has to be tested against all of these data and it is a challenge to find even one which is consistent with all observations. In fact we will show that the full range of observations can only be accomodated by additional components such as γ-ray point sources and also differences between local direct measurements and large-scale Galactic properties of cosmic rays.

II COSMIC RAY NUCLEONS

First we show results from CR composition which are relevant to the propagation of cosmic rays. For a given halo size (defined here as the z value at which the cosmic-ray density goes essentially to zero) the parameters of the diffusion/reacceleration model can be adjusted to fit the important secondary/primary ratios, illustrated in Fig 1 for a halo size of 4 kpc. In addition we can use the constraints on the halo size given by the radioactive CR species ^{10}Be and ^{26}Al, Fig 2. For details of Ulysses results on radioactive nuclei see [10–13]. Based on Ulysses ^{10}Be data, a range for the halo height of 4–10 kpc was derived in [6,14]. This is consistent with other analyses [15,16]. New results from the Advanced Composition Explorer satellite (ACE) will constrain the halo size better, but the above range is consistent with ACE results as presented in [17]. Other radioactive nuclei (^{36}Cl and ^{54}Mn) will provide further independent information; at present one can only say that they are consistent with the other nuclei. Having obtained sets of propagation parameters based on isotopic composition, we can proceed to use the model to study diffuse γ-rays.

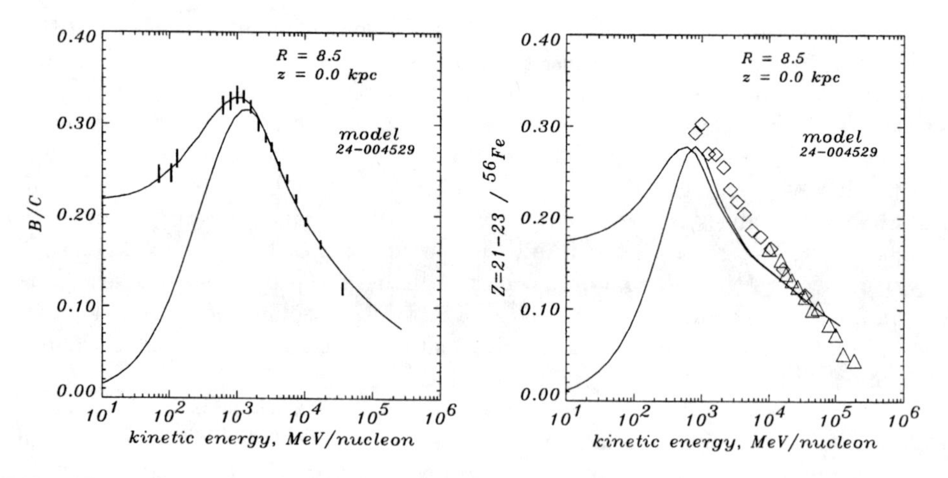

FIGURE 1. Cosmic-ray B/C and sub-Fe/Fe ratios for a diffusive halo model with reacceleration, halo height 4 kpc. For details of model and data see [14].

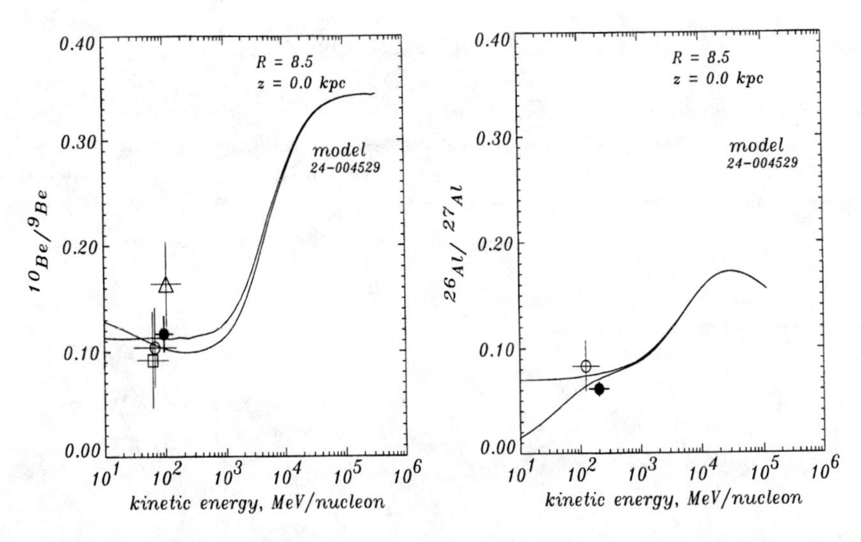

FIGURE 2. Cosmic-ray ^{10}Be/^{9}Be and ^{26}Al/ ^{27}Al ratio for the same model as used for Fig 1. For details of model and data see [14].

III GAMMA RAYS

Figure 3 shows the diffuse spectrum of the inner Galaxy for what we call a 'normal' or 'conventional' CR spectrum which is consistent with direct measurements of high energy electrons and synchrotron spectral indices (Figs 5, 6; see [7,8]). Clearly this model does not fit the γ-ray data at all well.

Consider first the well known problem of the high energy (> 1 GeV) EGRET excess [18]. One obvious solution is to invoke π^o-decay from a harder nucleon spectrum than observed in the heliosphere, which might for example be the case if the local nucleon spectrum were dominated by a local source which is not typical of the large-scale average. Then the local measurements would give essentially no information on the Galactic-scale spectrum. One can indeed fit the EGRET excess if the Galactic proton (and Helium) spectrum is harder than measured by about 0.3 in the index (Fig 3).

But there are two critical tests of this hypothesis provided by secondary antiprotons and positrons. It was shown in [19] that such a hard nucleon spectrum produces too many antiprotons. The new MASS91 measurements [20], which give the absolute antiproton spectrum from 3.7 to 24 GeV, have clinched this test, as shown in Fig 4. Quite independently, secondary positrons give a similar test, which the hard nucleon hypothesis equally fails (Fig 4). Again new data, this time from the HEAT experiment [21], give a good basis for this test. We conclude that there are significant problems if one wants to explain the GeV excess with π^o-decay. This illustrates the importance of considering all the observable consequences of any model. Of course it is anyway difficult to imagine such spectral variations of nucleons given the large diffusion region and isotropy of CR nucleons.

An alternative idea, first investigated in detail in [22], is inverse Compton (IC) from a hard electron spectrum. The point is that the electron spectrum we measure locally may not be representative of the large-scale Galactic spectrum due to the large spatial fluctuations which arise because of the large energy losses at high energies. What is measured directly may therefore depend only on the chance locations of the nearest electron sources, and the average interstellar spectrum could

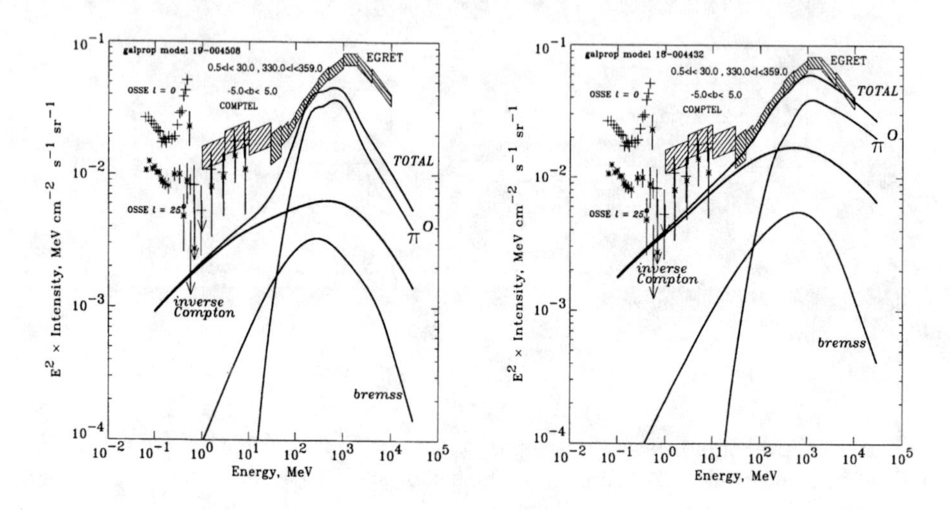

FIGURE 3. Gamma-ray spectrum of inner Galaxy for (left) 'conventional' CR spectra; (right) hard nucleon spectrum. Data: OSSE [4], COMPTEL [2], EGRET [3].

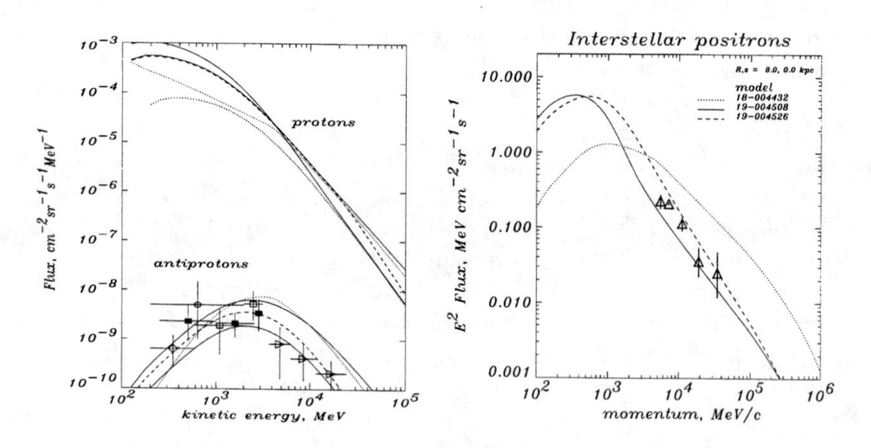

FIGURE 4. Secondary antiproton (left) and positron spectra (right) for a hard nucleon spectrum. Data: antiprotons [20], positrons [21].

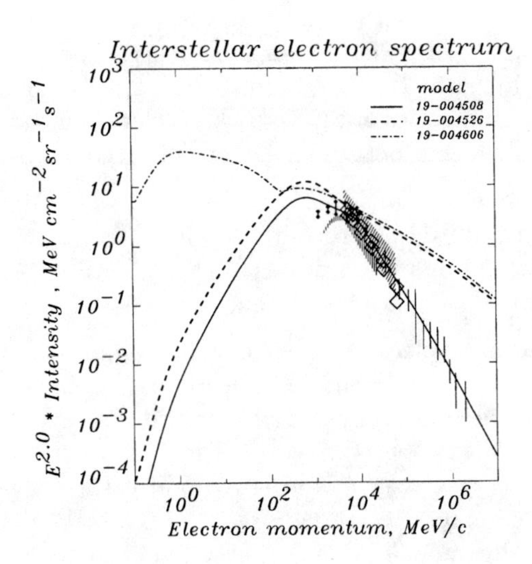

FIGURE 5. Electron spectra observed locally and for various models. Solid line: 'conventional' model, dashed line: injection spectrum 1.8, dash-dot line: spectrum reproducing low-energy γ-rays. For data see [7].

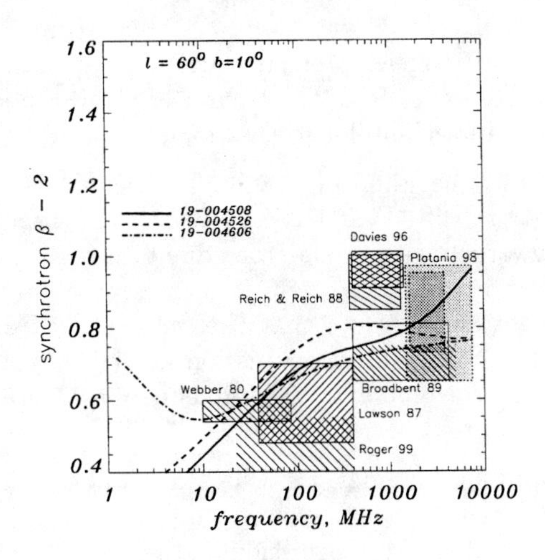

FIGURE 6. Synchrotron index for various electron spectra as in Fig 5. For data see [7,8].

be very different, in particular it could be much harder. An injection spectral index around 1.8 is required (Fig 5) and the corresponding γ-ray spectrum is shown in Fig 7. Note that modern theories of SNR shock acceleration can give hard electron injection spectra [23] so such a behaviour is not entirely unexpected.

To predict reliably the IC emission, we also need an updated model for the interstellar radiation field; we have recomputed it [7] using new information from IRAS, COBE, and stellar population models. There is still much scope for further improvement in the ISRF calculations however.

Note that for these hard electron spectra IC dominates above 1 GeV, and is everywhere a very significant contributor, while bremsstrahlung is relegated to third position in contrast to the more conventional picture (presented e.g. in [24]). Even if we can fit the inner Galaxy spectrum, the critical test is the spatial distribution: from Fig 8 one can see that it can indeed reproduce the longitude and latitude profiles. In fact it can reproduce latitude profile up to the Galactic pole (Fig 9) which is not the case for models with less IC. This can be seen as one proof of the importance of IC. But there is at least one problem associated with the hard electron spectrum hypothesis. A recent reanalysis of the full EGRET data for the Orion molecular clouds [25] determined the γ-ray emissivity of the gas, and this also shows the GeV excess, which would not expected since it should not involve IC. This could be a critical test. Perhaps the increased radiation field in the Orion star-forming region could boost the IC, and this ought to be investigated in detail.

An earlier analysis correlating EGRET high-latitude γ-rays with 408 MHz survey data [26] found evidence for IC with an $E^{-1.88}$ spectrum. This is very much in accord with the present models. More recently a study [27] which used a wavelet analysis to look for deviations from the Hunter et al. [18] model provided evidence for a γ-ray halo with a form similar to that expected from IC.

An effect which may be important at high latitudes is the enhancement due to the anisotropy of the ISRF and the fact that an observer in the plane sees preferentially downward-travelling electrons due to the kinematics of IC [28]. This can enhance the flux by as much as 40% for a large halo. Even in the plane it can have a significant effect. Note that the halo sizes considered here imply an increased contribution from Galactic emission at high latitudes, which will affect determinations of the isotropic extragalactic emission. More precise evaluation of these implications is in progress.

We mention finally low energies, for which a detailed account is given in [7,8]. Conventionally one invoked a soft electron injection, $E^{-2.1}$ or steeper, and this could then explain the 1–30 MeV emission as the sum of bremsstrahlung and IC. However it seems impossible to find an electron spectrum which reproduces the γ-rays without violating the synchrotron constraints, unless there is a very sharp upturn below 200 MeV; but even there it fails at to give the intensities measured by OSSE below 1 MeV. Therefore a source contribution appears to be the most likely explanation.

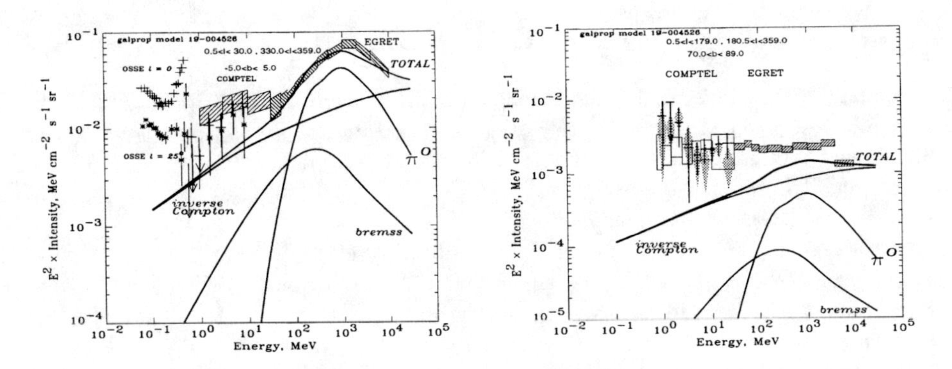

FIGURE 7. Gamma-ray spectrum for a hard electron injection spectrum. Left: inner Galaxy ; Right: high latitudes. Data as Fig 3.

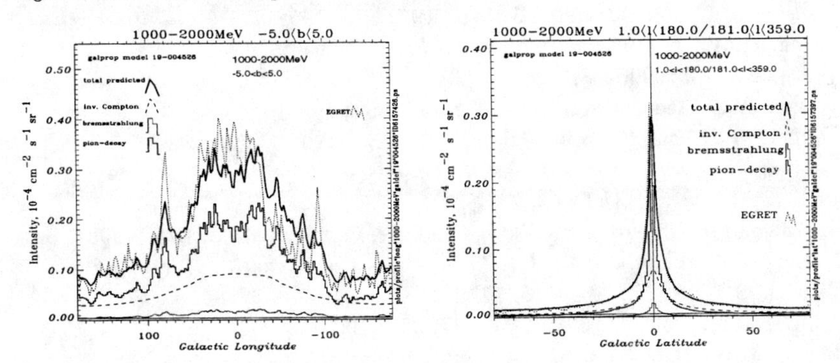

FIGURE 8. Gamma-ray profiles in the energy range 1–2 GeV for a model with a hard electron spectrum [7]. Dotted line: EGRET data, dashed line: inverse Compton, upper histogram: π^o-decay, lower histogram: bremsstrahlung, upper solid line: sum of components. Left: longitude profile, right: latitude profile.

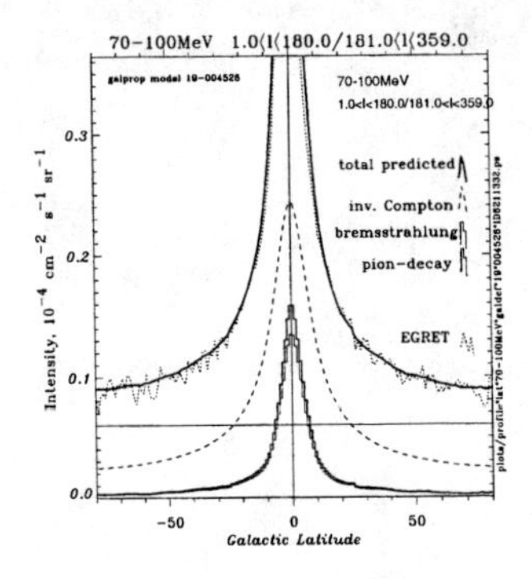

FIGURE 9. Gamma-ray profile at high latitudes, for the energy range 70–100 MeV [7]. Components as Fig 8; horizontal line: isotropic background.

REFERENCES

1. Hunter, S.D., Kinzer, R.L., and Strong, A.W. *AIP Conf. Proc.* **410**, 192 (1997).
2. Strong, A.W., et al., *Proc 3rd INTEGRAL Workshop*, in press, astro-ph/9811211.
3. Strong, A.W., and Mattox, J.R., *A&A* **308**, L21 (1996).
4. Kinzer, R.L., et al., *ApJ* **515**, 215, (1999).
5. Bloemen, H., et al., *these proceedings.*
6. Strong, A.W., and Moskalenko, I.V., *ApJ* **509**, 212 (1998).
7. Strong, A.W., Moskalenko, I.V., and Reimer, O., astro-ph/9811296.
8. Strong, A.W., and Moskalenko, I.V., *these proceedings.*
9. http://www.gamma.mpe-garching.mpg.de/~aws/aws.html
10. Connell, J.J., *these proceedings.*
11. Connell, J.J., *ApJ* **501**, L59 (1998).
12. Connell, J.J., DuVernois, M.A., and Simpson, J.A. *ApJ* **509**, L97 (1998).
13. Simpson, J.A., and Connell, J.J., *ApJ* **497**, L85 (1998).
14. Strong, A.W., and Moskalenko, I.V., *Proc. 26th Int. Cosmic Ray Conf.* **4**, 255 (1999); astro-ph/9906228.
15. Ptuskin, V.S., and Soutoul, A., *A&A* **337**, 859 (1998).
16. Webber, W.R., and Soutoul, A., *ApJ* **506**, 335 (1998).
17. Yanasak, N., et al., *Proc. 26th Int. Cosmic Ray Conference* **3**, 9 (1999).
18. Hunter, S.D., et al., *ApJ* **481**, 205 (1997).
19. Moskalenko, I.V., Strong, A.W., and Reimer, O., *A&A* **338**, L75 (1998).
20. Basini, G., et al., *Proc. 26th Int. Cosmic Ray Conference* **3**, 77 (1999).
21. Barwick, S.W., et al., *ApJ* **498**, 779 (1998).
22. Pohl, M., and Esposito, J.A., *ApJ* **507**, 327 (1998).
23. Baring, M.G., et al., *ApJ* **513**, 311 (1999).
24. Strong, A.W., and Moskalenko I.V., *AIP Conf. Proceedings* **410**, 1162 (1997).
25. Digel, S.W., et al., *ApJ* **520**, 196 (1999).
26. Chen, A., et al., *ApJ* **463**, 169 (1996).
27. Dixon, D.D., et al., *New Astronomy* **3(7)**, 539 (1998).
28. Moskalenko, I.V., and Strong, A.W., *ApJ* **528**, Jan 1 (2000), astro-ph/9811284.

Evidence for a discrete source contribution to low-energy continuum Galactic γ-rays

Andrew W. Strong* and Igor V. Moskalenko*†‡

*Max-Planck Institut für extraterrestrische Physik, 85740 Garching, Germany
† Institute for Nuclear Physics, M.V. Lomonosov Moscow State University, Moscow, Russia
‡ LHEA NASA/GSFC Code 660, Greenbelt, MD 20771, USA

Abstract. Models for the diffuse Galactic continuum emission and synchrotron radiation show that it is difficult to reproduce observations of both of these from the same population of cosmic-ray electrons. This indicates that an important contributor to the emission below 10 MeV could be an unresolved point-source population. We suggest that these could be Crab-like sources in the inner Galaxy. Alternatively a sharp upturn in the electron spectrum below 200 MeV is required.

I INTRODUCTION

Although 'diffuse' emission dominates the COMPTEL all-sky maps in the energy range 1–30 MeV, its origin is not yet firmly established; in fact it is not even clear whether it is truly diffuse in nature. This is in contrast to the situation at higher energies where the close correlation of the EGRET maps with HI and CO surveys establishes a major component as cosmic-ray interactions with interstellar gas. This paper discusses recent studies of the low-energy diffuse continuum emission based on the modelling approach described in [1]. The high energy (> 1 GeV) situation is addressed in [2,3]. The present work uses observational results reported in [4]; new imaging and spectral results from COMPTEL are presented in [5] but differences are not important for our conclusions.

II ELECTRONS, γ-RAYS AND SYNCHROTRON

Conventionally the low-energy γ-ray continuum spectrum has been explained by invoking a soft electron injection spectrum with index 2.1–2.4, and this could reproduce the 1–30 MeV emission as bremsstrahlung plus inverse Compton emission (see e.g. [6]). Fig 1 shows a range of electron spectra which result from propagation of injection spectral indices 2.0–2.4; the model is from [2]; in order to illustrate more

CP510, *The Fifth Compton Symposium,* edited by M. L. McConnell and J. M. Ryan

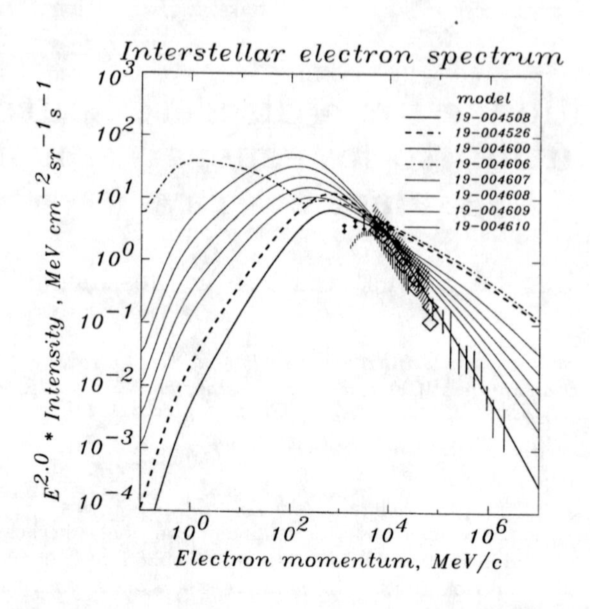

FIGURE 1. Electron spectrum after propagation for various electron injection spectra. Injection index 2.0 to 2.4 (narrow full lines). Also shown is a spectrum which reproduces the high-energy γ-ray excess (dashed line) and a spectrum with a sharp upturn below 200 MeV which can reproduce the low-energy γ-rays without violating synchrotron constraints (dash-dot line). The thick solid line is a spectrum consistent with both local measurements and synchrotron constraints. For the data compilation see [2].

clearly the effect these spectra are without reacceleration. The nucleon spectrum is consistent with local observations and is described in [2]. Fig 2 shows the inner Galaxy γ-ray spectrum for the same electron spectra. The best fit is evidently obtained for index 2.2–2.3.

A problem with this, which was noted earlier but has become clearer with more refined analyses, is the constraint from the observed Galactic synchrotron spectrum on the electron spectral index above 100 MeV. The synchrotron index is hard to measure because of baseline effects and thermal emission, but there has been a lot of new work in this area, in part because of interest in the cosmic microwave background. Fig 3 summarizes relevant measurements of the synchrotron index together with the predictions for the range of electron spectra in Fig 1. The new 22–408 MHz value from [7] is of particular importance here; it is consistent with that derived earlier in a detailed synchrotron modelling study [8]. The γ-rays fit best for an injection index 2.2–2.3, but the synchrotron index for 100–1000 MHz is then about 0.8 which is above the measured range. Although we illustrate this for just one family of spectra for a particular set of propagation parameters, it is clear that it covers the possible range of plausible spectra so that changing the propagation model would not alter the conclusion. Hence we are unable to find an electron spectrum which reproduces the γ-rays without violating the synchrotron constraints. If there were a very sharp upturn in the electron injection spectrum below 200 MeV, as illustrated in Fig 1, then we could explain the γ-rays as bremsstrahlung emission without violating the synchrotron constraints, but even then it would not reproduce the intensities below 1 MeV measured by OSSE [9].

III AN UNRESOLVED SOURCE POPULATION ?

In view of the problems with diffuse emission we suggest that an important component (at least 50%) of the γ-ray emission below 10 MeV originates in a population of unresolved point sources; it is clear that these must anyway dominate eventually as we go down in energy from γ-rays to hard X-rays (see e.g. [10]), so we propose the changeover occurs at MeV energies. For illustration we have tried adding (with arbitrary normalization) to the diffuse emission possible spectra for the unresolved population (Fig 4): a low-state Cyg X-1 type [11] appears too steep, but a Crab-like type ($E^{-2.1}$) would be satisfactory, and would require a few dozen Crab-like sources in the inner Galaxy. These would not be detectable as individual sources by COMPTEL and such a model not violate any observational constraints which we know of. In the examples in Fig 4 we have used the hard electron injection spectrum (index 1.8) required to fit the >1 GeV excess [2,3] so that with Crab-like sources we can finally reproduce the entire spectrum from 100 keV to 10 GeV.

This hypothesis has many observational consequences which can only be investigated by detailed modelling of source populations.

REFERENCES

1. Strong, A.W., and Moskalenko I.V., *ApJ* **509**, 212 (1998).
2. Strong, A.W., Moskalenko I.V., and Reimer, O., astro-ph/9811296 (1998).
3. Strong, A.W., Moskalenko I.V., and Reimer, O., *these proceedings.*
4. Strong, A.W. et al., *Proc 3rd INTEGRAL Workshop*, in press, astro-ph/9811211.
5. Bloemen, H., et al. *these proceedings.*
6. Strong, A.W., and Moskalenko I.V., *AIP Conf. Proceedings* **410**, 1162 (1997).
7. Roger, R.S., et al., *A&AS* **137**, 7 (1999).
8. Lawson, K.D., et al., *MNRAS* **225**, 307 (1987).
9. Kinzer, R.L., et al., *ApJ* **515** 215, (1999).
10. Valinia, A., and Marshall, F.E., *ApJ* **505**, 134 (1998).
11. McConnell, M., et al., *these proceedings.*
12. Strong, A.W., and Mattox, J. R., *A&A* **308**, L21 (1996).
13. Broadbent, A., Haslam, C.T.G, and Osborne, J.L., *MNRAS* **237**, 381 (1989).
14. Davies, R.D., et al., *MNRAS* **278**, 925 (1996).
15. Platania, P., et al.l *ApJ* **505**, 473 (1998).
16. Reich, P., and Reich, W., *A&A* **196**, 211 (1988).
17. Webber, W.R., Simpson, G.A., and Cane, H.V., *ApJ* **236** 448 (1980).

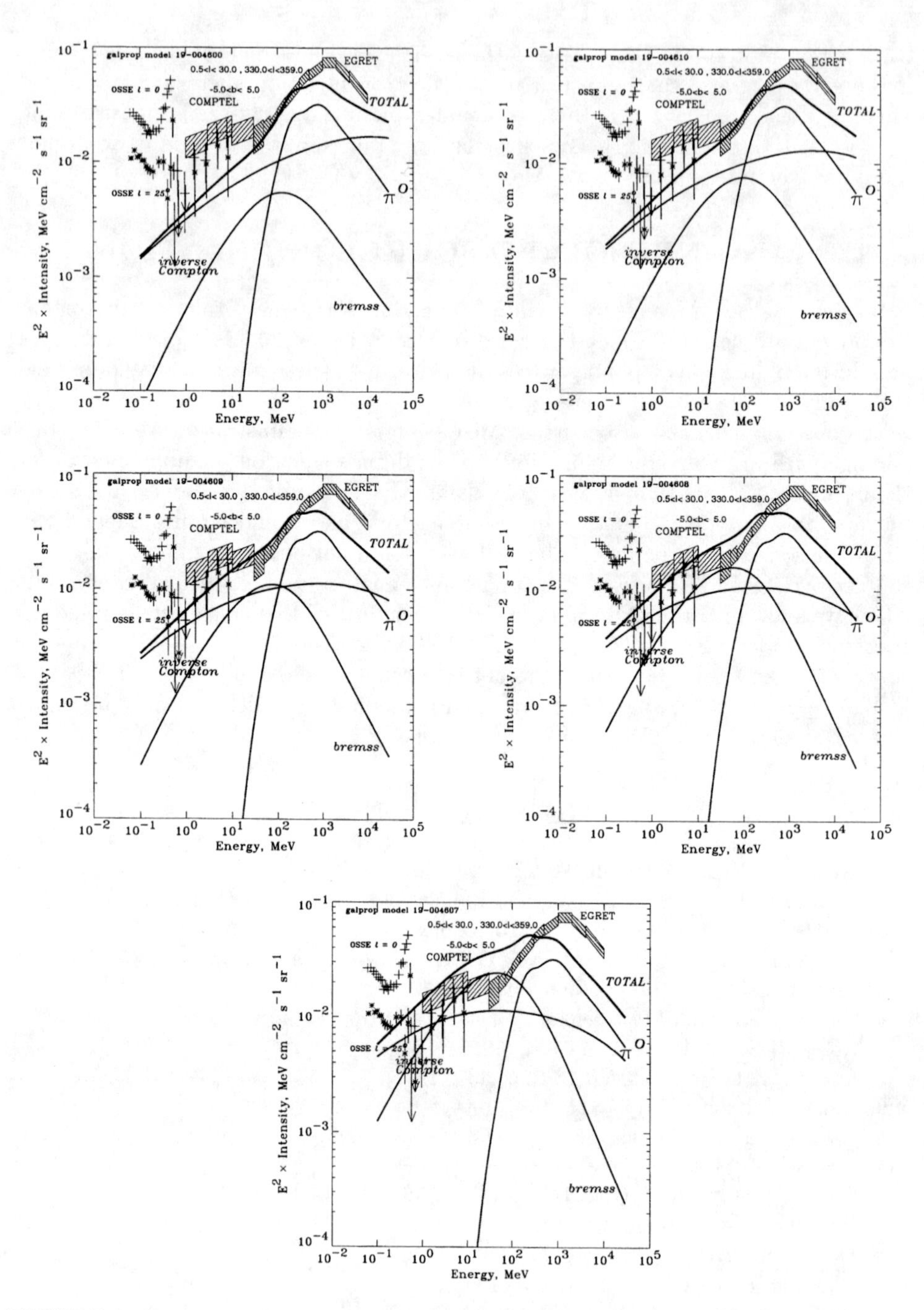

FIGURE 2. Gamma-ray spectrum of the inner Galaxy for various electron injection spectra. Injection index 2.0 to 2.4 (from top to bottom), no reacceleration. Data: [9,4,12] (for details see [2]).

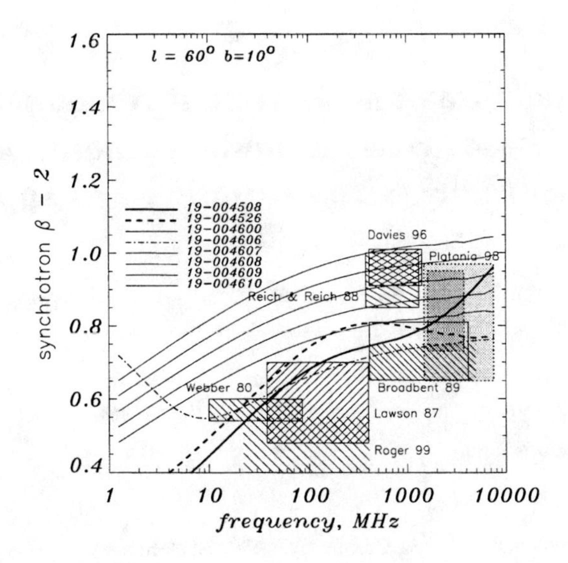

FIGURE 3. Synchrotron index for the electron injection spectra shown in Fig 1. Thin solid lines (from bottom to top): injection index 2.0 to 2.4. Data: [7,8,13–17] (for details see [2]).

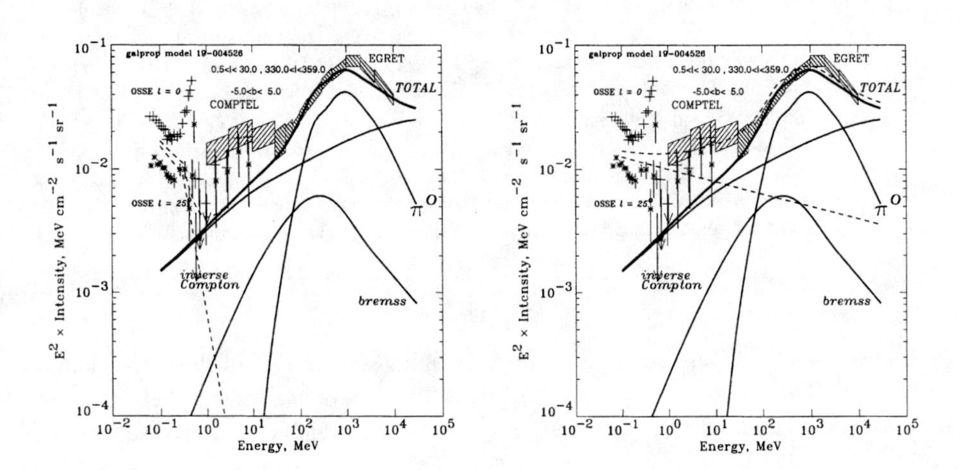

FIGURE 4. Gamma-ray spectrum of the inner Galaxy with possible unresolved source population components. The dashed lines show the assumed source contribution and the sum of source and diffuse components. The solid lines are the diffuse components alone. Left: Cyg X1- (low soft X-ray state) like source spectrum; Right: Crab-like source spectrum. The electron injection spectrum is hard (see text). Data as Fig 2.

The γ-Ray and Cosmic Ray Connection: *Ulysses* HET Secondary Radioisotope Measurements and Cosmic Ray Propagation

J. J. Connell and J. A. Simpson*

*Laboratory for Astrophysics and Space Research, The Enico Fermi Institute, The University of Chicago, Chicago, IL 60610, USA (*Also Department of Physics)*

Abstract. Cosmic rays constitute a super-thermal gas of charged particles magnetically confined within the Galaxy. Cosmic ray nuclei and electrons are a principal source of the diffuse γ-ray background in the Galaxy. Thus, understanding diffuse γ-ray production is directly related to understanding cosmic ray propagation in the Galaxy. While propagating though the interstellar medium (ISM), cosmic ray nuclei undergo nuclear spallation reactions, producing both stable and unstable secondary nuclei. Measurements of secondary radioisotopes are crucial tests of cosmic ray propagation. The abundances of some radioactive secondary isotopes (^{10}Be, ^{26}Al, ^{36}Cl, etc.) measure the average density of material cosmic rays traverse and relate to the confinement times of cosmic rays in the Galaxy. The abundances of electron capture isotopes and their daughter nuclei (for example, ^{49}V and ^{51}V) test the role of cosmic ray reacceleration. The *Ulysses* High Energy Telescope (HET) is a cosmic ray isotope spectrometer with sufficient mass resolution (~0.28 u at Fe) and collecting area to measure these rare isotopes. The latest HET measurements of the radioactive secondary cosmic ray isotopes are given, and the implications for cosmic ray propagation and diffuse γ-ray production are discussed.

INTRODUCTION

The Galactic diffuse γ-ray background constitutes over 90% of the photons in the γ-ray sky. The dominant source of Galactic diffuse γ-rays is cosmic ray electrons and nuclei. Cosmic ray electrons produce γ-rays by bremsstrahlung and inverse Compton interactions while cosmic ray nuclei produce γ-rays via nuclear collisions with interstellar material, mainly through π^0 production (1). A full understanding of the diffuse γ-ray background requires an understanding of cosmic rays. The ultimate goal is a model of cosmic ray production and propagation in the Galaxy that is consistent with all the available data, including cosmic ray measurements and γ-ray and radio observations. No such model is presently available.

Besides producing γ-rays during Galactic propagation, cosmic ray nuclei also produce secondary cosmic rays by nuclear interactions with the ISM. Measurements

CP510, *The Fifth Compton Symposium*, edited by M. L. McConnell and J. M. Ryan

of cosmic ray secondaries are thus a direct link between cosmic ray and γ-ray observations and theory. For example, the amount of material cosmic rays on average traverse during propagation is indicated directly by such cosmic ray measurements as the secondary-to-primary abundance ratios B/C, (Sc-Mn)/Fe ("sub-Fe/Fe"), p^-/p^+ and, to a lesser extent because of uncertainties in energy loss, e^+/e^-.

Unstable cosmic ray secondaries are an equally important test of cosmic ray propagation.

THE AVERAGE INTERSTELLAR DENSITY

The abundance of unstable secondaries observed in the cosmic rays results from the competition between production by nuclear spallation and loss through radioactive decay and escape from the Galaxy. The abundance of long lived nuclides reflect the average density of material cosmic rays traverse, ρ, with the average total amount of material constrained by the stable secondary-to-primary ratios (i.e. B/C). Particularly useful secondary radioisotopes include ^{10}Be, ^{26}Al and ^{36}Cl. Among the three isotopes, ^{26}Al is unique, with its 1.809 MeV γ-ray emission making possible the detection of ^{26}Al throughout the Galaxy as well as in cosmic rays.

The *Ulysses* spacecraft (a joint NASA and ESA mission) carries the University of Chicago High Energy Telescope (HET) described in detail in Simpson et al. (2). The HET clearly resolves individual isotopes through Fe (3), and easily resolves the isotopes of Be, Al and Cl. The curves in Figure 1A (4) show the $^{26}Al/^{27}Al$ ratio calculated using our model (5, 6) with interstellar densities of 0.20 to 0.45 atoms/cm^3. The data point is our *Ulysses* HET value with the vertical error bars reflecting the statistical (1 σ) uncertainties and the horizontal bars the energy interval of the measurement. The results for $^{10}Be/^{9}Be$ (7) and $^{36}Cl/Cl$ (8), summarized in Table 1, are similar. The calculated energy of the nuclei in interstellar space (before solar modulation), E_{IS}, is also shown (7).

Based on ^{10}Be, ^{26}Al and ^{36}Cl, the average density of material that cosmic rays traverse is ~0.25 atom/cm^3 which is significantly less than the ~1 atom/cm^3 accepted for the disk of the Galaxy. This may well reflect the extent to which cosmic rays propagate in a Galactic halo. There is substantial evidence that other galaxies containing cosmic ray electrons have magnetic halos with relativistic electrons, as evidenced by synchrotron emissions (e.g. 9). Using EGRET data, Dixon et al. (10) have reported an extensive Galactic halo distribution of γ-rays. Thus, it appears that both high energy cosmic ray nuclei and electrons escape from the Galactic disk to penetrate throughout the Galactic halo. If so, these isotopic abundances are the best constraints on the size of the halo (11).

While the average density of material cosmic rays traverse, ρ, is relatively well defined, the cosmic ray time-scale for propagation must be interpreted within the context of a given model. Our model is an approximation of the "leaky box", in which

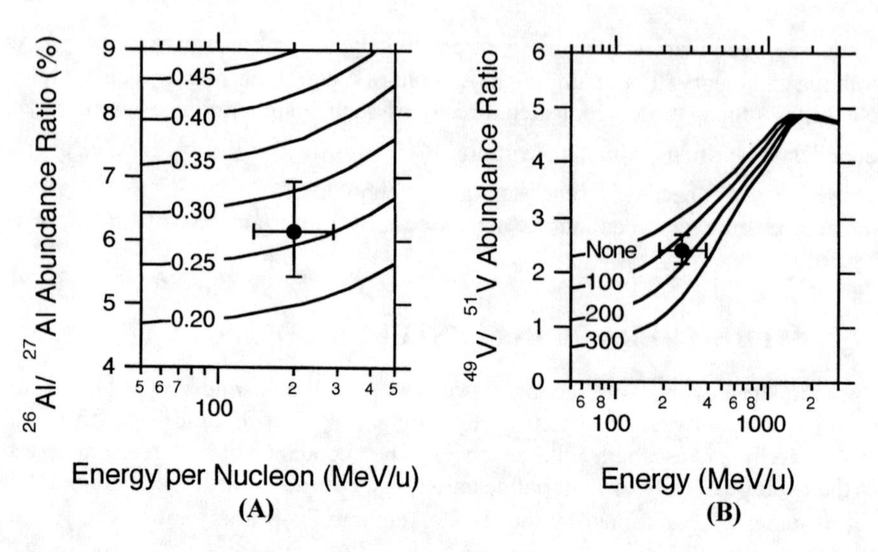

FIGURE 1. **(A)** Plot of calculated isotopic ratio ^{26}Al/ ^{27}Al versus energy per mass unit for various average interstellar densities labeled in atom/cm^3. Point with error bars is measured *Ulysses* HET ratio (4). **(B)** Model predictions for the isotopic abundance ratio ^{49}V/ ^{51}V with no reacceleration ("None"), and with 100, 200 and 300 MeV/u energy "boosts" as indicated. Point with error bars is the isotopic abundance ratio derived from *Ulysses* HET data (17).

TABLE 1. Summary of HET Radio-Chronometer Measurements

	^{10}Be/^{9}Be (7)	^{26}Al/^{27}Al (4)	^{36}Cl/Cl (8)
$\tau_{1/2}$ (yr)	1.6×10^6	8.7×10^5	3.08×10^5
E_{IS} (MeV/u)	523	603	677
ρ (atom/cm^3)	$0.187^{+0.034}_{-0.032}$	$0.26^{+0.05}_{-0.04}$	$0.35^{+0.15}_{-0.13}$
$\langle\lambda\rangle$ (g/cm^2)	6.84	7.45	7.96
T_{esc} (My)	$25.5^{+4.4}_{-4.6}$	$19.1^{+2.8}_{-3.3}$	$14.9^{+5.5}_{-6.4}$

the time-scale is understood as a confinement or escape time, T_{esc} (12). It is noteworthy that the densities found from the three radioisotopes (Table 1) are essentially consistent, despite the fact that each is predominantly a secondary of different primary nuclides.

REACCELERATION

One of the most elusive questions involving cosmic ray propagation in the Galaxy is the role and significance of cosmic ray reacceleration (13 and references therein). If, as is generally believed, cosmic rays with energies up to at least 10^{14} eV are accelerated by supernova shocks, then reacceleration must take place simply because the shocks must encounter some cosmic rays. Reacceleration can also arise from interstellar turbulence (e.g. 14). Whether reacceleration plays a major part in cosmic ray propagation, or is so insignificant that it can be ignored, is largely unknown. Strong reacceleration is ruled out (15) but weak reacceleration could explain the decrease in the secondary to primary elemental ratios, such as B/C, at low energy (13). To date, modeling efforts have been unable to break the degeneracy between reacceleration parameters and other physical parameters. The level of reacceleration is also crucial for understanding diffusive γ-ray production. For example, Moskalenko, et al. (1) find that it is more difficult to fit the observed γ-ray spectrum at high energy with reacceleration than without.

One crucial test of reacceleration is to measure the abundance of electron capture secondary isotopes in cosmic rays. Electron capture decay is strongly suppressed during cosmic ray propagation and is only possible by electron pick-up from the interstellar medium during propagation. This process is highly energy dependent, with low energy nuclei far more likely to pick-up an electron and decay than high energy nuclei (16). Thus, if cosmic rays experience significant reacceleration, the observed cosmic ray nuclei will have spent some time at lower energies during propagation, and electron capture isotopes will be less abundant than otherwise expected while their daughters will be correspondingly more abundant.

We have adopted the simplest possible approach to test for the effects of reacceleration (17). The output of the interstellar propagation model was "boosted" in energy (per nucleon) before being modulated. The pathlength was adjusted for each energy boost (100, 200 and 300 MeV/u) to agree with the secondary to primary ratios (B/C and sub-Fe/Fe) with no reacceleration. While this model is not physically correct, it is indicative of the significance of reacceleration.

Figure 1B shows the results of these propagation calculations for the $^{49}V/^{51}V$ isotopic abundance ratio (17). The curve marked "None" is with no reacceleration, while the curves marked 100, 200 and 300 are for calculations in which the cosmic rays are "boosted" by those energies (in MeV/u) before entering the solar system. ^{49}V decays solely by electron capture to ^{49}Ti ($\tau_{1/2}$ = 331 d) while ^{51}V is the daughter of the electron capture nuclide ^{51}Cr ($\tau_{1/2}$ = 27.7 d), so this ratio is particularly sensitive to reacceleration. (Note, the half-lives quoted are for neutral atoms—for a single pick-up electron the half-lives are slightly more than twice as long.) The point with error bars is the ratio derived from the HET data, 2.4 ± 0.3, corrected for energy intervals in the instrument and spectral shape. The V/Ti and Cr/V elemental abundance measurements (which are only weakly sensitive to reacceleration) are consistent with our model, giving added confidence to our result. The calculated ratios are insensitive to changes in pathlength distribution and interstellar density. Less clear are the uncertainties associated with energy loss from solar modulation: this is a subject of ongoing study.

Previously reported isotopic abundance ratios for vanadium using data from ISEE-3 (18) and *Voyager*-1 and -2 (19) were based on fits to mass distributions with unresolved peaks. Their values for $^{49}V/^{51}V$ ($1.6^{+0.6}_{-0.5}$ and $1.4^{+0.3}_{-0.2}$ respectively) have been interpreted (20) as indicating reacceleration or a lower energy loss from Solar modulation than in present theoretical models.

CONCLUSIONS

Based on ^{10}Be, ^{26}Al and ^{36}Cl, the average density of material that cosmic rays traverse is ~0.25 atom/cm^3. These measurements are an important constraint on the size of the Galactic magnetic halo. Future γ-ray measurements in the halo will be important for further testing the extent of cosmic ray propagation.

Based on our measurements of the electron capture secondary isotopes ^{49}V and ^{51}Cr and their daughters, ^{49}Ti and ^{51}V, we conclude that weak reacceleration of Galactic cosmic rays is indicated. Although our analysis shows evidence of reacceleration, it will be absolutely essential to study other electron capture nuclides as further tests before we can be certain of the significance of reacceleration in cosmic ray propagation. Future measurements of the diffuse γ-ray background will also be important for further constraining models of the propagation of Galactic cosmic rays.

ACKNOWLEDGMENTS

This work was supported in part under NASA Grant NGT-51300 with data prepared under NASA/JPL Contract 955432. Special thanks are due to C. Lopate for solar modulation parameters based on IMP-8 (NASA grant NAG 5-8032) and *Pioneer* 10 (NASA grant NAG 5-6472) data.

REFERENCES

1. Moskalenko, I. V., Strong, A. W. & Reimer, O., *A. & A. Lett.* **338**, L75 (1998).
2. Simpson, J. A., et al., *A. & A. Sup.* **92**, 365 (1992).
3. Connell, J. J. and Simpson, J. A., *Ap. J. Lett.* **475**, L61 (1997).
4. Simpson, J. A. and Connell, J. J., *Ap. J. Lett.* **497**, L85 (1998).
5. Garcia-Munoz, M., Simpson, J. A., Guzik, T. G., Wefel, J. P. and Margolis, S. H., *Ap. J. Sup.* **64**, 269 (1987).
6. DuVernois, M. A., Simpson, J. A. and Thayer, M. R., *A. & A. Sup.* **316**, 555 (1996).
7. Connell, J. J., *Ap. J. Lett.* **502**, L59 (1998).
8. Connell, J. J., DuVernois, M. A. and Simpson, J. A., *Ap. J. Lett.* **509**, L97 (1998).
9. Ekers, R. D. and Sancisi, R., *A. & A.* **54**, 973 (1977).
10. Dixon, D. D., et al., *New Astron.* **3**, 539 (1998).
11. Strong, A. W. and Moskalenko, I. V., *Ap. J.* **509**, 212 (1998).
12. Simpson, J. A. and Garcia-Munoz, M., *Space Sci. Rev.* **46**, 205 (1988).
13. Silberberg, R., Tsao, C. H. and Shapiro, M. M. in *Towards the Millennium in Astrophysics*, M. M. Shapiro, R. Silberberg and J. P. Wefel, eds., Singapore: World Scientific, 1998, p. 227.
14. Seo, E. S. and Ptuskin, V. S., *Ap. J.* **431**, 705 (1994).
15. Hayakawa, S., *Cosmic Ray Physics*, New York: Wiley, 1969.
16. Crawford, H. J., Ph.D. thesis, University of California at Berkeley (1979).
17. Connell, J. J. and Simpson, J. A., *Proc. of the Intern. Cosmic Ray Conf. (Salt Lake City)* **3**, 33 (1999).
18. Leske, R. A., *Ap. J.* **405**, 567 (1993).
19. Lukasiak, A., McDonald, F. B. and Webber, W. R., *Ap. J.* **488**, 454 (1997).
20. Soutoul, A. Legrain, R., Lukasiak, A., McDonald, F. B. and Webber, W. R., *A. & A. Lett.* **336**, L61 (1998).

ACTIVE GALAXIES

Multifrequency Observations of the Virgo Blazars 3C 273 and 3C 279 in CGRO Cycle 8

W. Collmar[1], S. Benlloch[2], J.E. Grove[3], R.C. Hartman[4], W.A. Heindl[5], A. Kraus[6], H. Teräsranta[7], M. Villata[8], K. Bennett[9], H. Bloemen[10], W.N. Johnson[3], T.P. Krichbaum[6], C.M. Raiteri[8], J. Ryan[11], G. Sobrito[8], V. Schönfelder[1], O.R. Williams[9], J. Wilms[2]

[1] *Max-Planck-Institut für extraterrestrische Physik, Postfach1603, 85740 Garching, Germany*
[2] *Institut für Astronomie und Astrophysik, Univ. of Tübingen, Tübingen, Germany*
[3] *Naval Research Lab., 4555 Overlook Av., SW, Washington, DC 20375-5352, USA*
[4] *NASA/Goddard Space Flight Center, Greenbelt, MD 20771*
[5] *Center for Astrophysics and Space Sciences, UCSD, La Jolla, CA, USA*
[6] *Max-Planck-Institut für Radioastronomie, D-53121 Bonn, Germany*
[7] *Metsähovi Radio Research Station, FIN-02540 Kylmälä, Finland*
[8] *Osservatorio Astronomico di Torino, I-10025 Pino Torinese, Italy*
[9] *Astrophysics Division, ESTEC, NL-2200 AG Noordwijk, The Netherlands*
[10] *SRON-Utrecht, Sorbonnelaan 2, NL-3584 CA Utrecht, The Netherlands*
[11] *Universtity of New Hampshire, Durham NH 03824-3525, USA*

Abstract. We report first observational results of multifrequency campaigns on the prominent Virgo blazars 3C 273 and 3C 279 which were carried out in January and February 1999. Both blazars are detected from radio to γ-ray energies. We present the measured X- to γ-ray spectra of both sources, and for 3C 279 we compare the 1999 broad-band (radio to γ-ray) spectrum to measured previous ones.

INTRODUCTION

We report on simultaneous multifrequency observations of the prominent Virgo blazars 3C 273 and 3C 279 during CGRO Cycle 8. Because both blazars are known γ-ray sources, which have been detected by the CGRO experiments several times before, we proposed for simultaneous CGRO (OSSE, COMPTEL) and RXTE high-energy observations. The prime goal was to simultaneously measure their high-energy spectra from about 2.5 keV to 30 MeV. Because of the shortage of spark chamber gas, the EGRET experiment is hardly available anymore and therefore was not requested in the proposals. After the proposed simultaneous high-energy observations were approved and scheduled, additional simultaneous observations

TABLE 1. Summary of the satellite observations of both Virgo blazars during the campaigns in 1999. The observation periods as well as the coverage in energy are given.

Source	Experiment	Obs. Period in '99	Energy Band
3C 273	EGRET	Jan. 20 - Feb. 2	30 MeV - 10 GeV
	COMPTEL	Jan. 5 - Feb. 2	750 keV - 30 MeV
	OSSE	Jan. 19 - Jan. 26	50 keV - ~1 MeV
	RXTE [a]	Jan 19, 26, Feb. 1	2.5 keV - ~200 keV
3C 279	EGRET[b]	Jan. 20 - Feb. 2	30 MeV - 10 GeV
	COMPTEL	Jan. 5 - Feb. 2	750 keV - 30 MeV
	OSSE	Jan. 5 - Jan. 19	50 keV - ~1 MeV
	OSSE[b]	Jan. 26 - Feb. 2	50 keV - ~1 MeV
	RXTE[a]	Jan 5, 12, 17	2.5 keV - ~200 keV

[a] 3 individual pointings of ~13 ksec each
[b] 3C 279 ToO observation; not reported here

were performed from ground-based observers extending the energy range of the campaigns to lower energies.

In this paper we report first observational results of the campaigns with emphasis on the X- and γ-ray part. In particular we present the measured X- to γ-ray spectra of both sources.

OBSERVATIONS

The multifrequency observations were carried out between 1999 January 5 and February 2. The observational strategy was that both blazars are within the COMPTEL field-of-view for the whole 4 weeks reaching the optimal sensitivity for the γ-ray observations, and OSSE observes simultaneously each source for 2 weeks. Within these two-week OSSE periods, three RXTE pointings were scheduled for each source covering simultaneously the X- and hard X-ray part of the spectrum and providing information on the X-ray variability. To supplement these high-energy observations both sources were simultaneously observed in different optical and radio bands. On 1999 January 15 the optical flux of 3C 279 reached a high level which – after some discussion – triggered target-of-opportunity (ToO) observations of this source. This resulted in a switch-on of EGRET which led to EGRET observations of both quasars during roughly the second half of the campaigns. Because of the 3C 279 ToO, the OSSE 3C 273 observation was stopped after one week, and OSSE observed 3C 279 again. These EGRET and OSSE ToO observations of 3C 279 are property of a different CGRO proposal and therefore their results are not reported here. The detailed high-energy observational log is given in Table 1.

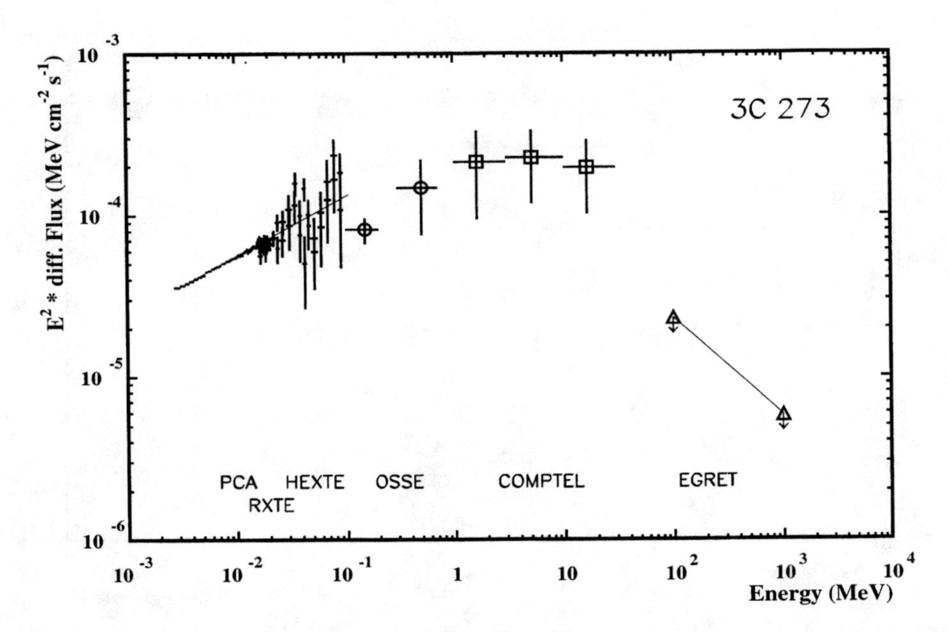

FIGURE 1. Simultaneous X-ray and γ-ray spectrum of 3C 273 as observed during the January/February 1999 campaign. The data points from the different experiments are derived from their total observation times during the campaign, which are given in Table 1. The RXTE (+) spectral points are derived from the observation sum of the three individual pointings, and are shown together with the best-fit power-law spectrum (solid line). EGRET did not detect the quasar at energies above 100 MeV providing an upper limit on the flux. For the upper limit line, drawn between 100 MeV and 1 GeV, a spectral power-law shape with photon index of 2.6 is assumed. The error bars are 1σ.

RESULTS

3C 273

3C 273 is significantly detected in all observed low-energy (radio, optical) bands. At high energies the quasar is significantly detected in X- and hard X-rays by RXTE from about 2.5 to $\sim$100 keV, showing a power-law spectrum with a photon index ($E^{-\alpha}$) α of 1.6. OSSE detects the source in the one-week observation at a significance level of $\sim 5\sigma$. Therefore its spectrum had to be rebinned severely to reach two significant spectral points. COMPTEL detects the blazar in the sum of the 4-week observation at the $\sim 5\sigma$ level. However, despite the COMPTEL detection at MeV-energies, EGRET – covering only half of the COMPTEL observation time – does not detect 3C 273 at energies above 100 MeV. The combined – more or less simultaneous – high-energy spectrum of 3C 273 is shown in Figure 1. The

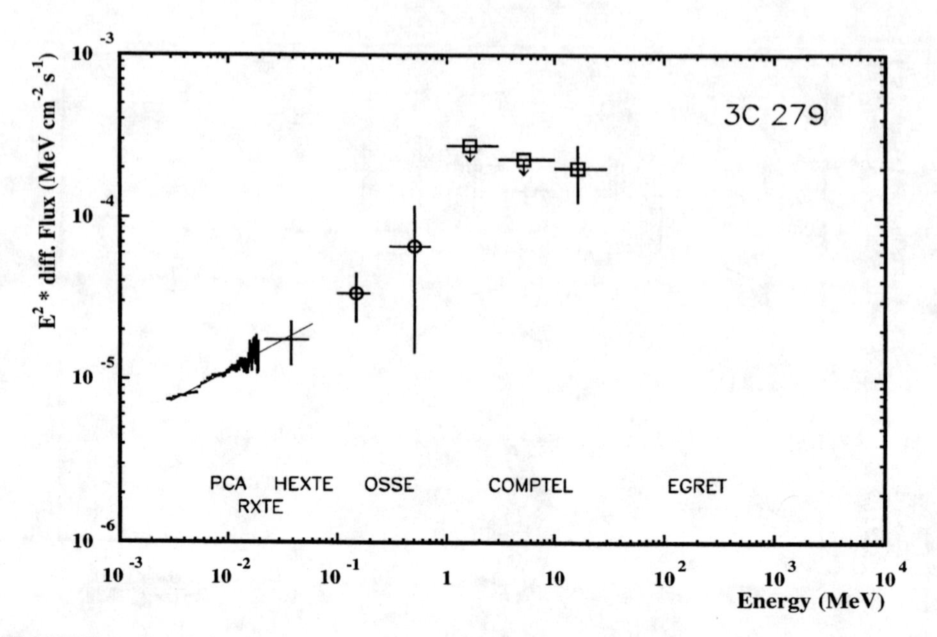

FIGURE 2. Quasi-simultaneous keV- to MeV-spectrum of the γ-ray blazar 3C 279 as observed in early 1999. The data points from the different experiments are derived from their total observations times during the campaign, which are given in Table 1. The RXTE (+) spectral points are derived from the observation sum of the three individual pointings, and are shown together with the best-fit power-law spectrum (solid line). The error bars are 1σ and the upper limits are 2σ. Simultaneous EGRET observations exist (Table 1), which will be reported elsewhere. The spectrum can be described by a single power-law shape from about 2.5 keV to 30 MeV.

well-known bending (e.g. [1], [2]) at MeV-energies is visible. The most surprising result however, is the non-detection by EGRET at high-energy γ-rays, despite the COMPTEL detection at MeV-energies. This requires a strong spectral turnover between 30 and 100 MeV, and might hint at different generation mechanisms for the MeV and >100 MeV photon populations.

3C 279

3C 279 is significantly detected at the radio and optical bands, showing strong time variability and flaring activity in the optical. At X-rays the blazar is significantly observed up to 20 keV by the RXTE/PCA, and is detected in hard X-rays between 20 and 50 keV by RXTE/HEXTE with a significance of about 5σ. The RXTE/PCA spectrum is well fitted by a single power-law model with an index α of 1.6. At higher energies the detection significances become marginal. OSSE

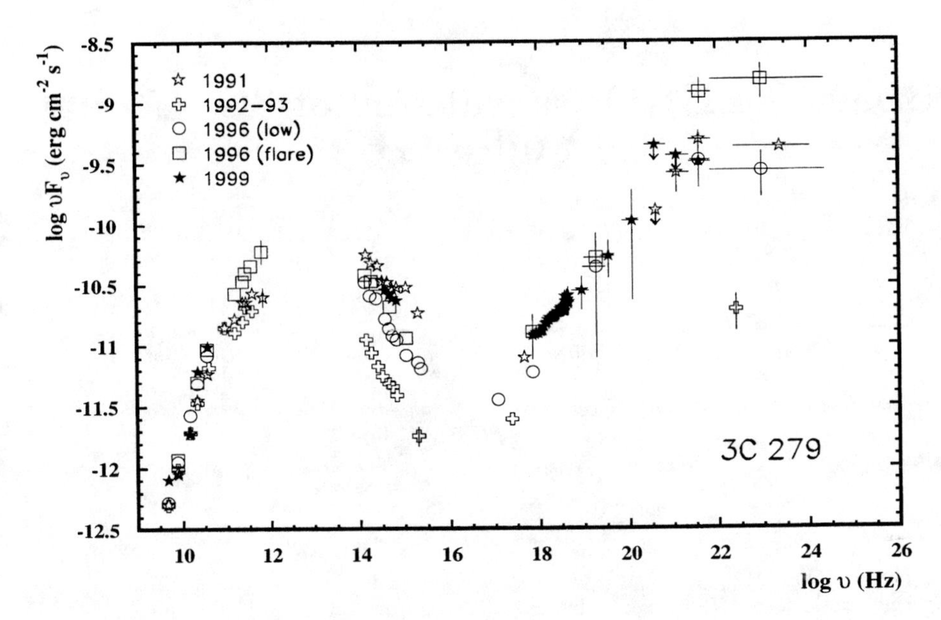

FIGURE 3. Broad-band spectrum of 3C 279 for different epochs. The 1999 results (filled stars) reported here fit nicely to the two previous measurements in 1991 and 1996 when 3C 279 was in a γ-ray high state. The spectral points of the earlier measurements are taken from [4].

found 3σ-evidence for the source only at their lower energies (near 100 keV), and COMPTEL – also at the 3σ level – only at their upper energies. At energies above 100 MeV EGRET has significantly detected 3C 279 ([3]), however these data are not reported here. The measured X- to γ-ray spectrum of 3C 279 is given in Figure 2. It shows that 3C 279 was observed in a bright γ-ray state. The flux at the COMPTEL upper energies is at the same level as measured during the two previous γ-ray high states in 1991 and 1996 (Figure 3). The spectral power-law shape measured from ~2.5 to 20 keV can – according to the current state of analysis – be extrapolated up to 30 MeV without any obvious breaks or bendings. This suggests that this part of the spectrum, which is considered to be non-thermal inverse-Compton radiation, is emitted by a single emission component or mechanism.

REFERENCES

1. Lichti, G.G., et al., *A&A* **298**, 711 (1995).
2. v. Montigny, C., et al., *ApJ* **483**, 161 (1997).
3. Hartman, R.C., *priv. comm.*, (1999).
4. Wehrle, A.E., et al., *ApJ* **497**, 178 (1998).

Space VLBI Observations of 3C 279 at 1.6 and 5 GHz

B.G. Piner[1], P.G. Edwards[2], A.E. Wehrle[1], H. Hirabayashi[2], J.E.J. Lovell[3], & S.C. Unwin[1]

[1]*Jet Propulsion Laboratory, California Institute of Technology, 4800 Oak Grove Dr., Pasadena, CA 91109*
[2]*Institute of Space and Astronautical Science, Sagamihara, Kanagawa 229-8510, Japan*
[3]*Australia Telescope National Facility, PO Box 76, Epping NSW 1710, Australia*

Abstract. We present the results of VLBI Space Observatory Programme (VSOP) observations of the gamma-ray blazar 3C 279 at 1.6 and 5 GHz from January 1998. The combination of the VSOP and VLBA-only images at these two frequencies maps the jet structure on scales from 1 to 100 mas. A spectral index map was made by combining the VSOP 1.6 GHz image with a matched-resolution VLBA-only image at 5 GHz from our VSOP observation on the following day. The spectral index map shows the core to have a highly inverted spectrum, with some areas having a spectral index approaching the limiting value for synchrotron self-absorbed radiation of +2.5 ($S \propto \nu^{\alpha}$). Gaussian model fits to the VSOP visibilities reveal high brightness temperatures ($> 10^{12}$ K) that are difficult to measure with ground-only arrays. An extensive error analysis was performed on the brightness temperature measurements. Most components did not have measurable brightness temperature upper limits, but lower limits were measured as high as 5×10^{12} K. This lower limit is significantly above both the nominal inverse Compton and equipartition brightness temperature limits. The derived Doppler factor in the case of the equipartition limit is at the upper end of the range of expected values for EGRET blazars.

I INTRODUCTION

The quasar 3C 279 (z=0.536) is one of the most intensively studied quasars for several reasons. It was the first radio source whose milliarcsecond scale structure was observed to exhibit superluminal motion [1]. It was the first blazar — and remains one of the brightest — detected in high-energy γ-rays by the EGRET instrument on the *Compton Gamma Ray Observatory* [2], which prompted several large multiwavelength studies of this source (e.g. [3]). For these reasons 3C 279 was chosen as part of a VSOP Key Science Project. We report in this paper on the VSOP observations of 3C 279 made during the 1st VSOP Announcement of Opportunity (AO1).

CP510, *The Fifth Compton Symposium*, edited by M. L. McConnell and J. M. Ryan

II IMAGES

The quasar 3C 279 was observed on 1998 January 9 at 1.6 GHz, and on 1998 January 10 at 5 GHz. The arrays were made up of the Japanese HALCA satellite and nine elements of the NRAO Very Long Baseline Array (VLBA) with the addition of the 70 m telescopes at Goldstone, California, U.S.A. and Tidbinbilla, Australia on January 9, and the 64 m telescope at Usuda, Japan on January 10. Figure 1 shows two images from the 5 GHz observation: the full-resolution space VLBI image and the VLBA-only image made from the same dataset with the space baselines removed. Figure 2 shows the full-resolution space VLBI and the VLBA-only images from the 1.6 GHz observation. Preliminary versions of these 5 and 1.6 GHz VSOP images have appeared in [4] and [5], respectively. The VSOP and VLBA-only images differ in scale by about a factor of four, showing the greatly increased resolution provided by the space baselines. The VSOP images presented in Figures 1 and 2 are the highest resolution images yet produced of 3C 279 at these frequencies. The 5 GHz VSOP image shows that on small angular scales 3C 279 is dominated by a double structure, consisting of the core and a bright jet component about 3 mas from the core along a position angle of $-115°$. This bright jet component is a well known superluminal feature denoted C4 [6]. The structure on slightly larger angular scales is quite different. The 5 GHz VLBA image and the 1.6 GHz VSOP image both show the double structure mentioned above, as well as a more extended jet to the southwest along a position angle of approximately $-140°$. This position angle is similar to that seen in older VLBI images as well as VLA and MERLIN images (e.g. [7]). The jet emission in the 1.6 GHz VSOP image is quite complex, and it appears that the jet may be limb brightened. However, we caution against over-interpreting these transverse features because the striping produced by the holes in the (u, v) plane coverage runs parallel to the jet. Future AO2 VSOP observations of 3C 279 at 1.6 GHz in which the baselines to the orbiting antenna have a different orientation in the (u, v) plane will allow a consistency check on these transverse structures. The 1.6 GHz VLBA image shows structure extending out to $\sim$ 100 mas from the core, all the way out to the smallest size scales sampled by the VLA images of de Pater & Perley [7].

III SPECTRAL INDEX MAP

Construction of spectral index maps is often hindered by the differing resolutions of the images at different frequencies; however, a powerful capability of the VSOP mission is that it can provide matched resolution images to ground-based images at higher frequencies. We have used this capability to produce a spectral index map from the 1.6 GHz VSOP image and the 5 GHz VLBA-only image; this spectral index map is shown in Figure 3. The core of 3C 279 has an inverted spectrum with steep spectral index gradients. Such inverted spectra are commonly interpreted as being due to self-absorption of the radio synchrotron emission. The calculated

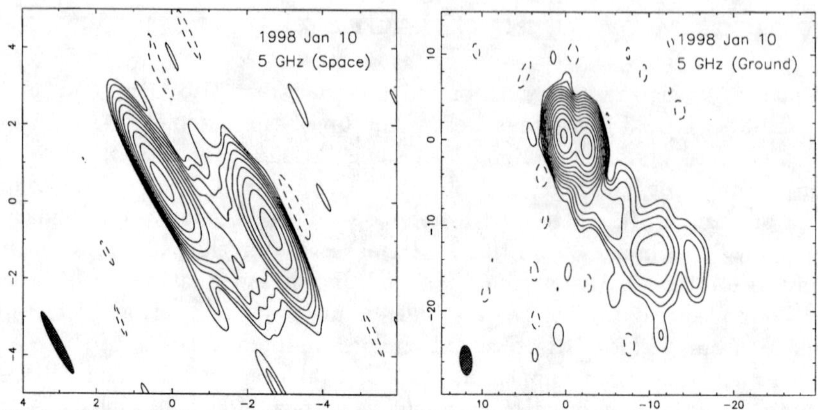

FIGURE 1. Space (left) and VLBA-only (right) 5 GHz images of 3C 279. The peak flux densities are 3.98 and 6.67 Jy beam^{-1}, the contour levels are 6.0 mJy beam^{-1} $\times$ 1,2,4,...512 and 1.4 mJy beam^{-1} $\times$ 1,2,4,...4096, and the beam sizes are 1.83 $\times$ 0.24 mas at 28° and 3.52 $\times$ 1.51 mas at 4° for the space and ground images respectively.

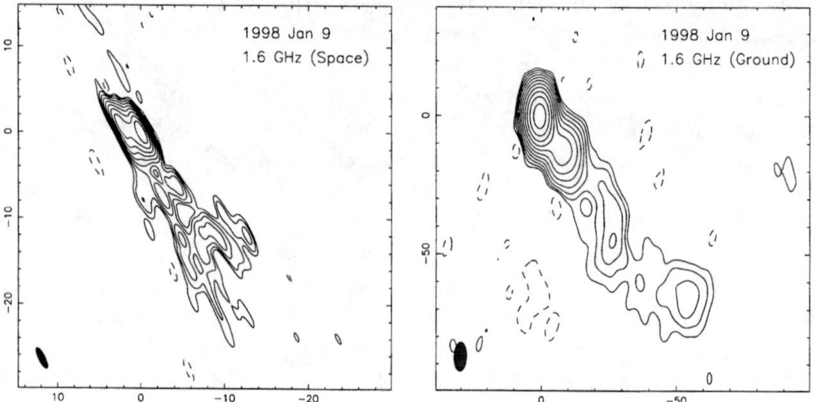

FIGURE 2. Space (left) and VLBA-only (right) 1.6 GHz images of 3C 279. The peak flux densities are 2.29 and 4.73 Jy beam^{-1}, the contour levels are 4.7 mJy beam^{-1} $\times$ 1,2,4,...256 and 2.9 mJy beam^{-1} $\times$ 1,2,4,...1024, and the beam sizes are 2.71 $\times$ 0.77 mas at 28° and 10.7 $\times$ 4.77 mas at $-2°$ for the space and ground images respectively.

spectral index in the core region ranges from $\sim$ 1.0 at the western edge to the theoretical limiting value for synchrotron self-absorption of 2.5 over a small region at the eastern edge. Spectral indices approaching this theoretical value are almost never seen; the flatter spectra are commonly interpreted as being due to an inhomogeneous source made up of a number of synchrotron components with differing turnover frequencies. The highly inverted spectrum at the eastern edge implies the detection of a homogeneous compact component in this region, which should be an efficient producer of inverse Compton gamma-rays.

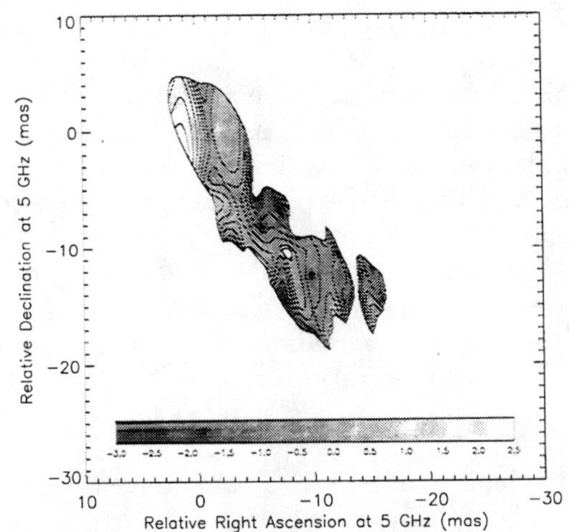

FIGURE 3. Spectral index map of 3C 279. The gray-scale color bar indicates the value of the spectral index ($S \propto \nu^{\alpha}$). Spectral index contours are also plotted at intervals of 0.5, from −2.5 to 2.5.

IV HIGH BRIGHTNESS TEMPERATURES

After fitting elliptical Gaussian components to the visibility data, we calculated their brightness temperatures from their modeled flux densities and sizes. Observed brightness temperatures are often used to calculate Doppler beaming factors by assuming an intrinsic brightness temperature and using the fact that $T_{B,obs} = \delta T_{B,int}$, where $T_{B,obs}$ is the observed source frame brightness temperature, $T_{B,int}$ is the intrinsic brightness temperature, and δ is the Doppler factor. The intrinsic brightness temperature depends on the physical mechanism imposing the limitation. Kellermann & Pauliny-Toth [8] showed that inverse Compton losses limit the intrinsic brightness temperatures to $\sim 5 \times 10^{11} - 1 \times 10^{12}$ K. Readhead [9] proposed that the limiting mechanism is equipartition of energy between the particles and magnetic field, and that intrinsic brightness temperatures are limited to $\sim 5{\times}10^{10}{-}1{\times}10^{11}$ K. Space VLBI observations have a major advantage over matched resolution ground-based observations at higher frequencies because the maximum observable brightness temperature depends on baseline length but not on observing frequency. The improvement gained by space VLBI covers the interesting transition region around 10^{12} K, the nominal inverse Compton brightness temperature limit [8]. An extensive error analysis was conducted on the model-fit brightness temperatures using the "Difwrap" program[1] coauthored by one of us (Lovell). For many components an acceptable fit was found with a component of zero area and infinite brightness temperature (in all cases caused by a valid fit with zero axial ratio at some position angle), and therefore it appears that many measured brightness temperatures, even

[1] http://halca.vsop.isas.ac.jp./survey/difwrap/

those measured by space VLBI, may have error bars that extend to infinity in the positive direction. Three components had minimum brightness temperatures over 10^{12} K, with the highest minimum brightness temperature of 4.9×10^{12} K being measured for component C4 in 1.6 GHz VSOP model fit. Our highest brightness temperature lower limit of $\sim 5 \times 10^{12}$ K implies Doppler factor lower limits of 5 and 50 for the inverse Compton and equipartition brightness temperature limits respectively. A Doppler factor of 50 is at the upper end of the Doppler factor distributions expected for flux-limited samples of flat-spectrum radio sources [10] and gamma-ray sources [11]. Bower & Backer [12] found similar values for the Doppler factor from their VSOP observations of NRAO 530 under these same two limiting conditions. If VSOP observations reveal a brightness temperature much higher than 5×10^{12} K, or many brightness temperatures around 5×10^{12} K, it may be difficult to reconcile the high Doppler factors implied by the equipartition brightness temperature limit with beaming statistics and with the relatively slow speeds measured in studies of apparent velocity distributions [13].

V ACKNOWLEDGMENTS

Part of the work described in this paper has been carried out at the Jet Propulsion Laboratory, California Institute of Technology, under contract with the National Aeronautics and Space Administration. A.E.W. acknowledges support from the NASA Long Term Space Astrophysics Program. We gratefully acknowledge the VSOP Project, which is led by the Japanese Institute of Space and Astronautical Science in cooperation with many organizations and radio telescopes around the world. The National Radio Astronomy Observatory is a facility of the National Science Foundation operated under cooperative agreement by Associated Universities, Inc.

REFERENCES

1. Whitney, A.R., et al., *Science*, **173**, 225 (1971).
2. Hartman, R.C., et al., *ApJ*, **385**, L1 (1992).
3. Wehrle, A.E., et al., *ApJ*, **497**, 178 (1998).
4. Hirabayashi, H., et al., *Adv. Sp. Res.*, in press.
5. Edwards, P.G., et al., *Astronomische Nachrichten*, in press.
6. Unwin, S.C., et al., *ApJ*, **340**, 117 (1989).
7. de Pater, I. & Perley, R.A., *ApJ*, **273**, 64 (1983).
8. Kellermann, K.I. & Pauliny-Toth, I.I.K., *ApJ*, **155**, L71 (1969).
9. Readhead, A.C.S., *ApJ*, **426**, 51 (1994).
10. Lister, M.L., & Marscher, A.P., *ApJ*, **476**, 572 (1997).
11. Lister, M.L., PhD thesis, Boston University (1998).
12. Bower, G.C. & Backer, D.C., *ApJ*, **507**, L117 (1998).
13. Vermeulen, R.C., *Proc. Natl. Acad. Sci.*, **92**, 11385 (1995).

*Beppo*SAX Observations of Mkn 421: clues on the particle acceleration ?

G. Fossati[1], A. Celotti[2], M. Chiaberge[2] and Y.H. Zhang[2]

[1] *UCSD/CASS, 9500 Gilman Drive, La Jolla, CA 92093-0424, U.S.A. — gfossati@ucsd.edu*
[2] *SISSA, via Beirut 2-4, 34014 Trieste, Italy — celotti, chiab, yhzhang@sissa.it*

Abstract. Mkn 421 was repeatedly observed with *Beppo*SAX in 1997–1998. We present highlights of the results of the thorough temporal and spectral analysis discussed by Fossati et al. (1999) and Maraschi et al. (1999), focusing on the flare of April 1998, which was simultaneously observed also at TeV energies. The detailed study of the flare in different energy bands reveals a few very important new results: (a) hard photons lag the soft ones by 2–3 ks –a behavior opposite to what is normally found in High energy peak BL Lacs X–ray spectra; (b) the flux decay of the flare can be intrinsically achromatic if a stationary underlying emission component is present. Moreover the spectral evolution during the flare has been followed by extracting X–ray spectra on few ks intervals, allowing to detect for the first time the peak of the synchrotron component shifting to higher energies during the rising phase, and then receding. The spectral analysis confirms the delay in the flare at the higher energies, as above a few keV the spectrum changes only after the peak of the outburst has occurred. The spectral and temporal information obtained challenge the simplest models currently adopted for the (synchrotron) emission and most importantly provide clues on the particle acceleration process. A theoretical picture accounting for all the observational constraints is discussed, where electrons are injected at low energies and then progressively accelerated during the development of the flare.

INTRODUCTION

Blazars are radio–loud AGNs characterized by strong variability, large and variable polarization, and high luminosity. The spectral energy distribution (SED) typically shows two broad peaks in a νF_ν representation (Fossati et al. 1998), with the emission up to X–rays thought to be due to synchrotron radiation from high energy electrons, while it is likely that γ-rays derive from the same electrons via inverse Compton (IC) scattering. In X–ray bright BL Lacs (HBL, from High-energy-peak-BL Lacs, Padovani & Giommi 1995) the synchrotron maximum occurs in the soft-X–ray band, and the IC emission extends in some cases to the TeV band.

Mkn 421 ($z = 0.031$) is the brightest HBL at X–ray and UV wavelengths and the first extragalactic source discovered at TeV energies (Punch et al. 1992), where dramatic variability has been observed (Gaidos et al. 1996).

CP510, *The Fifth Compton Symposium,* edited by M. L. McConnell and J. M. Ryan

THE 1998 X–RAY/TEV FLARE

In 1998 *Beppo*SAX observed Mkn 421 as part of a long lasting monitoring campaign (see also Takahashi in these Proceedings). *Beppo*SAX observations are dominated by an isolated flare (see Fig. 1), and one of the striking and important results is that in correspondence with the X–ray flare of April 21st a sharp TeV flare was detected by the Whipple Cherenkov Telescope (Figure 1). The peaks in the 0.1–0.5 keV, 4.0–6.0 keV and 2 TeV light curves are *simultaneous within one hour* (see Maraschi et al. 1999).

Here we will focus on the X–ray characteristics of the April 21st flare. We accumulated light curves for different energy bands. The post–flare light curves have been modeled with an exponential decay, superimposed to a steady emission. Four the main results:

<u>Decay Timescales:</u> the timescales range between 30 and 45$\times 10^3$ seconds, and *do not* show a clear (if any) relationship with the energy, rather suggesting that the post–flare spectral evolution can be *achromatic*. This result leads to reject the simplest possibility that the decay evolution is driven by the radiative cooling of emitting electrons (this simplest picture would produce a dependence of the timescale with energy, $\tau \sim E^{-1/2}$).

<u>Flaring/Steady components:</u> exponential decay fits require the presence of an underlying less variable component.

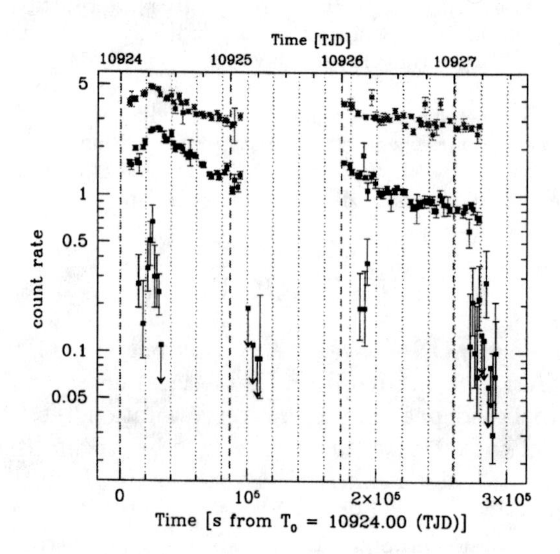

FIGURE 1. Light curves of Mkn 421 at TeV and X–ray energies, during the 1998 campaign. They are shown in order of increasing energy from bottom to top: Whipple E$\geq$ 2 TeV, MECS 4.0–6.0 keV and LECS 0.1–0.5 keV (both with 1500 s bins, multiplied by a factor 4 and 8, respectively). The count rate units are cts/s for *Beppo*SAX data, and cts/min for Whipple data.

Time Lag: the harder X–ray photons lag the soft X–ray ones. We performed a cross correlation analysis using the DCF (Edelson & Krolik 1988) and the MMD (Hufnagel & Bregman 1992) techniques and statistically determined the significance of the time lags using Monte Carlo simulations (Peterson et al. 1998). We refer to Zhang et al. (1999) for the relevant details of such analysis. The result is an average lag of $-2.7^{+1.9}_{-1.2}$ ks for DCF, and $-2.3^{+1.2}_{-0.7}$ ks for MMD (1 σ). This finding is opposite to what is normally found in the best monitored HBL X–ray spectra (e.g. Urry et al. 1993; Kohmura et al. 1994; Takahashi et al. 1996; Zhang et al. 1999) whose hard-to-soft behavior is usually interpreted in terms of cooling of the synchrotron emitting particles.

Rise vs. Fall: possible "asymmetry" of the rise/decay of the flare especially for the higher energy X–rays. The flare seems to be symmetric at the energies corresponding (roughly) to the synchrotron peak, while it might have a faster rise at higher energies. This could be connected to the observed hard–lag.

SPECTRAL VARIABILITY (1997 & 1998)

We accumulated spectra in sub-intervals, and developed an *intrinsically curved spectral model* to be able to estimate the position of the peak of the synchrotron component. In 1997 the source was in a lower brightness state, with a softer ($\Delta\alpha_{97,98} \simeq 0.4$) X–ray spectrum at all energies, and the peak energy 0.5 keV lower. There is a clear relation between the flux variability and the changes in the spectral parameters, both in 1997 and in 1998.

Synchrotron peak energy: the main new result is that we were *able to determine the energy of the peak of the synchrotron component* (with its error). We find a correlation between changes in the brightness and shifts of the peak position (e.g. Fig. 2). The source reveals a strikingly coherent spectral behavior between 1997 and 1998, and through a large flux variability (a factor 5 in the 0.1–10.0 keV band). The peak energies lie along a tight relation $E_{peak} \propto F^{0.55\pm0.05}$.

Hard Lag in 1998 spectra: the spectral analysis confirms the signature of the hard lag. A blow up of the 1998 flare interval is shown in Figure 3. The main remarkable features are: (a) the synchrotron peak shifts toward higher energy during the rise, and then decreases as soon as the flare is over. (b) The spectral index at 1 keV reflects exactly the same behavior, as expected being computed at the energy around which the peak is moving. (c) On the contrary, the spectral shape at 5 keV does not vary until a few ks after the peak of the flare, and only then –while the flux is decaying and the peak is already receding– there is a response with a significant hardening of the spectrum.

The fact that the spectral evolution at higher energies develops during the decay phase of the flare, produces a nice counter-clockwise loop in the α vs. Flux diagram, i.e. *opposite* way with respect to all the other known cases for HBLs (e.g. Sembay et al. 1993; Kohmura et al. 1994; Takahashi et al. 1996).

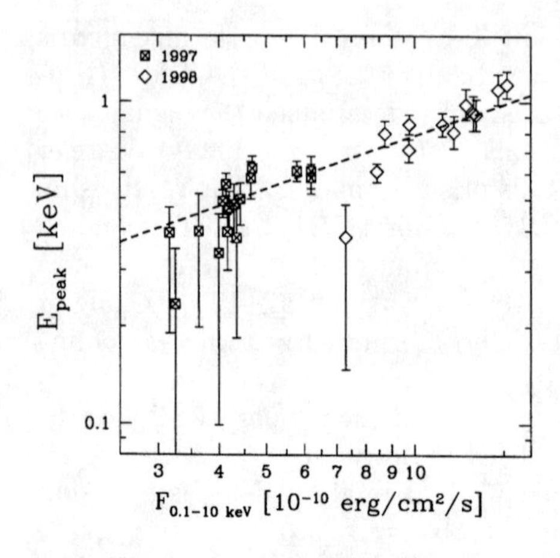

FIGURE 2. Synchrotron peak energy plotted versus "de-absorbed" 0.1–10.0 keV flux. The dashed line represents the best fitting power law, having a slope $\epsilon = 0.55$.

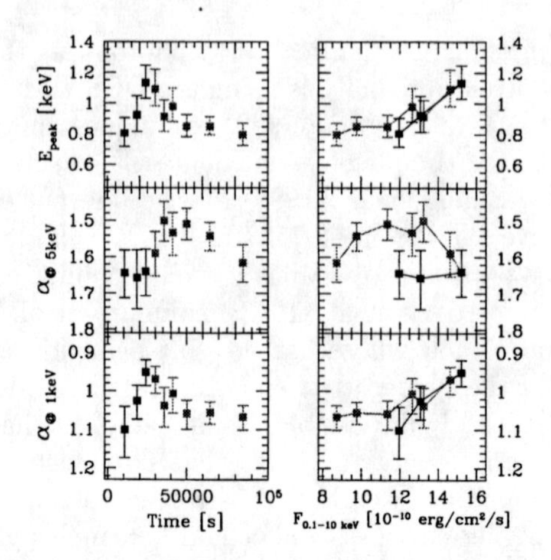

FIGURE 3. The photon spectral indices at 1 keV and at 5 keV, and energy of the peak of the synchrotron component, are plotted versus time and "de-absorbed" 0.1–10.0 keV flux.

PHYSICAL INTERPRETATION

Let us now focus on the possible interpretation of the two main results of this work: the *hard lag* and the *evolution of the synchrotron peak.*

The occurrence of the flare peak at different times for different energies is most likely related to the particle acceleration/heating process.

We therefore *introduced an acceleration term in the time dependent particle kinetic equation* within the model proposed by Chiaberge & Ghisellini (1999), which takes into account the cooling and escape terms and the role of delays in the received photons due to the travel time from different parts of the emitting volume.

The main constraints on the (parametric) form of the acceleration are: [A] particles have to be progressively accelerated from lower to higher energies within the flare rise timescale to produce the hard lag; [B] the emission in the LECS band from the highest energy particles (those radiating initially in the MECS band) should not exceed that from the lower energy ones, as after the peak no further increase of the (LECS) flux is observed; [C] the total decay timescale might be dominated by the achromatic crossing time effects, although the initial phase might be partly determined by the different cooling timescales.

It should be also noted that –within this scenario– the symmetry between the raise and decay of the softer energy light curve seems to suggest that at the same very energies where most of the power is released –possibly determined by the balance between the acceleration and cooling rates– the acceleration timescales are comparable to the region light crossing time.

If the timescales associated with this process are intrinsically linked to the typical size of the emitting region, we indeed expect the observed light curve to be symmetric where the bulk of power is concentrated, and an almost achromatic decay.

Indeed, within a single emission region scenario, we have been able to reproduce the sign and amount of lags, postulating that particle acceleration follows a simple law, and stops at the highest particle energies. The same model can account for the spectral evolution (shift of the synchrotron peak) during the flare.

CONCLUSIONS

These results provide us with several *temporal* and *spectral constraints* on any model. In particular, they could possibly be the *first direct signature of the ongoing acceleration process*, progressively "pumping" electrons from lower to higher energies. The measure of the delay provides a tight constraint on the timescale of the acceleration mechanisms.

A last crucial point is that our results support the possibility of the presence and role of quasi–stationary emission. The short-timescale, large-amplitude variability events could be attributed to the development of new individual flaring components (possibly maintaining a quasi-rigid shape), giving rise to a spectrum outshining a more slowly varying emission. The decomposition in these two components might allow to determine the nature and modality of the dissipation in relativistic jets.

REFERENCES

1. Chiaberge, M., Ghisellini, G. 1999, MNRAS, **306**, 551
2. Edelson, R. A., Krolik, J. H. 1988, ApJ, **333**, 646
3. Fossati, G., et al. 1998, MNRAS, **299**, 433
4. Fossati, G., et al. 1999, submitted to ApJ
5. Gaidos, J. A., et al. 1996, Nature, **383**, 319
6. Hufnagel, B. R., Bregman, J. N. 1992, ApJ, **386**, 473
7. Kohmura, Y., et al. 1994, PASJ, **46**, 131
8. Maraschi, L., et al. 1999, ApJLetters, in press
9. Padovani, P., and Giommi, P. 1995, ApJ, **444**, 567
10. Peterson, B. M., et al. 1998, PASP, **110**, 660
11. Punch, M., et al. 1992, Nature, **358**, 477
12. Sembay, S., et al. 1993, ApJ, **404**, 112
13. Takahashi, T., et al. 1996, ApJ, **470**, L89
14. Urry, C. M., et al. 1993, ApJ, **411**, 614
15. Zhang, Y. H., et al. 1999, ApJ, in press

GeV outbursts in Mrk 501

P. Sreekumar*, D. L. Bertsch**, S. D. Bloom†, R. C. Hartman**, Y. C. Lin††, R. Mukherjee‡, D. J. Thompson**

*ISRO Satellite Center, Bangalore, India

**NASA/Goddard Space Flight Center, Code 661, Greenbelt MD 20771

†Hampden-Sydney College, Hampden-Sydney, VA 23943

††W. W. Hansen Experimental Physics Laboratory, Stanford University, Stanford, CA

‡Barnard College and Columbia University, New York, NY 10027

Abstract. Mrk 501 is the third TeV blazar with a known GeV component. Previous multiwavelength campaigns on Mrk 501 showed well correlated outbursts at x-ray and TeV energies with no significant activity at GeV energies. We present here new evidence suggesting GeV outbursts in Mrk 501 when the spectrum appears to be extremely hard. However, this outburst appears uncorrelated with emission at x-ray energies. The resulting spectral energy distribution suggests a sharp cut off in the high-energy emission beyond a few hundred GeV.

I INTRODUCTION

Observations by the high-energy telescope EGRET on board the Compton Observatory have shown the presence of a class of active galactic nuclei called blazars that emit strongly in γ-rays (Mukherjee *et al.* 1997). Blazars are characterized by flat radio spectra ($\alpha > 0.5$) and rapid time variability at most wavelengths. The γ-ray luminosity in blazars often dominate the bolometric luminosity, especially during outbursts. The third EGRET catalog (Hartman *et al.* 1999) lists 66 active galactic nuclei, of which 43 are clearly classified as flat-spectrum radio quasars (FSRQ), 16 as BL Lac objects (XBLs=2 and RBLs=14) and 7 belonging to a less well-defined category of sources (intermediate spectrum/radio sources). The catalog covered the Observatory phases (1–4) and does not include Mrk 501, since it was clearly detected only during phase 5 observations.

CP510, *The Fifth Compton Symposium*, edited by M. L. McConnell and J. M. Ryan

The discovery of nearby XBLs at TeV energies has reinvigorated γ-ray studies of these objects. Mrk 421, at a z of 0.031 is the closest BL Lac object seen at GeV energies and was the first discovered TeV blazar (Punch 1992). Gaidos (1996) reported the discovery of extremely short bursts at TeV energies in this source with doubling times of the order of 1 hour or less. Such short variability timescales strongly constrain possible emission mechanisms. A 30-minute burst when the TeV flux increased by a factor of 25, suggests extremely small emission regions (a few light hours) if one uses light-travel time arguments. Observations show no clear evidence for spectral variability during the flare. Correlated variability at x-ray energies suggests strongly the predominance of an inverse Compton process that scatters the soft photons (synchrotron or direct/scattered accretion disk photons) to higher energies. Unlike the dramatic variability seen at x-ray and keV energies, there is only a weak indication of an increase in 100 MeV - 10 GeV flux measured by EGRET during an outburst (Macomb 1995). PKS 2155-304 and Mrk 501 are the two other TeV blazars that have also been detected at GeV energies. PKS 2155-304 was first detected at GeV energies in phase 4 of CGRO (Vestrand, Stacy & Sreekumar 1995). Recently, this source was detected at TeV energies (Chadwick *et al.* 1998) during a γ-ray/x-ray high state (Sreekumar & Vestrand 1997; Vestrand & Sreekumar 1999). At a z of 0.1, PKS 2155-304 is the most distant TeV source detected to date and is an ideal candidate to study the intergalactic infra-red photons using the absorption signatures in the high-energy spectrum.

Mrk 501 at a z of 0.033, is the second closest BL Lac object known. It was discovered at TeV energies (E>300 GeV) by the Whipple group (Quinn 1996; Catanese 1997) , the HEGRA Cherenkov telescope (Bradbury 1997; Aharonian 1997) and more recently by the Telescope Array Project (TAP) (Hayashida 1998). Mrk 501 also shows significant variability at x-ray, low-energy γ-ray and TeV energies. This source was detected for the first time in the 100 keV to 1 MeV range by OSSE, the resulting spectral energy distribution (SED) showing this emission to be most likely of synchrotron origin. This represents the largest extension of the synchrotron spectrum in any blazar to date. More importantly, the shift in the synchrotron cutoff energy from about 1 keV in the quiet phase to about 100 keV during the outburst, suggest an unprecedented increase in the maximum energy of the charged particle spectrum in the jet. Though initial analysis of the EGRET data showed no detection in the 100 MeV – 10 GeV range, recent analysis using observations in 1996 (Kataoka *et al.* 1999) reported a $\sim 4\sigma$ detection. In this paper, we present results on Mrk 501 showing convincing detection above 500 MeV and new evidence for γ-ray outbursts at GeV energies.

II OBSERVATIONS AND ANALYSIS

EGRET observations of Mrk 501 are listed in Table 1. The spark-chamber was operated in the narrow FOV mode during observations after phase 4. MeV–GeV detection of the source is evident in viewing periods (VP) 516.5 and 519.0. The

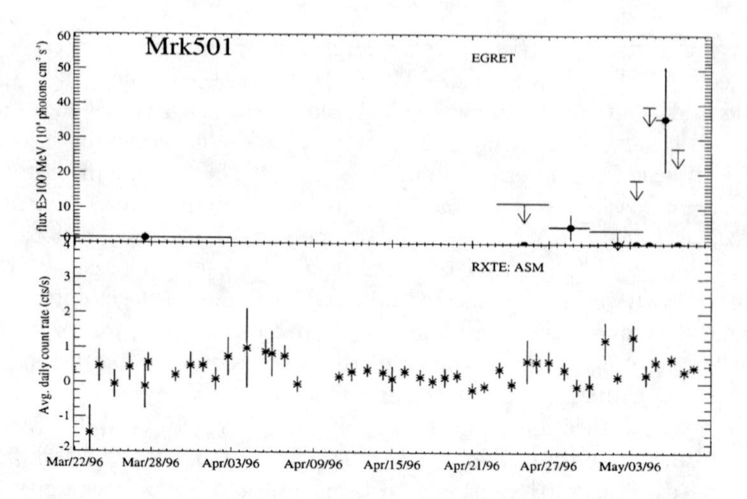

FIGURE 1. Time variability observed above 500 MeV (EGRET) and in the 2-12 keV band (ASM on RXTE). The γ-ray outbursts is uncorrelated with the x-ray emission.

strongest detection occurred during VP 519.0 (5.3σ above 500 MeV). A short time scale analysis (1-day) of the 2-week interval showed most of the source signal arrived within approximately a day (May 6 1996) (figure 1). Using photons that originate from within 2°of Mrk 501, a scatter plot of photon energy versus arrival time, showed 6 photons with energies >1 GeV in ~1-day interval about May 6th (figure 2). The archival EGRET data yielded the mean expected rate from that region of the sky of ~0.19 photons per day (from 6 weeks of exposure). This yields a <1e-6 Poisson probability for detecting 6 GeV photons from the direction of Mrk 501 during 1-day. The EGRET sky exposure was also examined on a sub-hour time scale to determine any significant variations that could simulate an outburst and none was found.

III RESULTS AND DISCUSSION

Careful analysis of EGRET data shows a clear detection of emission from Mrk 501 above 500 MeV where the much improved angular resolution makes the positional identification more certain. The unique synchrotron spectrum which sometimes extends to 100 keV implies that the inverse Compton emission peaks well beyond the EGRET energy range. This may explain why Mrk 501 was not detected as a strong GeV source during most EGRET observations even during x-ray/TeV outbursts. The new results presented here shows the first evidence for significant GeV emission from Mrk 501 that varies sharply over a time interval of 1-day or more. The GeV spectrum during this outburst (figure 3) is poorly determined (index = 1.1 ± 0.5)

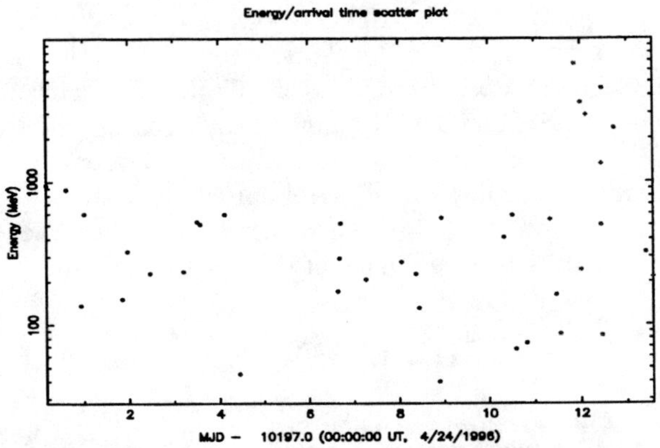

FIGURE 2. Photon energy vs. arrival time in the EGRET spark-chamber from a 2°region around Mrk 501. The GeV photons arrive mostly within a 24-hr interval on May 6 '99.

due to the limited statistics; however it shows the hardest known γ-ray spectrum in blazars. Previous correlated observations of Mrk 501 have shown nearly simultaneous outbursts at x-ray and TeV energies. Our search for multiwavelength data on Mrk 501 for May 6 yielded only 2–12 keV ASM data from the RXTE satellite. Figure 1 compares the ASM count rate with the EGRET measurements. No increase in the x-ray emission correlated with the GeV outburst is observed. Figure 3 shows the mean TeV spectrum (index = 2.47±0.07) published earlier by Aharonian *et al.* (1997). Though recent CAT results (Djannati-Ataï *et al.* 2000) have suggested changes in the TeV spectrum for different intensity states of Mrk 501, the GeV spectrum suggests a break at ~100 GeV. Alternately, using the observed correlation between x-ray and TeV emission from earlier outbursts, the ASM data can be used to set approximate upper limits on the TeV emission on May 6. The derived upper limit of 1×10^{-18} ergs/cm^2-s-keV at 0.5 TeV, also requires the GeV

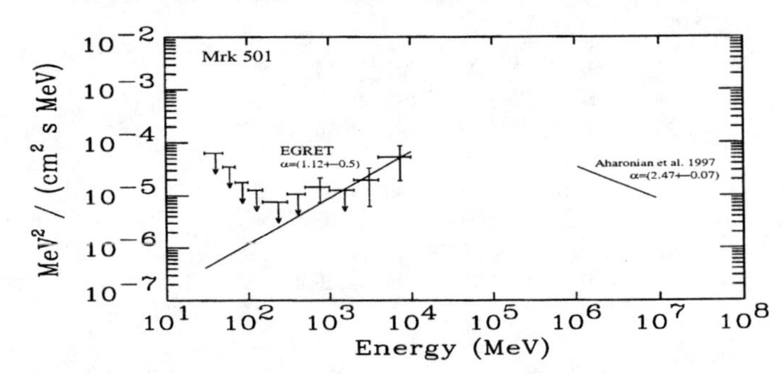

FIGURE 3. High-energy spectrum of Mrk 501. The non-simultaneous average TeV spectrum (Aharonian *et al.* 1997) suggests a break in the spectrum at ~ 100 GeV

spectrum to break sharply around ~100 GeV.

A likely scenario to explain the GeV outburst is a fresh injection of soft IR/optical photons that are Compton upscattered into the GeV range. However, preliminary analysis using standard SSC models (Bloom , private comm.) indicates difficulties in incorporating the derived TeV upper limit given the extremely hard GeV spectrum. It is unfortunate that no ground-based optical/IR data are available during the outburst in order to validate this. The exact nature of GeV outbursts in Mrk 501 maybe resolved only after the launch of the GLAST mission.

REFERENCES

1. Aharonian, F., *et al.* 1997, A&A, 327, L5
2. Bradbury, S.M., *et al.* 1997, A&A, 320, L5
3. Catanese, M., *et al.* 1997. ApJ, 487, L143
4. Chadwick, P., *et al.* 1998, ApJ, 503, 391
5. Djannati-Ataï, A., *et al.* 2000, AIP Conf. Proc. (this issue).
6. Gaidos, J.A., *et al.* 1996, Nature, 383, 319
7. Hartman, R.C., *et al.* 1999, ApJS, 123, 79
8. Hayashida, N., *et al.* 1998, astro-ph/9804043
9. Kataoka, J., *et al.* 1999, ApJ (in press).
10. Macomb, D.J., *et al.* 1995, ApJ, 449, L99
11. Mukherjee, R., *et al.* , 1997, ApJ, 490, 116
12. Punch, M., *et al.* 1992, Nature, 358, 477
13. Quinn,J., *et al.* 1996, ApJ, 456, L83
14. Sreekumar, P. & Vestrand, W.T. 1997, IAU Circular 6774
15. Vestrand, W.T., Stacy, G. & Sreekumar, P. 1995, ApJ, 454, L93
16. Vestrand, W.T., & Sreekumar, P. 1999, Astroparticle Phys. 11, 197

TABLE 1. EGRET Observations of Mrk 501 (E>0.5 GeV).

VP	Start	Stop	Flux	err	σ	Aspect
9.5[b]	09/12/91	09/19/91	<11		0.0	3.4°
201.0[b]	11/17/92	11/24/92	9	6	1.7	2.5°
202.0[b]	11/24/92	12/01/92	8	7	1.1	5.8°
516.5	03/21/96	04/03/96	1.5	1.2	2.0	3.1°
519.0	04/23/96	04/27/96	3.75	3,16	1.6	1.23°
	04/27/96	04/30/96	4.93	3.60	2.1	
	04/30/96	05/04/96	<3.88		0.0	
	05/04/96	05/05/96	<39.4		1.6	
	05/05/96	05/06/96	35.85	15.15	5.3	
	05/06/96	05/07/96	<27.3	0.0	1.3	
617.8	04/09/97	04/15/97	2.13	1.74	1.9	3.0°

[a]flux in 10^{-8} photons $(cm^2\text{-}s)^{-1}$; [b]E>100 MeV flux

a

TeV/X-ray observations of Mkn 501 during 1997 and 1998

H. Krawczynski*, F.A. Aharonian* for the HEGRA collaboration, R.M. Sambruna†, L. Chou†, P.S. Coppi‡, C.M. Urry§

*Max Planck Institut für Kernphysik, Heidelberg, Germany
†The Pennsylvania State University, State College, PA 16802
‡Yale University, New Haven, CT 06520-8101
§Space Telescope Science Institute, Baltimore, MD 21218

Abstract. The stereoscopic Cherenkov telescope system of HEGRA was used in 1997 and in 1998 to observe the BL Lac object Mkn 501 on a regular basis. After a period of exceptional activity in 1997 of continuous flares and a mean flux roughly one order of magnitude stronger than in the preceding years the source returned in 1998 to a state of moderate flux levels with only very sparse flaring activity. In this paper we summarize the TeV characteristics of Mkn 501 as deduced from the 1997 and 1998 HEGRA observations. Furthermore, we describe a joint HEGRA/RXTE campaign which we organized in June 1998.

INTRODUCTION

In the northern hemisphere two BL Lac objects have firmly been established as emitters of TeV γ-rays: Mkn 421 and Mkn 501. Due to the extreme conditions needed for the production of TeV photons, this TeV blazars can be revealing laboratories for the understanding of the nonthermal high energy emission of blazars in general. Furthermore, observations of TeV blazars contribute to the field of *Observational Cosmology*: the extinction of TeV photons due to pair production processes of the TeV γ-rays on photons of the Diffuse Extragalactic Background Radiation (DEBRA) [1,2] is expected to rapidly increase with γ-ray energy; the TeV energy spectra henceforth constrain the DEBRA intensity in the notoriously difficult wavelength region from 1μm-50μm. The DEBRA spectrum in this wavelength region is very interesting since it depends on the star formation history during the evolution of the universe and the cosmological parameters [3].

The stereoscopic HEGRA system of five identical Imaging Atmospheric Cherenkov Telescopes [4,5] is located on the Canary Island of La Palma (Spain) and is characterized by a large effective area of $\sim 10^5\,\mathrm{m}^2$ and excellent spectroscopic capabilities with an energy resolution better than 20% for individual TeV photons. The simultaneous observations of the Cherenkov light produced by air showers with several Cherenkov telescopes results in an unprecedented angular resolution of 0.1°

CP510, *The Fifth Compton Symposium*, edited by M. L. McConnell and J. M. Ryan

per individual γ-ray and a reduction of the background of hadronic cosmic rays by O(100) based on the analysis of the stereoscopic air shower images. The telescope system achieves an energy flux sensitivity of $\nu\,F_\nu$ at 1 TeV of 10^{-11} erg cm^{-2}s^{-1} (S/N = 5σ) for one hour of observation time which is ideally suited for the study of the rapid variability which is characteristic for blazars in general and especially for TeV blazars.

Multiwavelength observations of blazars promise to provide enough information to constrain the emission mechanism and the jet structure. In this context simultaneous TeV/X-ray observations are of special interest since both radiation components are believed to be produced by a population of high energy electrons producing synchrotron photons at longer wavelengths (X-rays) and Inverse Compton photons at shorter wavelengths (TeV γ-rays). In the following we first describe the TeV and TeV/X-ray observations; subsequently we discuss the results.

THE 1997/1998 TEV CHARACTERISTICS OF MKN 501

The HEGRA telescope system was used in 1997 and 1998 to extensively monitor the two TeV blazars Mkn 501 and Mkn 421. The 1997 Mkn 501 observations as well as the standard analysis tools and the methods to determine the systematic error on spectral estimates are described in [6,7]; interested readers can find a detailed account of the Mkn 421 observations in [8]. The results presented in this paper for Mkn 501 are based on 200 h of high quality, low zenith angle ($\theta < 30°$) data taken from March 16th, 1997 to Aug 25th, 1998.

The 1997/1998 light curve is shown in Fig. 1. During 1997 the source was extremely bright and variable. The integral fluxes above 1 TeV reached (during flares) 10 times the flux level of the Crab Nebula. Flux variability was found on time scales smaller than one day: from the flux measurements of different days shortest exponential flux increase/decrease time-constants of about 12 h were derived; on shorter time scales some evidence was found for substantial variability within several hours. For 63 days *diurnal* differential γ-ray spectra could be determined with a median accuracy of 0.2 in the 1-5 TeV photon index. In Fig. 2 (left side) an example of a diurnal differential spectrum is given (May 8th, 1997). After 1.9 h of observation time, the curvature of the spectrum can clearly be recognized and the 1-5 TeV

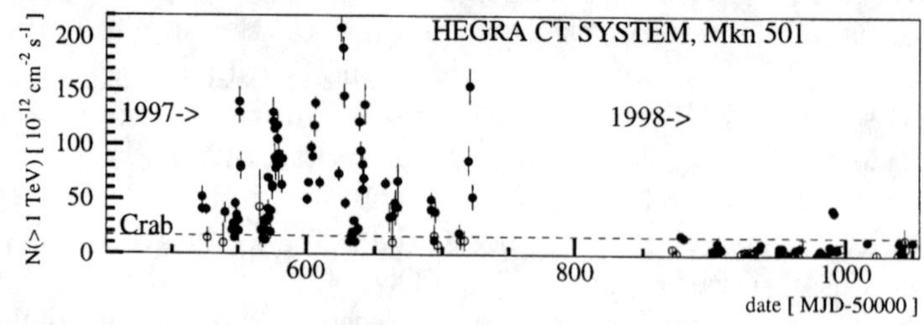

FIGURE 1. The 1997-1998 Mkn 501 light curve (1 day binning). MJD 50600 corresponds to June 1st, 1997.

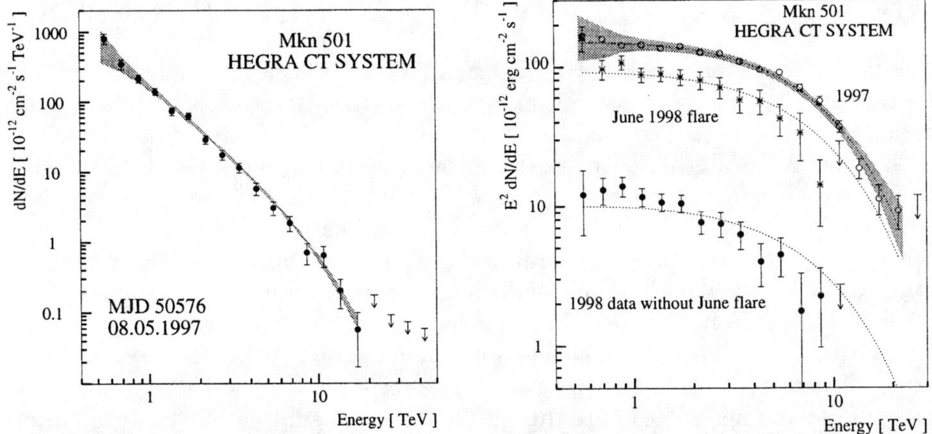

FIGURE 2. The left side shows a diurnal Mkn 501 energy spectrum taken with an integration time of 1.9 hours (data with $<30°$ zenith angle). The right side shows the Spectral Energy Distribution (E^2 dN/dE) averaged over 1997 (open symbols), for the 2 days of maximum flux during the June 1998 flare (asterisks), and for the rest of the 1998 data (full symbols). The systematic error on the curvature of the spectra is represented by the hatched region. Additionally, there is a 15% systematic error on the absolute energy scale. All flux upper limits are on a 2σ confidence level.

photon index could be determined to be $2.24 \pm 0.11_{stat} \pm 0.05_{syst}$. Remarkably, the HEGRA 1997 observations showed that the drastic flux variability was accompanied by at most modest variations in the TeV photon index. Only marginal evidence was found for spectral variability, i.e. two days (April 13th, 1997 and September 3rd, 1997) showed a spectrum harder than the time averaged spectrum on the $\sim 3\sigma$ confidence level.

The stability of the spectral shape justified a determination of the time averaged energy spectrum which is shown in Fig. 2 (right side). From 500 GeV to 24 TeV, the spectrum could be described by a power law model with an exponential cutoff: $dN/dE \propto (E/1\,\mathrm{TeV})^{-\beta} \exp(-E/E_0)$, with $\beta = 1.92 \pm 0.03_{stat} \pm 0.20_{syst}$, and $E_0 = (6.2 \pm 0.4_{stat} (-1.5 + 2.9)_{syst})$ TeV. Note that the errors on β and E_0 are strongly correlated. A dedicated χ^2-analysis yielded a very conservative lower limit on the maximum γ-ray energy in the signal of 16 TeV.

In 1998 the source showed very modest flux levels with only one flare with a flux clearly above the level of the Crab Nebula (June 26th/27th and 27th/28th). While the shape of the energy spectrum during the maximum of the June 1998 flare is rather similar to the shape of the 1997 spectrum, the time averaged 1998 energy spectrum (excluding the two days of the flare) seems to be noticeably softer (Fig. 2, right side). A fit of a power law model to the ratio of the two spectra gives $(dN/dE)_{1997}/(dN/dE)_{1998,lowstate} \propto E^{\beta}$ with $\beta = 0.36 \pm 0.10$.

HEGRA/RXTE OBSERVATION DURING JUNE 1998

In June 1998 we organized a HEGRA/RXTE multiwavelength campaign which was characterized by exactly simultaneous TeV/X-ray observations with long integration times per day for approximately two weeks [9]. The results of the campaign are summarized in Fig. 3. A strong flare was observed in both, the TeV and the X-ray bands. Mkn 501 showed very low TeV flux levels of about 1/10 the flux of the Crab Nebula. Only during a flare of approximately two days the TeV flux increased to about two times the Crab level. The X-ray flux varies synchronously; a time delay between the the TeV and the X-ray variability is surely smaller than one day. Remarkably, the TeV flux changed during the flare by a factor of $\sim$20 but the 10-20 keV X-ray flux changed only by a factor of $\sim$4. The TeV hardness ratios indicate that the TeV spectrum was already rather hard during the first days of observations, stayed hard during the flare, but substantially softened after the flare. However, the statistical significance for the softening after the flare is only about 2σ. In the X-ray band a simple correlation of the flux level and the spectral hardness is observed. Note that for the day of maximum TeV emission (June 26th/27th) we found evidence on the 0.5% chance probability level for a flux increase and subsequent decrease by a factor of 2 on a 1 h timescale.

DISCUSSION

The implications of the observations of the Mkn 501 spectrum on the DEBRA intensity have been discussed in detail in [7]. Here we focus on the evolution of the source from 1997 to 1998. Mkn 501 returned after 1997 – a year of high activity – to a state of moderate activity, both in the TeV and the X-Ray energy

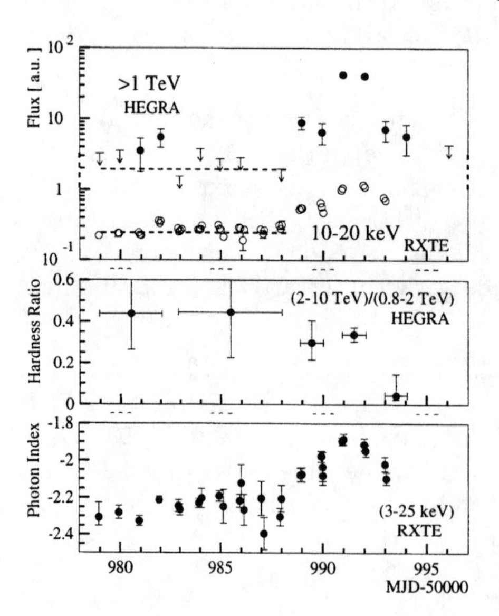

FIGURE 3. The temporal evolution of the TeV/X-ray fluxes and spectra during the RXTE/HEGRA campaign of June 1998. The upper panel shows the integral flux above 1 TeV as measured by HEGRA and the 10-20 keV count rate as determined from RXTE data. The dashed lines show the mean flux levels during the first 10 days of the campaign. The HEGRA flux upper limits are at the 1σ confidence level in order to facilitate the comparison with the 1σ error bars on the flux points. The second and third panels show the HEGRA hardness ratio and the RXTE 3-25 keV photon index (β from $dN/dE \propto E^{\beta}$), respectively.

bands. The decrease in flux is accompanied by a softer average TeV spectrum. Also the X-ray photon indices typically observed with RXTE are softer in 1998 than in 1997: between -1.7 and -2.1 were observed in 1997 [10], and between -1.9 to -2.3 during the observations of 1998 (this work). The BeppoSAX observations of April 1997 and April/May 1998 [11,12] show a similar trend. The softer 1998 X-ray and TeV spectra can easily be understood in terms of the high energy cutoff of accelerated particles moving towards lower energies. Certainly also the other parameters characterizing the production of the γ-ray radiation, as for example a softer spectrum of accelerated particles, could contribute to the evolution of the mean spectra. A decreasing bulk Lorentz factor of the emission region would account not only for a shift of the peaks of the X-ray and γ-ray spectral energy distributions towards lower energies, but also for the decrease of the mean flux level. Interestingly, the relation between the simultaneously observed X-ray and TeV fluxes did not change drastically (by much less than an order of magnitude), neither during 1997 [10] nor from 1997 to 1998. In most models, including Inverse Compton models, the TeV/X-ray flux relation is very sensitive to the model parameters as e.g. the magnetic field or the radius of the emission volume. It is thus puzzling that the source exhibits dramatic flares on a time scale of days or less while maintaining a rather constant TeV to X-ray flux relation on a time scale of two years. In Inverse Compton models this means that the magnetic field and/or the radius of the source have to stay very constant during all these flares or that they are connected to each other by a fixed relationship.

Acknowledgments
We thank the Instituto de Astrofísica de Canarias (IAC) for supplying excellent working conditions at La Palma. HEGRA is supported by the BMBF (Germany) and CYCIT (Spain).

REFERENCES

1. Gould J., Schréder G., *Phys. Rev. Lett.* **16**, 252 (1966).
2. Stecker F.W., De Jager O.C., Salamon M.H., *ApJ.* **390**, L49 (1992).
3. Primack J.R., Bullock J.S., Somerville R.S., Macminn D., *Astropart. Phys.* **11**, 93 (1999).
4. Daum A., Hermann G., Heß M., et al., *Astropart. Phys.* **8**, 1 (1998).
5. Konopelko A., Hemberger M., Aharonian F.A., et al., *Astropart. Phys.* **10**, 275 (1999).
6. Aharonian F.A., Akhperjanian A.G., Barrio J.A., et al., *A&A* **342**, 69 (1999).
7. Aharonian F.A., Akhperjanian A.G., Barrio J.A., et al., *A&A* **349**, 11 (1999).
8. Aharonian F.A., Akhperjanian A.G., Barrio J.A., et al., *A&A*, in press (1999).
9. Sambruna R.M., Aharonian F.A., Krawczynski H., et al., in preparation.
10. Krawczynski H., Coppi P.S., Maccarone T., Aharonian F.A., submitted to *A&A* (1999).
11. Pian E., Vacanti G., Tagliaferri G., et al., *ApJ* **492**, L17 (1998).
12. Pian E., Palazzi E., Chiappetti L., et al., *In: Procs. of the "BL Lac Phenomenon Workshop"*, Ed. L. Takalo, Turku, Finland, p. 180 (1998).

CAT VHE γ-ray Observations of Mkn 501 and Mkn 421, and implications on the blazar phenomenon

Arache Djannati-Ataï, for the CAT collaboration

Physique Corpusculaire et Cosmologie, Collège de France, Paris, France (IN2P3/CNRS)

Abstract. Two blazars, Mrk421 and Mrk501, have been detected at TeV energies by the CAT telescope operating in the French Pyrenées. The Very High Energy (VHE) emission observed from Mrk501 exhibits dramatically different behaviour in the quiescent and flaring states: the steep power-law spectrum seen in the low-activity state of 1998 contrasting with the harder, curved spectrum in the high-activity state of 1997. Mrk421 observations, on the other hand, are consistent with a power-law spectrum from the low-to-midrange intensities seen in 1998.

The VHE spectra, considered along with the low-energy (X-ray) behaviour of these sources, shed light on gamma-ray production and absorption mechanisms within blazar jets.

VHE ACTIVITY OF MRK 501 AND MRK 421 IN 1997-98

The two closest BL Lac objects, Mrk 421 and Mrk 501, have been observed with the CAT telescope (Cherenkov Array at Thémis), above a γ-ray detection threshold energy of 250 GeV. The telescope and its dedicated analysis method are fully described in [3] [8].

Fig. 1 shows the VHE light curves of the two blazars. Mrk 501 was in an unprecedentedly bright state during 1997, exhibiting flares during which it became up to $\sim$8 times brighter than the Crab Nebula. During the same period, Mrk 501 was also extremely active in the X-ray band, with its synchrotron peak lying above an exceptionally high energy of 100 keV (see e.g., [9]). In 1998, Mrk 501's mean VHE flux decreased markedly, with a few low-intensity flares in June. The X-rays showed also a much softer spectrum in the observations carried on by BeppoSAX [10].

Mrk 421, which was almost quiet in 1996-97, showed small bursts during the whole 1997-98 CAT observation period, resulting in a higher mean flux.

CP510, *The Fifth Compton Symposium,* edited by M. L. McConnell and J. M. Ryan

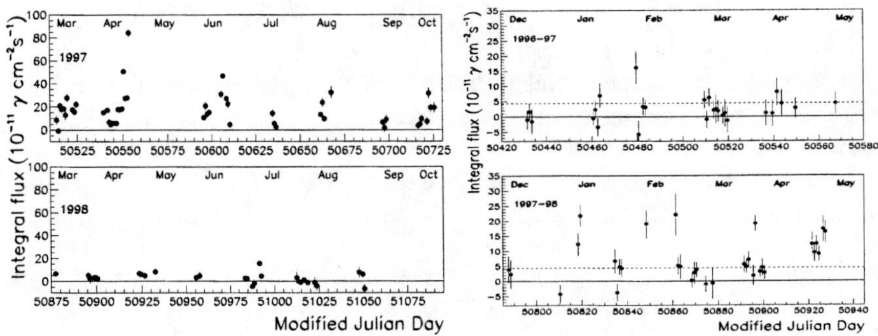

FIGURE 1. *(left)* Mrk 501 nightly integral flux above 250GeV in 1997 and 1998; *(right)* Mrk 421 nightly integral flux above 250 GeV between December 1996 and May 1998. The dashed line represents the mean flux over the two years. The abscissae for the two sources are different.

VHE SPECTRA OF MRK 501 AND MRK 421

Detailed analysis of the 1997 data (given in [4]) yields several spectral properties for Mrk 501. Its VHE spectral energy distribution (SED) shows a significant curvature and an emission extending above 10 TeV (Fig. 2*(left)*). The peak γ-energy is found to lie just above the CAT threshold. The study of hardness–intensity correlations for five different-level intensities (Fig. 2*(right)*) reveals the hardening of the VHE SED during flaring periods. As no significant variation in the curvature is seen (see [4]), the hardening effect is equivalent to the shift of the peak γ-energy towards higher energies with flux, ranging from 521 ± 82 GeV to 890 ± 77 GeV.

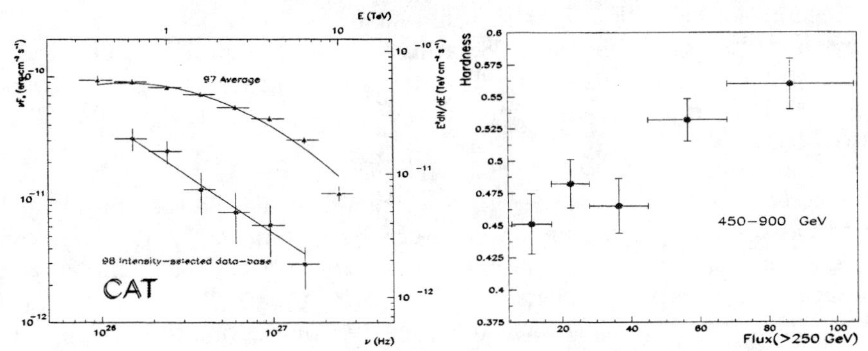

FIGURE 2. *(left)* Mrk 501 average VHE SED in 1997 between 330GeV and 13TeV, as compared to the 1998 SED from 520GeV to 8.25TeV (see Table 1); *(right)* Hardness-ratio ($HR = \frac{N_{E>900\text{ GeV}}}{N_{E>450\text{ GeV}}}$) *vs.* source intensity ($\Phi_{>250\text{ GeV}}$ in units of 10^{-11} cm^{-2}s^{-1}) from 1997 measurements.

In contrast with 1997, the spectrum of Mrk 501 in 1998 does not show any measurable curvature (Fig. 2*(left)*), and is much softer (see Table 1). This indicates that the VHE peak energy was well below the CAT threshold at that time. The comparison between the confidence contours of the spectral parameters for the two periods is shown in Fig. 3*(left)*. The incompatibility of the contours confirms, in particular, that in spite of larger statistical errors, the 1998 data cannot be fitted by the average spectral shape found in 1997 (see Table 1), as would have been expected in the absence of any spectral variability.

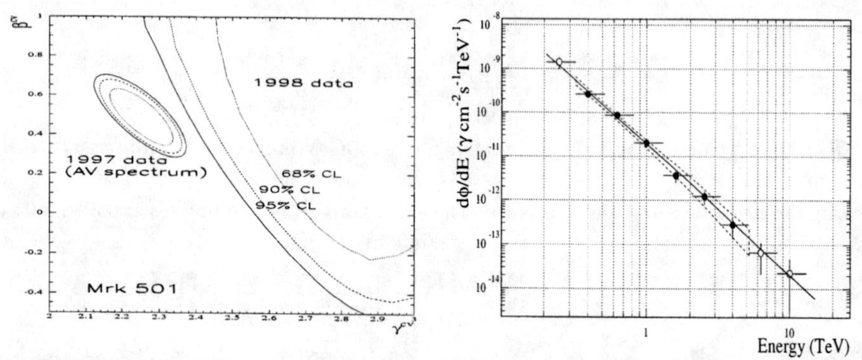

FIGURE 3. *(left)* 68%, 90% and 95% CL contours in the γ, β plane for the 1997 average and 1998 data sets (see Table 1); each contour is obtained by projecting the 3-dimension ellipsoid along the ϕ_0 axis. *(right)* Differential flux of Mrk 421 for the flaring periods in 1998. Only the energy bins shown with filled circles (from 330GeV to 5.2TeV) were used for estimation of spectral paremeters.

As mentioned earlier (Fig. 1), Mrk 421 was more active in 1998 than in 1997. Its energy spectrum derived for the 1998 flaring periods, at a mean level of ~1 Crab, resembles that of Mrk 501 in 1998. It is well-represented by a simple power law (Fig. 3) with a steep differential spectral index $\gamma = 2.96 \pm 0.13^{\text{stat}} \pm 0.05^{\text{syst}}$ [11] between 330 GeV and 5.2 TeV.

The absence of curvature in the VHE spectrum of Mrk 421 is confirmed by other ground-based telescopes, within a even-larger dynamic range [7]. However, different values are reported for the spectral index. The HEGRA collaboration [2] has published a very similar index for the 1997-98 period ($\gamma = 3.03 \pm 0.08^{\text{stat}} \pm 0.10^{\text{syst}}$), whereas the Whipple group [7] finds a harder spectrum during the 1995-96 flaring period, with $\gamma = 2.54 \pm 0.03^{\text{stat}} \pm 0.10^{\text{sys}}$. It is interesting to note that the mean flux level of the data used to derive this spectrum ($22.0 \times 10^{-11}\,\text{cm}^{-2}\text{s}^{-1}\text{TeV}^{-1}$) is about an order of magnitude higher than that of 1997-98 period used by CAT and HEGRA. This apparent discrepancy could well be understood in terms of intensity–hardness correlations, such as observed for Mrk 501 in 1997 by CAT. Similar evidence for Mrk 421 was in fact suggested in [14] on the basis of the Whipple 1995-96 low-flux data.

TABLE 1. Best fitted spectral parameters of Mrk 501 and Mrk 421 in 1997 and 1998 as obtained for a curved or power law shape. The average data-set contains the data of the whole year 1997. The differential flux are given as $d\Phi/dE_{TeV} = \phi_0 E_{TeV}^{-(\gamma+\beta \log_{10} E_{TeV})}$ in units of 10^{-11} cm^{-2}s^{-1}TeV^{-1}. If, according to the test statistic (λ), the curved shape hypothesis is favored, then the last two columns give the curvature term β and the peak-emission energy $E_{GeV}^{peak} = 10^{\frac{2-\gamma}{2\beta}}$. If not, β is quoted between brackets.

Data set	T [a]	λ	ϕ_0	γ	β	E_{GeV}^{peak}
Mrk 501 '97	57.2	61.5	5.19 ± 0.13	2.24 ± 0.04	0.50 ± 0.07	578 ± 98
Mrk 501 '98	13.0	0.09	1.25 ± 0.16	2.97 ± 0.20	(0.21 ± 0.73)	–
Mrk 421 '98	5.1	0.34	1.96 ± 0.20	2.96 ± 0.13	(0.28 ± 0.49)	–

[a] total time (h) ON-source.

DISCUSSION

In the last few years, much progress has been made towards a unifying scheme for gamma-ray bright blazars. Recent work by [5] and [6] based on a purely observational and a more theoretical approach, respectively, tend to lead to a continuous sequence of blazars, shown in (Fig. 4*(left)*). With increasing observed power, a decrease is seen in the frequencies of the synchrotron and inverse Compton (IC) peaks, together with an increase of the ratio of the powers of the high and low energy spectral components. In the underlying physical picture, an energetic electron beam propagating in the magnetized plasma jet produces X-rays through synchrotron radiation as well as VHE γ-rays through IC scattering of low-energy photons. The

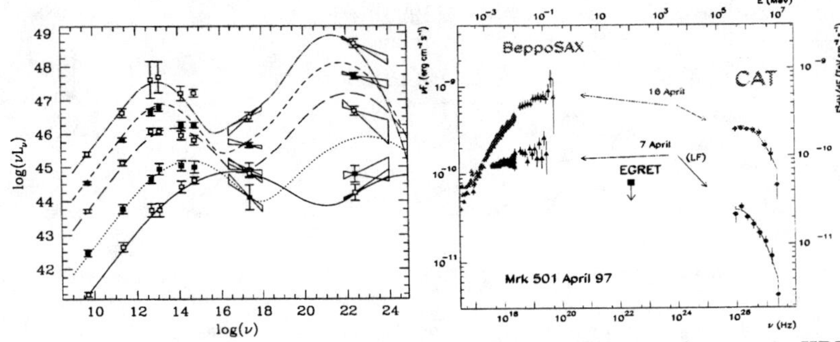

FIGURE 4. *(left)* The average SED's of blazars, showing the spectral sequence HBL, LBL, FSRQ as the synchrotron and IC frequencies decrease, while the luminosity increases [5]. *(right)* Mrk 501 X-ray and VHE spectra for April 7th and 16th showing their correlation. The EGRET upper limit corresponds to observations between April 9th and 15th. Such correlations have also been observed on Mrk 421 during coordinated observation campaigns (see e.g., [13]).

blazar sequence reflects then differences in terms of magnetic field, cooling rate, radiation fields, etc.

It is interesting to examine how the most extreme known blazars, Mkn 501 and Mkn 421, fit within this scheme. Since Mrk 501 and Mrk 421 lie at the the same redshift ($\sim$0.03), spectral differences between them must be intrinsic and not due, in particular, to absorption by the intergalactic infrared background radiation.

The SEDs of these sources show two bumps with correlated hard X-ray and γ-ray peak frequencies, as expected (see Fig. 4*(right)*). The hardness–intensity correlations, seen for both components of Mkn 501 by BeppoSAX and CAT in 1997 strongly suggest that the same particle population is responsible for emission in both energy-ranges, supporting the leptonic scenario used by [6]. With a mean flux much lower than that of 1997, the steeper and power-law shaped spectrum of Mkn 501 measured in 1998 is also consistent with this scenario. The lower-energy peaked synchrotron bump of Mkn 421, would imply that its IC peak is at a lower frequency than that of Mkn 501, and hence below CAT's threshold. Observations fully confirm this, also.

However, we note that the bolometric luminosity and the synchrotron peak frequency of Mkn 501 *both* increased in 1997. This is not consistent with the universal acceleration scheme proposed in [6].

One should keep in mind the fact that our understanding of blazars is based on time averaged data for which the timescales are, most probably, larger than those intrinsic to the sources. This might yield imprecise notions of *quiscent* or *flaring* states, which are in fact dependent on the sensitivity of the available instruments. Next generation telescopes, such as HESS, VERITAS or GLAST, should provide the opportunity of a deeper physical understanding of blazars.

REFERENCES

1. Aharonian, F.A., *et al*, *A&A* **349**, 11 (1999).
2. Aharonian, F.A., *et al*, *A&A* **350**, 757 (1999).
3. Barrau, A., *et al*, *Nucl. Instr. Meth.* A **416**, 278 (1998).
4. Djannati-Ataï, A., *et al*, *A&A* **350**, 17 (1999).
5. Fossati, G., Maraschi, L., Celotti, A., Comastri, A., & Ghisellini, G., *MNRAS* **299**, 433 (1998).
6. Ghisellini, G., Celotti, A., Fossati, G., Maraschi, L., & Comastri, A., *MNRAS* **301**, 451 (1998).
7. Krennrich, F., *et al*, *ApJ* **511**, 149 (1999).
8. Le Bohec, S., *et al*, *Nucl. Instr. Meth.* A **416**, 425 (1998).
9. Pian, E., *et al*, *ApJL* **492**, 17 (1998).
10. Pian, E., *et al*, *Proc. BL Lac Phenomenon, Turku, ed. Leo Takalo* (1998).
11. Piron, F., *et al*, *Proc. XXVI ICRC* **3**, 326 (Salt-Lake City, 1999).
12. Samuelson, F., *et al*, *ApJL* **501**, 17 (1998).
13. Takahashi, T., *et al*, *APh* **11**, 177 (1999).
14. Zweerink, J.A., *et al*, *ApJL* **490**, 141 (1997).

Microvariability in the southern gamma-ray blazar PKS 0537-441

Gustavo E. Romero*, J. A. Combi*, and S. A. Cellone†

*Instituto Argentino de Radioastronomía, C.C. 5, 1894 Villa Elisa, Argentina
†Observatorio Astronómico, Paseo del Bosque S/N, 1900 La Plata, Argentina

Abstract. The southern blazar PKS 0537-441 is one of the most variable active galactic nuclei. It has displayed extreme forms of variability in its emission from radio to gamma-ray wavelengths. Here we present results of new optical microvariability observations of this blazar in the V and R bands, obtained with high-quality CCD photometry made with a 2.15-m telescope at CASLEO, Argentina. We have found variability with amplitudes of more than 100% over timescales of ~ 2 days, similar to what has been observed in gamma-rays by EGRET. The spectral index, contrarily to what is expected from microlensing, showed changes of $\sim 27\%$ during the observing period, with a trend to steepen with decreasing flux. This seems to reflect an intrinsic mechanism behind the variability and, consequently, suggests that optical and gamma-ray emitting regions in this source can be co-extensive. If the most extreme manifestations of radio variability are also intrinsic, they might involve coherent emission processes. We have used our observations, along with EGRET and ROSAT data, to constrain the central black hole mass, which seems to be $\sim 8 \times 10^7\ M_{\odot}$. For such a mass the γ-ray variability is originated at ~ 290 gravitational radii from the central source.

I INTRODUCTION

The southern blazar PKS 0537-441 (at a redshift $z = 0.894$) is an extremely variable source across the entire electromagnetic spectrum. There are reports of strong and rapid variations in its flux density at radio [8,9], optical [5,11], and gamma-ray wavelengths [13,4]. The source has been claimed to have a foreground galaxy and microlensing-based variability [12,10], although this has been recently disputed by some authors [2,6]. In 1993 this blazar displayed one of the most extreme episodes of variability ever detected with changes up to 40% of the radio emission over timescales $\sim 10^4$ s [8]. Such variability, if produced in a synchrotron source, would imply brightness temperatures $T_{\rm b} \sim 10^{21}$ K, which are 9 orders of magnitude above the inverse Compton limit. The origin of this phenomenon, later observed in other sources, is unknown at present.

Here we present results of an optical monitoring campaign of this peculiar blazar

CP510, *The Fifth Compton Symposium,* edited by M. L. McConnell and J. M. Ryan

TABLE 1. Microvariability optical observations of PKS 0537-441

Band	Date	σ (mag)	t_v (hours)	Variable?	$C = \sigma_s/\sigma$	Y (%)
V	12/15-20/1998	0.0034	42.4	Yes	31.4	109.2
R	12/15-20/1998	0.0034	42.4	Yes	33.1	96.4

aimed to detect rapid variability in the R and V bands over timescales from minutes to several days, and we discuss the implications for the *gamma*-ray emission.

II OBSERVATIONS

The observations were carried out in December 1998 with the 2.15-m CASLEO telescope at San Juan, Argentina. The instrument was equipped with a cryogenically-cooled CCD camera with a Tek-1024 chip. Each field frame contained the blazar and several stars, six of which were used for calibration and control purposes following the procedures described in detail in Ref. [11].

The microvariability observations were made using Johnson's V and R filters with integration times of $\sim$ 1 min. Calibration frames (bias and flat field images) were taken each night according to standard procedures. The data reduction was performed through the IRAF software package running in a UNIX workstation.

III RESULTS

The results are summarized in Table 1. Observational errors, determined from the scatter of the control lightcurves, are of $\sim$ 0.003 mag. Blazar lightcurves are variable at both observing frequencies at a confidence level of more than 30 σ. In the V band, the variability amplitude, defined as

$$Y = \frac{100}{< D >}\sqrt{(D_{\rm max} - D_{\rm min})^2 - 2\sigma^2}, \quad (1)$$

where $D_{\rm max}$ and $D_{\rm min}$ are the maximum and the minimum in the differential lightcurve, reaches 109.24 %, whereas it results of 96.37 % in the R band. There is no time lag between the brightness changes at the different frequencies. The timescale of the major outburst is $t_v \sim \Delta F/(dF/dt) \approx 1.5 \times 10^5$ s. Within a single night, variations of about 30 % are observed in both bands.

The average spectral index is $\alpha_{RV} = -1.6$ ($S_\nu \propto \nu^\alpha$). Variability amplitudes of 27.44 % are present with a similar timescale to that observed in brightness. The index changes in the sense that it becomes steeper with decreasing flux density.

IV DISCUSSION

The observed variability in the spectral index at two close frequencies implies that gravitational microlensing (basically an acromatic phenomenon) cannot be the source of the optical variability in PKS 0537-441. Models involving changes in the viewing angle of superluminal components are also discarded by the same reason. Other extrinsic effects that could be effective in the radio band, as interstellar scintillation, are not present at optical frequencies. We conclude, then, that the variability is intrinsic to the source. The correlation between the flux level and the continuum slope in the sense that the continuum is steeper when the source is fainter agrees with the picture of synchrotron radiation originated in a shocked jet [7].

If no beaming is introduced in the source, our data provides an upper limit to the central black hole mass, which is constrained by a gravitational radius not larger than 2.4×10^{15} cm to be $< 2.7 \times 10^9\ M_\odot$. A more accurate determination can be obtained from the γ-ray variability data combined with the ROSAT X-ray information.

Since PKS 0537-441 is a strong γ-ray emitter, the optical depth to pair production for a γ-ray of energy E created at a height x above the center of the X-ray emitting accretion disk that surrounds the central black hole and that is propagating outward along the x-axis (i.e. towards the observer) should be $\tau_{\gamma\gamma} \sim 1$.

The optical depth is

$$\tau_{\gamma\gamma} = \int_x^\infty \alpha_{\gamma\gamma}(E, X)dX, \tag{2}$$

where $\alpha_{\gamma\gamma}$ is the $\gamma - \gamma$ absorption coefficient. Following Ref. [1] we compute $\tau_{\gamma\gamma}$ for a two-temperature accretion disk model with an intensity distribution given by $I(E,R) \propto E^{-\alpha}R^{-w}$, with R being the radial distance over the disk ($R_{\rm min} \leq R \leq R_{\rm max}$). For a 2-T disk model with $R_{\rm min} = 6R_{\rm g}$ (Schwarzschild geometry), $R_{\rm max} = 100R_{\rm g}$, and $w = 3$, we get, using the reported ROSAT flux $F_{\rm 1KeV} = 0.79$ μJy, the index $\alpha_{\rm X} = 2.1$ [14], and cosmological parameters $q_0 = 1/2$ and $h = 0.7$,

$$\tau_{\gamma\gamma} \approx 2 \times 10^{28} \left(\frac{M}{M_\odot}\right)^{-1} \left(\frac{x}{R_{\rm g}}\right)^{-7.2}, \tag{3}$$

where we have considered 1-GeV photons.

The variability observed by EGRET in the γ-ray emission [4] imposes a constraint to the size of the γ-spheres at $E > 100$ MeV:

$$x_\gamma \leq ct_{\gamma,\rm v}\frac{\delta}{1+z} \quad \text{cm}, \tag{4}$$

where δ is the Doppler factor and $t_{\gamma,\rm v}$ is the timescale of the γ-ray variability. Adopting the Doppler factor $\delta \sim 3.8$ estimated by Fan et al. in Ref. [3], we obtain $x_\gamma \leq 3.5 \times 10^{15}$ cm, and then, demanding that $\tau_{\gamma\gamma} \sim 1$, we get:

$$M_{\rm Schw} \approx 3 \times 10^7 \quad M_\odot. \tag{5}$$

In a similar way, if we consider a Kerr black hole in our disk model, the mass results:

$$M_{\rm Kerr} \approx 8.3 \times 10^7 \quad M_\odot. \tag{6}$$

These estimates are consistent with the value derived by Fan et al., which is insensitive to the kind of black hole involved (they obtain 5.8×10^7 $M_\odot$). For a mass of $\sim 8 \times 10^7$ $M_\odot$ the observed γ-ray variability must be produced at a distance of ~ 290 gravitational radii from the central source.

Two points are noteworthy. First, the mass of the hole seems to be relatively small compared to those determined in other not-so-variable objects like M87 and 3C273. This could be a generic feature of extremely variable AGNs, as it is also suggested by the estimates of Fan et al. [3] for several strongly variable γ-ray blazars. And secondly, the size scales of the optical and γ-ray regions are similar, indicating that they are possibly generated by the same population of particles. Future simultaneous optical and γ-ray observations of PKS 0537-441 could be used to set this question.

Finally, we notice that if the ultra-rapid radio variability observed by Romero et al. in 1993 [8] is also intrinsic to the source, then coherent emission processes should be occurring in the innermost region of this object, otherwise violations of the inverse Compton limit seems to be unavoidable.

Acknowledgments. G.E.R. and J.A.C. are very grateful to the organizers of 5th Compton Symposium for financial support that make possible their participation. The work described in this paper has been supported by Fundación Antorchas, CONICET, and the agency ANPCT (PICT 03-04881).

REFERENCES

1. Becker P.A., Kafatos M., ApJ **453**, 83 (1995)
2. Falomo R., Melnick J., Tanzi E.G., A&A **255**, L17 (1992)
3. Fan J.H., Xie G.Z., Bacon R, A&AS **136**, 13 (1999)
4. Hartman R.C. , ASP Conf. Ser. **110**, 33 (1996)
5. Heidt J., Wagner S.J., A&A **305**, 42 (1996)
6. Lewis G.F., Williams L.L.R., MNRAS **287**, 155 (1997)
7. Marscher A.P., Perugia University Obs. Publ. **3**, 81 (1997)
8. Romero G.E., Combi J.A., Colomb, F.R., A&A **288**, 731 (1994)
9. Romero G.E., Benaglia P., Combi J.A., A&A **301**, 33 (1995)
10. Romero G.E., Surpi G., Vucetich H., A&A **301**, 641 (1995)
11. Romero G.E., Cellone S.A., Combi J.A., A&AS **135**, 477 (1999)
12. Stickel M, Fried J.W., Khür H., A&A **206**, L30 (1988)
13. Thompson D.J., et al., ApJ **410**, 87 (1993)
14. Treves A. et al., ApJ **406**, 447 (1993)

Evidence for emission in the MeV band from GRO J1837+59 and QSO 1739+522

O.R. Williams[4], K. Bennett[4], R. Much[4], V. Schönfelder[1], W. Collmar[1], H. Steinle[1], H. Bloemen[2], W. Hermsen[2], and J. Ryan[3]

[1] *Max-Planck-Institut für Extraterrestrische Physik, P.O. Box 1603, 85740 Garching, Germany*
[2] *SRON-Utrecht, Sorbonnelaan 2, NL-3584 CA Utrecht, The Netherlands*
[3] *Space Science Center, Univ. of New Hampshire, Durham NH 03824, U.S.A.*
[4] *Astrophysics Division, Space Science Department of ESA, ESTEC, 2200 AG Noordwijk, NL*

Abstract. Emission in the MeV band from the region containing the bright unidentifed EGRET source GRO J1837+59 and the steep-spectrum EGRET blazar QSO 1739+522 was first reported by COMPTEL following an observation in November 1992. During this observation the emission was consistent with a single point source, designated GRO J1753+57. However, the location of GRO J1753+57 was not consistent with either GRO J1837+59 or QSO 1739+522 and identifying its counterpart in other wavelengths proved difficult. Moreover, subsequent observations suggested that the source could not in fact be described as a single point source and explanations involving extended or multiple sources were invoked.

We present an analysis of recent obervations of this region, which confirm that the emission cannot arise from a single source, but can be modelled as a combination of emission from both GRO J1837+59 and QSO 1739+522. The spectrum and time variability derived for these two sources are discussed and compared to observations in neighbouring bands.

I INTRODUCTION

Emission in the MeV band from the region containing the bright unidentifed EGRET source GRO J1837+59 and the steep-spectrum EGRET blazar QSO 1739+522 was first reported by COMPTEL during an observation in November 1992. The emission was consistent with a single, bright, variable point source which was designated GRO J1753+57 [1]. However, it was not possible to convincing identify a candidate at other wavebands within the 3σ location contour, despite a thorough search and follow-up optical observations [2].

Subsequent observations of the region showed evidence that the emission was not in fact consistent with a single source. Hypothesised explanations for the emission ranged from a diffuse Galactic source related to high-velocity clouds [3] and emission from unresolved point sources [4].

CP510, *The Fifth Compton Symposium*, edited by M. L. McConnell and J. M. Ryan

This paper investigates the hypothesis that all the observed emission from this region can be explained by the known γ-ray sources GRO J1837+59 and QSO 1739+522. To do this, we model the observed signal in terms of emission from these two sources and then compare the spectra and light-curves so obtained with EGRET observations of this region. The region has been observed 28 times by COMPTEL during CGRO mission phases I-VI between 30-May-91 and 15-Apr-97. Data from the viewing period (VP) 22 between 05-Mar-92 and 19-Mar-92 is omitted because COMPTEL coverage was intermittent due to problems with the onboard tape drives.

II MAXIMUM LIKELIHOOD-RATIO MAPS

Binned datasets in the energy intervals 0.75-1.0 MeV, 1.0-3.0 MeV, 3.0-10.0 MeV and 10.0-30.0 MeV have been separately created for each of the individual observations. These datasets were analysed using a modified version of the maximum likelihood-ratio method [5] to give a single maximum likelihood-ratio map (MLM) for each energy range.

Fig.1a shows a contour plot in Galactic coordinates of the MLM of the 1.0-3.0 MeV range. There is a significant feature in the vicinity of QSO 1739+522 and GRO J1837+59. The bright γ-ray source Cygnus X-1 and the Galactic diffuse emission are included in the background model. We note that the detailed results obtained are dependent on the models assumed for any nearby diffuse emission and will result in an increased uncertainty in fluxes extracted from these maps. In this paper, however, the reported uncertainties are purely statistical.

The likelihood analysis has been repeated including terms for GRO J1837+59 and QSO 1739+522 in the model fitted to the background. Summed MLMs were then generated using backgrounds including first GRO J1837+59 (Fig.1b), then QSO 1739+522 (Fig.1c), and finally both (Fig.1d). Including a source at the location of QSO 1739+522 in the background the remaing excess is centred on GRO J1837+59 with a significance of 5σ (Fig.1b). Similarly, including a source at the location of GRO J1837+59 in the background gives rise to an excess in the region of QSO 1739+522 of 4σ (Fig.1c). Finally, including both QSO 1739+522 and GRO J1837+59 in the background results in a map with no significant excess ($< 2\sigma$).

III SPECTRA

Fig.2 shows the EGRET spectra of GRO J1837+59 and QSO 1739+522 using Phase I-IV data [7] compared with the COMPTEL data. The data covers the same viewing periods except that the EGRET data includes VP 22 for which COMPTEL data is not available.

Extrapolation of the best-fit EGRET data suggests that QSO 1739+522 should be visible in the COMPTEL energy range. The detection in the 1-3 MeV band and the upper limits in the other COMPTEL bands are entirely consistent with

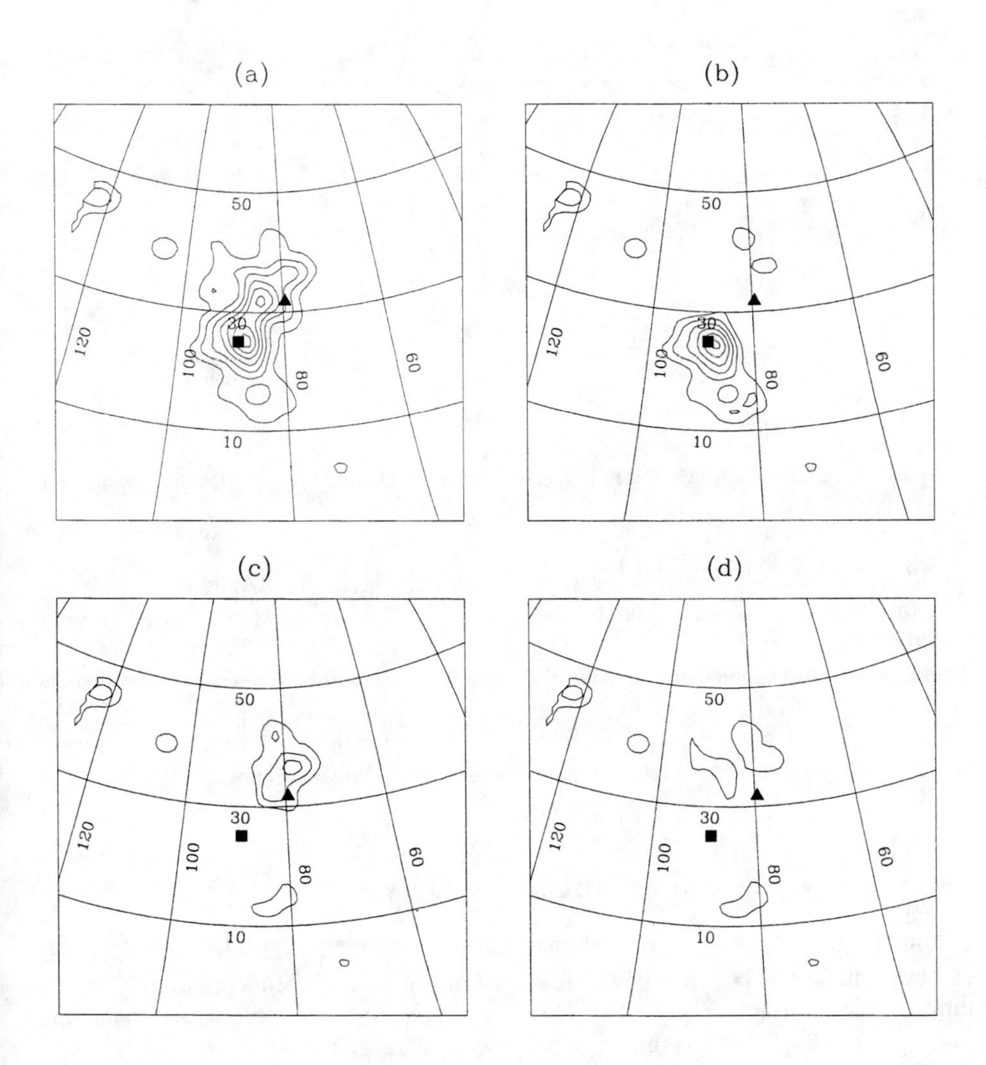

FIGURE 1. MLMs for the period 30-May-91 to 27-May-98 in the 1-3 MeV band: (a) Neither GRO J1837+59 nor QSO 1739+522 in background (b) GRO J1837+59 in background (c) QSO 1739+522 in background (d) QSO 1739+522 and GRO J1837+59 in background. GRO J1837+59 in indicated by a triangle, QSO 1739+522 by a square. Likelihood values start at 10 with steps of 5.

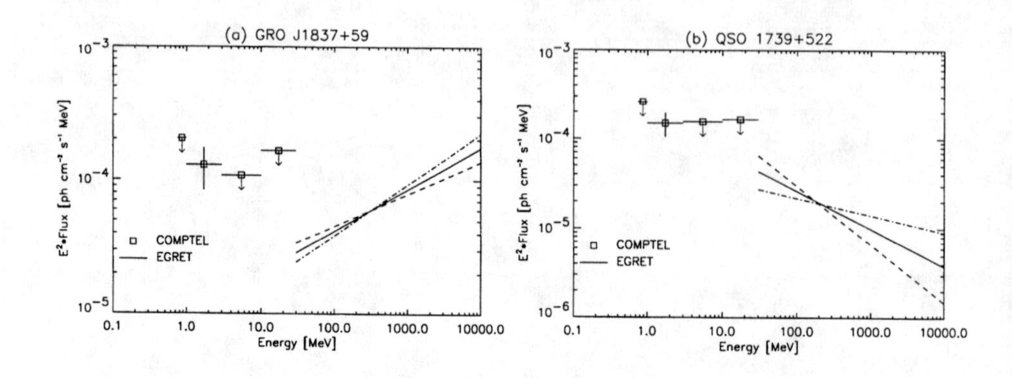

FIGURE 2. EGRET and COMPTEL spectra of (a) GRO J1837+59 (b) QSO 1739+522 for Phase I-IV.

the EGRET data. In contrast, extrapolation of the GRO J1837+59 data suggests that this source should not be detected by COMPTEL, unless there is either a very sudden break in the slope or the spectral shape in the EGRET band is variable. There is some evidence in the EGRET data for a variable spectral slope [6] and the EGRET data from individual observations are compatible with the COMPTEL data e.g. VP 201 when COMPTEL measures a significant signal from this direction [4]. If this explanation is correct then the source must be variable in the COMPTEL band.

IV LIGHT-CURVES

The emission ascribed to GRO J1837+59 and QSO 1739+522 in the individual observations in the 1-3 MeV band are shown in Fig.3. The points are mainly upper limits. The emission from QSO 1739+522 is entirely consistent with a constant flux ($\chi^2/n = 31/27$) while there is a suggestion of variability in the output of GRO J1837+59 ($\chi^2/n = 55/27$) at the 3 σ level.

V DISCUSSION

In the presence of either several near-by point sources or of extended sources the unambigous interpretation of COMPTEL data is difficult. However, the explanation of the signal described in this paper by GRO J1837+59 and QSO 1739+522 is plausible. Firstly, EGRET data suggests that QSO 1739+522 should be detected by COMPTEL and therefore any model of emission from this region in the MeV band must include that source. Secondly, including QSO 1739+522 in the subtracted background gives rise to a single point-like excess at the location of GRO

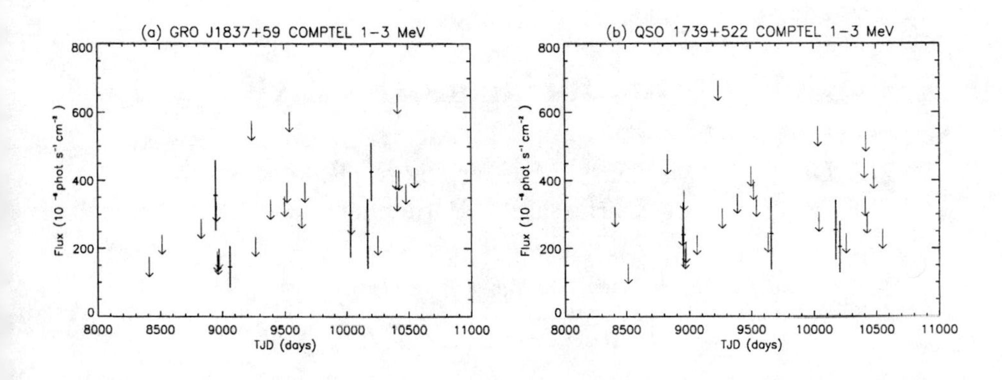

FIGURE 3. Light-curves of (a) GRO J1837+59 (b) QSO 1739+522.

J1837+59, which is the brightest unidentified EGRET source. A detection of GRO J1837+59 by COMPTEL is consistent with the EGRET data only if the source is variable. In support of this we note that EGRET has seen variations in flux, and possibly slope, while COMPTEL sees suggestions of flux variability.

REFERENCES

1. Williams, O.R. et al.,1995, A&A 297, 21.
2. Carramiana, A., et al. 1996, A&AS 120, 595.
3. Blom, J.J., et al,.1997, Proc. 2nd Integral Workshop, C. Winkler, T.J.-L. Courvoisier, Ph. Durouchoux (eds.), ESA SP-382, 119.
4. Williams, O.R. et al., 1997, AIP Conf. Proc. 410, 1243.
5. Bloemen, H., et al. 1994, A&AS 92, 419.
6. Nolan,. P.J., et al. 1996, ApJ, 459, 100.
7. Hartman, R.C. et al. 1999, ApJS, 123, 79.

BATSE Monitoring of BLAZARS

A.J. Dean [*], A. Malizia [*†], M.J. Westmore [*], R. Gurriaran [*],
F. Lei [*], L. Bassani [‡], J.B. Stephen [‡]

[*] *Department of Physics, The University, Southampton, UK*
[‡] *Istituto TeSRE/CNR, Via Gobetti 101, Bologna, ITALY*
[†] *BeppoSAX Science Data Centre, Rome, ITALY*

Abstract. Since the time coverage of BATSE is nearly complete over 8 years, this instrument offers a unique opportunity to monitor highly variable sources such as blazars in a relatively unexplored energy band (20-100 keV). Here we present some examples of this capability which is particularly useful during period of flaring activity. Four years of monitoring of the QSO 71.07 in the hard X-ray band is presented which indicates enhanced hard X-ray emission during an optical flare. BATSE also allowed the extraction of spectra during bright states. These data indicate that the spectrum steepens when the source is brighter. Results are also presented for the BL Lacs MKN501 and PKS 2005-489 during high energy flares, first detected by other instruments. The potential of using BATSE to study sources over long periods of time and to follow the emission before and after flares is clearly demonstrated.

INTRODUCTION

The BATSE all sky capability coupled with its ability to monitor detected sources over a long time period makes it ideal for the study of Blazars. Also, for this class of objects, BATSE's limited sensitivity is compensated by the source brightness increase during periods of flaring activity that is a common feature of all these objects. Here we demonstrate this capability with observations of both QSO's and BL Lac objects studied by BATSE, and highlight the potential of this instrument not only to monitor continuously the source flux behaviour, but also to study possible spectral changes in a relatively unexplored energy range (20-100 keV).

4C 71.07

4C 71.07 (0836+710),is a high red-shift (z = 2.172) QSO recently detected at high energies by means of OSSE and BATSE observations (1); it is probably the farthest object so far detected in hard X-rays. The long term (3-year) light curve is shown in figure 1, each bin corresponding to 60 days integration time. The crosses superimposed on the BATSE data correspond to optical (R band) measurements where the optical minimum has been set to m_R=17.4 (2), while the two vertical lines indicate the period

CP510, *The Fifth Compton Symposium,* edited by M. L. McConnell and J. M. Ryan

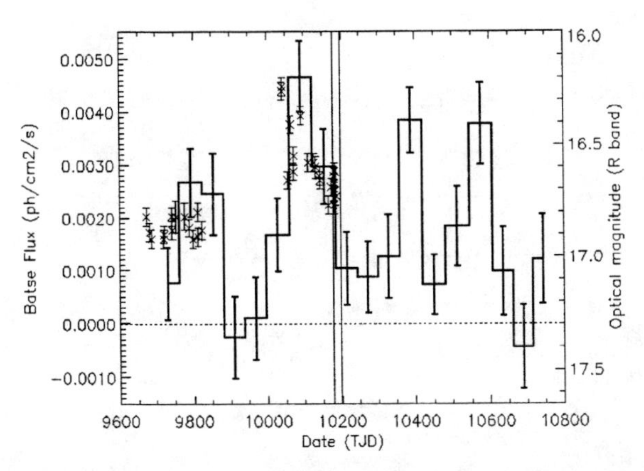

FIGURE 1. The BATSE 2-monthly flux measurements of 4C71.07 in comparison with the optical R magnitude. The period of the OSSE observations is indicated by the vertical lines.

of an OSSE observation. A few flare-like events are visible around April 1995 (TJD 9750-9900), January 1996 (TJD 10000-10200), November 1996 (TJD10350--10400) and May 1997 (TJD10598--10600) separated by periods of low flux when the source is below the detection limit of BATSE. Of particular interest is the second flare which corresponds to a period of optical flaring activity monitored in the R band (3): the source was at its historical maximum (m_R=16.1) on November 20, 1995 (TJD10042) about 55 days before the BATSE peak on January 14, 1996 (TJD10097). The optical lightcurve is highly structured with a second peak on December 8, 1995 and a third on January 7, 1996; both these maxima are characterised by a lower intensity in R than the first. BATSE data, having much lower statistical significance than the optical monitoring, is not able to resolve these structures but they are indicative of a similar broad structure shifted in time by about 50 days, similar to that observed in a previous flare (4).

The source reached a 20-100 keV flux of 2.9 10^{-10} erg cm^{-2} s^{-1} during the peak but was back to its mean flux about 3 months later. BATSE spectra relative to 60 days around the January 1996 peak (TJD 10063-10123) and around an OSSE observation in April 1996 were also extracted and analysed over the 20-500 keV energy range. The source was detected at 9.5σ and 4σ significance respectively. The data were fitted with a single power law which in both cases resulted in a satisfactory fit with photon indices of α = 2.3 ± 0.4 (χ^2=0.93 for 12 dof) and 1.3 ± 0.5 (χ^2=0.63 for 12 dof). The two BATSE measurements are shown in figure 2. Interestingly the latter data set has a spectral index similar to that obtained by OSSE (1.1 ± 0.3) although the intensity is slightly higher (as is usual when comparing BATSE and OSSE measurements (1). The former data set, however, is best represented a steeper photon index, in contrast with the typical *'flatter when brighter'* rule of Blazar behaviour.

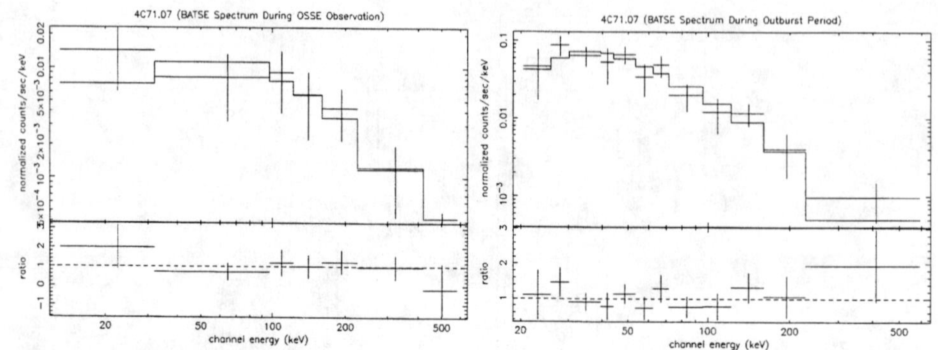

FIGURE 2. The BATSE spectra of 4C71.07 corresponding to 60-day intervals around the OSSE observation (*left*) and during the outburst (*right*).

This behaviour, however, is also seen in PKS 0528+134 (5) and so, although the evidence is marginal, it is an important observational constraint.

MKN501

According to the RXTE all sky monitor, the high X-ray activity of the source started in March 1997 and lasted until at least October of the same year (6). In figure 3 we show the BATSE light curve, which show the source increasing in brightness at the same time as detected by RXTE, indicating that the low energy and high energy X-rays are correlated in intensity. During April 1997 the source was monitored at various wavelengths, but the only continuous coverage was that provided by BATSE while

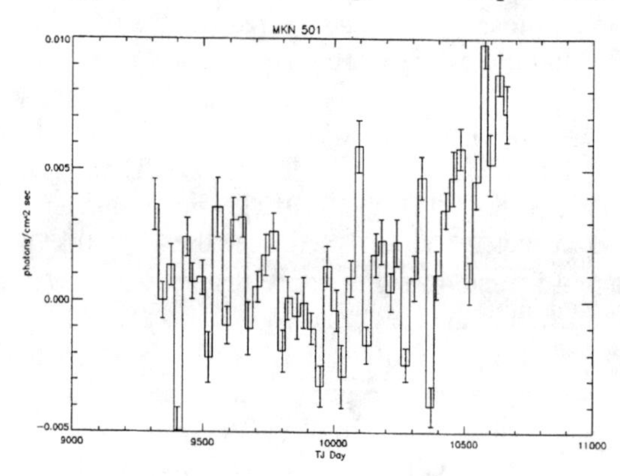

FIGURE 3. The BATSE lightcurve of Mrk 501 from 10th August 1993 until 8th September 1997. A general increase in actvity is seen from the end of February 1997 (TJD 10500) in agreement with the RXTE observations.

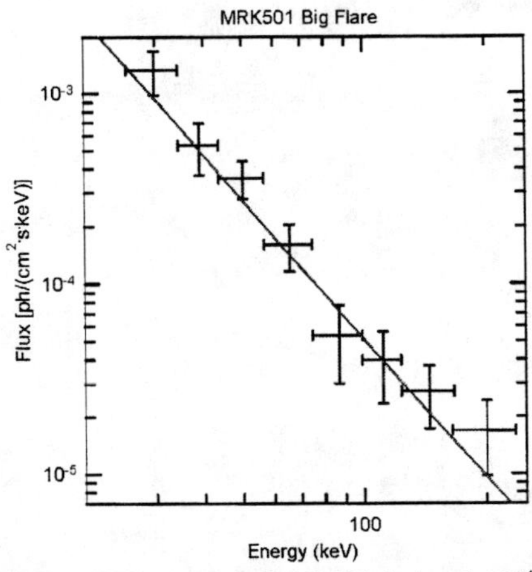

FIGURE 4. The BATSE spectrum of Mrk 501 during the 11th April 1997 flare.

both OSSE and SAX performed pointed observations. In particular SAX observed the source on April 7th ,11th and 16th, at which times the PDS spectrum was consistent with a single power law of photon index between 1.8 and 2.0. A spectrum was obtained from BATSE (figure 4) contemporaneous with the 11th April 1997 flare. The spectrum is well fitted by a single power law of photon index 2.4 ± 0.3, consistent with the SAX /PDS results (7).

PKS 2005-489

Over the period 11th - 28th October 1998, the RXTE all sky monitor detected an increase in flux from the BL Lac object PKS 2005-489 by a factor of 2-3 times its average intensity from 1996-98 (IAU Circ 7041, 1998). This activity triggered a BeppoSAX observation at the beginning of November 1998 (IAU Circ. 7055,1998). The source was bright and clearly visible in the PDS up to 100 keV, with a preliminary analysis indicating an average flux of 1.7 10^{-10} erg cm^{-2} s^{-1} in the 10-100 keV band, and a power-law of photon index 2.21 ± 0.02 well describing the data above 2 keV. Follow-up observations with RXTE indicated that the peak of emission occurred in November 1998. Prompted by these observations and the high reported flux, we searched the BATSE database for emission from this source. The source was visible (at 5σ) only during the period of RXTE and SAX reported activity i.e. between TJD 1110 and 1151 as shown in figure 5, at a flux level of 7.4 ± 1.3 10^{-10} erg cm^{-2} s^{-1}, and in particular reached a maximum at the same time as the RXTE peak.

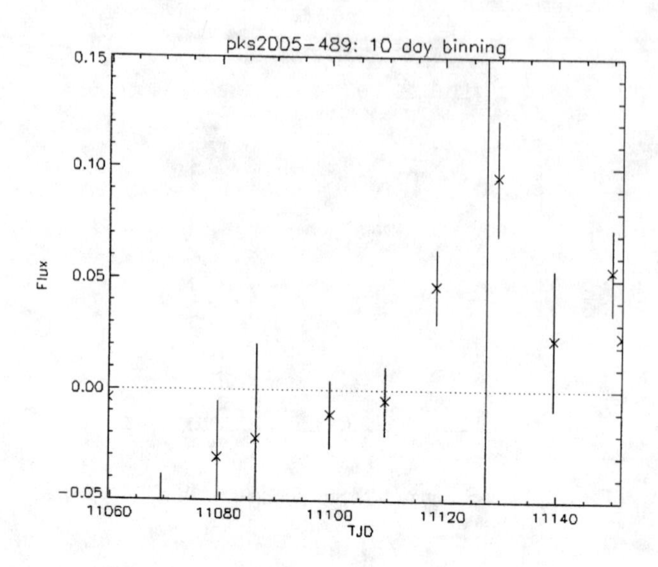

FIGURE 5. The BATSE lightcurve of PKS 2005-489 over the period around the RXTE and BeppoSAX reported activity. The vertical line at TJD 11127.5 corresponds to the 10th November flare.

REFERENCES

1. Malizia, A. *et al*, *Astrophys. J., submitted*
2. Von Linde, J. *et al*, *Astron. & Astrophys.* **267**, L23, (1993)
3. Raiteri, C.M. *et al.*, *Astron. & Astrophys.* **127**, 445, (1998)
4. Otterbein, K. *et al.*, *Astron. & Astrophys.* **334**, 489, (1998)
5. Ghisellini, G. *et al*, *Astron. & Astrophys. In press (astro-ph 9906165)*, (1999)
6. Remillard, R.A. and Levine, M.L. *in Proceedings of the conference on All Sky X-ray observations in the Next Decade (astro-ph 9707338)*, (1997)
7. Pian, E. *et al.*, *Astrophys. J.* **492**, L17, (1998)

Some Aspects of the Radio Emission of EGRET-Detected Blazars

Y.C.Lin[1], D.L.Bertsch[2], S.D.Bloom[3], J.A.Esposito[2], R.C.Hartman[2], S.D.Hunter[2], D.A.Kniffen[3], G.Kanbach[4], H.A.Mayer-Hasselwander[4], P.F.Michelson[1], R.Mukherjee[5], A.Mücke[6], P.L.Nolan[1], M.K.Pohl[7], O.L.Reimer[4], and D.J.Thompson[2]

[1] *W. W. Hansen Exp. Phys. Lab., Stanford University, Stanford, CA 94305 USA*
[2] *Code 661, Goddard Space Flight Center, Greenbelt, MD 20771 USA*
[3] *Dept. of Phys. and Astronomy, Hampden-Sydney College, Hampden-Sydney, VA 23943 USA*
[4] *Max-Planck-Institut für Extraterrestrische Physik, Giessenbachstr, D-85748 Garching Germany*
[5] *Barnard College and Columbia University, New York, NY 10027 USA*
[6] *Dept. of Phys. and Mathematical Phys., Univ. of Adelaide, Adelaide, SA 5005 Australia*
[7] *Institut für Theoretische Physik 4, Ruhr-Universität Bochum, 44780 Bochum Germany*

Abstract. It has long been recognized that the high-latitude EGRET sources can be identified with blazars of significant radio emission. Many aspects of the relation between high-energy gamma-ray emission and radio emission of EGRET-detected blazars remain uncertain. In this paper, we use the results of the recently published Third EGRET Source Catalog to examine in more detail to what extent the EGRET flux and the radio flux are correlated. In particular we examine the correlation (or the lack of it) in flux level, spectral shape, temporal variation, and detection limit. Many significant previous studies in these areas are also evaluated.

INTRODUCTION

Ever since EGRET began in 1991 to detect extragalactic objects generally referred to as blazars, the radio emission of such EGRET sources has been found to be closely related to the detected gamma radiation [1–3]. Over the years, many studies have been carried out to investigate the question of radio-γ-ray connection. In this paper, we examine and summarize such results for blazars.

FLUX CORRELATION

Among the EGRET-detected blazars, there are cases when strong γ-ray sources are found to have strong radio fluxes as well, as pointed out by, e.g., Mattox et al.

CP510, *The Fifth Compton Symposium,* edited by M. L. McConnell and J. M. Ryan

[4]. But the true nature of the flux correlation between radio and γ-ray emission is more complicated than a simple one-to-one correspondence.

One-to-One Flux Correlation

Mücke et al. [5] have made a thorough and comprehensive study and found no statistically significant one-to-one correlation between radio flux and EGRET flux. This study provides in-depth analyses on this correlation question. It produces a negative result. Unless future data can sustain a claim otherwise, it is advisable that the radio flux and the EGRET flux should not be regarded as being proportional to each other or having a one-to-one relationship.

A Possible Correlation Pattern

The radio flux and the high-energy γ-ray flux of EGRET-detected blazars could be correlated in some other ways. In Figure 1, we plot the radio fluxes at 5 GHz versus the EGRET fluxes for E > 100 MeV in individual viewing periods with the EGRET measurement significance $\sqrt{\mathrm{TS}} > 3.0$. The radio fluxes are taken from the NED database, one value for each source. The EGRET fluxes are those listed in the Third EGRET Catalog [3]. The radio fluxes and the EGRET fluxes are

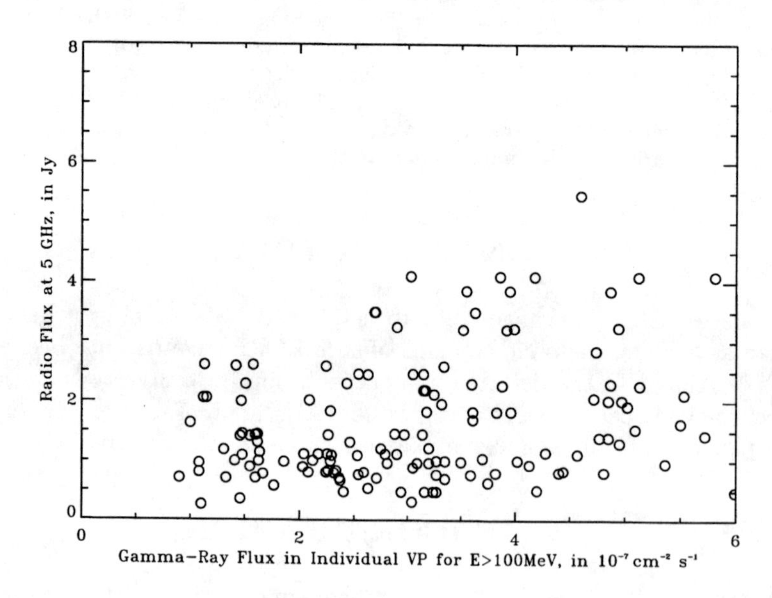

FIGURE 1. Radio flux density at 5 GHz vs EGRET flux for E>100 MeV in individual viewing periods with $\sqrt{TS} > 3.0$.

not simultaneous data. One can see that the data points in Figure 1 occupy the lower right half of the graph. The EGRET flux limit at $\sim 1.0 \times 10^{-7}$ cm^{-2} s^{-1} for E > 100 MeV reflects the EGRET sensitivity. Beyond the radio flux of $\sim$ 2 Jy, the minimum detected EGRET flux for a source seems to increase with the corresponding radio flux, or at least the EGRET flux level seems more likely to become higher when the radio flux increases in Figure 1. But five of the EGRET-detected blazars, off scale in radio fluxes in Figure 1, do not follow this pattern: 3C 273, 3C 279, 3C 454.3, PKS 0521−365, and PKS 1830−210. These are all very strong radio sources. Their EGRET fluxes are much lower than what this pattern would indicate. At this time, we do not know whether such prominent sources form a true subclass of EGRET-detected blazars or this pattern will disappear under more extensive observations. Furthermore, the variability of radio fluxes, which can easily rearrange the data points in Figure 1, is not considered here. Maybe future high-energy γ-ray missions like the GLAST telescope [6] can confirm or disprove this pattern.

Flux Correlation during Radio Flares

Valtaoja et al. [7] have published a result on the correlation of EGRET flux with radio flux during radio flares. They have found that during a flare the EGRET flux is correlated with the *increase* in the radio flux at the time of the EGRET measurement, but not to the size of the flare. This study is based on about thirty simultaneous measurements between EGRET observations and Metsähovi 22 GHz monitoring data. The statistical significance is thus not very high, but this is a very interesting result. It is directly related to the radio state at the time of EGRET detection. See the section "RADIO STATE FOR Γ-RAY EMISSION."

SPECTRUM CORRELATION

Both the radio spectral indices and the high-energy γ-ray spectral indices of the EGRET-detected blazars extend over large ranges of values. It would be interesting to see if the spectral indices in these two wavebands are correlated in some way. We have calculated the two-point radio spectral indices with data taken from the NED database between 2.7 and 5 GHz, 5 and 31 GHz, and 5 and 90 GHz, for EGRET-detected blazars. The two radio measurements for each spectral index calculation are required to be in the same radio catalog and simultaneous data are used whenever available. The EGRET spectral indices are taken from the Third EGRET Catalog [3]. No correlation whatsoever can be seen between the radio spectral indices and the EGRET spectral indices. It may seem that, although the radio and high-energy γ-ray bands are closely related to each other, the beam of particles that produces one band is unlikely to be the same one that produces the other band. These two bands of radiation are likely to be related to each other at a deeper level of the radiation mechanism.

RADIO AND EGRET FLUX LIMITS

When the EGRET Team first tried to search for counterparts in radio sources for the high-latitude EGRET sources, the radio flux was restricted to $>\sim$ 1 Jy, later changed to $>\sim$ 0.5 Jy, at 5 GHz in order to reduce the number of source candidates [1]. This has created an uncertainty as to whether the unidentified high-latitude EGRET sources are actually radio sources with fluxes lower than this artificial search limit. To answer this question, Sreekumar et al. [8], Nolan et al. [9], Dingus et al. [10], and Lin et al. [11] have devoted special attention to search for counterparts for the unidentified high-latitude ($|b| > 10°$) EGRET sources in the Second EGRET Catalog [2] among radio sources with fluxes as low as 0.3 Jy at 5 GHz or even lower. Only one possible identification was found in this way. It now appears certain that the radio flux limit of $\sim$ 0.5 Jy at 5 GHz is an instrinsic property of the EGRET-detected blazars for the EGRET detection limit of $\sim 1.0 \times 10^{-7}$ cm^{-2} s^{-1} for E > 100 MeV in one viewing period. It is true that some of the EGRET-detected blazars do have radio fluxes below 0.5 Jy at 5 GHz [3]. Furthermore, some of the radio sources with fluxes above 0.5 Jy at 5 GHz could be historically much weaker. But it seems that to find many more γ-ray-emitting blazars with radio fluxes lower than $\sim$ 0.5 Jy at 5 GHz, the γ-ray detection limit would have to be much lower than what EGRET can provide.

RADIO STATE FOR Γ-RAY EMISSION

From the studies of Reich et al. [12], Mücke et al. [13], Valtaoja et al. [14], Pohl et al. [15], Lähteenmäki et al. [16], and Marscher et al. [17,18], opinions now all seem to converge to the picture that: (1) higher the radio activities are, more often high-energy γ-rays are detected; (2) high-energy γ-rays are most likely detected when the source is in the rising phase of a radio flare; (3) it is moderately likely when the radio flux is in a high-flux stage; (4) it is least likely when the source is in the declining phase of a flare. We must also mention, as described above, that the high-energy γ-ray flux is moderately correlated with the *increase* of radio flux at the time of the EGRET measurement, but not to the flare size itself [7]. This picture represents the current understanding of the radio state when an EGRET flux is detected. It points to the possibility that the high-energy γ-rays as detected by EGRET are most likely emitted in flares and the durations of γ-ray flares are much shorter than the radio flares. But it does not preclude the possibility that low-level continuous fluxes of high-energy γ-rays may also exist in blazars.

RADIO MORPHOLOGY

Recently Piner and Kingham [19] published their VLBI study of six EGRET blazars and a number of blazars not detected by EGRET for comparison. Based on their observations, they indicate that the γ-ray flares do not necessarily correlate with component ejections, (component ejections during γ-ray flares have been reported before; see e.g. Wehrle et al. [20]), the γ-ray blazars do not preferentially belong either to the population with misaligned jets or to the population without misaligned jets, and the γ-ray blazars are not found to be more strongly beamed than those which have not been detected by EGRET. In an ongoing VLBA monitoring program by Marscher et al. [17,18], with a large sample size and a long observation history, it has been found that about 50% of the observed radio flares are correlated with EGRET detections; the lack of detections in the other 50% can be explained with paucity of EGRET observations and brevity of γ-ray flares. Marscher et al. [17,18] also indicate that EGRET-detected blazars do show evidence of being more strongly beamed than those not detected by EGRET. This is at variance with what Piner and Kingham [19] find in this beaming question. But as pointed out by Piner [21], the measured average speed of EGRET sources, at 6 $h^{-1}c$, by Piner and Kingham [19], is very similar to the value obtained by Marscher et al. [17,18]. The difference lies in the choice of objects for the sample of blazars not detected by EGRET. Marscher's sample [17,18] contains more recent results. We can perhaps draw a tentative conclusion for the beaming question at this time that EGRET-detected blazars are indeed more strongly beamed on the average than those not detected by EGRET.

REFERENCES

1. Fichtel, C. E., *ApJS* **94**, 551 (1994).
2. Thompson, D. J., et al., *ApJS* **101**, 259 (1995).
3. Hartman, R. C., et al., *ApJS* **123**, 79 (1999).
4. Mattox, J. R., et al., *ApJ* **481**, 95 (1997).
5. Mücke, A., et al., *A&A* **320**, 33 (1997).
6. Michelson, P. F., these proceedings.
7. Valtaoja, E., et al., *Proc. Heidelberg Work. on γ-Ray Emitt. AGN*, p.121 (1996).
8. Sreekumar, P., et al., *ApJ* **464**, 628 (1996).
9. Nolan, P. L., et al., *ApJ* **459**, 100 (1996).
10. Dingus, B. L., et al., *ApJ* **467**, 589 (1996).
11. Lin, Y. C., et al., *ApJS* **105**, 331, (1996).
12. Reich, W., et al., *A&A* **273**, 65 (1993).
13. Mücke, A., et al., *A&AS* **120**, 541 (1996).
14. Valtaoja, E., et al., *A&AS* **120**, 491 (1996).
15. Pohl, M., et al., *A.S.P. CONF. SER.* **110**, p.268 (1996).
16. Lähteenmäki, A., et al., these proceedings.
17. Marscher, A. P., et al., these proceedings.
18. Marchenko, S. G., et al., these proceedings.
19. Piner, B. G., and Kingham, K. A., *ApJ* **507**, 706 (1998).
20. Wehrle, A. E., et al., *Astroparticle Phys.* **11**, 169 (1999).
21. Piner, B. G., private communication.

Spectral properties of gamma-ray detected blazars from 5 to 37 GHz

Harri Teräsranta

Metsähovi Radio Observatory, Metsähovintie 114, FIN-02540 Kylmälä, Finland

Abstract. Radio monitoring data from the Metsähovi Radio Observatory and the University of Michigan Radio Astronomical Observatory are used to study if the spectral indexes of the mean flux between used frequencies are different between γ-ray detected and not detected AGN. The γ-ray detected AGN seem to have a flatter spectral index towards the higher frequencies.

INTRODUCTION

Decades long monitoring data from Metsähovi Radio Observatory at 22 and 37 GHz and from the University of Michigan Radio Astronomical Observatory at 5, 8 and 14.5 GHz has been used in studying the mean spectral indexes of AGN detected at γ-rays by the EGRET and for reference a similar group of flat spectrum radio sources which have not been detected at γ so far. The purpose of this study is to find criteria for identifying the still unknown sources detected by EGRET. As the pointing accuracy of EGRET is only in the order of one degree, in the error box remain many sources, which could be considered as potential radio counterparts to the γ-ray emission. The flat spectrum (from 1.4 or 2.7 to 5 GHz) of the source together with its strong radio flux combined with earlier blazar type behaviour have been the criteria for source identification. The spectral index α is calculated in this study according to Eq.1.

$$\alpha = \frac{log(\frac{S_2}{S_1})}{log(\frac{f_2}{f_1})} \tag{1}$$

where, S_1 is the flux density at the lower frequency f_1 and S_2 is the flux at the higher frequency f_2. Simultaneous observations with the γ-detections would give best results in source identifications, at least in the mm-region [1], but as the γ-ray observations were done many years ago, another approach must be tried.

CP510, *The Fifth Compton Symposium,* edited by M. L. McConnell and J. M. Ryan

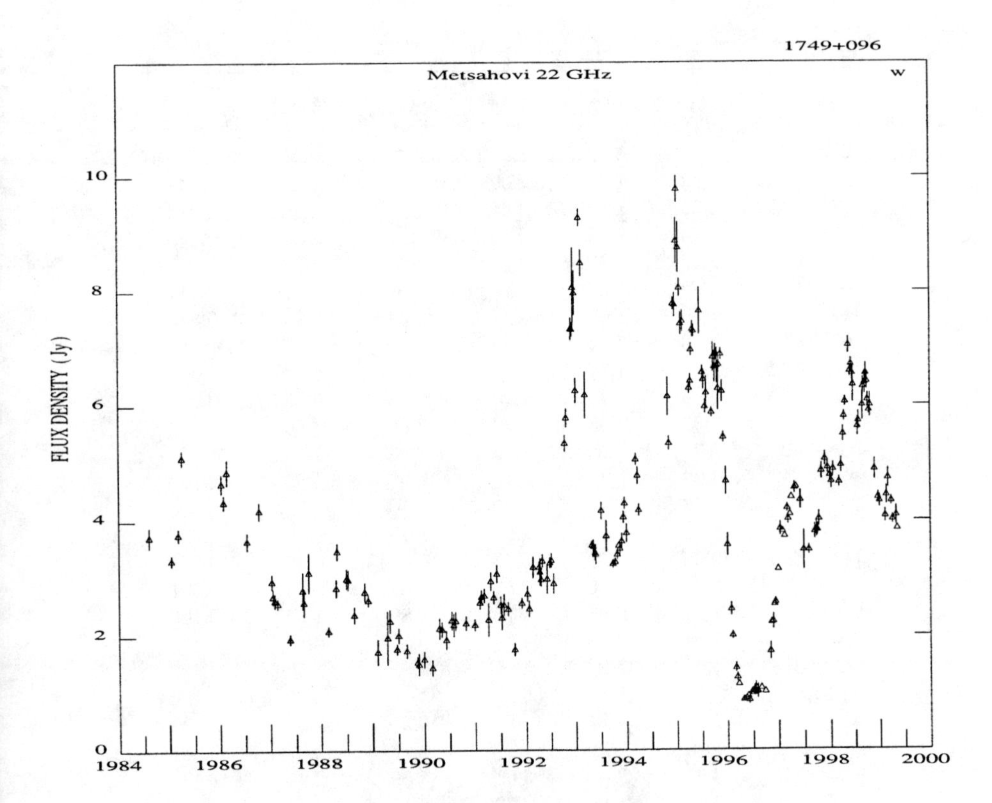

FIGURE 1. The flux density of 1749+096 at 22 GHz as measured at Metsähovi.

THE SAMPLES

In this study 31 γ-ray detected sources from the third EGRET cataloque [2], which had been observed over a long period of time with both telescopes, were chosen (Table 1). For reference a group of 27 sources with about equal sampling at radio frequencies was chosen (Table 2). The Michigan data used was monthly averaged and the Metsähovi data is weekly averaged. After this first smoothing of data, a simple mean value for the flux was calculated at all frequencies.

The data is averaged in two parts mainly to the possibility of too dence sampling during outbursts, which would overestimate the mean values. In ultimate cases the flux of a source can vary with over a factor of 10, as seen in figure 1, where is shown the flux density of 1749+096 at 22 GHz. The Michigan data is quite evenly sampled throughout the year, as the Metsähovi data suffers from random breaks due to other telescope users.

TABLE 1. Mean flux values at 5-37 GHz for the γ-detected sources

Source	Sm(5) (Jy)	Sm(8) (Jy)	Sm(14) (Jy)	Sm(22) (Jy)	Sm(37) (Jy)	z
0202+149	2.75	2.92	2.63	2.41	2.24	0.405
0219+428	2.18	1.53	1.09	1.10	1.07	0.444
0234+285	2.69	2.89	2.66	2.51	2.58	1.213
0235+164	1.75	2.13	2.04	2.17	2.34	0.940
0336-019	2.22	2.37	2.36	2.45	2.36	0.852
OA 129	3.69	4.27	4.40	4.67	5.07	0.915
0440-00	1.59	1.68	1.43	1.09	0.98	0.844
0458-020	2.98	3.12	3.20	2.45	2.31	2.286
0528+134	3.81	4.76	5.11	6.01	5.87	2.060
0716+714	0.58	0.82	0.70	0.94	0.76	0.3
0735+178	2.31	2.55	2.32	2.06	2.36	0.424
0804+499	1.00	1.18	1.23	1.36	1.36	1.43
0836+710	2.09	1.98	1.79	1.90	1.55	2.172
OJ 287	2.68	3.45	4.05	3.70	3.82	0.306
0954+55	1.95	1.69	1.30	1.10	0.90	0.901
0954+658	0.66	0.78	0.73	0.69	0.86	0.368
Mark 421	0.73	0.71	0.61	0.61	0.55	0.031
4C 29.45	1.57	1.75	1.78	1.89	1.67	0.729
ON 231	1.13	1.19	1.03	0.83	0.83	0.102
1222+216	1.91	2.19	1.85	1.81	1.58	0.435
3C 273	37.33	38.24	36.11	34.52	30.26	0.158
3C 279	11.39	13.76	14.87	18.31	17.51	0.538
1510-089	2.44	2.78	2.91	2.78	2.87	0.361
1611+343	3.32	3.77	3.65	3.59	2.94	1.401
4C 38.41	2.64	2.72	2.39	2.18	2.12	1.814
1730-130	6.81	9.01	8.94	10.07	9.15	0.902
1739+522	1.93	2.11	2.06	1.84	1.67	1.375
1741-03	2.99	3.95	4.55	4.91	3.69	1.054
BL Lac	3.57	3.91	3.60	3.57	3.23	0.069
CTA 102	4.08	3.64	3.09	3.40	3.24	1.037
3C 454.3	12.33	12.60	10.43	9.62	8.58	0.859

RESULTS

For the 5 mean fluxes at specific frequencies all combinations of spectral indexes (α) were calculated. Mean values of these spectral indexes and their error estimates were then calculated for both groups (Table 3). As most of these sources, also from the refrence list, are violently variable, the mean value for the flux should be a clear improvement to using a single epoch flux.

Most of the EGRET-detected AGN have allready been in both Metsähovi and Michigan monitoring samples prior their detection at γ. The Metsähovi monitoring sample has been formed mainly from flat spectrum sources (α between 1.4 or 2.7 and 5 GHz being close to 0), likewise to the sample for optical variability from three decades ago for the Rosemary Hill group [3]. Clearly this flat spectrum behaviour

TABLE 2. Mean flux values at 5-37 GHz for the reference group

Source	Sm(5) (Jy)	Sm(8) (Jy)	Sm(14) (Jy)	Sm(22) (Jy)	Sm(37) (Jy)	z
III ZW 2	0.29	0.49	0.88	0.83	0.85	0.090
0109+224	0.55	0.75	0.73	1.05	1.05	
DA 55	1.54	1.98	2.18	2.48	2.37	0.859
3C 84	41.20	39.12	33.52	29.09	25.52	0.018
NRAO 140	1.74	1.63	1.45	1.61	1.48	1.259
NRAO 150	4.36	4.55	4.17	4.03	3.62	
3C 120	3.57	3.65	3.08	3.10	2.44	0.033
DA 193	5.81	6.70	5.32	4.83	4.16	2.365
0736+017	1.64	1.86	1.77	1.76	1.61	0.191
OI 090.4	1.17	1.35	1.35	1.40	1.21	0.660
0814+425	1.07	1.17	1.08	1.10	0.84	0.258
4C 39.25	8.03	9.31	7.82	8.92	7.26	0.698
OL 093	2.97	3.61	3.94	3.80	3.68	0.888
1308+326	2.22	2.79	2.88	2.78	2.46	0.997
1413+135	0.99	1.52	1.93	1.94	2.00	0.247
4C 14.60	1.06	1.20	1.01	0.84	0.68	0.605
OS 562	1.28	1.43	1.63	1.54	1.43	0.751
3C 345	8.86	10.11	10.15	9.42	8.83	0.595
1749+096	1.87	2.69	3.25	4.12	3.99	0.320
1803+784	2.49	2.85	2.80	2.26	1.98	0.68
3C 371	1.83	1.88	1.72	1.55	1.41	0.050
1928+738	3.76	3.89	3.71	3.31	2.45	0.302
2005+40	3.46	3.83	3.17	2.48	2.05	1.736
OX 057	8.74	8.68	6.31	4.92	3.26	1.932
2145+067	3.77	7.43	8.32	8.09	6.96	0.999
2201+315	3.05	3.58	3.49	2.96	2.60	0.298
3C 446	4.80	5.24	5.25	5.58	5.22	1.404

has been of importance, as all of those EGRET-detected sources seem to belong to this group.

From the Metsähovi monitoring data it was earlier seen, that the γ-ray emission is seen only when the source is in a higher state of activity, or in a rising phase [1]. From Table 3 it can be conluded, that the differences in spectral indexes between the γ-detected and non-detected AGN are in most cases not significant. From the lower part of the spectrum, 5 to 8 and 5 to 14 GHz the non-γ sources tend to have a more inverted spectrum, even the flux is rising on average for both groups. At the higher end of our spectral coverage, 14-37 and 22-37 GHz spectral indexes tend to be flatter for the γ-detected sources, even both groups have a falling flux on average. Thereby, for identification of γ-ray sources, a better indicative than the lower frequency (5-8) spectral index would be the high frequency (20-40GHz) spectral index.

TABLE 3. Mean spectral indexes between all frequency pairs, their standard errors and the chance of the means being from the same population

$f_1 - f_2$	$\alpha(\gamma)$	σ	$\alpha(ref)$	σ	P
5-8	0.18	0.05	0.33	0.07	0.09
5-14	0.02	0.04	0.14	0.06	0.10
5-22	0.01	0.04	0.08	0.05	0.28
5-37	-0.02	0.03	0.00	0.04	0.69
8-14	-0.11	0.03	-0.03	0.06	0.23
8-22	-0.07	0.03	-0.04	0.05	0.60
8-37	-0.09	0.03	-0.11	0.04	0.69
14-22	-0.03	0.05	-0.06	0.06	0.70
14-37	-0.08	0.03	-0.17	0.04	0.08
22-37	-0.12	0.04	-0.25	0.04	0.03

FUTURE CONSIDERATIONS

This method of trying to predict γ-ray visibility of sources from their continuum properties at 5-37 GHz frequencies has several drawbacks. First we see the combined radiation coming from a large area: core, jet and diffuse radiation from outer regions. At lower frequencies, 5 and 8 GHz it is impossible to extract the outburst components, as the outbursts tend to last for many years and thus the quiet levels are not reached before the next outburst starts. If the γ-ray flares are coming from the very base of the jet, close to the core, then higher radio frequency radiation, above 20 GHz could have a connection with this. An extension of this study with similar demands on the sampling to higher frequencies, 90 and 230 GHz would need weekly observations. For identification of still unknown γ-ray sources, it is proposed that the sources within error boxes that show flatter spectra towards higher frequencies, 20-40 GHz, should be more propable than the ones peaking clearly at lower frequencies (5-8 GHz). Future observations with AGILE and GLAST will prove if this is true, as their sensitivity and pointing accuracy is far better than with EGRET.

ACKNOWLEDGEMENTS

This research has made use of data from the University of Michigan Radio Orservatory which is supported by funds from the University of Michigan.

REFERENCES

1. Valtaoja E. & Teräsranta H. *A&A* **297**, L13, (1995).
2. Hartman R.C., Bertsch D.L., Bloom S.D., et al. *ApJS* **123**, 79–202, (1999).
3. Smith A. G. *Astron. Soc. Pac. Conf. Ser.*, Vol **110**, p 3–16, (1996).

Comparison of Epochs of Ejection of Superluminal Components with the Gamma-Ray Light Curves of EGRET Blazars

S. G. Marchenko[1,2], A. P. Marscher[1], J. R. Mattox[1]
J. Hallum[1], A. E. Wehrle[3], S. D. Bloom[3]

[1]Institute for Astrophysical Research, Boston University, U.S.A.
[2]Astronomical Institute, St. Petersburg State University, Russia
[3]Infrared Processing and Analysis Center, U.S.A.

Abstract. We compare the extrapolated epochs of zero separation of apparent superluminal components from the cores in γ-ray bright blazars with the EGRET light curves. In 10 out of 20 superluminal components with sufficient γ-ray data, the epochs of zero separation coincide with times of global or local maxima of the γ-ray light curves within the 1-σ uncertainties of the epochs of zero separation. If we focus only on the γ-ray peaks, nine out of ten with sufficient VLBA data are accompanied by superluminal ejections. The γ-ray flares are therefore associated with energetic events in the jets rather than with minor episodes of enhanced particle acceleration.

OBSERVATIONS

We have completed a program of monitoring of the milliarcsecond-scale structure of γ-ray bright blazars with the VLBA at 22 and 43 GHz during the period from November 1993 to July 1997. (Only a subset of the sample was observed at each epoch). The sample consists of 30 quasars and 12 BL Lac objects. We have determined velocities of jet components in 33 sources, 20 of which have multiple moving knots, and compare the epochs of zero separation from the (presumed stationary) core with the γ-ray light curves obtained from the 3rd EGRET catalog (1) in order to determine whether γ-ray flares are associated with major energetic disturbances that propagate down the jet.

Fig. 1 presents the γ-ray light curves (the flux is in units of 10^{-8} photons cm^{-2} s^{-1}, error bars correspond to 1-σ, and points without error bars are upper limits)

CP510, *The Fifth Compton Symposium,* edited by M. L. McConnell and J. M. Ryan

TABLE 1. Epochs of Zero Separation and Times of γ-ray Peaks

Name	z	T_γ	T_0	$\beta_{app}h$
0336-019	0.852	1995.27	1995.28 ±0.02	6.11±0.06
0440-003	0.844	1994.63	1994.66±0.19	8.5±0.8
0528+134	2.06	1994.21	1994.2±0.4	27±15
0716+714	?	1995.13	1995.1±0.1	—
0836+710	2.17	1992.20	1992.2±0.2	13.3±2.0
1156+295	0.729	1993.02	1992.8±0.2	4.3±1.0
1622-253	0.786	1995.73	1995.7±0.2	17.5±1.1
1622-297	0.815	1995.46	1995.2±0.2	16.0±0.8
1730-130	0.902	1995.46	1995.3±0.2	6.1±2.1
1908-201	?	1995.73	1995.9±0.1	—

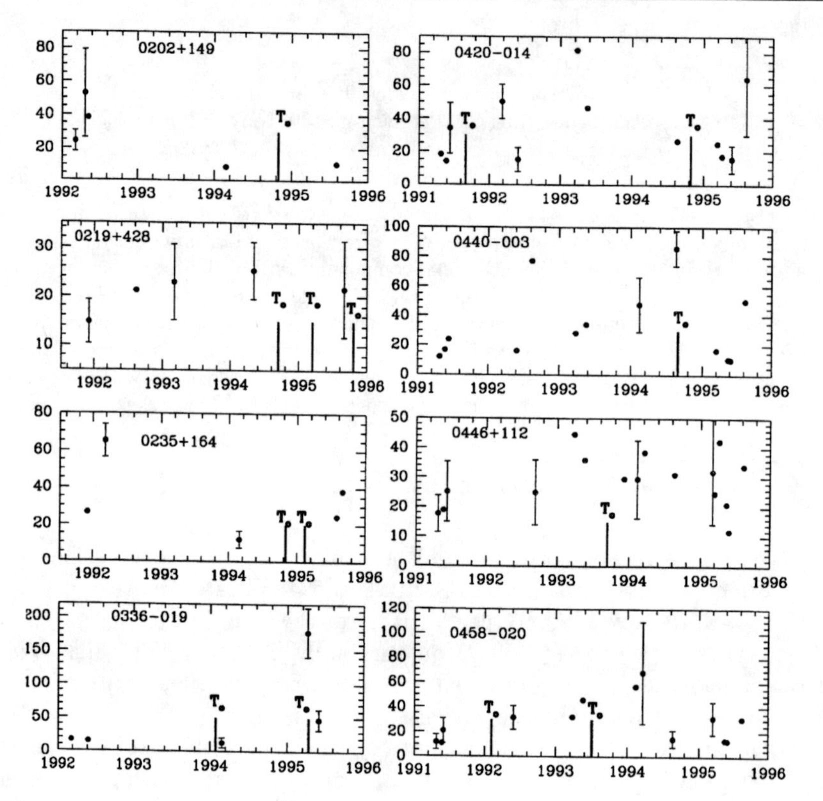

FIGURE 1. EGRET γ-ray light curves of sources with 43 GHz VLBA data near times of peaks in the light curves. Vertical lines marked with "T_0" indicate epochs of zero separation from the core on the VLBA images, as extrapolated from the motions observed at later epochs.

for 27 sources, where the times marked "T_0" correspond to the epochs of zero separation of superluminal jet components from the cores. However, the epochs of zero separation are contemporaneous with γ-ray observations for only 19 sources (26 components). For six of the 26 superluminal components, the γ-ray fluxes near the zero-separation epochs were only uninteresting (i.e., too high to determine whether there was a flare) upper limits to the γ-ray fluxes. In ten of the remaining cases, the epochs of zero separation coincided with the times of global or local maxima of the γ-ray light curves within the 1-σ uncertainties of the zero-separation epochs. These results are presented in Table 1, where z is the redshift, T_γ is the time of the global or local maximum of the γ-ray light curve, T_0 is the epoch of zero separation, and $\beta_{app}h$ is the apparent speed in light units for $H_0 = 100h$ km s^{-1} Mpc^{-1}, q_0=0.1, and Λ=0.

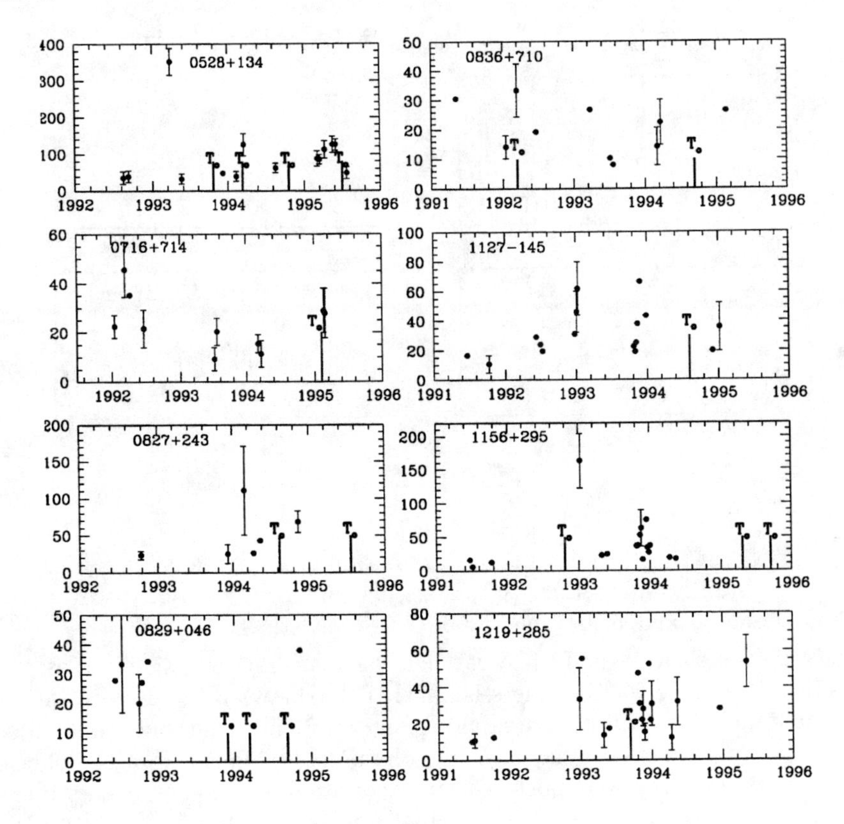

FIGURE 1. cont.

DISCUSSION

It is very striking that, despite the sparse character of the γ-ray light curves and uncertainties of epochs of zero separation of about 0.1–0.2 yr, in 50% of the superluminal components, the epoch of ejection corresponded to a high γ-ray flux, and that 90% of the γ-ray flares were accompanied by superluminal ejections.

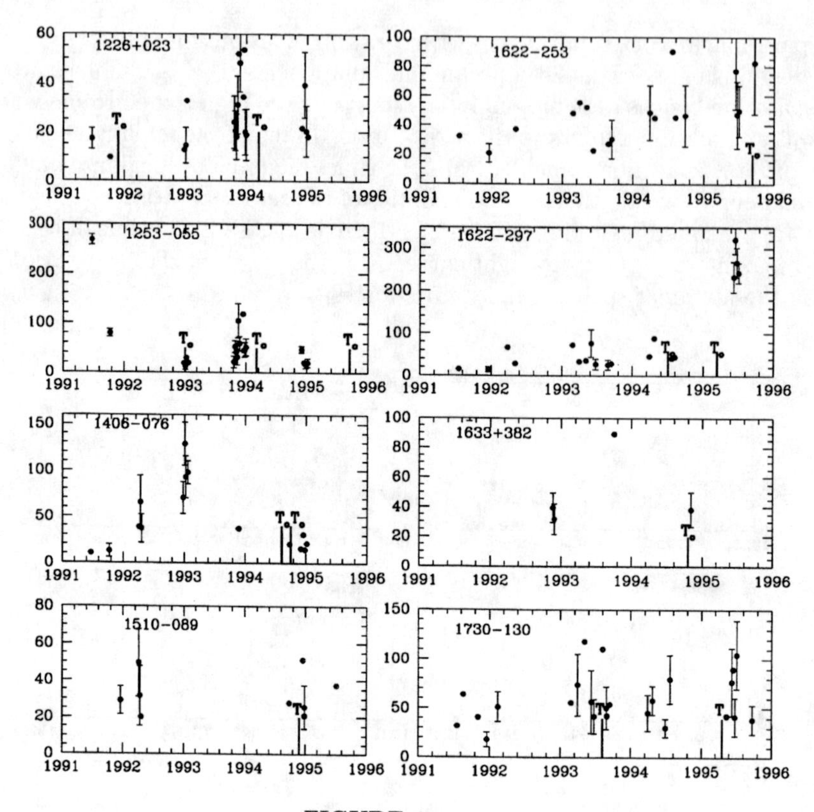

FIGURE 1. cont.

This implies that the γ-ray high states are not, in general, merely very short-term peaks of constantly fluctuating γ-ray emission, but rather major disturbances in the Lorentz factor and/or flow energy of the relativistic jet.

Since the time coverage of the γ-ray light curves is rather sparse in most cases, we cannot determine whether there is a time delay between the peak in the γ-ray light curve and the ejection of a superluminal component, although we can state that we see no evidence for such a delay: the average differences of time between the γ-ray flux maxima and epochs of zero separation for all components in Table 1 is very close to zero ($\langle T_\gamma - T_0 \rangle$=0.05±0.13 yr).

In the remaining 50% of the superluminal components, the ejection was not accompanied by a high γ-ray flux. Such non-correspondence can be explained if the time scale of a high γ-ray state is of the same order as the typical 1-σ uncertainties of the epoch of zero separation. We are in the process of carrying out numerical similations to determine the statistical significance of our result.

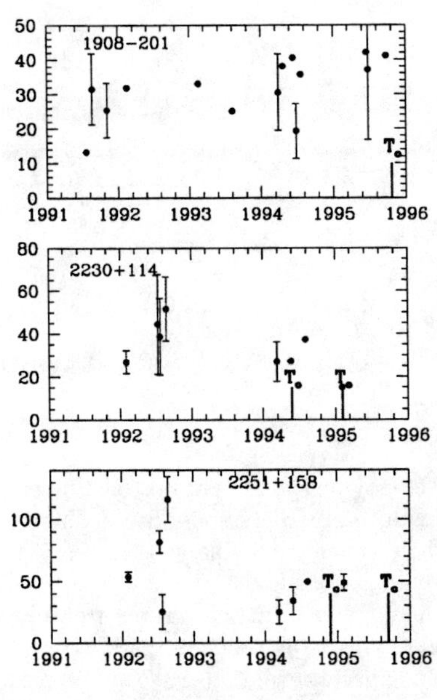

FIGURE 1. cont.

ACKNOWLEDGMENTS

This research was supported in part by a number of NASA guest investigator grants (RXTE and CGRO).

REFERENCES

1. Hartman, R. C., et al., *ApJS*, **123**, 79 (1999).

Multi-Epoch VLBA Observations of Gamma-Ray Bright Blazars

Alan P. Marscher[1], Svetlana G. Marchenko[1,2], John R. Mattox[1]
Jeremy Hallum[1], Ann E. Wehrle[3], Steven D. Bloom[3]

[1]Institute for Astrophysical Research, Boston University, U.S.A.
[2]Astronomical Institute, St. Petersburg State University, Russia
[3]Infrared Processing and Analysis Center, U.S.A.

Abstract. We report some of the statistical results from our multi-epoch VLBA study of motions in the jets of 42 γ-ray bright blazars. The apparent velocities are high, which agrees with the expectations of inverse Compton models in which the γ rays are more highly beamed than is the radio synchrotron radiation. We find evidence that the apparent speeds are higher farther from the core, which can be explained best by bending from a viewing angle near zero to one closer to $\sin^{-1}(\Gamma^{-1})$, rather than by increasing Lorentz factors. In 25 of the sources in the sample, there is at least one non-core component that appears to be stationary during our observations. We offer explanations for these results.

OBSERVATIONS AND RESULTS

We have completed a program of monitoring of the milliarcsecond-scale structure of γ-ray bright blazars with the VLBA at 22 and 43 GHz during the period from November 1993 to July 1997. The sample consists of 30 quasars and 12 BL Lac objects (only a subset of which was observed at each epoch). We have determined velocities of jet components in 33 sources, 20 of which have multiple moving knots. To convert from proper motions to velocities, we assume a standard Friedmann cosmology with H_o=65 km s^{-1} Mpc^{-1}, q_o=0.1, and Λ=0.

The plot shown in Figure 1 (bottom panel) reveals a positive correlation of apparent velocity with redshift. It appears that the correlation is stronger for BL Lac objects (solid circles) than for quasars (open circles). However, the number and range of redshifts of the BL Lac objects in the sample are both small, hence the statistical significance of this trend is not high.

The distribution of apparent velocities (Fig. 1, top panel) has a global maximum at values of 10–12c, with a long tail extending to very high speeds. In fact, this is

CP510, *The Fifth Compton Symposium,* edited by M. L. McConnell and J. M. Ryan

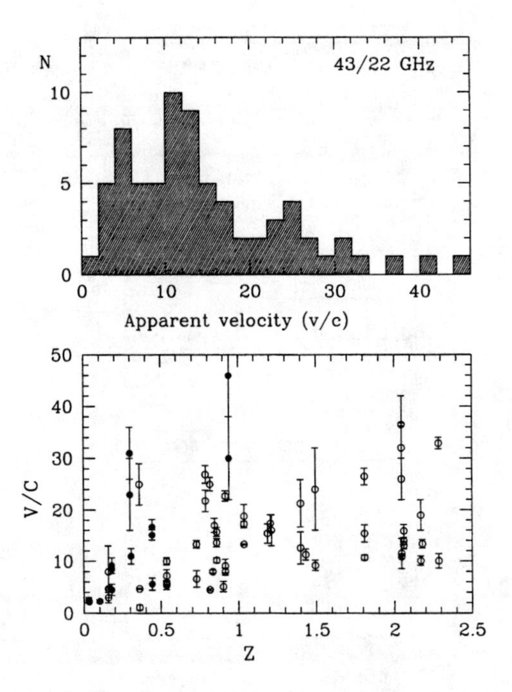

FIGURE 1. *Top panel:* Histogram of apparent velocities of jet components with proper motions detected (33 sources). *Bottom panel:* Correlation plot of apparent velocities vs. redshift. BL Lac objects are denoted by filled circles, quasars by open circles.

the first large sample to contain many sources with apparent speeds exceeding $20c$.

Although our time base is not sufficiently long to determine with confidence whether individual components accelerate, we can decide statistically whether proper motions are higher or lower farther from the core. In order to compare motions in different sources, we must deproject the distances along the jet, which requires an estimate of the angle between the jet axis and the line of sight. The Lorentz factor Γ must be greater than the apparent speed in light units ($\beta_{\rm app}$). Furthermore, the apparent motion is maximized at $\beta_{\rm app} \approx \Gamma$ when the angle between the velocity vector and the line of sight equals $\sin^{-1}(\Gamma^{-1})$. Hence, we can perform a crude deprojection by assuming that the orientation angle equals $\sin^{-1}(\beta_{\rm app}^{-1})$. Still, since there are a number of sources in which different components have different proper motions, there is no unambiguous way to determine the orientation angle. In Figure 2 we plot apparent velocity vs. distance from core for three cases: (1) (top panel) constant Lorentz factor within each source, equal to the maximum ($11c$) of the distribution of apparent speeds (cf. Fig. 1, top panel) — this serves mainly to illustrate the typical deprojected distances in the sources; (2) (middle panel) constant Lorentz factor within each source, equal to the average apparent velocity of the components in that source (with error bars corresponding to the range of apparent velocities in each source); and (3) (bottom panel) Lorentz factor of each component equal to its apparent

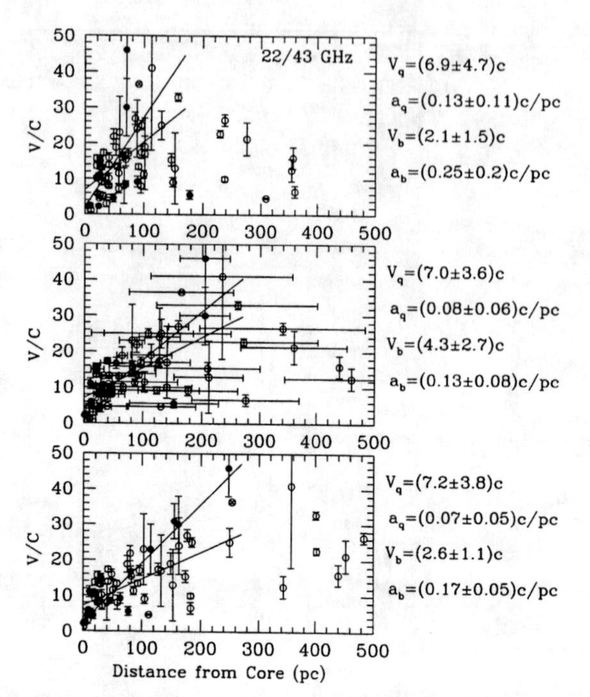

FIGURE 2. Apparent velocities vs. deprojected distance from the core, following (in the three different panels) various schemes to estimate the angle between the jet and the line of sight (see text). BL Lac objects are denoted by green circles, quasars by red circles. The lines are root-mean-square fits to the data points for each class of object. At right are indicated the intercept (V) and the slope (a) for both quasars (subscript "q") and BL Lac objects (subscript "b").

speed, which allows different Lorentz factors and angles for different components in the same source. Other possible deprojection schemes can be devised, but all give the same general result: For both quasars (open circles) and BL Lac objects (solid circles), the apparent velocities are faster farther from the core up to a deprojected distance of at least 100–200 pc. The apparent acceleration is lower for quasars, however their mean apparent velocities close to the core are approximately two times faster.

We measure no significant motion relative to the core for 37 non-core components in 25 of the sources. The distribution of distances from the core of these stationary components is shown in Figure 3. The top panel presents angular distances, while in the bottom panel we convert these to linear distances, deprojected by assuming that the angle between the jet axis and the line of sight equals the value of $\sin^{-1}(\beta_{\rm app}^{-1})$ of the moving component closest in position to the stationary component. Using this scheme, we find two preferred ranges of distances from the cores: 60–80 pc and 140–160 pc. The distribution therefore appears not to be random, and could even be periodic.

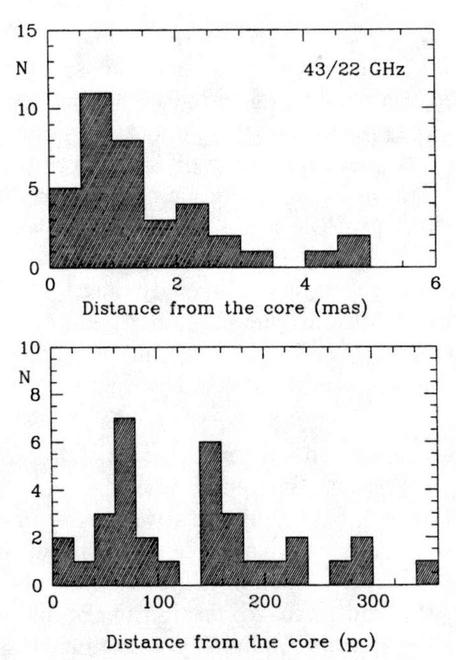

FIGURE 3. Distances from the core of components with no significant motion (relative to the core) during our observations. *Top panel:* Angular distances in mas. *Bottom panel:* Linear distances in pc (sources without known redshifts have been deleted).

DISCUSSION

The maximum of the apparent velocity distribution of about 11c and the long high-velocity tail contrasts sharply with the preliminary apparent velocity distribution of the Caltech-Jodrell flat-spectrum (CJF) sample (1), for which most of the speeds lie in the 0–8c range (after adjustment to our adopted values of H_o and q_o). The high speeds are expected for a sample selected for γ-ray brightness if the source of the high-energy emission is inverse Compton scattering. In the synchrotron self-Compton case (seed photons from the jet), both the synchrotron and the inverse Compton emission are beamed by a factor of $\delta^{2+m+\alpha}$, where δ is the Doppler beaming factor, and m is zero for steady-state emission and unity for emission from a single, evolving component. However, the value of the spectral index α (flux density $F_\nu \propto \nu^{-\alpha}$) is considerably steeper (~ 1) in the γ-ray regime than in the radio (~ 0). If the seed photons are from unbeamed (relative to the observer) sources outside the jet that are scattered by relativistic electrons inside the jet, these photons are blueshifted in the frame of the jet plasma and an additional effective Doppler boosting occurs, such that the γ-ray flux density is proportional to $\delta^{4+\alpha}$. In either case, the higher Doppler beaming for the γ-rays causes the γ-ray bright blazars to have higher mean Lorentz factors and therefore higher apparent velocities than sources that are not so γ-ray bright.

Our result that the apparent velocities are higher farther from the core is intriguing. While it might imply acceleration of the jet on the tens and hundreds of parsec scale, as in the original Blandford & Rees (2) jet model, the γ-ray emission, which originates within a parsec or less of the central engine, must be highly beamed to avoid high opacities to pair production off the X-rays (e.g., ref. 3). We consider it more likely that the effect is caused by bending. For a given Lorentz factor, the strong relativistic beaming of the γ-ray blazars favors very small viewing angles, close to zero. Projection effects should accentuate bends, and indeed the jets of many of the blazars in our sample are strongly bent. If the viewing angle is near zero in the most compact regions, then the only possible bending is toward larger angles, which [up to an angle $\sin^{-1}(\Gamma^{-1})$] results in higher apparent velocities, as observed. The bending required is only a few degrees over ~ 100 pc, and hence would be quite gradual if a source were viewed at a right angle to the jet.

The presence of stationary components is expected according to hydrodynamical simulations, because of pressure differences with the external medium. This is the result of pressure imbalances caused by variations either in the external medium or in the energy (or velocity) at the injection point at the throat of the jet (4,5). The stationary features, which are standing shocks, are spaced roughly periodically, which would explain the multi-peaked distribution of distances from the core. In fact, in these models the core itself might be a stationary shock.

ACKNOWLEDGMENTS

This research was funded by the NASA Gamma Ray Observatory Guest Investigator Program under grants NAG5-2508, NAG5-3829, and NAG5-7323. The VLBA is a facility of the National Radio Astronomy Observatory (NRAO), which is operated by Associated Universities Inc., under cooperative agreement with the National Science Foundation.

REFERENCES

1. Pearson, T. J., *et al.*, "The Caltech-Jodrell Bank VLBI Surveys," in *IAU Colloq. 164: Radio Emission from Galactic and Extragalactic Compact Sources*, ed. J. A. Zensus, G. B. Taylor, J. M. Wrobel, *ASP Conf. Ser.*, **144**, San Francisco: Astron. Soc. Pacific, 17–24 (1998).
2. Blandford, R. D., & Rees, M. J., *MNRAS*, **169**, 395–415 (1974).
3. Wehrle, A. E., it et al., *ApJ*, **497**, 178–187 (1998).
4. Daly, R. A., & Marscher, A. P., *ApJ*, **334**, 539–551 (1988).
5. Gómez, J. L., Martí, J. M., Marscher, A. P. et al., *ApJ*, **482**, L33–L36 (1997).

Space VLBI Observations of Gamma-Ray Sources

P.G. Edwards and H. Hirabayashi

Institute of Space and Astronautical Science, Sagamihara, Kanagawa 229-8510, Japan

Abstract. The radio astronomy satellite HALCA, launched in February 1997, is the main element of the VLBI Space Observatory Programme (VSOP) which enables images to be made at 1.6 GHz and 5 GHz with resolutions up to three times higher than can be obtained from ground VLBI. Most VSOP observations are of compact, radio-loud flat-spectrum extragalactic sources, with many of the identified CGRO gamma-ray sources already having been observed. In this paper VSOP observations of gamma-ray sources are reviewed.

I INTRODUCTION

The technique of Very Long Baseline Interferometry (VLBI) utilises widely spaced radio telescopes simultaneously observing the same celestial radio source. At each telescope, the radio-astronomical signal is mixed with a highly stable signal from a local frequency standard and recorded on magnetic tape. These data are then coherently combined at a correlator to synthesize a telescope with dimensions equal to the largest projected baselines between the telescopes.

At 5 GHz, the limit to the angular resolution that can be achieved by ground-based VLBI, imposed by the diameter of the Earth, it is about 1 milliarcsecond (mas). However, many radio sources contain components that remain unresolved at this resolution. The feasibility of extending VLBI to space was demonstrated in a series of experiments using a TDRSS satellite [1,2]. These results and other studies paved the way for the development of the satellite MUSES-B by the Institute of Space and Astronautical Science (ISAS).

The 830 kg MUSES-B satellite was launched on 12 February 1997 and the satellite then renamed HALCA (the Highly Advanced Laboratory for Communications and Astronomy). The 8 metre diameter main reflector, a gold-coated molybdenum mesh, was deployed two weeks after the launch. HALCA's orbit takes it from a perigee height above the Earth's surface of 560 km to an apogee height of 21,400 km, with the orbital period being 6.3 hours. The elliptical orbit enables a wide range of baseline lengths and orientations to be sampled, enabling imaging VLBI observations at 1.6 GHz (λ18cm) and 5 GHz (λ6cm) on baselines over three times longer

CP510, *The Fifth Compton Symposium,* edited by M. L. McConnell and J. M. Ryan

than those achievable on Earth [3].

HALCA detects left-circular polarized radio signals, and VSOP observations are made using two 16 MHz channels of two-bit sampled data. Correlated flux densities of at least ~100 mJy are required in order for sources to be detected on baselines to HALCA. Observing is further constrained by a number of conditions. A two-way link between HALCA and one of five dedicated ground tracking stations is required for VSOP observations: the tracking station up-links a suitably doppler-shifted reference signal to the satellite, and the satellite down-links the digitized and formatted science data in real time. HALCA is powered by solar panels and so no observing is possible when the Sun is eclipsed by the Earth or Moon. Furthermore, HALCA cannot observe sources within 70° of the sun as the main antenna shadows the solar panels at these angles.

II VSOP OBSERVATIONS

Approximately 30% of HALCA's in-orbit time is used for observing projects selected by international peer-review from open proposals submitted by the astronomical community in response to Announcements of Opportunity. This part of the mission's scientific programme constitutes the General Observing Time (GOT).

A further 15% of the in-orbit time will be devoted to a mission-led systematic survey of Active Galactic Nuclei (AGN): the VSOP Survey Program. Survey observations are in general of shorter duration and made with fewer telescopes than the GOT observations. Nevertheless, the Survey will provide a large complete sample of data on sub–milli-arcsecond radio structures for studying the statistical properties of AGN and planning future VLBI (and space VLBI) observations.

The remainder of the in-orbit time is used for maneuvering between sources, calibration and testing of the satellite systems. Further details of the VSOP observations are available from the VSOP website: http://www.vsop.isas.ac.jp.

III RESULTS

All but a small fraction of VSOP observations are made of extragalactic sources, however two of the galactic sources that have been observed are associated with gamma-ray sources. The Vela pulsar has been observed at 1.6 GHz in the speckle limit of interstellar scattering as a continuation of studies which inferred the size of the pulsar's emission region as being 500 km [4]. More recently, a long 5 GHz VSOP observation was made of the binary system LSI +61°303 during one of its regular non-thermal radio outbursts. This massive X-ray binary system has been associated with 3EG J0241+6104, although the identification is not yet firmly established [5].

The nearest extragalactic source observed by HALCA and ground-radio telescopes is Cen A, which is classified as a firmly identified AGN in the third EGRET catalog [5]. The 5 GHz VSOP observations to date have barely detected Cen A on the baselines to the satellite, probably because the core is significantly self-absorbed at this frequency, as shown by ground VLBI observations [6].

The TeV gamma-ray sources Mkn 421 and Mkn 501 have also been the subject of VSOP observations. Piner et al. combined the 5 GHz VSOP observation of Mkn 421 with a series of ground-based VLBI observations to monitor the motion of components in the parsec-scale radio jet, and found that the jet component speeds are sub-luminal [7]. The moderate VLBI component speeds and high inferred Doppler factor from observations of TeV variability [8] can be reconciled if a small angle to the line-of-sight is assumed for the jet, or the Doppler factor is assumed to change between the 10^{-4} pc-scale TeV emitting region and the pc-scale radio jet, or the component motions do not reflect the real bulk flow speed [7].

VSOP images of Mkn 501 have been presented by Giovannini et al. [9,10]. The 1.6 GHz images indicate that the inner jet is centrally brightened at its beginning but becomes extended and limb brightened at ~8 mas from the core, possibly as a result of interactions between the jet and the surrounding medium. Giovannini et al. report a possible component speed from a comparison of two VSOP observations eight months apart of $5.2\,c$ ($H_0 = 65\,\mathrm{km\,s^{-1}\,Mpc^{-1}}$) [9]. This is in contrast with the speed of $2.5\pm0.3\,c$ derived from three epoch 22 GHz Very Long Baseline Array (VLBA) observations by Marscher [11], although insufficient details are given to know whether or not the quoted speeds refer to the same jet component. Changing component speeds, and differing speed for different components within the same source, have been reported in other sources (e.g. [12]).

The synthesized beam for VSOP observations is usually quite elliptical, with the highest resolution along the projected major axis of HALCA's orbit. As HALCA's orbit precesses, the position angle of the highest resolution changes with time. VSOP observations of 3C273 in December 1997 were made while in the highest resolution was perpendicular to the jet direction, and Lobanov et al. have transversely resolved the jet, revealing it to be significantly edge brightened. The jet appears to contain several emitting components, possibly relating to shocks and plasma instabilities [13].

One of the strengths of VSOP observations is the ability to make matched-resolution images with space VLBI at one frequency and ground VLBI at a frequency three times higher (but with baselines three times shorter). A good illustration of this is given by Piner et al. for 3C279, where the combination of a VSOP 1.6 GHz observation and a VLBA 5 GHZ observation enabled a spectral index map to be produced, showing strong synchrotron self-absorption in the core and the loss of synchrotron emitting electrons along the jet [16].

A VSOP observation of BL Lac was made in December 1997 [14], several months after a well-studied outburst from this source (see, e.g., IAUCs 6693, 6700, 6702, 6703, 6705, 6708). No new features (cf higher resolution ground VLBI studies [15]) are visible in the 5 GHz VSOP image, however the observation provides a useful first epoch for comparison with later observations to study the source evolution after such an outburst. By changing the weighting scheme used in imaging the data, the jet can be traced out as far as 20 mas from the core. The jet bends

$\sim$4 mas from the core but remains straight, albeit becoming more extended, out to 20 mas.

In contrast, the EGRET source 1156+293 exhibits much larger bends. The 1.6 GHz VSOP image from an In-Orbit Checkout observation in June 1997 clearly shows a large bend several milli-arcseconds from the core [3]. (Although the projected bend is large, the deprojected bend may be quite modest if the jet is closely aligned with our line of sight.) Another dramatic bend occurs between the parsec-scale and the kiloparsec-scale [17].

IV DISCUSSION

In a quantitative study of the suggestion that jet-bending may be an important factor in determining whether a source is observed at gamma-ray energies, Tingay et al. reached the tentative conclusion from a study of published parsec- and kiloparsec-scale radio images that the parsec-scale jets in gamma-ray–quiet AGN appear to have more and larger bends than do the parsec-scale–jets of EGRET-identified AGN [18]. The need for caution in the interpretation of this result, advocated by Tingay et al., is borne out by the fact that previously published images of 1156+295 had not revealed the bend close to the core seen in the VSOP image. VSOP and higher-frequency ground VLBI observations will enable this hypothesis to be studied in more detail.

Another reported difference between gamma-ray–loud and gamma-ray–quiet sources is in their brightness temperature distributions. Moellenbrock et al. found that EGRET-detected sources tend as a class to have higher brightness temperatures [19]. The measurement of brightness temperature depends on the component flux density and maximum baseline lengths sampled during the observation. There is no explicit dependence on frequency, although there is an implicit (albeit minor for the flat-spectrum sources of interest here) variation in component flux densities with frequency. Thus VSOP observations, with longer baselines than ground observations, are ideally placed to probe the brightness temperature distributions of gamma-ray–loud and gamma-ray–quiet sources.

There is a brightness temperature upper limit of $\sim 10^{12}$ K for incoherent synchrotron emission in the rest frame of the emitting plasma, which happens to coincide with the maximum brightness temperatures that can be measured from ground VLBI. Above this limit the energy losses due to inverse Compton scattering become catastrophic and the source rapidly 'cools' to below 10^{12} K [20]. However, as the brightness temperature observed in our reference frame differs from its value in the emitted frame by the factor $\delta/(1+z)$, where δ is the Doppler factor of the component in the rest frame of the radio source core and z is the source redshift. VSOP observations of brightness temperatures in excess of the inverse Compton limit enable constraints to be placed on the Doppler factors of sources (e.g. 1730−130 [21], 1921−293 [22]).

V THE FUTURE

The successes of the VSOP mission suggest a bright future for space VLBI. Proposed future missions include RadioAstron, a Russian mission with a 10 m diameter solid-panel main-reflector, and ARISE, a more ambitious US mission employing a 25 m inflatable antenna and operating up to 86 GHz. Planning of a follow-up mission to HALCA, currently dubbed VSOP-2, is underway at ISAS, with the project aiming to observe up to 43 GHz [23]. Of these, VSOP-2 appears the most likely to be operational during the lifetime of proposed gamma-ray missions such as GLAST, with good prospects for coordinated multiwavelength observations spanning the electromagnetic spectrum.

VI ACKNOWLEDGMENTS

The VSOP Project is led by ISAS in cooperation with many organizations and radio telescopes around the world. David Murphy is thanked for comments on the manuscript.

REFERENCES

1. Levy, G.S., et al., *Science*, **234**, 187 (1986).
2. Linfield, R.P., et al., *ApJ*, **358**, 350 (1990).
3. Hirabayashi, H., et al., *Science*, **281**, 1825 (1998) and erratum **282**, 1995 (1998).
4. Gwinn, C.R., et al., *ApJ*, **483**, L53 (1997).
5. Hartman, R. et al., *ApJS*, **123**, 79 1999.
6. Tingay, S.J., et al., *AJ*, **115**, 960 (1998).
7. Piner, B.G., et al., *ApJ*, in press.
8. Gaidos, J.A., et al., *Nature*, **383**, 319 (1996).
9. Giovannini, G., et al., *Adv. Sp. Res.*, in press.
10. Giovannini, G., et al., in *BL Lac Phenomena*, ed. L. Takalo, 439 (1999).
11. Marscher, A.P., *Astropart. Phys.* **11**, 19 (1999).
12. Zensus, J.A., et al., *ARAA*, **35**, 607 (1997).
13. Lobanov, A.P., et al., *Adv. Sp. Res.*, in press.
14. Okayasu, R., et al., *Adv. Sp. Res.*, in press.
15. Denn, G.R. and Mutel, R.L., *ASP Conf. Series*, **144**, 169 (1998).
16. Piner, B.G., et al., *these proceedings*.
17. McHardy, I., et al., *MNRAS* **246**, 305 (1990).
18. Tingay, S.J., et al., *ApJ*, **500**, 673 (1998).
19. Moellenbrock, G.A., et al., *AJ*, **111**, 2174 (1996).
20. Kellermann, K.I. & Pauliny-Toth, I.I.K., *ApJ*, **155**, L71 (1969).
21. Bower, G.C. and Backer, D.C., *ApJ*, **507**, L117 (1998).
22. Shen, Z.-Q., et al., *PASJ*, **51**, 513 (1999).
23. Hirabayashi, H., *Adv. Sp. Res.*, in press.

The radio/gamma-ray connection in AGNs

A. Lähteenmäki*, E. Valtaoja† and M. Tornikoski*

*Metsähovi Radio Observatory, FIN-02540 Kylmälä, Finland
†Tuorla Observatory, FIN-21500 Piikkiö, Finland

Abstract. We have continued the comparison of the Metsähovi total flux density radio variations and EGRET gamma-ray observations from Cycles 1–4. This work compliments the arguments presented at the 4th Compton Symposium. We have also attempted to remove the effect that different radio and gamma-ray spectral indices have on the observed properties. We discuss the relationship between the two frequency regions and the consequences it has on the source modeling. It seems that the radio emission and the gamma-rays originate within the same shocked area of the relativistic jet and that the gamma-rays are most likely produced by the synchrotron self-Compton (SSC) mechanism.

I UPDATED METSÄHOVI AND EGRET RESULTS

The third EGRET catalog [1] has provided new gamma-ray detections of active galactic nuclei (AGNs), as well as some improvements to the data already published. We have studied the connection between radio and gamma-ray emission based on earlier EGRET catalogs, and concluded that the most probable gamma-ray AGN to be detected is a highly polarized quasar (HPQ) with an ongoing and rising high frequency radio flare and a large associated variability brightness temperature, i.e. Doppler boosting factor [2]. We have now updated our radio database of Metsähovi 22 and 37 GHz observations with the third EGRET catalog (Cycles 1, 2, 3 and 4) results, thus reaching over 400 individual EGRET pointings. The average gamma-ray flux was obtained for over 30 sources. The properties of each individual radio flare were modeled by the method described in [5], and then compared with the gamma-ray data. The statistical correlations support the SSC model and suggest that the gamma-ray emission originates within the same shocked regions of radio jets which produce the synchrotron flares at radio and higher frequencies.

Averages. Figure 1 shows the variability brightness temperature $T_{b,var}$ distribution of detected and nondetected sources, based on the average gamma-ray flux P1234, or upper limits in the case of nondetections, and the highest ever observed $T_{b,var}$. The results are similar to previous work since the amount of new average

CP510, *The Fifth Compton Symposium,* edited by M. L. McConnell and J. M. Ryan

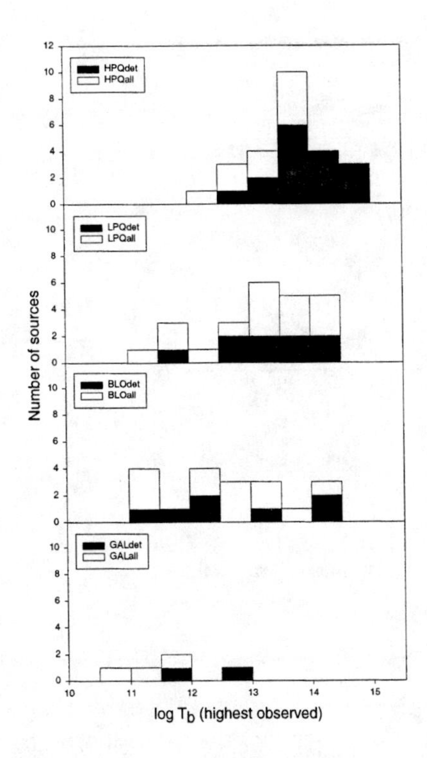

FIGURE 1. Variability brightness distributions of detected and nondetected sources.

gamma-ray data was small. The HPQs are more likely to be detected than any other AGNs. The detection probability depends only on $T_{b,var}$ ($P_{Kruskall-Wallis} = 0.0249$). This implies that the sources with largest amounts of Doppler boosting are also strong gamma-ray emitters. BL Lac objects (BLOs) have a low detection rate, possibly due to more modest values of $T_{b,var}$ [2,3]. Other source classes shown are low polarization quasars (LPQs) and radio galaxies (GALs). The average gamma-ray flux of the detected sources depends mostly on the Lorentz factor of the source ($P_{Spearman} = 0.013$, notice that $D = D(\Gamma, \theta)$). It also possibly depends on the average radio flux ($P_{Spearman} = 0.082$). The correlation with $T_{b,var}$ was much lower ($P_{Spearman} = 0.159$).

Individual pointings. The number of individual EGRET pointings was increased by over 100 %, totaling in 415 pointings. The correlation between the radio and gamma-ray parameters was also significantly increased. The strength of the gamma-ray emission depends on $T_{b,var}$ ($P_{Spearman} = 0.002$), the maximum radio flare flux ($P_{Spearman} = 0.005$), the radio flux at the time of the pointing

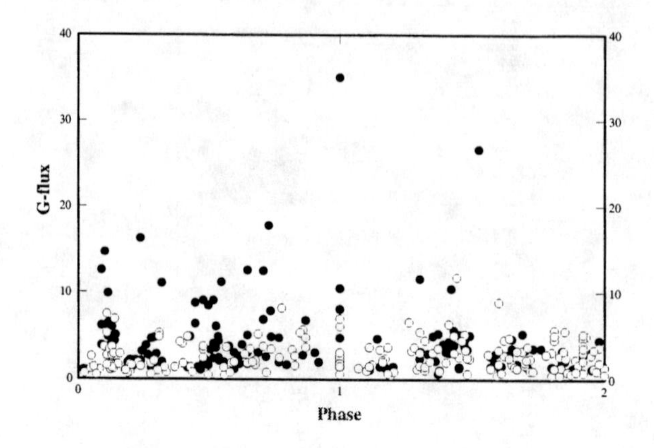

FIGURE 2. Observed gamma-ray flux vs. radio flare phase. Filled, detections; open, upper limits

($P_{Spearman}$ = 0.010) and the phase of the radio flare, i.e. is it rising or falling ($P_{Spearman}$ = 0.010, Fig. 2). Interestingly, the detection probability does not seem to depend on any of the radio properties. This is explained by Fig. 2. At first sight there seems to be an equal number of detections both in the rising part of the flare and in the falling part. In fact, there are 78 detections and 86 nondetections in the rising phase ($0 \longrightarrow 1$), and 103 detections and 109 nondetections in the falling phase ($1 \longrightarrow 2$). However, if one takes a look at only those pointins with gamma-ray flux exceeding, e.g., 7 $photon/cm^2/s$, a clear trend emerges. It seems that in some sources a small amount of gamma-ray emission is present at all times. Once the 'basic' gamma-ray flux level is removed, the detection probabilities on $T_{b,var}$, the maximum radio flare flux, the radio flux at the time of the pointing and the phase emerge again. We have consistently used the strongest radio model flare at the time of each pointing. Assuming, like the results suggest, that gamma-rays are emitted at the beginning of radio flares, we find that at least 34 detections in the falling phase of the radio flare actually belong to another flare component rising at the same time. Then there would be 112 detections and 86 nondetections in the rising phase, and 69 detections and 109 nondetections in the falling phase. The problem of fast or overlapping radio flares —and thus difficult to model— remains, causing the apparent large number of detections in the falling phases of radio flares.

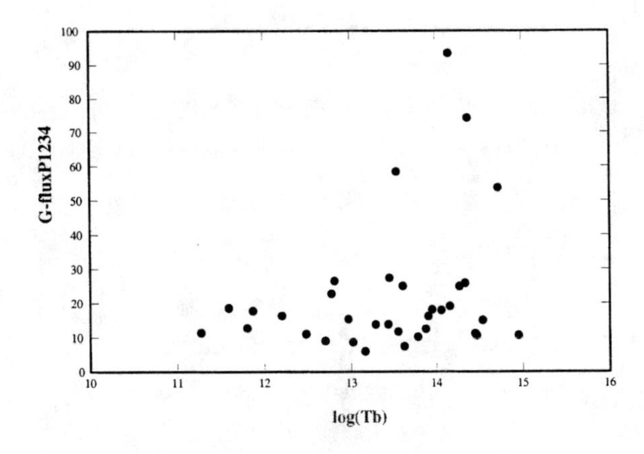

FIGURE 3. Average gamma-ray flux vs. brightness temperature.

II THE EFFECT OF DIFFERENT RADIO AND GAMMA-RAY SPECTRAL INDICES

The radio variations should be comparable to gamma-ray variations through IR variations (the inverse Compton seed photons for gamma-rays). Also, both frequencies are affected by the same amount of Doppler boosting. Instead we find that the sources with the highest estimated Doppler factors are also the strongest gamma-ray emitters (Fig. 3). This may be due to different spectral indices in the radio and gamma-ray regions. The frequency shift caused by Doppler boosting leads to a situation where the radio and the gamma-ray fluxes vary by different amounts. This effect can be removed by correcting the observed gamma-ray flux. It is not possible to reliably define the turnover peaks of the radio spectra and thus obtain corrected radio fluxes in a straightforward way. Instead we have assumed that the radio spectral index $\alpha_r = 0$ (defined by $S \propto \nu^{-\alpha}$) which is generally true for this type of sources. The gamma-ray spectra, however, are relatively straight, so obtaining corrected fluxes is not a problem —even though they still may be slightly uncertain due to possible curvature of the spectra. The spectral indices (usually $\alpha_g \approx 2$) for gamma-ray sources are listed in [1].

The corrected gamma-ray flux (in the source frame) S_{source}, was computed as $S_{source} = S_{obs}(\frac{D}{1+z})^{-\alpha_g}$ where S_{obs} is the observed gamma-ray flux, z is the redshift

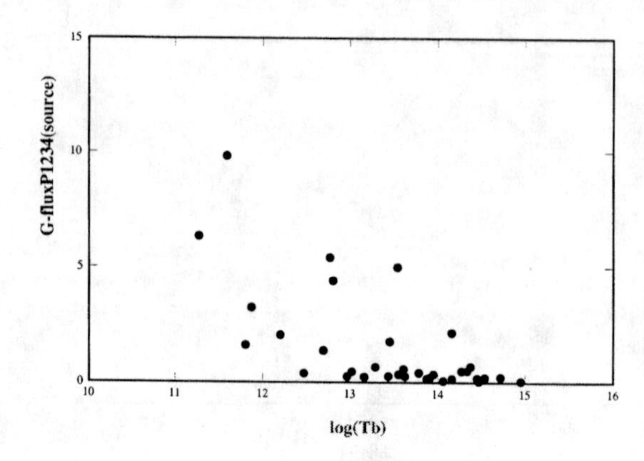

FIGURE 4. Corrected average gamma-ray flux vs. brightness temperature.

and D is the Doppler factor calculated from total flux density variations [3], for each source (average flux) or pointing separately. The fact that the sources with the highest Doppler factors are also the strongest gamma-ray emitters can be rather well explained by the effect of different radio and gamma-ray spectral indices (Fig. 4). This also strengthens the case for SSC models, in which the co-spatial components should be boosted by similar amounts.

We also plotted the corrected observed gamma-ray flux vs. the radio flux at the time of the EGRET pointing. Some sources appear to be rather strong gamma-ray emitters even when they are in a relatively quiescent radio state. These results will be more closely reviewed in [4].

REFERENCES

1. Hartman, R. C., et al., *ApJS*, 123, 79 (1999)
2. Lähteenmäki, A., Teräsranta, H., Wiik, K., and Valtaoja, E., *Proc. 4th Compton Symposium*, eds. C. D. Dermer, M. S. Strickman, & J. D. Kurfess (New York, AIP), p. 1452 (1997).
3. Lähteenmäki, A., and Valtaoja, E., *ApJ*, 521, 493 (1999).
4. Lähteenmäki, A., and Valtaoja, E., *ApJ*, in preparation (2000).
5. Valtaoja, E., Lähteenmäki A., Teräsranta, H., and Lainela, M., *ApJS*, 120, 95 (1999).

Millimeter-wave behavior of EGRET-detected and non-EGRET-detected AGNs

Merja Tornikoski and Anne Lähteenmäki

Metsähovi Radio Observatory, Metsähovintie 114, FIN–02540 Kylmälä, Finland.
E–mail: Merja.Tornikoski@Hut.Fi

Abstract. We present results from an analysis of the radio properties of two samples of sources: 1. EGRET-detected AGNs and 2. relatively bright and densely sampled southern AGNs not detected with EGRET. The radio properties of the detected and non-detected sources show similarity. When analysing the millimeter flux curves it turns out that most of the EGRET-detections were made when the source was going through the initial stages of a radio flare. These results suggest that many sources that exhibit blazar-like behavior in the millimeter domain could be detected at gamma energies if observed during the initial growth stages of a strong radio flare.

INTRODUCTION

We have studied long time series of high radio frequency (22 to 230 GHz) data of two samples of sources:

1. Southern or equatorial EGRET-detected AGNs listed in the Third EGRET Catalog of High-Energy Gamma-Ray Sources [1].
2. A comparison sample of southern sources with relatively long and densely sampled data sets at 90 GHz and which at some point have had a flux density >2 Jy at 90 GHz.

The sources are listed in Table 1. The first sample consists of 19 sources, and the second one of 16 sources. We have compared these two samples to each other to see if there are any fundamental differences in the high radio frequency behavior between the two samples.

ANALYSIS

We have calculated spectral indices and variability indices for both the EGRET-detected and non-detected sources to see if there are significant differences in the

CP510, *The Fifth Compton Symposium,* edited by M. L. McConnell and J. M. Ryan

TABLE 1. The two samples.

EGRET-detected sources		Comparison sample	
0208−512	0336−019	0003−066	0007+106
0420−014	0446+112	0048−097	0238−084
0454−234	0454−463	0332−403	0402−362
0458−020	0521−365	0438−436	0605−085
0537−441	1127−145	0607−157	0637−752
1334−127	1406−076	1921−293	1954−388
1510−089	1514−241	2223−052	2227−088
1606+106	1622−297	2243−123	2345−167
1730−130	1741−038		
1830−210			

high radio frequency properties between these two samples of sources.

The spectral indices were calculated between 37 and 90 GHz (for the sources for which also 37 GHz data were available), and between 90 and 230 GHz. We calculated the spectral indices using three different flux values for each source: 1) the average fluxes from the complete time series; 2) the minimum values, assuming that this is the quiescent flux, i.e. the base level when no radio flares are seen; 3) the maximum values, i.e. the flux during the peak of the strongest flare in each source.

No statistical differences were found between the EGRET-detected and the non-detected sources when comparing any of the above mentioned spectral indices.

We also calculated the fractional variability index $\Delta S = (S_{max} - S_{min})/S_{min}$ at 90 GHz for each source, and at 37 GHz for all the sources for which also 37 GHz were available. Again, no statistical differences were found between the two samples of sources.

SIMULTANEITY OF RADIO AND GAMMA FLARES

We have examined visually all of the radio data streams and compared the epochs of the EGRET pointings / gamma-detections to the behavior in the radio flux curves. For the sources with very densely sampled data sets we used the method described in [9] to decompose the radio flares into exponential flare components.

A closer look at some of the sources

PKS 0454−234. Figure 1. This source has been in the EGRET field of view 8 times before the end of Cycle 4. The only detection was made when the millimeter flux curves also showed a rapid flare. Using the method described in [9] we find that there are two superposed flare components in the millimeter flare at the time of the EGRET detection, first of which is probaly still growing and the other one is in its initial growth stage at the time of the EGRET observation.

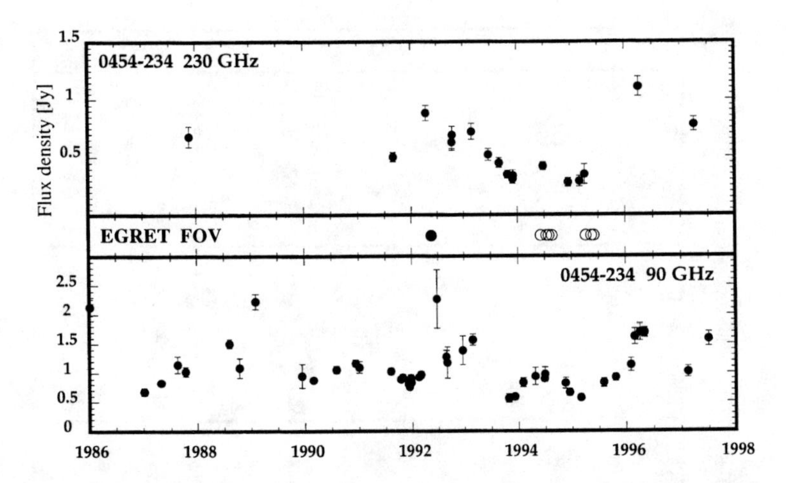

FIGURE 1. PKS 0454−234 at 90 GHz and 230 GHz. The data are our group's SEST observations, published until mid-1994 in [7], complemented by IRAM data [3], [4], [5], [2]. The epochs when the source was in the EGRET field of view are marked with circles in the center panel. The filled circle denotes a significant detection.

PKS 1510−089. Figure 2: The only significant EGRET-detection was made when the radio flux curve shows a quiet state. When decomposing the flares into components using the method described in [9] we find, however, that there are two superposed flare components which at this very epoch are at their initial growth stages.

PKS 1921−293. Figure 3: This source has not been detected by EGRET, even though its extreme properties (strong radio flux, rapid variability) would make it a good candidate for gamma-detections. Since 1990 the source has been in a relatively active state in the mm-domain, with several rapid radio flares. It has only seldom been in the EGRET field of view, and these epochs coincide with steady or (rapidly) decreasing radio flux rather than growing/peaking radio flares.

The case of PKS 2255−282

This source, not included in our present sample, underwent a very strong millimeter-wave outburst in mid-1997, after which we initiated a multifrequency radio observing campaign for the source. Not long after that, in January 1998, the source was also seen flaring at gamma-ray energies. Historically the source was below the detection threshold of EGRET, although it has been in the EGRET field of view several times.

For PKS 2255−282 one can assume that the high activity in the radio domain was correlated with the activity observed at the gamma-ray energies. (The multi-

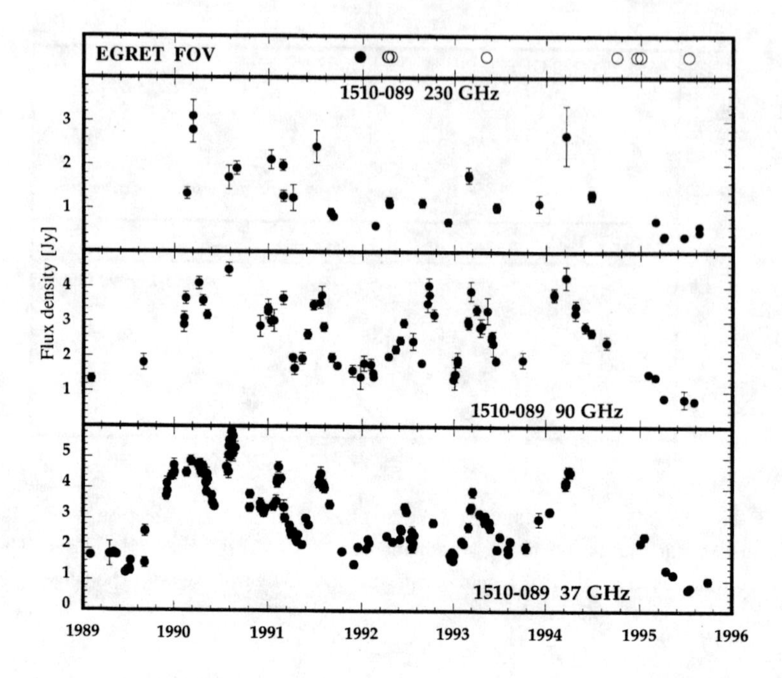

FIGURE 2. PKS 1510−089 at 37, 90 and 230 GHz. The 37 GHz data are from Metsähovi [6]. The millimeter-data are our group's SEST observations, published until mid-1994 in [7], complemented by IRAM data [3], [4], [5], [2]. The epochs when the source was in the EGRET field of view are marked with circles in the top panel. The filled circle denotes a significant detection.

frequency behavior of this source is discussed in [8]). Even though the source has always been relatively bright at millimeter wavelenghts, the prominent outburst was a sign of a period of major acitivity in the radio and the high energy domains. The structure of the radio flare leads us to believe that the major flare was also composed of 2–3 superposed flare components.

CONCLUSIONS

There were no statistical differences between the two samples of sources when comparing the various spectral indices and the fractional variability indices. The EGRET-detected sources exhibit no more extreme spectral properties in the radio domain than the other well-sampled strong and variable sources. In many cases, the EGRET-detections were made when the source was going through the initial stages of a radio flare. For many of the sources with no EGRET-detections, the source was in the EGRET field of view during a relatively quiet stage in the radio

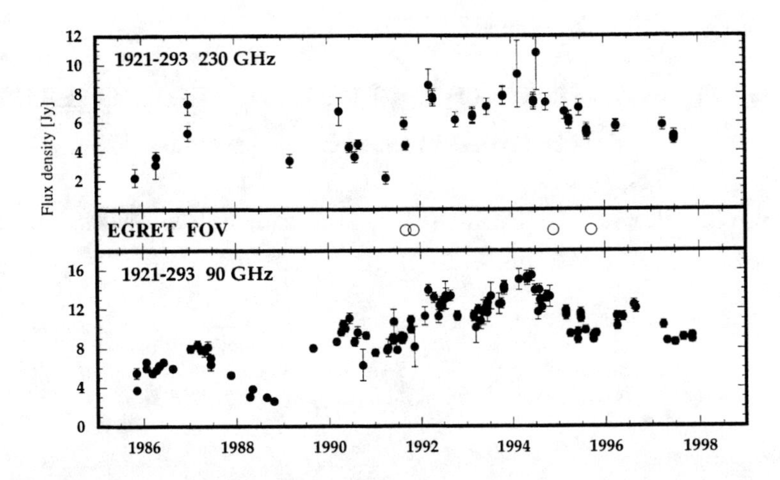

FIGURE 3. PKS 1921−293 at 90 GHz and 230 GHz. The data are our group's SEST observations, published until mid-1994 in [7], complemented by IRAM data [3], [4], [5], [2]. The epochs when the source was in the EGRET field of view are marked with circles in the center panel.

domain. (Please see the paper by Lähteenmäki, Valtaoja & Tornikoski in these proceedings for additional discussion.)

The radio properties of the detected and non-detected sources show similarity, and new sources are being detected with repeated gamma observations, often simultaneously with a radio flare. This suggests that most of the sources that exhibit blazar-like behavior in the millimeter domain could be detected with high-energy gamma instruments if observed during the initial growth stages of a strong radio flare.

REFERENCES

1. Hartman R. C. et al., *ApJS* **123**, 79 (1999).
2. Reuter H.-P. et al., *A&AS* **122**, 271 (1997).
3. Steppe H. et al., *A&AS* **75**, 317 (1988).
4. Steppe H. et al., *A&AS* **96**, 441 (1992).
5. Steppe H. et al., *A&AS* **102**, 611 (1993).
6. Teräsranta H. et al., *A&AS* **132**, 305 (1998).
7. Tornikoski M. et al., *A&AS* **116**, 157 (1996).
8. Tornikoski M. et al., *AJ* **118**, 1161 (1999).
9. Valtaoja E. et al., *ApJS* **120**, 95 (1999).

Broad Band Properties of Radio-Loud Emission Line AGNs

P. Grandi[1], G.G.C. Palumbo[2,3], P. Giommi[4], G. Malaguti[3], L. Maraschi[5], C.M. Urry[6], E. Piconcelli[2,3]

[1] *IAS/CNR, Roma, Italy;* [2] *University of Bologna, Italy;*
[3] *ITesre/CNR Bologna, Italy;* [4] *BeppoSAX Science Data Center, Roma, Italy;*
[5] *Osservatorio Astronomico di Brera, Milano, Italy;* [6] *STscI, Baltimore, USA*

Abstract. Recent BeppoSAX observations of bright radio galaxies have shown that they have a considerable variety of spectral properties and important differences with respect to Seyfert galaxies. Reprocessing features (Fe line and reflection hump) are not always present and generally weak. We suggest two possible scenarios: either radio-loud objects have disk X-ray emission diluted by the jet, or their X-ray reprocessing material subtends a smaller solid angle because of different accretion disk geometry and physical characteristics.

INTRODUCTION

The most basic classification of Active Galactic Nuclei (AGN) consists in dividing them in two classes: radio-quiet and radio-loud. This represents not only an observational distinction but a basic physical difference whose basis is still not understood. Radio morphologies (lobes, jets etc.) are obviously the product of some physical mechanisms at work in the nuclei of radio-loud objects. The primary engine has been identified with spinning massive black hole fed by thick, and hot, accretion flow [1]; Blandford [2] and Meier [3] have explicitly identified the black hole spin as a possible physical parameter responsible for the radio-loud and radio-quiet dichotomy. If the hole rotates faster, it is more efficient in producing the jets observed in radio-loud objects.

X-ray photons seem to be the best probe to investigate the radio loudness issue as they are produced (and reprocessed) in the inner regions of AGNs (< 1 pc) where accretion occurs. X-ray observations of radio-loud objects therefore provide unique data to test the model hypotheses. BeppoSAX is a broad band (0.1 - 150 keV) X-ray satellite [4] which has ideal characteristics for this purpose. Here we present the on-going BeppoSAX analysis of bright radio-loud emission line AGNs and discuss the preliminary results of a comparison between the Broad-Line Radio Galaxies (BLRG) of our sample and 12 Seyfert 1 galaxies also observed by BeppoSAX [5].

CP510, *The Fifth Compton Symposium,* edited by M. L. McConnell and J. M. Ryan

Note that, within the Unified Schemes for AGNs [6], the Broad-Line Radio Galaxies are considered the radio-loud counterpart of the Seyfert 1s.

THE SAMPLE AND THE DATA

Our sample consists of 6 Broad-Line Radio Galaxies. As indicated in Table, 4 BLRGs display Fanaroff-Riley II (FRII) radio morphology and half show superluminal motions. For comparison the Narrow-Line Radio Galaxy (NLRG) Centaurus A was also included, for which variability can be checked as there were two separate observations, and the quasar 3C 273.

Models more complex than power laws have been tested in order to account for all the observed spectral features under the assumption that spectral components represent physical situations similar to the ones observed in Seyfert galaxies. Seyfert models explain observed data in the framework of accretion on a black hole from a cold physically thin but optically thick disk with a hot corona above it [7,8]. The role of the corona is to transform into X-ray photons, via inverse Compton scattering, the UV photons generated by the disk. Downscattered X-rays, in turn, hit the disk and are reprocessed, generating an iron line and reflection hump above 10 keV which are both observed Seyfert features.

The BLRG data of the present sample are generally well fit by a multicomponent spectrum although not all features are always present in each source (Table 1).

TABLE 1. BeppoSAX Sample of Radio-Loud AGN: X-ray Spectral Features

Radio Galaxy	Optical/Radio[a] type	$N_H > N_H^{Gal}$	Soft Excess	Fe Line	Refl. Comp.	Cutoff	Lum[b]
PKS2152-69	BLRG/FRI-FRII	No	No	No	No	No	0.2
Pictor A	BLRG/FRII	No	No	No	No	No	1.0
3C120	BLRG/FRI(S)	Yes	Yes	Yes	Yes	Yes	2.4
3C111	BLRG/FRII(S)	Yes	No	Yes (?)	No	No	2.9
3C390.3	BLRG/FRII(S)	Yes	No	Yes	Yes	No	3.3
3C382	BLRG/FRII	No	Yes	Yes	Yes	Yes	9.4
Cen A	NLRG/FRI	Yes	Yes*	Yes	No(?)	Yes	0.05
3C273	Quasar/(S)	No	Yes	Yes	No	Yes	81

[a] – S = superluminal source
[b] – Luminosity (2-10 keV) corrected for absorption ($\times 10^{44}$ erg cm^{-2} sec^{-1})
* – In Cen A, the soft excess is extended thermal emission

In order to better present the results, the continuum, soft excess, cold absorber, iron line and reflection component are discussed separately.

Continuum emission – The BLRG average spectral slope is $\Gamma^{BLRG} = 1.73$ (rms dispersion $\sigma = 0.08$) consistent with that observed in the radio-quiet AGNs ($\langle\Gamma^{Sey1}\rangle = 1.85$, $\sigma = 0.22$).

In two BLRGS, 3C120 and 3C382, we detected a steepening of the spectrum at high energies, which was modeled with an exponential cutoff. In two cases, 3C111 and 3C390.3, we could estimate a 2σ lower limit $E_{cutoff} > 90$ keV. A steepening of the high energy spectrum was detected also in the NLRG CenA [9] and in the quasar 3C273 [10,11]. The cut-off energies of our radio-loud AGNs fall in the same energy range of Seyfert 1s ($E^{Sey1}_{cutoff} = 237$ keV, $\sigma_{rms} = 150$ keV).

Soft Excess – Soft excesses are present in 3C120 and 3C382, the only two BLRGs with strong UV bumps [12,13]. As in the case of 3C273, which also shows a deviation from the X-ray continuum power law below 1 keV [14], the observed soft excess might be the hard tail of the accretion disk thermal emission.

Cold Absorber – Half the sample objects show an absorption column in excess of the galactic one as found for the NLRG Centaurus A. On the other hand a warm absorber, typical Seyfert 1 signature, has never been detected. This suggests that radio galaxy nuclei are probably embedded in a different environment than radio-quiet ones. An historical study of the column density changes in 3C390.3 further supports this result. The long-time (years) variability of the intrinsic N_H does not appear to be correlated to the flux intensity at 1 keV (see fig. 2 in [15]). It is possible that variations in the geometry of the absorber rather than changes of its ionization state (as expected in the case of a warm absorber) are responsible for the long term variability of N_H. This might also imply that in radio galaxies the absorber is more distant.

Iron Line – In general the iron lines in our sample are weak and in some sources were not detected. In 3C111 an iron line is only marginally detected and in Pictor A it is absent. For PKS2152-69 the BeppoSAX exposure time was too short to allow the detection of even a strong feature. While in the Seyfert 1 sample the iron line is always detected [5], in BLRGs the reprocessed feature is detected in only half the sources. In addition, in BLRGs the iron line EW are significantly smaller than in Seyfert 1s (see also fig.1): $\langle EW^{BLRG} \rangle = 71$ eV, $\sigma^{BLRG}_{rms} = 45$ eV; $\langle EW^{Sey1} \rangle = 175$ eV, $\sigma^{Sey1}_{rms} = 52$ eV.

In an effort to understand the reason for this dichotomy, we investigated the iron line variability in Centaurus A. Centaurus A is a powerful X-ray source ($F_{2-10\ \mathrm{keV}} \sim 3 \times 10^{-10}$ erg cm^{-2} sec^{-1}) and was observed twice by BeppoSAX. During the second pointing the nuclear flux of the source was 25% lower. Independent of the brightness of the point-like nuclear component, its spectral shape did not change significantly. In contrast, the iron line flux was more intense when the source was weaker (see fig. 2 in [9]). This result seems to indicate a delay between the continuum variation and the corresponding excitation of the line emitting material. In the accretion disk geometry this implies that such line emitting material cannot be located close to the primary X-ray source. Similar conclusions were reached with Ginga and ASCA data for 3C390.3 [16].

A comparison with previous Ginga and ASCA data could also be drawn for 3C120 [17] and 3C382 [18]. In both radio galaxies, BeppoSAX measured a continuum flux higher than ASCA and Ginga, but significantly weaker Fe equivalent width

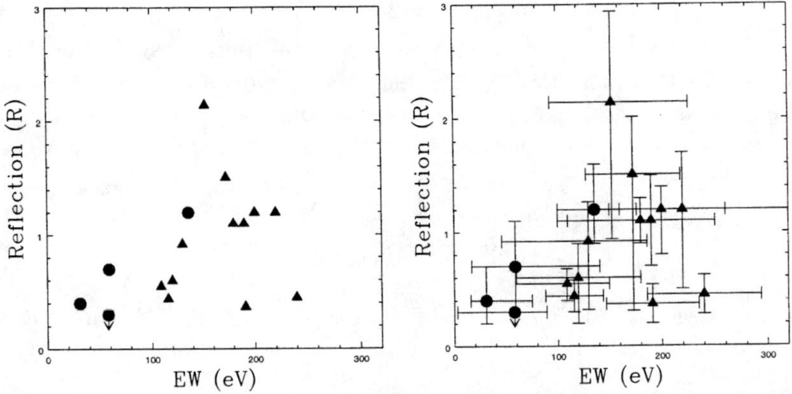

FIGURE 1. (*left panel*) – The amount of reflected radiation (Reflection) is plotted as a function of the strength of the iron line (EW) for the BLRG sample (circles). For comparison the Seyfert 1 (triangles) sample of Matt (1999) is also shown. Note that the radio-loud AGN are characterized by generally weaker iron lines and reflection components. (*right panel*) – Same plot with error bars

($EW^{SAX}_{3C120} \sim 60$ eV and $EW^{SAX}_{3C382} \sim 30$ eV). However these sources show Seyfert 1 spectral characteristics (huge UV bump, a soft excess and broad iron line [17,19]) and therefore one cannot exclude that Fe is produced close to the central black hole. The diluted EW could then be due to an increase in jet intensity, rather than a temporal delay.

Reflection Component – All BLRGs with a detected Fe line show a reflection hump. There is indication that weak reflection corresponds to a weak Fe line (Fig. 1). If proven, this would imply that the line is generated in the same material which reflects and reprocess the X-ray continuum.

DISCUSSION

BeppoSAX observations of BLRGs, exploiting the full broad band, have allowed to point out several important differences between radio-loud and radio-quiet AGNs. Moreover, for the first time, it has been established that BLRG Fe line equivalent widths are significantly weaker than the one observed in Seyfert 1 galaxies.

For some BLRGs, the data suggest that the cold matter responsible for the Fe line emission is physically far from the primary X-ray source. One possible interpretation (suggested by Chen and Halpern [20]) assumes that the accretion flow is cold thin optically thick [21] at large radii and hot and geometrically thick [22,23] close to the hole. Such a description of the accretion flow has the advantage that it would subtend a smaller solid angle than the corresponding cold disk in Seyfert 1s.

On the other hand, BLRGs with soft excess, strong UV bump and broad Fe lines (i.e., Seyfert 1 like spectra) could have cold material quite close to the primary X-ray source. In this case, the reprocessing gas should be a cold thin disk [21] also in the innermost part. The weakness of the iron line might then be produced by the Doppler-enhanced non-thermal jet continuum that dilutes the Seyfert-like emission.

REFERENCES

1. Rees M.J., Begelman M.C., Blandford R.D., Phinney E.S., 1982, Nature, 295, 17
2. Blandford, R. D., 1990 in Active Galactic Nuclei, Saas-Fee Advanved Course, pag. 264
3. Meier D. L., 1999, New Astronomy Reviews in press (astro-ph/9908283)
4. Scarsi, L. 1993, A&AS, 97, 371
5. Matt, G., 1999, in *X-ray Astronomy '99: Stellar Endpoints, AGN and the Diffuse Background*, Bologna, September 6-10
6. Urry C.M. and Padovani P., 1995, PASP, 107,803
7. Haardt, F., Maraschi, L., 1991, ApJ, 380, L51
8. Haardt F., Maraschi L., 1993, ApJ, 413, 507
9. Grandi, P., et al. 1998, in Advances in Space Research, proceedings of 32nd COSPAR Symposium, in press (astro-ph/9811468)
10. Haardt F., et al. 1998, A&A, 340, 35
11. Grandi P. et al., in preparation
12. Maraschi, L., et al., 1991, ApJ, 368, 13
13. Tadhunter, C. N., et al., 1986, MNRAS, 219, 55
14. Grandi P., et al., 1997, A&A, 325, L17
15. Grandi, P., et al. 1999, A&A, 343, 40
16. Wozniak, P. R., et al. 1988, MNRAS, 299, 449
17. Grandi, P., et al. 1997, ApJ, 487, 636
18. Nandra, K. & Pounds, K. A. 1994, MNRAS, 268, 405
19. Reynolds C. S., 1997, MNRAS 286, 513
20. Chen, K. and Halpern, J. P., 1989, ApJ, 344, 115
21. Shakura, N. I., Sunyaev, R. A. , 1973, A&A, 24, 337
22. Shapiro, S. L., Lightman A. P., Eardley D. M., 1976, ApJ, 204, 187
23. Narayan, R., Mahadevan, R. and Quataert, E. 1998, *The Theory of Black Hole Accretion Disk*, eds M.A. Abramowicz, G. Bjornsson, J.E. Pringle, pag.148

An Unbiased Hard X-Ray Survey of the Nearest Seyfert 2 Galaxies

M.J. Westmore*, R. Gurriaran*, A.J. Dean*, F. Lei*, A. Malizia*, L. Bassani†, J.B. Stephen†, B.A. Harmon‡, AND M.L. McCollough§

*University of Southampton, Southampton, England, SO17 1BJ
†ITeSRE/C. N. R., Via P. Gobetti 101, 40129 Bologna, Italy
‡NASA/Marshall Space Flight Center, Huntsville, AL 35812
§Universities Space Research Association, Huntsville, AL 35805

Abstract. The BATSE Earth Occultation data is being successfully used to study the temporal and spectral properties of an unbiased sample of Seyfert 2 galaxies in the 20-100 keV energy range over a period of 8 years. The sample used is the enlarged Maiolino and Reike sample restricted in distance to those (46) located within 42 Mpc from Earth. Of the 20 sources so far analysed 12 were detected at greater than the 5 sigma level; some of these are new detections in the hard X-ray band. Preliminary results of this study are presented including example light-curves and spectra. To enlarge the scope of our study we also present a comparison of our data with data obtained by BeppoSAX. The implications of these results to the Unified Model of AGN, and the synthesis of the Cosmic Diffuse Background are discussed.

INTRODUCTION:

The Burst and Transient Source Experiment (BATSE) on board The Compton Gamma Ray Observatory (CGRO) has viewed the entire sky all of the time since 1991. This has provided a unique database where the 20-100 keV emission from *any* high energy source in the sky can be studied. For the purposes of this study the BATSE Earth Occultation technique [5] was applied to study the 20-100 keV emission from an unbiased sample of Seyfert 2 galaxies. This paper reviews the current status of this study; including examples of the results that are being obtained.

Our parent sample of Seyfert galaxies is that of Maiolino and Reike [8] (hereafter MR) complemented with NGC 1808, which was missed in the original version of the sample, and 18 new Seyfert's listed by Ho et al. [6]. The MR sample includes all Seyfert's identified spectroscopically within the Revised Shapeley-Ames catalog which in turn is limited in B magnitude of the host galaxy ($B_T < 13.4$). We then applied a cut in redshift such that all Seyfert 2 galaxies with z < 0.007 (or equivalently distance < 42 Mpc with $H_0 = 50$ km/s/Mpc) are included; this results in 46 Seyfert 2 galaxies tabulated in table 1.

CP510, *The Fifth Compton Symposium*, edited by M. L. McConnell and J. M. Ryan

TABLE 1. An unbiased sample of the nearest 46 Seyfert 2 galaxies

Circinus	NGC 185	NGC 3982	NGC 4639	NGC 5505	NGC 7410
ESO 428-G014	NGC 2273	NGC 4138	NGC 4698	NGC 5506	NGC 7465
NGC 1058	NGC 2655	NGC 4258	NGC 4941	NGC 5643	NGC 7496
NGC 1068	NGC 3079	NGC 4395	NGC 4945	NGC 5953	NGC 7582
NGC 1365	NGC 3185	NGC 4472	NGC 5033	NGC 6221	NGC 7590
NGC 1386	NGC 3254	NGC 4565	NGC 5128	NGC 6300	NGC 7743
NGC 1433	NGC 3486	NGC 4579	NGC 5194	NGC 676	
NGC 1808	NGC 3941	NGC 4594	NGC 5275	NGC 7314	

The unbiased nature of the sample will allow qualitative and quantitative statements to be made on the nature of the high energy characteristics of Seyfert 2 galaxies. These results will also have important implications on the Unified Model of AGN (hereafter UM) as well as on the synthesis of the Hard X-ray Cosmic Diffuse Background (hereafter CDB).

Our collaboration is also developing a fully physical model of the time varying systematic components of the background signal detected by the LAD's. This model considers the material and geometric composition of the entire CGRO spacecraft and the radiation environment in which operates. By considering all the possible and relevant nuclear interactions that *could* take place, a Monte-Carlo approach is being applied to accurately predict the time varying components of the background signal as seen by the LAD's. By subtracting the predicted curvature of the background from the BATSE database before the standard occultation analysis is begun, we are essentially *Flat Fielding* [4] the data. A study of a low-luminosity sample of objects will provide a good tool to optimise this technique. The results presented here were obtained with the use of the standard occultation techniques and without the application of this flat-fielding technique.

DATA ANALYSIS AND RESULTS:

The BATSE Earth Occultation technique relies on the principal fact that a characteristic step will be seen in the count rate detected by the LAD's whenever a high energy source is occulted by the Earth's limb. The size of this step is measured for each occultation of an object of interest and so a count rate history is generated. This algorithm is applied independently to BATSE's 16 energy channels and so continuum spectra may also be generated. This raw count rate history is then cleaned in two ways. Firstly the effects of contaminating sources are removed by removing all step measurements due to the target source that occur at times when a known contaminating source is also within a certain angular distance of the Earth's limb. Secondly, the time taken for a full occultation is dependent on the relative position of the source, the Earth's limb, and CGRO [5]. Consequently, step measurements in the count rate history for which this has become a dominant and detrimental effect on the accuracy of the measuring process are removed.

TABLE 2. Data analysed so far and preliminary results

Source name	% analysed	Significance	Comment
Cen A	95%	115	
Circinus	95%	21	Possible contamination
NGC 1365	95%	14	Possible contamination
NGC 3185	95%	14	Probably NGC 3227
NGC 4258	95%	11	Possible contamination
NGC 4579	95%	10	
NGC 4945	95%	23	
NGC 5194	95%	9	
NGC 5506	95%	24	
NGC 7465	95%	16	
NGC 7582	95%	12	Possible contamination
NGC 4395	60%	6	

Data for 20 sources has been analysed so far. Twelve sources are seen at greater than the 5 sigma level; table 2 lists these detections. Col. [1] lists the source name; col. [2] lists the percentage of the total 8 year database analysed to date for that source; col. [3] lists the significance of the detection. At present there remains the possibility that for some of the detections tabulated, we are actually seeing a nearby source rather than the target intended. We intend to resolve this ambiguity in due course. The sources for which this may be true have a pertinent comment entered in col. [4].

Comparison with BeppoSAX: Circinus is a well known highly absorbed Seyfert 2 galaxy. The Spectra shown in figure 1 consists of two observations. The plotted points with error bars are the differential flux measured with BATSE and the solid histogram is the best fit absorbed power law model to a BeppoSAX observation made on 13th March 1998 [9]. Seyfert 2 nuclei are known to be highly variable sources. Our BATSE measurement and the BeppoSAX observation were made with vastly different integration times and so it may be concluded that the two observations show good agreement.

New Hard X-ray Detections: Of the 20 sources analysed so far, to our knowledge, 5 are new hard X-ray detections; NGC 4258, NGC 4395, NGC 4579, NGC 5194, and NGC 7465. Figures 2 & 3 display examples (for NGC 5194) of the two principal data products available from this work; figure 2 shows the BATSE spectrum with best fit power law as the solid histogram whilst figure 3 shows the count rate history.

UM & CDB; A Diagnostic Diagram: Almost all variants of the UM predict that there exists a large population of highly absorbed Seyfert nuclei [1]. It has been shown [3] that a dominant contribution to the CDB could be due to a population of highly absorbed Seyfert nuclei if the distribution of N_H within Seyfert 2's is be peaked at a higher value than is currently believed.

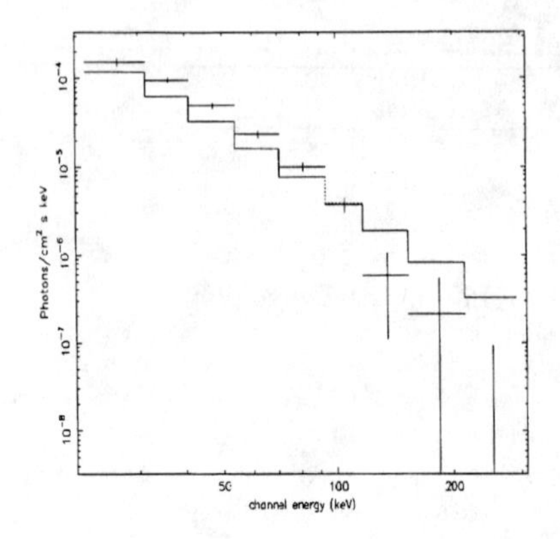

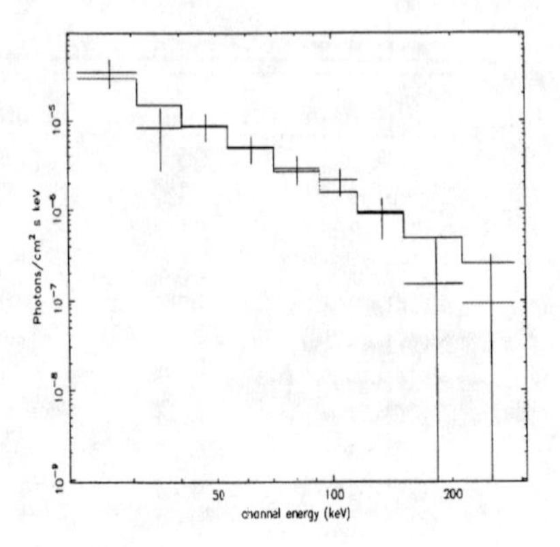

FIGURE 1. BATSE spectral data points and BeppoSAX best fit absorbed power law for Circinus

FIGURE 2.BATSE spectral data points and best fitpower law for new detection NGC 5194

Bassani et al. [2] found, via the use of *diagnostic diagrams*, that there is good evidence that the distribution of N_H within Seyfert 2 nuclei is indeed peaked at a much higher value ($\text{Log}(N_H) \approx 24$) than previously believed. Figure 4 plots the ratio of the 20-100 to 2-10 keV flux against the *Thickness Parameter* of Bassani et al. [2]. This is the ratio of the 2-10 keV flux normalised to an isotropic indicator; the optical [O III]. The 2-10 keV flux has not been corrected for intrinsic absorption whilst the [O III] has been corrected with the Balmer decrement. The diagram plots data from the literature and the BATSE detections discussed in this paper. The BATSE detections have been subdivided to highlight the position in the diagram of the new hard X-ray sources; however, all the BATSE detections are plotted with our BATSE derived fluxes complemented with 2-10 kev and [OIII] fluxes taken from ADS literature sources. The diagram is partitioned (crudely) into regions that should reflect the varying values of N_H found in Seyfert 2 nuclei. From the less absorbed objects to the right of the F[2-10]/[OIII]=1 line up to the highly absorbed objects with F[2-10]/[OIII]<1. Immediately with this diagram one can draw conclusions about the nature of our BATSE detections. For example one may conclude that Circinus and NGC 7582 are highly absorbed objects of the type that Bassani et al. [2] posits to be representative of the median Seyfert 2 nucleus. Conversely, one may conclude that NGC 4579 and NGC 5506 are less absorbed objects. Once this study is complete the unbiased nature of the sample will allow a more sophisticated and fruitful study of the nature of obscuration in Seyfert 2 nuclei.

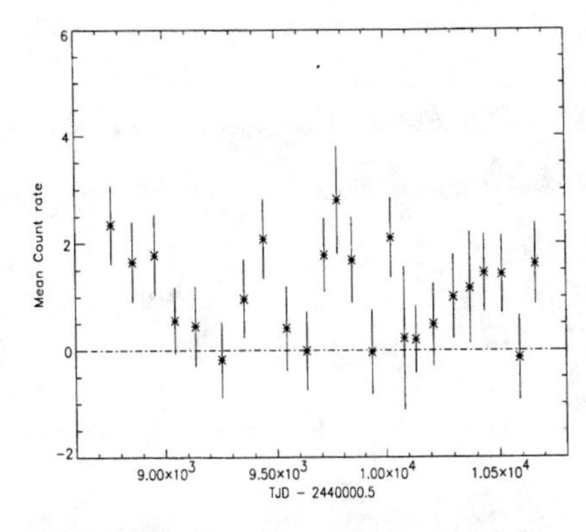

FIGURE 3. Count rate history for new detection; NGC 5194

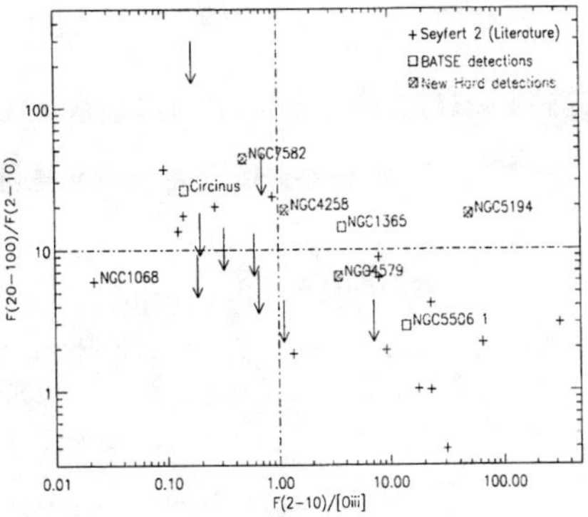

FIGURE 4. A Diagnostic diagram for Seyfert 2 galaxies

SUMMARY AND FUTURE WORK:

We have begun to exploit the unique BATSE database by applying the standard techniques to a low-luminosity sample of objects. Currently 20 sources out of the 46 in our sample have been analysed with a ($> 5\sigma$) detection efficiency of 60%.

This collaboration is also committed to developing and modernising the standard occultation techniques by introducing the Southampton Flat-Fielding technique as well as developing direct all-sky imaging software. The development of the all-sky imaging suite will allow this collaboration to produce the first complete all-sky survey in the hard X-ray range since the HEOA 1 A4 experiment [7].

REFERENCES

1. Antonucci, R.R.J., 1993, *ARA&A*, **31**, 473-521
2. Bassani et al., 1999, *ApJS*, **121**, 4735
3. Comastri, A., Setti, G., Zamorani, G., Hasinger, G. 1995, *A&A*, **296**, 1-12
4. Gurriaran, R., Lei, F., Dean, A.J., Westmore, M.J., Harmon, B.A., 1999, *In Preparation*
5. Harmon, B.A., *et al. The Compton Observatory Science Workshop,* pp. 69-75, Annapolis, Maryland, Sept. 23-25, 1991
6. Ho, L.C., Filipenko, V., Sargent, W.L.W. 1997, *ApJS*, **112**, 315
7. Levine, A.M., et al. 1984, *ApJS*, **54**, 581-617
8. Maiolino, R., Reike, G.H. 1995, *ApJ*, **454**, 95-105
9. Matt et al. 1999, *A&A*, **341**, L39-L42

Limits on MeV Emission from Active Galaxies Measured with COMPTEL

J. G. Stacy[1,2], S. C. Kappadath[1], W. Collmar[3], V. Schoenfelder[3], H. Steinle[3], A. Strong[3], H. Bloemen[4], W. Hermsen[4], J. M. Ryan[5], and O. R. Williams[6]

[1]*Louisiana State University, Baton Rouge, LA 70803*
[2]*Southern University, Baton Rouge, LA 70813*
[3]*Max Planck Institute for Extraterrestrial Physics, Garching, Germany*
[4]*SRON-Utrecht, Utrecht, The Netherlands*
[5]*Space Science Center, University of New Hampshire, Durham, NH 03824*
[5]*Astrophysics Division, ESA/ESTEC, Noordwijk, The Netherlands*

Abstract. We describe our program to provide cumulative flux limits in the COMPTEL energy range for a large sample of high-energy active galactic nuclei (AGN) of general interest. The First COMPTEL Source Catalogue [1,2] will contain cumulative two-sigma upper limits to the time-averaged MeV-emission measured with COMPTEL for 142 AGN and other unidentified gamma-ray sources detected at high Galactic latitudes ($|b| > 10^o$). These limits were derived using composite COMPTEL all-sky maximum-likelihood maps for the 4.5-year period covering Phases 1 through 4 of the CGRO mission (1991-1995). The composite all-sky maps were produced from standard-processing COMPTEL datasets for individual CGRO viewing periods, in four standard energy bins spanning the sensitive range of COMPTEL (0.75-1, 1-3, 3-10, and 10-30 MeV). From these maps we have extracted statistical likelihoods, significances of potential source detections, and associated fluxes, errors, or upper limits, for an extensive list of target objects. In the choice of candidate objects, emphasis was placed on known or suspected gamma-ray sources, particularly those detected in neighboring energy bands to COMPTEL by the CGRO/EGRET and OSSE instruments. These limits will be used to assess the possible contribution of AGN to the diffuse gamma-ray background measured with COMPTEL in the MeV regime.

INTRODUCTION

The Imaging Compton Telescope (COMPTEL) aboard the Compton Gamma Ray Observatory (CGRO) is sensitive to medium-energy gamma radiation from 0.8 to 30 MeV. As a wide-field, imaging instrument COMPTEL has carried out the first comprehensive survey of the sky at MeV-energies [1-3]. The medium-energy gamma-ray regime is of prime importance in the study of the broadband properties of a number of classes of astrophysical sources. For active galactic nuclei (AGN), in particular, the power per natural logarithmic frequency interval (νF_ν) is known to peak in the MeV region of the spectrum. Spectral breaks are typically required to join

CP510, *The Fifth Compton Symposium,* edited by M. L. McConnell and J. M. Ryan

observations spanning several decades in energy around the MeV gamma-ray band [4]. To more fully characterize the broad-band properties of high-energy AGN and related sources, we have undertaken a systematic search through COMPTEL composite all-sky maps for evidence of MeV emission from known or suspected gamma-ray sources at high Galactic latitude ($|b|>10^o$). Here we describe our analysis methods and general results. A full listing of the upper limits obtained will be presented in the First COMPTEL Source Catalogue [1,2].

DATA PROCESSING AND ANALYSIS

The limits on MeV-emission from AGN presented at this conference were derived using composite COMPTEL all-sky maximum-likelihood maps for the 4.5-year period covering Phases 1 through 4 of the CGRO mission (1991-1995). The composite all-sky maps were produced from standard-processing COMPTEL maximum-likelihood datasets for individual CGRO viewing periods, in four standard energy bins spanning the sensitive range of COMPTEL (0.75-1, 1-3, 3-10, and 10-30 MeV). In particular, the significance of the summed log-likelihood ratio in a given all-sky map, corresponding to a confidence level in the detection of possible source emission, is computed separately for each pixel of a composite skymap, based on the number of individual input maps contributing to that pixel. Event selections and data processing followed COMPTEL team-standard methods [5]. A more detailed description of the data-processing procedure used to obtain the composite all-sky maps can be found in [6].

CANDIDATE TARGET SELECTION

An extensive list of target objects lying at high Galactic latitudes ($|b|>10^o$), consisting of AGN and other known but unidentified gamma-ray sources has been compiled for this study. In the choice of candidate objects, emphasis was placed on known or suspected gamma-ray sources, particularly those detected in neighboring energy bands to COMPTEL by the CGRO/EGRET and OSSE instruments [7-11]. The 142 objects selected are summarized by general category in Table 1 below.

TABLE 1. Categories of High-Latitude Target Objects.

Object Type	Number	References
Seyfert Galaxies	25	[7]
OSSE-Identified Blazars	2	[8]
Identified EGRET AGN	58	[9-11]
Unidentified EGRET Sources	57	[9-11]
TOTAL	142	

The vast majority of objects in the candidate source list consist of known high-energy gamma-ray blazars, and of an approximately equal number of unidentified EGRET sources previously detected at high Galactic latitudes [9-12].

UPPER LIMITS TO MEV EMISSION

Time-averaged cumulative flux limits in the COMPTEL energy range have been obtained for the 142 candidate objects summarized in Table 1. Space constraints preclude a full listing of these upper limits in the present report. For these results we refer the reader to the First COMPTEL Source Catalogue [1,2]. In this reference the full results are presented in tabular form, containing the following information: the object name in standard coordinate format; the object position in both right ascension and declination, and Galactic longitude and latitude; the measured COMPTEL two-sigma upper limits in four energy bands (0.75 to 1, 1 to 3, 3 to 10, and 10 to 30 MeV); the object "type" (SY for Seyfert galaxy, from the target list of [7]; O for OSSE-detected blazar [8], 1EG for the First EGRET Catalog [9], 2EG for the Second EGRET Catalog [10], or 2EGS for the Supplement to the Second EGRET Catalog [11]); and, other common names or identifiers for the object.

DISCUSSION

In general, the flux limits described here demonstrate that COMPTEL does *not* detect cumulative time-averaged MeV-emission from the vast majority of high-energy gamma-ray blazars detected with EGRET, nor from other well-known classes of x-ray sources such as Seyfert galaxies. This result is consistent with similar COMPTEL analyses for individual CGRO viewing periods [13] and separate long-term time-averaged source searches using the full COMPTEL database [14].

The limits presented here also bear on the question of the origin of the cosmic diffuse gamma-ray (CDG) background in the COMPTEL energy range [15-18]. While recent reports suggest that AGN may not be a major contributor [19] to the diffuse gamma-ray background measured in the EGRET energy band [20], the highly variable nature of blazar emission complicates such analyses [21]. The spectrum of the CDG clearly breaks in the MeV region observed by COMPTEL, and is likely due to a superposition of unresolved sources. Candidate object classes range from Type Ia supernovae [22], to hard x-ray emitters such as Seyfert galaxies [23] or "off-axis" jet sources such as Cen-A [24] around 1 MeV, to possible "MeV Blazars" in the ~1-10 MeV range [25], to flaring high-energy blazars in the 10-30 MeV region and above [26]. The present results suggest that flaring gamma-ray blazars are not a dominant contributor to the CDG observed by COMPTEL.

ONGOING WORK

We continue to assess the significance of a small number of marginal detections with COMPTEL of MeV emission from high-latitude sources. These apparent detections may in fact be due to confusion with other nearby known point sources (e.g., the quasar PKS 0528+134 with the Crab pulsar), or to the presence of extended, rather than point-like, features in the composite skymaps [13]. In parallel, we also continue to cross-check the present results against those obtained using more recent all-sky maximum likelihood maps [13,14].

Finally, the primary scientific focus of our ongoing analysis is to place firm quantitative limits on the contribution of active galaxies to the cosmic diffuse gamma-ray background in the COMPTEL energy range.

ACKNOWLEDGMENTS

The COMPTEL project is supported by the German government through DARA grant 50 QV 90968, by NASA under contract NAS 5-26645, and by the Netherlands Organization for Scientific Research (NOW). JGS acknowledges partial support for this work under NASA grants NAG5-7355 and NAG5-8100.

REFERENCES

1. Schoenfelder, V., et al., A&A Suppl. Ser., in press (1999).
2. Schoenfelder, V., et al., these proceedings.
3. Schoenfelder, V., et al., ApJS 86, 657 (1993).
4. Collmar, W., et al., A&A 328, 33 (1997).
5. Diehl, R., "COMPTEL Data Analysis Standards," COMPTEL Internal Report COM-MO-DRG-MGM-231.9 (3 March 1996).
6. Stacy, J. G., et al., in Fourth Compton Symposium, AIP Conf. Proc. 410, eds. C. D. Dermer and J. D. Kurfess (AIP: New York), 1356 (1997).
7. Maisack, M., et al., A&A 298,400 (1995).
8. McNaron-Brown, K., et al., ApJ 451, 575 (1995).
9. Fichtel, C. E., et al., ApJS 94, 551 (1994)(=1EG).
10. Thompson, D. J., et al., ApJS 101, 259 (1995)(=2EG).
11. Thompson, D. J., et al., ApJS, 107, 227 (1996)(=2EGS).
12. Hartman, R. C., et al., ApJS 123, 79 (1999)(=3EG).
13. Blom, J. J., "COMPTEL High-latitude Gamma-ray Sources," Ph.D. dissertation, University of Leiden (1997).
14. Collmar, W., et al., these proceedings.
15. Kappadath, S. C., et al., A&A Suppl. Ser. 120, C619 (1996).
16. Kappadath, S. C., "Measurement of the Cosmic Diffuse Gamma-Ray Spectrum from 800 keV to 30 MeV," Ph.D. dissertation, University of New Hampshire (1998).
17. Weidenspointner, G., "The Origin of the Cosmic Gamma-Ray Background in the COMPTEL Energy Range," Ph.D. dissertation, Technische Universitaet Muenchen (1999).
18. Weidenspointner, G., et al., these proceedings.
19. Chiang, J., and Mukherjee, R., ApJ 496, 752 (1998).
20. Sreekumar, P., et al., ApJ 494, 523 (1998).
21. Stecker, F., and Salamon, M., ApJ 464, 600 (1996).
22. The, L., et al., ApJ 403, 32 (1993).
23. Zdziarski, A. A., MNRAS 281, L9 (1996).
24. Steinle, H., et al., A&A 330, 97 (1998).
25. Bloemen, H., et al., A&A 293, L1 (1995).
26. Sreekumar, P., Stecker, F., and Kappadath, S. C., in Fourth Compton Symposium, AIP Conf. Proc. 410, eds. C. D. Dermer and J. D. Kurfess (AIP: New York), 344 (1997).

VHE Gamma Ray Observations of Southern Hemisphere AGNs

P. M. Chadwick, K. Lyons, T. J. L. McComb, K. J. Orford, J. L. Osborne, S. M. Rayner, S. E. Shaw, and K. E. Turver

Department of Physics, Rochester Building, Science Laboratories, University of Durham, Durham, DH1 3LE, U.K.

Abstract. A range of AGNs visible from the Southern hemisphere has been observed with the University of Durham Mark 6 very high energy gamma ray telescope. Results of the observations of PKS 2155–304, 1ES 0323+022, PKS 0829+046, 1ES 1101–232, Cen A, PKS 1514–24, RXJ 10578–275, 1ES 2316–423, PKS 2005–489 and PKS 0548–322 are presented.

INTRODUCTION

One of the most unexpected results in high energy astrophysics in the last decade has been the discovery of high energy and very high energy (VHE) emission from active galactic nuclei (AGNs). The EGRET detector on board the *Compton Gamma Ray Observatory* established that BL Lacs (predominantly radio selected) and flat-spectrum radio sources are strong high energy gamma ray emitters, while X-ray selected BL Lacs (XBLs) have been identified as a source of VHE gamma rays.

The Durham AGN dataset consists of observations of 10 AGNs made with the Mark 6 telescope from 1996 to 1998. The discovery of VHE gamma rays from PKS 2155–304 has already been reported; this is the most distant BL Lac yet detected at these energies [1]. Here we describe observations of PKS 0548–322, PKS 2005–489, 1ES 0323+022, PKS 0829+046, RXJ 10578–275, 1ES 1101–232, Cen A, PKS 1514–24, and 1ES 2316–423, covering a range of classes of AGN. The typical energy threshold for these observations is $\sim$ 300 to 400 GeV. This is $\sim$ 5 times lower than the typical threshold of the CANGAROO telescope, which has also been used to observe Southern hemisphere AGNs [2].

CP510, *The Fifth Compton Symposium,* edited by M. L. McConnell and J. M. Ryan

OBSERVATIONS

Current VHE γ-ray observations of AGNs support the idea that it is the XBLs which are the most promising sources of VHE emission, as suggested by Stecker et al. [3]. The nine AGNs which are discussed in this paper comprise five XBLs, two RBLs, one intermediate class object and one close radio galaxy (Cen A) which has been detected previously as a VHE γ-ray source. While RBLs are thought to be less promising as VHE γ-ray sources than XBLs, observations in the VHE range will help to confirm the fundamental differences between the XBLs and RBLs. VHE γ-ray observations of BL Lacs have, in general, concentrated on the closest objects, but we have sought to extend the current redshift limit of $z = 0.117$ by observing more distant AGNs. With an energy threshold of $\sim$ 300 GeV, the Mark 6 Telescope is well-suited to this task. In the case of one XBL, 1ES 1101–232, the VHE γ-ray observations were made nearly contemporaneously with *BeppoSAX* observations.

The selection criteria applied to these data used a standard set of criteria developed from our successful observations of PKS 2155–304, and include allowance for the variation of image parameters with event size. They are routinely applied to data from all objects recorded at zenith angles less than 45°, which is the case for all the observations reported here.

RESULTS

The dataset for each source has been tested for the presence of gamma ray signals. The flux limits from the nine AGNs are summarised in Table 1. They are all 3 σ flux limits, based on the maximum likelihood ratio test [4,5]. The threshold energy for the observations has been estimated on the basis of preliminary simulations, and is in the range 300 to 400 GeV for these objects, depending on the object's elevation. The collecting areas which have been assumed, again from simulations, are 5.5×10^8 cm^2 at an energy threshold of 300 GeV and 1.0×10^9 cm^2 at an energy threshold of 400 GeV. These are subject to systematic errors estimated to be $\sim$ 50%. We have assumed that our current selection procedures retain $\sim$ 20% of the γ-ray signal, which is subject to a systematic error of $\sim$ 60 %.

We have also searched our dataset for γ-ray emission on timescales of $\sim$ 1 day. The search for enhanced emission has been conducted by calculating the on-source excess after the application of our selection criteria for the pairs of on/off observations recorded during an individual night. A typical observation comprising 6 on/off pairs of observations (1.5 hours of on-source observations) yields a flux limit of $\sim 1 \times 10^{-10}$ cm^{-2} s^{-1} at 300 GeV. Conversely, had any of the objects on which we report here produced a 15-minute flare similar to that seen from Mrk 421 with the Whipple telescope on 1996 May 7 [6], it would have been detected with the Mark 6 telescope at a significance of around 7 σ. There is no evidence for any flaring activity.

TABLE 1. Flux limits (3σ) for observations of active galactic nuclei made with the University of Durham Mark 6 Telescope. Also shown are the predictions of the Stecker et al. model [3].

Object	Estimated Threshold (GeV)	Flux Limit ($\times 10^{-11}$ cm^{-2} s^{-1})	Predicted Flux ($\times 10^{-11}$ cm^{-2} s^{-1})
Cen A	300	5.2	
PKS 0829+046	400	4.7	
PKS 1514–24	300	3.7	
1ES 2316–423	300	4.5	0.15
1ES 1101–232	300	3.7	2.0
RXJ 10578–275	300	8.2	0.33
1ES 0323+022	400	3.7	0.40
PKS 2005–489	400	0.79	0.51
PKS 0548–322	300	2.4	1.3

DISCUSSION

Whilst the interpretation of VHE upper limits from BL Lacs is complicated by the lack of a complete theory of VHE γ-ray emission from AGNs, Stecker et al. [3] have predicted the TeV fluxes from a range of objects, three of which (1ES 0323+022, PKS 0548–322, and PKS 2005–489), are included in the present work. The expected fluxes from the other XBLs included in this paper may be estimated on the basis of the work of Stecker et al. [3,8] using the simple relation $\nu_x F_x \sim \nu_\gamma F_\gamma$ and the published X-ray fluxes. We estimate that the 300 GeV fluxes of 1ES 1101–232, 1ES 2316–423 and RXJ 10578–275 would be 2.0×10^{-11} cm^{-2} s^{-1}, 1.5×10^{-12} cm^{-2} s^{-1}, and 3.3×10^{-12} cm^{-2} s^{-1} respectively, taking into account photon-photon absorption using the recent determination of γ-ray opacity by Stecker [8]. All these suggested fluxes are lower than the flux limits reported here. However, the lack of contemporaneous X-ray measurements in the case of most of our observations limits the usefulness of these predictions and emphasises the importance of simultaneous X-ray and γ-ray observations and multiwavelength campaigns. In the case of the RBLs, an extended observation of PKS 1514–24, a close RBL, lends support to the suggestion that RBLs are not strong VHE γ-ray emitters.

Our observations of Cen A were made when it was in an X-ray low state, in contrast to the earlier VHE detection of Cen A reported by Grindlay et al. [7], which was made when Cen A was in X-ray outburst. Further VHE γ-ray observations during an X-ray high state would be desirable.

CONCLUSIONS

The Durham University Mark 6 Telescope has been used to make observations of 9 close AGNs: 1ES 0323+022 (XBL, $z = 0.147$), PKS 0548–322 (XBL, $z = 0.069$), PKS 0829+046 (RBL, $z = 0.18$), RXJ 10578–275 (XBL, $z = 0.092$) 1ES 1101–

232 (XBL, $z = 0.186$), Cen A (low luminosity radio galaxy, $z = 0.0089$), PKS 1514–24 (RBL, $z = 0.049$), PKS 2005–489 (XBL, $z = 0.071$) and 1ES 2316–423 (transitional BL Lac, $z = 0.055$). We find no evidence for either steady or flaring emission of γ-rays above 300 – 400 GeV in any of these sources. The flux limits are in excess of the fluxes predicted on the basis of the simple model of Stecker et al. [3]. The flux limits derived for 1ES 0323+022 (3.7×10^{-11} cm^{-2} s^{-1}), PKS 0548–322 (2.4×10^{-11} cm^{-2} s^{-1}), and PKS 2005–489 (0.79×10^{-11} cm^{-2} s^{-1}) are not in conflict with the specific predictions of [3] (4.0×10^{-12} cm^{-2} s^{-1}, 1.3×10^{-12} cm^{-2} s^{-1}, and 0.51×10^{-12} cm^{-2} s^{-1} respectively).

We are grateful to the UK Particle Physics and Astronomy Research Council for support of the project and the University of Sydney for the lease of the Narrabri site. The Mark 6 telescope was designed and constructed with the assistance of the staff of the Physics Department, University of Durham. The efforts of Mrs. S. E. Hilton and Mr. K. Tindale are acknowledged with gratitude. We would like to thank Anna Wolter for providing us with information about *BeppoSAX* observations of 1ES 1011–232 in advance of publication. This paper uses quick look results provided by the ASM/*RXTE* team and uses the NASA/IPAC Extragalactic database (NED), which is operated by the Jet Propulsion Laboratory, Caltech, under contract with the National Aeronautics and Space Administration.

REFERENCES

1. Chadwick, P. M., et al., *Ap. J.*, **513**, 161 (1999).
2. Roberts, M. D., et al., *astro-ph/9902008*, (1999).
3. Stecker, F. W., de Jager, O. C., & Salamon, M. H., *Ap. J.*, **473**, L75 (1996).
4. Gibson, A. I., et al., *Proc. Intl. Workshop on Very High Energy Gamma Ray Astro.*, Bombay: Tata Institute, ed. P. V. Ramana Murthy & T. C. Weekes, 97 (1982).
5. Li, T. P., & Ma, Y. Q., *Ap. J.*, **272**, 317 (1983).
6. Gaidos, J. A., et al., *Nature*, **383**, 319 (1996).
7. Grindlay, J. E. et al., *Ap. J.*, **197**, L9 (1975).
8. Stecker, F. W., *astro-ph/9812286*, (1998).

Blast-Wave Physics Model for Blazars

Charles D. Dermer

Naval Research Laboratory, Code 7653, Washington, DC 20375-5352

Abstract. A blast-wave physics model is applied to observations of the spectral energy distributions of blazars. Numerical simulation results showing instantaneous and time-averaged spectra for blast-wave evolution are presented for an idealized case with different values of the initial Lorentz factors of the outflow. The results are compared with the trend of blazar properties ranging from flat spectrum radio quasars to high-frequency peaked BL Lac objects. The long-term averaged spectra and short-term flaring spectra calculated from temporally evolving blast waves are in qualitative agreement with the different spectral states observed from BL Lac objects.

I INTRODUCTION

Multiwavelength observations show that the typical spectral energy distributions (SEDs) of blazars are characterized by two broadband peaks in the νF_ν spectrum, consisting of a lower energy peak in the 10^{13}-10^{17} Hz range and a second peak at gamma-ray energies [1,2]. The properties of the lower energy component are consistent with a nonthermal synchrotron origin from nonthermal electrons in relativistic bulk plasma. Blazars can be divided into three categories on the basis of their overall SEDs: flat spectrum radio quasars (FSRQs), low energy-peaked BL Lacs (LBLs), and high energy-peaked BL Lacs (HBLs). The photon energy of the peak of the νF_ν synchrotron component in FSRQs and LBLs occurs at radio/IR and IR/optical energies, respectively, and is in the UV/X-ray regime in HBLs. The absolute luminosity declines with increasing peak frequency of the synchrotron component, and the ratios of the Compton to synchrotron fluxes are, on average, smaller in the LBLs and HBLs than in the FSRQs [3].

Correlated observations indicated that the higher energy component is probably due to Compton scattering by blazar jet electrons [4]. The soft photons that are scattered to produce this peak could be internal synchrotron photons [5,6], accretion disk radiation that enters the jet directly [7] or after being scattered by surrounding BLR clouds and circumnuclear debris [8], or reflected jet synchrotron radiation [9]. Much effort has gone into modeling the overall SEDs of blazars by specifying the magnetic field, bulk Lorentz factor, and nonthermal electron distribution (e.g., [10]). Long-term monitoring observations and multiwavelength campaigns are providing information on how the SEDs of blazars vary with time, and this offers a new avenue

CP510, *The Fifth Compton Symposium,* edited by M. L. McConnell and J. M. Ryan
2000 American Institute of Physics 1-56396-932-7

to understand the blazar system. In this paper, we apply blast-wave physics to blazar models in order to explain their temporal evolution. We describe the model in Section 2, present numerical results in Section 3, and briefly discuss these results in terms of the blazar phenomenology summarized above.

II THE MODEL

In the standard model for blazars, a supermassive black hole ejects a collimated jet of plasma traveling at relativistic speeds. The outflowing plasma carries directed kinetic energy that can be converted to radiation. This conversion can occur when nonthermal electrons and protons are injected into the comoving plasma frame either by sweeping up particles from the surrounding medium [12], or when two plasmoids, moving with different bulk Lorentz factors, collide [13]. Here we consider only the case where the bulk plasma sweeps up particles from a surrounding medium through an external shock. The energy injected in the form of nonthermal protons and electrons comes at the expense of the bulk internal motion of the plasma, causing the plasmoid to decelerate.

For the simulations shown here, we consider idealized conditions where the medium is uniform with density $n = 1$ cm^{-3}. Even given this extreme simplification, there are several other parameters that must be specified for a complete blast-wave model. One is the energy E_0 released in an injection event, which is chosen to be 10^{54} ergs per 4π sr. For simplicity, we calculate the SEDs for uncollimated outflows. Beaming of the emission can introduce an additional set of parameters that will not affect the issues considered here. We examine intial Lorentz factors $\Gamma = 10, 30$, and 100. The fraction ϵ_e of energy transferred from the nonthermal protons to the electrons is set equal to 0.5. The electrons are injected with a Lorentz factor $p = 2.5$, with a maximum electron Lorentz factor $\gamma_{\rm max} = 4 \times 10^7/[B({\rm G})]^{1/2}$, where B is the mean magnetic field in the comoving fluid frame. (The minimum Lorentz factor of the injected electrons is implied by p, $\gamma_{\rm max}$, ϵ_e, and the condition that there is no particle escape.) The magnetic field is given in terms of a fraction ϵ_B of the equipartition energy density of the downstream nonthermal particles [11]. We let $\epsilon_B = 10^{-2}$ here.

The numerical simulation model calculates synchrotron and synchrotron self-Compton (SSC) emission, and treats the blast wave deceleration self-consistently. It follows the evolution of the electron spectrum through synchrotron, SSC, and adiabatic energy losses. The high-energy radiation is attenuated through $\gamma\gamma$ pair production attenuation, though the reinjected pairs are not included in the calculation. The external Compton scattering process is also not treated here.

III NUMERICAL RESULTS

Fig. 1 shows calculations using the parameter set quoted in the previous section. As the blast wave expands, it becomes energized as it sweeps up material from the

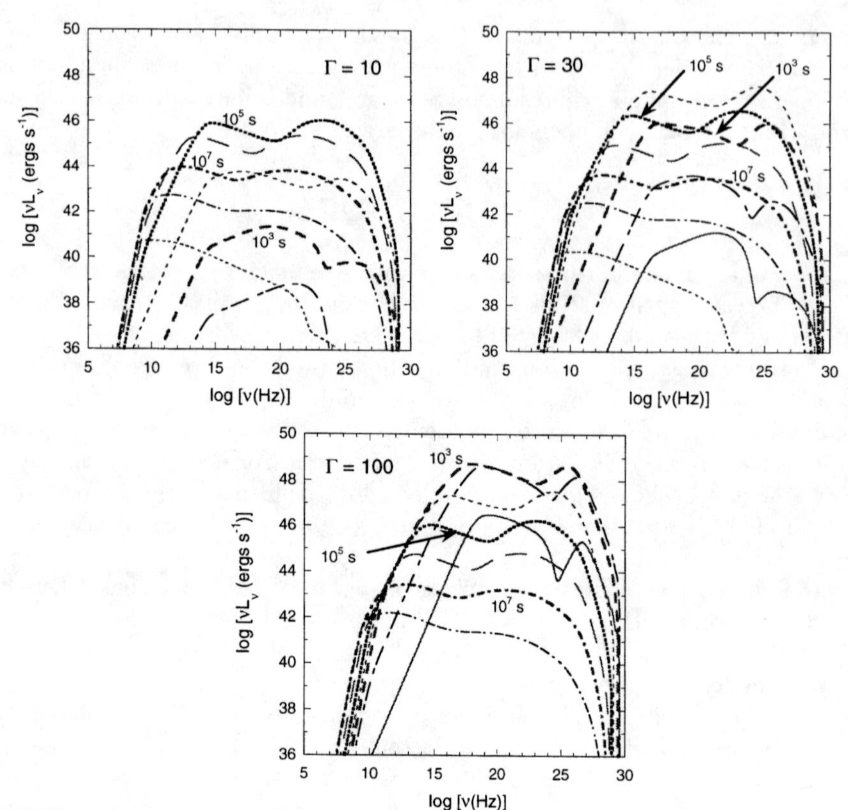

FIGURE 1. Calculations of the instantaneous νL_ν spectra that would be observed when outflowing plasma, with a directional energy release of 10^{54} ergs/4π sr, decelerates and radiates by encountering a surrounding uniform medium with density $n = 1$ cm^{-3}. Only the bulk Lorentz factors differ between the three panels (see text for parameters). Spectra are calculated at observing times differing by factors of 10, with spectra calculated at 10^3 s, 10^5 s, and 10^7 s indicated.

external medium. This causes the increase in the flux at early times. The blast wave begins to decelerate after it has swept up an amount of relativistic inertia equal to the baryon mass in the outflowing plasma. This causes the overall νL_ν flux to decrease and the photon energies of the peaks of the νL_ν spectrum to decay to lower energies.

The generic two-component form of the SED of blazars is easily reproduced by this model. For the parameters chosen, the SSC component has a flux that is comparable to the synchrotron component. The effect of increasing Γ is to cause the duration $t_{\rm dec}$ of the prompt luminous phase, during which the blast wave has not yet experienced significant deceleration, to decrease, and to cause the peak νL_ν luminosity Π to reach much larger values. In fact, $t_{\rm dec} \propto (E_0/n)^{1/3}\Gamma^{-8/3}$ [14] and $\Pi \propto (E_0^2 n)^{1/3}\Gamma^{8/3}$ [15]. The photon energy $E_{\rm pk}$ of the peak of the νL_ν synchrotron

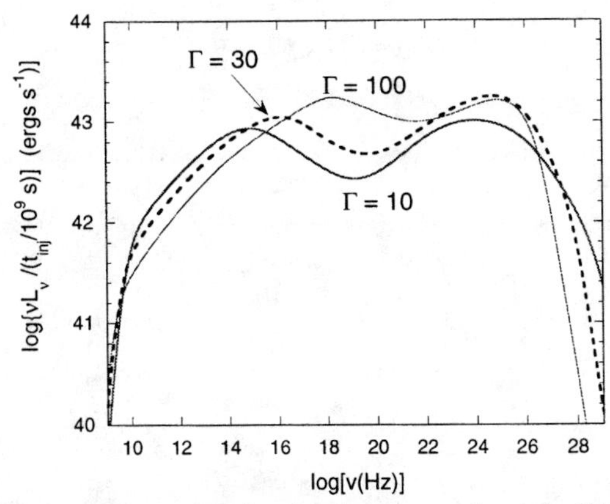

FIGURE 2. Long-term average spectra for the three cases shown in Fig. 1. This figure shows the average luminosity measured over an observing interval of 10^9 s for a single injection event with directional energy release of $10^{54}/(4\pi)$ ergs sr^{-1}. The average power would be multiplied by the factor $t_{\rm inj}/10^9$ s for an identical ejection event every $t_{\rm inj}$ s.

spectrum at the time when the emission is brightest varies as $E_{\rm pk} \propto n^{1/2}\Gamma^4$. These relations can be checked against the Fig. 1 calculations.

Fig. 2 shows time-averaged SEDs obtained by integrating the curves in Fig. 1 over an observing interval of 10^9 s. Because the change in the duration of the prompt phase compensates the change in power output to give a constant total energy, the long-term averaged power is roughly constant. However, $E_{\rm pk}$ still varies $\propto \Gamma^4$ because this is the photon energy where the bulk of the power is radiated. Fig. 3 compares the time-averaged SEDs for $\Gamma = 30$ and 100 with the νL_ν SEDs emitted near the times when the power output is maximum. We have also renormalized the long-term average spectrum as if the central engine ejected a plasmoid every 10^5 s; the $\Gamma = 100$ curves are offset for clarity. As can be seen, the instantaneous spectra near the peak power output has a harder low-energy spectrum and is radiated over a narrower waveband than the time-averaged spectrum.

IV COMPARISON WITH BLAZAR OBSERVATIONS

Because the initial Lorentz factor Γ has the most dramatic effect upon spectra, we ask whether a single parameter family consisting of Γ could explain the trend of FSRQs, LBLs, and HBL properties. The peak frequency of the νL_ν spectra is strongly dependent on Γ, going roughly as Γ^4. The ~ 4 order-of-magnitude increase in the peak frequency in the FSRQ→LBL→HBL sequence of SEDs might therefore be expected to be due to an increase by a factor of ~ 10 in Γ. The trend for the measured bolometric luminosities to decline in the FSRQ→LBL→HBL sequence, however, would mean that the time-averaged jet power $L_{\rm jet}$ must decrease with increasing Lorentz factor. Thus, one must postulate an *ad hoc* relationship $\Gamma \propto 1/L_{\rm jet}$ to account for the observed trend in BL Lac properties. In contrast, this

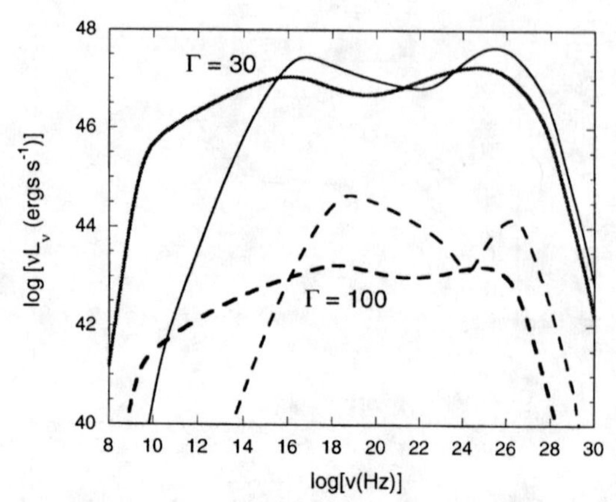

FIGURE 3. Thick curves show the long-term average spectra from Fig. 2 for the cases $\Gamma = 30$ and $\Gamma = 100$. Thin curves show flaring spectra representing peak νL_ν spectra from Fig. 1, corresponding to $t = 10^4$ and $t = 10^2$ s cases for $\Gamma = 30$ and $\Gamma = 100$, respectively. The $\Gamma = 100$ curves have been offset by a factor 10^4 for clarity.

model implies flaring SEDs that are harder than the long-term average SED, and which evolve to longer wavelengths with time, in qualitative agreement with the behavior of the spectral states observed from Mrk 501 [16]. Future work must consider whether the implied *ad hoc* relationship has a physical basis.

This work is supported by the Office of Naval Research. I thank J. Chiang for use of the blazar code.

REFERENCES

1. von Montigny, C. *et al.*, *Astrophys. J.* **440**, 525 (1995).
2. Sambruna, R., Maraschi, L., and Urry, C. M., *Astrophys. J.* **463**, 444 (1996).
3. Fosatti, G., *et al.*, *MNRAS* **299**, 433 (1998).
4. Macomb, D. J., *et al.*, *Astrophys. J.* **449**, L99 (1995).
5. Maraschi, L., Ghisellini, G., and Celotti, A., *Astrophys. J.* **397**, L5 (1992).
6. Bloom, S. D., and Marscher, A. P., *Astrophys. J.* **461**, 657 (1996).
7. Dermer, C. D., and Schlickeiser, R., *Astrophys. J.* **416**, 458 (1993).
8. Sikora, M., Begelman, M. C., and Rees, M. J., *Astrophys. J.* **421**, 153 (1994).
9. Ghisellini, G., and Madau, P., *MNRAS* **280**, 67 (1996).
10. Böttcher, M. 1999, *Astrophys. J.*, **515**, L21 (1999).
11. Chiang, J., and Dermer, C. D., *Astrophys. J.* **512**, 699 (1999).
12. Dermer, C. D., *Astropar. Ph.* **11**, 1 (1999).
13. Ghisellini, G., astro-ph/9906111 (1999).
14. Mészáros, P., and Rees, M. J., *Astrophys. J.* **405**, 278 (1993).
15. Dermer, C. D., Chiang, J., and Böttcher, M., *Astrophys. J.* **513**, 656 (1999).
16. Pian, E., *et al.* *Astrophys. J.* **492**, L17 (1998).

Physics of Relativistic Jets in Blazars

Fumio Takahara*

*Department of Earth and Space Science, Osaka University
Machikaneyama 1-1, Toyonaka, Osaka 560-0043, Japan

Abstract. Gamma-ray emission from blazars provides us with important diagnostics on the size and beaming factor of the emitting region as well as the energy densities of magnetic field, relativistic electrons and various radiation components. Through a simple analysis of multiwavelength observations, I argue that relativistic jets are kinetic power dominated and are composed mainly of electron-positron pairs and that particle acceleration in shock waves conforms very well with observations. I also discuss implications on the formation and bulk acceleration of relativistic jets.

INTRODUCTION

EGRET on board the CGRO satellite has identified more than 50 gamma-ray blazars [1,2]. Observational properties of blazars such as rapid time variability, strong optical polarization, superluminal expansion and strong gamma-ray emission are theoretically understood in terms of a relativistic jet viewed from close to the line of sight [3]. The broad band spectra of blazars are well interpreted in terms of the synchrotron emission and inverse Compton scattering by a population of relativistic electrons, although various kinds of target photons for inverse Compton scattering are possible [4–7].

Since both synchrotron and Compton components are observed, we can estimate the size and beaming factor of the emission region as well as the energy densities of magnetic field, relativistic electrons and various radiation components. The energetics of relativistic jets thus obtained then gives important implications on theoretical mechanisms of particle acceleration as well as the formation and bulk acceleration of relativistic jets [8–10]. After briefly summarizing the basic scenario, I estimate energy densities of various components and discuss theoretical implications. I also discuss several theoretical issues to be fully explored in near future.

BASIC SCENARIO

There are two general restrictions on the modeling of the gamma-ray emission from blazars as discussed by [4]. While the size of the emitting region should be

CP510, *The Fifth Compton Symposium*, edited by M. L. McConnell and J. M. Ryan

small enough to explain the observed rapid time variabilities, it should be large enough to avoid the absorption of gamma-rays due to pair creation by a collision off soft photons. These restrictions prove the necessity of the relativistic beaming, the degree of which is measured by the beaming factor defined by $\delta = 1/(\Gamma(1 - \beta_\Gamma \cos\theta))$, where β_Γ and Γ are the velocity in units of the light velocity and the Lorentz factor of the jet whose direction makes an angle θ to the line of sight. Assuming a spherical emission region of a radius R, for simplicity, the resultant lower limit of δ turns out to be around $5 \sim 10$, which is consistent with the value inferred from other observations such as superluminal expansion. If we adopt $\delta = 10$ as a fiducial value, the typical size of the emission region becomes 0.003pc which corresponds to $100r_g$ for the black hole mass of $3 \times 10^8 M_\odot$, where r_g is the Schwarzschild radius. Thus, the gamma-ray emission probes deep inner part of the relativistic jets, one or two orders of magnitude deeper than the radio VLBI observations.

This size is understood in a very simple way. Suppose that jets are produced in the vicinity of the central black hole, $\sim 10r_g$ with $\Gamma \sim 10$. Time varying Γ on the time scale of $10r_g/c$ causes internal shocks at a typical distance of $\Gamma^2 \times 10r_g \approx 10^3 r_g$. Since the opening angle of the jet is roughly Γ^{-1}, the size of the shocked region becomes about $100r_g$. Since the observed variability time scale of emission is shortend by another factor of $\delta \approx \Gamma$, eventually the time scale of $10r_g/c$ is recovered.

Shocks naturally produce a population of relativistic electrons with a power law spetrum by the diffusive acceleration machanism. The canonical power law index is 2 when the effect of radiative cooling is neglected. Radiative cooling makes the index of high energy electrons steeper by 1. The break energy is determined by the balance between particle escape and cooling, while the maximum energy is determined by the balance between acceleration and cooling. These features are fully consistent with the observed emission spectra of blazars if the shock is at least mildly relativistic [8,9].

PHYSICAL STATE OF EMITTING REGION

The observed isotropic luminosity of each component $L_{i,\rm obs}$ is related to the comoving energy density of photons u_i as

$$L_{i,\rm obs} = 4\pi R^2 c u_i \delta^4. \tag{1}$$

The ratio of the luminosity of synchrotron self-Compton component to synchrotron luminosity is given by

$$\frac{L_{\rm SSC,obs}}{L_{\rm syn,obs}} = \frac{u_{\rm SSC}}{u_{\rm syn}} = \frac{f u_{\rm syn}}{u_{\rm mag}}, \tag{2}$$

where $f < 1$ denotes the effect of Klein-Nishina reduction; not all synchrotron photons can become effective target photons of inverse Compton scattering. Similar formulas can be written for other components, too. Typical observations suggest that $L_{\rm SSC,obs}$ is comparable to $L_{\rm syn,obs}$ in magnitude so that $u_{\rm mag}$ is comparable to or less than $u_{\rm syn}$. If we assume the values of R and δ, we can estimate the magnetic field strength which turns out to be rather weak [11].

Moreover, energy spectrum of relativistic electrons can be inferred from the observed emission spectra. As a typical electron spectrum which is predicted by diffusive shock acceleration and can reproduce observations rather well, we adopt the double power law form,

$$n(\gamma) = \begin{cases} K\gamma^{-2}, & \text{for } \gamma_{\rm min} < \gamma < \gamma_{\rm br} \\ K\gamma_{\rm br}\gamma^{-3}, & \text{for } \gamma_{\rm br} < \gamma < \gamma_{\rm max}. \end{cases} \tag{3}$$

Corresponding number and energy densities of electrons are given by

$$n_{\rm rel} = \frac{K}{\gamma_{\rm min}} \tag{4}$$

and

$$u_{\rm rel} = K m_{\rm e} c^2 \ln(\gamma_{\rm br}/\gamma_{\rm min}), \tag{5}$$

respectively.

Then, the energy density of synchrotron photons is given by

$$u_{\rm syn} = \frac{R\sigma_{\rm T}\gamma_{\rm br}u_{\rm rel}u_{\rm mag}}{m_{\rm e}c^2}\frac{\ln(\gamma_{\rm max}/\gamma_{\rm br})}{\ln(\gamma_{\rm br}/\gamma_{\rm min})}. \tag{6}$$

Since $\gamma_{\rm br}$ can be estimated from the observed spectral break, we can estimate $u_{\rm rel}$ as well. From a theoretical point of view, it is more instructive to introduce escape velocity of electrons $\beta_{\rm esc}$, noting that $\gamma_{\rm br}$ should be determined by the balance between radiative cooling and escape; in diffusive shock acceleration escape velocity is basically identified with shock velocity. We obtain

$$\gamma_{\rm br} = \beta_{\rm esc}\frac{m_{\rm e}c^2}{R\sigma_{\rm T}u_{\rm soft}}, \tag{7}$$

where

$$u_{\rm soft} = u_{\rm mag} + f u_{\rm syn} + u_{\rm ext}. \tag{8}$$

Thus, we obtain

$$u_{\rm rel} = \frac{u_{\rm soft}u_{\rm syn}}{\beta_{\rm esc}u_{\rm mag}}\frac{\ln(\gamma_{\rm br}/\gamma_{\rm min})}{\ln(\gamma_{\rm max}/\gamma_{\rm br})}. \tag{9}$$

For high luminosity blazars, external Compton scattering dominates over other components, which means that $u_{\rm rel}$ is larger than $u_{\rm mag}$. For low lumonosity blazars,

$\gamma_{\rm br}$ is near $\gamma_{\rm max}$, which means the same inequality. Thus, the jets turn out to be particle dominated; the energy density of relativistic electrons is larger than that of magnetic field by one or two orders of magnitude [8,9].

The kinetic power of relativistic electrons and Poynting power are estimated by

$$L_{\rm rel,kin} = \pi R^2 c u_{\rm rel} \Gamma^2 = \frac{L_{\rm obs}}{4\beta_{\rm esc}\Gamma^2} \frac{\ln(\gamma_{\rm br}/\gamma_{\rm min})}{\ln(\gamma_{\rm max}/\gamma_{\rm br})}. \tag{10}$$

and

$$L_{\rm mag,kin} = \pi R^2 c u_{\rm mag} \Gamma^2 \tag{11}$$

It should be noted that the above kinetic power is only for relativistic electrons. If the composition of jets is electron-proton, unless $\gamma_{\rm min}$ is extremely large, the true kinetic power would be much larger, exceeding the Eddignton power and the observed large scale power of extended radio sources. Then, the composition of jets is likely to be electron-positron, in accord with recent reports in favor of electron-positron pair dominance based on other methods [12–14].

DISCUSSION

The above results raise several important theoretical issues on the production and bulk acceleration of relativistic jets. First, Poynting power is insufficient to accelerate jets up to the bulk Lorentz factor of 10 so that other mechanisms should be reconsidered. Historically, the difficulties with radiative acceleration have led to magnetic acceleration. Possibilities of combined thermal and radiative mechanisms should be pursued.

Second, we should find mechanisms for the production of electron-positron pairs. Recently we have proposed a possible mechanism for the electron-positron outflow from hot accretion disks [15]. Within a disk, hard photons can produce pairs which can escape from the disk before annihilation through their own gas pressure, thus a large part of accretion power can be converted into the kinetic power of pairs.

Third, one potential problem with this model is pair annihilation near the disk as was argued by [16,17]. Careful treatment about the reduction of annihilation cross section for mildly relativistic pairs and effects of pair production by trapped high energy photons should be examined to have a final outcome.

Final issue is the radiation drag and associated bulk Compton radiation. At least for some high luminosity objects, external Compton component is necessary. The external photon field works to decelerate jets through Compton scattering by bulk motion. The resultant soft X-ray component is yet identified and its upper limit places a strong constraint on the external photon field [10]. One possible way out may be that its energy density is fairly constant with distance during the bulk acceleration stage.

CONCLUSIONS

I have shown that multiwavelength spectra from blazars can well be interpreted in term of shock acceleration of relativistic electrons in the relativistic jets. I have estimated the energetics of relativistic jets and found several important results. Most important feature is that the jet is particle dominated and that a large kinetic power of relativistic electrons suggests that the jet is comprised of electron-positron pairs rather than a usual electron-proton plasma. I have discussed sveral potential problems with this picture, which should be worked out in near future.

Acknowledgements

This work is supported in part by the Scientific Research Fund of the Ministry of Education, Science and Culture under Grant No. 11640236.

REFERENCES

1. Mukherjee, R. et al., *Ap.J.* **490**, 116-135 (1997).
2. von Montigny, C. et al., *Ap.J.* **440**, 525-553 (1995).
3. Blandford, R. D. and Konigl, A., *Ap.J.* **232**, 34-48 (1979).
4. Maraschi, L., Ghisellini, G. and Celotti, A., *Ap.J.* **397**, L5-L9 (1992).
5. Sikora, M., Begelman, M. C. and Rees, M. J., *Ap.J.* **421**, 153-162 (1994).
6. Dermer, D. C. and Schlickeiser, R., *Ap.J.* **416**, 458-484 (1993).
7. Inoue, S. and Takahara, F., *Ap.J.* **463**, 555-564 (1996).
8. Takahara, F., *Towards a Major Atmospheric Cerenkov Detector III*, ed. T. Kifune, Tokyo: Universal Academy Press, 1994, pp.131-137.
9. Takahara, F., *Relativistic Jets in AGNs*, eds. M. Ostrowski, M. Sikora, G. Madejski and M. Begelman, Crakow: Jagellonian University, 1997, pp. 253-261.
10. Sikora, M. et al., *Ap.J.* **484**, 108-117 (1997).
11. Kubo, H. et al., *Ap.J* **504**, 693-701 (1998).
12. Wardle, J. F. C. et al., *Nature* **395**, 457-461 (1998).
13. Reynolds, C. S. et al., *M.N.R.A.S.* **283**, 873-880 (1996).
14. Hirotani, K. et al., *P.A.S.J.* **51**, 263-267 (1999).
15. Yamasaki, T., Takahara, F. and Kusunose, M., *Ap.J* **523**, L21-L24 (1999).
16. Ghisellini, G.et al., *M.N.R.A.S.* **258**, 776-786 (1992).
17. Blandford, R. D. and Levinson, A., *Ap.J.* **441**, 79-95 (1995).

Spectral Variability of Blazars

Markus Böttcher*[1]

*Space Physics and Astronomy Department; Rice University, MS 108
6100 S. Main Street; Houston, TX 77005 - 1892, USA

Abstract. Broadband spectral characteristics of different subclasses of γ-ray blazars in different γ-ray intensity states are contrasted and discussed in the framework of currently popular leptonic jet models. Differences between quasars and BL-Lacs may be understood in terms of the dominance of different radiation mechanisms in the gamma-ray regime. Spectral variability patterns of different blazar subclasses appear to be significantly different and require different intrinsic mechanisms causing gamma-ray flares. As examples, recent results of long-term multiwavelength monitoring of PKS 0528+134 and Mrk 501 are presented.

INTRODUCTION

Recent high-energy detections and simultaneous broadband observations of blazars, determining their spectra and spectral variability, are posing strong constraints on currently popular jet models of blazars. 66 blazars have been detected by EGRET at energies above 100 MeV [1]. Most EGRET-detected blazars exhibit rapid variability [2], in some cases on intraday and even sub-hour (e. g., [3]) timescales, where generally the most rapid variations are observed at the highest photon frequencies. The broadband spectra of blazars consist of at least two clearly distinct spectral components. The first one extends in the case of flat-spectrum radio quasars (FSRQs) from radio to optical/UV frequencies, in the case of HBLs up to soft and even hard X-rays, and is consistent with non-thermal synchrotron radiation from ultrarelativistic electrons. The second one peaks at several MeV – a few GeV in most quasars, while in the case of some HBLs the γ-ray peak appears to be located at TeV energies.

In this paper, I am focusing on the interpretation of blazar broadband spectra by leptonic jet models [5–8]. In Section 2, I will give a description of the model and discuss the different γ-ray production mechanisms. In Section 3, I will review recent progress in understanding intrinsic differences between different blazar classes. In Section 4, I will discuss how this may be related to broadband spectral variability of individual blazars.

[1] Chandra Fellow

CP510, *The Fifth Compton Symposium,* edited by M. L. McConnell and J. M. Ryan

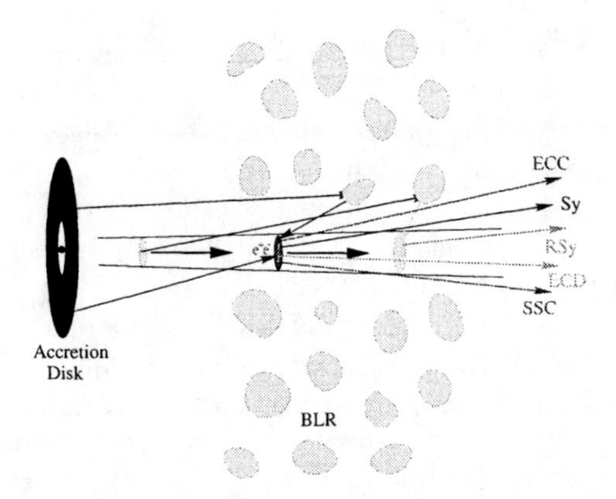

FIGURE 1. Illustration of the model geometry and the relevant γ radiation mechanisms for leptonic jet models.

MODEL DESCRIPTION AND RADIATION MECHANISMS

The basic geometry of leptonic blazar jet models is illustrated in Fig. 1. At the center of the AGN, an accretion disk around a supermassive, probably rotating, black hole is powering a relativistic jet. Along this pre-existing jet structure, occasionally blobs of ultrarelativistic electrons are ejected at relativistic bulk velocity.

The electrons are emitting synchrotron radiation, which will be observable at IR – UV or even X-ray frequencies, and hard X-rays and γ-rays via Compton scattering processes. Possible target photon fields for Compton scattering are the synchrotron photons produced within the jet (the SSC process, [5,9,10]), the UV – soft X-ray emission from the disk — either entering the jet directly (the ECD [External Comptonization of Direct disk radiation] process; [6,11]) or after reprocessing at the broad line regions or other circumnuclear material (the ECC [External Comptonization of radiation from Clouds] process; [7,12,13]), or jet synchrotron radiation reflected at the broad line regions (the RSy [Reflected Synchrotron] mechanism; [14–16]).

The relative importance of these components may be estimated by comparing the energy densities of the respective target photon fields. Simple analytical estimates of these photon densities may be found, e. g., in [17].

TRENDS BETWEEN DIFFERENT BLAZAR CLASSES

There appears to be a continuous sequence in the broadband spectral properties of blazars, ranging from FSRQs over LBLs to HBLs [18]. While in FSRQs the synchrotron and γ-ray peaks are typically located at infrared and MeV – GeV energies, they are shifted towards higher frequencies in BL Lacs, occurring at medium to even hard X-rays and at multi-GeV – TeV energies in some HBLs. The bolometric luminosity of FSRQs is — at least during γ-ray high states — strongly dominated by the γ-ray emission, while in HBLs the relative power outputs in synchrotron and γ-ray emission are comparable.

Detailed modeling of several blazars has indicated that this sequence appears to be related to the relative contribution of the external Comptonization mechanisms ECD and ECC to the γ-ray spectrum. While most FSRQs are successfully modelled with external Comptonization models (e. g., [13,19–21]), the broadband spectra of HBLs are consistent with pure SSC models (e. g., [22–24]).

A physical interpretation of this sequence in the framework of a unified jet model for blazars was given in [25]. If the average energy of electrons, γ_e, is determined by the balance of an energy-independent acceleration rate $\dot{\gamma}_{acc}$ and radiative losses, $\dot{\gamma}_{rad} \approx -(4/3)\, c\sigma_T\, (u'/m_e c^2)\, \gamma^2$, where u' is the total target photon density u', then the average electron energy will be $\gamma_e \propto (\dot{\gamma}_{acc}/u')^{1/2}$. If one assumes that the properties determining the acceleration rate of relativistic electrons do not vary significantly between different blazar subclasses, then an increasing energy density of the external radiation field will obviously lead to a stronger radiation component due to external Comptonization, but also to a decreasing average electron energy γ_e, implying that the peak frequencies of both spectral components are displaced towards lower frequencies.

SPECTRAL VARIABILITY OF BLAZARS

Between flaring and non-flaring states, blazars show very distinct spectral variability. FSRQs often show spectral hardening of their γ-ray spectra during γ-ray flares (e. g., [26,29,30]), and the flaring amplitude in γ-rays is generally larger than in all other wavelength bands. The concept of multi-component γ-ray spectra of quasars [26] offers a plausible explanation for this spectral variability due to the different beaming patterns of different radiation mechanisms [27]. This has been applied to PKS 0528+134 in [28] and [20].

The results of [20] indicate that γ-ray flaring states of PKS 0528+134 are consistent with an increasing bulk Lorentz factor Γ of ejected jet material, while at the same time the low-energy cutoff γ_1 of the electron distribution injected into the jet is lowered. This is in agreement with the physical picture that due to an increasing Γ, the quasi-isotropic external photon field is more strongly Lorentz boosted into the blob rest frame, leading to stronger external Compton losses, implying a lower value of γ_1. The external Compton γ-ray components depend much more strongly on the bulk Lorentz factor than the synchrotron and SSC components do. This leads naturally to a hardening of the γ-ray spectrum, if the SSC mechanisms plays an important or even dominant role in the X-ray — soft γ-ray regime, while external Comptonization is the dominant radiation mechanism at higher γ-ray energies.

While this flaring mechanism is plausible for FSRQs, short-timescale, correlated X-ray and γ-ray flares of the HBLs Mrk 421 and Mrk 501 [22,23] and synchrotron flares of other HBLs (e. g., PKS 2155-304, [31,32]) have been explained successfully in the context of SSC models where flares are related to an increase of the maximum

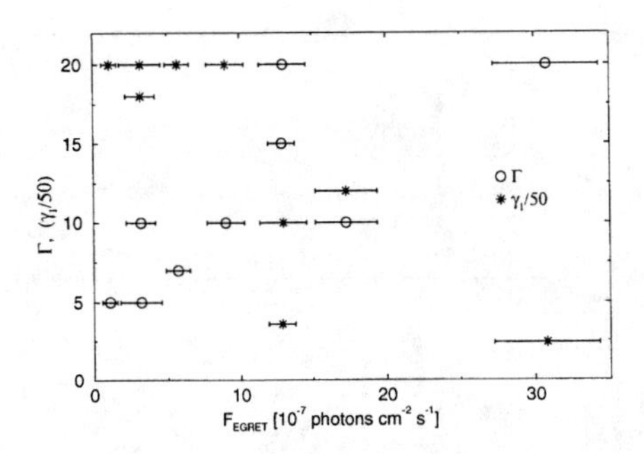

FIGURE 2. The dependence of the fit parameters Γ and γ_1 (low-energy cut-off of the electron distribution) on the EGRET flux for fits to simultaneous broadband spectra of PKS 0528+134 (see [20]).

electron energy, γ_2, and a hardening of the electron spectrum.

Comparing detailed spectral fits to weekly averaged broadband spectra of Mrk 501 [24] over a period of 6 months, we have found that TeV and hard X-ray high states on intermediate timescales are consistent with a hardening of the electron spectrum (decreasing spectral index) and an increasing number density of high-energy electrons, while the value of γ_2 has only minor influence on the weekly averaged spectra. Fig. 3 shows how the spectral index of the injected electron distribution and the density of high-energy electrons resulting from our fits are varying in comparison to the RXTE ASM, BATSE, and HEGRA 1.5 TeV light curves.

These variability studies seem to indicate that due to the different dominant γ radiation mechanisms in quasars and HBLs also the physics of γ-ray flares and extended high states is considerably different. While in FSRQs the γ-ray emission and its flaring behavior appears to be dominated by conditions of the external radiation field, this influence is unimportant in the case of HBLs where emission lines are very weak or absent, implying that the BLR might be very dilute, leading to a very weak external radiation field, which becomes negligible compared to the synchrotron radiation field intrinsic to the jet.

REFERENCES

1. Hartman, R. C., et al., *ApJS*, **123**, 79 (1999a).
2. Mukherjee, R., et al., *ApJ*, **490**, 116 (1997).
3. Gaidos, J. A., et al., *Nature*, **383**, 319 (1996).
4. Mannheim, K., *A&A*, **269**, 67 (1993).
5. Marscher, A. P., & Gear, W. K., *ApJ*, **298**, 114 (1985).
6. Dermer, C. D., Schlickeiser, R., & Mastichiadis, A., *A&A*, **256**, L27 (1992).
7. Sikora, M., Begelman, MM. C., & Rees, M. J, *ApJ*, **421**, 153 (1994).
8. Böttcher, M., Mause, H., & Schlickeiser, R., *A&A*, **324**, 395 (1997).

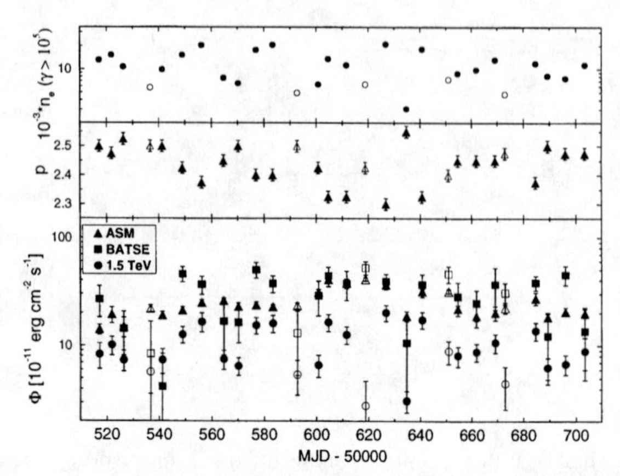

FIGURE 3. Temporal variation of the fit parameters $n_e(\gamma > 10^5)$ (density of high-energy electrons) and p (spectral index of injected electron distribution) compared to the weekly averaged light curves from RXTE ASM, BATSE and HEGRA (see [24]) for Mrk 501.

9. Maraschi, L., Ghisellini, G., & Celotti, A., *ApJ*, **397**, L5 (1992).
10. Bloom, S. D., & Marscher, A. P., *ApJ*, **461**, 657 (1996).
11. Dermer, C. D., & Schlickeiser, R., *ApJ*, **416**, 458 (1993).
12. Blandford, R. D., & Levinson, A., *ApJ*, **441**, 79 (1995).
13. Dermer, C. D., Sturner, S. J., & Schlickeiser, R., *ApJS*, **109**, 103 (1997).
14. Ghisellini, G., & Madau, P., *MNRAS*, **280**, 67 (1996).
15. Bednarek, W., *A&A*, **342**, 69 (1998).
16. Böttcher, M., & Dermer, C. D., *ApJ*, **501**, L51 (1998).
17. Böttcher,M., in "Towards a Major Atmospheric Cherenkov Detector VI", GeV – TeV Gamma-Ray Astrophysics Workshop, *AIP proc.*, in press (1999)
18. Fossati, G., et al., *MNRAS*, **289**, 136 (1997).
19. Sambruna, R., et al., *ApJ*, **474**, 639 (1997).
20. Mukherjee, R., et al., *ApJ*, **527**, in press (1999).
21. Hartman, R. C., et al., in preparation (1999b).
22. Mastichiadis, A., & Kirk, J. G., *A&A*, **320**, 19 (1997).
23. Pian, E., et al., *ApJ*, **492**, L17 (1998).
24. Petry, D., et al., *ApJ*, submitted (1999).
25. Ghisellini, G., et al., *MNRAS*, **301**, 451 (1998).
26. Collmar, W., et al., *A&A*, **328**, 33 (1997).
27. Dermer, C. D., *ApJ*, **446**, L63 (1995).
28. Böttcher, M., & Collmar, W., *A&A*, **329**, L57 (1998).
29. Hartman, R. C., et al., *ApJ*, **461**, 698 (1996).
30. Wehrle, A. E., et al., *ApJ*, **497**, 178 (1998).
31. Georganopoulos, M., & Marscher, A. P., *ApJ*, **506**, L11 (1998).
32. Kataoka, J., et al. *ApJ*, in press (1999).

Obscuration model of Variability in AGN

B. Czerny*, A. Abrassart†, S. Collin-Souffrin† and A.-M. Dumont†

*Copernicus Astronomical Center, Bartycka 18, 00-716 Warsaw, Poland
†DAEC, Observatoire de Meudon, F-92195 France

Abstract. There are strong suggestions that the disk-like accretion flow onto massive black hole in AGN is disrupted in its innermost part (10-100 Rg), possibly due to the radiation pressure instability. It may form a hot optically thin quasi spherical (ADAF) flow surrounded by or containing denser clouds due to the disruption of the disk. Such clouds might be optically thick, with a Thompson depth of order of 10 or more. Within the frame of this cloud scenario [1,2], obscuration events are expected and the effect would be seen as a variability. We consider expected random variability due to statistical dispersion in location of clouds along the line of sight for a constant covering factor. We discuss a simple analytical toy model which provides us with the estimates of the mean spectral properties and variability amplitude of AGN, and we support them with radiative transfer computations done with the use of TITAN code of [3] and NOAR code of [4].

INTRODUCTION

The variability of radio quiet AGN has been established since the early EXOSAT observations. However, the nature of this variability, observed in the optical, UV and X-ray band is not clear.

The emission of radiation is caused by accretion of surrounding gas onto a central supermassive black hole. The observed variability may be therefore directly related to the variable rate of energy dissipation in the accretion flow. However, it is also possible that the observed variability does not represent any significant changes in the flow. Such an 'illusion of variability' may be created if we do not have a full direct view of the nucleus. We explore this possibility in some detail.

Clumpy accretion flow has been suggested by various authors in a physical context of gas thermal instabilities or strong magnetic field (e.g. [5–7]). Here we follow a specific accretion flow pattern described by [1].

We assume that the cold disk flow is disrupted at the distance of 10 - 100 $R_{\rm Schw}$ from the black hole. The resulting clumps of cool material are large and optically very thick for electron scattering, and they become isotropically distributed around the central black hole. Hot plasma responsible for hard X-ray emission forms still

CP510, *The Fifth Compton Symposium,* edited by M. L. McConnell and J. M. Ryan

closer to a black hole, perhaps due to cloud collision. We do not discuss the dynamics of cloud formation but we concentrate on the description of the radiation produced by such a system. We consider the radiative transfer within the clouds and radiative interaction between the clouds and a hot phase, and we relate the radiation flux and spectra to the variations in the cloud distribution.

VARIABILITY MECHANISM

Within the frame of the cloud scenario, our line of sight to the hot X-ray emitting plasma is partially blocked by the surrounding optically thick clouds. Variations in the cloud distribution lead to two types of phenomena: slower variations due to the systematic change in a total number of clouds and fastest variations due to the random rearrangement of the clouds without any change of their total number. We concentrate on this second type of variability.

The amplitude of such a variability is determined by the number of clouds N surrounding a black hole at any given moment and the mean covering factor C of the cloud distribution $\left(\frac{\delta L_X}{L_X}\right)_{obs} = \frac{C\sqrt{2/N}}{1-C}$.

Such variations do not reflect any deep changes in the hot plasma itself so they are expected to happen without the change in the hard X-ray slope of the plasma emission. However, the UV variability amplitude caused by the same mechanism as well as some variations in hard X-rays due to the presence of the X-ray reflection depend in general on the cloud properties like X-ray albedo and radiative losses through the unilluminated dark sides of clouds. We use very detailed radiative transfer in X-ray heated optically thick clouds in order to estimate those quantities. We develop a complementary toy model describing the energetics of the entire hot plasma/cloud system for easy use to estimate the model parameters from the observed variability amplitudes.

RADIATIVE TRANSFER SOLUTIONS FOR MEAN SPECTRA

Two codes are used iteratively in order to compute a mean spectrum emitted by the clouds distribution. TITAN [3] is designed to solve the radiative transfer within an optically thick medium, including computations of the ionization state of the gas and its opacities. NOAR [4] is a Monte Carlo code designed to follow the hard X-ray photons using Monte Carlo method.

The result of the numerical computation of a single mean spectrum is shown in Figure 1. The cloud distribution was assumed to be spherical, with the covering factor $C = 0.9$, all clouds being located at a single radius. The hot medium in this computation was replaced by a point like source of a primary emission, with flux normalizations fixed through specification of the ionization parameter ξ. However, the multiple scattering of photons of different clouds was included.

We see that the broad band spectrum clearly consists of two basic components but there are also detailed spectral features in UV and soft X-ray band in addition to hard X-ray iron K_{α} line.

TOY MODEL AND VARIABILITY AMPLITUDES

In our toy model we replace the radiative transfer computations with analytical description of the probabilities of the X-ray and UV photon fate. X-ray photons can be reflected by bright sides of the clouds, can escape from the central region towards an observer or can be absorbed and provide new UV photons as well as energy for the emission from the dark sides. UV photon can also escape, can be reflected and can be upscattered to an X-ray photon by a hot plasma. All those probabilities are determined by four model parameters: covering factor C, probability of Compton upscattering γ, X-ray albedo a and efficiency of dark side emission β_d. The condition of compensating for the system energy losses with Compton upscattering relate those four quantities to the Compton amplification factor. The variability amplitude depends also on the number of clouds, N.

Such a model allow us to calculate all basic properties of the stationary model,

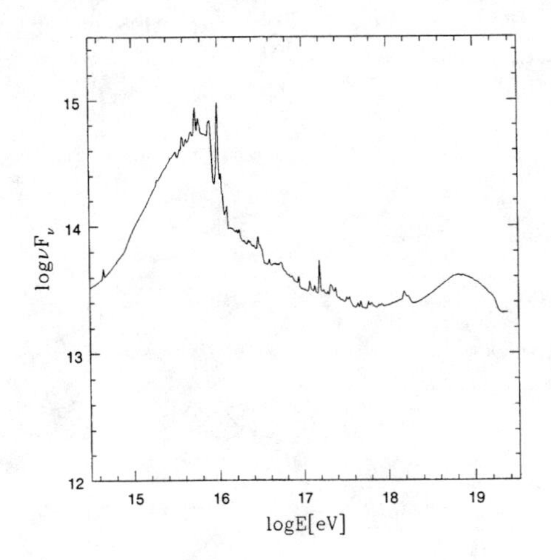

FIGURE 1. The exemplary spectrum calculated with the coupled codes TITAN and NOAR for the following parameters of the shell: number density log$n$ = 14, column density log N_H = 26, covering factor $C = 0.9$. The incident primary radiation was assumed to be a power law extending from 1 eV to 100 keV, with energy index $\alpha = 1$, and the ionization parameter $xi = 300$. The size of the central source was neglected.

like the observed ratio of the X-ray luminosity to the UV luminosity, the intrinsic ratio of those two quantities as seen by the clouds, the slope of the hard X-ray emission and the variability amplitudes in UV and X-ray band.

In particular, we can determine the ratio of the normalized variability amplitudes in X-ray and UV band predicted by our model $R = \left(\frac{\delta L_{UV}}{L_{UV}}\right)_{obs} / \left(\frac{\delta L_X}{L_X}\right)_{obs}$. In Figure 2 we show the dependence of this ratio on the covering factor C and the efficiency of the dark side energy loss by the clouds. The dependence on other model parameters was reduced by assuming the X-ray albedo $a = 0.5$ supported by numerical results and the Compton amplification factor $A = 4$ which well describes the mean hard X-ray spectral slope.

We see that if the clouds are very opaque (β_b negligible) the normalized amplitude ratio is always equal 1 within the frame of our model. Significant dark side energy losses reduce the UV amplitude since they add a constant contribution to UV flux.

DISCUSSION

The presented model well reproduces large observed variability amplitudes if the covering factor is close to 1. It also explains why large variability amplitudes are not necessarily accompanied by the change of the slope of the hard X-ray emission coming from comptonizing hot plasma. In order to check whether the model requires unacceptable values of the parameters we confront the model with the data in the following way.

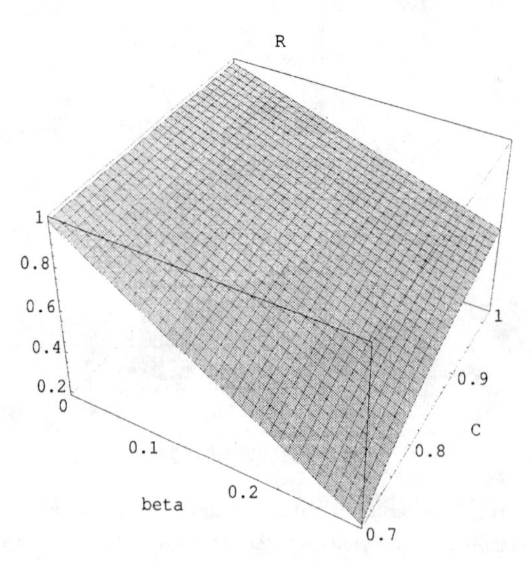

FIGURE 2. The dependence of the ratio of the normalized variability amplitude in UV and in X-rays on the dark side loss eficciency β_d and covering factor C; other parameters: X-ray albedo $a = 0.5$, Compton amplification factor $A = 4$.

TABLE 1. Toy model parameters for AGN.

Object	rms_{UV}	rms_X	C	N	γ	β_d
NGC 3516	0.333	0.357	0.90	1180	0.047	0.04
NGC 7469	0.167	0.167	0.90	5400	0.044	0.00
NGC 4151	0.009	0.024	0.90	2610	0.095	0.39
NGC 5548	0.222	0.222	0.90	3050	0.044	0.00

We apply our toy model to observed variability of four Seyfert 1 galaxies extensively monitored in UV and X-ray band (see Table 1). The variability amplitudes are taken from [8–12], and we estimated the mean X-ray to UV luminosity ratio as 1/3 in all objects. We assumed the X-ray albedo $a = 0.5$ and the Compton amplification factor as $A = 4$. We calculated the remaining model parameters: C, N, γ, β_d.

All four objects are consistent with the model, having relative UV amplitude smaller than the relative X-ray amplitude. As expected, the covering factor (determined by the luminosity ratio) is large and the probability of Compton upscattering is low either due to small optical depth of the hot plasma or due to small radial extension of the hot plasma. The required dark side losses are comparable to the value of 0.20 obtained from the numerical solution of the radiative transfer within a cloud (see Figure 1). Therefore cloud scenario offers an attractive explanation of the observed variability of AGN if further observations will confirm that the slope of the direct Compton component and its high energy cut-off do not vary.

REFERENCES

1. Collin-Souffrin S., Czerny, B., Dumont, A.-M., and Życki, P.T., *A&A* **314**, 393 (1996)
2. Czerny B., and Dumont A.-M., *A&A*, **338**, 386 (1998)
3. Dumont A.-M., Abrassart A., and Collin S., *A&A*, (1999) (submitted)
4. Abrassart A., (1999) (in preparation)
5. Celotti, A, Fabian, A.C., and Rees, M., *MNRAS*, **255**, 419 (1992)
6. Krolik, J.H., *ApJ*, **498**, L13 (1998)
7. Torricelli-Ciamponi, G., and Courvoisier, T.J.-L., *A&A*, **335**, 881 (1998)
8. Goad M.R., Koratkar A.P., Axon D.J., Korista K.T., O'Brien P.T.O, 1999, *ApJ*, **512**, L95
9. Edelson R., and Nandra K., *ApJ*, **514**, 682 (1999)
10. Nandra, K., Clavel, J., Edelson, R.A., George, I.M., Malkan, M.A., Mushotzky, R.F., Peterson, B.M., Turner, T.J., *ApJ*, **505**, 594 (1998)
11. Edelson, R. et al. *ApJ*, **470**, 364 (1996)
12. Clavel et al. *ApJ*, **393**, 113 (1992)

Time Variability of Emission from Blazars and Electron Acceleration

M. Kusunose*, F. Takahara†, and H. Li††

*Kwansei Gakuin University, Nishinomiya 662-8501, Japan
†Osaka University, Osaka 560-0043, Japan
††Los Alamos National Laboratory, Los Alamos, NM 87545

Abstract. Recent observations of blazars by X- and gamma-rays show that blazars exhibit rapid time variations. We use the synchrotron-self-Compton (SSC) model to simulate the time variations of the energy spectra of radiation and electrons; we solve the kinetic equations of electrons and photons simultaneously, including a simple model of electron acceleration which mimics the shock acceleration. Electrons escaping from the acceleration region are *injected* into a cooling region where electrons emit radio through very high energy gamma-rays via SSC. The time evolution of the energy spectra of electrons and photons are shown, when low energy electrons ($\gamma \sim 2$) are supplied in the acceleration region and accelerated up to $\gamma \sim 10^7$. Then TeV gamma-rays are emitted as observed from Mrk 421 and Mrk 501.

INTRODUCTION

High energy emission from blazars [4,6,8,10,14,16] is generally attributed to relativistically moving jets or blobs from the nucleus of galaxies [2,3,7,12,15,17]. The emission mechanisms of blazars are thought to be synchrotron emission and inverse Compton scattering, while there are models which assume a soft-photon supply by accretion disks [15]. Hadronic processes are also assumed in some models [11].

Though most models are focused on steady state emission from blazars, time variability is often observed from blazars by X- and gamma-rays [8]. Models with time-dependent calculations are recently presented by [9,13]. Mastichiadis and Kirk [13] calculated the time evolution of the photon spectrum when electrons obeying a power law with an exponential cutoff are injected. On the other hand, Kirk et al. [9] included electron acceleration, electron transport, and synchrotron emission, though Compton scattering was not included. Because very high energy gamma-ray emission is an important feature of blazars, the effects of Compton scattering need to be included in theoretical models. We then have developed a numerical code to calculate the time evolution of electron and photon energy-spectra including Compton scattering. Below we show the time evolution of electrons and photons with a simple model for electron acceleration.

CP510, *The Fifth Compton Symposium*, edited by M. L. McConnell and J. M. Ryan

MODEL

We assume that the emission (cooling) region is a spherical blob (radius R) moving at a relativistic speed with Doppler factor $\mathcal{D}$. The blob contains an acceleration region, which is presumably a shock region. In the acceleration region which is assumed to be a disk with radius R and thickness $R_{\rm acc}$, electrons are accelerated on timescale $t_{\rm acc}$ and escape on timescale $t_{e,\rm esc}$. Escaping electrons are injected into the cooling region. Electrons suffer from cooling losses in both the acceleration and cooling regions.

The kinetic equation describing the time evolution of electrons is given by

$$\frac{\partial N(\gamma)}{\partial t} = -\frac{\partial}{\partial \gamma}\left\{\left[\left(\frac{d\gamma}{dt}\right)_{\rm acc} - \left(\frac{d\gamma}{dt}\right)_{\rm loss}\right] N(\gamma)\right\} - \frac{N(\gamma)}{t_{e,\rm esc}} + Q(\gamma)\,, \tag{1}$$

where γ is the Lorentz factor of electrons and $N(\gamma)d\gamma$ is the number density of electrons. We assume that monochromatic electrons with Lorentz factor γ_0 are injected in the acceleration region, i.e., $Q(\gamma) = Q_0\,\delta(\gamma - \gamma_0)$; in numerical results shown below, we assume $\gamma_0 = 2$. Electrons are then accelerated and lose energy by synchrotron radiation and Compton scattering [the energy loss rate is denoted by $(d\gamma/dt)_{\rm loss}$]. The acceleration term is approximated by $(d\gamma/dt)_{\rm acc} = \gamma/t_{\rm acc}$. In the framework of diffusive shock acceleration, e.g. [1,5], $t_{\rm acc}$ can be approximated as

$$t_{\rm acc} = 20\lambda(\gamma)c/(3v_s^2) \sim 3.79 \times 10^{-6}(0.1{\rm G}/B)\xi\,\gamma \quad {\rm sec}, \tag{2}$$

where $v_s \approx c$ is the shock speed with c being the light speed, B is the strength of magnetic fields, and $\lambda(\gamma) = \gamma m_e c^2 \xi/(eB)$ is the mean free path assumed to be proportional to the electron Larmor radius with ξ being a parameter, m_e the electron mass, and e the electron charge. For the convenience of numerical calculations, we assume $t_{\rm acc}$ does not depend on γ:

$$t_{\rm acc} = 3.79 \times 10(0.1{\rm G}/B)(\gamma_f/10^7)\xi \quad {\rm sec}, \tag{3}$$

where γ_f is assumed to be a characteristic Lorentz factor of relativistic electrons and used as a parameter; we set $\gamma_f = 10^7$ throughout this paper. Although realistic acceleration time for the smaller values of γ should be correspondingly shorter, we make this choice because we mainly concern about the electrons with the large values of γ. We make sure that the resultant spectrum is that expected in diffusive shock acceleration by choosing $t_{e,\rm esc} = t_{\rm acc}$ in the acceleration region. The electron spectrum in the cooling region is calculated by Eq. (1), with $(d\gamma/dt)_{\rm loss}$ dropped. Also $Q(\gamma)$ is replaced by the escaping electrons from the acceleration region and $t_{e,\rm esc}$ is set to be $2R/c$.

The photon spectrum is calculated by

$$\frac{\partial n_{\rm ph}(\epsilon)}{\partial t} = \dot{n}_{\rm C}(\epsilon) + \dot{n}_{\rm em}(\epsilon) - \dot{n}_{\rm abs}(\epsilon) - \frac{n_{\rm ph}(\epsilon)}{t_{\gamma,\rm esc}}\,, \tag{4}$$

where $n_{\rm ph}(\epsilon)d\epsilon$ is the photon number density. The term $\dot{n}_{\rm C}(\epsilon)$ denotes Compton scattering. Photon production and self-absorption by synchrotron radiation are included in $\dot{n}_{\rm em}(\epsilon)$ and $\dot{n}_{\rm abs}(\epsilon)$, respectively. External photon sources are not included. The rate of photon escape is estimated as $n_{\rm ph}(\epsilon)/t_{\gamma,\rm esc}$. We set $t_{\gamma,\rm esc} = R_{\rm acc}/c$ and R/c in the acceleration and cooling regions, respectively.

NUMERICAL RESULTS

In the following we assume $\mathcal{D} = 10$, $R/c = 5 \times 10^5$ sec in the blob frame, and $B = 0.1$ G. Electrons are injected at the rate of 0.1 electrons cm^{-3} s^{-1} in the acceleration region. The acceleration timescale is set $t_{\rm acc} = 2 \times 10^4$ sec in the blob frame. We show in Figure 1 the time evolution of the electron spectra for both the acceleration and cooling regions, when the blob is empty at $t = 0$. [Because of numerical errors, the power-law index of the electron spectrum in the acceleration region slightly deviates from -2.] The increase in $\gamma_{\rm max}$ is observed. In Figure 2, the time evolution of the emission spectra in the observer's frame is shown. The evolution of the Compton component is delayed from that of the synchrotron component, because the time required for photons to fill the blob is roughly R/c.

In this particular simulation, the energy density of electors is much larger than those of photons and the escaping electrons carry more energy than emitted photons. It is also found that the ratio of the energy densities of Compton photons and synchrotron photons is about 0.7, while the ratio of the energy densities of synchrotron photons to magnetic fields is about 9.

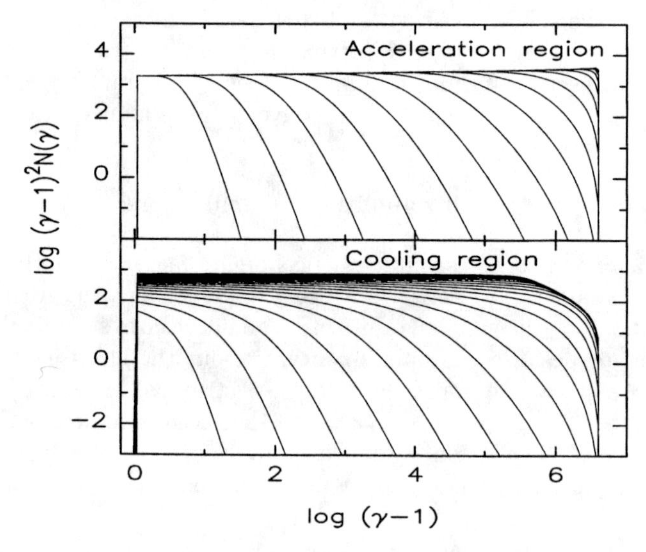

FIGURE 1. Time evolution of electron spectra for $t = 0 - R/c$ with equally spaced time span $0.05R/c$.

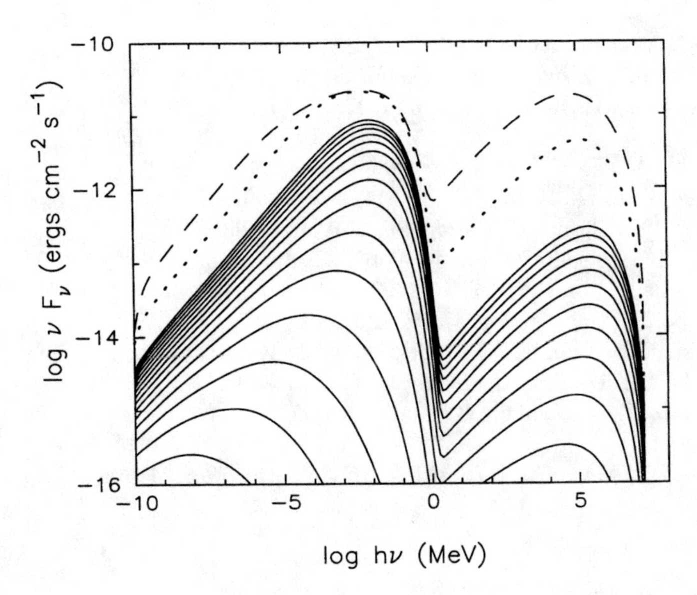

FIGURE 2. Time evolution of the photon spectra (*solid curves*) corresponding to Figure 1. The dotted curve is for $t = 2R/c$ and the dashed curve is for $t = 10R/c$, which is in a steady state.

SUMMARY

We showed the time evolution of electron and photon spectra when electron acceleration occurs in an acceleration region and the accelerated electrons are then injected into a cooling region. The maximum Lorentz factor becomes large enough to emit TeV gamma-rays, depending on the acceleration timescale. With our code, it is possible to simulate flares, which are often observed by X- and gamma rays, by temporarily changing the acceleration timescales, magnetic fields, etc., which will be presented in future work.

M.K. and F.T. have been partially supported by Scientific Research Grants (M.K.: Nos. 09223219 and 10117215; F.T.: Nos. 09640323, 10117210, and 11640236) from the Ministry of Education, Science, Sports and Culture of Japan.

REFERENCES

1. Blandford, R. D., and Eichler, D., *Phys. Rep.*, **154**, 1 (1987).
2. Blandford, R. D., and Königle, A., *Astrophys. J.*, **232**, 34 (1979).
3. Blandford, R. D., and Rees, M. J., *Pittsburgh Conf. on BL Lac Objects*, ed. A. M. Wolfe (Pittsburgh: Univ. Pittsburgh Press), 328 (1978).
4. Catanese, M. et al., *Astrophys. J. Letters*, **487**, 143 (1997).
5. Druly, L. O.' C., *Rep. Prog. Phys.*, **46**, 973 (1983).

6. Gaidos, J. A. et al., *Nature*, **383**, 319 (1996).
7. Inoue, S., and Takahara, F., *Astrophys. J.*, **463**, 555 (1996).
8. Kataoka, J. et al., *Astrophys. J.*, **514**, 138 (1999).
9. Kirk, J. G., Rieger, F. M., and Mastichiadis, A., *Astron. & Astrophys.*, **333**, 452 (1998).
10. Krennrich, F. et al., *Astrophys. J.*, **511**, 149 (1999).
11. Mannheim, K., *Astron. & Astrophys.*, **269**, 67 (1993).
12. Maraschi, L., Ghisellini, G., and Celotti, A., *Astrophys. J. Letters*, **397**, L5 (1992).
13. Mastichiadis, A., and Kirk, J. G. 1997, *Astron. & Astrophys.*, **320**, 19 (1997).
14. Mukherjee, R. et al., *Astrophys. J.*, **490**, 116 (1997).
15. Sikora, M., Begelman, M. C., and Rees, M. J., *Astrophys. J.*, **421**, 153 (1994).
16. Takahashi, T., Tashiro, M., Madejski, G., Kubo, H., Kamae, T., Kataoka, J., Kii, T., Makino, F., Makishima, K., and Yamasaki, N., *Astrophys. J. Letters*, **470**, L89 (1996).
17. Ulrich, M.-H., Maraschi, L., and Urry, C. M., *Ann. Rev. Astron. & Astrophys.*, **35**, 445 (1997).

Model for the Redshift and Luminosity Distributions of Gamma Ray Blazars

Charles D. Dermer* and Stanley P. Davis†

*Naval Research Laboratory, Code 7653, Washington, DC 20375-5352
†Lincoln University, and NASA/Goddard Space Flight Center

Abstract. A simple model is used to fit the redshift and luminosity distributions of gamma-ray blazars observed with EGRET. The model consists of collimated relativistic bulk plasma outflow, with radiation beamed along the jet axis due to Doppler boosting. The EGRET detection sensitivity is used to assess source detectability. The best-fit Doppler factors and intrinsic source luminosities are obtained from fits to the BL Lac object (BL) and flat spectrum radio quasar (FSRQ) distributions, assuming that the source density distributions follows the star formation rate history of the universe. Predictions for the detectability of blazars with GLAST are made. We argue that GLAST will detect a much larger fraction of BLs in its sample than found by EGRET in consequence of the weaker cosmological effects on detectability of the nearer BLs.

I INTRODUCTION AND SAMPLE

Population studies of gamma-ray blazars are difficult because of the unknown beaming patterns and Doppler factors of the jets. An accurate determination of the space density of these sources is required to identify parent populations [1] and to assess the formation history of supermassive black-hole jet sources and their contribution to the diffuse extragalactic gamma-ray background. We perform such a study here by fitting a simple blazar model to the observed redshifts and peak luminosities of gamma-ray blazars in the Third EGRET catalog [2].

The sample consists of 60 high-confidence gamma-ray blazars, consisting of 14 BLs and 46 FSRQs. We exclude sources within 10° of the Galactic plane, and use source catalogs [3,4] to establish BL identifications. Table 1 lists the sources from the Third EGRET catalog [2] used in this study and their classifications. The gamma-ray blazar sample, binned by redshift and grouped into BLs and FSRQs, is plotted in Fig. 1. Note the very different distributions of BLs and FSRQs. The BLs are less numerous and closer, with an average redshift $z \sim 0.5$, whereas the FSRQ distribution peaks at $z \sim 1$, and shows a tail reaching to $z \sim 2.3$. The average peak powers of the FSRQs are $\sim$ 1-2 orders of magnitude greater than those of the BLs (Fig. 2a). Fig. 2b shows that there are about 5 times as many FSRQs as BLs per unit peak power flux in the EGRET range ($\gtrsim 10^{-10}$ ergs cm^{-2} s^{-1}).

CP510, *The Fifth Compton Symposium,* edited by M. L. McConnell and J. M. Ryan

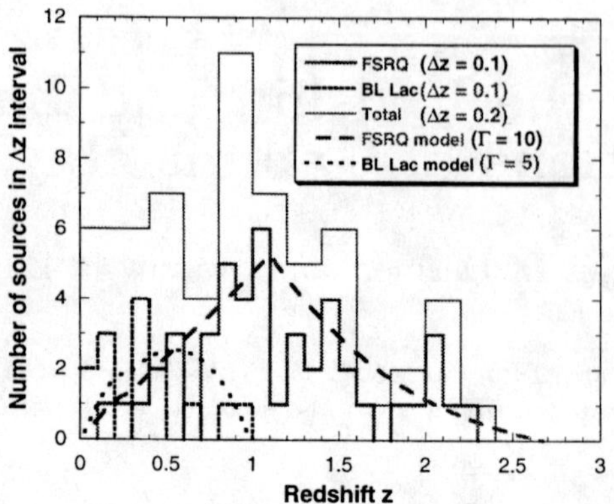

FIGURE 1. Redshift distribution of EGRET blazars (histograms) from the Third EGRET Catalog [2]. Curves show model fits assuming comoving density proportional to the star formation rate (SFR; see Fig. 3b). Fit parameters for the FSRQs are $\Gamma = 10$ and comoving luminosity $L = 3 \times 10^{44}$ ergs s^{-1}; fit parameters for the BL Lacs are $\Gamma = 5$ and $L = 5 \times 10^{44}$ ergs s^{-1}.

II THE MODEL

We employ a simplified version of the standard model for blazars described in Ref. [5]. A relativistically moving plasmoid emits a power-law radiation spectrum that is Doppler boosted by the effects of the relativistic motion. For the calculations shown here, we use the beaming factor where the flux density $S \propto \mathcal{D}^{3+\alpha}$, where $\mathcal{D} = [\Gamma(1 - \beta\cos\theta)]^{-1}$. Here Γ is the bulk Lorentz factor, $\beta = \sqrt{1 - \Gamma^{-2}}$, α is the energy spectral index of the radiation, and θ is the angle between the jet and line-of-sight directions. Continuous outflow scenarios and external Compton scattering

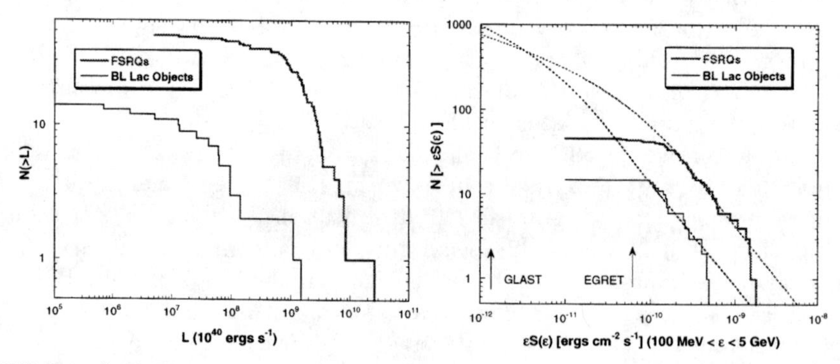

FIGURE 2. (left) Distribution of observed peak luminosities of FSRQs and BL Lac objects as measured in the EGRET energy range. (right) Histogram showing the measured peak power flux size distribution of EGRET blazars, the fit to the size distribution with the model, and predictions for gamma-ray blazar detectibility with GLAST for a sensitivity threshold of 1.3×10^{-12} ergs cm^{-2} s^{-1}.

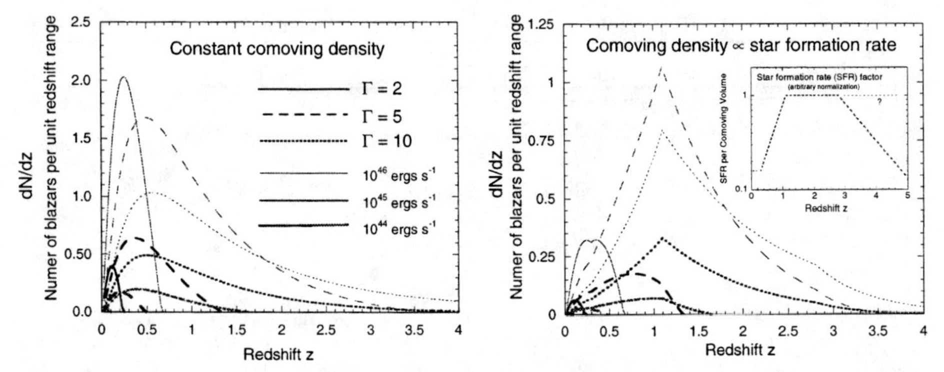

FIGURE 3. Fig. 3a (left). Model redshift distributions of EGRET blazars, assuming constant comoving density of blazars and an $\Omega_m = 0.3$, $\Omega_\Lambda = 0.5$ cosmology with a Hubble constant $H_0 = 65 \text{ km s}^{-1} \text{ Mpc}^{-1}$. Fig. 3b (right). Same as Fig. 3a, but for a comoving density proportional to the function [7] shown in the inset, which is a fit to the SFR derived from faint galaxy counts in the Hubble Deep Field [8].

processes produce different beaming patterns [6], not considered here.

Fig. 3 shows the expected number of blazars detected per unit redshift above the EGRET sensitivity threshold of 6.4×10^{-11}, for randomly oriented blazars with different values of Γ and comoving luminosity L (see legend). We let $\alpha = 1$ in this calculation. Models for constant comoving density of blazars and a comoving density proportional to the SFR factor are shown in Figs. 3a and 3b, respectively. The inset in Fig. 3b shows the SFR factor that is applied to the constant comoving density case to obtain the distribution of detected blazars shown in Figs. 3b and 1. Our approach differs from that of Chiang and Mukherjee [9] by considering density rather than luminosity evolution. We note that models with different underlying assumptions could give considerably different results for the number of blazars that GLAST might detect (e.g., [10]).

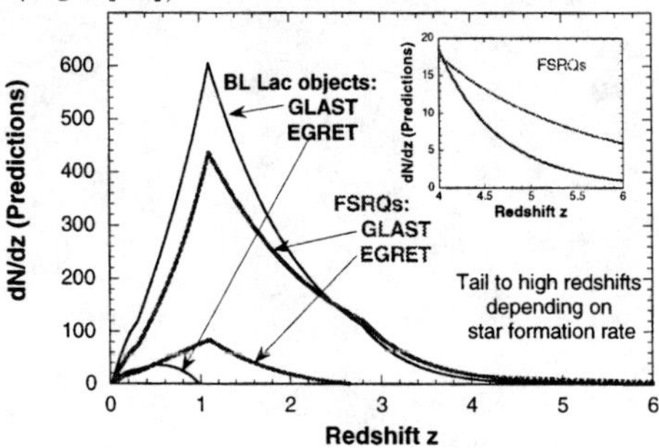

FIGURE 4. Predictions for the number of FSRQs and BLs to be detected with GLAST, assuming that the blazar rate follows the SFR history of the universe. Inset shows the range of high-redshift blazars implied by uncertainties in the SFR at large redshift (see inset in Fig. 3b).

TABLE 1. Sample of High-Confidence Gamma-Ray Blazars used in Analysis

Catalog Name	F(peak)[a]	ΔF	Redshift z	Other Name	Classification[b]
3EG J0204+1458	52.8	26.4	0.405		
3EG J0210-505	134.1	24.9	1.003		
3EG J0222+4253	25.3	5.80	0.444	3C 66A X	
3EG J0237+1635	65.1	8.80	0.940	AO 0235+164	X
3EG J0340-0201	177.6	36.6	0.852		
3EG J0412-1853	49.5	16.1	1.536		
3EG J0422-0102	64.2	34.2	0.915		
3EG J0442-0033	85.9	12.0	0.844	NRAO 190	
3EG J0450+1105	109.5	19.4	1.207	PKS 0446+112	
3EG J0456-2338	14.7	4.20	1.009	PKS 0454-234	
3EG J0458-4635	22.8	7.40	0.8580		
3EG J0459+0544	34.0	18.0	1.106		
3EG J0500-0159	68.2	41.3	2.286		
3EG J0530+1323	351.4	36.8	2.060	PKS 0528+134	
3EG J0540-4402	91.1	14.6	0.894	0537-441	R
3EG J0721+7120	45.7	11.1	0.30	0716+714	R
3EG J0737+1721	29.3	9.90	0.424		X
3EG J0743+5447	42.1	8.30	0.723		
3EG J0828+0508	35.5	16.3	0.180		X
3EG J0829+2413	111.0	60.1	2.046		
3EG J0845+7049	33.4	9.00	2.172		
3EG J0852-1216	44.4	11.6	0.566		
3EG J0853+1941	15.8	6.90	0.306	OJ 287	X
3EG J0952+5501	47.2	15.5	0.901		
3EG J0958+6533	18.0	9.40	0.368		R
3EG J1104+3809	27.1	6.90	0.031	Mkn 421	X
3EG J1200+2847	163.2	40.7	0.729		
3EG J1222+2841	53.6	14.1	0.102	W Comae	X
3EG J1224+2118	48.1	15.3	0.435		
3EG J1229+0210	48.3	11.3	0.158	3C 273	
3EG J1230-0247	15.5	4.10	1.045		
3EG J1246-0651	44.1	29.6	1.286		
3EG J1255-0549	267.3	10.7	0.538	3C 279	
3EG J1329+170	33.1	19.3	2.084		
3EG J1339-1419	20.2	11.6	0.539		
3EG J1409-0745	128.4	23.4	1.494		
3EG J1429-42	55.3	16.3	1.522		
3EG J1512-0849	49.4	18.3	0.361		
3EG J1605+1553	42.0	12.3	0.357	QSO 1604+159	R
3EG J1608+1055	62.4	13.0	1.226		
3EG J1614+3424	68.9	15.3	1.401		
3EG J1625-2955	258.9	15.3	0.815		
3EG J1626-2519	82.5	35.0	0.786		
3EG J1635+3813	107.5	9.60	1.814		
3EG J1727+0429	30.2	18.8	0.296		
3EG J1733-1313	104.8	34.7	0.902		
3EG J1738+5203	44.9	26.9	1.375		
3EG J1744-0310	48.7	19.6	1.054		

TABLE 1. (cont.) Sample of High-Confidence Gamma-Ray Blazars Used in Analysis

Catalog Name	F(peak)[a]	ΔF	Redshift z	Other Name	Classification[b]
3EG J1935-4022	93.9	31.4	0.966		
3EG J1937-1529	55.0	18.6	1.657		
3EG J2025-0744	74.5	13.4	1.388		
3EG J2036+1132	35.9	15.0	0.601		R
3EG J2055-4716	35.0	20.9	1.489	QSO 2052-474	
3EG J2158-3023	30.4	7.70	0.116	PKS 2155-304	X
3EG J2202+4217	39.9	11.6	0.069	BL Lac	X
3EG J2232+1147	51.6	15.0	1.037	CTA 102	
3EG J2254+1601	116.1	18.4	0.859	3C 454.3	
3EG J2321-0328	38.2	10.1	1.411		
3EG J2358+4604	42.8	20.3	1.992		
3EG J2359+2041	26.3	9.00	1.066		

[a] F: peak flux ($E > 100$ MeV) in units of 10^{-8} photons cm^{-2} s^{-1}.

[b] X: X-ray selected BL Lac object if detected by Einstein Slew Survey or HEA0; R: radio-selected BL Lac; no entry: FSRQ.

III PREDICTIONS

The redshift distributions for the constant comoving density model (Fig. 3a) does not give a good fit to the data, especially for the FSRQs: it peaks at too small a redshift and has a high redshift tail not seen in the data (compare Figs. 1 and 3a). The model with blazar density proportional to SFR gives a reasonable fit to the data (see Fig. 1) and the observed power flux distribution (Fig. 2b). By extrapolating to the flux threshold for GLAST, we see from Fig. 2b that GLAST should detect $\sim$ 2000 blazars and, moreover, a larger fraction ($\sim$ 50% rather than the $\sim$ 25% observed with EGRET) should be BLs. Thus we predict a flowering of BL Lac objects with GLAST. As these are nearly all X-ray selected (or "high-frequency peaked") BLs, many of which will not have been previously identified, this will provide a rich new class of targets for TeV observatories.

REFERENCES

1. Urry, C. M., and Padovani, P., *Proc. Astron. Soc. Pacific*, **107**, 803 (1995).
2. Hartman, R. C. *et al.*, *Astrophys. J. Supp.* **123**, 79 (1999).
3. Padovani, P. and Giommi, P., *MNRAS* **277**, 1477 (1995).
4. Perlman, E. S. *et al.*, *Astrophys. J. Supp.* **104**, 251 (1996).
5. Dermer, C. D., and Gehrels, N., *Astrophys. J.* **447**, 103 (1995).
6. Dermer, C., Sturner, S., and Schlickeiser, R., *Astrophys. J. Supp.*, **109**, 103 (1997).
7. Böttcher, M., and Dermer, C. D., *Astrophys. J.* **529**, (2000) in press (astro-ph/9812059).
8. Madau, P., Pozzetti, L., and Dickinson, M., *Astrophys. J.* **498**, 106 (1998).
9. Chiang, J., and Mukherjee, R., *Astrophys. J.* **496**, 752 (1998).
10. Stecker, F. W., & Salamon, M. H., *Astrophys. J.*, **464**, 600.

A Novel Mechanism of the Formation of Electron-Positron Outflow from Hot Accretion Disks

Tatsuya Yamasaki*[1], Fumio Takahara* and Masaaki Kusunose†

*Department of Earth and Space Science, Graduate School of Science, Osaka University, Toyonaka, Osaka 560-0043, Japan
†Department of Physics, School of Science, Kwansei Gakuin University, Nishinomiya 662-8501, Japan

Abstract. We propose a mechanism of the relativistic jet formation in active galactic nuclei and galactic black hole binaries. We consider the ejection of electron-positron pairs produced in the two-temperature accretion disks, by solving the pair momentum equation in the one-zone approximation, in which we assume that the electron-positron component can escape independently of the electron-proton one which forms a hydrostatic atmosphere. The results show that, in the inner region of the disks, when the mass accretion rate becomes larger than about a tenth of the Eddington rate, most of the viscously dissipated energy is converted into the thermal and kinetic energy of the ejected electron-positron pairs. The produced pairs are accelerated in the vertical direction by its own gas pressure rather than by the radiative force. Thus, this mechanism is successful in extracting accretion power to form powerful electron-positron jets as suggested by recent observations.

INTRODUCTION

Relativistic jets are observed in active galactic nuclei and galactic black hole candidates. Although it is still unknown how these relativistic jets are formed, their physical properties become rather clear by recent observations (e.g., [1], [2], [3], [4]). They suggest that some of the jets are so energetic as their power amounts to about 10 % of the Eddington luminosity of the central objects. Further, the jets are mainly composed of electron-positron pairs rather than electron-proton plasmas, at least within a parsec scale distance from the central black holes ([1], [4], [5], [6]). Thus the relativistic jets are most likely powered by accretion disks around the central black holes and formed by the ejection of electron-positron pairs produced in the accretion disks. We investigate the pair production processes in the accretion disks, coupled with the pair ejection by solving the disk structure equations, including

---

1) Research Fellow of the Japan Society for the Promotion of Science

CP510, *The Fifth Compton Symposium,* edited by M. L. McConnell and J. M. Ryan
 1-56396-932-7/00/$17.00

the vertical momentum equation of the electron-positron pairs, within the one-zone approximation, and investigate whether the above jet formation mechanism does operate or not.

CALCULATIONS

Solving the disk structure equations, we assume that i) the disk is steady, axisymmetric and Keplerian one, ii) the disk is optically thin and two temperature; energy exchange between electron-positron and proton are only via the Coulomb coupling, iii) the electron-proton component stays in the disk and the electron-positron pair component is allowed to move vertically without any friction with the former one, iv) advective cooling of the proton can be neglected. We use the cylindrical coordinates (r, ϕ, z) with the z-axis being perpendicular to the disk plane. We do not consider detailed vertical structure of the disk and adopt a one-zone approximation.

Then, the one-zone approximated equations for the electron-positron pairs are written as,

$$\frac{\rho_{\pm} u_z}{H} = 2(\dot{n}_{\rm P} - \dot{n}_{\rm A}) m_{\rm e}, \tag{1}$$

$$\frac{\rho_{\pm} u_z^2}{H} = \frac{p_{\pm}}{H} + \rho_{\pm} \frac{\sigma_{\rm T} F}{c m_{\rm e}} - \rho_{\pm} H \Omega_{\rm K}^2. \tag{2}$$

and

$$\rho_{\pm} u_z \left(U_{\pm} + \frac{p_{\pm}}{\rho_{\pm}} \right) = (\Lambda_{\rm ie} - \Lambda^-) H + 2(\dot{n}_{\rm P} - \dot{n}_{\rm A}) m_{\rm e} \frac{u_z^2}{2} H. \tag{3}$$

Equation (1) is the continuity equation, where $m_{\rm e}$, u_z and $\rho_{\pm}$ are the electron mass, the z-component of the velocity of the pairs and the mass density of the pairs (defined as $\rho_{\pm} \equiv 2 m_{\rm e} n_+$ by using positron number density n_+), respectively; H is the scale height of the disk estimated by $H = c_{\rm s}/\Omega_{\rm K}$, where $c_{\rm s}$ and $\Omega_{\rm K}$ are the isothermal sound speed and the Keplerian frequency, respectively; $\dot{n}_{\rm P}$ and $\dot{n}_{\rm A}$ are the pair production rate and annihilation rate per unit volume, respectively. Equation (2) is the vertical component of the momentum equation, where F, $p_{\pm}$, c, and $\sigma_{\rm T}$ are the radiation flux, the pair pressure, the speed of light, and the Thomson cross section, respectively. Equation (3) is the energy equation, where $U_{\pm}$, $\Lambda_{\rm ie}$ and Λ^- are the internal energy of the electron-positron pairs per unit mass, the energy transfer rate from protons to electron-positron pairs by the Coulomb coupling and the radiative cooling rate per unit volume, respectively.

We can solve these equations by coupling with the vertically integrated equations for the protons:

$$\dot{M} r^2 \Omega_{\rm K} \left[1 - \left(\frac{3 r_{\rm g}}{r} \right)^{1/2} \right] = -2\pi r^2 \tau_{r\phi} 2H, \tag{4}$$

$$\left(r \frac{\partial \Omega_{\rm K}}{\partial r} \right) \tau_{r\phi} H = \Lambda_{\rm ie} H, \tag{5}$$

where $\dot{M}$, $r_{\rm g}$ and $\tau_{r\phi}$ are the mass accretion rate, the Schwarzschild radius and the $r\phi$-component of the viscous-stress tensor, respectively; and the radiative transfer equation:

$$\frac{F}{H} = \Lambda^{-} - 2(\dot{n}_{\rm P} - \dot{n}_{\rm A}) m_{\rm e} c^2. \tag{6}$$

RESULTS

The results are as follows ([7]). There are two solutions throughout the whole disk, when $\dot{M}_*$, the mass accretion rate normalized by the Eddington one, is less than a critical value, $\dot{M}_{*,\rm crit}$. The upper high pair density branch is unphysical, because the proton temperature is higher that the virial temperature ([8]). Hereafter, we consider only the lower pair density branch. When $\dot{M}_*$ is larger than $\dot{M}_{*,\rm crit}$, there is no solution around $r \sim 5 r_{\rm g}$, where pair annihilation and ejection

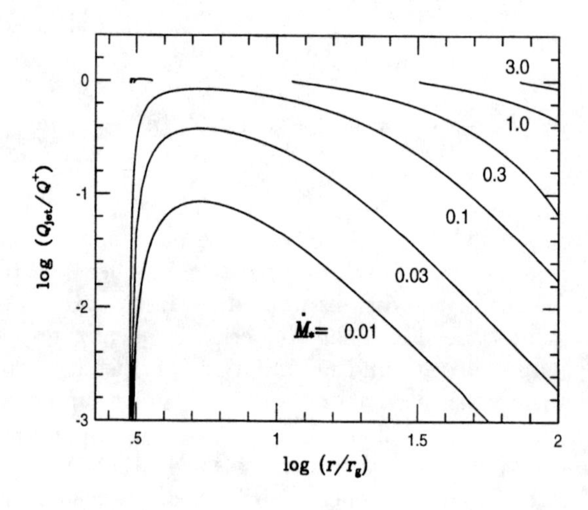

FIGURE 1. Radial distribution of $Q_{\rm jet}/Q^+$, where $Q_{\rm jet}$ is the energy extracted from a disk by pairs and Q^+ is the viscously dissipated energy, for various values of $\dot{M}_*$. Only the low pair density solutions are shown.

cannot cope with pair production, because of the high photon density ([8]). In our calculations, the value of $\dot{M}_{*,\mathrm{crit}}$ is 0.189. This is about ten times as large as the value obtained in the case without pair ejection. Our results show that the effect of pair ejection raises the value of $\dot{M}_{*,\mathrm{crit}}$, because it supplements the shortage of annihilation and balances with pair production. These results agree with those of previous work taking into account the effects of pair escape by giving the escape velocity *a priori* ([9], [10]).

Figure 1 shows the radial distributions of Q_{jet}, the vertically integrated power extracted from the disks by the outflow of pairs, to Q^{+}, the vertically integrated viscous dissipation rate. The latter is defined as the left-hand side of equation (3), and the former is defined as $[\Lambda_{\mathrm{ie}} - \Lambda^{-} + 2(\dot{n}_{\mathrm{P}} - \dot{n}_{\mathrm{A}})m_{\mathrm{e}}c^2]H$, which include the rest mass energy of the ejected pairs. As the accretion rate becomes large, this fraction becomes large, and when $\dot{M}_{*}$ becomes as large as 0.1, the outflowing energy carried by the escaping pairs exceeds the radiated energy. The ratio of the power of outflow to the Eddington luminosity, $4\pi \int_{r_{\mathrm{in}}}^{100r_g} Q_{\mathrm{jet}} r dr / L_{\mathrm{Edd}}$ have its maximum value 0.136 when $\dot{M}_{*} = \dot{M}_{*,\mathrm{crit}}$, so that about half of the accretion power is transferred to the electron-positron outflow power.

Figure 2 shows the ratio of F_{rad}, the force acting on electron-positron pairs by radiative force, to F_{gas}, the force acting on them by their own pressure. The former is defined by the second term in the right-hand side of equation (2), and the latter is defined by the first term in the right-hand side of equation (2). It shows that electron-positron pairs are ejected by the gas pressure gradient force rather than the radiative force.

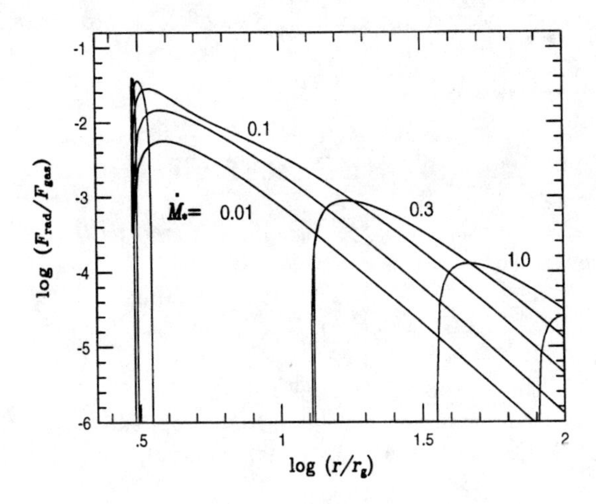

FIGURE 2. Radial distribution of $F_{\mathrm{rad}}/F_{\mathrm{gas}}$ for the low pair density solutions, where F_{rad} is the force acting on electron-positron pairs by their own pressure and F_{gas} is the force acting on them by the radiative force, for various values of $\dot{M}_{*}$.

CONCLUDING REMARKS

Our results show that roughly half of the viscously dissipated energy is transferred to the thermal and the kinetic energies of the ejected pairs when the normalized mass accretion rate $\dot{M}_*$ is as large as 0.1. The other half of the dissipated energy is radiated from the disk. That is, accretion disks can efficiently transform the dissipated energy to the outflow energy. The present pair outflow formation model is plausible to explain the fact that the ejected jets have energy at least comparable to that radiated from the accretion disks. This is owing to another result that the pairs produced in accretion disks are ejected by their own gas pressure rather than the radiative force. This result means that the outflow energy is not supplied by the radiation field, but by the thermal energy of the electron-positron pairs which comes from protons through Coulomb collisions. It solves the difficulty explaining this observational fact in models that the pairs are produced and accelerated outside of the disks (In these models, most of the dissipated energy is inevitably lost by escape of photons).

More detailed formulation including special relativistic effects, the vertical structure of disks and processes above disk will be promising to further understanding the observed properties of jets, for instance, the jet power. The time-dependent behavior of jets is another interesting topic, because it might be related to the radio and X-ray observations of Galactic superluminal sources.

ACKNOWLEDGMENTS

This work was supported in part by a Grant-in-Aid from the Ministry of Education, Science, Sports and Culture of Japan (6293, T.Y.; 10117210, 09640323, F.T.; 09223219, 10117215, M.K.).

REFERENCES

1. Takahara, F., *Towards a Major Atmospheric Cerenkov Detector III*, Tokyo: Universal Academy, 1995, p. 131
2. Sikora, M., Madejski, G., Moderski, R., & Poutanen, J., *ApJ*, **484**, 108 (1997).
3. Kubo, H., Takahashi, T., Madejski, G., Tashiro, M., Inoue, S., & Takahara, F., *ApJ*, **504**, 693 (1998).
4. Takahara, F., these proceedings (1999).
5. Reynolds, C. S., Fabian, A. C., Celotti, A., & Rees, M. J., *MNRAS*, **283**, 873 (1996).
6. Wardle, J. F. C., Homan, D. C., Ojha, R., & Roberts, D. H., *Nature*, **395**, 457 (1998).
7. Yamasaki, T., Kusunose, M., and Takahara, F., *ApJ*, **523**, L21 (1999).
8. Kusunose, M., & Takahara, F., *PASJ*, **40**, 435 (1988).
9. White, T. R., & Lightman, A. P., *ApJ*, **340**, 1024 (1989).
10. Misra, R., & Melia, F., *ApJ*, **449**, 813 (1995).

Measuring Black Hole Masses in X-ray Bright Galactic Nuclei

Insu Yi* and Stephen P. Boughn†

*Korea Institute for Advanced Study, 207-43 Cheongryangri, Dongdaemun, Seoul 130-012, Korea; iyi@kias.re.kr
†Astronomy Department, Haverford College, Haverford, PA 19041, USA; sboughn@haverford.edu

Abstract. X-ray and radio emission from X-ray bright galactic nuclei such as LINERS and low luminosity Seyferts is likely to arise from hot two-temperature, advection-dominated accretion flows. The observed hard X-ray and high frequency radio luminosities could be used to estimate the central supermassive black hole masses, which has been tested in some cases with available mass estimates. Large scale and small scale jet activities appear to be an emission process additional to the accretion flow itself for which the critically rotating black hole spin is likely to be responsible. Detection of X-ray bright, low luminosity galactic nuclei in the gamma-ray range could further test the ADAF models of such sources.

INTRODUCTION

Rotating accretion flows with low radiative efficiencies, advection-dominated accretion flows (ADAFs), have been widely applied to a variety of astrophysical systems including Galactic X-ray transients, active galactic nuclei, and low luminosity galactic nuclei (see e.g. [13] and references therein). The radiative efficiency is low since the dissipated gravitational binding energy or accretion energy is not efficiently radiated away but stored within the flows and advected inward [7,9,5]. Since the radiative cooling processes in the optically thin ADAFs are well understood, the resulting radiation spectra could be used as a diagnostic tool to probe the plasma conditions around black holes. As a consequence, the black hole mass and the accretion rate could be estimated. In ADAFs, due to low radiative cooling efficiencies of electrons and ineffective ion-electron Coulomb coupling, ions are nearly virialized [9] while the electrons' energy balance is mainly determined by the balance between the ion-electron energy transfer and the cooling of electrons.

We introduce the following usual physical scalings; mass $m \equiv M/M_\odot$, radius $r = R/R_s$ ($R_s = 2GM/c^2 = 2.95 \times 10^5 m\ cm$), accretion rate $\dot{m} = \dot{M}/\dot{M}_{Edd}$ ($\dot{M}_{Edd} = L_{Edd}/0.1c^2 = 1.39 \times 10^{18} m\ g/s$ where L_{Edd} is the Eddington luminosity). The equipartition magnetic field's pressure $B^2/8\pi = (1-\beta)P_{gas}$ with $\beta = 0.5$ where β is the ratio of magnetic to total pressure P_{gas} is the gas pressure. The viscosity

CP510, *The Fifth Compton Symposium,* edited by M. L. McConnell and J. M. Ryan

parameter $\alpha \lesssim 1$ and the $\beta \lesssim 0.5$ are the two major uncertainties along with some possible non-Coulomb heating and coupling for electrons. Various radiative cooling channels for electrons give rise to distinct spectral components [9,13]. The synchrotron cooling is responsible for spectral emission components in radio, IR, or optical/UV depending on the mass m and accretion rate $\dot{m}$. The Compton cooling contributes to optical/UV/soft X-ray emission. The bremsstrahlung cooling emits X-ray and soft gamma-ray photons in the KeV range.

ADAFs exist when accretion rates fall below a certain critical rate $\dot{M}_{crit} = \dot{m}_{crit}\dot{M}_{Edd}$. Such a critical rate arises because there exists a maximum accretion rate above which heating could be balanced by radiative cooling without any necessity of advective cooling [12,9]. In the two-temperature ADAFs, the critical rate $\dot{m}_{crit} \approx 0.3\alpha^2$. In ADAFs, the radiative luminosity $L_{ADAF} = \eta_{ADAF}\dot{M}c^2$ where $\eta_{ADAF} = \eta_{eff} \times 0.2\dot{m}\alpha^{-2} \propto \dot{m} \propto \dot{M}/M$ [9], which is much lower than the efficiency for the thin disk $\eta_{eff} \sim 0.1$.

LOW LUMINOSITY X-RAY BRIGHT GALACTIC NUCLEI

ADAFs may exist in relatively dim galactic nuclei including the Galactic center source Sgr A^*. We define the following scaling relevant for galactic nuclei containing supermassive black holes; $m_7 = m/10^7$, $\dot{m}_{-3} = \dot{m}/10^{-3}$, and $R_s = 2GM/c^2 = 3 \times 10^{12} m_7\ cm$. For a thin disk accretion flow, the luminosity $L = \eta\dot{M}c^2$ with the efficiency $\eta = \eta_{eff} \sim 0.1$. A high luminosity from a thin disk coupled with the low emission temperature $T_{disk} \sim 6 \times 10^6 m_7^{-1/5} \dot{m}_{-3}^{3/10} r^{-3/4} K$ is not suitable for low luminosity, hard X-ray sources. The disk luminosity $L_{disk} \sim 1 \times 10^{42} m_7 \dot{m}_{-3} erg/s$ mainly contributes to optical/UV/soft X-ray. The hard X-ray and radio emission require separate emission components such as an optically thin corona and radio jets. The jet power could be related to the black hole spin parameter through $L_{jet} \sim 1 \times 10^{42} \bar{a}^2 \eta_{jet} m_7 \dot{m}_{-3} erg/s$ where $\bar{a}$ is the black hole spin parameter and η_{jet} is the uncertain jet radiative efficiency.

In contrast, ADAFs with equipartition strength $\beta \sim 0.5$ magnetic fields $B \sim 1 \times 10^4 m_7^{-1/2} \dot{m}_{-3}^{1/2} r^{-5/4} G$ can account for radio and X-ray emission in terms of the emission from hot electrons with the electron scattering depth $\tau_{es} \sim 5 \times 10^{-2} \dot{m}_{-3}$ and the characteristic electron temperature $T_e \sim 5 \times 10^9\ K$. ADAFs are most interesting for mass accretion rates below $\dot{m}_{crit} = \dot{M}_{crit}/\dot{M}_{Edd} \approx 0.3\alpha^2 \sim 10^{-3} - 10^{-2}$. Radio emission comes from the synchrotron emission with the characteristic synchrotron emission frequency [14,15],

$$\nu_{sync} \sim 1 \times 10^{12} m_7^{-1/2} \dot{m}_{-3}^{1/2} r^{-5/4} T_{e9}^2\ Hz \tag{1}$$

where $T_{e9} = T_e/10^9 K \sim 5$. The radio luminosity with the inverted spectra

$$L_R \sim \nu L_\nu^{sync} \sim 2 \times 10^{32} x_{M3}^{8/5} T_{e9}^{21/5} m_7^{6/5} \dot{m}_{-3}^{4/5} \nu_{10}^{7/5}\ erg/s \tag{2}$$

where $x_{M3} = x_M/10^3$ is a dimensionless synchrotron self-absorption parameter and $\nu_{10} = \nu/10^{10}\ Hz$. ADAFs' angular radio size is

$$\theta(\nu) \sim 2m_7^{3/5}\dot{m}_{-3}^{2/5}\nu_{10}^{-4/5}(D/10Mpc)^{-1}\ \mu as \quad (3)$$

which suggests that the radio sources at high frequencies $\gtrsim 10GHz$ are very compact and practically point-like.

Hard X-rays in ADAFs are from bremsstrahlung and multiple Compton scattering. For $\dot{m}_{-3} \lesssim 1$, X-ray emission is dominated by bremsstrahlung with $L_x \sim L_x^{brem} \propto m\dot{m}^2$. Since the radio luminosity $L_R \propto m^{8/5}\dot{m}^{6/5}$,

$$L_R \propto mL_x^{3/5}. \quad (4)$$

For $\dot{m}_{-3} \gtrsim 1$, $L_x^{Compt} \propto \dot{m}^{7/5+N}$ with $N \geq 2$ which comes from the average number of electron scatterings. In this case,

$$L_R \propto mL_x^{6/5(N+1)}. \quad (5)$$

Yi & Boughn [14,15] have demonstrated the radio/X-ray luminosity relation for ADAFs using $L_x = L_x(2-10keV)$;

$$L_R \sim 10^{36}m_7(\nu/15GHz)^{7/5}(L_x/10^{40}erg/s)^x\ erg/s \quad (6)$$

where $x \sim 1/5$ for $\dot{m} \lesssim 10^{-3}$ and $x \sim 1/10$ for $\dot{m} > 10^{-3}$ or $L_{R,adv}/L_{x,adv} \propto mL_{x,adv}^{-1}$.

ADAFs are likely to drive jets/outflows [8,1]. If jets are powered by black hole's spin energy,

$$L_{R,jet}/L_{R,adv} \sim 4 \times 10^5\bar{a}^2\eta_{jet}m_7^{-1/5}\dot{m}_{-3}^{1/5} \quad (7)$$

and $L_{R,jet}/L_{x,adv} \propto \bar{a}^2 mL_{x,adv}^{-1}$ are expected. That is, $L_{R,jet} \gg L_{R,adv}$ for $\bar{a} \gg 2\times10^{-3}\eta_{jet}^{-1/2}$. If this power contributes to the radio emission, the observed emission inevitably becomes large scale jets/outflows with much higher luminosites. Such emission components should be able to be distinguished from the compact ADAF radio sources. Characteristic ADAF emission spectra are determined primarily by $\dot{m}$ and weakly affected by the black hole mass M. Any combinations among L_x, L_R, and M give useful information on the nature of emission from galactic nuclei.

Galactic Center Source Sgr A^*: Galactic center radio source Sgr A^* is a prime candidate for an ADAF [12,10,6]. Several spectral fitting studies based on the observed M and the estimated $\dot{m}$ have shown that the observed emission seen from radio to hard X-ray is indeed well accounted for by an ADAF [6,10]. *NGC 4258:* NGC 4258 almost certainly has a central black hole with mass $M = 3.5 \pm 0.1 \times 10^7 M_\odot$. Although a good spectral fit from radio to X-rays has been suggested, the non-detection of this source at a radio frequency of 22 GHz has put a severe constraint on the possible ADAF models. The non-detection of the core at 22 GHz could imply that $\dot{m} \sim 10^{-2}$ and the outer extent of the ADAF $r \sim 30$ [4].

Such a constraint is highly suspect due to a possibility of strong variabilities in low luminosity galactic nuclei [11].

X-ray Bright Galactic Nuclei: Yi & Boughn [14,15] have applied the ADAF model to a small sample of X-ray bright galactic nuclei with black hole mass estimates. Since ADAFs are most relevant for low luminosity, hard X-ray sources, faint, hard X-ray galactic nuclei are likely candidates for ADAFs [3,14,15]. ADAFs are likely for massive black holes accreting at accretion rates $\lesssim 10^{-2} \dot{M}_{Edd}$ for which hard X-ray and inverted spectrum radio emission from compact core are expected. Black hole masses could be estimated using observed radio/X-ray luminosities (i.e. eq. (6)). ADAF sources could, however, contain jets/outflows which can contribute to radio emission. Depending on the level of radio activities and existence of extended radio emission features, galactic sources could be classified [14,15].

X-ray bright galactic nuclei (XBGN) are defined as galactic nuclei with X-ray luminosities in the range $10^{40} \lesssim L_x \lesssim 10^{42} erg/s$ which is sub-luminous compared with the more powerful active galactic nuclei (AGN). Most of XBGN are expected to overlap with emission line galaxies with $L_x \sim 10^{39} - 10^{42} erg/s$. However, some of the low luminosity Seyferts with $L_x \gtrsim 10^{42} erg/s$ could belong to the XBGN. For $L_x \sim 10^{41} erg/s$,

$$L_R \sim 4 \times 10^{36} (M_{BH}/3 \times 10^7 M_\odot) \; erg/s \tag{8}$$

at 20 GHz with the characteristic inverted radio spectrum $I_\nu \propto \nu^{2/5}$. If X-ray and radio are indeed from ADAFs, the black hole masses can be estimated.

Yi and Boughn [14,15] proposed the source classification based on the known black hole masses and the ADAF radio/X-ray luminosity relation. Adopting Sgr A*, NGC 4258, NGC 1068, NGC 1316, NGC 4261, and NGC 4594 as fiducial sources, XBGN are classified into radio-loud XBGN and radio-quiet XBGN. The former show that the observed radio luminosity $L_{R,obs} \sim L_{R,jet} \gg L_{R,adv}$ where $L_{R,jet}$ and $L_{R,adv}$ are the expected radio jet luminosity and ADAF radio luminosity, respectively. These sources are expected to show extended radio emission, unlikely to have strongly inverted radio spectra, and may have compact ADAF radio emission from compact cores separate from the extended emission components. The latter show that $L_{R,obs} \sim L_{R,adv}$ and that the dominating emission components are compact cores with inverted spectra.

ADAFs are prone to outflows or jets [7,8,1] although a self-consistent inflow/outflow solution has not been found yet (see [1] and references therein). It remains to be seen if a self-consistent inflow/outflow solution can account for compact and extended jet-like emission components in XBGN. The large radio powers in extended radio sources are clearly distinguished from compact radio cares, which suggests that the core radio components could indeed be ADAF sources and the radio lobes are powered by spinning black holes [12,15].

CONCLUSIONS

A simple ADAF radio/X-ray luminosity relation could ccount for the observed luminosity ratios in some XBGN provided that large scale jet powers are removed. The comparison between the observed ratios and the predicted ratios results in some quite reliable black holes mass estimates [15]. When the dynamical mass estimates are available, the ADAF estimates are consistent with the existing ones.

ADAFs emission could be highly variable. The issue of steady vs. non-steady ADAFs is directly related to the observed variabilities in ADAF candidate sources which show occasional non-detections. A clear understanding of the variabilities in XBGN is necessary in order to verify the existence of ADAFs in these sources.

The driving mechanism for large scale powerful jets remains yet to be understood. The small scale outflows perhaps with less collimation could be distinct from the large scale jets. It is tempting to speculate that the large scale jets derive their powers mainly from spinning black holes while the small scale outflows are driven by high entropy ADAFs.

The hard X-ray spectra coupled with the inverted radio spectra are a clear evidence for the ADAFs. The existence of ADAFs in low luminosity XBGN could be further tested by detecting gamma-rays near $\sim 1MeV$. The ADAFs predict a cutoff from bremsstrahlung emission. In contrast, luminous X-ray binaries or medium mass black hole sources would have much softer spectra truncated well below $\sim 1MeV$ (e.g. [2]).

REFERENCES

1. Blandford, R. D. & Begelman, M., *MNRAS* **303**, L1 (1999).
2. Colbert, E. J. M. & Mushotzky, R. F., *ApJ* **519**, 89 (1999).
3. Fabian, A. C. & Rees, M. J., *MNRAS* **277**, L5 (1995).
4. Gammie, C. F., Narayan, R., & Blandford, R. D., *ApJ* **516**, 177 (1999).
5. Ichimaru, S., *ApJ* **214**, 840 (1977).
6. Narayan, R. et al., *ApJ* **492**, 554 (1998).
7. Narayan, R. & Yi, I., *ApJ* **428**, L13 (1994).
8. Narayan, R. & Yi, I., *ApJ* **444**, 231 (1995).
9. Narayan, R. & Yi, I., *ApJ* **452**, 710 (1995).
10. Narayan, R., Yi, I., & Mahadevan, R., *Nature* **374**, 623 (1995).
11. Ptak, A. et al., *ApJ* **501**, L37 (1998).
12. Rees, M. J., Begelman, M. C., Blandford, R. D., & Phinney, E. S., *Nature* **295**, 17 (1982).
13. Yi, I., in *Astrophysical Discs-An EC Summer School*, ed. J. A. Sellwood and J. Goodman, Astronomical Society of the Pacific Conference Series, 1999, Vol. 160, p279.
14. Yi I. and Boughn S. P., *ApJ* **499**, 198 (1998).
15. Yi I. and Boughn S. P., *ApJ* **515**, 576 (1999).

On the X-ray heated skin of Accretion Disks

Sergei Nayakshin

Laboratory for High Energy Astrophysics, NASA Goddard Space Flight Center.

Abstract. We present a simple analytical formula for the Thomson depth of the X-ray heated skin of accretion disks valid at any radius and for a broad range of spectral indices of the incident X-rays, arbitrary geometry of the X-ray source, accretion rates and black hole masses. We expect that this formula may find useful applications in studies of geometry of the inner part of accretion flows around compact objects, and in several other astrophysically important problems, such as the recently observed X-ray "Baldwin" effect (i.e., monotonic decrease of Fe line's equivalent width with the X-ray luminosity of AGN), the problem of missing Lyman edge in AGN, and line and continuum variability studies in accretion disks around compact objects.

Introduction

X-ray illumination of an accretion disk surface is a problem of general astrophysical interest. Since the X-ray heating of the disk atmosphere changes energy and ionization balances there, the spectra emitted by X-ray illuminated accretion disks in any wavelength may be quite different from those resulting from non-illuminated disks. It has been known for many years (e.g., Basko, Sunyaev, & Titarchuk 1974) that X-ray illumination leads to formation of a hot (and often completely ionized) X-ray "skin" above the illuminated material. Unfortunately, due to numerical difficulties, most of the previous studies of X-ray illumination had to rely on a constant density assumption for the illuminated gas (e.g., Ross & Fabian 1993; Życki et al. 1994; Ross, Fabian & Brandt 1996), in which case the completely ionized skin forms only for very large ionization parameters (e.g., Ross, Fabian & Young 1999).

Recently, Nayakshin, Kazanas & Kallman (1999; hereafter paper I) have shown that if the assumption of the constant density is relaxed, then the temperature and ionization structure of the illuminated material is determined by the thermal ionization instability, and that the X-ray heated skin always forms on the top of the disk (see also Raymond 1993; Ko & Kallman 1994; Różańska & Czerny 1996). Due to the presence of the hot skin, the resulting reflected spectra are quite different from those obtained with the constant density assumption. In this paper, we present an

CP510, *The Fifth Compton Symposium*, edited by M. L. McConnell and J. M. Ryan

approximate expression for the Thomson depth of the hot skin which allows one to qualitatively understand effects of the thermal instability on the reflected spectra.

Approximate hydrostatic balance

We assume that X-rays are incident on the surface of an accretion disk whose structure is given by the standard accretion disk theory. As explained in paper I, even Fe atoms are completely ionized in the X-ray skin if the X-ray flux, F_x, is comparable with or larger than the disk intrinsic flux, F_d, and if the illuminating X-rays have relatively hard spectra, i.e., photon index $\Gamma \lesssim 2$. Under those conditions, the Compton temperature, T_c, is close to $\sim$ few keV. The Thomson depth of the hot layer, τ_1, is obtained by integrating $d\tau_1 = \sigma_T n_e(z) dz$, from $z = z_b$ to infinity, where $n_e(z)$ is the electron density; z is the vertical coordinate; $z_b \simeq H$ gives the location of the bottom of the ionized skin; and H is the disk scale height. Note that a simpler estimate of τ_1 can be obtained if one assumes that the gas temperature is equal to the Compton one, and that the density law follows a Gaussian law (see Kallman & White 1989).

One can show (Nayakshin 1999, paper II hereafter) that, for incident angles not too far from normal, the vertical integration of $d\tau_1$ leads to

$$\tau_1 \simeq 6.9\, T_1 \frac{F_x}{F_d} G^{1/2}(r)\, \dot{m}\,(1-f) \tag{1}$$

$$G(r) \equiv \frac{2^{16}}{27} [1-(3/r)^{1/2}]^2\, r^{-3}\,,$$

where $T_1 = kT_c/1$ keV, $r = R/R_s$, $R_s = 2GM/c^2$ is the Schwarzschild radius, $\dot{m}$ is the dimensionless accretion rate ($\dot{m} = 1$ corresponds to Eddington luminosity for the accretion disk) and $0 \leq f < 1$ is the coronal dissipation parameter (e.g., Svensson & Zdziarski 1994). The maximum of function $G(r)$ occurs at $r = 16/3$ where it is equal to unity. This treatment tacitly assumes that the ionizing radiation does not change with the depth into the illuminated material. The following simple modification takes into account the decrease in the value of τ_1 due to spectral reprocessing (paper II):

$$\tau_h = \frac{\tau_1}{1+\tau_1/3}\,, \tag{2}$$

A comparison of our analytical results with values of τ_h numerically calculated in paper I shows that the deviation of our approximate expression from "exact" results is less than $\sim 20\%$ (see Fig. 1 in paper II).

"Lamp Post" Model

As an application of our methods, we choose to analyze the model where the X-ray source is located at some height h_x above the black hole (the "lamp post model" hereafter; Reynolds & Begelman 1997; Reynolds et al. 1999). In this paper we will not discuss the region within the innermost stable orbit for the reason that the properties of the accretion flow there are not well constrained. Further, we will assume a non-rotating black hole and that all the X-rays are produced within the central source and neglect all relativistic effects. Let us define η_x as the ratio of the total X-ray luminosity to the integrated disk luminosity of the source, i.e., $\eta_x \equiv L_x/L_d$. Using equation (1), one obtains

$$\tau_1 \simeq 27.2 \frac{h_x}{R_s} \eta_x T_1 \left[1 + (h_x/R)^2\right]^{-3/2} r^{-3/2} \dot{m} \tag{3}$$

As discussed in paper I, the local Compton temperature depends on the cosine of the X-ray incidence angle, μ_i, and the ratio $F_{\rm x}/F_{\rm d}$ approximately as:

$$T_c \simeq T_x \left[1 + \mu_i \sqrt{3} \frac{F_{\rm x} + F_{\rm d}}{F_{\rm x}}\right]^{-1} . \tag{4}$$

Figure (1) shows the Thomson depth of the skin as a function of radius for several values of $\dot{m}$. The parameters in Figure (1) are chosen to be: (a) $\Gamma = 1.9$, $\eta_x = 1$ ("X-ray strong" case); (b) $\Gamma = 1.9$, $\eta_x = 0.1$ ("X-ray weak" case), and (c) $\Gamma = 2.3$, $\eta_x = 1$, and $h_x = 6R_s$ for all cases.

Figure (2) shows the angle-averaged reflected spectra and temperature profiles of the hot layer for the three cases just considered and the accretion rate equal to the Eddington value ($\dot{m} = 1$) for $r = 10$. For hard X-ray spectra, the local emissivity of the line is negligible when $\tau_{\rm h} \gtrsim 1$. Thus, the strength of the iron line will be decreasing with $\dot{m}$ for the X-ray strong case (Fig. 1a) when $L_x \gtrsim$ few percent of the Eddington value. The skin is thickest for smaller radii, and therefore the broad iron line component will decrease first. The narrow line component (emitted farther away from the black hole) will also decrease with X-ray luminosity, but considerably slower than the broad line. This is qualitatively consistent with Figure (3) of Nandra et al. (1997), suggesting an explanation to the X-ray Baldwin effect.

The X-ray weak case (Figs. 1b & 2b) is different in two respects. Firstly, the Thomson depth of the hot skin is smaller at a given accretion rate compared with the X-ray strong case. Most importantly, however, the skin is not "that hot". Namely, the skin temperature is only ~ 0.3 keV in the inner disk. For that reason, it turns out that the iron line centroid energy is close to 6.7 keV, and a very deep absorption edge appears. This edge is in fact much stronger than the one resulting from a neutral material. Hence it is possible that this relatively cold X-ray heated skin can be unambiguously detected in spectra of real AGN. Similarly, the soft incident X-ray spectrum leads to a relatively cool skin because $T_x \sim 1$ keV only (see Figs. 1c & 2c). As in the case $\eta_x = 0.1$, a strong absorption edge is observed. In addition, the 6.7 keV Fe line is stronger, with EW of 65 eV. Therefore, such skin may also be detectable if it exists.

Comparing the value of $\tau_{\rm h}$ resulting from analytical formulae with those seen in Fig. (2b), one notes that deviations are as large as $\sim 50\%$. This relatively large

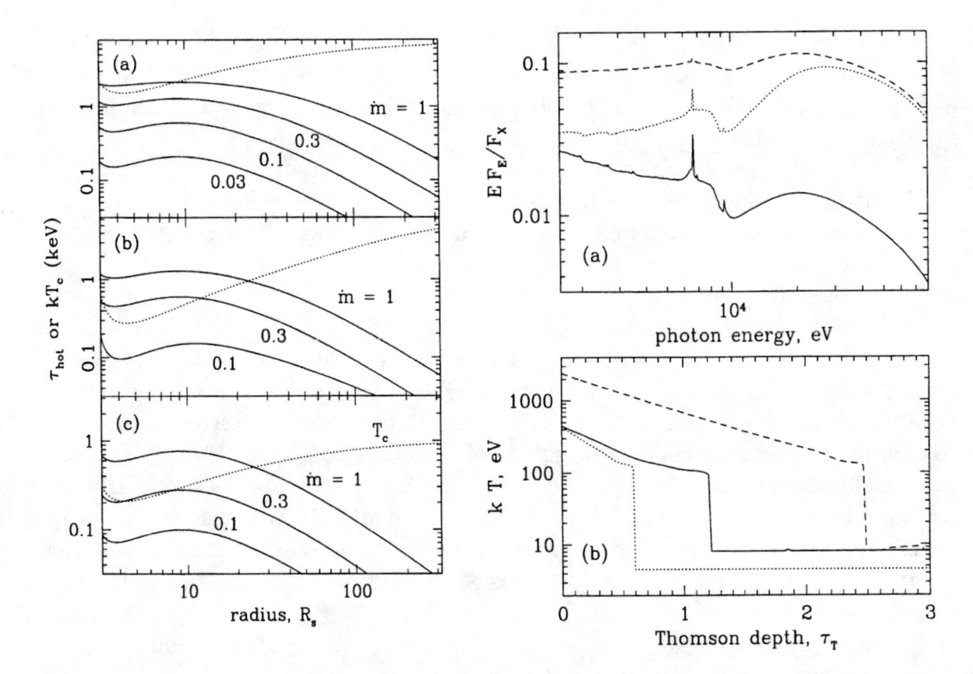

FIGURE 1. (LEFT) Thomson depth (solid curves) of the hot skin as a function of radius for different accretion rates (whose values are shown next to corresponding curves). The Compton temperature is shown by the dotted curve.

FIGURE 2. (RIGHT) Reflected spectra (a) and temperature profile (b) computed for the lamp post geometry at $r = 10$ with accretion rate $\dot{m} = 1$. The dashed, dotted and solid curves correspond to models shown in Fig. (1) [a], [b] and [c], correspondingly. The solid curve in [a] was scaled down by factor of 2 for clarity.

deviations exemplify the fact that our equations (1) and (2) are good approximations only to cases with strong X-ray flux and hard incident spectra (i.e.,$F_x \gtrsim F_d$ and $\Gamma \lesssim 2$). When one of these two conditions is not satisfied, our results can only be used as an order of magnitude estimate of τ_h.

Summary

We derived an approximate expression for the Thomson depth of the hot completely ionized skin on the top of an accretion disk illuminated by X-rays. Our results are weakly dependent on the a priori unknown α-viscosity parameter (because it only enters through the boundary conditions). Under certain conditions ($F_x \gtrsim F_d$ and $\Gamma \lesssim 2$), the X-ray heated skin may act as a perfect mirror for photons with energies below $\sim$ 30 keV (see paper I). Because the X-ray heated skin is thickest in the inner part of an accretion disk, the observed absence or deficit of the relativistically broadened line and other reprocessing features in some systems (e.g., Życki , Done & Smith 1997, 1998), which was interpreted as a possible evidence for a disruption of the cold disk for small radii, may equally well mean that the "cold" disk is still present up to the innermost radius, but the skin effectively shields it from the X-ray illuminating flux. Further, the hot skin in the inner part of accretion disk also explains why iron Kα Equivalent Width decreases with luminosity in AGN (see Nandra 1997).

It is interesting, however, that the presence of the skin becomes apparent in systems that have $F_x \lesssim F_d$ or $\Gamma \gtrsim 2$ (i.e., many of the NLS1 Galaxies), because the Fe atoms may be not completely stripped of their electrons and thus produce strong ionized edges and lines. We believe these predictions should be testable observationally with current X-ray missions such as Chandra, Astro-E and XMM. Also note that for a patchy corona model of accretion disks (e.g., Haardt, Maraschi & Ghisellini 1994), the Thomson depth of the hot skin is always larger than the one found here, since the ratio F_x/F_d is larger. Finally, the presence of the ionized skin is important not only for the X-rays, but for other wavelengths as well (e.g., in studies of Lyman edge of accretion disks, correlation of optical/UV light curves with X-rays).

The author acknowledges support from NAS/NRC Associateship and many useful discussions with D. Kazanas and T. Kallman.

REFERENCES

Basko, M.M., Sunyaev, R.A., & Titarchuk, L.G. 1974, A&A, 31, 249
Kallman, T.R., & White, N. E. 1989, *ApJ*, 341, 955.
Ko, Y-K, & Kallman, T.R. 1994, *ApJ*, 431, 273.
Krolik, J.H., McKee, C.F., & Tarter, C.B. 1981, *ApJ*, 249, 422.
Lightman, A.P., & White, T.R. 1988, *ApJ*, 335, 57.
Nandra, K., et al. 1997, *ApJL*, 488, L91.
Nayakshin, S., Kazanas, D., & Kallman, T. 1999 (paper I), submitted to ApJ (astro-ph/9909359).
Nayakshin, S. 1999 (paper II), to appear in ApJ
Raymond, J.C. 1993, *ApJ*, 412, 267.
Reynolds, C.S., & Begelman, M.C. 1997, *ApJ*, 488, 109.
Reynolds, C.S., Young, A.J., Begelman, M.C., & Fabian, A.C. 1999, *ApJ*, 514, 164.
Ross, R.R., & Fabian, A.C. 1993, *MNRAS*, 261, 74.
Ross, R.R., Fabian, A.C., & Brandt, W.N. 1996, *MNRAS*, 278, 1082.
Ross, R.R., Fabian, A.C., & Young, A.J. 1999, *MNRAS*, 306, 461.
Różańska , A., & Czerny, P.T. 1996, Acta Astron., 46, 233
Svensson, R. & Zdziarski, A. A. 1994, *ApJ*, 436, 599.
Życki , P.T., Krolik, J.H., Zdziarski, A.A., & Kallman, T.R. 1994, *ApJ*, 437, 597.
Życki , P.T., Done, C., & Smith, D.A. 1997, *ApJL*, 488, L113.
Życki , P.T., Done, C., & Smith, D.A. 1997, *ApJL*, 496, L25.

CLUSTERS OF GALAXIES

Hard X-Ray Observations of a Sample of Clusters of Galaxies

L. Bassani[1], E. Caroli[1], G. DiCocco[1], G. Malaguti[1], J.B. Stephen[1]
A. Malizia[(2,3)], M.J. Westmore[2], A.J.Dean[2]

(1) Istituto TeSRE/CNR, Via Gobetti 101, Bologna, ITALY
(2) Dept. of Physics, The University, Southampton, UK
(3) BeppoSAX SDC, Rome, ITALY

Abstract. The BeppoSAX public archive has been searched for hard X-ray data obtained with the PDS instruments on observations of clusters of Galaxies . Of the fourteen objects analysed with the standard procedure, eight are detected with a significance > 5σ in the 13-100 keV range, while for the remaining objects a marginal detection (3 < σ < 5) can be claimed. A preliminary spectral analysis of source data having more than 10σ detection indicates that at least in some cases a second component is present besides the thermal bremsstrahlung emission that dominates at low energy. However, the lack of imaging capability in the PDS prevents discrimination between an origin of this radiation in terms of an AGN or diffuse cluster emission. The potential of the INTEGRAL instruments to detect this high energy emission in Clusters of Galaxies is briefly discussed.

INTRODUCTION

To date, observations of high energy (> 10 keV) emission from clusters of galaxies have been limited to only a few objects and any positive results have not always been confirmed by subsequent observations. For example, the Coma cluster detections with the POKER (1) and MIFRASO (2) balloon-borne instruments were followed by negative results with HEAO A-4 (3) and OSSE (4) respectively. The better sensitivity of new high energy instruments, such as the BeppoSAX PDS, has recently allowed an unambiguous detection to be made of the high energy emission from Coma (5,6), but whether this is typical of clusters in general has still to be determined.

In order to investigate this issue, and also in view of future high energy missions such as INTEGRAL, we have analysed the PDS data from a number of clusters of galaxies observed by BeppoSAX using the publicly available data. The data reduction was performed using the SAXDAS software package using standard screening criteria.

CP510, *The Fifth Compton Symposium,* edited by M. L. McConnell and J. M. Ryan

TABLE 1. Data Analysis Results

Name	Observation Date	Integration Time (s)	Count rate (counts/sec)	20-100 keV Flux (ergs cm^{-2} s^{-1})
Abell 33	23 Nov 96	3.35 10^4	0.1200 ± 3.70 10^{-2}	< 2.30 10^{-11}
Perseus	19 Sep 96	3.84 10^4	1.3790 ± 2.80 10^{-2}	4.85 10^{-11}
Abell 3266	24 Mar 98	3.19 10^4	0.2433 ± 3.94 10^{-2}	1.10 10^{-11}
Abell 496	05 Mar 98	4.17 10^4	0.1165 ± 3.33 10^{-2}	< 1.20 10^{-11}
Virgo	14 Jul 96	1.21 10^4	0.2668 ± 6.24 10^{-2}	2.58 10^{-11}
Coma-1	28 Dec 97	3.17 10^4	0.9400 ± 3.95 10^{-2}	4.03 10^{-11}
Coma-2	19 Jan 98	1.11 10^4	0.9739 ± 6.80 10^{-2}	4.40 10^{-11}
Coma Total	-	4.28 10^4	0.9348 ± 3.40 10^{-2}	4.10 10^{-11}
Abell 2029	4 Feb 98	1.79 10^4	0.1850 ± 5.10 10^{-2}	< 3.50 10^{-11}
Abell 2142	26 Aug 97	4.55 10^4	0.4232 ± 3.22 10^{-2}	2.11 10^{-11}
Abell 3627	01 Mar 97	2.42 10^4	0.3258 ± 4.60 10^{-2}	1.52 10^{-11}
Abell 2163-1	06 Feb 98	2.05 10^4	0.1907 ± 4.95 10^{-2}	0.70 10^{-11}
Abell 2163-2	21 Feb 98	2.25 10^4	0.2000 ± 4.74 10^{-2}	1.51 10^{-11}
Abell 2163 (Tot)	-	4.30 10^4	0.2057 ± 3.33 10^{-2}	1.23 10^{-11}
Abell 2199	21 Apr 97	4.29 10^4	0.1485 ± 3.42 10^{-2}	1.15 10^{-11}
Abell 2256	11 Feb 98	2.38 10^4	0.1336 ± 4.52 10^{-2}	1.00 10^{-11}
Abell 2319	16 May 97	1.96 10^4	0.5924 ± 4.86 10^{-2}	7.89 10^{-11}
Abell 3667	13 May 98	3.65 10^4	0.2100 ± 3.47 10^{-2}	2.40 10^{-11}

RESULTS

Table 1 shows the results obtained of this analysis, indicating that high energy emission (> 13 keV) is detected at greater than 5σ significance from 8 of the 14 sources, strongly indicating both that radiation at these energies is typical of this class of objects and that we now have the instrumental sensitivity to be able to measure it. The only major drawback of the SAX/PDS is the angular resolution which, at 1.3° FWHM, is not sufficient to exclude contamination from other sources in the field of view. Indeed, by checking the BeppoSAX MECS images as well as searching in public archives for possible 2-10 keV emitting sources (i.e. those most likely to contaminate the PDS at high energies) within the instrument field of view, we find that this possibility exists for 7 of the 14 clusters: Perseus, Abell 33, 2029, 2142, 3627, 3666 and Virgo. In most cases the candidate source of contamination is an AGN, likely to be part of the cluster itself.

Figure 1 shows the spectra obtained by PDS for the Coma cluster and Abell 2142 regions using a bremsstrahlung model fit to the data. In both cases, the bremsstrahlung temperature was fixed at a value in line with previous 2-10 keV observations (8.2 keV for Coma (6) and 8.8 keV for Abell 2142 (7)). It is evident that not only is the source visible in each case, but that there is a high energy excess above the thermal emission.

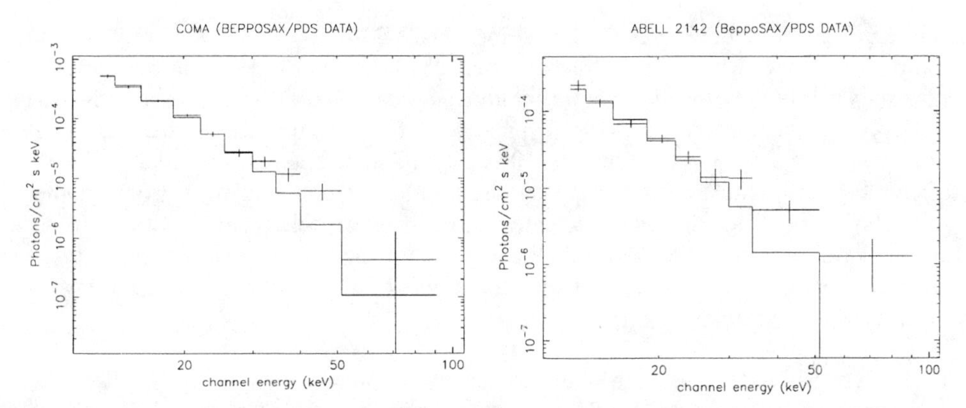

FIGURE 1. The spectra of the COMA *(left)* and Abell 2142 clusters as measured with SAX/PDS.

In table 2 we show the results of the spectral analysis of the four sources with PDS detections above 10σ significance. The 4 models correspond to:

Model I	a bremsstrahlung fit with the temperature fixed at a value corresponding to previous 2-10 keV measurements (6,7,8).
Model II	as above but with the addition of a power law component
Model III	as Model I but allowing the temperature to be a free parameter
Model IV	as Model III but with the addition of a power law component

TABLE 2. Spectral Analysis Results for 4 Models

		Model I	Model II	Model III	Model IV
Perseus	kT	6.5	6.5	7.4±0.5	6.6±0.7
	α	-	$1.7^{+0.9}_{-0.7}$	-	1.5±1.3
	χ^2/dof	50.3/13	10.5/11	30.3/12	10.4/10
Coma	kT	8.2	8.2	$9.4^{+1.0}_{-0.9}$	$7.9^{+1.9}_{-7.0}$
	α	-	$3.0^{+0.6}_{-2.1}$	-	2.8*
	χ^2/dof	27.4/13	18.1/11	21.5/12	18.0/10
Abell 2142	kT	8.8	8.8	$13.5^{+5.7}_{-3.4}$	-
	α	-	1.6*	-	-
	χ^2/dof	16.3/13	5.8/11	10.1/12	-
Abell 2319	kT	9.2	9.2	$10.2^{+2.9}_{-2.1}$	-
	α	-	1.7*	-	-
	χ^2/dof	5.6/13	4.8/11	5.1/12	-

* = not constrained

It is clear that for Perseus, Coma and Abell 2142 the data require an extra component to the thermal fit, while in Abell 2319 the high energy emission can be explained purely in terms of the bremsstrahlung spectrum (see also (9) for a discussion of BeppoSAX data of Abell 2319). Typically we find that when a hard tail exists, it contributes about 50% to the total 20 -100 keV emission. The hard tail evident from 3 of the 4 clusters can be interpreted in terms either of contamination from another source(s) in the field of view or to interactions between relativistic electrons (6) or cosmic rays (10) and the CMB; in the latter case the emission is expected to be diffuse. In order to discriminate between these possibilities there are two requirements for future observations:

1) High sensitivity so as to be able to determine precisely the shape of the non-thermal excess.
2) Imaging capability so as to be able to identify the precise location of the high energy emission.

The INTEGRAL mission, meeting both these requirements, and also having a wide field of view will be well suited to this type of observation, even in terms of serendipitous source studies. We estimate, for example that 4 or 5 of the 14 sources listed in Table 1 will be visible at more than 5σ in a 10^5 second observation with IBIS. The case of Abell 2142 is representative of this type of object: apart from the cluster emission the MECS image (see figure 2, *left*) shows a source at position (RA 15^h 59^m $19^s.2$, DEC +27° 03'18''.8) identified as a Seyfert galaxy (QSO 1552+272). An even more spectacular example of possible contamination at high energies is given by Abell 3667 where a galaxy at (RA 20^h 11^m $56^s.5$, DEC -56° 43'54''.7) is quite bright in the BeppoSAX MECS image (figure 2 *right*); although this galaxy is not classified as active, its brightness at X-ray energies is suspicious and it could be a major source of high energy photons.

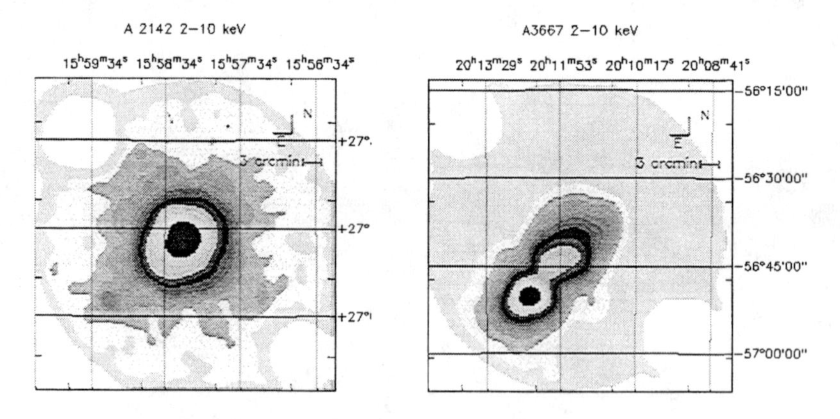

FIGURE 2. The BeppoSAX MECS images of the Abell 2142 region *(left)*, showing a possible confusion source at the lower left. The corresponding image of the Abell 3667 cluster *(right)* reveals that the cluster emission (*lower left source*) is the brightest in this energy range

A deeper analysis of the MECS spectra of these two sources could provide further information as to their relative intensities at high energies. This would still, however, rely on an extrapolation of the spectra, and clearly what is required for a definitive answer to the source of the high energy emission is a deep INTEGRAL exposure of these two sky regions.

REFERENCES

1. Bazzano, A. *et al*, Astrophys. J. **279**, 515 (1984)
2. Bazzano, A. *et al*, Astrophys. J. **362**, L51 (1990)
3. Rephaeli, Y., and Gruber, D., Astrophys. J. **333**, 133 (1988)
4. Rephaeli, Y., Ulmer, M. and Gruber, D., Astrophys. J. **429**, 554 (1994)
5. Rephaeli, Y., Gruber, D. and Blanco, P., Astrophys. J. **511**, L21 (1999)
6. Fusco Femiano, R. *et al*, Astrophys. J., **513**, L21 (1999)
7. Markevitch, M. et al, Astrophys. J. **503**, 77 (1998)
8. Arnaud, K.A. *et al*, Astrophys. J. **436,** L67 (1994)
9. Kaastra, J.S. *et al*, Astrophys. J. Letts. **519** L119 (1999)
10. Lieu *et al*, Astrophys. J. Letts **510**, L25 (1999)

Hard X-Ray Emission from Clusters of Galaxies

V.A.Dogiel

P.N.Lebedev Physical Institute, Leninskii pr. 53, 117924 Moscow, Russia

Abstract. We analyze the origin of the flux of hard X-rays from the Coma cluster observed with the Beppo-SAX telescope. It is often assumed that this flux is produced by nonthermal particles in-situ accelerated in the Coma halo. We show, however, that in the case of in-situ acceleration the hard X-ray flux may be generated by thermal particles whose spectrum is distorted by the acceleration. This mimic the effect of two-temperature distribution of the bremsstrahlung X-ray emission. A power-law non-thermal spectrum is formed at energies much higher than the energy range observed by Beppo-SAX.

I INTRODUCTION

The EUV and radioemission of the Coma cluster coming from an extended halo (size $\sim$ 1 Mpc) are generated by nonthermal electrons which are in-situ accelerated in the halo medium (see e.g. [1,3]). The energy of these electrons is about hundred MeV – several GeV. Recently the Beppo-SAX telescope found a hard X-ray flux from Coma whose intensity is higher than expected from the thermal distribution [2]. Enßlin et al. [1] assumed that this emission is due to bremsstrahlung losses of suprathermal electrons accelerated by turbulence within the medium. This means that electrons with energies slightly higher than kT are accelerated in the Coma halo.

The other explanation was analyzed by Fusco-Femiano et al. [2]. They attempted to fit their results with two temperatures, but ruled out this interpretation, since one of these temperatures was 8 keV, but the other was unbelievably high, $>$ 40 keV.

Nevertheless we show below that the thermal interpretation of the Beppo-SAX results may be correct. The point is that the effect of acceleration is not limited by generation of the nonthermal spectrum only. This process changes significantly the equilibrium distribution of thermal particles. We show that the hard X-ray emission may be thermal in that sense that it is produced by electrons whose spectrum is formed by Coulomb collisions (not by acceleration processes).

CP510, *The Fifth Compton Symposium,* edited by M. L. McConnell and J. M. Ryan

II THE EQUATION FOR ACCELERATED PARTICLES

In the case of in-situ acceleration in the halo the flux of accelerated particles is determined by their injection from the thermal pool by Coulomb collisions. Since the rate of injection is fixed by Coulomb collisions establishing the thermal spectrum of background particles, we have a nice possibility to estimate parameters of acceleration mechanism needed to produce the number of nonthermal particles in the halo. A natural source for particle acceleration in cosmic space is an exchange of energy between particles and waves which are excited due to various kinds of instabilities. This leads to a slow stochastic acceleration which is described as diffusion in the momentum space. For many cases the diffusion coefficient $\alpha(p)$ has the form (see e.g. [5])

$$\alpha(p) = \alpha_0 p^2 \,. \tag{1}$$

To estimate the number of the accelerated particles we have to solve an equation describing the background spectrum as well as the spectrum of accelerated particles. If the acceleration takes place everywhere in the volume and we can neglect processes of particle spatial propagation, then the equation for background and accelerated particles converges to

$$\frac{df}{d\tau} - \frac{1}{u^2}\frac{\partial}{\partial u}\left(\left(\frac{1}{u} + u^2\alpha(u)\right)\frac{\partial f}{\partial u} + f\right) = 0 \,, \tag{2}$$

where $\alpha(u)$ describes processes of stochastic particle acceleration (momentum diffusion) τ is the dimensionless time, $\tau = t \cdot \nu_0$, ν_0 is the characteristic frequency of Coulomb collisions, and u is the dimensionless particle velocity

$$u = \frac{v}{\sqrt{kT/m}} = \sqrt{\frac{2E}{kT}} \,. \tag{3}$$

Here we assume (that is rather naturally) that only a small part of background particles is accelerated, i.e. $\alpha_0 \ll \nu_0$.

Thus the background particles in our case are under the influence of acceleration gains and of ionization losses. Their energy variations at $E > kT$ are given by

$$\frac{dE}{dt} = \alpha_0 E - \nu_0 E \left(\frac{kT}{E}\right)^{3/2} \tag{4}$$

It equals zero at the energy

$$E_{inj} = kT \left(\frac{\nu_0}{\alpha_0}\right)^{2/3} \tag{5}$$

Only above this energy ($E > E_{inj}$) where $dE/dt > 0$ the acceleration increases the particle energy continuously in time. As a result a flux of run-away particles from collisional region into the region of nonthermal particles is formed. Below this energy ($E < E_{inj}$) the particle spectrum is formed by Coulomb collisions.

It seems that the problem is characterized by the energy E_{inj} only. Then the particle spectrum below E_{inj} can be described by Maxwellian spectrum with a single temperature T, and above this energy a power-law spectrum is formed by the acceleration mechanism. However, as it was shown by Gurevich (1960) the run-away flux into the region of acceleration changes the distribution function not only in the region above E_{inj} but also below it, in the range $E_M \ll E \ll E_{inj}$. The function can be described as Maxwellian only in the range $E \ll E_M$. We determine the value E_M below.

This distortion of the distribution function in the region $E_M \ll E \ll E_{inj}$, where collisions play still a significant role, can be in principle interpreted as an appearance of the "second" effective temperature which is higher than the gas temperature T.

III DISTORTION OF THE MAXWELLIAN SPECTRUM

The problem of particle acceleration from background gas for a spatially uniform case of stochastic acceleration was analyzed by Gurevich [4]. He showed that a flux of run-away particles along the spectrum is generated by the acceleration mechanism. The value of this flux S can be written as

$$S(u) = S_0 \sqrt{\frac{2}{\pi}} \int_0^u x^2 \exp\left(-\frac{x^2}{2}\right) dx \,, \tag{6}$$

where the constant S_0 is

$$S_0 = \sqrt{\frac{2}{\pi}} n_0 \exp\left(-\int_0^\infty \frac{x\,dx}{1 + \alpha_0 x^5/\nu_0}\right) , \tag{7}$$

here n_0 is the density of background particles.

The analysis of Eq.(6) shows that the flux value is almost zero at small u and it increases to $S_0 = const$ at large u. The important point is that the flux reachs the value $S \simeq S_0$ at the energies $E \simeq E_M \ll E_{inj}$. This means that we have the two (not one) energies which characterize the problem. Therefore at small kinetic energies $E = kTu^2/2$ determined by

$$E < E_M = kT(\nu_0/\alpha_0)^{2/5} \,, \tag{8}$$

where the flux is small enough the distribution function is almost equilibrium and can be presented by the thermal (Maxwellian) spectrum with the temperature T,

$$F_M \simeq n_0 \frac{2}{\sqrt{\pi (kT)^3}} \sqrt{E} \exp\left(-\frac{E}{kT}\right) . \tag{9}$$

In the energy range

$$E_M \ll E \ll E_{inj} = kT(\nu_0/\alpha_0)^{2/3} \tag{10}$$

the spectrum is also determined by collisions, but (since the flux S in this region is not negligible) this collisional distribution function is strongly (exponentially) distorted and cannot be described the Maxewllian distribution. There are substantial deviations from the Maxwellian distribution in the energy range $E \gg E_M = E_{inj}\alpha_0^{4/15}$, because of the constant particle flux escaping from the collisional region. If one tries to reproduce this part of the spectrum by thermal one, he should assume another effective temperature of these particles $T^* \gg T$.

Only in the energy range $E \gg E_{inj}$ the collisions are ineffective and a nonthermal power-law spectrum of accelerated particles is formed there

$$F_{ac} = C_2(\tau)E^{-1} \tag{11}$$

The solution of the distribution function in the energy range $E_M \ll E \ll E_{inj}$ can be written in the form

$$F = n_0 \frac{2}{\sqrt{\pi(kT)^3}}\sqrt{E}\exp\left(-\int_0^\infty \frac{xdx}{1+\alpha_0 x^5/\nu_0}\right)\cdot \left(\exp\left(\int_{\sqrt{2E/kT}}^\infty \frac{xdx}{1+\alpha_0 x^5/\nu_0}\right) - 1\right). \tag{12}$$

IV BREMSSTRAHLUNG RADIATION

Using the spectrum (12) we calculated the intensity of X-ray photons from the Coma cluster. The soft part of the spectrum is determined by the temperature and the density of the background gas. Therefore this part of the spectrum can be used to estimate the Coma medium parameters: the density $n_0 \simeq 3 \cdot 10^{-3}$ cm^{-3} and the temperature $T \sim 8$ keV. The high energy range of the spectrum depends on the acceleration parameter α. Any variations of α does not change the Maxwellian distribution of background particles at $E < E_M$ but at $E > E_M$ the intensity of electrons and the flux of bremsstrahlung photons are sensible to this parameter. From the Beppo-SAX data we derive that $\alpha_0/\nu_0 \simeq 10^{-3}$. The frequency of Coulomb collisions for the parameters of the Coma halo is $\nu_0 \sim 10^{-12}$s, therefore the characteristic time of particle acceleration equals $\alpha_0^{-1} \sim 10^{15}$s. Then we obtain that

$$E_M \sim 100 \text{ keV, and } E_{inj} \sim 800 \text{ keV}. \tag{13}$$

If we take into account that the energy of bremsstrahlung photons is less than the energy of parent electrons, then we expect the thermal bremsstrahlung emission at energies below $30-50$ keV whose spectrum corresponds to the temperature T, nonthermal bremsstrahlung emission at energies above $300-400$ keV, and in the range $30-400$ keV the emission is generated by thermal particles, whose spectrum is distorted by the run-away flux that can be interpreted as a thermal bremsstrahlung emission at the temperature $T^* \gg T$.

ACKNOWLEDGMENTS

VAD thanks Dr.L.Feretti for her useful comments. He is also grateful to Dr.T.Enßlin who sent a copy of his paper before its publication.

This work was partly supported by the Government of Russia through the Russian Foundation of Fundamental Researches under the grant No. 98-02-16248.

REFERENCES

1. Enßlin, T. A., Lieu, R., & Bierman, P. L., *A&A*, **344**, 409 (1999). Giovannini, G., Feretti, L., Venturi, T., Kim, K.-T., & Kronberg, P. P. 1993, *ApJ*, **406**, 399 (1993)
2. Fusco-Femiano, R., Dal Fiume, D., Feretti, L., Giovannini, G., Grandi, P., Matt, G., Molendi, S., & Santangelo, A., *ApJ*, **513**, L21 (1999).
3.
4. Gurevich, A. V., *Sov.Phys JETP*, **38**, 1150 (1960).
5. Toptygin, I. N., *Cosmic Rays in Interplanetary Magnetic Fields*, Reidel, Amsterdam (1985).

DIFFUSE COSMIC GAMMA-RAY EMISSION

Extragalactic Gamma-Ray Emission: CGRO results

P. Sreekumar

ISRO Satellite Center, Bangalore, India

Abstract. The first all-sky survey above 1 MeV by the COMPTEL and EGRET instruments on board the Compton Gamma Ray Observatory has provided new results on the extragalactic diffuse emission. The extragalactic emission over the energy range 1 MeV – 100 GeV is well described by a broken power law spectrum, the spectrum being harder above 30 MeV. No large scale spatial anisotropy or changes in the energy spectrum are observed in the deduced extragalactic emission. The most likely explanation for the origin of this extragalactic emission, is that it arises primarily from unresolved γ-ray-emitting blazars at least above 10 MeV. The consistency of the average γ-ray blazar spectrum with the derived extragalactic diffuse spectrum strongly argues in favor of such an origin. The origin of the emission in the 1–30 MeV remains less well understood. Additional contributions from sources with jets not necessarily aligned close to our line-of-sight are also examined. Future gamma-ray missions such as GLAST with enhanced spectral capabilities, are expected to provide the next major advances in resolving the origin of the extragalactic γ-ray background.

I INTRODUCTION

The study of the cosmic diffuse γ-ray emission has always been challenging as a result of limited flux sensitivity, poor angular resolution, and limited sky coverage of earlier γ-ray detectors. The investigation has been further hindered by significant diffuse γ-ray emission arising from our own galaxy as a result of cosmic-ray interactions with the interstellar matter and radiation (Hunter *et al.* 1997). Furthermore, there exists no observable (at least by current capabilities) spatial or temporal signature characteristic of the diffuse extragalactic background that distinguishes it from other radiation. Since its launch in 1991, one of the most important goals of the Compton Observatory has been to carry out a detailed study of the cosmic γ-ray emission. The two wide field-of-view instruments COMPTEL and EGRET on CGRO have carried out the first all-sky survey of the sky from 1 MeV to 100 GeV. Detailed descriptions on the design and performance of the COMPTEL instrument is given by Schönfelder *et al.* 1993 and of the EGRET instrument by Thompson *et al.* 1993 and Esposito *et al.* 1999. Significant achievements in the area of extragalactic emission studies were made possible primarily

CP510, *The Fifth Compton Symposium,* edited by M. L. McConnell and J. M. Ryan

due to the low-instrumental background of EGRET and the good understanding of the instrumental background of COMPTEL. Here we will discuss only briefly the main results on the cosmic diffuse emission from CGRO and some recent findings on the origin of the emission. A more detailed discussion is included in the review by Sreekumar, Stecker & Kappadath (1997).

If we examine the situation prior to the launch of CGRO, at low/medium energies (< 10 MeV), the most surprising result was the detection of an *MeV bump* around 2 MeV. This emission was well in excess of the extrapolated hard X-ray diffuse continuum emission (Trombka *et al.* 1977). Interestingly, many other experiments had also independently confirmed the existence of the *MeV bump*. Although it was pointed out very early on by Fishman (1972) that cosmic-ray induced radioactivity could contribute significantly in the 1-10 MeV range, various theories of cosmic origin were proposed to explain the bump. Above 30 MeV, the OSO-3 (Kraushaar *et al.* 1972) and the SAS-2 (Fichtel *et al.* 1975) experiments confirmed the existence of an extragalactic emission at high energies. The final SAS-2 results reported by Thompson & Fichtel (1982) provided the first spectral measurements of the high-energy extragalactic spectrum. The spectrum was best fit using a power-law with an index of -(2.4±0.4). No results were available on the intensity or spectrum beyond 200 MeV.

II SUMMARY OF CGRO RESULTS

With the launch of CGRO in 1991, significant advances were made with respect to the spectrum and spatial distribution of the extragalactic γ-ray emission (figure 1). We summarize below some of the important results.

A 1 – 30 MeV

• no Evidence for an *MeV bump*. It has been understood as mostly arising from cosmic-ray induced background in the instrument (Kappadath *et al.* 1996)
• a power law (index =-2.2±0.2) best fits the spectrum in the 1–30 MeV range (Kappadath *et al.* 1998; Weidenspointer 1999)
• the derived spectrum is consistent with the extrapolation of the hard X-ray diffuse spectrum near the lower end of the COMPTEL spectrum (~ 1 MeV) and with the EGRET diffuse spectrum at the high end (~ 30 MeV).

B 30 MeV – 100 GeV

• spectral measurements were extended to 100 GeV (Sreekumar *et al.* 1998)
• smooth single power-law fit (index=-2.01±0.03) that is harder than the average SAS-2 spectrum (though within error bars)

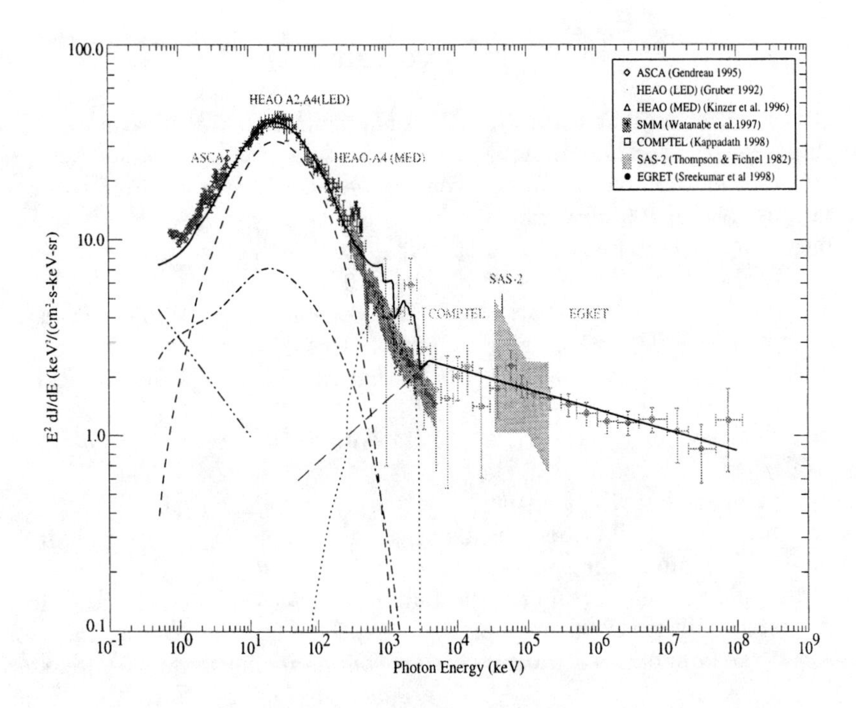

FIGURE 1. The extragalactic γ-ray spectrum

• near-isotropy (galactic plane and Galactic center excluded in the study) of the diffuse emission intensity and spectrum throughout the sky.
• evidence for an excess Galactic component (Dixon *et al.* 1998), in the inner Galaxy that most probably arises from unaccounted inverse Compton emission.

III ORIGIN OF THE EXTRAGALACTIC EMISSION

The key to understanding the origin of the extragalactic emission may lie in the realization that it may be composed of a number of different components with different origins. These components in turn may dominate the observed emission only in specific energy ranges. In the discussion to follow, we will broadly distinguish between the two clear scenarios: truly diffuse emission of cosmological origin and emission from unresolved point sources of γ-rays.

A Diffuse origin

High-energy processes that results in γ-ray emission are discussed comprehensively by Stecker (1971). These include

- matter-antimatter annihilation in a baryon-symmetric universe (Stecker, Morgan & Bredekamp 1971) (1–100 MeV range)
- γ-ray lines from Type Ia SNR (decay of ^{56}Ni $\rightarrow^{56}$Co $\rightarrow^{56}$Fe and from ^{26}Al, ^{44}Ti, and ^{60}Co (The *et al.* 1993, Watanabe *et al.* 99)(1–10 MeV range)
- annihilation of supersymmetric particles (Silk & Srednicki 1984; Rudaz & Stecker 1991; Kamionkowski 1995) (>1 GeV)
- decay of 500–1000 GeV Higgs bosons (Bhattacharjee, Shafi & Stecker, 1998) (>1 GeV)
- primordial black hole evaporation (Page & Hawking 1976, McGibbon *et al.* 1991)

Most of these models predict continuum or line contributions that are not easily separable from other contributions, considering our current detector capabilities. The characteristic bump in the neutral pion decay spectrum led to the pre-CGRO belief that this process could produce the bulk of the *MeV bump* emission. In a baryon-symmetric universe the neutral pions are expected from nucleon-antinucleon annihilation at the boundaries of galaxy clusters containing matter and antimatter. The results from COMPTEL and SMM (Watanabe 1997) suggests that contributions from this process has to be weaker than previously estimated. In the 1–10 MeV range the emission from various SNR decay lines, smeared by integration over various redshifts, could account for the observed diffuse emission. γ-rays from primordial black hole (PBH) evaporation integrated over all redshifts could produce a diffuse background. Contribution from the PBH is predicted to have a steeper spectrum than that observed for the diffuse emission above 30 MeV. The final moments of evaporation of PBHs is expected to produce a bursts of γ-rays lasting a few microseconds. However EGRET observations has yielded null results (Fichtel *et al.* 1996), suggesting weaker contribution (if any) to the diffuse emission.

B Unresolved sources

The observed diffuse emission can also arise from unresolved point sources. Table 1 summarizes the list of high-energy γ-ray sources from the 3rd EGRET catalog (Hartman *et al.* 1999). Blazars form the dominant class of identified sources. Possible source classes include

- active galactic nuclei (Bignami *et al.* 1982; Stecker *et al.* 1993; 1996; Chiang *et al.* 1995; Chiang & Mukherjee 1998, Mücke & Pohl 1999)
- normal galaxies (Strong *et al.* 1976)
- clusters of galaxies (Dar & Shaviv 1995; Erlykin *et al.* 1996 ; Colafrancesco & Blasi 1998)
- FIR galaxies

With the exception of contribution from AGNs, the predicted contributions from the other source classes all fall below 10%. Detailed discussion of the AGN contribution has been covered in the earlier review (Sreekumar, Stecker & Kappadath 1997). With the recent work of Chiang & Mukherjee (1998) and Mücke & Pohl (1999), it appears that not all of the observed diffuse emission above 10 MeV can be accounted from blazar emission. We briefly discuss a new alternative.

1 *Contribution from off-axis emission in blazar jets*

The nearest giant radio galaxy, Centaurus A has been observed to emit high-energy radiation upto at least 1 GeV (Sreekumar *et al.* 1999). Studies at lower energies have shown the Cen A jet to be offset from our line-of-sight by almost 70°. This is the first observational evidence for emission upto 1 GeV from a large-inclination jet. The OSSE experiment on CGRO had reported early on the detection of Cen A, however, OSSE has no imaging capability to confirm the source position. COMPTEL's error region for Cen A unfortunately also contained a nearby BL Lac object MS1312-423 (Steinle *et al.* 1998). The strongest evidence for the identification of the source with Cen A comes from the recent EGRET result where the signal is centered on Cen A, with the BL Lac object 1312-423 well outside the 95% confidence contour (Sreekumar *et al.* 1999). Finally, the smooth continuation of the spectrum from 50 keV to 1 GeV strongly argues that OSSE, COMPTEL and EGRET have detected the same source (figure 2).

Cen A is the closest AGN and the derived γ-ray luminosity is small ($\sim 10^{-5}$ weaker) compared to the typical γ-ray blazar. If Cen A represents inclined-jet AGN with emission extending at least upto 1 GeV, its spectrum being steeper (index=-2.40 ± 0.28) than the γ-ray background, (index=2.10 ± 0.03), suggests very limited contribution to the γ-ray background above 10 MeV from large-inclination jet sources. The spectral energy distribution of Cen A shows peak power around 1 MeV or below, and hence such sources could be significant contributors to the extragalactic γ-ray background around 1 MeV (Steinle *et al.* 1998).

TABLE 1. Summary of third EGRET catalog (Hartman *et al.* 1999)

Source class	*Number* $> 5\sigma$	other references
blazars	67 ($94 > 4\sigma$)	Mukherjee *et al.* 1997
pulsars	6	Thompson *et al.* 1997
unidentified	170	Hartman *et al.* 1999
OTHER		
normal galaxy	1 (LMC)	Sreekumar *et al.* 1992
radio galaxy	1 (Cen A)	Sreekumar *et al.* 1999
X-ray binary	1 (Cen X-3)	Vestrand,Sreekumar,Mori 1997

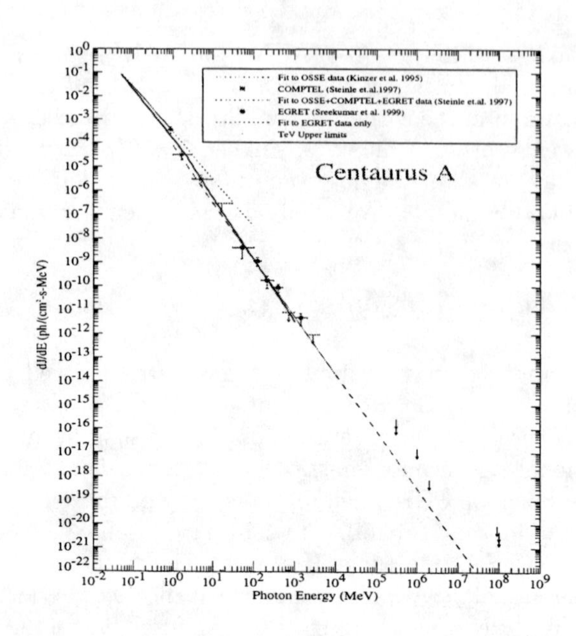

FIGURE 2. Cen A differential photon spectrum from CGRO observations

Recently, Weferling and Schickeiser (1999) discussed contributions from blazars whose jets are pointed away from our line-of-sight. The scattered energy decreases rapidly with increasing jet angle w.r.t. the observer (figure 3). However these sources have fluxes below the EGRET sensitivity due to intrinsic low luminosity and large distances. Assuming Cen A represents an extreme case of a large-inclination jet from a nearby source, γ-ray emission from jets at intermediate inclination angles can be significant. With the number density of radio-loud FRI source population almost 1000 times larger than that of FSRQs and BL Lac objects, the contribution from AGNs jets that are not nearly-aligned to the observer can be significant. The future GLAST mission with its significantly improved γ-ray sensitivity should detect a large number of intermediate-inclination jet sources at distances well beyond that of Cen A. This will enable a realistic determination of the diffuse emission contribution from AGNs.

IV CONCLUSION

CGRO observations by COMPTEL and EGRET show the presence of an extragalactic diffuse emission above 1 MeV. The previous measurements of an *MeV bump* has been attributed mostly to instrumental background contributions arising from cosmic-ray interactions. A power-law spectrum (index -2.2±0.2) well repre-

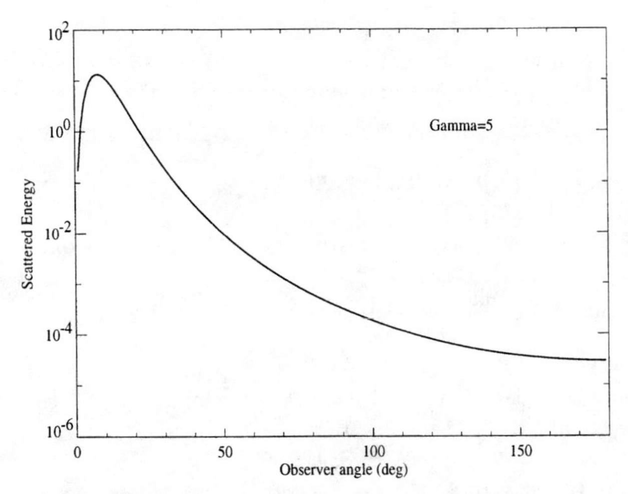

FIGURE 3. Observed emission as a function of jet orientation w.r.t observer (Weferling & Schickeiser (1999))

sents the 1–30 MeV emission while a flatter power-law spectrum with an index of -(2.1±0.03) characterizes the emission from 30 MeV to almost 100 GeV. The precise origin of the extragalactic γ-ray emission is not fully resolved, though observations suggest strongly that a large fraction of the emission arises from unresolved AGN emission.

If Cen A is indeed a misaligned blazar (Bailey *et al.* 1986) this provides fresh evidence for >100 MeV emission from radio-loud AGN with jets at large inclination angles. Assuming a unification model for AGN, and increasing high-energy emission with decreasing inclination angles (Dermer, Sturner & Schlickeiser 1997; Weferling & Schlickeiser 1999), the detection of more distant radio-loud AGN with intermediate-inclination angle jets can be expected. Though the intrinsic luminosity is lower than other on-axis sources, the significantly larger space density of radio-loud FRI sources, points to a new unresolved source class that could contribute to the extragalactic γ-ray background.

REFERENCES

1. Bailey, W., *et al.*, 1986, Nature, 322, 150
2. Bhattacharjee, P., Shafi, & Stecker, F.W., 1998, PRL,
3. Bignami, G. F., *et al.*, 1979, ApJ, 232, 649
4. Chiang, J., *et al.*, 1995, ApJ, 452, 156
5. Chiang, J., & Mukherjee, R., 1998, ApJ, 496, 752
6. Colafrancesco, S., & Blasi, P., 1998, Astroparticle Phys., 9, 227
7. Comastri, A., Setti, G., Zamorani, G., & Hasinger, G., 1995, A&A, 296, 1

8. Dar, A., & Shaviv, N., 1995, PRL, 75, 3052
9. Dermer, C.D. Sturner, S.J. & Schlickeiser, R., 1997, ApJS, 109, 103
10. Dixon, D. D., *et al.*, 1998, New Astronomy, 3, 539
11. Erlykin, A.D., Osborne, J.L., Wolfendale, A. W., & Zhang, L., 1996, A&AS, 120, 623
12. Esposito, J.A., *et al.* 1999, APJS, 123, 203
13. Fichtel, C.E., *et al.* 1975, ApJ, 198, 163
14. Fichtel, C.E., *et al.* 1996, Proc. of 3rd Huntsville Symp., AIP Conf. Proc. 384, 368
15. Fishman, G., 1972, ApJ, 171, 163
16. Hartman, R.C., *et al.* 1999, ApJS, 123, 79
17. Hunter, S.D. *et al.* 1997, ApJ, 481, 205
18. Kaminonkowski, M. 1995, The Gamma Ray Sky with Compton GRO and SIGMA, ed. M. Signore, P. Salati, & G. Vedrenne (Dordrecht: Kluwer), 113
19. Kanbach, G., et al., 1988, Space Sci. Rev., 49, 69
20. Kappadath, S. C., *et al.*, 1996, A&AS, 120, 619
21. Kappadath, S. C.,1998, Ph.D. Thesis, Univ. of New Hampshire
22. Kraushaar, W.L., *et al.* 1972, ApJ, 177, 341
23. Mücke, A., & Pohl, M., 1999, Proc. BL Lac Phenomenon, Turku, Finland, 217
24. Mukherjee, R., *et al.*, 1997, ApJ, 490, 116
25. Page, D. N., & Hawking, S. W., 1976, ApJ, 206, 1
26. Rudaz, S., & Stecker, F. W., 1991, ApJ, 368, 40
27. Schönfelder, V., *et al.*, 1993, ApJS, 86, 657
28. Silk, J., & Srednicki, M., 1984, PRL, 53, 624
29. Sreekumar, P. *et al.* 1992, ApJ, 400, L67
30. Sreekumar, P., Stecker, F. W., & Kappadath, S. C., 1997, Proc. of the 4th Comp. Symp, AIP Conf. Proc. 410, 344
31. Sreekumar, P. *et al.* 1998, ApJ, 494, 523
32. Sreekumar, P. *et al.* 1999, Astroparticle Phys., 11, 221
33. Stecker, F. W. 1971, *Cosmic Gamma Rays*, Mono Book Co., Baltimore
34. Stecker, F. W., & Salamon, M. H., 1996a, PRL, 76, 3878
35. Stecker, F. W., & Salamon, M. H., 1996b, ApJ, 464, 600
36. Stecker, F. W., Salamon, M. H., & Malkan, M., 1993, ApJ 410, L71
37. Stecker, F. W., Morgan, D. L., Bredekamp, J., 1971, PRL, 27, 1469.
38. Steinle, H., *et al.*, 1998, A&A,330, 97
39. Strong, A. W., Wolfendale, A. W., & Worrall, D. M., 1976, J. Phys. a, 9, 1553
40. The, L.-S., Leising, M. D., & Clayton, D. D., 1993, ApJ, 403, 32
41. Thompson, D.J., *et al.* 1993, ApJS, 86, 629
42. Vestrand, W. T., Sreekumar, P., & Mori, M., 1997, ApJL, 483, L49
43. Watanabe, K., *et al.* 1997,in AIP Conf. Proc. 410, 1223
44. Watanabe, K., *et al.* 1999, ApJ, 516, 285
45. Weferling & Schlickeiser, R., 1999, A&A, 344, 744
46. Weidenspointer, G. 1999, Ph.D. Thesis, Max-Planck Inst. für Extraterretrial Physik.

The cosmic diffuse gamma-ray background measured with COMPTEL

G. Weidenspointner*, M. Varendorff*, S.C. Kappadath¶, K. Bennett||, H. Bloemen‡, R.Diehl*, W. Hermsen‡, G.G. Lichti*, J. Ryan†, V. Schönfelder*

* *Max-Planck-Institut für extraterrestrische Physik, Postfach 1603, 85740 Garching, Germany*
† *Space Science Center, University of New Hampshire, Durham, NH 03824, USA*
‡ *SRON-Utrecht, 3584 CA Utrecht, The Netherlands*
|| *Astrophysics Division, ESTEC, 2200 AG Noordwijk, The Netherlands*
¶ *Louisiana State University, Baton Rouge, Louisiana, USA*

Abstract. We report a refined analysis of the cosmic diffuse gamma-ray background (hereafter CDG) in the energy range 0.8–30 MeV with the Compton telescope COMPTEL onboard the Compton Gamma-Ray Observatory. We have identified all major instrumental-background lines, included the results of a detailed study of the instantaneous instrumental continuum-background characteristics, and used all available COMPTEL data at high galactic latitudes.

The new "whole-sky" average CDG spectrum again shows no evidence for an MeV-bump, merges smoothly with the spectra at higher and lower energies, and is consistent with a transition from a softer to a harder component around a few MeV. This spectrum is consistent with previous COMPTEL results. In addition, comparison of the CDG intensity from various regions of the sky allows us to place limits on the large-scale anisotropy of the CDG in selected energy bands. Upper limits on the relative deviations from isotropy consistent with the data at the 95% confidence limit range from about 24% to about 45% on scales of a few steradian.

INTRODUCTION

One of the principal goals of the COMPTEL mission [6] is the study of the cosmic diffuse gamma-ray background (CDG). Early COMPTEL results on the CDG spectrum have been reported in, e.g., [2,3], an alternative analysis has been performed by [1]. Advances in our understanding of the instrumental background, such as the identification of all major instrumental-background lines and a detailed study of the characteristics of the instantaneous continuum-background, resulted in a refined analysis of the CDG spectrum from 0.8–30 MeV [12,13]. In addition, comparison of the CDG intensity from various regions of the sky allows us to place limits on the large-scale anisotropy of the CDG in selected energy bands.

CP510, *The Fifth Compton Symposium,* edited by M. L. McConnell and J. M. Ryan

Throughout this work the term CDG refers to the total gamma-ray intensity from high galactic latitudes (including a possible galactic high-latitude component, the contribution of resolved point sources being insignificant).

DATA ANALYSIS

The quest for the CDG in the MeV range is severely impeded by the low intensity of the signal in the presence of a much higher instrumental background. As is described in more detail elsewhere (e.g. [3,13,15]), the CDG intensity is determined in this analysis approach first by fitting the time-of-flight (ToF) distribution to determine the ToF forward-peak event rate. The ToF forward-peak is composed of proper double-scattered events from the sky and various prompt and long-lived background components that have to be accounted for, as outlined below. The residual ToF forward-peak event rate after subtraction of these instrumental-background components is then attributed to the CDG intensity.

Prompt background components are instantaneously produced in proton and neutron interactions within the instrument material and follow the local cosmic-ray intensity (see e.g. [5]). The intensity of incident cosmic-ray particles is monitored with the instrument's charged-particle shields, so-called veto domes. A detailed study of the characteristics of prompt background components confirmed that their variation with veto rate is linear to a good approximation. When linearly extrapolated to low veto rates, however, the event rate of prompt components does not vanish at veto rate zero, contrary to the findings by [3], but at a small yet significant positive veto-rate value, referred to as veto-rate offset [13,15]. Most likely, this veto-rate offset is an artifact of small non-linearities in the correlation between prompt-background event rate and veto rate. Taking into account this offset in our analysis elevates each point in the spectrum.

Long-lived background components arise from the decay of activated radioactive isotopes. The major components of the COMPTEL instrumental-line background can be attributed to eight individual isotopes, namely ^{2}D, ^{22}Na, ^{24}Na, ^{28}Al, ^{40}K, ^{52}Mn, ^{57}Ni, and ^{208}Tl (see [14] and references therein). Subtraction of these lines, which has to be performed as a function of veto rate before the veto-rate extrapolation [14,15], is essential to determine the CDG intensity below about 4 MeV.

In our analysis COMPTEL was used as a pointed collimator, the field-of-view restricted to a half-opening angle of 40° [3]. To determine the CDG intensity above 4.3 MeV we used all observations which were pointed at galactic latitudes $|b^{\mathrm{II}}| > 30°$ from the beginning of the mission in 1991 to early 1997, since at these energies all instrumental background appears to be prompt. Although we have identified and subtracted all major instrumental lines we limit the data sets used to determine the CDG intensity below 4.3 MeV, where long-lived instrumental background is present, to those where the activity of (long-lived) instrumental lines is the lowest (1992–1993). The instrument's response was determined from Monte Carlo simulations of a diffuse isotropic power-law source with slopes -2.5 and -2.0 for energies below

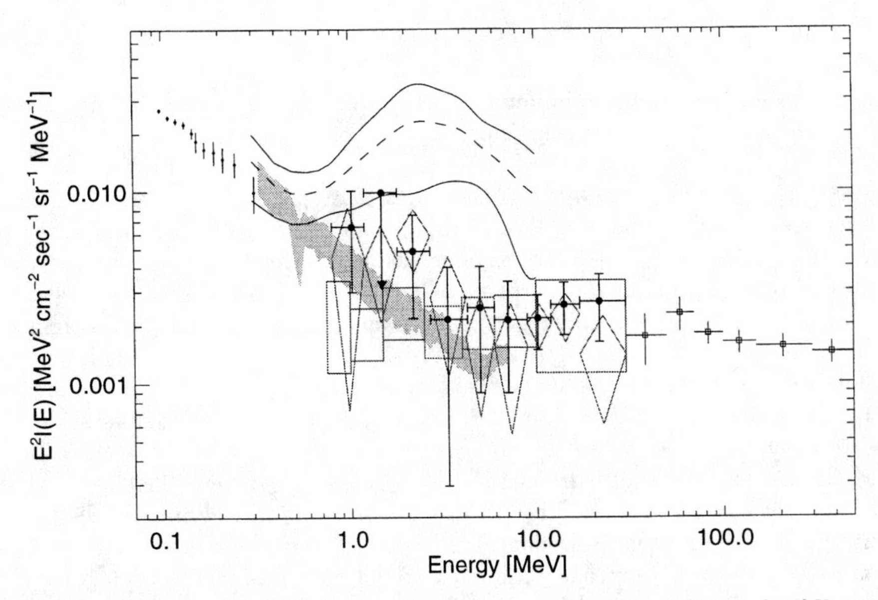

FIGURE 1. A comparison of various results on the spectrum of the CDG. The hard X-ray data are from [4], the grey-shaded band depicts the preliminary result of [11], the high-energy data are from [7]. The COMPTEL results are from [1] (boxes), [3] (diamonds) and this work (data points, the arrow of the 2σ upper limit reaches down to the intensity value). The historic 'MeV-bump' is illustrated by the result of [9].

and above 4.3 MeV, respectively.

RESULTS AND DISCUSSION

Our refined measurement of the CDG spectrum from 0.8–30 MeV is depicted in Fig. 1. The error bars give the total error, which is the linear sum of the total statistical error and the individual systematic errors. The new "whole-sky" average CDG spectrum is consistent with earlier results obtained with the veto-rate extrapolation technique [3], in which the isotopes ^{52}Mn, ^{57}Ni and ^{208}Tl were not accounted for, which was based on an extrapolation to veto rate zero, and which was restricted to the Virgo and South-Galactic Pole observations only. The refined CDG spectrum again shows no evidence for an MeV-bump, merges smoothly with the spectra at lower [4] and higher [7] energies, and is consistent with a transition from a softer to a harder component around about 5 MeV. Because of the size of the total error bars a broken power-law fit is statistically not preferred to a single power-law fit (slope -2.20 ± 0.22), albeit yielding a smaller χ^2_ν. Above 2.7 MeV our spectrum is also consistent with the preliminary results of an alternative, imaging approach to determine the CDG spectrum with COMPTEL from maximum likelihood all-sky model fitting [1]. Below 2.7 MeV there may be a small discrepancy between the results of the extrapolation technique and the alternative approach. Finally, our spectrum is consistent with a preliminary analysis of SMM data [11].

In addition to investigations of the spectrum of CDG, COMPTEL is the first instrument that allows us to address the CDG isotropy on large angular scales. The large-scale isotropy was investigated in a number of energy bands by comparing intensities measured for selected sky regions at intermediate and high galactic latitudes. No significant deviation of the CDG intensity from isotropy could be found within the statistical uncertainty, consistent with a cosmological origin of the CDG. Assuming that the CDG intensity is, indeed, isotropic, upper limits on the relative deviations from isotropy consistent with the data at the 95% confidence level were derived, which range from about 24% to about 45% on angular scales of a few steradian.

Over the last years, the measurements at MeV energies by COMPTEL and SMM, and at higher energies by EGRET, have significantly improved our knowledge of the extragalactic gamma-ray emission. The implications of these new results for the origin of the high-energy extragalactic background have repeatedly been reviewed (see e.g. [8] and references therein). Briefly, the extragalactic background appears to result from the superposed emission of various classes of unresolved point sources. Truly-diffuse components, such as the annihilation of matter and anti-matter in a baryon-symmetric universe, seem to be insignificant. A variety of possible source populations for the CDG at MeV energies have been proposed, such as various classes of blazars, in particular so-called misaligned blazars or radio-loud Seyfert galaxies and MeV-blazars, as well as Type Ia supernovae. The contributions of each of these populations to the CDG intensity are still uncertain.

REFERENCES

1. Bloemen H., et al., *Proc. of 3rd INTEGRAL Workshop*, in press.
2. Kappadath S.C., et al., *A&AS* **120**, C619 (1996).
3. Kappadath, S.C., Ph.D. Thesis, University of New Hampshire, USA (1998).
4. Kinzer R., et al., *ApJ* **475**, 361 (1997).
5. Ryan J., et al., *1997 Conf. on the High-Energy Bgd. in Space (IEEE)*, 13 (1997).
6. Schönfelder V., et al., *ApJ Suppl. Ser.* **86**, 657 (1993).
7. Sreekumar P., et al., *ApJ* **494**, 523 (1998).
8. Sreekumar P., these proceedings.
9. Trombka J.I., et al., *ApJ* **212**, 925 (1977).
10. Varendorff M., et al., *Proc. of the 4th Compton Symposium* (**AIP 410**), 1577 (1997).
11. Watanabe K., et al., *Proc. of the 4th Compton Symposium* (**AIP 410**), 1577 (1997).
12. Weidenspointner G., et al., *Proc. of 3rd INTEGRAL Workshop*, in press.
13. Weidenspointner G., Dissertation, Technical University Munich, Germany (1999).
14. Weidenspointner G., et al., these proceedings.
15. Weidenspointner G., et al., *A&A*, in preparation.

The MeV Cosmic Gamma-ray Background Measured with SMM

K. Watanabe[1,2], M. D. Leising[3], G. H. Share[4] and R. L. Kinzer[4]

[1] *NASA/Goddard Space Flight Center, Code 661.0, Greenbelt, MD 20771,*
[2] *Universities Space Research Association,*
[3] *Department of Physics and Astronomy, Clemson University, Clemson, SC 29634-1911,*
[4] *E. O. Hulburt Center for Space Research Naval Research Laboratory, Washington D.C. 20375*

Abstract. Given the Solar Maximum Mission (SMM) Gamma-Ray Spectrometer's (GRS) nine years of exposure and large field of view, its data contain a tremendously significant signal from the isotropic cosmic γ-ray background (CGB) in the energy range 0.3 – 8.0 MeV. We have extracted this signal by modeling its modulation by the Earth's motion through the GRS field of view, along with several other background components, such as from the SAA and Earth Albedo γ rays. We can quantify the success of the technique and evaluate possible systematic errors because we have many independent measurements of the CGB, which should be constant in time, and because all known other background components have narrow lines that should not be present in the CGB. We thus obtain the definitive measurement of the CGB in this energy range to date. We compare the CGB spectrum with line emission from iron production in thermonuclear supernovae, and conclude that some other source(s) probably dominate the 1 MeV region.

I INTRODUCTION

The isotropic cosmic gamma-ray background (CGB) from a few hundred keV to several MeV carries important information about numerous astrophysical phenomena. The thermal spectra of AGN cutoff at the low end of this range, and the higher-energy nonthermal spectra of Blazars apparently turnover at energies below a few MeV. The evolution of these sources with cosmological redshift is in principle discernable from a detailed measurement of the CGB. Cosmic star formation up to redshift of a few could be monitored, through the ^{56}Co lines from thermonuclear supernovae (a secondary indicator of star formation because of binary evolution delays.) Universal matter-antimatter symmetry should be revealed most clearly at a few MeV from proton/antiproton annihilation at redshifts of 10–100.

Absolute measurements at 1 MeV are notoriously difficult; most instruments have used some modulation of a celestial signal on a timescale over which the variations of the large instrumental backgrounds are small. Here we use the Solar Maximum

CP510, *The Fifth Compton Symposium,* edited by M. L. McConnell and J. M. Ryan

Mission (SMM; 1980–1989) Gamma-Ray Spectrometer (GRS) with its 130° field of view and a reasonably well-matched occulter, the Earth, to provide the modulation of a large signal. Complications arise because the occulter is itself a strong source of gamma rays, and because the modulation occurs over the orbital period–a period shared by the variations of the dominant backgrounds. Still we have the advantages of long time coverage and detailed spectral information. We can make significant independent measurements at hundreds of times at a few hundred energies. Thus we can define a successful measurement: the spectrum should be independent of time, and it should contain no sharp spectra features such as narrow lines. These characteristics are not shared by any of the known background effects.

II DATA ANALYSIS

We select the 5–6 non-SAA orbits for each of 3000 days, and at each energy we fit the SMM/GRS count rate with a model which is a linear combination of (a) *a constant*, (b) *the expected variation Earth albedo gamma rays*, (c) *free decay of SAA-induced radioactivity*, (d) *the prompt effects of local cosmic rays*, (e) *decay of continuing cosmic-ray-induced radioactivity*, and (f) *the expected variation of the count rate due to the sky emission.* The constant term simply represents the many strong background components that vary on times longer than one day. Components (b) through (e) require a great deal of study of the data to establish adequate models. We have explored over one hundred combinations of possible models. In the end we favor empirical determinations of these components over first-principles models. In particular, the total count rate in the detectors and shields at energies above 10 MeV provides an excellent monitor of the cosmic ray environment. The radioactive decay terms are simple exponentials, but the integrated doses and decay lifetimes are determined from the data. The CGB signal term, (f), determined each day is the total flux from that part of the sky that the Earth transits in that day.

Having chosen the basic model components, we have a many-parameter, nonlinear fit to perform tens of thousands of times. Because of the frequent strong correlations among the fit parameters, this is impossible to do in a simple automated fashion. We first establish, at each energy, the values of the nonlinear parameters, such as radioactive lifetimes and assumed powerlaw dependences of detector count rates on measured parameters. For example, we assume the 1 MeV count rate is proportional to the >10 MeV rate to some power determined from the data. We do this with a subset of the nonlinear fits that appear to "work", having no unphysical parameters or strong correlations. These nonlinear model parameters, generally strong functions of energy, are then fixed, and we perform the six-term linear fits with all coefficients constrained to be positive and bounded. The upper bounds are interpolated from brute-force six-dimensional maps of χ^2 at ten energies, which presumable identify the true global minima in this space, with generous amounts added. These upper bounds never explicitly exclude any previously measured CGB intensity values, e.g., [9], but they do prevent the fits from getting stuck in local minima.

III RESULTS

These fits provide our basic results, the intensities of each of the six components at 400 energies for each of 3000 days. We average these over various times to study the spectra, or over various energy bands to study the time dependences. Fig. 1 shows the mission-averaged spectra of the six components. The CGB spectrum above $\sim$ 4 MeV has a (negative) "leakage" of the atmospheric albedo (as do others.) The correlation of the atmospheric component with others is large, and we are unable to completely remove it from the CGB signal. Because of its unique shape, we are able to remove it afterward, and because its slope is so flat, it is irrelevant below 2 MeV. For our final spectrum, we add our measured atmospheric spectrum to the CGB spectrum in an amount necessary to eliminate the 4.44 MeV line.

Apart from this, all known background features appear only in the component where they are supposed to be present, as identified by their narrow-line signatures. Interestingly, the galactic 1.8 MeV line emission appears entirely in the prompt cosmic ray coefficient, where it is added to a prompt 1.8 MeV background line from ^{27}Al spallation. This is because the >10 MeV rate used to monitor cosmic rays contains a galactic photon signal as well, and the galactic 0.3–8.0 MeV counts are apportioned to this term by the fit. We can restore the galactic counts to the CGB term by using a McIlwain-L-parameter model of the cosmic ray intensity, but then the fits are generally not as good, and we want to remove the galactic signal anyway. There are, of course, also CGB counts in the >10 MeV rate, but as the CGB spectrum is so steep compared to the galaxy and backgrounds, only a small fraction (we estimate $\leq 6\%$) of the CGB counts are credited to this term.

Collecting the counts in broad energy bands over times of days to months reveals long-term variations, due to the solar cycle and SMM altitude changes, in all components – except the CGB. There are, however, variations in the CGB rates on the orbital precession timescale, 53 days. The amplitude of these variations is 10–20% (depending on energy), much larger than the statistical uncertainties, and so these variations define our estimate of the total uncertainties in the measurements. We use the RMS deviations of the 30-day intensities to plot uncertainties of our final spectrum. We find no evidence for any annual modulation of the CGB signal, as we would expect if it were still contaminated with the galactic signal.

Our confidence in our CGB spectrum generally decreases with energy. At 300–500 keV the result is robust against a wide variety of changes in the model components and analysis methods. Above 2 MeV there is a great deal more dependence on technique, and the possible systematic errors are larger at higher energies.

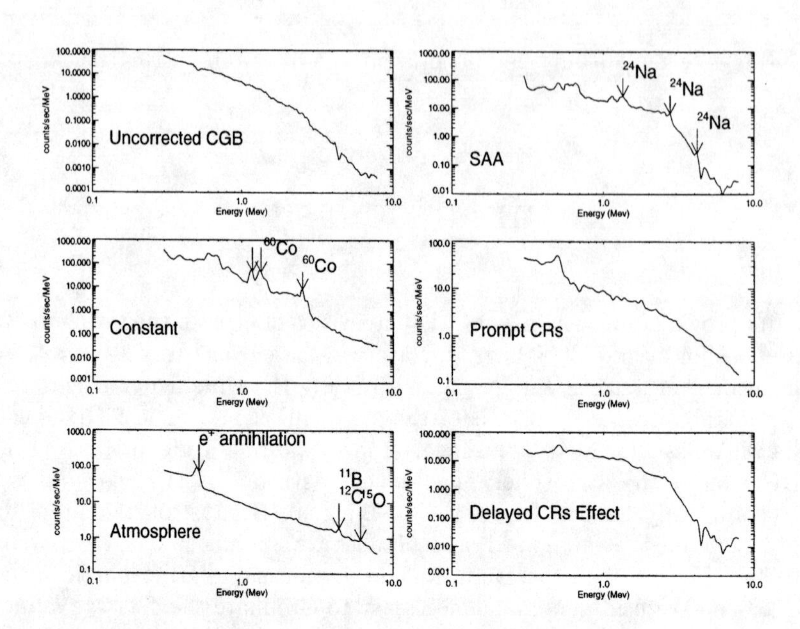

FIGURE 1. Energy spectra of the model terms averaged over nine years.

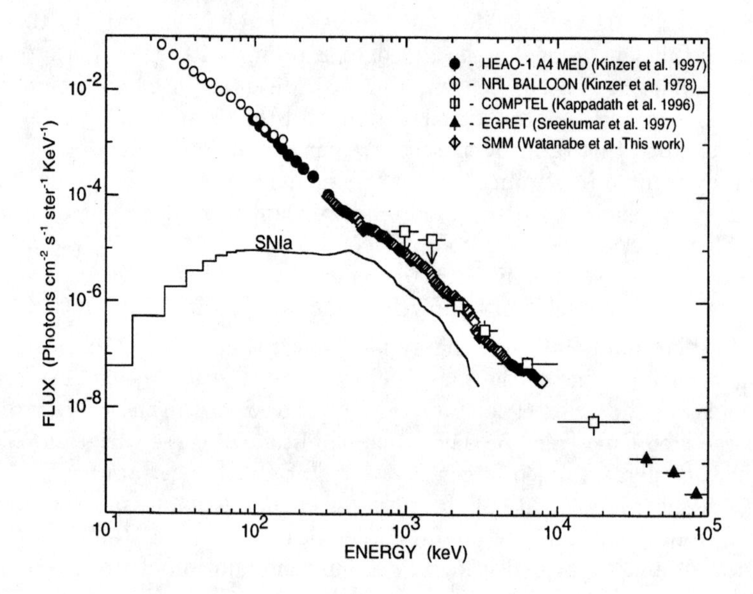

FIGURE 2. The CGB spectrum with previous measurements. The Cosmological SNIa contribution to the CGB is also shown for comparison.

IV CONCLUSION

In Figure 2 we plot the SMM CGB spectrum, corrected for atmospheric contamination and the GRS instrument response, with selected other measurements ([4], [3], [2] and [7]) for comparison. Our measurement is in good agreement with other hard X-ray experiments, but is incompatible with early measurements of the "bump" at $\geq$1 MeV. It is slightly lower than a COMPTEL measurement ([2]) at a few MeV. The SMM/GRS spectrum appears to smoothly connect the hard X-ray measurements with the presumably more reliable measurements at $\geq$10 MeV. The isotropic background at energy 10 keV $\leq$ E $\leq$ 0.5 MeV is thought to be dominated by unresolved Seyfert galaxies ([11]). Typical nearby Seyfert galaxies' spectra cutoff sharply above $\sim$ 0.1 Mev ([1]), but we see no hint of a such a cutoff in our (redshift integrated) spectrum. Unresolved SNIa might contribute to a large fraction of the CGB in this energy range([8], [10]), but we see no sign of the predicted ^{56}Co line edges. This is consistent with the most recent estimate that SNIa make only 25% of the CGB flux at 1 MeV, and with the current idea that more of the nucleosynthesis, and therefore iron production, took place at redshift near unity than in the very recent past. That we find such a smooth spectrum throughout this range, where so many different spectra are thought to meet, suggests that some other continuum source(s) probably dominate(s) this range.

REFERENCES

1. Cameron, R. A., et al. 1993, in Proceedings of the First Compton Observatory Symposium, p. 478-482
2. Kappadath, S. C., et al. 1996, *Astronomy & Astrophysics.* **120**, 619
3. Kinzer, R. L., et al. 1978,*Astrophysical Journal.* **222**, 370
4. Kinzer, R. L., et al. 1997, *Astrophysical Journal.* **475**, 361
5. Letaw, J. R.,et al. 1989, *Journal of Geophysical Research.* **94**, 1211
6. Share, G. H., et al. 1989, High-energy radiation background in space, AIP, p. 266-277
7. Sreekumar, P., et al. 1998,*Astrophysical Journal.* **494**, 523
8. The, L.-S., et al. 1993, *Astrophysical Journal.* **403**, 32
9. Trombka, J. I., et al. 1977, *Astrophysical Journal.* **212**, 925
10. Watanabe, K.,et al.1999, *Astrophysical Journal.* **516**,285
11. Zdziarski, A. A., et al. 1995, *Astrophysical Journal.* **438**, L63

UNIDENTIFIED SOURCES

EGRET/COMPTEL Observations Of An Unusual, Steep-Spectrum Gamma-Ray Source

D. J. Thompson*, D. L. Bertsch*, R.C. Hartman*, W. Collmar†, W. N. Johnson§

**NASA/GSFC*
†MPE
§NRL

Abstract. During analysis of sources below the threshold of the third EGRET catalog, we have discovered a source, named GRO J1400-3956 based on the best position, with a remarkably steep spectrum. Archival analysis of COMPTEL data shows that the spectrum must have a strong turn-over in the energy range between COMPTEL and EGRET. The EGRET data show some evidence of time variability, suggesting an AGN, but the spectral change of slope is larger than that seen for most gamma-ray blazars. The sharp cutoff resembles the high-energy spectral breaks seen in some gamma-ray pulsars. There have as yet been no OSSE observations of this source.

INTRODUCTION

The construction of the third EGRET catalog (1) selected sources based on their statistical significance for the energy range E > 100 MeV. After the catalog was completed, we investigated the possibility that some interesting steep-spectrum sources might have fallen below the catalog significance threshold. The strongest example of such a source is the one reported here. Its statistical significance in the summed maps from Phases 1-4 of the CGRO mission is highest in the energy band 50-70 MeV (over 6 σ), while falling below 4 σ in the E > 100 MeV energy range.

The position of this source is shown in the likelihood map of Figure 1. This map is constructed by combining likelihood maps from the 50-70, 70-100, and 100-150 MeV bands, where the source significance was highest. We define the best position by the centroid of the 95% confidence contour, and the uncertainty as the radius of this contour, as shown in the table below.

TABLE 1. Source Characteristics

Name	Gal. Long.	Gal. Lat.	R.A.	Dec.	95% radius
GRO J1400-3956	317.1°	21.0°	210.2°	-39.9°	1.3°

CP510, *The Fifth Compton Symposium,* edited by M. L. McConnell and J. M. Ryan

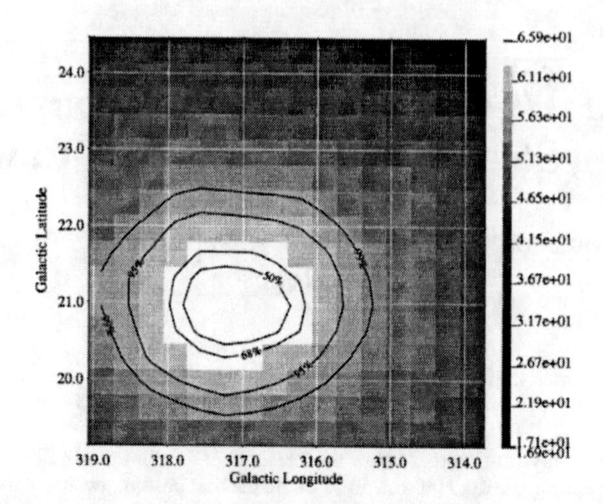

FIGURE 1. EGRET likelihood map for GRO J1400-3956 . The map represents a combination of maps for 50-70, 70-100, and 100-150 MeV for the summed Phases 1-4 data set. Third EGRET catalog sources were modeled (1).

SIMULTANEOUS COMPTEL OBSERVATIONS

Because COMPTEL and EGRET are co-aligned on the Compton Observatory, and COMPTEL has a larger field of view than EGRET, GRO J1400-3956 was observable by COMPTEL in the same viewing periods as EGRET (primarily VP 0120, 0230, 0270, 2070, 2080, 2150, 2170, 3160, and 4240). For the map combining all the Phases 1-4 data, COMPTEL found little evidence for a source at the position identified by EGRET. Below 10 MeV, COMPTEL found only upper limits. In the 10-30 MeV band, COMPTEL's indication of a source was at the 2 σ level.

SEARCH FOR COUNTERPARTS

At a Galactic Latitude of 21°, this source could be either Galactic or extragalactic. The Princeton pulsar catalog shows no pulsars within the error contours. The deeper pulsar survey now underway at Parkes covers this part of the sky and might offer new possibilities. The NASA Extragalactic Database (NED) shows many objects within this relatively large error box, but none that are obvious candidates to be the gamma-ray source: 15 galaxies, 8 IR sources, 8 weak radio sources, and several other objects. The brightest radio source, PKS 1402-388, has a 5 GHz flux density of only 0.25 Jy and a steep radio spectrum (-0.7), unlike the EGRET-detected blazars, which are typically brighter and have flat radio spectra.

One positional coincidence found in NED is with a gamma-ray burst, 3B940703B, whose error box is centered on celestial coordinates J1401-3911, well within the EGRET 95% error contour. Because the time of this burst was not during one of the

EGRET/COMPTEL pointings toward this direction, and both the EGRET and BATSE error boxes are relatively large, this is probably just a chance alignment.

Because this source is only about 8° from the core of radio galaxy Cen A, which has extended radio lobes, we checked for possible alignment of the source with the radio lobes. The new source is not aligned with the radio jets.

ENERGY SPECTRUM

The EGRET detection of GRO J1400-3956 was strong enough in several energy bands to construct a spectrum. The data points and upper limits are consistent with a single power law with number index 3.41 ± 0.34. For every energy bin above 500 MeV, the likelihood test statistic is 0, showing no hint of emission at higher energies. This is one of the steepest source spectra seen by EGRET (see the paper by Bertsch et al. at this conference). If this spectrum extended unbroken into the COMPTEL band, it would be a bright COMPTEL source. The fact that COMPTEL has little evidence of the source indicates a strong change of slope.

In Figure 2, the COMPTEL limits and the one 2 σ excess (10-30 MeV) are combined with the EGRET data. In this case, the spectrum has been multiplied by E^2, giving the equivalent of a power per logarithmic energy interval. The dramatic change of spectral slope is obvious. The dotted line shows the extrapolation of the EGRET spectrum to lower energies. In order to be consistent with the data between 10 and 100 MeV, the change of slope must occur near 50 MeV (the EGRET 50-70 MeV point lies above the fitted line). Taking the slope above 50 MeV as the 3.4 index seen for

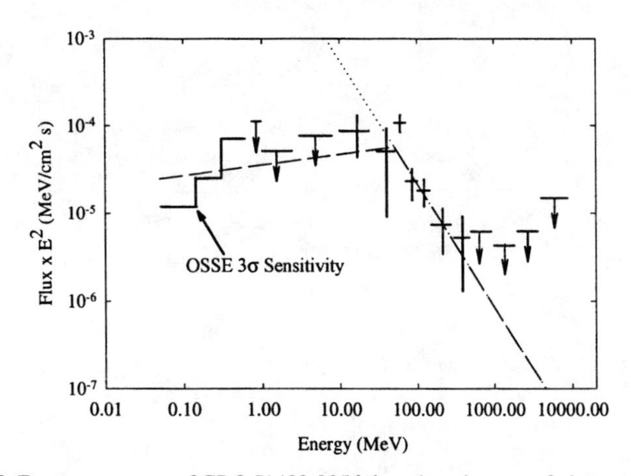

FIGURE 2. Energy spectrum of GRO J1400-3956, based on the sum of observations in Phases 1-4. Dotted line: power law fit to the EGRET data. Dashed line: broken power law consistent with the COMPTEL upper limits for the same observations. The histogram shows the OSSE 3 σ continuum sensitivity, although no OSSE observations have yet been done.

the EGRET data alone, the slope needed below 50 MeV in order to be consistent with the COMPTEL upper limits is 1.8, a change of 1.6 in index.

If the dashed line is the true spectrum, then OSSE should be able to detect the source. Conversely, an OSSE upper limit could further constrain the flattening seen below 50 MeV.

TIME VARIABILITY SEARCH

The EGRET data were examined for time variability, using the 50-70 MeV band where the source is the brightest. The data show some, though not overwhelming, evidence of time variability of the source (Figure 3). In terms of the known EGRET sources, this behavior is more characteristic of AGN than of pulsars, although none of the known EGRET blazars show spectra as steep as this source.

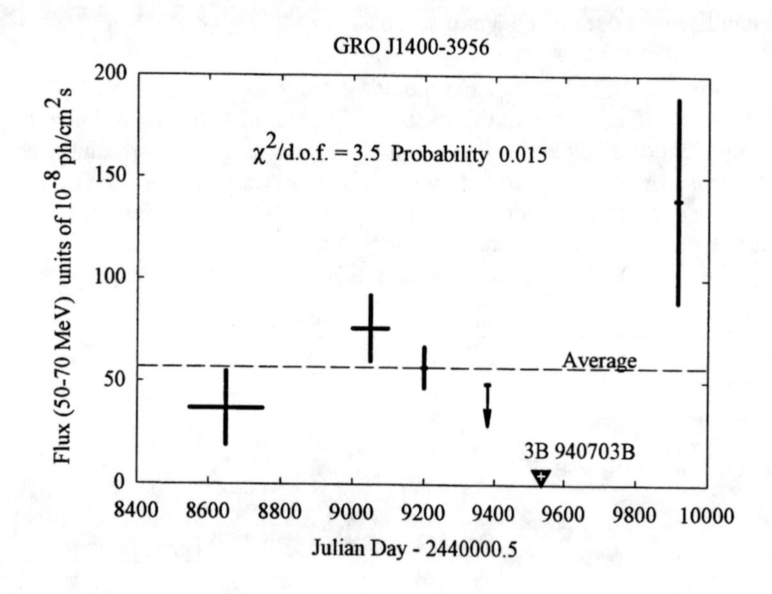

FIGURE 3. EGRET 50-70 MeV flux from GRO J1400-3956 during different observations. The dashed line shows the average from all EGRET data. The time of 3B940703B, from the same direction, is also shown.

SUMMARY

The distinguishing feature of GRO J1400-3956 is the spectral shape with a strong change of slope near 50 MeV. Among the known gamma-ray sources, this feature is unique. Such a strong change of slope is not expected in most physical models involving accelerated particles, unless there is some sort of cutoff in the particle spectrum. The slope change most resembles the pulsar cutoffs seen in the GeV range for Vela and Geminga (2,3) or the spectral changes seen near 1 MeV for some "MeV-peaked" blazars.

This source represents a unique combination of features: a strong change in spectral slope near 50 MeV, a suggestion of time variability, and a lack of pulsar or blazar radio counterparts. Whether it represents an unusual example of a known class of gamma-ray sources or something entirely different remains an open question.

We continue to study GRO J1400-3956 with the Compton Observatory in two ways:

- There have been a number of COMPTEL observations of this sky region since the last useful EGRET observation (VP4240), because the source is often with the COMPTEL field of view during observations of Cen A or PSR B1509-58, both of which have been frequent COMPTEL targets. Preliminary analysis of these later COMPTEL observations has not yielded a strong detection, but the work is ongoing. These data might help clarify the spectrum and/or the possibility of time variability.

- A CGRO Cycle 9 proposal for additional COMPTEL observations and the first OSSE observations has been accepted. As shown in Figure 2, the OSSE data should either provide a detection or a further constraint on the spectral shape. The tentative scheduling shows the source being observed during the early part of 2001. Perhaps these observations will turn up a new gamma-ray surprise for the new Millennium.

In the longer run, the peak of the luminosity appearing in the 50 MeV range suggests that this will be a good candidate for observations with GLAST. With its much larger sensitivity and better angular resolution, GLAST should have the capability of shedding more light on this intriguing source.

REFERENCES

1. Hartman, R.C. et al., *ApJS* **123**, 279-202 (1999).

2. Kanbach, G. et al., *A&A* **289**, 855-867 (1994).

3. Mayer-Hasselwander, H.A. et al., *ApJ* **421,** 276-283 (1994).

X-ray and γ-ray Observations of the COS-B Field 2CG 075+00

R. Mukherjee*, E. Gotthelf†, D. Stern*, M. Tavani†+

* *Dept. of Physics & Astronomy, Barnard College & Columbia University, NY, NY 10027*
† *Columbia Astrophysics Lab., Columbia University, NY, NY 10027*
+ *IFCTR-CNR, via Bassini 15, I-20133 Milano, Italy*

Abstract. We present a summary of γ-ray and X-ray observations of the intriguing COS-B field, 2CG 075+00, in order to search for potential counterparts. The third EGRET (3EG) catalog shows that the COS-B emission corresponds to at least two localized γ-ray sources, 3EG 2016+3657 and 3EG J2021+3716. We present analyses of archival X-ray fields which overlap error boxes of both the EGRET sources.

INTRODUCTION

The EGRET (Energetic Gamma Ray Experiment Telescope) instrument on the Compton Gamma Ray Observatory (CGRO) has surveyed the γ-ray sky at energies above 100 MeV, detecting more than 270 point sources [4]. Of these, a large fraction ($\sim$ 60%) remain unidentified, with no convincing counterparts at other wavelengths. Some of these unidentified sources were previously observed with the COS-B satellite, which carried out one of the first surveys of the γ-ray sky [14]. Surprisingly, only two of the unidentified COS-B sources have been subsequently associated with EGRET sources, and both are pulsars namely, Geminga [1], and 2CG 342-02 (PSR B1706-44) [15]. The nature of the unidentified γ-ray sources remains a long-standing mystery of high energy astrophysics.

In this article we re-visit the region containing the unidentified COS-B source 2CG 075+00, located in the Cygnus region, for which a significant amount of archival γ-ray (EGRET) and X-ray (ASCA & ROSAT) data have accumulated. Previous attempts to locate the origin of the high energy emission have been frought with frustration, as the position associated with 2CG 075+00 in the second EGRET catalog [16], 2EG J2019+3719, has shifted significantly in the third ERGET catalog, and split between two nearby sources, 3EG 2016+3657 and 3EG 2021+3716. Fortunately, both revised EGRET error boxes have overlapping archival ASCA and ROSAT observations.

CP510, *The Fifth Compton Symposium,* edited by M. L. McConnell and J. M. Ryan

I THE γ-RAY OBSERVATIONS

2CG 075+00, first observed by COS-B, is located in the Galactic plane, at $l = 75°$, $b = 0.0°$. The second COS-B catalog indicates an error radius of $\sim 1.0°$ for the source and notes that the source structure could possibly be interpreted as extended features [14]. The integrated γ-ray flux from the source was given to be 1.3×10^{-8} ph cm^{-2} s,$^{-1}$ although no spectral information was available.

Since its launch in 1991, EGRET has observed the error circle of 2CG 075+00 several times. Spatial analysis of the EGRET fields is performed by comparing the observed γ-ray map to that expected from a model of the diffuse Galactic and extragalactic radiation [5,12]. A maximum likelihood method is used to determine the source location and flux as a function of energy [7].

In the second EGRET (2EG) catalog, 2CG 075+00 was weakly detected as 2EG J2019+3719. Reanalysis of the region using a larger data set for the third EGRET calolog (3EG) revealed two sources, 3EG 2016+3657 and 3EG 2021+3716, located 0.8° away from the initial EGRET source. These more accurate, revised positions were derived from a likelihood analysis of the EGRET data for energies > 100 MeV for the combined Phase 1 through Cycle 4 data (1991-1995) [4]. The two source positions have errors of 33′ and 18′, respectively, at the 95 % contour.

Figure 1 shows the light curves of 3EG 2016+3657 and 3EG 2021+3716. The horizontal bars on the individual data points denote the extent of the viewing period for that observation. Fluxes have been plotted for all detections greater than 2σ. For detections below 2σ, upper limits at the 95% confidence level are shown. The flux levels of both the sources are roughly constant over the period of the EGRET observations, in contrast to that observed in blazars.

II THE X-RAY OBSERVATIONS

Archival ROSAT and ASCA observations were available for both 3EG 2016+3657 and 3EG 2021+3716. The fields were of interest historically due to the presence

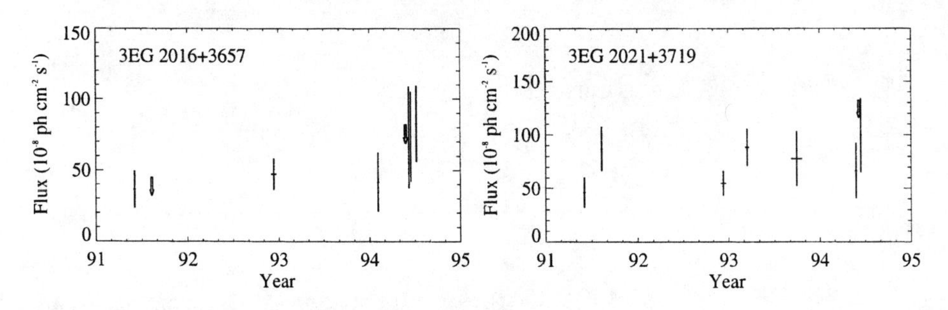

FIGURE 1. EGRET light curves for (a) 3EG 2016+3657 and (b) 3EG 2021+3716 from 1991 to 1995. 2σ upper limits are shown as downward arrows.

of the 2CG source, as well as several other X-ray sources known to exist in the region. Two adjacent observations with ROSAT and ASCA fall nicely on the two 3EG error boxes.

We present data acquired with the ROSAT PSPC (Position Sensitive Proportional Counter) and the ASCA Gas Imaging Spectrometer (GIS) which allow complementary broad-band X-ray data in the 0.2 – 10 keV range with arcmin spatial resolution and moderate energy resolution. The PSPC 1° radius field-of-view is about twice that of the GIS. All data was obtained from the HEASARC archive at Goddard Space Flight Center and edited using the latest standard processing for each mission.

We created ROSAT and ASCA images of the region containing 3EG J2016+3657 and 3EG J2021+3716 by co-adding exposure corrected sky maps from each mission (see Fig. 2). These images are centered on the position of the earlier second EGRET catalog source, 2EG 2019+3716. However, the PSPC image size is large enough to include the 95 % error contours of both the 3EG sources, the positions of which are indicated with crosses. Note that the ASCA images are not centered on the EGRET positions, and only part of the 95 % error contour of 3EG 2016+3657 is covered by the ASCA observation.

The ROSAT maps were examined to search for a possible X-ray counterpart to the two 3EG γ-ray sources. The detected positions of X-ray sources in the ROSAT field are numbered in the image and are tabulated in Table 1. Several of these source are well known and were the target of the X-ray study.

The ASCA images were, similarly, searched for corresponding X-ray counterparts. No point sources were found in the ASCA image within the 95 % contour of 3EG

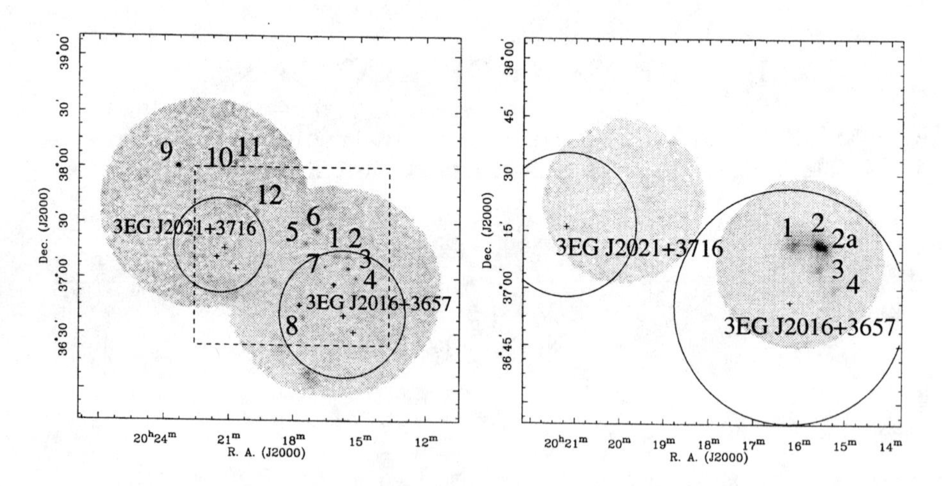

FIGURE 2. ROSAT (left) and ASCA (right) images of 3EG 2016+3657 and 3EG 2019+3719. The circles correspond to the 95% contours for the EGRET sources. The dashed rectangle in ROSAT image corresponds to the size of the ASCA image.

TABLE 1. X-ray sources in the ROSAT field of 3EG 2016+3657

Number[a]	Source Name	RA	Dec	Count Rate[b]	Other sources
1	CTB87	20 16 09.67	+37 12 17.5	24 ± 0.7	4C+37.57
2	1RXP J201534+3	20 15 34.57	+37 11 08.9	24 ± 0.7	1WGA J2015.5+3, 2E2013.7+370
3	2E2013.7+3655	20 15 38.52	+37 04 45.0	30 ± 0.7	
4	No counterparts			9 ± 0.6	
5	246	20 17 29.71	+37 18 31.3	3 ± 0.4	PPM 74637, SAO 69765
6	2E2015.1+3715	20 16 59.56	+37 25 18.6		1RXS J201700.4, HD193077 PPM 84624, SAO 69755
7	1WGA J2016.6+3	20 16 37.70	+37 05 53.8		
8	1WGA J2017.5+3	20 17 34.6	+36 38 06.6		1RXP J201736+3
9	1WGA J2023.3+3	20 23 21.70	+38 00 03.7		1RXP J202322+3
10	No counterparts				
11	1WGA J2020.7+3	20 20 43.30	+38 02 00.8		1RXP J202042+3
12	1WGA J2019.7+3	20 19 44.4	+37 35 44.8		

(a) Identifying number in ROSAT image (Fig. 2). (b) Source counts (ASCA) extracted from a 3 arcmin diameter aperture, in the 2-10 keV energy band and are background subtracted.

J2021+3716. The ASCA image of 3EG 2016+3657 revealed 5 point sources, as indicated with numbers in Fig. 2 (right). Source numbers 1, 2, 3 and 4 correspond to ROSAT sources of the same numbers in Fig. 2 (left). Source number 1 is coincident with the supernova remnant (SNR), CTB 87 (G74.9+1.2), that has a flat radio spectrum, with spectral index 0.2 ± 0.2. Table 1 gives the ASCA count rate for the four sources corresponding to the ROSAT sources. To measure the source count rate we extracted photons using a $2'$ radius aperture and estimated the background contribution using a large annulus away from the other source following the method described in [3]. Source 2a in the ASCA image has no counterpart in the ROSAT image. We get an ASCA count rate of $15 \pm 0.4 \times 10^{-3}$ for this source. Source 4 in the ROSAT and ASCA images appears to have no counterparts at other wavelengths. Further work on these sources is currently in progress, and will be presented elsewhere [9].

III SUMMARY

We present a high energy study of the revised EGRET position of the intriguing COS-B field, 2CG 075+00, in order to search for possible X-ray counterparts. Neither of the two EGRET sources, 3EG 2016+3657 and 3EG 2021+3716, exhibit any significant evidence of variability, unlike for the typical EGRET blazar observation. No potential spectrally flat, radio-loud AGN counterparts exist for these sources.

In the past, efforts to identify the COS-B sources have included systematic multiwavelength observations. The field of 2CG 075+00 was mapped with the Effelsberg radio telescope at several frequencies [10], but no convincing counterparts were obtained.

It is interesting that no prominent X-ray source is in the gamma-ray error boxes considered here. Isolated gamma-ray pulsars at the distance of a few hundred parsecs might be consistent with both 3EG sources. Our study of archival X-ray (ASCA and ROSAT) data yields several faint sources within the error boxes of the two 3EG sources. The region contains tracers of star formation, several Wolf–Rayet stars, and OB associations. We notice the presence of the SNR CTB87 in the field of 3EG 2016+3657, a fact potentially quite important in light of the gamma-ray source/SNR associations noticed in previous investigations [13]. However, with the present data, given the large error boxes, it is not possible to argue in favor of any one source as the plausible counterpart to the EGRET sources.

Gamma-ray production from SNRs, Wolf-Rayet and OB associations is expected in several theoretical models, and our observations are a step towards the identification of a class of non-blazar unidentified gamma-ray sources near the Galactic plane. This subject was extensively investigated in the past for COS-B sources [8,17], and recently for EGRET sources [11,2,6]. Clearly, for the 3EG sources considered here, we need more refined gamma-ray positions and estensive monitoring (possibly by AGILE and GLAST) to establish their ultimate nature.

E.V.G's research is supported by NASA LTSA grant NAG5-7935. D.S. acknowledges support from the Hughes Grant at Barnard College.

REFERENCES

1. Bertsch, D. L., et al. 1992, Nature, 357, 306.
2. Esposito, J. A., et al. 1996, ApJ, 461, 820.
3. Gotthelf, E. V. & Kaspi, V. 1998, ApJ, 497, L29.
4. Hartman, R. C., et al. 1999, ApJS, 123, 79.
5. Hunter, S. D., et al. 1997, ApJ, 481, 205.
6. Kaaret, P. & Cottam, J., 1996, ApJ, 462, 35L.
7. Mattox, J. R., et al. 1996, 461, 396.
8. Montmerle, T. 1979, ApJ, 231, 95.
9. Mukherjee, R., Gotthelf, E. V., & Tavani, M., in prep.
10. Ozël, M. E., et al. 1988, A&A, 200, 195.
11. Romero, G. E., Benaglia, P., & Torres, D. F. 1999, A&A, 348, 868.
12. Sreekumar, P., et al. 1998, ApJ, 494, 523.
13. Sturner, S. J. & Dermer, C. D. 1995, A&A, 293, L17.
14. Swanenburg, B. N., et al. 1981, ApJ, 243, L69.
15. Thompson, D. J., et al. 1992, Nature, 359, 615.
16. Thompson, D. J., et al. 1995, ApJS, 102, 259.
17. Völk, H. J. & Forman, M. 1982, ApJ, 253, 188.

Multiwavelength studies of the peculiar gamma-ray source 3EG J1835+5918

O. Reimer[1], K.T.S. Brazier[2], A. Carramiñana[3], G. Kanbach[1], P.L.Nolan[4], D.J. Thompson[5]

[1] *Max-Planck-Institut für extraterrestrische Physik, 85740 Garching, Germany*
[2] *University of Durham, DH1 3LE Durham, England*
[3] *Instituto Nacional de Astrofísica Optica y Electrónica, Tonantzintla, México*
[4] *W. W. Hansen Experimental Physics Lab, Stanford University, CA 94395 Stanford, USA*
[5] *LHEA Code 660, NASA GSFC, MD 20771, Greenbelt, USA*

Abstract. The source 3EG J1835+5918 was discovered early in the CGRO mission by EGRET as a bright unidentified γ-ray source outside the galactic plane. Especially remarkable, it has not been possible to identify this object with any known counterpart in any other wavelengths band since then. Analyzing our recent ROSAT HRI observation, for the first time we are able to suggest X-ray counterparts of 3EG J1835+5918. The discovered X-ray sources were subject of deep optical investigations in order to reveal their nature and conclude on the possibility of being counterparts for this peculiar γ-ray source.

GAMMA-RAY OBSERVATIONS

EGRET observations of the unidentified γ-ray source 3EG J1835+5918 above 100 MeV in CGRO observation cycles 1 to 4 are covered in the Third EGRET catalog [1]. Moreover, 3EG J1835+5918 has been reported as a GeV γ-ray emitter [2], [3]. In order to obtain the most comprehensive data base on 3EG J1835+5918, we expanded the analysis up to the most recent EGRET observations (CGRO cycle 7). Viewing periods with 3EG J1835+5918 in the field of view were examined separately at energies above 100 MeV and above 1 GeV. As reported earlier [4], 3EG J1835+5918 was only seen by EGRET at large off-axis angle early in the mission, resulting in the indication of flux variability. The most recent variability study of EGRET sources above 100 MeV [5] restricts the off-axis location of any γ-ray source to be within 25°. Considering only nine periods matching this criterion, 3EG J1835+5918 was found to be constant within statistics. In order to acknowledge this approach, we label observations with up to 25° off-axis location different than observations outside 25°, see fig.1. The flux of 3EG J1835+5918 during the observatins in cycle 7 (13-27 January 1998, aspect angle 5°) can be evaluated by

CP510, *The Fifth Compton Symposium,* edited by M. L. McConnell and J. M. Ryan

considering a similar on-axis observation of Geminga during 7-21 July 1998. If we assume that the EGRET sensitivity has not changed appreciably between these observations and that Geminga remains a stable emitter in γ-rays as previously observed, we can derive an normalization for the flux of 3EG J1835+5918 in cycle 7. Figure 1 shows the resulting flux history of 3EG J1835+59 above 100 MeV throughout the EGRET mission.

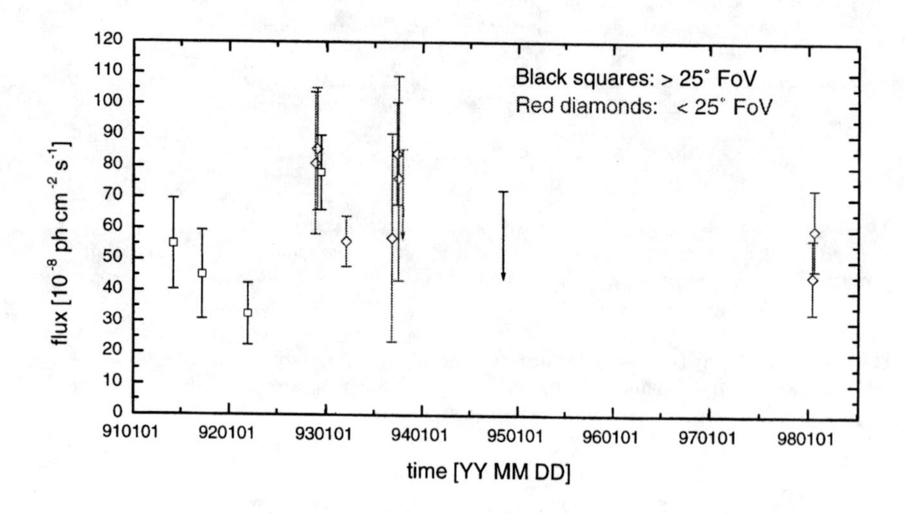

FIGURE 1. Flux history of EGRET observations on 3EG J1835+5918

The high-energy γ-ray spectrum is determined from EGRET observation of CGRO cycle 1 to 4. The power law spectral index is about -1.7 between 70 MeV and 4 GeV. Striking similarities to the γ-ray spectra of identified pulsars like Geminga and Vela can be seen in fig.2: the hard power law spectral index, a high-energy spectral cut-off or turnover and a low energy spectral softening.

The γ-ray source location is determined separately above 1 GeV using observations from cycle 1 to 7. Its precision (68% and 95% source location within a few arcminutes) allows us to cover the complete γ-ray error box with only one ROSAT HRI pointing. The γ-ray source confidence contours and the ROSAT HRI photon density is shown in fig.3.

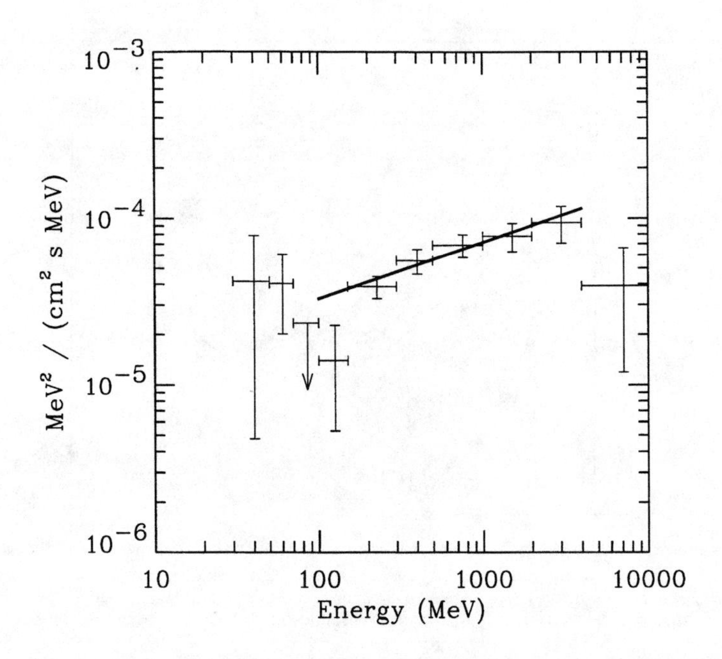

FIGURE 2. High energy gamma-ray spectrum of 3EG J1835+5918

X-RAY OBSERVATIONS

With the 60 ksec ROSAT High Resolution Imager observation from December 1997/January 1998, the only previous HRI X-ray exposure of this source could be increased by a factor of 12. For the first time, we discovered point sources at X-ray energies between 0.1 and 2.4 keV. The sources are all faint with typical HRI count rates of 1-3 $10^{-3}s^{-1}$. Two of the ten discovered sources are not in positional agreement with the determined > 1 GeV γ-ray source location contour, and therefore not considered as counterpart candidates. Using only ROSAT HRI data at this time, no spectral information on the discovered X-ray point sources is available.

OPTICAL OBSERVATIONS

The discovered X-ray sources were subject of optical identification campaigns at the 2.12m telescope of the Observatorio Astrofísico Guillermo Haro (Cananea, México). A detailed description of the optical observations on 3EG J1835+5918 is presented elsewere in these proceedings [6].

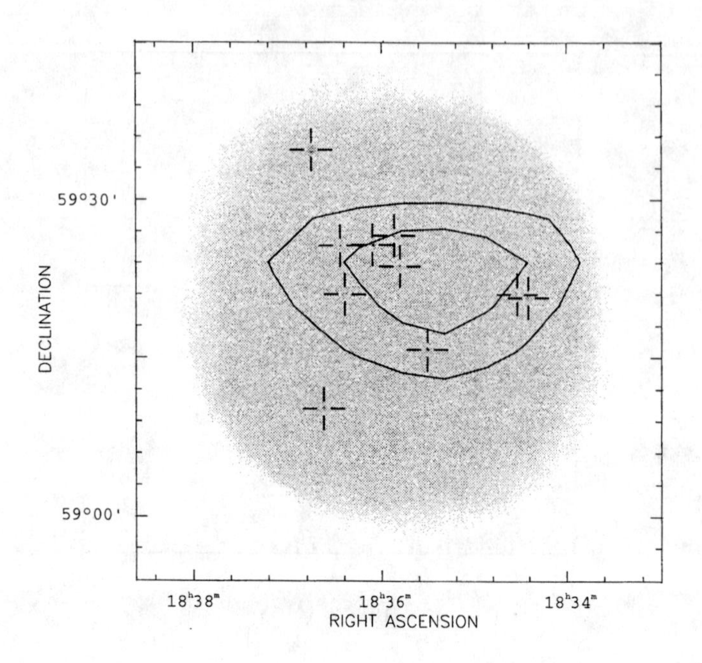

FIGURE 3. EGRET > 1 GeV source location contours (68% and 95%) overlaid on the 60 ksec ROSAT HRI image

RADIO OBSERVATIONS

Deep searches at radio wavelengths (770 MHz) at the position of 2EG J1835+59 have not detected any object above 0.5 Jy [7]. This result is in agreement with the correlation study between unidentified EGRET sources and catalogued flat-spectrum radio sources using the Green Bank 4.85 MHz and Parkes-MIT-NRAO 4.85 MHz surveys, which also did not find any counterpart for 2EG J1835+59 [8].

SUMMARY & CONCLUSIONS

The brightest unidentified EGRET source at high galactic latitudes was subject of a multifrequency identification campaign. For the first time, counterparts in X-rays are suggested. The optical identification of the X-ray counterparts has been finished for the brighter sources [6], resulting in the elimination of four of the viable eight X-ray candidates. The eliminated X-ray sources are identified with stars or distant galaxies unlikely to be the γ-ray source. Spectra for the fainter candidates will have to be obtained at larger telescopes. This is currently in progress at the 6m telescope of the Special Astrophysical Observatory (Zelenchuk, Russia). The pulsar-like spectrum in γ-rays, the high-galactic latitude source location and the lack of any blazar class object or flat spectrum radio source would suggest a nearby radio-quiet pulsar. Such pulsars are predicted [9] to exist among the unidentified γ-ray sources seen by EGRET. Perhaps the first ones were already found within the γCygni supernova remnant [10], the CTA1 SNR [11], with GeV J1417-6100 [12], and 2EG J0635+0521 [13]. We will conclude on the nature of this enigmatic γ-ray source when we will have completed the optical identifications of the remaining four weak X-ray sources coincident with 3EG J1835+5918.

REFERENCES

1. Hartman, R.C. et al., *ApJS* **123**, 79 (1999)
2. Lamb, R.C. and Macomb, D.J., *ApJ* **488**, 872 (1997)
3. Reimer, O., Dingus, B. and Nolan, P.L., Proc. 25th ICRC, Vol.3, 97 (1997)
4. Nolan, P.L. et al., AIP Conf. Proc. 304, eds. Fichtel, C.E., Gehrels, N. and Norris, J.P., 361, (1994)
5. Tompkins, W., Ph.D. thesis, Stanford University, March 1999
6. Carramiãna, A. et al., these proceedings
7. Nice, D.J. and Sayer, R.W., *ApJ* **476**, 261 (1997)
8. Mattox, J.R., Schachter, J., Molnar, L., Hartman, R.C. and Patnaik, A.R., *ApJ* **481**, 95 (1997)
9. Yadigaroglu, I.-A. and Romani, R.W. *ApJ* **449**, 211 (1995)
10. Brazier, K.T.S., Kanbach, G., Carramiñana, A., Guichard, J., & Merck, M., *MNRAS* **281**, 1033 (1996)
11. Brazier, K.T.S., Reimer, O., Kanbach, G. & Carramiñana, A., *MNRAS* **295**, 819 (1998)
12. Roberts, M.S.E., Romani, R.W., Johnston, S., Green, A.J. *ApJ* **515**, 712 (1999)
13. Kaaret, P., Pirano, S., Halpern, J, Eracleous, M., *ApJ* **523**, 197 (1999)

Optical studies of potential counterparts for unidentified EGRET sources

A. Carramiñana[1], V. Chavushyan[1,2], S. Zharikov[2], O. Reimer[3], K.T.S. Brazier[4]

[1] *Instituto Nacional de Astrofísica Optica y Electrónica, Tonantzintla, México*
[2] *Special Astrophysical Observatory, Zelenchuk, Russia*
[3] *Max Planck Institut für Extraterrestrische Physik, Garching, Germany*
[4] *University of Durham, United Kingdom*

Abstract. We present optical observations of candidate X-ray counterparts for the EGRET sources 3EG J2020+4017, 3EG J0010+7309 and 3EG J1835+5918. Preliminary spectroscopic observations of the late-type star coincident with the X-ray counterpart of 3EG J2020+4017, the γ-Cygni source, show no evidence of binariety. In the case of 3EG J0010+7309, the CTA-1 source, we performed deep optical imaging, finding a red $V \sim 22$ magnitude object inside the ROSAT contour. Finally, a detailed analysis of optical observations of several X-ray sources within the EGRET error box of 3EG J1835+5918 is presented.

THE IDENTIFICATION PROGRAMME

The Third EGRET catalog contains 170 unidentified of high-energy γ-ray sources. The distribution of these objects in the sky indicates a strong Galactic component. Unidentified EGRET sources are concentrated towards the Galactic Plane and Galactic Centre, in a similar way than pulsars, the only identified Galactic population of high-energy γ-ray emitters. The identification of Geminga, the second brightest source in the sky in the range $E_\gamma > 100\,\mathrm{MeV}$, as a radio-quiet pulsar ([5]) opened the possibility that some of these sources could be of the same nature. In fact, it is expected that at least a fraction of the unidentified EGRET sources *are* Geminga-type pulsars.

Our identification programme follows the steps which led to the identification of Geminga and is aimed at identifying the best radio-silent candidates through observations at different spectral bands, mainly X-rays and optical. Apart from our data, we rely on the public radio data, like those from the Green-Bank survey at 4.85 GHz ([4]), through the study presented in [6] regarding the identifications

[1)] Sponsored by CONACyT research grant 211290-5-25539E

CP510, *The Fifth Compton Symposium,* edited by M. L. McConnell and J. M. Ryan

of γ-ray sources with radio-loud flat-spectrum sources. None of the sources studied has a strong association with radio-loud flat-spectrum sources.

We start selecting bright, non-variable γ-ray sources, and compute an improved positional error-box using photons with energies above 1 GeV, taking advantage of the narrower EGRET point-spread function at GeV energies. We then proceed to identify X-ray point sources from X-ray images of the ≥ 1 GeV error-box, which then are targets of optical follow-up observations. Up to now ROSAT archival data and pointed observations have been used for the X-ray analysis.

Optical observations of the ROSAT X-ray candidate counterparts of the given unidentified EGRET source are useful to identify the best of these candidates. Radio-quiet AGNs, Seyfert, starburst, normal galaxies, binary and flaring stars can be X-ray sources without expected detectable γ-ray emission. Finding no likely optical counterpart for a given X-ray source strengthens the case for the association between the X-ray and γ-ray sources. And at some level, faint blue optical objects can also be considered neutron star candidates on their own right. The ultimate goal of our programme is finding pulsations in X-ray data and test the γ-ray data for their presence.

Optical observations presented here have been carried out with the 2.12m telescope of the Observatorio Astrofísico Guillermo Haro (Cananea, Sonora) and with the 6m telescope of the Special Astrophysical Observatory (Zelenchuk, Russia). We present results for: (i) 3EG J2020+4017 (γ-Cygni); (ii) 3EG J0010+7309 (CTA-1) and, (iii) 3EG J1835+5918.

3EG J2020+4017: THE γ-CYGNI SOURCE

The multiwavelength analysis of this EGRET and COS-B source was presented in [1]. A single X-ray point source was found within the γ-ray (≥ 1 GeV photons) error-box of 3EG J2020+4026, the region also showing extended X-ray emission, reminiscent of SN remnants. Inside the X-ray $\lesssim 10''$ ROSAT-HRI error box is located a 14.5 magnitude star, spectroscopically identified as K0V, late-type star, with no signs of chromospheric activity and therefore ruled out as γ-ray source and the X-ray source.

We performed spectroscopic observations using the Boller & Chivens spectrograph at the 2.12m Cananea telescope on September 27 and 29, 1997. The purpose of these was to test the K0V star for orbital motion looking at the λ5890 NaI absorption line, with a sampling of 0.43Å/pixel. Preliminary analysis shows no sign of doppler orbital displacements. Although more systematic observations -considering the expected ranges of periods and velocity amplitudes- might be of interest, we conclude unlikely that the K0V star might have a neutron star companion. Imaging observations of the star are underway to detect fainter objects inside the ROSAT error box.

3EG J0010+7309 = CTA-1

The multiwavelength analysis of this EGRET source, previously associated with a radio-l;oud AGN, was presented in [2]. The improved X-ray contours ruled out the association with the AGN and a single ROSAT source, RX J0007.0+7302 is included inside the 95% confidence γ-ray contour. Independent analysis by [8] concludes this X-ray source to be a plerion. BVRI observations with the 2.12m inside the X-ray $\lesssim$ 10" ROSAT-HRI error box of RX J0007.0+7302 showed no optical counterpart down to $R \gtrsim 23$.

We performed observations with the 6m telescope at SAO in November 19, 1997, under poor seeing conditions. BVR images (figure shows the R image) clearly show a point source, of R magnitude between 22 and 23, which in fact can be seen in the original Cananea R and I images shown in Fig 4 of [2], just below the detection level for the 2m images. The 6m R image has a detection limit of about 24.5 magnitudes and only the cited object is seen above that threshold.

Further observations under good -photometric- conditions were performed this year (to be reported in detail elsewhere, [3]). The visual magnitude is estimated as V=22.59 ± 0.09, with the object showing clear red colours (V-R=0.92±0.10). The magnitude limit of these new images is around 25th magnitude. The colours are more consistent with a faint dwarf star or a late-type galaxy than with a neutron star, weakening its possible association with the X-ray and γ-ray source(s). Spectroscopic identification might require a 8-10m class telescope.

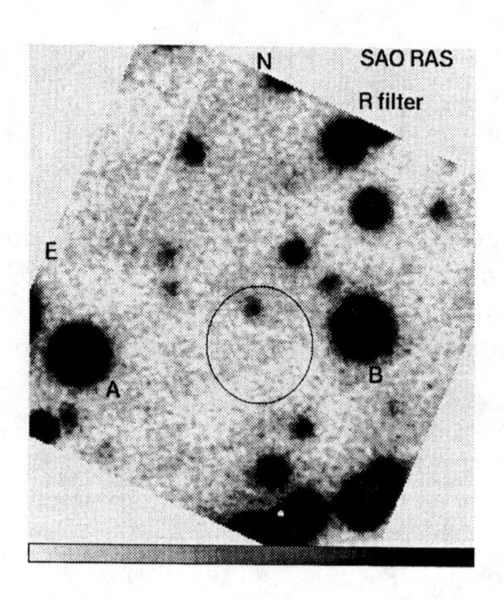

FIGURE 1. SAO 6m R image of the ROSAT error box of RX J0007.0+7302.

3EG J1835+5919

HRI observations of this bright EGRET source were performed during ROSAT AO8. Our analysis found ten X-ray point sources, which we labelled as X1 to X10 and targetted for optical observations. Although the improved 95% confidence GeV error-box excludes X1 and X10 we did include them in the optical study. A more complete multiwavelength analysis of 3EG J1835+5918 is presented elsewhere in these proceedings ([7]).

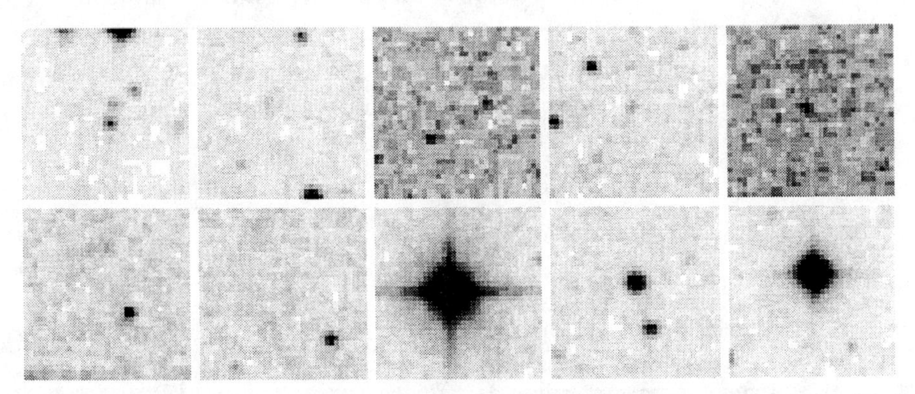

FIGURE 2. POSS 1×1 images centered in the positions of the ten X-ray sources found in the 3EG J1835+5919 region. Top images correspond, from left to right, to X1 to X5 and bottom images to X6 to X10

POSS 1×1 arcmin images of these sources are shown in Figure 2. From optical examination of these images the optical 2m follow-up strategy was decided. We summarize our results for each object:

- X1: this X-ray source is practically ruled out by positional arguments. Two objects are consistent with the X-ray position, and the spectrum obtained for the closest to the ROSAT position indicates a line emitting quasar, with $z \simeq 0.466$. We propose identifying this (radio-quiet) AGN with the X-ray source and ruling it out as counterpart for the γ-ray source.

- X2: a very faint object is seen. VRI imaging was performed, with preliminary analysis giving a non-detection in V ($m_V > 20.2$) and marginal detection in R ($m_R \simeq 20.5 \pm 0.7$). No conclusion is made on this object.

- X3: two objects lie just outside the ROSAT error box. VRI imaging performed, giving non detections inside the ROSAT contour. No conclusion is made.

- X4: very faint object inside error-box. VRI imaging performed, with weak detections ($m_V \simeq 19.0 \pm 0.2$ and $m_R \simeq 19.5 \pm 0.2$), suggesting a red object. Spectroscopy attempted with null results. No conclusion is made.

- X5: single object inside error-box, bright enough for spectroscopy. The spectrum shows an emitting line quasar at $z \simeq 1.865$. We propose identifying this (radio-quiet) AGN with the X-ray source and ruling it out as counterpart for the γ-ray source.

- X6: single object at the edge of the error-box, bright enough for spectroscopy. The spectrum indicates a late type star, probably a M5V star.

- X7: nothing inside error box; VRI imaging and spectroscopy of nearby object was performed. The S/N of the spectrum doesn't allow us to identify the object. No conclusion is made.

- X8: single very bright object at X-ray positions ($m_V \sim 11$). The spectrum obtained indicates a G dwarf star. We propose its identification as the X-ray source and ruling it out as the γ-ray source.

- X9: two fairly bright objects inside error box. Both spectra correspond to late type stars, M-type for the brightest and G for the faintest. The positional coincidence with the M star makes the identification likely, ruling it out as the γ-ray source.

- X10: practically ruled out by positional arguments. A single very bright object is at X-ray positions and its spectrum indicates a K5V star. We rule out the X-ray source as the γ-ray source.

In short we consider the association of any of X1, X5, X6, X8, X9 and X10 with the γ-ray source unlikely. We are left with four candidates (X2, X3, X4 and X7) which deserve further optical study. With the exception of X7, conclusive observations almost certainly require observations with a $>$ 4m telescope.

REFERENCES

1. Brazier K.T.S., Kanbach G., Carramiñana A., Guichard J., & Merck M., *MNRAS* **281**, 1033 (1996).
2. Brazier K.T.S., Reimer O., Kanbach G. & Carramiñana A., *MNRAS* **295**, 819 (1998).
3. Carramiãna el al. in preparation
4. Condon JJ, Broderick JJ & Seielstad GA, AJ 97, 1064 (1989)
5. Halpern J.P. & Holt S., Nature 357, 222(1992)
6. Mattox J.R., Schachter J, Molnar L, Hartman RC & Patnaik AR, *ApJ* **481**, 95 (1997)
7. Reimer et al., these proceedings
8. Slane P. et al., ApJ 485, 221 (1997)
9. Yadigaroglu I.A. & Romani R., ApJ 449, 211 (1995)

A Systematic Search for Short-term Variability of EGRET Sources

P. M. Wallace[1], D. L. Bertsch[2], S. D. Bloom[3], N. J. Griffis[1], S. D. Hunter[2], D. A. Kniffen[4], D. J. Thompson[2]

[1]*Department of Physics and Astronomy, Berry College, Mt. Berry, GA 30149*
[2]*NASA/Goddard Space Flight Center, Greenbelt, MD 20771*
[3]*IPAC, JPL/Caltech, MS 100-22, Pasadena, CA 91125*
[4]*Department of Physics and Astronomy, Hampden-Sydney College, Hampden-Sydney, VA 23943 and NASA HQ, Washington, DC 20546*

Abstract. The 3rd EGRET Catalog contains 170 unidentified high-energy (E>100 MeV) gamma-ray sources, and there is great interest in the nature of these sources. One means of determining sources class is the study of flux variability on time scales of days; pulsars are believed to be stable on these scales while blazars are known to be highly variable. In addition, previous work has led to the discovery of 2CG 135+01 and GRO J1838-04, candidates for a new high-energy gamma-ray source class. These sources display transient behavior but cannot be associated with any known blazars. These considerations have led us to conduct a systematic search for short-term variability in EGRET data, covering all viewing periods through cycle 4. Three unidentified sources show some evidence of variability on short time scales; the source displaying the most convincing variability, 3EG J2006-2321, is not easily identified as a blazar.

INTRODUCTION

There are 271 sources listed in the 3rd EGRET Catalog of High-energy Gamma-ray Sources[1]. Besides one solar flare, the Large Magellanic Cloud, and a possible association with a radio galaxy (Cen A), the identified sources are distributed among two established classes of high-energy gamma rays: pulsars and radio-loud blazars. Pulsars are believed to not vary in gamma-ray output over time scales of one or two days, while blazars are known to be highly variable. While many instances of blazar flares have been reported, no comprehensive survey of EGRET data has been performed. It is the purpose of this study to conduct a systematic search for short-term variability in EGRET data from cycles 1-4. This paper focuses on the unidentified 3EG sources.

DATA & ANALYSIS

All unidentified 3EG sources are examined across all viewing periods (VP's) for evidence of variability. The VP's are broken down into one-day intervals and intensity

CP510, *The Fifth Compton Symposium,* edited by M. L. McConnell and J. M. Ryan

maps are generated for each day. With such short intervals, the statistics are extremely limiting; therefore this study is sensitive to only the strongest changes in gamma-ray output. Only those light curves with at least one 4σ one-day detection are considered for close analysis.

The remaining light curves are analyzed using the variability index V. If Q is the probability of obtaining a value of χ^2 equal to or greater than the empirical χ^2 from an intrinsically nonvariable source, then $V \equiv -\log Q$. All curves with $V \leq 1.0$ are considered to be not variable. The curves are also inspected for evidence of flaring; those that display such evidence are modeled by Monte Carlo methods in order to determine the probability of finding such a flare from an intrinsically stable source. Three unidentified 3EG sources displayed $V \geq 1.0$ and/or evidence of flaring. They are discussed below.

3EG J1410-6151

During the first four days of VP 14.0, the flux of 3EG J1410-6151 fell from $(5.4\pm1.5) \times 10^{-6}$ photons $cm^{-2}s^{-1}$ to below EGRET's sensitivity where it remained for the rest of the 14-day period; this is suggestive of flaring behavior. (See Figure 1.) Monte Carlo simulation gives a probability of 0.0007 that the fluctuation found in this VP is produced by a nonvariable source. It has been suggested[2] that 3EG J1410-6103 ($l = 312.18$, $b = -0.35$) is associated with SNR G312.4-0.4, which falls just outside the 68% error contour. It should be noted that although this VP occurred early in EGRET's life when its sensitivity was high, the source is 27° off-axis.

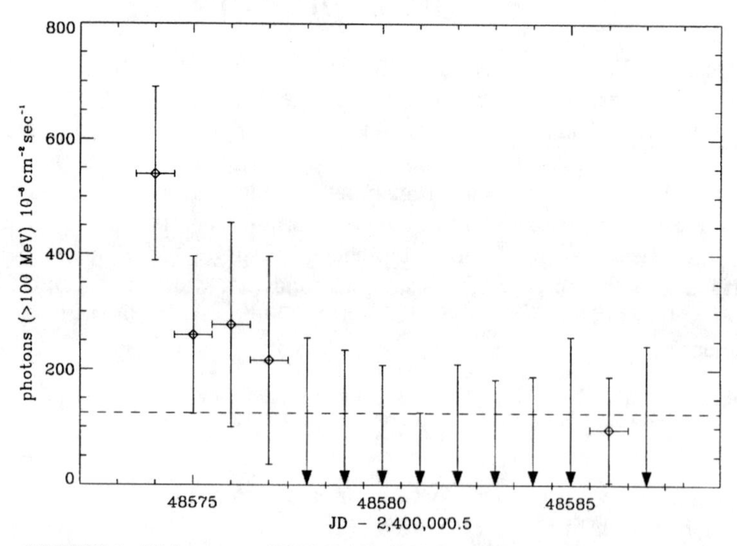

FIGURE 1. Light curve of 3EG J1410-6151 from VP 14.0. $V = 1.47$.

3EG J1746-2851

As this source is unidentified, strong, and coincident with the Galactic Center, it has been studied in some detail[3]. However, until now its short-term variability has not been examined, and there is some evidence of variability in VP's 16.0 and 429.0. 3EG J1746-2851 sits in the most densely-packed region of the high-energy gamma-ray sky; there are ten sources listed in the 3rd EGRET Catalog within 10° of the Galactic Center. Given the broad EGRET PSF, source confusion is a serious problem. However, while 3EG J1746-2861 appears to fluctuate during two different VP's, no other sources in confused regions display any evidence of short-term variability.

The light curve of 3EG J1746-2851 during VP 16.0 is shown in Figure 2. The three strongest one-day detections fall on days 7-9 of the two-week VP, during which the aspect was 20°. The peak detection has a significance of 4.3σ and is flanked by detections of 3.9σ and 3.1σ. The variability index is 2.09, corresponding to a probability of 0.008 that these data are consistent with a nonvariable source. Monte Carlo analysis is more restrictive. This source and the seven others within a 7° radius were modeled and there is a probability of 0.0004 that a three-day fluctuation of this or greater significance will occur in a 14-day period given intrinsically nonvariable sources.

3EG J1746-2851 also shows evidence of variability in VP 429.0. During this pointing the aspect is only 6° and $V = 3.0$. The peak flux is $(6.4\pm1.7) \times 10^{-6}$ photons cm^{-2} s^{-1} and on two days the source is not detected at all, but there is no evidence of flaring.

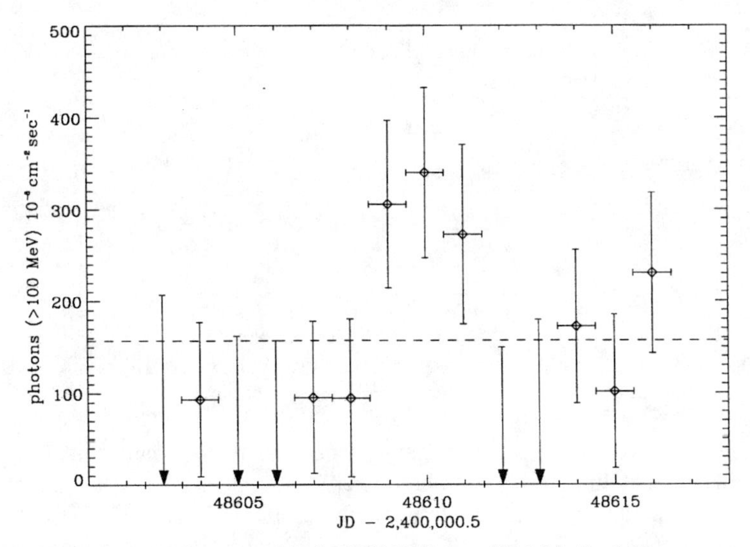

FIGURE 2. Light curve of 3EG J1746-2851 from VP 16.0. $V = 2.09$.

3EG J2006-2321

This source shows strong variability in VP 13.1, during which it was 13° from the instrument axis. The light curve shows evidence of flaring and is shown in Figure 3. The variability index for this curve is 3.18; the Monte Carlo probability that the source is nonvariable is 0.0006. 3EG J2006-2321 is well-isolated and lies 26° off the Galactic Plane, free of the bright galactic diffuse radiation; thus the claim of variability is strengthened.

The combination of large $|b|$ and flaring behavior suggests an association with the blazar class of AGN. However, all of the 66 3EG sources identified as AGN are associated with loud spectrally flat radio sources; for 3EG 2006-2321 no such association can be easily made. The best candidate is the radio source PMN J2005-2310 (260 mJy at 4.85 GHz, α_r not known), for which the probability of association[4] with 3EG J2006-2321 is only 0.015. If this source is of extragalactic origin, then it is unlike other EGRET AGN; of the 10 AGN with peak flux above 10^{-6} photons cm^{-2} s^{-1}, none are weaker than 1.0 Jy at 4.85 GHz.

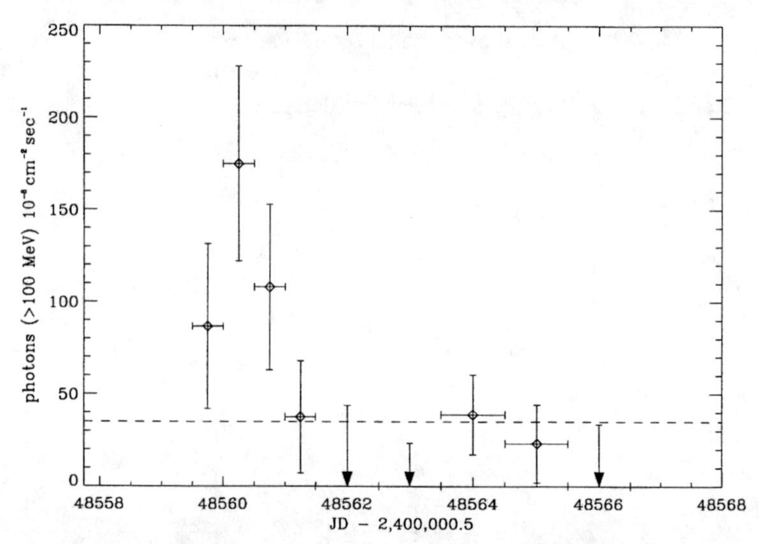

FIGURE 3. Light curve of 3EG J2006-2321 from VP 13.1. $V = 3.18$. The first 4 points represent 12-hour integration times; the final 5 represent 24-hour integration times.

Recently, two other sources have been found to share this combination of variability, peak flux above 10^{-6} photons cm^{-2} s^{-1}, and lack of easy association with a radio-loud spectrally flat counterpart: 2CG 135+01 and GRO J1838-04[5,6]. An association of 2CG 135+01 (3EG J0241+6103) with the radio source GT 0236+610 has been suggested but not confirmed; GT 0236+610 itself is associated with the massive binary system LS I +61°303. To date there are not plausible counterparts, galactic or extragalactic, for GRO J1838-04 (3EG J1837-0423). 3EG J2006-2321, along with these two sources, may be representative of a new class of high-energy gamma-ray emitters. However, there are some differences among these three sources; unlike 3EG J2006-2321, the other two sources are very close to the Galactic Plane. Also, while 2CG 135+01 is found by the present study to be variable on short time scales, GRO 1838-04 is not; it is a very bright but steady source in VP 423.0.

Further study of possible association of this source with PMN J2005-2310 is underway.

CONCLUSION

The survey of EGRET data from cycles 1-4 finds three unidentified sources that display some evidence of short-term variability; large statistical errors ensure that we detect only the strongest variations. Of these three, only 3EG J2006-2321 is strongly variable. If this source is an AGN, its radio characteristics are unlike those of other bright (peak flux > 10^{-6} photons cm^{-2} s^{-1}) EGRET blazars. If it is not extragalactic in origin, it may, with 2CG 135+01 and GRO J1838-04, represent a new class of high-energy gamma-ray emitters. Study of this source continues.

REFERENCES

1. Hartman, R. C. et al. 1999 ApJS, 123, 79
2. Sturner. S. J. & Dermer, C. D. 1995, A&A, 293, L17
3. Mayer-Hasselwander, H. A. et al. 1998, A&A 335,161
4. Mattox, J. R. 1999, these proceedings
5 Kniffen, D. A. et al. 1997, ApJ, 486, 126
6. Tavani, M. et al. 1998, ApJ, 497, L89

Spectral Modeling of the EGRET 3EG Gamma Ray Sources Near the Galactic Plane

D.L. Bertsch[1], R.C. Hartman[1], S.D.Hunter[1], D.J. Thompson[1], Y.C. Lin[2], D.A. Kniffen[3], G. Kanbach[4], H.A. Mayer-Hasselwander[4], O. Reimer[4], and P. Sreekumar[5]

1. *Code 661, NASA/Goddard Space Flight Center, Greenbelt, MD 20771 USA*
2. *W. W. Hansen Exp. Phys. Lab., Stanford Univ., Stanford, CA 94305 USA*
3. *Code S, NASA Headquarters, 300 E. St. SW, Washington, DC 20024*
4. *Max-Planck Institut fur Extraterrestrische Physik, Giessenbachstr. D-85748, Garching, Germany*
5. *ISRO Satellite Center, Bangalore, India*

Abstract. The third EGRET catalog lists 84 sources within 10° of the Galactic Plane. Five of these are well-known spin-powered pulsars, 2 and possibly 3 others are blazars, and the remaining 74 are classified as unidentified, although 6 of these are likely to be artifacts of nearby strong sources. Several of the remaining 68 unidentified sources have been noted as having positional agreement with supernovae remnants and OB associations. Others may be radio-quiet pulsars like Geminga, and still others may belong to a totally new class of sources. The question of the energy spectral distributions of these sources is an important clue to their identification. In this paper, the spectra of the sources within 10° of Galactic Plane are fit with three different functional forms; a single power law, two power laws, and a power law with an exponential cutoff. Where possible, the best fit is selected with statistical tests. Twelve, and possibly an additional 5 sources, are found to have spectra that are fit by a breaking power law or by the power law with exponential cutoff function.

INTRODUCTION

The gamma ray sources near the Galactic Plane are likely to be from a more than one class of objects that are associated with our Galaxy. The spectral properties of these sources may offer a distinction between different source mechanisms. The five known pulsars for example exhibit relatively hard spectra and all except the Crab break at high energies. This paper examines the spectral properties of the EGRET third catalog (Hartman et al. 1999) sources within ±10° of the Galactic plane.

ANALYSIS

Each source in the EGRET Third Catalog (Hartman et al., 1999) was fitted with three functional forms, a single power law, two matching power laws, and a power law, modified by an exponential cut-off as shown in the following equations;

$$\frac{\partial J}{\partial E}(E, K, E_0, \lambda) = K\left(\frac{E}{E_0}\right)^{-\lambda} \tag{1}$$

CP510, *The Fifth Compton Symposium,* edited by M. L. McConnell and J. M. Ryan

$$\frac{\partial J}{\partial E}(E,K,\lambda_1,\lambda_2)=K\left(\frac{E}{1000\,MeV}\right)^{-\lambda_1} \quad for \quad E\leq 1000\,MeV$$
$$=K\left(\frac{E}{1000\,MeV}\right)^{-\lambda_2} \quad for \quad E\geq 1000\,MeV \tag{2}$$

$$\frac{\partial J}{\partial E}(E,K,\lambda,E_f)=K\left(\frac{E}{300\,MeV}\right)^{-\lambda}\exp(-E/E_f) \tag{3}$$

In eq. 1, E_0 was set to the value determined by the EGRET Spectral program to minimize the correlation between the two other fit parameters. The location of the break energy in eq. 2 was set to 1000 MeV to keep the number of parameters at a minimum since at best, there are only 10 energy points available. With each fit, a reduced χ^2 was obtained, and an F-Test was done to see if there is statistical justification in using either of the forms in eqs. 2 or 3 rather than a simple power law to fit the observed spectral data. In the F-Test, a value of $P < 0.05$ is generally taken as the point where the more complicated fit is warranted. Summed data sets for Phases 1 through 4 of the CGRO (Compton Gamma Ray Observatory) mission were used to maximize the statistics since the sources in the Galactic plane region of the sky do not show strong variability.

RESULTS

Six sources that are listed in the third EGRET catalog are thought to be artifacts from residual emission in the wings of the PSF (Point Spread Function) from Vela (see Hartman et al., 1999). These were removed from consideration, and just the remaining 78 sources were modeled. The photon energy spectra of the majority of these 78 sources were found to be best represented by a simple power law whose index is given in the catalog paper (Hartman et al., 1999). However, 28 of these were judged to be too limited statistically at the extremes of EGRET's energy range to have a meaningful measure of a departure from a simple power law spectrum.

The F-Test analysis of the change in χ^2 between a simple power law and the more complex forms (eqs. 2 and 3) indicated that 12 sources have complex spectra. Another 5 sources that are weak statistically also may exhibit a curving or breaking form. Table 1 lists the results of the two-power law modeling of these 17 sources. The column labeled "Spectral Category" indicates by the "2P" designation the 5 sources that are best fit by a breaking power law. These are also shown in bold in Table 1. The spectra of sources designated by "PE" are better described by power-law with an exponential cutoff. The sources labeled "SL" (italicized) are statistically limited and either model fits them reasonably well. The parameters of eq. 2 and their

Table 1. Two-Power-Law Fits to the EGRET Sources Within 10° of the Galactic Plane

Name	Type	Galactic Long. deg.	Galactic Lat. deg.	Spectral Sigif. σ	Spectral Cat.	Coefficient 10^{-11} cm^{-2} s^{-1}	Index-1	Index-2	Red. χ^2
3EG_J0617+2238	**U**	**189.00**	**3.05**	**17.4**	**2P**	**5.89 ± 0.70**	**-1.79± 0.09**	**-2.65 ± 0.24**	**0.91**
3EG_J1710-4439	**P**	**343.10**	**-2.69**	**21.4**	**2P**	**16.71 ± 9.08**	**-1.69 ± 0.07**	**-2.26 ± 0.13**	**0.73**
3EG_J1736-2908	**U**	**358.79**	**1.56**	**5.8**	**2P**	**7.15 ± 1.67**	**-1.49 ± 0.23**	**-5.80 ± 1.39**	**0.92**
3EG_J1746-2851	**U**	**0.11**	**-0.04**	**17.5**	**2P**	**13.29 ± 2.83**	**-1.20 ± 0.27**	**-2.31 ± 0.26**	**2.64**
3EG_J2021+3716	**U**	**75.58**	**0.33**	**10.3**	**2P**	**10.19 ± 1.44**	**-1.23 ± 0.15**	**-3.39 ± 0.36**	**0.55**
3EG_J0633+1751	P	195.13	4.27	76.4	PE	49.3 ± 14.1	-1.38 ± 0.07	-2.54 ± 0.16	7.51
3EG_J0834-4511	P	263.55	-2.79	73.8	PE	107.6 ± 40.2	-1.48 ± 0.08	-2.50 ± 0.19	13.46
3EG_J1655-4554	U	340.48	-1.61	5.2	PE	5.23 ± 1.37	-1.44 ± 0.26	-6.41 ± 1.96	0.38
3EG_J1741-2050	U	6.44	5.00	6.6	PE	3.68 ± 0.67	-1.71 ± 0.16	-3.77 ± 0.59	0.62
3EG_J2020+4017	U	78.05	2.08	21.0	PE	13.63 ± 2.16	-1.87 ± 0.10	-2.71 ± 0.34	2.60
3EG_J2027+3429	U	74.08	-2.36	5.8	PE	2.55 ± 0.75	-2.02 ± 0.18	-20 ± 20	0.73
3EG_J2033+4118	U	80.27	0.73	11.8	PE	7.72 ± 2.50	-1.60 ± 0.26	-3.85 ± 1.29	2.12
3EG_J0634+0521	*U*	*206.18*	*-1.41*	*4.6*	*SL*	*1.80 ± 0.67*	*-1.56 ± 0.37*	*-3.74 ± 1.54*	*0.55*
3EG_J0747-3412	*U*	*249.35*	*-4.48*	*3.5*	*SL*	*2.23 ± 0.66*	*-1.77 ± 0.23*	*-21 ± 21*	*0.29*
3EG_J1316-5244	*U*	*306.85*	*9.93*	*5.7*	*SL*	*1.43 ± 0.34*	*-2.26 ± 0.14*	*-19 ± 19*	*0.50*
3EG_J1810-1032	*U*	*18.81*	*4.23*	*4.9*	*SL*	*2.69 ± 0.57*	*-1.98 ± 0.15*	*-4.85 ± 1.36*	*0.42*
3EG_J2206+6602	*a*	*107.23*	*8.34*	*5.2*	*SL*	*2.04 ± 0.58*	*-1.99 ± 0.20*	*-5.66 ± 2.29*	*0.34*

Table 2. Power-Law-With-Exponential-Cutoff Fits to Sources Within 10° of the Galactic Plane

Name	Type	Galactic Long. deg.	Galactic Lat. deg.	Spectral Sigif. σ	Spectral Cat.	Coefficient 10^{-11} cm^{-2} s^{-1}	Index	*e*-Folding Energy MeV	Red. χ^2
3EG_J0633+1751	**P**	**195.13**	**4.27**	**76.4**	**PE**	**37.14 ± 1.69**	**-1.29 ± 0.06**	**2770 ± 472**	**4.24**
3EG_J0834-4511	**P**	**263.55**	**-2.79**	**73.8**	**PE**	**80.61 ± 4.14**	**-1.45 ± 0.06**	**3807 ± 925**	**7.92**
3EG_J1655-4554	**U**	**340.48**	**-1.61**	**5.2**	**PE**	**10.10 ± 4.36**	**-0.27 ± 0.63**	**299 ± 111**	**0.30**
3EG_J1741-2050	**U**	**6.44**	**5.00**	**6.6**	**PE**	**4.97 ± 0.80**	**-1.19 ± 0.24**	**692 ± 187**	**0.36**
3EG_J2020+4017	**U**	**78.05**	**2.08**	**21.0**	**PE**	**14.70 ± 1.69**	**-1.78 ± 0.13**	**2804 ± 1428**	**2.28**
3EG_J2027+3429	**U**	**74.08**	**-2.36**	**5.8**	**PE**	**14.08 ± 7.73**	**-0.81 ± 0.47**	**234 ± 85**	**0.33**
3EG_J2033+4118	**U**	**80.27**	**0.73**	**11.8**	**PE**	**15.07 ± 6.87**	**-0.56 ± 0.62**	**370 ± 172**	**1.69**
3EG_J0617+2238	U	189.00	3.05	17.4	2P	5.97 ± 0.70	-1.68 ± 0.15	2226 ± 1003	1.27
3EG_J1710-4439	P	343.10	-2.69	21.4	2P	12.14 ± 0.78	-1.75 ± 0.07	8320 ± 3914	1.20
3EG_J1736-2908	U	358.79	1.56	5.8	2P	10.42 ± 3.90	-0.69 ± 0.58	407 ± 172	1.02
3EG_J1746-2851	U	0.11	-0.04	17.5	2P	7.02 ± 1.74	-1.12 ± 0.37	629 ± 154	0.61
3EG_J2021+3716	U	75.58	0.33	10.3	2P	7.94 ± 1.28	-0.63 ± 0.30		
3EG_J0634+0521	*U*	*206.18*	*-1.41*	*4.6*	*SL*	*23.3 ± 11.3*	*-0.85 ± 0.78*	*550 ± 385*	*0.48*
3EG_J0747-3412	*U*	*249.35*	*-4.48*	*3.5*	*SL*	*60.7 ± 46.5*	*-0.71 ± 0.86*	*300 ± 193*	*0.36*
3EG_J1316-5244	*U*	*306.85*	*9.93*	*5.7*	*SL*	*26.5 ± 12.4*	*2.20 ± 0.38*	*1201 ± 1635*	*0.91*
3EG_J1810-1032	*U*	*18.81*	*4.23*	*4.9*	*SL*	*65.5 ± 16.3*	*-1.30 ± 0.27*	*473 ± 139*	*0.22*
3EG_J2206+6602	*a*	*107.23*	*8.34*	*5.2*	*SL*	*67.6 ± 28.8*	*-1.10 ± 0.43*	*341 ± 135*	*0.19*

uncertainties along with the reduced χ^2 value of the fit are given in Table 1.

Table 2 summarizes the fits using a power-law with an exponential cutoff form to the same 17 sources. There are 7 sources that are fitted best by this model. The fit parameters of eq. 3 are tabulated here.

Figure 1 compares the F-Test probability with the source significance. It is evident that most of the sources are near the significance threshold of 5σ required of sources near the plane for inclusion in the third EGRET catalog. Some of the sources here are below the cutoff. They exceeded the Catalog threshold in either one viewing period or in some combination of viewing periods, but are not as strong in the Phase 1 through 4 data used here. Three pulsars, Vela, Geminga, and PSR 1706-44 (3EG-J1710-4439) below the dotted line and the Crab pulsar above the dotted line have the four highest significance levels in figure 1. Discounting these four sources, the remaining points have a source significance distribution that is similar to the points above the line (power-law spectra). In other words, there is not a significant bias for strong sources to have non-power law spectra.

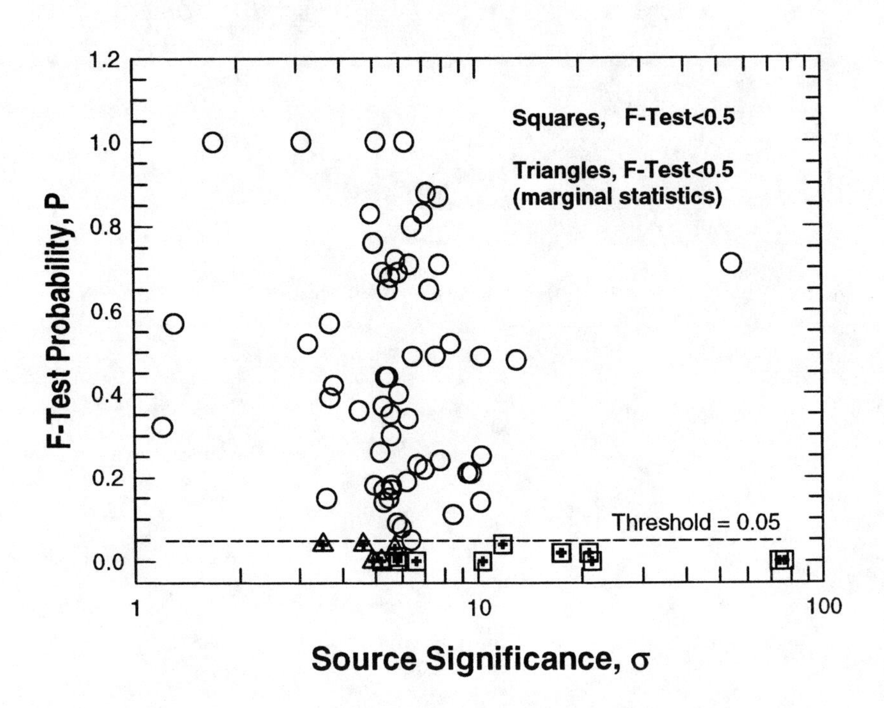

FIGURE 1. F-Test probability as a function of source significance. Points below the threshold line of P = 0.05 are the sources that have spectra that are modeled best by a breaking power law or a power law with an exponential cutoff. The distribution with significance is similar for the sources above and below the dotted line if the four highest points (pulsars) are ignored.

CONCLUSIONS

Among the sources within 10° of the Galactic Plane, at least 12 have spectral features that break at high energies. Three of these are known pulsars as noted above. The remaining 9 may be from a distinct class and perhaps are pulsar candidates themselves. EGRET will not be able to add significantly to the statistics on any of these sources, and it remains for the next generation gamma ray telescope, GLAST, to better determine their spectral features.

REFERENCES

Hartman et al., 1999, ApJS, 123, 79.

On the nature of the galactic population of 3EG sources

Gustavo E. Romero*, Paula Benaglia*, and Diego F. Torres†

*Instituto Argentino de Radioastronomía, C.C. 5, 1894 Villa Elisa, Argentina
†Departamento de Física, Universidad Nacional de La Plata, C.C. 64, 1900 La Plata, Argentina

Abstract. We present the results of a study on the possible association of unidentified gamma-ray sources in the 3EG Catalog with different types of galactic objects: Wolf-Rayet (WR) and Of stars, supernova remnants (SNRs), and OB star associations (considered as pulsar tracers). We have made simulations of large numbers of galactic populations of gamma-ray sources in order to weight the statistical significance of the positional coincidences. We have found that 6 EGRET detections are coincident with WR stars, 4 with Of stars, 22 with SNRs, and 26 with OB associations. The probability that all the SNR and OB coincidences were the pure effect of chance is negligible ($< 10^{-5}$ and $< 10^{-3}$, respectively). The statistical support for the association of massive stars with EGRET sources is not compelling (probabilities $\sim 10^{-2} - 10^{-3}$). However, we find that there are a posteriori arguments to support at least three candidates for gamma-ray production in star systems with strong stellar winds: WR 140, 142, and Cyg OB2 No.5.

I INTRODUCTION

The publication of the Third EGRET (3EG) catalog of high-energy gamma-ray sources [1] provides new and valuable elements to deepen the quest for the origin of the unidentified γ-ray sources. The new catalog lists 271 point sources, including 170 detections with no conclusive counterparts. Of the unidentified sources, 74 are located at $|b| < 10^o$ (this number can be extended to 81 if we include sources with their 95 % confidence contours reaching latitudes $|b| < 10^o$). This means that the number of possible galactic unidentified sources is now nearly doubled respect to the 2EG catalog.

With the aim of finding the positional coincidence between unidentified 3EG sources at $|b| < 10^o$ and different populations of galactic objects, we have developed a computer code that determines the angular distance between two points in the sky, taking into account the positional uncertainties in each of them. The code can be used to obtain a list of γ-ray sources with error boxes (95 % confidence contours given by the 3EG catalog) overlapping extended (like SNRs or OB associations) or point-like objects.

CP510, *The Fifth Compton Symposium,* edited by M. L. McConnell and J. M. Ryan

II RESULTS

We ran the code with the 81 unidentified EGRET sources at galactic latitudes $|b| < 10^o$ and complete lists of Wolf-Rayet (WR) stars [4], Of stars [2], SNRs [3], and OB associations [7]. We have found that 6 γ-ray sources of the 3EG catalog are positionally coincident with WR stars, 4 with Of stars, 22 with SNRs, and 26 with OB associations.

In order to estimate the statistical significance of these coincidences, we have simulated a large number of sets of EGRET sources, retaining for each simulated position the original uncertainty in its galactic coordinates. We have generated 1500 populations (a larger number do not significantly modify the results) of 81 γ-ray sources through rotations on the celestial sphere. Since we simulated a galactic source population and not arbitrary sets at $|b| < 10^o$, we imposed that the new distribution (i.e. each of the simulated sets) retained the form of the actual histogram in latitude of the unidentified 3EG sources, with 1° or 2°-binning. The histogram, for 1°-binning, is shown in Fig. 1.

The unidentified 3EG sources have, additionally, a non-uniform distribution in galactic longitude, showing a concentration towards the galactic center (see Fig. 1). This is basically an a posteriori result: a galactic population is not necessary compelled to follow such a distribution. If we impose on the simulated sets the observed distribution in longitude, probabilities of pure chance association will result naturally raised. When doing the simulations we have considered both cases: we made simulations of any kind of possible galactic populations adopting just constrains in b, and we made also simulations under the stronger hypothesis that just populations with the observed concentration in longitude are possible. These latter results provide an absolute upper limit for the pure chance overlapping probabilities between EGRET sources and the different types of known objects.

We estimated the level of positional coincidence between each simulated set and the different galactic populations under consideration. From these results we obtained an average expected value of chance associations and a corresponding standard deviation. The probability that the observed association level had happened

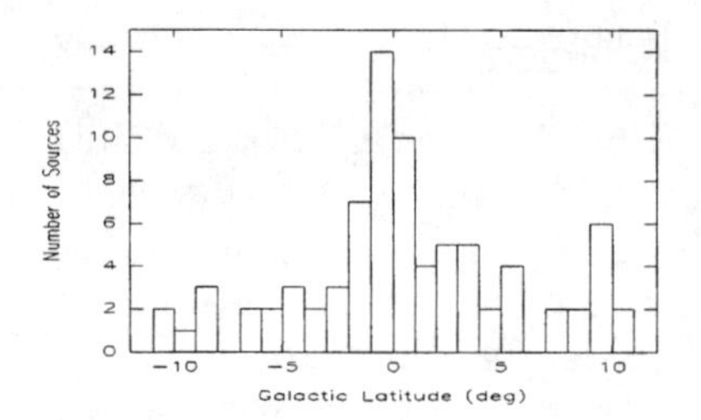

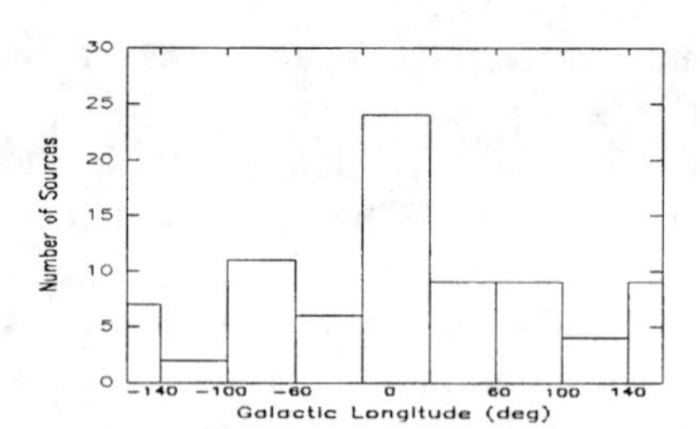

FIGURE 1. (Upper) Distribution in galactic latitude of the 81 unidentified EGRET sources with positions at $|b| < 10^o$ (within errors); (lower) distribution in galactic longitude of the same sources.

TABLE 1. Statistical results obtained from simulations with a random distribution in galactic longitude.

Object type	Actual coincidence	Simulated 1°-bin	Probability 1°-bin	Simulated 2°-bin	Probability 2°-bin
WR	6	2.3 ± 1.4	8.3×10^{-3}	2.1 ± 1.4	5.8×10^{-3}
Of	4	1.2 ± 1.1	1.5×10^{-2}	1.1 ± 1.0	5.9×10^{-3}
Assoc.OB	26	12.7 ± 3.1	1.2×10^{-5}	12.5 ± 3.1	9.8×10^{-6}
SNR	22	7.8 ± 2.5	1.6×10^{-8}	7.0 ± 2.4	5.4×10^{-10}

TABLE 2. Simulations constrained in both galactic latitude and longitude. The indicated binning interval is the binning of the preserved histogram in galactic longitude.

Object type	Actual coincidence	Simulated 20°-bin	Probability 20°-bin	Simulated 40°-bin	Probability 40°-bin
WR	6	4.2 ± 1.8	1.3×10^{-1}	3.5 ± 1.7	8.3×10^{-2}
Of	4	2.0 ± 1.3	2.9×10^{-1}	1.7 ± 1.2	5.4×10^{-2}
Assoc.OB	26	18.7 ± 3.1	8.4×10^{-3}	16.4 ± 3.1	1.0×10^{-3}
SNR	22	11.1 ± 2.6	2.2×10^{-5}	10.6 ± 2.6	1.1×10^{-5}

by chance was then evaluated assuming a Gaussian distribution of the outputs. The results of this study are shown in Tables 1 and 2.

From Table 1, it can be seen that there is a strong statistical correlation between unidentified γ-ray sources of the 3EG catalog and SNRs (at $\sim 6\sigma$ level) as well as with OB associations (at $\sim 4\sigma$ level). We also find that there is a marginally significant correlation with stars ($\sim 3\sigma$). Remarkably, the probability of a pure chance association for SNRs is as low as 5.4×10^{-10} for 2°-binning and 1.6×10^{-8} for 1°-binning. For the stars, we obtain probabilities in the range $10^{-2} - 10^{-3}$, which are suggestive but not overwhelming.

When simulations constrained to follow the observed distribution of unidentified sources in both b and l are considered, we see that SNRs still present quite negligible probabilities for pure chance overlapping with 3EG sources ($\sim 10^{-5}$), whereas OB associations show marginally significant probabilities ($\sim 10^{-3}$). The case for physical association of 3EG sources with stars is not compelling in this case, and must be supported by further a posteriori evidence.

III DISCUSSION

Our results for SNRs and OB associations confirm the correlations observed in the 2EG catalog in Refs. [9] and [5]. The case for possible association of unidentified EGRET sources with WR stars was previously presented (using data from the 2EG catalog) by Kaul and Mitra [6]. They proposed, on the basis of positional correlation, that 8 of these sources could be produced by WR stars. Their analysis of the possible chance occurrence of these associations, which was purely analytic, yielded an a priori expectation of $\sim 10^{-4}$. But their results are notably modified when the 3EG catalog is considered. Most of the possible associations claimed by Kaul & Mitra are no longer viable ones. Just WR stars 37-39, 138, and 142 of their list stay after our analysis. The a posteriori analysis of our results show that three stars are of special interest as possible counterparts of EGRET sources: WR 140, WR 142, and Cyg OB2 No.5 (see Ref. [8] for a detailed discussion).

IV CONCLUSIONS

The main conclusion to be drawn is that there seems to exist more than a single population of galactic γ-ray sources. Pulsars constitute a well established class of sources, and there is no doubt that under certain conditions some SNRs are also responsible for significant γ-ray emission in the EGRET scope. Isolated and binary early-type stars are likely to present high-energy radiation strong enough to be detected by EGRET in some special cases. We propose that, in addition to the well-known WR stars 140 and 142, the Cyg OB2 No. 5 binary system could be a strong γ-ray source, the first one to be detected involving no WR stars.

Acknowledgments. G.E.R. is grateful to the organizers of 5th Compton Symposium for financial support that make possible his participation. This research has been supported by Fundación Antorchas, CONICET, and the agency ANPCT (PICT 03-04881).

REFERENCES

1. Hartman R.C., et al., ApJS **123**, 79 (1999)
2. Cruz-González C., et al. , Rev. Mex. Astron. Astrofis. **1**, 211 (1974)
3. Green D.A., A Catalog of Galactic Supernova Remnants, Mullard Radio Astronomy Observatory, Cambridge, UK (1998)
4. van der Hucht K.A., et al., A&A **199**, 217 (1988)
5. Kaaret P., Cottam J., ApJ **462**, L35 (1996)
6. Kaul R.K., Mitra A.K, Proceedings of the Fourth Compton Symposium, C. D. Dermer, M. S. Strickman, and J. D. Kurfess Eds., AIP, New York, p.1271 (1997)
7. Mel'nik A.M., Efremov Yu.N., Astron. Lett. **21**, 10 (1995)
8. Romero G.E., Benaglia P., Torres D.F., A&A **348**, 868 (1999)
9. Sturner S.J., Dermer C.D., Mattox J.R. A&AS **120**, 445 (1996)

GAMMA-RAY BURSTS

The 4.5 $\pm$0.5 Soft Gamma Repeaters in Review

K. Hurley

UC Berkeley
Space Sciences Laboratory
Berkeley, CA 94720-7450

Abstract. Four Soft Gamma Repeaters (SGRs) have now been identified with certainty, and a fifth has possibly been detected. I will review their X-ray and gamma-ray properties in both outburst and quiescence. The magnetar model accounts fairly well for the observations of SGR1806-20 and SGR1900+14, but data are still lacking for SGR1627-41 and SGR0525-66. The locations of the SGRs with respect to their supernova remnants suggest that they are high velocity objects.

I INTRODUCTION

The Soft Gamma Repeaters are sources of short, soft-spectrum ($\leq$ 100 keV) bursts with super-Eddington luminosities. They undergo sporadic, unpredictable periods of activity, sometimes quite intense, which last for days to months, often followed by long periods (up to years or decades) during which no bursts are emitted. Very rarely, perhaps every 20 years, they emit long duration *giant flares* which are thousands of times more energetic than the bursts, with hard spectra ($\sim$ MeV). The SGRs are quiescent, and in some cases periodic, 1-10 keV soft X-ray sources as well. They all appear to be associated with supernova remnants, and a good working hypothesis is that they are all *magnetars* , i.e. highly magnetized neutron stars for which the magnetic field energy dominates all other sources, including rotation [6,29]. Figure 1 shows the time histories of bursts from SGR1900+14, and figure 2 shows a typical energy spectrum.

In this paper, I will mainly review the radio, X-ray, and gamma-ray properties of the SGRs in outburst and in quiescence, and indicate how the magnetar model accounts for these properties.

II SGR1806-20

Kulkarni and Frail [22] suggested that this SGR was associated with the Galactic supernova remnant (SNR) G10.0-0.3, based on its localization to a $\sim$ 400 arcmin.^{2}

CP510, *The Fifth Compton Symposium,* edited by M. L. McConnell and J. M. Ryan

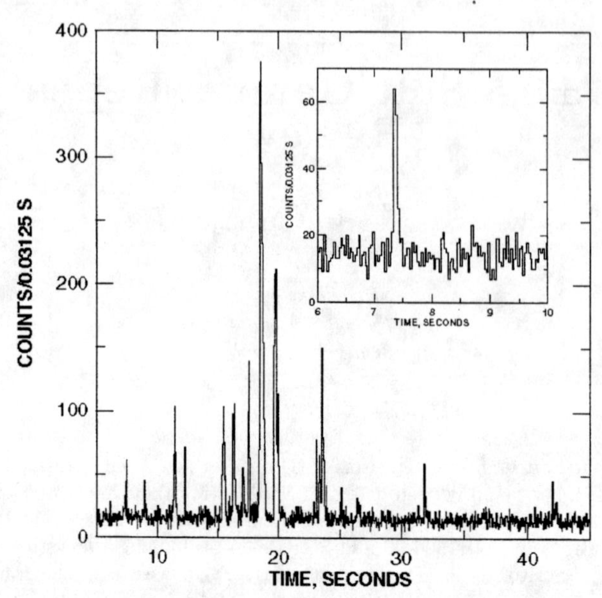

FIGURE 1. From Hurley et al. (1999a). Inset: a typical burst from SGR1900+14 as observed in the 25-150 keV range by *Ulysses* . Main figure: bursts during a period of intense activity.

error box by the old interplanetary network (IPN) [1]. This was confirmed when ASCA observed and imaged the source *in outburst* , localizing it to a 1' error circle [25]. A quiescent soft X-ray source was also detected by Cooke [5] using the ROSAT HRI. Based on more recent observations, Kouveliotou et al. [20] have found that the quiescent source is periodic (P=7.48 s) and is spinning down rapidly ($\dot{P}=2.8\times10^{-11}$s/s). If this spindown is interpreted as being due entirely to magnetic dipole radiation, the implied field strength is B=8 $\times10^{14}$G. The 2-10 keV X-ray luminosity of the source is 2×10^{35}erg/s, and the low energy X-ray spectrum may be fit by a power law with index 2.2.

The SNR G10.0-0.3 has a non-thermal core, and Frail et al. [8] have detected changes in the radio contours of the core on $\sim$ year timescales. Van Kerkwijk et al. [30] have found an unusual star at the center of this core, which they identify as a luminous blue variable (LBV). The presence of this object has been a mystery up to now, because it was thought that the SGRs were single neutron stars. Recent work from the 3rd IPN has shed some light on this issue [13]. Figure 3 shows the location of the SGR superimposed on the radio contours of the SNR. It can be seen that the SGR is in fact offset from the LBV. The LBV may be powering the non-thermal core of the SNR, and causing the changes in the radio contours. It is also possible that the SGR progenitor was once bound to the LBV, but that it became unbound when it exploded as a supernova. A transverse velocity of $\sim$100 km/s would then be required to explain the displacement between the two. Alternatively, it is possible that the apparent SGR-SNR association is due to a chance alignment of these two

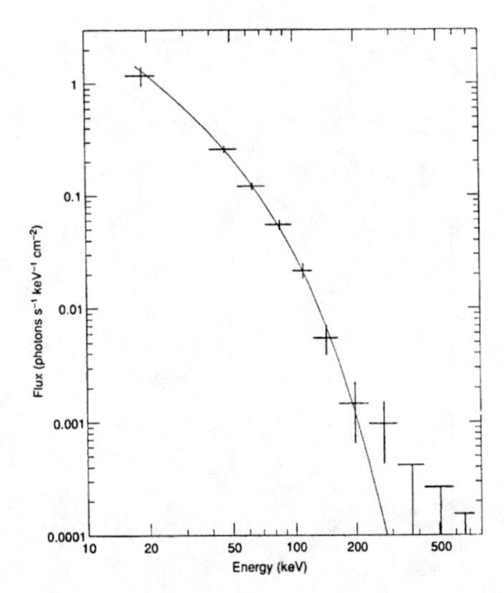

FIGURE 2. Reprinted by permission from Nature (Kouveliotou et al. 1993) copyright 1993 Macmillan Magazines Ltd.. Typical spectrum of a burst from SGR1900+14 as observed by BATSE. The spectrum is fit here with an optically thin thermal bremsstrahlung function, with kT=39 keV.

objects along the line of sight.

III SGR1900+14

SGR1900+14 was discovered by Mazets et al. [23] when it burst 3 times in two days. A precise localization by the IPN [12] showed that this source lay just outside the Galactic SNR G42.8+0.6, with an implied proper motion >1000 km/s. The SGR is associated with a quiescent soft X-ray source [32,14,21]. The quiescent source has a period 5.16 s, and a period derivative 6.1 $\times 10^{-11}$s/s; again, assuming purely dipole radiation, B $\sim 8 \times 10^{14}$G. The 2-10 keV luminosity is 3 $\times 10^{34}$erg/s, and the spectrum may be fit with a power law of index 2.2.

On 1998 August 27, the SGR emitted a giant flare which was probably the most intense burst ever detected at Earth [15]. Its luminosity was 2 $\times 10^{43}$ erg/s in >25 keV X-rays, or $10^5 L_E$ (the Eddington luminosity). The time history of this burst clearly displayed the 5.16 s periodicity of the quiescent source (figure 4). The magnetic field strength required to contain the electrons responsible for the X-ray emission is $> 10^{14}$G; this constitutes an independent argument for the presence of strong fields in SGRs. From measurements of the ionospheric disturbance which this burst caused, Inan et al. [18] have estimated that there must have been one order of magnitude more energy in 3-10 keV X-rays than in >25 keV X-rays, bringing the total energy to $\sim 4 \times 10^{44}$erg. Frail et al. [9] detected a transient radio source

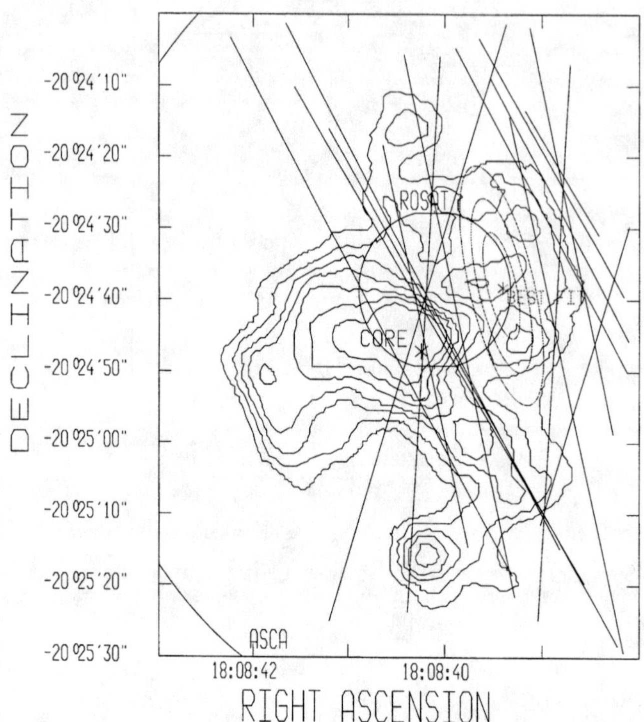

FIGURE 3. From Hurley et al. (1999b). Eight IPN annuli (lines), and the 1, 2, and 3 σ equivalent confidence contours (ellipses) for SGR1806-20. The best fit position and the position of the non-thermal core are indicated. The ASCA error circle is just visible in the lower left and upper left hand corners (Murakami et al. 1994). The ROSAT PSPC error circle is at the center; its radius is 11" (Cooke 1993). The 3.6 cm radio contours of G10.0-0.3 are also shown, from Vasisht et al. (1995).

with the VLA at the SGR position following the giant flare. This is the only case where a radio point source is present at an SGR position.

IV SGR0525-66

This SGR was discovered when it emitted the giant flare of 1979 March 5 [3,10]. It was localized by the IPN to a 0.1 arcmin^2 error box within the N49 supernova remnant [7]. For an LMC distance of 55 kpc, this burst had a luminosity of 5×10^{44} erg/s in X-rays >50 keV, or $2 \times 10^6 L_E$; the total energy emitted was $\sim 7 \times 10^{44}$erg in >50 keV X-rays. The time history displayed a clear 8 s periodicity [2]. Paczynski [26] was the first to suggest a strongly magnetized neutron star as the origin of this burst. Although the source remained active through 1983 [11], it has not been observed to burst since then.

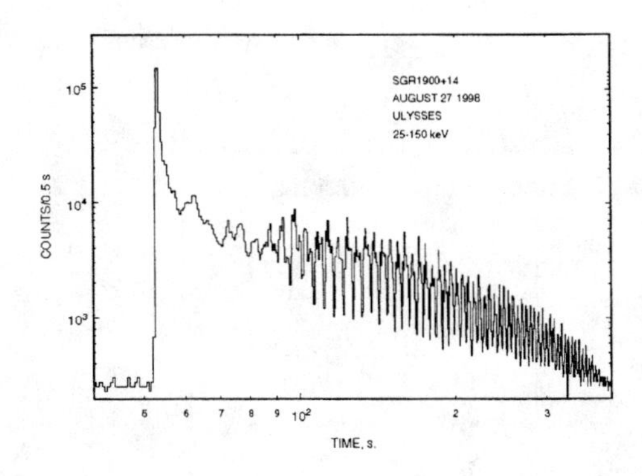

FIGURE 4. Reprinted by permission from Nature (Hurley et al. 1999d) copyright 1999 Macmillan Magazines Ltd.. The *Ulysses* 25-150 keV time history of the 1998 August 27 giant flare from SGR1900+14. Note the 5.16 s periodicity.

Rothschild et al. [27] found a quiescent soft X-ray point source in the SGR error box with a ROSAT HRI observation. As no energy spectra are obtained from the HRI, the soft X-ray luminosity can only be estimated by assuming various spectral shapes. The 0.1-2.4 keV luminosity is in the range $10^{36} - 10^{37}$ erg/s, depending on the assumed spectrum. No periodicity was detected in this observation, but the upper limit to the pulsed fraction is only 66%. If the age of the N49 SNR is taken to be 5 kyr [31], the implied transverse velocity of the SGR is several thousand km/s. *Chandra* observations of the SNR are scheduled, and are bound to reveal more about this interesting object.

V SGR1627-41

SGR1627-41 burst about 100 times in June-July 1998, and has not been observed to burst since then. During that period, observations by BATSE [34], *Ulysses* [16], KONUS-*Wind* [24], and RXTE [28] led to a precise source localization. The SGR lies near the SNR G337.0-0.1, at a distance of $\sim$ 11 kpc. The implied transverse velocity of the SGR is in the range 200 - 2000 km/s. Although no giant flare has been observed from this source, there is a KONUS-*Wind* observation of an extremely energetic event [24]. The luminosity and total energy of the burst in the >15 keV range were $\sim 8 \times 10^{43}$erg/s and $\sim 3 \times 10^{42}$erg/s, respectively.

Like the other SGRs, this one also appears to be a quiescent soft X-ray source. *BeppoSAX* observations revealed a variable source with spectral index 2.1 and luminosity $\sim 10^{35}$erg/s [35]. Although the *BeppoSAX* observations gave weak evidence for a possible 6.4 s periodicity, this was not confirmed in later ASCA observations of the source with better statistics [17].

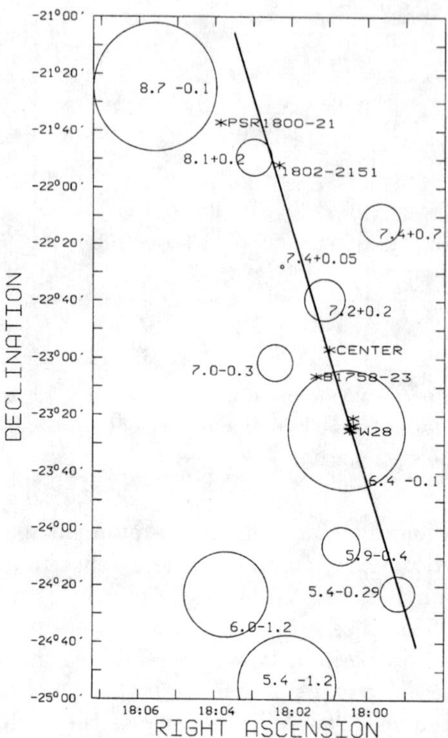

FIGURE 5. From Cline et al. (1999). IPN error box for SGR1801-23 (the lines are too closely spaced to distinguish). The center is indicated with an asterisk. Circles give the approximate locations of confirmed and suspected SNRs; the radii have been taken as half the size given in the catalogs. Asterisks give the positions of ROSAT X-ray sources, and two pulsars, PSR1800-21 and B1758-23, probably associated with SNRs 8.7-0.1 and 6.4-0.1. Coordinates are J2000.

VI SGR1801-23

The latest SGR to be discovered is 1801-23 [4]. It was observed to burst just twice, on June 29, 1997, by *Ulysses* , BATSE, and KONUS-*Wind.* The burst spectra were soft, and could be fit by an optically thin thermal bremsstrahlung function with a kT of $\sim$ 25 keV. The time histories were short. In both respects, then, the source properties resemble those of the other SGRs. However, because only two bursts were observed, and they occurred on the same day, the IPN localization is not very precise. The error box is 3.8 $^{\circ}$ long, and has an area of $\sim$ 80 arcmin^2. The source lies in the general direction of the Galactic center, and the error box crosses numerous possible counterparts (figure 5). The source would have a super-Eddington luminosity for any distance $>$ 250 pc; at the approximate distance of the Galactic center, its luminosity would be 1200L$_E$. At present, the best hypothesis is that this source is indeed an SGR; recall that SGR1900+14 was similarly detected when it burst 3 times in two days, and it remained quiescent for many years. Like SGR1900+14, the identification of SGR1801-23 may have to await a new period of bursting activity.

Table 1 summarizes the essential properties of the SGRs.

TABLE 1. Essential properties of the SGRs

SGR	Super-Eddington Bursts?	Giant Flare?	Periodicity Observed in Burst?	Quiescent Soft X-ray Source? (erg/s)	Periodicity in Quiescent Source?	$\dot{P}$ 10^{-11} s/s
1806-20	1000×	No	No	2×10^{35}	7.47 s	2.8
1900+14	1000×	270898	5.16 s	3×10^{34}	5.16 s	6.1
0525-66	20000×	050379	8 s	10^{36-37}	No	—
1627-41	400000×	No	No	10^{35}	6.4 s?	—
1801-23	?	No	No	?	—	—

VII THE MAGNETAR MODEL

Briefly, the magnetar model [6,29] explains the short, soft bursts by localized cracking on the neutron star surface, with excitation of Alfven waves which accelerate electrons. Every 20–100 y, a massive, global crustquake takes place. Regions of the neutron star with magnetic fields of opposite polarity suddenly encounter one another, resulting in magnetic field annihilation and energization of the magnetosphere, giving rise to a giant flare. Magnetars are thought to be born in ~ 1 out of 10 supernova explosions, and remain active for perhaps 10,000 y. Thus there should be about 10 active magnetars in the Galaxy at any given time. So far, we have found 4.5 ± 0.5. Stay tuned for more!

We are grateful for JPL support of *Ulysses* operations under Contract 958056 and to NASA for support of the IPN under NAG5-7810.

REFERENCES

1. Atteia, J.-L. et al., *Ap. J.* **320**, L105 (1987).
2. Barat, C. et al.,*Astron.Astrophys.* **79**, L24 (1979).
3. Cline, T. et al., *Ap. J.* **237**, L1 (1980).
4. Cline, T. et al., *Ap. J.*, accepted, astro-ph/9909054, (1999).
5. Cooke, B., *Nature* **366**, 413 (1993).
6. Duncan, R., and Thompson, C., *Ap. J.* **392**, L9 (1992).
7. Evans, W. D. et al., *Ap. J.* **237**, L7 (1980).
8. Frail, D. et al., *Ap. J.* **480**, L129 (1997).
9. Frail, D. et al., *Nature* **398**, 127 (1999).
10. Golenetskii, S. et al., *Sov. Astron. Lett.* **5**, 340 (1979).
11. Golenetskii, S. et al., *Sov. Astron. Lett.* **13(3)**, 166 (1987).

12. Hurley, K. et al., *Ap. J.* **510**, L107 (1999a).
13. Hurley, K. et al., *Ap. J.* **523**, L37 (1999b).
14. Hurley, K. et al., *Ap. J.* **510**, L111 (1999c).
15. Hurley, K. et al., *Nature* **397**, 41 (1999d).
16. Hurley, K. et al., *Ap. J.* . **519**, L143 (1999e).
17. Hurley, K. et al., *Ap. J.*, submitted, astro-ph/9909355 (1999f).
18. Inan, U. et al., *GRL*, in press (1999).
19. Kouveliotou, C. et al., *Nature* **362**, 728 (1993).
20. Kouveliotou, C. et al., *Nature* **393**, 235 (1998).
21. Kouveliotou, C. et al., *Ap. J.* **510**, L115 (1999).
22. Kulkarni, S., and Frail, D., *Nature* **365**, 33 (1993).
23. Mazets, E. et al., *Sov. Astron. Lett.* **5(6)**, 343 (1979).
24. Mazets, E. et al., *Ap. J.* **519**, L151 (1999).
25. Murakami, T. et al., *Nature* **368**, 127 (1994).
26. Paczynski, B., *Acta Astronomica* **42**, 145 (1992).
27. Rothschild, R. et al., *Nature* **368**, 432 (1994).
28. Smith, D. et al., *Ap. J.* **519**, L147 (1999).
29. Thompson, C., and Duncan, R., *Mon. Not. R. Astron. Soc.* **275**, 255 (1995).
30. van Kerkwijk, M. et al., *Ap. J.* **444**, L33 (1995).
31. Vancura, O. et al., *Ap. J.* **394**, 158 (1992).
32. Vasisht, G. et al., *Ap. J.* **431**, L35 (1994).
33. Vasisht, G. et al., *Ap. J.* **440**, L65 (1995).
34. Woods, P. et al., *Ap. J.* **519**, L139 (1999a).
35. Woods, P. et al., *Ap. J.* **519**, 139 (1999b).

Gamma Ray Bursts and Afterglow

Re'em Sari

Theoretical Astrophysics 130-33, California institute of Technology, Pasadena CA 91125

Abstract. The origin of GRBs have been a mystery for almost 30 years. Their sources emit a huge amount of energy on short time scales and the process involves extreme relativistic motion with bulk Lorentz factor of at least a few hundred. In the last two years, "afterglow", emission in X-ray, optical, IR, and radio was detected. The afterglow can be measured up to months and even years after the few seconds GRB. We review the theory for the γ-rays emission and the afterglow and show that it is strongly supported by observations. A recent detection of optical emission simultaneous with the GRB, well agrees with theoretical predictions and further constrains the free parameters of the models. We discuss the evidence that some of the bursts are jets, and discuss the prospects of polarization measurements.

I THE GENERIC PICTURE

The phenomena of GRBs was discovered almost thirty years ago, by the Vela defense satellites [1]. Today, the biggest catalog of GRBs [2] is due to the instrument BATSE on board of the Compton Gamma-ray Observatory. BATSE observes about one burst per day and more than two thousands bursts have been observed by now. The spectrum of GRBs is well described by a broken powerlaw, and usually peaks between 100keV-400keV [3]. In strong bursts, high energy powerlaw tails extending up to 200 MeV were seen and the most extreme case had a few GeV photons detected. On the average the high energy tail is characterized by $\nu F_\nu \sim \nu^{-0.25}$. The duration of the GRBs varies significantly, mainly between a few milliseconds to a few hundred seconds. One of the striking properties of GRBs is their erratic temporal structure. While only a few burst are smooth, most of them vary over a time scale δt which is much shorter than the burst's duration t. In many bursts the ratio $N \equiv t/\delta t \geq 100$.

The cosmological distance to the bursts (now well established due to detection of redshifts), combined with the large fluence observed at earth implies that the energy released in the event is huge, with a record of 3×10^{54}erg (GRB 990123, see e.g. [4]). This huge energy, together with the short variability time scale, places the GRBs phenomena as the most extreme in the universe.

The extreme characteristics of GRBs lead to a paradox, so called the "compactness problem". If one assumes that an energy of 10^{52}erg made of photons,

CP510, *The Fifth Compton Symposium*, edited by M. L. McConnell and J. M. Ryan

distributed according to the GRBs spectrum, is released in a small volume of linear dimensions $R \leq c\delta t$ then the optical depth to pair creation is $\tau \sim 10^{15}$. If that was true, all the photons would have interacted to create pairs and thermalize. However, the observed spectrum of GRBs is highly non-thermal!

The only known solution to the "compactness problem" is relativistic motion. If the emission site is moving relativistically, with a Lorentz factor γ, toward the observer, then the optical depth is reduced, compared to the stationary estimate, due to two effects: First, the size of the source can be bigger by a factor of γ^2. This will still produce variability over a short time scale given by $\delta T = R/\gamma^2 c$ since not all the source is seen as the radiation for a relativistically moving object is beamed. Second, the photons in the local frame are softer by a factor of γ, and therefore only a small fraction of them, at the high energy tail, have enough energy to create pairs. The combination of these two effects reduces the optical depth by a factor of $\sim \gamma^{6.5}$. Therefore, the optical depth is reduced below unity, and the "compactness problem" is solved, if the Lorentz factor is larger than about a hundred.

This solution led to a three stage generic scenario for GRBs. First, a compact source releases about 10^{52} erg, in a small volume of space and on a short time scale. This large concentration of energy expands due to its own pressure. If the rest mass that contaminates the site is not too large, $\leq 10^{-5} M_{\odot}$, this will result in relativistic expansion with $\gamma > 100$. Finally, at a large enough radius, the kinetic energy of the expanding material is converted to internal energy an radiated, mainly in γ-rays. At this stage the system is optically thin and high energy photons can escape.

In this talk we will concentrate mainly on the third stage. We will assume that a relativistic flow with a high Lorentz factor exists, carrying more than 10^{52} erg as kinetic energy, and discuss how this flow may produce the γ-ray photons as well as the afterglow. This presentation will be short in equations, stressing the main qualitative ideas.

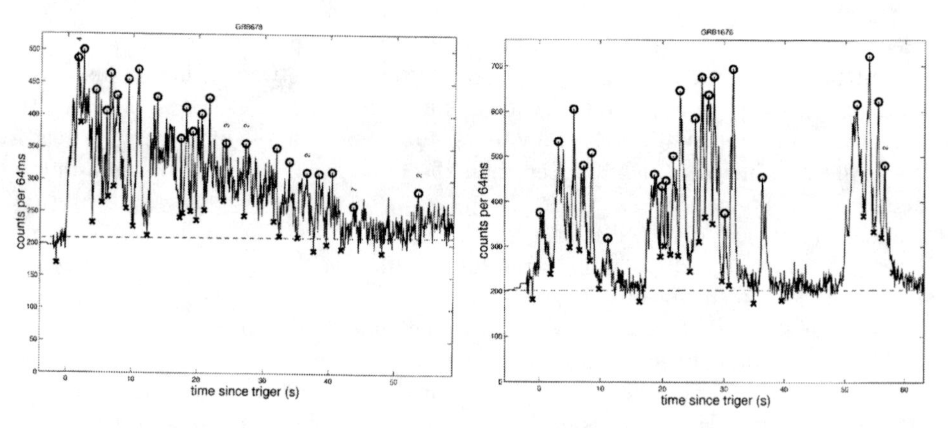

FIGURE 1. Two variable bursts, measured by BATSE. The dashed line is the background level.

II INTERNAL VS. EXTERNAL SHOCKS

Assume a flow carrying 10^{52} erg as kinetic energy. In order for this to produce photons, the kinetic energy must be converted back into internal energy and radiated away. The flow must therefore, at least partially, slow down. Two scenarios were proposed for this deceleration: external shocks [5] and internal shocks [6,7]. In the external shocks scenario, the relativistic material is running into some (external) ambient medium, probably the interstellar medium or a wind that was emitted earlier by the progenitor. In the internal shocks scenario the inner engine is assumed to emit an irregular flow, that consists of many shells, that travel with a variety of Lorentz factors and therefore colliding into each other and thermalizing some of their kinetic energy.

The property that proved to be very useful in constraining these two possibilities is the variability observed in many of the bursts. In the external shocks scenario, this variability is attributed to irregularities in the surrounding medium, e.g., clouds. Each time the ejecta runs into a higher density environment, it produces a peak. In the internal shocks scenario, the source has to emit many shells, and when every two of them collide, a peak is produced. External shocks require a complicated surrounding with a relatively simple source that explodes once, while internal shocks require a more complicated source that will explode many times to produce several shells. Due to these very different requirements from the source, the question of internal or external shocks is of fundamental importance in understanding the basic nature of the phenomena.

The size of the clouds in which the ejecta runs into, in the external shocks scenario, has to be very small to produce peaks that are narrower than the duration of the burst [8]. Sari & Piran [9] gave the following argument: The size of the clouds has to be smaller than $R/N\gamma$ to produce peaks that are narrower by a factor of N than the duration of the burst. The number of clouds should be smaller than N otherwise pulses arriving from different clouds will overlap and the amplitude of the variability will be reduced. Finally the observable area of the ejecta, due to relativistic beaming is $(R/\gamma)^2$. The maximal efficiency of the external shocks scenario is therefore given by (cloud area) $\times$(number of clouds)/(shell area)$\leq 1/N \sim 1\%$. Since in many bursts $N > 100$, external shocks have a sever efficiency problem, when constructed to produce highly variable bursts. Other predictions of external shock are also inconsistent with the observed temporal profile [10]. Moreover, the density ratio between the cloud and the surrounding has to be huge, of the order of $\gamma N^2 \sim 10^6$, in order that the ejecta will be slowed down mainly by the dense clouds rather than by the low density uniform medium.

Internal shocks do not suffer from these problems. The variability can be produced even without breaking the spherical symmetry. Detailed calculations show that the observed temporal structure coming from internal shocks, closely follows the operation of the inner engine that generated the shells [11]. In this scenario, the source must be variable on time scales shorter than a second and last for as long as 100 seconds, just as the bursts themselves.

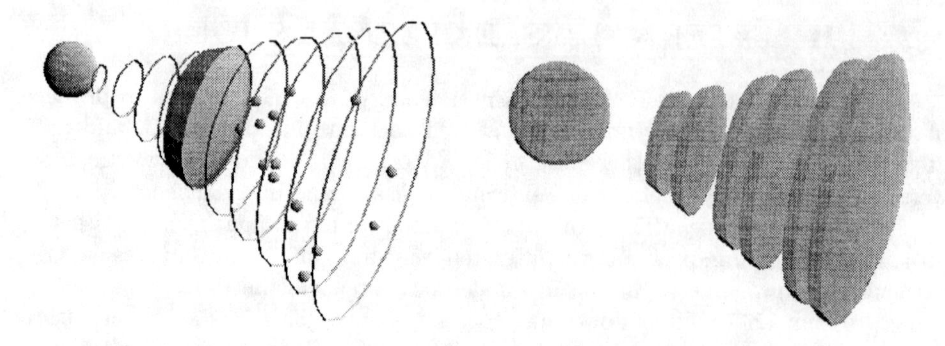

FIGURE 2. Producing variability by external shocks (left) or internal shocks (right).

III THE AFTERGLOW REVOLUTION

The study of γ-ray bursts was revolutionized when the Italian Dutch satellite BeppoSAX delivered arcminutes positioning of some GRBs, within a few hours time scale. This enabled other ground and space instruments to monitor the relatively narrow error box. Emission in X-ray, infrared, optical and radio, so called "afterglow" was observed by now for about a dozen of bursts. The study of GRBs, that was up to then collimated to a narrow energy band, immediately turned into a multi wavelength astronomy field. Due to the transient nature of the afterglow, a major part of the game is to observe the GRBs field early enough, when the afterglow is still bright. Within the first day, the optical emission is usually brighter than 20th magnitude and therefore small telescopes can play an important role in measuring the lightcurve. A large worldwide collaboration is observing these events and the data is submitted to an impressive Global-Coordinate-Network [12] in real time, allowing other observatories to react accordingly.

The observed afterglow usually shows a power law decay $t^{-\alpha}$ in the optical and X-ray where a typical value is $\alpha \cong 1.2$. Some afterglows show a steeper decline with $\alpha \cong -2$. On the radio wavelength, the flux seems to rise on timescale of weeks and then decay with a similar powerlaw. In some cases the radio flux was observed for about a year following the few seconds GRB.

The Afterglow was predicted well before it was observed [13–16] After the internal shocks produced the GRB, the shell interacts with the surrounding medium and decelerates. The emission shifts into lower and lower frequencies. Excitingly, the afterglow theory is relatively simple. It deals with the emission on timescale much longer than those of the GRBs. The details of the complex initial conditions are therefore forgotten and the description depends on a small number of parameters, such as the total energy and the external density.

The basic model assumes that electrons are accelerated by the shock into a powerlaw distribution $N(\gamma_e) \sim \gamma_e^{-p}$ for $\gamma_e > \gamma_m$. The lower cutoff of this distribution is assumed to be a fixed fraction of equipartition. It is also assumed that a considerable magnetic field is being built behind the shock, it is again characterized by a certain fraction ϵ_B of equipartition. The relativistic electrons then emit synchrotron radiation which is the observed afterglow. The broad band spectrum of such emission was given by Sari, Piran & Narayan [17].

At each instant, there are three characteristic frequencies: (I) ν_m which is the synchrotron frequency of the minimal energy electron, having a Lorentz factor γ_m. (II) The cooling time of an electron is inverse proportional to its Lorentz factor γ_e. Therefore, electrons with a Lorentz factor higher than a critical Lorentz factor $\gamma_e > \gamma_c$ can cool on the dynamical timescale of the system. This characteristic Lorentz factor corresponds to the "cooling frequency" ν_c. (III) below some critical frequency ν_a the flux is self absorbed and is given by the Rayleigh-Jeans portion of a black body spectrum. The broad band spectrum of the well studied GRB 970508 [18] is in very good agreement with the theoretical picture.

The evolution of this spectrum as a function of time depends on the hydrodynamics. The simplest, which also well describes the data, is the adiabatic model with a constant density surrounding medium. The rest mass collected by the shock at radius R is about $R^3\rho$. On the average, the particles move with a Lorentz factor of γ^2 in the observer frame and therefore the total energy is given by $E \sim \gamma^2 R^3 \rho c^2$. Assuming that the radiated energy is negligible compared to the flow energy, we

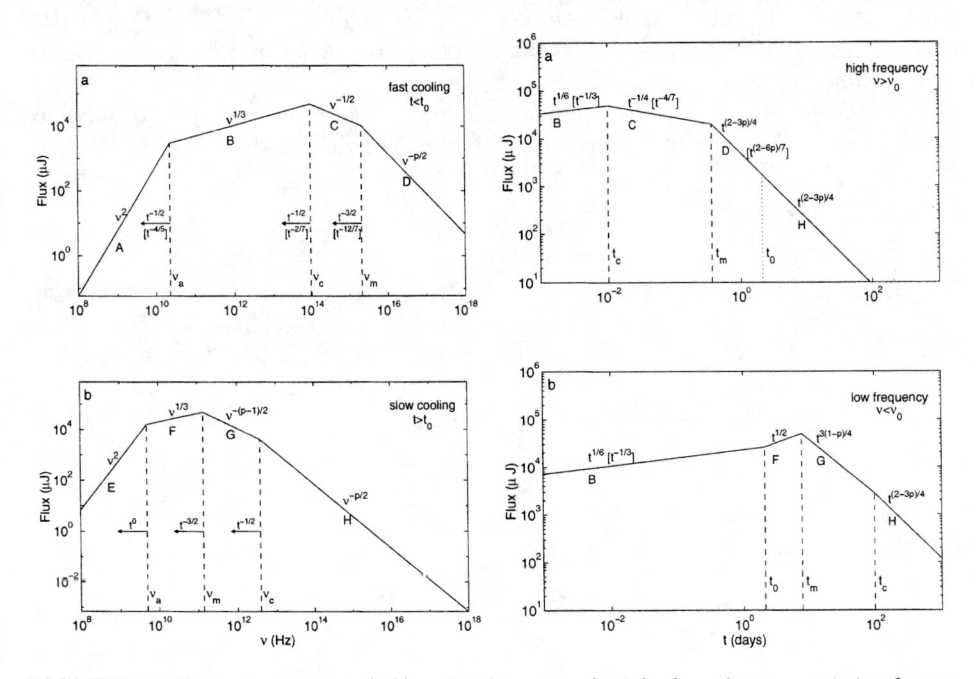

FIGURE 3. Theoretical spectra (left) and light curves (right) of synchrotron emission from a powerlaw distribution of electrons. $p = 2.2 - 2.4$ fits well the observed spectra and lightcurves.

obtain that $\gamma \sim R^{-3/2}$ or in terms of the observer time, $t = R/\gamma^2 c$, we get $\gamma \sim t^{-3/8}$. If on the other hand the density drops as R^{-2} (as is expected if the surrounding is a wind produced earlier by the progenitor of the burst) we get $\gamma \sim t^{-1/4}$. These simple scaling laws lead to the spectrum evolution as given in the above figure.

Given the above hydrodynamic evolution, one can construct light curves at any given frequency. These will also consist of power laws, changing from one power law to the other once the break frequencies sweep through the observed band. These power laws are in fair agreement with afterglow observations.

IV JETS AND BEAMING

The hydrodynamic evolution described above, assumed spherical symmetry. Scenarios in which the ejecta is not spherical but has a limited solid angle $\Omega = \pi\theta_0^2$ are usually called "jets". These "jets", should not be confused with the relativistic beaming of the radiation. The term "jet" corresponds to the physical shape of the outflow, and is created by the inner engine. In contrast, the relativistic beaming is a special relativity effect and has to do only with the fact that the ejecta is moving with relativistic Lorentz factor γ. The relativistic beaming allows an observer to see only a small angular extent, of size $1/\gamma$ centered around the line of sight.

The question of "jets" has two important implications. First, the true total energy emitted by the source is smaller by a factor of $\Omega/4\pi \sim \theta_0^2/4$ than if the ejecta was spherical. Second, the event rate must be bigger by the same factor to account for the observed rate.

Interestingly, due to the relativistic beaming (which is independent of jets) we are only able to see an angular extent of $1/\gamma < 0.01$ during the GRB itself where

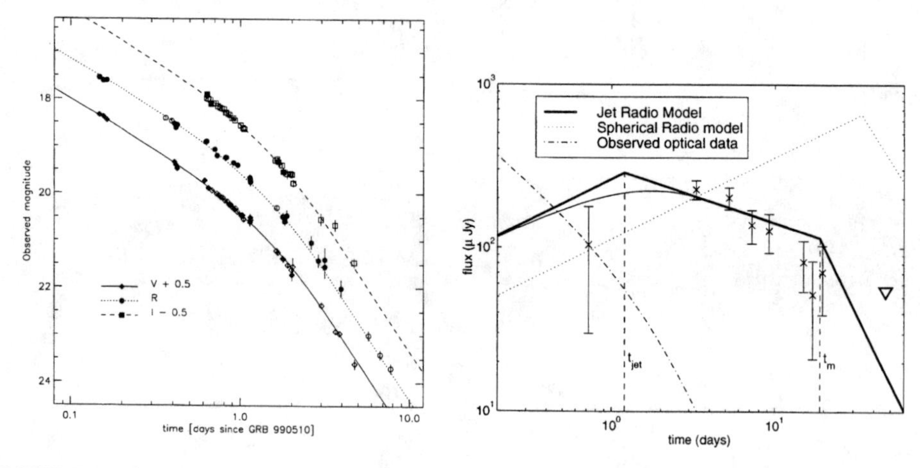

FIGURE 4. GRB 990510, the best evidence for a "jet": an achromatic break in optical and radio at $t_{jet} \cong 1.2$days implying $\theta_0 \cong 0.08$.

the Lorentz factor $\gamma > 100$. Therefore, we cannot distinguish a jet from spherical ejecta. Therefore, given the bursts only, the event rate and the energy in each GRBs are unknown to about four orders of magnitude! However, as γ decreases it will eventually fall below the inverse opening angle of the jet. The observer will appreciate that some of the sphere is missing from the fact that less radiation is observed. This effect, will produce a significant break, steepening the lightcurve decay by a factor of $\gamma^2 \sim t^{-3/4}$. The transition should occur when $1/\gamma = \theta_0$ and it therefore provides an indication for the jet's opening angle. Additionally, Rhoads [19] has shown that at about the same time, the physical size of the jet will begin to increase so that $\theta(t) \sim 1/\gamma$. Taking this effect into account, the break is even more significant and the decay is proportional to $t^{-p} \sim t^{-2.2} - t^{-2.4}$.

Evidence of a break from a shallow to a steep power law was seen in GRB 990123 [4,20], unfortunately the break was observed only in one optical band while data on other bands were ambiguous. A very clear break was seen in GRB 990510 [21,22] simultaneously in all optical bands and in radio. In GRB 990123 and GRB 990510 the transition times were about 2.1days and 1.2days reducing the isotropic energy estimate by a factor of $\sim$ 200 and $\sim$ 300 respectively.

Sari, Piran, & Halpern [23] have noted that the observed decays in GRBs afterglow that do not show a break are either of a shallow slope of $\sim t^{-1.2}$ or a very steep slope of $\sim t^{-2}$. They argued that the rapidly decaying bursts are those in which the ejecta was a narrow jet and the break in the light curve was before the first observations. Interestingly, evidence for jets are found when the inferred energy (without taking jets into account) is the largest. This implies that the jets account for a considerable fraction of the wide luminosity distribution seen in GRBs, and the true energy distribution is less wide than it seems to be.

V THE OPTICAL FLASH & THE RADIO FLARE

An exiting event this year was the first detection of a bright (9th magnitude) optical emission simultaneous with GRB 990123 [24]. Theoretical prediction for such a flash was recently given in detail by Sari & Piran [25,26] and was earlier suggested as a possibility by Mészáros & Rees [16]. During the first few tens of seconds, the evolution of the Lorentz factor as a function of time is not self similar. There are two shocks: a forward shock going into the surrounding medium and a reverse shock going into the expanding shell. The hydrodynamic details were discussed in [27]. During the initial stage, the internal energy stored behind the shocked surrounding matter and the shocked ejecta is comparable. However, the temperature of the shocked ejecta is much lower, typically by a factor of $\gamma \sim 10^2$. This results in an additional emission component with a typical frequency lower by a factor of $\gamma^2 \sim 10^4$, which, for typical parameters, falls in the optical regime. Contrary to the "standard" late afterglow, this emission is very sensitive to the initial Lorentz factor.

The optical flash of GRB 990123 peaked around 60 seconds after the burst trigger.

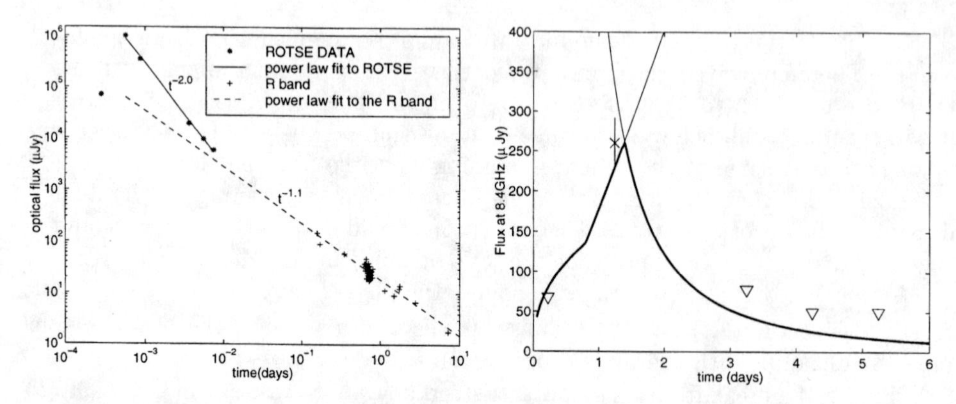

FIGURE 5. GRB990123: Optical (left) date fits theoretical prediction. Radio "flare" seen a day after the burst agrees with theory scaling of optical data (heavy solid line).

The observed optical properties of this event are well described by the emission from the reverse shock that initially decelerates the ejecta, provided the initial Lorentz factor is about 200 [28,29]. It takes tens of seconds for the reverse shock to sweep through the ejecta and produce the bright flash. Later, the shocked hot matter expands adiabatically and the emission quickly shifts to lower frequencies and considerably weakens.

Another new ingredient that was found in GRB 990123 is a radio flare [30]. Contrary to all other afterglows, where the radio peaks around few weeks and then decays slowly, this burst had a fast rising flare, peaking around a day and decaying quickly. Within a day the emission from the adiabatically cooling ejecta, that produced the 60s optical flash shifts into the radio frequencies [28]. The optical flash and the radio flare are therefore related. The fact that the "usual" forward shock radio emission did not show up later, on a timescale of weeks, is in agreement with the interpretation of this burst as a "jet" which causes the emission to considerably weaken by the time the frequency arrives to the radio.

VI POLARIZATION - A PROMISING TOOL

An exciting possibility to further constrain the models and obtain a direct proof to the geometrical picture of "jets" is to measure polarization. Gruzinov & Waxman and Medvedev & Loeb [31,32] considered the emission from spherical ejecta which by symmetry should produce no polarization on the average. Polarization is more natural if the ejecta is a "jet" and the observer is not directed at its very center [33–35] since the spherical symmetry is broken. For simplicity, lets assume that the direction of the magnetic field behind the shock is larger in the shock plane (the results are more general, unless the magnetic field has no preferred direction). The synchrotron polarization from each part of the fireball, which is perpendicular to

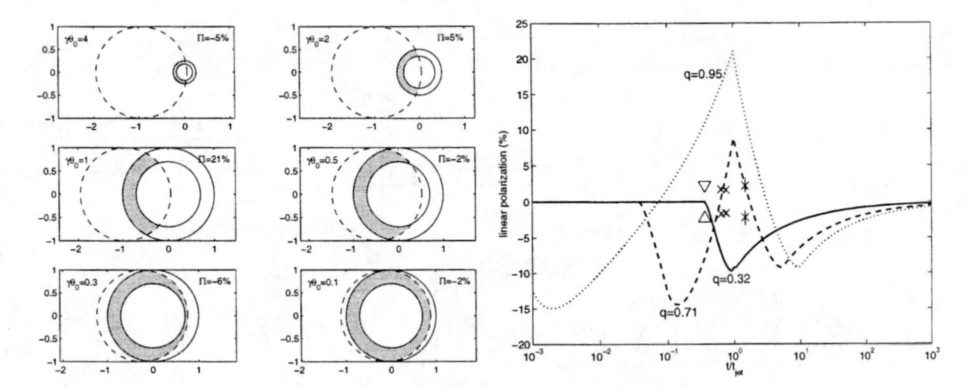

FIGURE 6. Left: Evolution of the observed ring (gray) and the physical jet (dash). Right: observed and theoretical polarization lightcurve.

the magnetic field, is directed radially.

As long as the relativistic beaming of size $1/\gamma$ is narrower than the physical size of the jet θ_0, one is able to see a full ring and therefore the radial polarization averages to zero. As the flow decelerates, the relativistic beaming $1/\gamma$ becomes comparable to θ_0 and only a part of the ring is visible. Net polarization is then observed. Note that due to the radial direction of the polarization from each fluid element, the total polarization is maximal when a quarter or three quarters of the ring are missing (or radiate less efficiently) and vanishes for a full or half ring. The observed polarization when more than half of the ring is missing is perpendicular to the direction when less than half of it is missing.

At late stages the jet expands and since the offset of the observer from the physical center of the jet is constant, spherical symmetry is regained. The vanishing and re-occurrence of significant parts of the ring results in a unique prediction: there should be three peaks of polarization, with the polarization position angle during the middle peak rotated by 90° with respect to the other two peaks. In case that the observer is very close to the center, more than half of the ring is always observed, and therefore only a single direction of polarization is expected. Few possible polarization light curve are presented in the figure.

VII WHAT DID WE LEARN ABOUT THE SOURCE?

(i) Internal shocks imply that the source is variable on $\lesssim$ 1s timescales but lasts for tens of seconds. (ii) The event rate is probably higher than observed by about a factor of a hundred since some events are narrow jets. This translates to one event per 10^5 years per galaxy (iii) The environment of at least some bursts well agrees with ordinary ISM densities. These bursts do not occur in their galaxies' halo. (iv) measurements of optical flashes and radio flares probe the ejecta material, allowing us to measure how many baryons reside in the explosion cite. GRB 990123 has $\gamma_0 \sim 200$. (v) Taking jets into account, the total energy involved can be "only" 10^{52}erg even in the most extreme case.

REFERENCES

1. Klebesadel, R. W.; Strong, I. B. & Olson, Roy A. 1973, ApJ, 182, 85
2. Paciesas, W. S. et al. 1999, ApJS, 122, 465
3. Band, D. et al., 1993, ApJ, 413, 281
4. Kulkarni, S. R., et al. 1999, Nature 398, 389
5. Mészáros, P., & Rees, M. J. 1993, ApJ, 405, 278
6. Narayan, R., Paczyński, B. & Piran, T. 1992, ApJ, 395, 83
7. Rees M. J. & Mészáros P. 1994, ApJ, 430, L93
8. Fenimore, E. E., Madras, C., & Nayakshine, S. 1996, ApJ, 473, 998
9. Sari, R., & Piran T. 1997, ApJ, 485, 270
10. Ramirez-Ruiz, E., & Fenimore, E. E. 1999, A&A, 138, 521
11. Kobayashi, S., Piran, T., & Sari, R. 1997, ApJ, 490, 92
12. Barthelmy, S.D., et al. 1994, in "Proceeding of the Second Huntsville Workshop"; eds. G.Fishman, J.Brainerd, K.Hurley; 307; 643
13. Paczyński, B. & Rhoads, J. 1993, ApJ, 418, L5
14. Katz, J. I., 1994, ApJ, 422, 248
15. Vietri, M. 1997, ApJ, 478, L9
16. Mészáros, P., & Rees M. J. 1997, ApJ, 476, 232
17. Sari, R., Piran, T. & Narayan, R. 1998, 497, L17
18. Galama, T. J. et al. 1998, ApJ, 500, 101
19. Rhoads, J. E. 1999, ApJ, 525, 737
20. Fruchter, A. S., et al., 1999, ApJ, 519, L13
21. Stanek, K. Z., Garnavich, P. M., Kaluzny, J., Pych, W. & Thompson, I. 1999, ApJ, 522, L39
22. Harrison F. A., et al. 1999, ApJ, 1999, 523, L121
23. Sari, R., Piran, T., & Halpern, J. 1999, ApJ, 1999, 519, L17
24. Akerlof, C. et al., 1999, Nature. 398, 400
25. Sari, R., & Piran T. 1999a, A&AS, 138, 537
26. Sari, R., & Piran T. 1999b, ApJ, 520, 641
27. Sari, R., & Piran T. 1995, ApJ, 455, L143
28. Sari, R., & Piran T. 1999c, ApJ, 517, L109
29. Mészáros, P., & Rees M. J. 1999, MNRAS, 306, L39
30. Kulkarni, S. R., et al. 1999, ApJ, 522, L97
31. Gruzinov A., & Waxman E., 1999, ApJ, 511, 852
32. Medvedev, M. V., & Loeb A., 1999, astro-ph/9904363
33. Gruzinov A. 1999, ApJ, 525, L29
34. Ghisellini, G., & Lazzati, D., 1999, MNRAS, 309, L7
35. Sari, R. 1999, ApJ, 524, L43

Modeling the Fe Kα Emission Line in GRB Afterglows

Markus Böttcher*[1]

*Space Physics and Astronomy Department; Rice University, MS 108
6100 S. Main Street; Houston, TX 77005 - 1892; USA

Abstract. The time and angle dependent yield of fluorescence line and continuum emission from a dense torus around a cosmological gamma-ray burst (GRB) source is simulated, taking into account photoionization, collisional ionization, recombination, and electron heating and cooling due to various processes. A model calculation to reproduce the Fe Kα line emission observed in the X-ray afterglow of GRB 970508 indicates that $\sim 10^{-4} M_{\odot}$ of iron must be concentrated in a metal-enriched region of $R \lesssim 10^{-3}$ pc of extent.

INTRODUCTION

The recent marginal detection of a redshifted iron Kα emission line in the X-ray afterglow of GRB 970508 [1] has stimulated vital interest in the processes of photoionization and fluorescence line emission in gamma-ray burst (GRB) environments. X-ray absorption features and fluorescence line emission from the environments of cosmological GRBs have been investigated by several authors [2–4]. Motivated by suggestions that GRBs are caused by the death of a very massive star [5–7] and are therefore likely to be embedded in the dense gaseous environment of a star-forming region, in [4,3] the influence of a dense, quasi-isotropic GRB environment on the observable radiation, in terms of X-ray absorption features and fluorescence line emission has been investigated. However, the results of [8,3] clearly show that while a temporally varying Fe K absorption edge might be detectable, the luminosity and duration of the Fe Kα line observed in GRB 970508 [1], if real, is inconsistent with a quasi-isotropic environment.

A plausible way to solve this problem is the assumption a dense torus surrounding the GRB source, which could be produced by anisotropic ejecta of the burst progenitor (e. g., [6,7]). An anisotropic geometry has recently been considered in [9]. The processes considered by these authors include fluorescence line emission following multiple photoionization events, recombination and fluorescence line emission following electron-collisional ionization in a dense, hot plasma, and fluorescence line

[1] Chandra Fellow

CP510, *The Fifth Compton Symposium,* edited by M. L. McConnell and J. M. Ryan

emission in the course of reflection of the GRB radiation off a dense, highly opaque medium, in which most of the iron remains in a low ionization state. They find that the multiple-ionization – recombination scenario has problems due to the high electron temperature and implied long recombination times expected under the conditions where this process might be dominant, while their "thermal emission" and "reflection" scenarios appear more promising. Both processes require a very dense, highly opaque torus illuminated and heated by the GRB radiation. Thus, under these conditions, both processes might be important to a certain extent.

More recently, in [10], the radiation transport and reflection processes in two fundamentally different GRB and afterglow geometries (quasi-isotropic shell and evacuated funnel) have been investigated using XSTAR, under the assumption that at all times the material illuminated by the GRB may be regarded as being in thermal and ionization equilibrium. They find that a scattering funnel model can reproduce the observed iron line equivalent width of GRB 970508. However, their underlying GRB continuum appears to decay much more rapidly than the observed $F_X \propto t^{-1.2}$ [11], so that the iron line luminosity reproduced by their calculation is actually two orders of magnitude lower than the observed one.

In this paper, I report on first results of a detailed numerical study of the relevant processes in a dense torus illuminated by GRB radiation and hydrodynamically interacting with the relativistic blast wave triggered by the GRB. The physics included in the numerical study is outlined in Section 2. In Section 3, general considerations and a model calculation reproducing the (marginally) detected Fe Kα line in the X-ray afterglow of GRB 970508 are discussed.

MODEL ASSUMPTIONS AND COMPUTATIONAL SCHEME

In [3], we have investigated the problem of time-dependent photoionization, photoelectric absorption and fluorescence line emission in the case of an isotropic, moderately dense GRB environment, where recombination and electron-collisional effects were negligible. In the situation investigated in this paper, such effects have to be considered carefully. Furthermore, the numerical problem is no longer isotropic. The code used in [3] has thus been modified in order to account for the anisotropy of the GRB environment.

The geometry assumed to treat this problem is illustrated in Fig. 1. The center of the coordinate system is the center of the GRB explosion. The GRB source is surrounded by a torus of dense material (particle density n_T), at a distance r_T from the center of the explosion. The radius of the cross-section of the torus is denoted a. The burst source and the torus are embedded in dilute ISM of density n_{ISM} which extends out to a radius r_{ISM} from the center of the explosion.

The ISM and the torus are illuminated by the time-dependent radiation field of a GRB, represented by the analytical expressions of [12]. The numerical scheme used to treat the time-dependent radiation transport problem is basically the same as

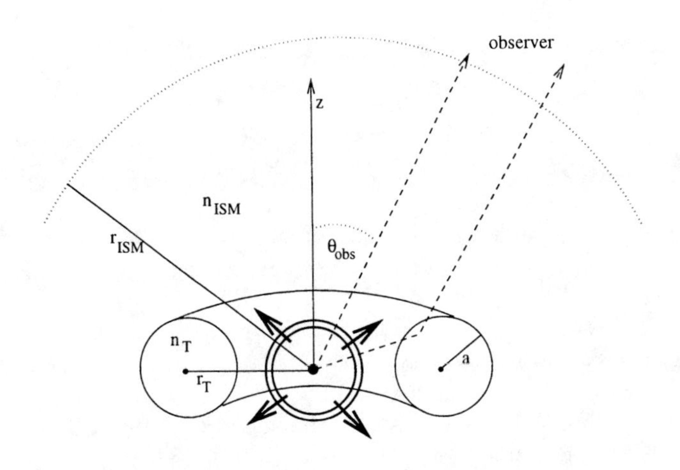

FIGURE 1. Sketch of the model geometry. See text for description.

described in [3], except that now the environment is anisotropic and more processes are included. To account for the anisotropy of the CBM, the expressions of [12] need to be modified because the blast wave will be decelerated much more efficiently in the dense torus than in the dilute ISM. When the blast wave begins to interact with the torus, it will produce an extremely short, extremely luminous flash of very high-energy radiation. The duration of this flash, as it would be measured by an observer located within or behind the torus, is $\Delta t_t \approx 1.1 \cdot 10^{-5}\,\mathrm{s}\, E_{54}/(n_{10}\,\Gamma_{300}^4\, r_{15}^2)$, where $E_{54} = E_0/(10^{54}\,\mathrm{erg})$, $n_{10} = n_T/(10^{10}\,\mathrm{cm}^{-3})$, $\Gamma_{300} = \Gamma/300$, and $r_{15} = r_1/(10^{15}\,\mathrm{cm})$.

As the parts of the blast wave interacting with the torus are decelerated to subrelativistic velocities almost instantaneously, the material of the torus will be energized via shock heating. This is taken into account using basic energy and momentum conservation arguments.

Line and continuum emission resulting from atomic processes in the ISM or the torus is assumed to be emitted isotropically at each point. The time delay (due to the light travel time difference) of such radiation reaching the observer from directions misaligned with respect to the line of sight to the GRB source, is properly taken into account. The output spectra and light curves are sampled under different viewing angles θ_{obs} with respect to the symmetry axis of the torus.

In addition to the processes of photoelectric absorption, photoionization and fluorescence line emission following photoionization events, which had been included already in [3], now radiative and dielectronic recombination [13,14], electron-collisional ionization [14], electron heating and cooling due to bremsstrahlung, Compton scattering and Coulomb scattering, and continuum emission due to radiative recombination and bremsstrahlung emission are taken into account. 133 strong UV and X-ray lines due to radiative transitions following recombination into excited states of H, He, C, O, Ne, Mg, Si, S, Ca, Fe, and Ni have been included using the line energies and branching ratios given in [15].

MODELING THE IRON LINE OF GRB 970508

GRB 970508 was a moderately bright burst with a peak flux of $\Phi_p \approx 3.4 \cdot 10^{-7}$ erg cm s^{-2} and a duration of $t_\gamma \approx 15$ s in the energy band of the GRBM on board the BeppoSAX satellite (40 – 700 keV). The X-ray flux measured by the WFC exhibits a power-law decay ($F_\nu(t) \propto t^{-\chi}$) up to $t \sim 6 \cdot 10^4$ s after the burst with index $\chi = 1.17 \pm 0.1$, before a secondary outburst at X-ray energies occurs [11]. [1] have recently reported the marginal detection of a possible Fe Kα line with a line flux of $\Phi_L = (5 \pm 2) \cdot 10^{-5}$ cm^{-2} s^{-1} at the likely redshift of the burst, $2.0 \cdot 10^4$ s – $5.6 \cdot 10^4$ s after the burst trigger time.

When estimating the efficiency of reprocessing the illuminating GRB flux into Fe Kα line flux and thus estimating the amount of mass required to produce the observed fluorescence line, it is important to take the effects of anisotropy of the CBM into account, as described in the previous section. If the density anisotropy due to the torus is strong, the assumption of the dense torus being illuminated and photoionized by radiation with the observed GRB characteristics will obviously yield a completely unrealistic picture.

The duration of the observed line emission is most probably dominated by light-travel time effects rather than the intrinsic time scale of one of the physical process involved, if the GRB and its afterglow are related to a relativistically expanding blast wave. A conservative estimate of the mass required to produce the observed iron line flux is based on the result of [3] that out to a radius of $R \sim 10^{20} \, (n_T[\text{cm}^{-3}])^{-1/3}$ cm iron in the CBM will be completely ionized. Taking into account the large number of Auger electrons ejected following photoionization of iron in low ionization states, a fiducial number of Kα line photons in the energy range 6.4 – 6.7 keV emitted in the course of complete ionization of an iron atom is ~ 5. Thus, for a line luminosity $L_{K\alpha} = 10^{44} L_{44}$erg s^{-1} emitted over a time scale $\Delta t_L = 10^5 \, t_5$ s, a total of $N_{Fe} = 2 \cdot 10^{56} \, L_{44} \, t_5 / f$ iron atoms is needed, where $f \geq 1$ is a correction factor accounting for the enhancement of the efficiency of line emission due to recombination and electron-collisional ionization. This yields a required mass of iron of $M_{Fe} = 0.16 \, (L_{44} \, t_5 / f) \, M_\odot$. The line flux measured in the afterglow of GRB 970508 corresponds to an intrinsic, isotropic luminosity of $L_{44} \approx 6$, which leads to a significantly higher mass estimate than found in [1] and [9], unles $f \gg 1$.

Following [12], the spectral and temporal properties of GRB 970508 can be reasonably well reproduced assuming $E_0 = 4.5 \cdot 10^{53}$ erg, $\Gamma_0 = 100$, $n_{ISM} = 4.8 \cdot 10^5$ cm^{-3}, $q = 8.15 \cdot 10^{-5}$, $g = 1.6$, and photon spectral indices $\alpha_{le} = 2/3$ and $\alpha_{he} = 2.1$ below and above the peak, respectively.

Fig. 2 shows the results of a model simulation assuming an intrinsically anisotropic blast wave, which is a factor of 100 less energetic in the direction of

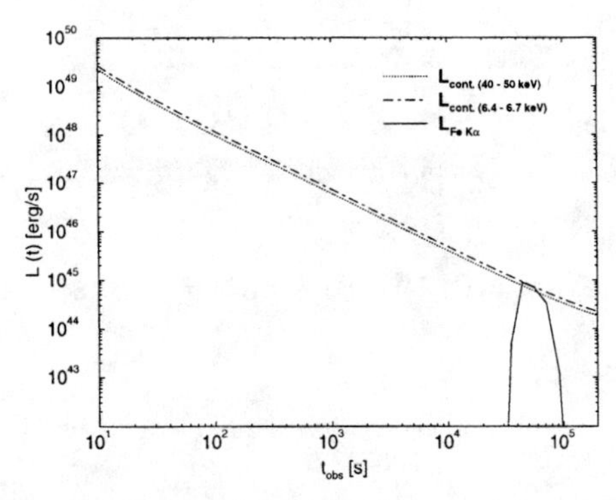

FIGURE 2. Temporal evolution of the afterglow continuum and the apparent isotropic luminosity in the Fe Kα line from a torus illuminated by an intrinsically isotropic GRB, as observed along the symmetry axis. Parameters: see text

the torus than along the evacuated funnel. The torus is assumed to have a total mass of $1.26\,M_\odot$, a density of 10^{11} cm^{-3}, and $r_T = 7 \cdot 10^{-4}$ pc, $a = 2 \cdot 10^{-4}$ pc. The material in the torus is assumed to have a two-fold iron overabundance with respect to solar-system abundances, which corresponds to a total of $8 \cdot 10^{-5}\,M_\odot$ of iron. The observed iron line intensity is very well reproduced with this model calculation.

REFERENCES

1. Piro, L., et al., *ApJ*, **514**, L73 (1999).
2. Mészáros, P., & Rees, M.J., *MNRAS*, **299**, L10 (1998).
3. Böttcher, M., et al., *A&A*, **343**, 111 (1999).
4. Ghisellini, G., et al., *ApJ*, **517**, 168 (1999).
5. Woosley, S. E., *ApJ*, **405**, 273 (1993).
6. Paczyński, B., *ApJ*, **494**, L45 (1998).
7. Stella, L., & Vietri, M., *ApJ*, **507**, L45 (1998).
8. Böttcher, M., et al., *A&AS*, proc. of "Gamma-Ray Bursts in the Afterglow Era", in press (1998).
9. Lazzati,, D., et al., *MNRAS*, **304**, L31 (1999).
10. Weth, C., et al., *ApJ*, submitted (astro-ph/9908243) (1999).
11. Piro, L., et al., *A&A*, **331**, L41 (1998).
12. Dermer, C. D., et al., *ApJ*, **513**, 656 (1999).
13. Aldrovandi, S. M. V., & Péquignot, D., *A&A*, **25**, 137 (1973).
14. Shull, J. M., & van Steenberg, M., *ApJS*, **48**, 95 (1982).
15. Kato, T., *ApJS*, **30**, 397, (1976).

Asymmetric subpeaks in short duration bursts

P.N. Bhat*, Varsha Gupta† and Patrick Das Gupta†

* *Tata Institute of Fundamental Research, Homi Bhabha Road, Colaba Mumbai 400 005*
† *Department of Physics and Astrophysics, University of Delhi, Delhi 110 007*

Abstract. Subpeaks in 65 short duration gamma ray bursts belonging to the 3B catalogue have been identified and are fitted with lognormal functions as most of them are of FRED kind. Characterising the symmetry of a subpeak by the ratio of rise and decay time, we find that statistically the first peak of a burst tends to be more asymmetric than the subsequent ones. In this work we have also studied the correlations between various parameters like duty cycle, hardness ratio, rapidity index etc. that characterise a burst.

INTRODUCTION

Following BeppoSAX observations, it is not only evident that GRBs are extra-galactic but also that a broad class of external shock models correctly describe the observed afterglow light-curves at various wavelengths [1]. However, when it comes to the basic mechanism that leads to such a wide variety of burst time-history in gamma rays, there is considerable uncertainty. It is therefore not surprising that a lot of exciting work is being done to study the underlying mechanism that gives rise to burst subpeaks [2] [3] [4] [5]. In what follows, we subject a subsample of bursts listed in the 3B catalogue, consisting of 65 short duration GRBs (i.e. $T_{90} < 2$s) with complete time-histories that include pre- and post- backgrounds, to several investigations.

ANALYSIS OF TIME-PROFILE DATA

To study the subpeaks quantitatively, we develop a systematic procedure to identify them unequivocally and then, since most peaks in a burst exhibit fast rise and a slow decay, we fit each of them with a lognormal function. The *modus operandi* for peak detection and fitting may be described briefly as follows: (i) The pre- and post- burst background counts are given a quadratic fit $bfit$ using a second order polynomial, and this fit is then extrapolated to the burst region. As some portion of the GRB could lie outside the T_{90} region, the burst extremities

CP510, *The Fifth Compton Symposium,* edited by M. L. McConnell and J. M. Ryan

are defined by noting the points of intersection between $bfit$ and the time-profile, at a resolution in which latter appears smooth. The instants at which first and the last intersections take place are termed as Tl and Tu, respectively, so that the duration of a burst may be defined as the difference $Tu - Tl$. (ii) A background time history BB of same duration and resolution as that of the burst time history SS is picked up from the post-burst background region, and then both on SS as well as on BB the following identical operations are employed: A first differential is taken for both, and the indication of peak presence in SS (BB) is spotted when the first differential for SS (BB) goes from positive to negative. A 1-dimensional low pass filter is applied to both the time histories successively, until BB shows a peak of 0 or 1. The peaks seen in SS at this stage are taken to be the actual signal peaks [6] [7]. The above exercise is repeated at various time resolutions, and at a certain optimum resolution, a given GRB shows maximum number of peaks, latter being subsequently accepted as genuine subpeaks of the burst. (iii) For a peak that begins at the kth time bin and ends at the lth, one estimates the following parameters - mean (α), standard deviation (β) and the number of photons (N): $\alpha = (\sum_{j=k}^{l} j\ f_j)/N$, $\beta^2 = (\sum_{j=k}^{l} (j-\alpha)^2 f_j)/(N-1)$, and $N = \sum_{j=k}^{l} f_j$, where $f_j = SS_j - bfit_j$. A subpeak is considered to be significant if the corresponding value of N is larger than the sum total of Poissonian fluctuation in the photon counts in each time bin. (iv) Each significant peak is fitted with a lognormal function so that a peak that starts from the kth bin and contains N signal photons is associated with the following theoretical photon count:

$$C_j = \begin{cases} \frac{N}{(j-k)\sqrt{2\pi}\sigma} \exp\left[-(\log(j-k)-\mu)^2/2\sigma^2\right] & j > k, \\ 0 & j \leq k. \end{cases} \quad (1)$$

The parameters (μ,σ) are related to the estimated mean and variances (α, β^2) from photon counts by: $\alpha = k + \exp(\mu + \sigma^2/2)$, $\beta^2 = \exp(2\mu + \sigma^2)[\exp(\sigma^2) - 1]$ Fitting of all significant peaks in a given burst is obtained by minimizing the reduced χ^2 determined using the observed count SS and the entire fit CC which is a sum of $bfit$ and M distinct lognormal functions, M being the total number of subpeaks in the burst. The minimization is carried out iteratively to arrive at the best fit parameters as well as corresponding uncertainties. Except in the case of the GRB with trigger no.2614, fits for all bursts closely follow the observed time-histories.

CHARACTERISATION OF BURST FEATURES

The GRBs, in this paper, have been characterised by the following quantities:

1. Burst Complexity Index (CI): This is defined as the number of peaks detected in a burst using the procedure described in the earlier section. An average value of 2.5 shows that a burst on an average has 2-3 peaks. Fig.1 provides a distribution of CI for the sample considered in this paper.

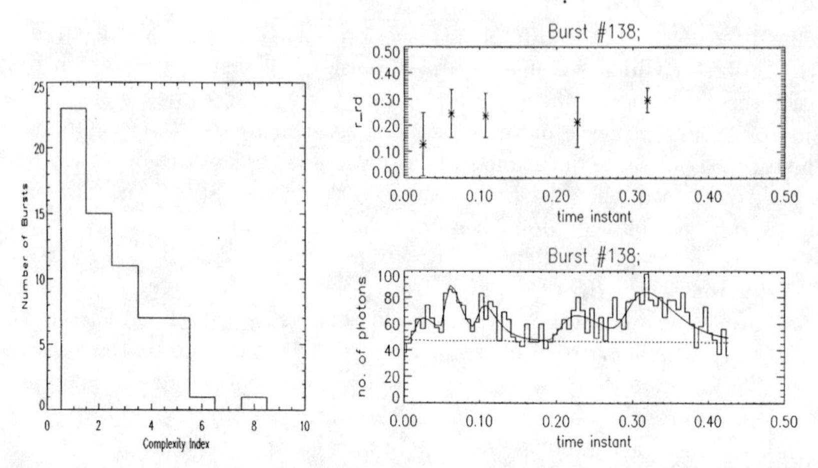

FIGURE 1. Left box shows the distribution of complexity index for 65 short duration GRBs. Right portion shows the time history of the GRB (trigger no.138) with fitted lognormal functions.On the top of it we show a plot of r_{rd} along with 1σ error against peak position - the straight line fit is obtained from χ^2 minimization

2. Rise and Decay times: Rise time t_r (decay time t_d) of a peak in a GRB is taken to be the time interval during which the value of the fitted lognormal function increases (decreases) from 5% (95%) to 95% (5%) of its maximum value. The lowest values of rise and decay times observed in the sample of 64 bursts are 1.15 ± 0.35 ms and 1.66 ± 0.73 ms. The maximum values are 90 ± 0.8 ms and 1.48 ± 0.42s .

3. Peak symmetry parameter, r_{rd} : We may quantify the symmetry of a peak using the parameter r_{rd} defined as: $r_{rd} \equiv t_r/t_d$. Since, most GRBs lack redshift information, use of r_{rd} to characterise peak symmetry has the merit of being independent of the burst distance because of the manifest cancellation of stretching in the rise and decay times due to cosmological expansion. For the sample used in our analysis, r_{rd} ranges from ~ 0.02 (highly asymmetric peak) to ~ 0.8 (almost symmetric case). An example of plots of r_{rd} versus peak position for individual GRBs is shown in fig.1 along with the burst time history. In many cases, we find that the first peak tends to be more asymmetric i.e. corresponds to a lower value of r_{rd} than the subsequent ones. To study this feature further, we fit these plots with straight lines after minimizing the corresponding χ^2. The slopes of all straight line fits are positive, barring 13 cases (out of a total of 41), suggesting that there is a tendency of subpeaks evolving towards increasing symmetry in a burst.

We also study the cumulative behaviour of r_{rd} as a function of peak position for all the 161 subpeaks belonging to 64 bursts (trigger no.2614 has been excluded) by making use of weighted correlation. The degree of correlation is

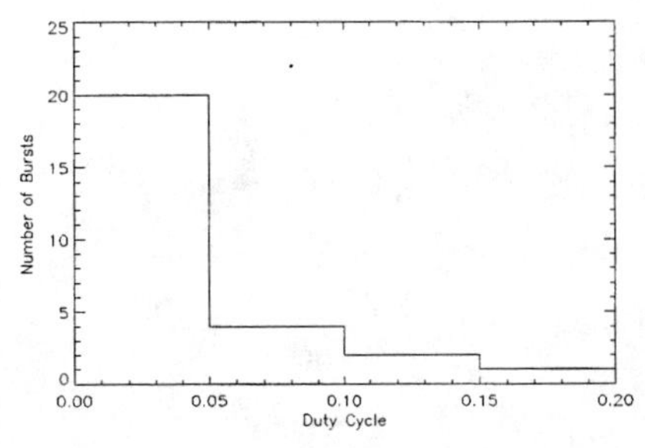

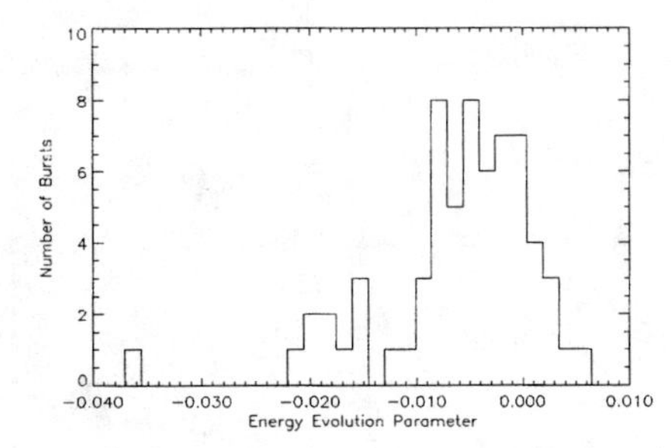

FIGURE 2. Left box shows the histogram of duty cycle for bursts having three or more subpeaks and in right box is shown the distribution of Energy Evolution Parameter for all the 65 bursts

0.17 corresponding to a chance probability of 0.03. A straight line fit to the scatter plot of r_{rd} as a function of peak position indicates a positive slope of 0.23, reinforcing the conclusion that subsequent peaks in a burst tend to be statistically more symmetric.

4. Duty Cycle (D): Motivated by the work of Sari and Piran [3] we define the duty cycle D of a burst to be the ratio of full width half maxima (FWHM) of the lognormal function fitted to the narrowest peak and the total duration of the burst. In this work, we estimate the duty cycle for GRBs with CI exceeding 2. From the distribution of D, presented in fig.2, it is evident that for most GRBs, the duty cycle lies between ~ 0.01 to ~ 0.05. The corresponding average value of D is 0.046.

5. Energy Evolution Parameter (EEP): For a given energy channel, the centroid (t_{cen}) of the time profile of a burst is obtained using,$t_{cen} = (\sum SS_i t_i)/\sum SS_i$ The mean slope of the straight line connecting the centroids with the channel number for the two brightest detectors is defined to be the EEP associated with the burst. A negative value of EEP would imply that high energy photons from the burst arrive at the detector earlier than the corresponding lower energy photons, suggesting a hard-to-soft evolution. Most GRBs show EEP to be negative, the sample average being -6×10^{-3}, as inferred from fig.2. Only for $\sim 14\%$ of the total sample of short duration GRBs, EEP is found to be positive. This is similar to the spectral evolution properties of long duration bursts [8].

CONCLUSIONS

The observed trend of r_{rd} increasing with peak position in a burst might be of significance to GRB models. If multiple shells are responsible for different subpeaks (as in internal shock models [3] [4] [5]) in a burst, it is plausible that the explanation of subsequent peaks being more symmetric lies in later shells suffering less deceleration due to sweeping away of ambient ISM by previous shells [9].

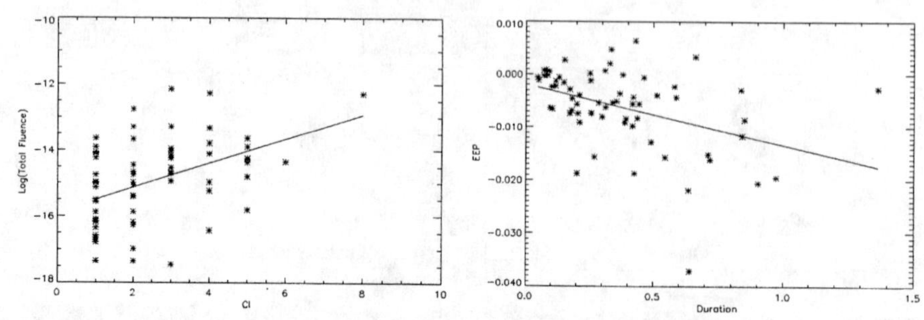

FIGURE 3. Left box shows the scatter plot of total fluence against complexity index of bursts. In the right box, EEP is plotted against duration.A least square fit is given to both the diagrams..

Correlation studies among various quantities characterising a burst show some interesting features. We find that both HR as well as CI are strongly correlated with total fluence. Probability that the observed correlation is purely due to chance is less than 2×10^{-4} in each case. Figure 3 shows a plot of total fluence versus CI. Hence, brighter bursts tend be harder and more complex, which is somewhat expected from energetics. The other strong correlations observed are between EEP and duration (see Fig.3) as well as among CI and duty cycle (for each pair, the degree of correlation is found to be negative with chance probability being less than 10^{-3}). This leads one to infer that soft photons tend to arrive at the detector with more delay from longer GRBs in the sample, while bursts with greater number of subpeaks often contain relatively very narrow subpeaks. It is interesting to note that we did not find any significant correlation between CI and duration.

The short duration GRBs exhibit spectral evolution similar to that of longer ones [8]. The more commonly observed hard-to-soft evolution has natural explanations in the context of expanding fire ball models. There still lies the challenge of modelling the soft-to-hard evolution seen in a smaller number of bursts ($\sim 14\%$).

REFERENCES

1. Tsvi Piran, Phys. Rep. **314**, 575-667 (1999).
2. E.E.Fenimore, C.D.Madras and S.Nayakshin, Astrophys.Jour. **473** ,998 -1012 (1996).
3. Re'em Sari and Tsvi Piran, Mon.Not.Astron.Soc. **287**, 110-116 (1997).
4. Shiho Kobayashi, Tsvi Piran and Re'em Sari, Astrophys.Jour. **490**, 92-98 (1997).
5. E.F.Fenimore, E. Ramirez-Ruiz and Bobing Wu, Astrophys.Jour. **518**, L73-L76 (1999).
6. Bhat, P. N. et al. AIP Conference Proceedings 307, Gamma Ray Bursts, Second Workshop, Huntsville, Al October 1993, pp. 953-957.
7. Bhat, P. N. et al. AIP Conference Proceedings 384,Gamma Ray Bursts, Third Huntsville Symposium, Huntsville, Al October 1995, pp. 197-201.
8. David L. Band, Astrophys.Jour. **486**, 928-937 (1997).
9. Varsha Gupta, Patrick Das Gupta and P.N. Bhat, in preparation.

First Results of a Study of TeV Emission from GRBs in Milagrito

R.S. Miller[8,6], R. Atkins[1], W. Benbow[2], D. Berley[3,10], M.L. Chen[3,11], D.G. Coyne[2], B.L. Dingus[1], D.E. Dorfan[2], R.W. Ellsworth[5], D. Evans[3], A. Falcone[6], L. Fleysher[7], R. Fleysher[7], G. Gisler[8], J.A. Goodman[3], T.J. Haines[8], C.M. Hoffman[8], S. Hugenberger[4], L.A. Kelley[2], I. Leonor[4], M. McConnell[6], J.F. McCullough[2], J. E. McEnery[1], A.I. Mincer[7], M.F. Morales[2], P. Nemethy[7], J.M. Ryan[6], B. Shen[9], A. Shoup[4], C. Sinnis[8], A.J. Smith[9], G.W. Sullivan[3], T. Tumer[9], K. Wang[9], M.O. Wascko[9], S. Westerhoff[2], D.A. Williams[2], T. Yang[2], G.B. Yodh[4]

(1) University of Utah, Salt Lake City, UT 84112, USA
(2) University of California, Santa Cruz, CA 95064, USA
(3) University of Maryland, College Park, MD 20742, USA
(4) University of California, Irvine, CA 92697, USA
(5) George Mason University, Fairfax, VA 22030, USA
(6) University of New Hampshire, Durham, NH 03824, USA
(7) New York University, New York, NY 10003, USA
(8) Los Alamos National Laboratory, Los Alamos, NM 87545, USA
(9) University of California, Riverside, CA 92521, USA
(10) Permanent Address: National Science Foundation, Arlington, VA ,22230, USA
(11) Now at Brookhaven National Laboratory, Upton, NY 11973, USA

Abstract. Milagrito, a detector sensitive to γ-rays at TeV energies, monitored the northern sky during the period February 1997 through May 1998. With a large field of view and high duty cycle, this instrument was used to perform a search for TeV counterparts to γ-ray bursts. Within the Milagrito field of view 54 γ-ray bursts at keV energies were observed by the Burst And Transient Satellite Experiment (BATSE) aboard the Compton Gamma-Ray Observatory. This paper describes the results of a preliminary analysis to search for TeV emission correlated with BATSE detected bursts. Milagrito detected an excess of events coincident both spatially and temporally with GRB 970417a, with chance probability 2.8×10^{-5} within the BATSE error radius. No other significant correlations were detected. Since 54 bursts were examined the chance probability of observing an excess with this significance in any of these bursts is 1.5×10^{-3}. The statistical aspects and physical implications of this result are discussed.

CP510, *The Fifth Compton Symposium,* edited by M. L. McConnell and J. M. Ryan

I OBSERVATIONS AND ANALYSIS

Milagro, a new type of TeV γ-ray observatory sensitive at energies above 100 GeV, with a field of view of over one steradian and a high duty cycle, began operation in February 1999, near Los Alamos, NM. A predecessor of Milagro, Milagrito [5], operated from February 1997 to May 1998. During this time interval, 54 γ-ray bursts (GRBs) detected by BATSE [1] were within Milagrito's field of view (less than 45° zenith angle).

A search was conducted in the Milagrito data for an excess of events above the cosmic-ray background coincident with each of these γ-ray bursts. For each burst, a circular search region was defined by the BATSE 90% confidence interval, which incorporates both the statistical and systematic position errors [2]. The size of this 90% confidence interval ranged from 4° to 26° for the 54 GRBs in the sample. The search region was tiled with an array of overlapping 1.6° radius bins centered on a $0.2° \times 0.2°$grid. This radius was derived from the measured angular resolution of Milagrito and was selected prior to the search. The number of events falling within each of the 1.6° bins was tallied for the duration of the burst reported by BATSE. This duration is defined as the time required for BATSE to accumulate 90% of the γ-rays(T90). T90 ranged from 0.1 seconds to 195 seconds for the 54 bursts examined.

The angular distribution of background events on the sky was characterized using two hours of data surrounding each burst. This distribution was normalized to the number of events detected by Milagrito over the entire sky during the T90 interval (N_{T90}). The resulting background data were also binned in 1.6° bins spaced 0.2° apart. The Poisson probability that the excess of events in each 1.6° bin was due to a background fluctuation was calculated and the bin with lowest probability was taken as the candidate position of a TeV γ-ray counterpart to the BATSE burst. The background and signal counts in this bin were used to calculate a fluence or fluence upper limit for each burst.

II RESULTS

The flux sensitivity of Milagrito to γ-ray bursts depends on the zenith angle and duration of the burst, and on the instrument conditions at the time. During the lifetime of the Milagrito detector, data were taken with three different water depths (0.9 m, 1.5 m and 2.0 m). In addition, for the period February 1997 through the end of March 1997 a considerable amount of snow collected on the cover of the pond. Detector simulations were used to obtain effective area as a function of zenith angle for an assumed $E^{-2.0}$ spectrum for each of these configurations . These were then used to calculate flux upper limits for each burst. Flux upper limits in the range $10^{-6} - 10^{-8}$ $\gamma/cm^2/s$ were obtained for 53 of the 54 bursts in the sample.

Of the 54 bursts one, GRB 970417a, shows a substantial excess above background in the Milagrito data. The BATSE detection of this burst is a weak burst with a

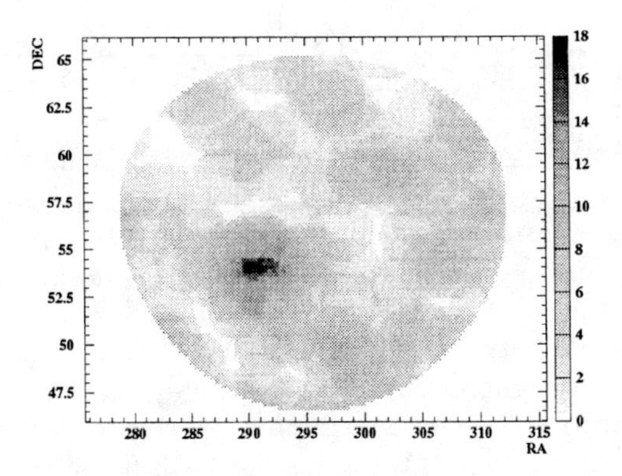

FIGURE 1. Number of events recorded by Milagrito during T90 in the BATSE 90% error radius for GRB 970417, each bin contains the number of events detected by Milagrito within a 1.6 degree radius.

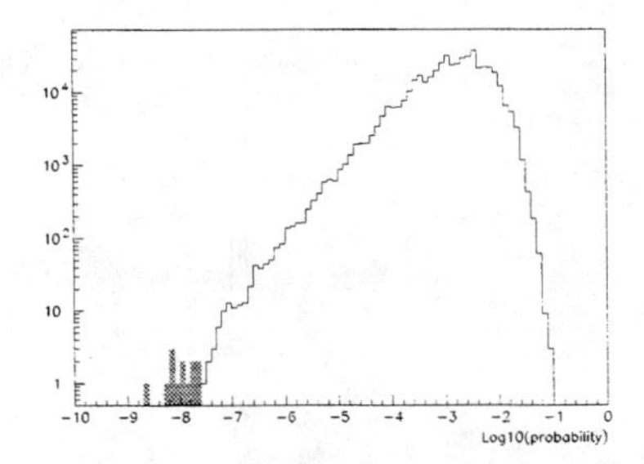

FIGURE 2. The distribution of minimum probabilities for the ensemble of simulated data-sets for GRB 970417a.

fluence in the 50–300 keV energy range of 1.5×10^{-7} ergs/cm^2 and T90 of 7.9 seconds. BATSE determined the burst position to be RA = 295.66°, DEC = 55.77°. The 90% positional uncertainty was 9.4° . The 1.6° radius bin with the largest excess in the Milagrito data is centered at RA = 289.89° and DEC = 54.0°, corresponding to a zenith angle of 21°. This position is 3.8° away from the position reported by BATSE; well within the BATSE 1-sigma position error 6.2°. The uncertainty in the position of the TeV candidate was determined by Monte-Carlo simulations to be approximately 0.5°. Figure 1 shows the number of counts in this search region for the array of 1.6° bins. The bin with the largest excess has 18 events with an expected background of 3.46 ± 0.11. The Poisson probability for observing a signal at least this large due to a background fluctuation is 2.89×10^{-8}.

To obtain the significance of this result one must account for the size of the search region. The probability of obtaining the observed significance anywhere within the entire search region was determined by Monte Carlo simulations. A set of simulated signal maps was made by randomly drawing N_{T90} events from the background distribution. Each map was searched, as before, for a significant excess within the search region defined by BATSE. The probability of the observation in the actual data being due to a fluctuation in the background, after accounting for the size of the search region, is given by the ratio of the number of simulated data sets with probability less than that observed for the actual data to the total number of simulated data sets. The distribution of the probabilities for 4.65×10^6 simulated data sets is shown in figure 2; thirteen of which had Poisson probability less than 2.89×10^{-8}. We therefore find that the chance probability of such a detection within the entire 9.4° search region for GRB 970417a to be 2.8×10^{-5}. The probabilities for each of the other 53 bursts in the sample were obtained using the same method, the distribution of these probabilities, after correcting for the size of the search region, is shown in figure 3. The histogram on the left, plotted on a log-linear scale, illustrates the significance of the excess for GRB 970417a relative to the rest of the sample. The histogram on the right of this figure, plotted on a linear scale is flat, as expected. 54 bursts were examined. Therefore the chance probability of observing such a significant excess due to fluctuations in the background for any of these bursts is 1.5×10^{-3}.

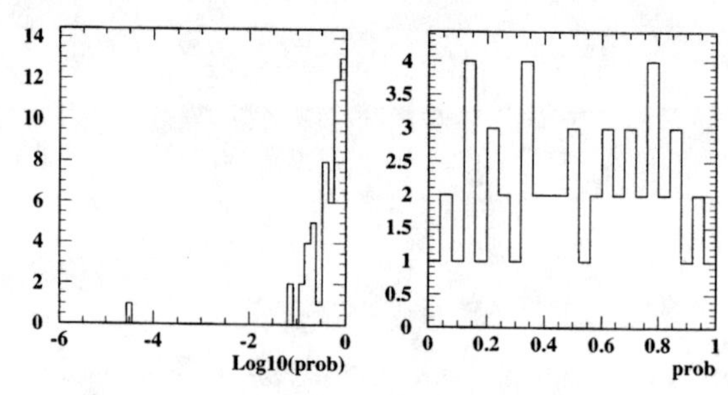

FIGURE 3. The distribution of probabilities, corrected for the size of the search region for the 54 GRBs in the sample, both plots show the same data with a linear and logarithmic scale for the x-axis

Although the initial search was limited to T90, for GRB 970417a longer time intervals were also examined. To allow for the positional uncertainty of the excess observed by Milagrito, the radius of the search bin was increased to 2.2° for this search. A search for TeV γ–rays integrated over long time intervals of one hour, two hours and a day after the GRB start time did not show any significant excess. Lightcurves where the data are binned in intervals of one second and of T90 (7.9 s) are shown in figure 4. A preliminary analysis reveals no statistically compelling evidence for TeV afterflares.

III CONCLUSION

An excess of events with chance probability 2.8×10^{-5} coincident both spatially and temporally with the BATSE emission for GRB 970417a was observed by Milagrito. The chance probability that an excess of this significance would be observed

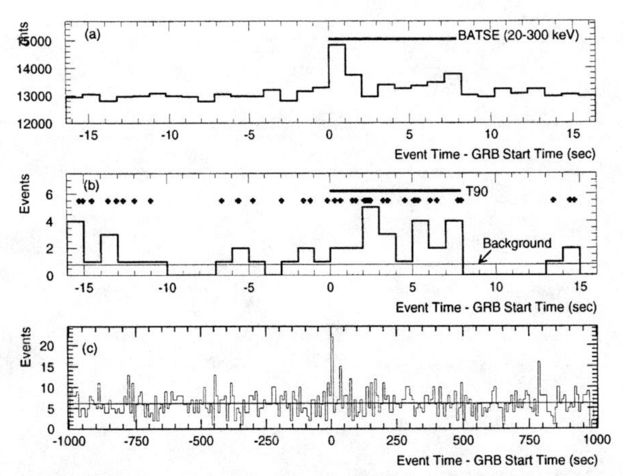

FIGURE 4. GRB 970417a: (a) The BATSE lightcurve, (b) Milagrito data within a 2.2° radius of GRB 970417a integrated in 1 second bins, the crosses indicate the arrival times of the events and (c) integrated in bins of 7.9 seconds (T90) for 2000 seconds

from the entire sample of 54 bursts is 1.5×10^{-3}. Preliminary analysis suggests that the spectrum must extend with no cutoff to at least a few hundred GeV.

If the observed excess from GRB 970417a is not a fluctuation of the background, then a new class of γ-ray bursts bright at TeV energies may have been observed. A search for other coincidences with BATSE, to verify this result, will be continued with the current instrument, Milagro, which has increased sensitivity to TeV γ-ray bursts.

ACKNOWLEDGMENTS

This research was supported in part by the National Science Foundation, the U.S. Department of Energy Office of High Energy Physics, the U.S. Department of Energy Office of Nuclear Physics, Los Alamos National Laboratory, the University of California, the Institute of Geophysics and Planetary Physics, The Research Corporation, and the California Space Institute.

REFERENCES

1. W. S. Paciesas et al., (Astro-Ph-9903205) (1999)
2. M. S. Briggs et al., *Astrophys. J. Supp.* **122(2)**, 503 (1999)
3. M.H. Salamon and F. W. Stecker, *Astrophys. J.* **493**, 547 (1998).
4. J. R. Primack et al, *Astroparticle Physics* **11**, 93 (1999).
5. R. Atkins et al., *Nucl. Inst. and Methods* (1999) (submitted).

SOLAR FLARES

High-Energy Solar Flare Observations at the Y2K Maximum

A. Gordon Emslie

Department of Physics, UAH, Huntsville, AL 35899

Abstract. Solar flares afford an opportunity to observe processes associated with the acceleration and propagation of high-energy particles at a level of detail not accessible in any other astrophysical source. I will review some key results from previous high-energy solar flare observations, including those from the *Compton* Gamma-Ray Observatory, and the problems that they pose for our understanding of energy release and particle acceleration processes in the astrophysical environment. I will then discuss a program of high-energy observations to be carried out during the upcoming 2000-2001 solar maximum that is aimed at addressing and resolving these issues. A key element in this observational program is the High Energy Solar Spectroscopic Imager (HESSI) spacecraft, which will provide imaging spectroscopic observations with spatial, temporal, and energy resolutions commensurate with the physical processes believed to be operating, and will in addition provide the first true gamma-ray spectroscopy of an astrophysical source.

INTRODUCTION

At a mere 8 light *minutes* distance, the Sun offers us a unique opportunity to examine astrophysical processes at close range. The high photon flux and fine spatial resolution (1 arc-second $\simeq$ 725 km) permits the study of these processes at a level of detail totally inaccessible in cosmic sources.

The long-term goals of high-energy solar flare research are to understand the processes that accelerate high-energy particles (electrons, protons, neutrons, and heavier ions are all involved to various degrees) and ultimately to reconcile these processes with a primary energy release mechanism. It is well-established that the primary source of flare energy resides in current-carrying magnetic fields, for two reasons:

- the observed energy released in a flare is of the same order of magnitude as the stored magnetic energy $(B^2/8\pi)V$ (where B is the magnetic field strength [Gauss] and V the flare volume [cm^3]); and
- no other energy source operating at the sub-nuclear temperatures appropriate to the outer solar atmosphere contains nearly enough energy to power a flare.

CP510, *The Fifth Compton Symposium,* edited by M. L. McConnell and J. M. Ryan

However, as will be elaborated on below, extracting energy from a magnetic field on a short timescale is highly problematic – the resistive diffusion timescale $\tau_r \simeq (4\pi L^2/\eta c^2)$ (where L is the scale length, c the speed of light, and $\eta \sim 10^{-7} T^{-3/2}$ is the resistivity [T being the plasma temperature]) is millions of years for typical solar parameters ($L \simeq 10^9$ cm, $T \simeq 10^7$ K). This timescale is many orders of magnitude longer than observed flare timescales, which range from seconds to hours. As a result, most "pure" theoretical modeling efforts must invoke either some form of "anomalous" process to increase η or a highly contrived geometry (to decrease L). However, the true nature of the energy release process is revealed not so much by such *ad hoc* approaches, but rather by a detailed study of the radiation signatures that are produced during the flare. In this article, we will show how observations of high-energy emissions can be used to answer important questions such as

- where is the particle acceleration region?
- what are the energy spectra and angular distributions of the accelerated particles?
- how do the energetic particles gain/lose energy during their acceleration and propagation?

Answers to these questions represent important constraints for modelers of the energy storage and release processes.

PERTINENT OBSERVATIONS

Observations that provide direct information on high-energy particles in solar flares include hard X-rays (photon energy $\epsilon \gtrsim 10$ keV) and gamma-rays ($\epsilon \gtrsim 500$ keV). Observations of thermal emissions in lower-energy photon bands are less pertinent, since they correspond to a situation where the entropy has increased to the point of obscuring important information (although such observations can provide essential context information – see below). Hard X-ray data includes the photon spectrum, the spatial distribution of the emission, flux vs. time profiles and, in rare instances, the polarization and directivity of the radiation field. Together, these constitute *complete knowledge* of the emitted radiation field, albeit at a level that can still be significantly limited by photon statistics. In gamma-rays, on the other hand, we have no spatial or polarization data (either existing or available in the forseeable future) and are instead limited to studies of temporal variations and spectral forms. Solar flare gamma-ray spectra consist of continua due to both bremsstrahlung and pion production, and nuclear de-excitation lines at a few MeV [12].

RESULTS FROM PREVIOUS OBSERVATIONS

Before discussing the exciting prospects for new observations at the upcoming solar maximum, it is worth reviewing some key results from previous observations.

Acceleration of Large Electron Fluxes

Hard X-ray observations of large solar flares reveal the emission of some 10^{32} photons s^{-1} at hard X-ray energies – note that this is a *directly observed* value, with little latitude for modification. If the electrons that produce these hard X-rays are considered "energetic," in the sense that their characteristic energy is much greater than the thermal energy of the electrons in the target with which they interact, then one can readily show (e.g., [13]) that only $\sim 10^{-5}$ of the electron energy is used to produce X-rays, with the rest of the energy dissipated in Coulomb collisions with ambient electrons. Therefore the observed hard X-ray fluence corresponds to a much larger rate of electron acceleration, viz. $\sim 10^{37}$ s^{-1}.

An apparently straightforward way of accelerating these electrons invokes large-scale direct electric fields [7]. Fields of modest strength $\simeq 10^{-4}$ V cm^{-1} (comparable to the Dreicer field that constitutes the threshold field for runaway of a significant fraction of the electron distribution), operating on characteristic scales of 10^9 cm, can readily produce electrons of the required 100 keV energies. However, practical application of such a mechanism requires that we satisfactorily resolve two critical issues:

- *Replenishment*
  The total number of electrons in a coronal loop is of order $nV \simeq 10^{10}$ cm^{-3} $\times$ 10^{27} cm^3 $\simeq 10^{37}$. (Here n is the plasma density and V the volume of the loop.) An acceleration rate of 10^{37} s^{-1} would therefore deplete the entire loop in about a second or so, unless there is some mechanism to replenish the source with fresh (or recycled) electrons.

- *Electrodynamics*
  The current I associated with the rate of electron acceleration is of order 10^{37} s^{-1} $\times$ 10^{-19} C $\simeq 10^{18}$ A. The steady-state magnetic field associated with such a current propagating in a $r \simeq 10^7$ m wide channel is $\mu_o I/2\pi r \simeq 10^4$ Tesla, or 10^8 Gauss. Magnetic fields of this magnitude are simply not observed on the Sun, even in flares. Worse, the associated energy content in the field is $(B^2/8\pi) \times V \simeq 10^{42}$ erg, some 10 orders of magnitude greater than the energy in the electrons that supposedly create it. This implies that a steady-state current of this magnitude would never be reached. Indeed, when one considers that the inductance of the loop is $\sim \mu_o V^{1/3} \simeq 10$ H, the voltage required to "switch on" a current of 10^{18} A in a timescale of the order of a few seconds (the rise time of a typical hard X-ray burst) is $\simeq 10^{19}$ Volts, an absurd value.

We thus see that the size of the inferred electron acceleration rate poses some real, fundamental, problems for acceleration mechanisms. A model which resolves both of the problems above has been proposed [5]. It involves the fragmentation of the flare volume into many separate current channels, with counterstreaming currents in adjacent channels. This reduces the current in any given channel and so both the magnetic field generated and voltage necessary to "switch on" the current, and provides replenishment though a closed circuit involving adjacent current channels. However, in order to keep the magnetic fields and electric potentials within acceptable limits, some 10^{14} (!!) separate channels are required. The simultaneous energization of such a large number of separate elements is a powerful constraint that has caused significant doubt to be cast on the viability of large-scale electric fields as an electron acceleration mechanism in solar flares. Instead, some researchers now strongly favor some form of stochastic acceleration model, such as resonant acceleration by cascading ensembles of magnetohydrodynamic waves [9], which result in the accelerated particles having a nearly isotropic distribution and so avoid unacceptably large directed currents.

Proton Acceleration

Using data from the OSSE experiment on the *Compton* Gamma-Ray Observatory, Murphy et al. [10] have reported observations of > 1 MeV flux in the flare of 1991 June 4 that persists for many hours after the onset of the event. Such a continued release of magnetic energy cannot readily be accounted for through trapping of an ensemble of previously-acclerated particles; instead long-term continuous acceleration is required. Mechanisms to produce such a long-term acceleration of protons, without any concomitant acceleration of hard X-ray-producing electrons, have not yet been put forward.

In addition, Ramaty et al. [11] (see also [4]) have analyzed *Solar Maximum Mission* observations of the ^{20}Ne gamma-ray line at 1.634 MeV. This line has a relatively low excitation threshold and so provides information on the proton spectrum at energies $\simeq 1$ MeV. Such observations suggest that the proton spectrum continues its steep form down to ~ 1 MeV energies and therefore that the energy content in protons rivals that in ~ 10 keV electrons.

Relative Timing of Hard X-ray Emissions

Aschwanden, Schwartz, & Alt [1] have shown that flare emission in different channels of the *Compton* Observatory BATSE instrument show slight, yet systematic, delays at low energies. Furthermore, these energy dependence of these time delays is consistent with simple time-of-flight propagation delays. This result not only implies a very prompt and localized acceleration process; it also allow us to set bounds on the distance between the acceleration site and the region (e.g., the chromosphere) where the bulk of the hard X-ray emission is produced. In a study of almost 3000 events, they report an average time delay between emissions in the 25-50 keV, and 50-100 keV, channels of 26.4 ± 1.5 ms, implying a propagation distance of 7300 ± 800 km. Note that in order for electrons of this energy to propagate such a distance without attenuation by Coulomb collisions, the ambient density must be $\ll 10^{11}$ cm^{-3}.

THRUSTS FOR THE NEW SOLAR MAXIMUM

The 2000-2001 solar maximum will see an unprecedented flotilla of spacecraft observatories aimed at the Sun. This, plus the coordinated use of ground-based observatories (e.g., optical, radio, neutron monitors) will permit us to address many of the issues highlighted above.

The "flagship" mission during this period will be the High Energy Solar Spectroscopic Imager (HESSI) mission [3], selected by NASA as a Small-Class Explorer and scheduled for a July 2000 launch. HESSI consists of a set of nine pairs of absorbing grids, which provide spatial information via a rotating modulation collimator technique. The spatial resolutions of the various grid pairs range from 2.3 arc seconds to 3 arc minutes. Beneath each set of collimators are situated cryogenically-cooled high-purity Germanium detectors, capable of providing spectral resolution of order 2 keV across the energy range from a few keV to several MeV. A complete set of spatial Fourier components is obtained each half-rotation of the spacecraft, so that a full image in each energy band can be obtained in this 2 second time period, limited only by photon statistics. Meaningful spatial information can also be obtained on much shorter ($\sim$ 50 milliseconds) timescales, the duration of one modulation cycle in the finest grid pair. HESSI will contribute fundamental new information in several key areas, as follows:

Hard X-Ray Imaging Spectroscopy

The length, time and energy resolutions achievable by the HESSI instrument are not only better than those of previous instruments by large factors, they are also for the first time commensurate with scales of physical interest to the electron-proton bremsstrahlung process that produces hard X-rays in flares. For example, consider the passage of a 20 keV electron through a medium of ambient density 10^{11} cm^{-3}, a typical value for the flaring corona. In about 200 ms, the electron travels the first 2.3 arc seconds ($\simeq$ 1500 km), during which its energy degrades from 20 keV to $\sim$ 16 keV. In the next 300 ms or so, it travels the next 2.3 arc seconds, degrading to $\sim$ 10 keV in the process. Thus changes in the hard X-ray spectrum in the $10-20$ keV range should be apparent on resolvable size and time scales.

Observation of the location of the predominant hard X-ray source will permit an identification of the spatial *location* of the energy release site, so observing directly the relative location of the acceleration region and the region of peak hard X-ray production, a quantity heretofore determined only *indirectly* (cf. discussion above [1]). Observation of the spatial *structure* of the hard X-ray emission will address the issue of source fragmentation referred to above in connection with large-scale electric field acceleration models. In addition, the ability to resolve hard X-ray structure on size scales smaller than the collisional mean free path of the electrons responsible permits a very straightforward computation of the electron spectrum as a function of position, since the emission can now be considered "thin-target"

[2]. Comparison of the hard X-ray spectra at different locations within the source will permit an *empirical* determination of the energy loss (or gain) rates associated with the bremsstrahlung-producing electrons. Consider the continuity equation for electron number, viz.

$$\frac{\partial N(E,t)}{\partial t} + \frac{\partial}{\partial E}\left[N(E,t)\,\frac{dE}{dt}\right] = 0 \tag{1}$$

and replace, in a steady-state analysis, the time derivative by $v(\partial/\partial z)$, where z is the distance along the electron's path. Solving for dE/dz gives

$$\frac{dE}{dz} = -\frac{1}{N(E,z)\,v(E)}\int_{E^*}^{E} v\,\frac{\partial N(E,z)}{\partial z}\,dE, \tag{2}$$

where E^* is such that $dE/dz = 0$ for $E = E^*$. This result gives the energy loss rate dE/dz as a function of energy E (and possibly also position z) from a series of measurements of the electron number spectrum $N(E)$ at different positions z, which are readily inferred from the hard X-ray spectra at these positions using standard inversion formulae [2].

These inferred energy loss/gain rates will help establish (or negate) the viablility of acceleration and transport models. For example, models of electron energization that invoke acceleration by large-scale electric fields make very definitive predictions as to the evolution of the electron spectrum with position and time, and the have the easily-tested property that dE/dz is independent of E. Modification to the accelerated electron spectrum through Coulomb collisions corresponds to $dE/dz \sim -n/E$, where n is the local density. The $1/E$ form is a testable result, and, if indeed consistent with observations, yields the ambient density n as a parameter.

Gamma-ray Line Spectroscopy

HESSI will produce the first observations of the *shape* of the nuclear gamma-ray lines emitted during flares. The spectral resolution of the Germanium detectors is well below the width of every significant nuclear line emitted in solar flares (except the deuterium formation line at 2.223 MeV, which is formed by capture of very slow neutrons on ambient protons and therefore has a very narrow width commensurate with the speed of these neutrons). Thus HESSI observations will constitute the first true *spectroscopic* observations of an astrophysical source, with line shapes and shifts defining the energy and angular distributions of the accelerated particles (e.g., protons, α-particles), as well as the abundance of the ambient nuclei. It will be of considerable interest, for example, to observe the change in the shape of these spectral lines during long-duration events such as that of 1991 June 4 (cf. [10]). A wealth of theoretical predictions regarding line shapes exists, and awaits the HESSI observations to test their predictions.

Hard X-ray Polarization

The layout of the detectors in the HESSI spacecraft offers a serendipitous opportunity to observe linear polarization of the hard X-ray emission. Although not a primary goal of the HESSI mission, this additional capability was recognized in the design stage, and has been enhanced by the introduction of a small beryllium passive scatterer in the detector tray. The geometry is such that five of the nine Germanium detectors will be exposed to photons scattered off the beryllium block. These scattered photons will be detected in the *lower* segments of the detectors. Low energy photons impingent directly on the detectors are stopped in the upper segments and do not propagate through to the lower segments. Thus, any low-energy photons detected by the lower segments *must* have arrived via the beryllium scatterer, and the variation in counts from detector to detector provides information on the degree and orientation of the incoming polarization vector. The rotation of the spacecraft (at 15 rpm) will reduce uncertainties due to differences in detector sensitivity. It is estimated that polarization signals in excess of few percent (in a large flare) will be detectable by HESSI.

There is no shortage of model predictions for the polarization signal expected (e.g., [8] and reference therein). Attempts have been made over the last 30 years to observe hard X-ray polarization from flares, but have been inconclusive for various reasons (see [8]). The prospect of observing this important quantity is therefore an intriguing one. Not only would a significant result (either positive or null) have important implications for the angular distibution of electrons accelerated in flares, but one should also note philosophically that this observation completes the information content (arrival time and direction, energy, and polarization) of the radiation field.

Complementary Observations

Notwithstanding the pioneering observations that HESSI will make, it cannot do everything. The scientific return from the next solar maximum will depend on a synthesis of the HESSI observations with data from a variety of other experiments. These include:

Context Observations

The physical context in which the processes of particle acceleration and transport occur can be revealed to only a limited extent by HESSI. More detailed knowl edge of the magnetic geometry can be obtained through the use of ground-based magnetographs and gyrosynchrotron radio measurements. The plasma environment (density, temperature, ...) in which the acceleration occurs can be revealed through soft X-ray and EUV measurements from spacecraft such as Yohkoh and SoHO. Optical observations reveal not only the sites of particle precipitation, but optical polarization data in lines such as Hα can be used to infer the flux of protons in the $\lesssim$1 MeV energy range [6].

Other High-Energy Observations

The effective detector area of HESSI is $\simeq 100$ cm^2. By comparison, the BATSE experiment on *Compton* has an effective area of order 5000 cm^2 (depending somewhat on the orientation of the spacecraft with respect to the Sun). The vastly greater number of photons detected by BATSE therefore permits studies of fine temporal fluctuations that HESSI cannot resolve. Previous BATSE results have suggested fundamental energy release (and particle acceleration) timescales of order several hundred milliseconds, and future such observations, coupled with the fine spatial resolution and exquisite spectral detail to be obtained with HESSI, will constitute a data set from which truly meaningful conclusions can be drawn.

SUMMARY

The proximity of the Sun makes it a unique laboratory in which to investigate processes of broader astrophysical interest. Observations from past instruments, including those on the *Compton* Gamma-Ray Observatory, have shown that there are significant issues that need to be resolved in understanding these processes. The HESSI spacecraft, in conjunction with various other space- and ground-based observatories, will provide data during the next solar maximum that will permit truly significant strides toward this goal to be made.

REFERENCES

1. Aschwanden, M., Schwartz, R..A., & Alt, *Ap.J.* **447**, 923 (1995).
2. Brown, J.C., *Solar Phys.* **18**, 489 (1971).
3. Dennis, B.R., et al., *SPIE* **2804**, 228 (1996).
4. Emslie, A.G., Brown, J.C., & MacKinnon, A.L. *Ap. J.* **485**, 430 (1997).
5. Emslie, A.G., & Hénoux, J.-C., *Ap.J.* **446**, 371 (1995).
6. Emslie, A.G., Miller, J.A., Hénoux, J.-C., Vogt, E., & Sahal-Bréchot, S., *Ap. J*, submitted (2000).
7. Holman, G.D., & Benka, S.G., *Ap.J.* **400**, L79 (1992).
8. Leach, J., Emslie, A.G., & Petrosian, V., *Solar Phys.*, **96**, 331 (1985).
9. Miller, J.A., *Ap.J.*, **491**, 939 (1997)
10. Murphy, R.J. et al., *Ap. J.* **490**, 883 (1997).
11. Ramaty, R., Mandzhavidze, N., Kozlovsky, B., & Murphy, R.J., *Ap. J.* **455**, 193 (1995).
12. Share, G.H., & Murphy, R.J., *Ap. J* **452**, 933 (1995).
13. Tandberg-Hanssen, E.A., & Emslie, A.G., *The Physics of Solar Flares*, Cambridge University Press : Cambridge (1986).

Accelerated-Particle Spectral Variability in the 1991 June 11 Solar Flare

R. J. Murphy and G. H. Share

E. O. Hulburt Center for Space Research
Naval Research Laboratory
Washington, DC 20375 USA

Abstract. The X12 solar flare that occurred on 1991 June 11 was well observed by the instruments on *CGRO*. Here we present observations of this flare obtained with OSSE. OSSE detected nuclear γ-ray lines, 2.223 MeV neutron-capture radiation, 0.511 MeV positron-annihilation radiation, >16 MeV γ rays and neutrons. We apply three techniques of measuring the accelerated-particle energy spectrum. Since each technique is sensitive to a different range of particle energies, comparison of the derived spectral indexes provides information about the shape of the particle spectrum over a broad energy range. We show that as the flare progressed, the particle spectrum clearly evolved. There were intervals when the particle spectrum was consistent with an unbroken power law from ~1 to 100 MeV. But there were also intervals when the spectrum deviated significantly from a simple, unbroken power law through this energy range.

INTRODUCTION

The 1991 June 11 solar flare was one of six large flares originating from AR6659. The GOES X12 flare was located at N31W17 solar heliographic coordinates and the soft X-ray flux peaked at ~2:09 UT (~7740 s UT). It was well-observed by the instruments on *CGRO*. Due to saturation of the anti-coincidence counter, the EGRET spark chamber could not observe the peak of emission and began monitoring after 3:26 UT. It detected γ-ray emission up to a GeV lasting for at least 8 hours after the peak (1) with a spectrum suggesting a pion origin. Data from the EGRET/TASC obtained during the peak of the flare showed evidence for spectral evolution (2). COMPTEL data also suggested that the >30 MeV ion spectrum hardened as the flare progressed (3). Nuclear lines were detected with *GRANAT*/PHEBUS (4).

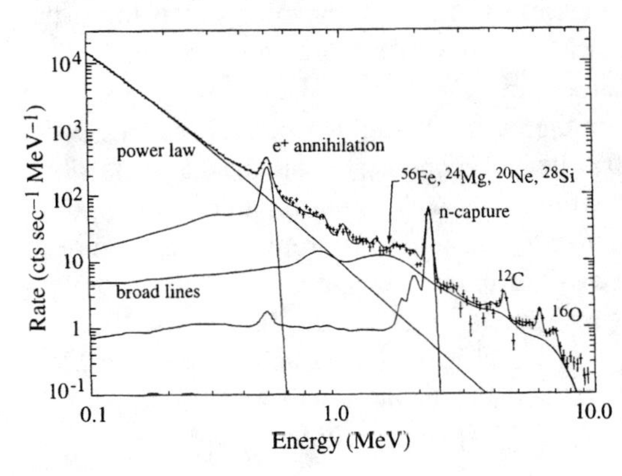

Figure 1. 1991 June 11 solar flare OSSE count spectrum from one detector.

OSSE covers γ-ray energies from 0.050 to 200 MeV and is sensitive to high-energy neutrons. Its good energy resolution allows spectroscopic study of γ-ray lines and its large area provides excellent sensitivity for weak emission. At the time of the June 11

CP510, *The Fifth Compton Symposium,* edited by M. L. McConnell and J. M. Ryan
2000 American Institute of Physics 1-56396-932-7

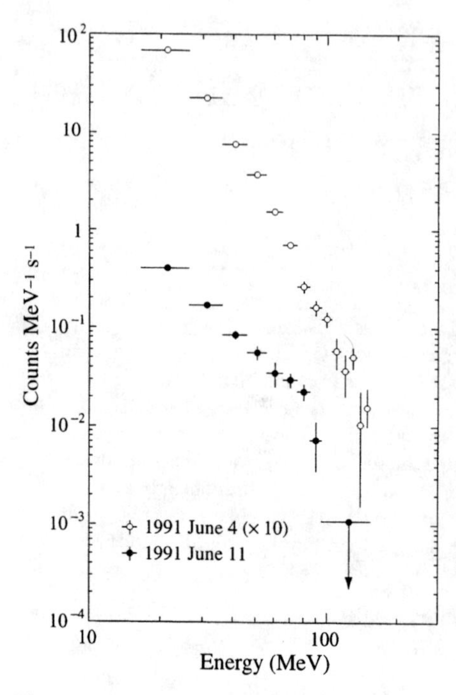

Figure 2. OSSE >16 MeV γ-ray count spectrum of the 1991 June 11 (solid) and 1991 June 4 (open) flares.

flare, OSSE was observing Cygnus when it received a BATSE solar trigger. OSSE immediately slewed to the Sun and observed for ~3000 seconds until satellite night and during the following orbits.

Figure 1 shows the OSSE 0.1–10 MeV count spectrum accumulated after the peak of the June 11 ~MeV emission. Also shown is a fit to the data with a model consisting of narrow and broad γ-ray lines and a power law representing electron bremsstrahlung. We note the excess of the data over the model above ~7.5 MeV implying that there is emission continuing above 10 MeV not accounted for by a simple extrapolation of the power law from lower energies. OSSE also observed these high-energy γ rays in addition to neutrons from this flare. OSSE separates the γ-ray and neutron contributions using measured pulse shapes and Figure 2 shows the OSSE >16 MeV count spectrum attributed to γ-rays from this flare. Shown for comparison is the spectrum observed by OSSE from the 1991 June 4 flare. The June 11 spectrum is significantly harder, suggesting that some of this emission may be from pion decay rather than from electron bremsstrahlung as in the June 4 flare.

Figure 3 shows the derived time profiles of the >2 MeV narrow nuclear lines, the 2.223 MeV neutron-capture line, the 0.511 MeV positron-annihilation line, and the >16 MeV emission. (The relative normalizations are arbitrary for display purposes.) The nuclear lines, the neutron-capture line, and the positron-annihilation line peak at ~7450 s UT. The >16 MeV emission peaks ~850 s later. A secondary peak in the 0.511 MeV line at this time supports the claim that much of this >16 MeV emission is from pion decay rather than electron bremsstrahlung. Four emission intervals are defined in Figure 3. Interval I (7126–7650 s UT) covers the peak of the nuclear lines and electron bremsstrahlung. Interval II (7650–7913 s UT) corresponds roughly to the "Interphase" defined by Dunphy *et al.* (2) for this flare. Interval III (7913–8634 s UT) covers the >16 MeV emission peak. Interval IV (8634–9420 s UT) covers the decay phase.

MEASUREMENT OF SPECTRAL HARDNESS

The cross sections for the processes responsible for γ-ray line emission can have significantly different energy dependencies. Ratios of line fluences are therefore sensitive to the steepness of the accelerated-particle energy spectrum. Depending on where the

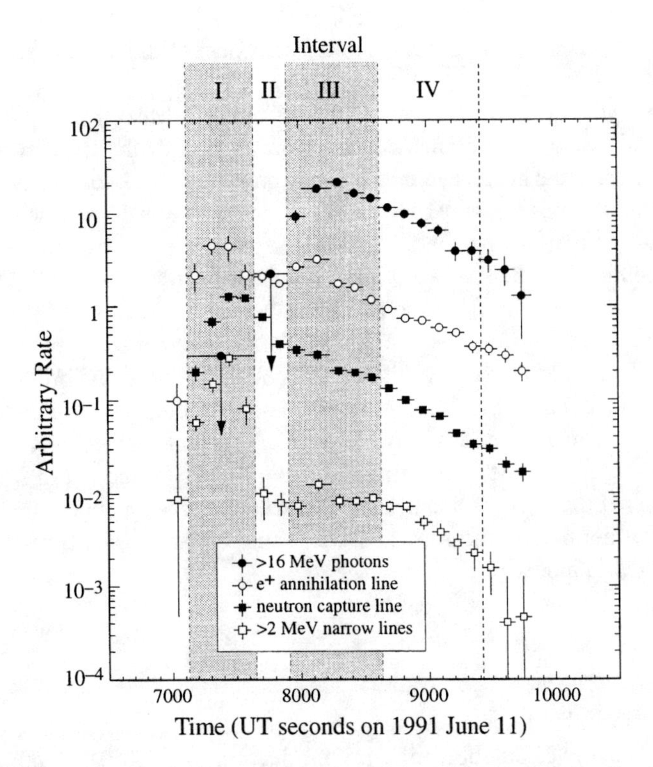

Figure 3. >16 MeV, 0.511 MeV line, 2.223 MeV line and >2 MeV narrow line fluxes.

specific cross sections peak, the ratios provide this information within relatively narrow energy bands. We use three ratios which together cover particle energies from ~1 to greater than several hundred MeV.

The first technique uses the **6.13-to-1.63 MeV line flux ratio**. The thresholds for the production of the $^{16}O^{*6.13}$ and $^{20}Ne^{*1.63}$ excited nuclei are ~2 and 8 MeV nucleon^{-1}, respectively, making the flux ratio quite sensitive to the particle spectral hardness. Both cross sections are falling rapidly by ~20 MeV nucleon^{-1}. As a result, this line ratio is sensitive to the steepness of the particle spectrum in the ~2–20 MeV nucleon^{-1} energy range.

The second technique uses the **0.511-to-4.44 MeV line ratio**. Positrons result from the decay of radioactive nuclei produced by interactions of ~10–50 MeV nucleon^{-1} particles. If the particle spectrum extends to several hundred MeV nucleon^{-1}, positrons can also result from the decay of pions produced. The positrons slow down in the solar atmosphere and annihilate with electrons to produce the 0.511 MeV line. On the other hand, the $^{12}C^{*4.44}$ excited nucleus is produced by ~10–30 MeV nucleon^{-1} particles. Therefore, this line ratio is sensitive to the steepness of the particle spectrum from ~10 to 50 MeV nucleon^{-1} or as much as several hundred MeV nucleon^{-1} if pions are produced. We assume a positronium fraction of 67%.

The third technique uses the **2.223-to-4.44 MeV line ratio**. Neutrons responsible for the 2.223 MeV neutron-capture line are produced primarily by 50–100 MeV nucleon^{-1}

particles. This ratio is therefore sensitive to the steepness of the particle spectrum in the 10–100 MeV nucleon^{-1} range.

Deexcitation γ rays such as those at 1.63, 4.44 and 6.13 MeV from ^{20}Ne, ^{12}C and ^{16}O are produced essentially instantaneously after production of the excited nuclei. Their fluxes therefore trace the nuclear interaction rate and so the 6.13-to-1.63 MeV line ratio provides an instantaneous measure of the particle spectral index. However, both the 0.511 MeV positron-annihilation and the 2.223 MeV neutron-capture lines are delayed. The 0.511 MeV line delay is due the decay time of the radioactive positron emitters and the subsequent slowing-down time of the positrons. Murphy and Ramaty (5) showed that for the 1981 June 21 *Solar Maximum Mission (SMM)* flare the slowing-down time was less than 16 s. The mean radioactive decay time, however, is on the order of a hundred seconds (6,7). The delay of the 2.223 MeV line is due to the slowing-down time of the neutrons to thermal energies and is also on the order of a hundred seconds (8,9). Because of these delays, spectral indexes determined using the 0.511 and 2.223 MeV lines are meaningful only if the data are accumulated over a long enough interval such as intervals I, III and IV defined for the June 11 flare. We note that interval II may be too short to provide a reliable index.

RESULTS

Power-law spectral indexes of the accelerated particles can be derived from the measured line ratios using calculations of γ-ray production (e.g., 5,10). Comparison of the indexes determined with the three techniques can then provide information about the overall shape of the particle spectrum from ~2 to as much as several hundred MeV.

Figure 4 shows spectral indexes derived by applying the three techniques to the June 11 data accumulated during the four Intervals of Figure 3. Technique 1 could not be used during the peak of the ~MeV emission (Interval I) due to saturation effects caused by the intense flux. Each Interval is discussed, starting with the two intervals (I and III) covering the peaks of the ~MeV and the >16 MeV emissions.

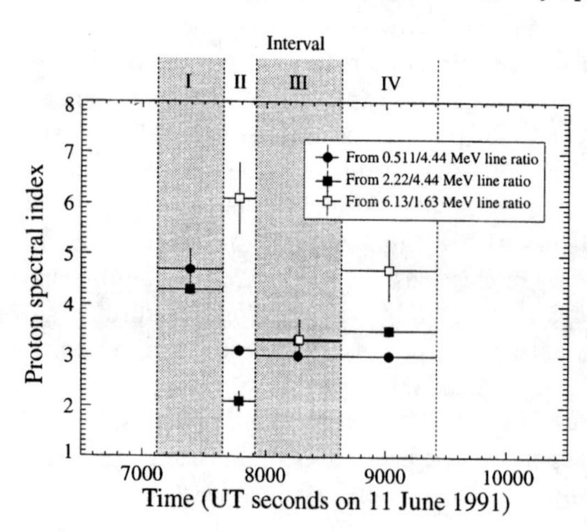

Figure 4. Accelerated-particle spectral indexes derived by the three techniques for the four intervals defined in Figure 3.

Interval I (peak of ~MeV γ-ray emission). The two available techniques give consistent spectral indexes, implying that the spectrum during this interval is an unbroken power law from ~10 to ~100 MeV. The derived index of ~4.5 is consistent with the index of ~4 derived by Dunphy *et al.* (2) for this interval.

Interval III (peak of >16 MeV γ-ray emission). All three techniques give nearly consistent indexes, implying that the spectrum is essentially an unbroken power law from ~2 to >100 MeV. The index (~3.2) is harder than that of Interval I and is consistent with the index of 3.35 ± 0.10 derived by Dunphy *et al.* (2) for this interval. Such a harder index is expected since the >16 MeV emission is probably from pion decay.

Interval II (Interphase). The indexes derived with the three techniques are *not* consistent, suggesting that the particle spectrum is *not* an unbroken power law during this interval. There appears to be a low-energy component with a soft spectrum and a high-energy component with a hard spectrum.

Interval IV (decay phase). The indexes derived with the three techniques again do not agree, suggesting that the spectrum again is not an unbroken power law. The hard, high-energy component may be that seen by EGRET late in the flare.

CONCLUSION

We applied three techniques of determining the steepness of the accelerated-particle energy spectrum, each sensitive to a different energy range, to OSSE data from the 1991 June 11 solar flare. We showed that the particle spectrum evolved as the flare progressed. During both the peak of the ~MeV emission and the peak of the >16 MeV emission, the spectrum was consistent with an unbroken power law from a few MeV to 100 MeV. The spectral indexes during these two intervals were ~4.5 and ~3.2, respectively. During the interval between these two emission peaks and during the decay phase, the spectrum deviated from a simple power law, showing a steeper low-energy component.

The peak of the >16 MeV emission was not coincident with that of the ~MeV line emission but occurred ~850 seconds later. The energy spectrum of this high-energy emission was very hard, suggestive of pion-decay radiation rather than electron bremsstrahlung. An enhancement of the 0.511 MeV positron annihilation radiation at the time of the >16 MeV peak supports this conclusion (see Figure 3). The arrival time of the neutrons (not shown here) was delayed relative to the ~MeV line radiation as compared to other flares such as the 1991 June 4 flare. A significant fraction of the neutrons were therefore probably produced during this delayed high-energy episode. This second acceleration episode is similar to the 1982 June 3 and 1984 April 24 flares observed with *SMM*.

REFERENCES

1. Kanbach, G., *et al.*, A&AS, **97**, 349 (1993).
2. Dunphy, P. P., *et al.* submitted to Solar Physics (1999).
3. Rank, G., PhD thesis, Max Plank Inst. (1995).
4. Trottet, G., *et al.*, A&AS, **97**, 337 (1993).
5. Murphy, R. J., and Ramaty, R., Adv. Sp. Res, **4**, #7, 127 (1984).
6. Kozlovsky, B., *et al.*, ApJ, **316**, 801 (1987).
7. Murphy, R. J., PhD thesis, Univ. of Md. (1985).
8. Prince, T. A., *et al.*, Proc. 18th Internat. Cosmic Ray Conf., **4**, 79 (1983).
9. Hua, X.-M., and Lingenfelter, R. E., Solar Physics, **107**, 351 (1987).
10. Ramaty, R., *et al.*, in AIP Conf. Proc. 374, 172 (1996).

Energetic Proton Spectra in the 11 June 1991 Solar Flare

C. A. Young, M. B. Arndt, A. Connors, M. McConnell, G. Rank, J. M. Ryan, R. Suleiman, and V. Schönfelder[†]

University of New Hampshire,Space Science Center
[†]Max-Planck Institut fur Extraterrestriche Physik

Abstract. We have studied a subset of the 11 June 1991 solar flare γ-ray data that we believe arise from soft proton or ion spectra. Using data from the COMPTEL instrument on the Compton Observatory we discuss the gamma-ray intensities at 2.223 MeV, 4-7 MeV, and 8-30 MeV in terms of the parent proton spectrum responsible for the emission.

INTRODUCTION

Flares and intervals within flares have been observed to have γ-ray spectra that are produced by soft proton spectra. Share and Murphy (1995) found evidence of soft proton spectra in several flares. During 1991 COMPTEL was used to measure the γ-ray spectra of several flares including some that exhibited features associated with soft proton spectra.

The 11 June 1991 flare observed with the COMPTEL instrument aboard CGRO shows several indicators of a soft proton spectrum. During a middle phase of the flare (Dunphy et al. 1999), there is low 4-7 MeV emission, no significant emission above 10 MeV, but a strong 2.2 MeV line (Rank et al. 1997). In addition, the 2.2 MeV line has a long (> 300 s) decay time. These two facts point toward an excess of lower energy neutrons and thus a softer parent proton spectrum (Ryan et al. 1996). Similarly, the 24 October 1991 flare shows a paucity of 4-7 MeV emission. There was a strong 2.2 MeV line and an even stronger ^{20}Ne line (Suleiman 1995). This ^{20}Ne line and other low threshold lines seen along with the 2.2 MeV line indicate low energy neutrons and thus a soft proton spectrum. The 11 June 1991 measurements were performed in COMPTEL's telescope mode, whereas the 24 October 1991 measurements were made in the burst mode.

COMPTEL has two modes that operate simultaneously. The telescope mode is a Compton imaging telescope that operates in the energy range of 0.75 MeV - 30 MeV. It consists of two detector planes. The top is a low-Z liquid scintillator that Compton scatters an incident photon to the bottom detector plane that is a

CP510, *The Fifth Compton Symposium,* edited by M. L. McConnell and J. M. Ryan
 1-56396-932-7/00/$17.00

set of high-Z NaI detectors. The second mode is the burst spectrometer and it operates from 0.1-1.1 MeV and 0.6-10 MeV, using two of the NaI detectors in the lower (D2) modules. The ideal analysis of the flares includes both data sets. The telescope mode provides spectral and imaging data so it has a high signal-to-noise ratio but a small effective area. The burst mode provides spectral information in an overlapping energy band with a lower signal-to-noise ratio but a much greater effective area. This provides two data sets that contain comparable information but with different systematics. In an earlier study (Young et al. 1999), we discussed the spectral deconvolution and its use with the burst mode data emphasizing the 24 October flare. Here, we discuss the analysis of the 11 June flare telescope data.

DISCUSSION

For analysis of the 11 June 1991 flare data, we divided the flare into 3 phases. Phase I covers 7030 to 7600 seconds (1:57:10 - 2:06:40 UT), Phase II is from 7600 to 8000 seconds (2:06:40 - 2:13:20 UT) and Phase III is from 8000 to 10400 seconds (2:13:20 - 2:53:20 UT). These phases differ slightly from those used by Dunphy et al. (1999). Fig. 1 shows the flare lightcurve uncorrected for background and deadtime and Fig. 2 shows the flux time profiles for the 2.2 MeV, 4-7 MeV, and 8-30 MeV energy ranges (Rank 1996) with background and continuum subtracted.

As seen in Fig. 2, during Phase I there is a larger flux in the 4-7 MeV range than in 2.2 MeV line with a significant 8-30 MeV component. This is consistent with a high-energy population of ions. During the transition from Phase I to Phase II the relative amounts of 4-7 MeV and 2.2 MeV emission are reversed with an order of magnitude more 2.2 MeV emission. This would normally indicate a very hard ion spectrum. This, however, is inconsistent with the lack of significant 8-30 MeV emission. Alternately, the spectrum could have been produced by a softer population of ions. This would better explain the extended 2.2 MeV emission which has a decay time of >300 s. Aside from extended production, a long decay time requires a low hydrogen density which would imply a smaller penetration depth for the neutrons. The neutrons would penetrate deeper if they were more energetic (Hua and Lingenfelter 1987).

The three plots in Fig. 3 show the deconvolved photon spectra (using the Maximum Entropy Method (Gull and Skilling 1991) and a Monte Carlo simulated effective area (Schönfelde et al. 1991)) for the three phases. We have included with each a best fit powerlaw for the bremsstrahlung continuum. The powerlaw indices were based on a single measurement from the PHEBUS instrument by Trottet et al. (1993). The normalizations where based on the COMPTEL data at 600 keV and above 10 MeV. The 2.2 MeV emission is the most prominent feature in all three phases. The measurements with EGRET-TASC (Dunphy et al. 1999) and COMPTEL (Rank 1996) both show spectral hardening from Phase I to Phase III. Also, both indicate two independent ion populations in Phase I and Phase III due to the low value of the nuclear line flux in Phase II.

FUTURE WORK

Several important features will be added to this analysis. The initial broad-banded work is based on the work of Rank (1997). Finer energy binning and an improved deconvolution method will enable a focus on individual lines. Improved power law fits based on BATSE (Rank 1999) and OSSE will provide better continuum subtraction. These improvements will allow a comparison of photon spectra from both the telescope and burst mode. The time intervals from the TASC analysis will be closer matched to allow a better comparison of the 2 data sets. This provides two independent measurements in the same energy band. Then using a nuclear γ-ray template model, the proton spectrum can be analyzed along with the associated low energy neutrons (Young and Ryan 1997).

ACKNOWLEDGMENTS

Using this work is supported by NASA under contract NASS-26645, by the German government through DARA grant 50 QV 90968 and the Netherlands Organization for Scientific Research (NWO).

REFERENCES

1. Dunphy et al., *Solar Physics* **187**, 45 (1999).
2. Gull, S. and Skilling, J., *Quantified Maximum Entropy User's Manual*, MEDC: Meldreth (1991).
3. Hua and Lingenfelter, R., *Solar Physics* **107**, 351 (1987).
4. Murphy et al., *ApJ* **358**, 259 (1990).
5. Rank, G. et al., *28th AAS Solar Physics Division*, Bozeman, Montana (1997).
6. Rank, G., *PhD Thesis*, der Technischen Universität München (1996).
7. Rank, G. et al., work in preparation (1999).
8. Schönfelder et al., *ApJS* **86**, 657 (1991).
9. Share, G.H. and Murphy, R.J., *ApJ* **452**, 993 (1995).
10. Suleiman, R., *Masters Thesis*, University of New Hampshire (1995).
11. Trottet,G. et al., *Astron. Astroph. Suppl.* **97**, 337 (1993).
12. Young, C.A. and Ryan, J.M., *25th ICRC*, Durban, South Africa (1997).
13. Young, C.A. et al., *26th ICRC*, Salt Lake City, Utah, USA (1999).

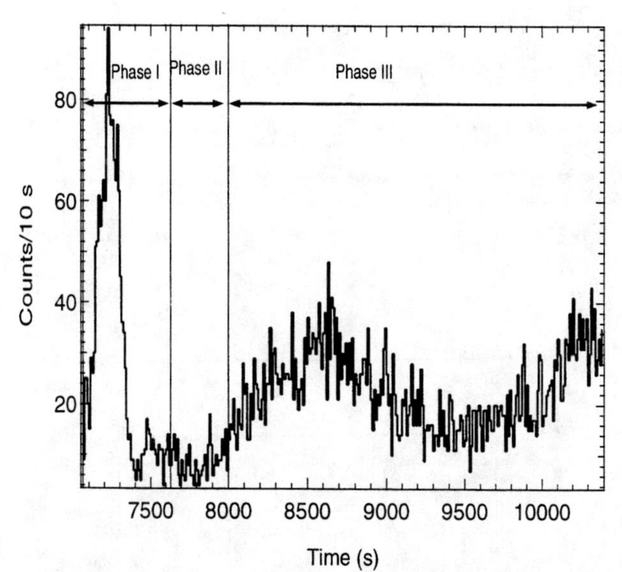

FIGURE 1. The 11 June 1991 solar flare uncorrected light curve showing the 3 phases. The odd shape is an effect of the deadtime from the middle of Phase I until a few hundred seconds into Phase III. This is due too the large soft x-ray flux that impinged upon the veto domes of COMPTEL.

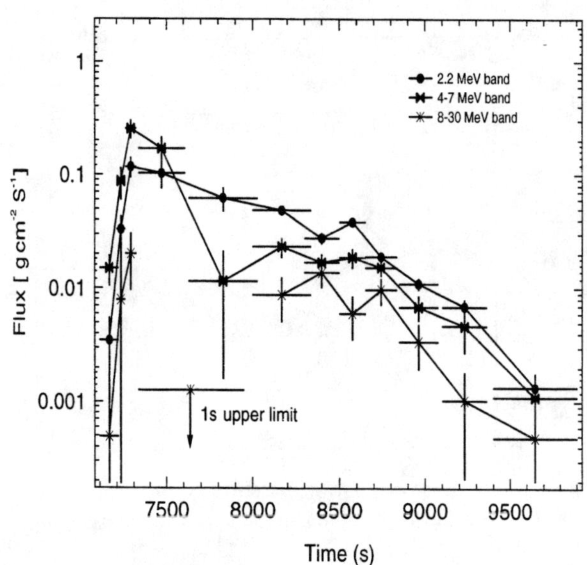

FIGURE 2. The 11 June 1991 solar flare flux time histories in the 2.2, 4-7, and 8-30 MeV energy bands with subtraction of background and continuum (based on Trottet et al. 1993).

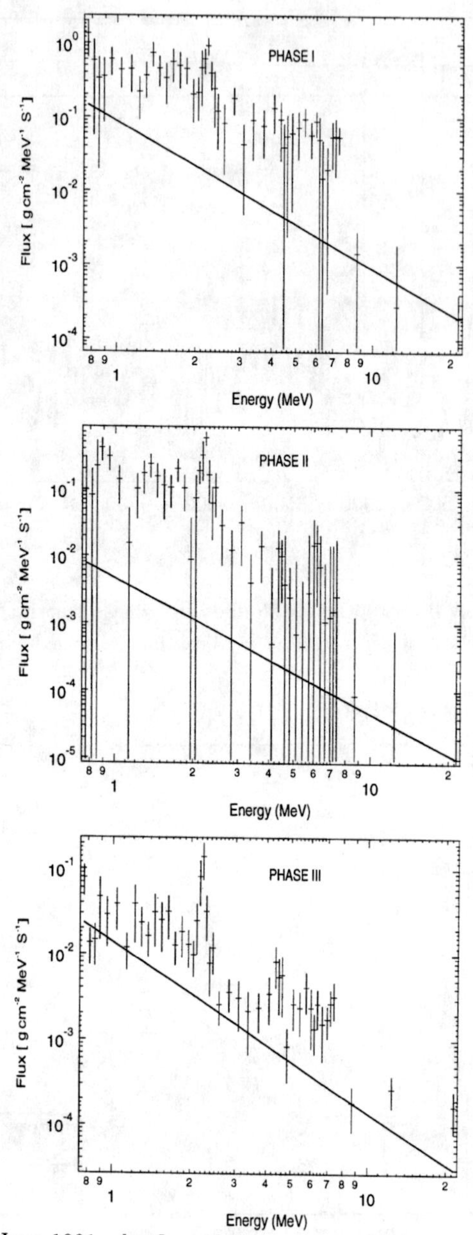

FIGURE 3. The 11 June 1991 solar flare photon spectra with bremsstrahlung power law fit for the 3 time phases.

Gamma Ray Measurements of the 1991 November 15 Solar Flare

Martina B. Arndt[1], Kevin Bennett[2], Alanna Connors[1], Mark McConnell[1], Gerhard Rank[3], James M. Ryan[1], Volker Schönfelder[3], Raid Suleiman[1], and C. Alex Young[1]

[1]*Space Science Center, University of New Hampshire, Durham, NH 03824*
[2]*Astrophysics Division of ESA/Estec, NL-2200 AG Noordwijk, The Netherlands*
[3]*Max-Plank Institute für Extraterrestrische Physik, D-8046, Garching Germany*

Abstract. The 1991 November 15 X1.5 flare was a well observed solar event. Comprehensive data from ground-based observatories and spacecraft provide the basis for a contextual interpretation of gamma-ray spectra from the Compton Gamma Ray Observatory (CGRO). In particular, spectral, spatial, and temporal data at several energies are necessary to understand the particle dynamics and the acceleration mechanism(s) within this flare. X-ray images, radio, Ca XIX data and magnetograms provide morphological information on the acceleration region [4,5], while gamma-ray spectral data provide information on the parent ion spectrum. Furthermore, time profiles in hard X-rays and gamma-rays provide valuable information on temporal characteristics of the energetic particles. We report the results of our analysis of the evolution of this flare as a function of energy (~25 keV - 2.5 MeV) and time. These results, together with other high energy data (e.g. from experiments on Yohkoh, Ulysses, and PVO) may assist in identifying and understanding the acceleration mechanism(s) taking place in this event.

INTRODUCTION

The 1991 November 15 X1.5 solar flare was a well observed event. It occurred near disk center at S13W19 (~23° heliocentric angle) in active region NOAA 6919. X-ray emission commenced ~2235:00 UT, peaked ~2237:30 UT, and lasted on the order of 5 minutes. High energy observations of this event between ~25 keV and 10 MeV were made by instruments on Yohkoh [6], Ulysses [6], Pioneer Venus Orbiter [8], and the Compton Gamma Ray Observatory (CGRO). In this work we use data from the BATSE [1] and COMPTEL [11] instruments on CGRO.

Four of the eight BATSE detectors were exposed to the event, with 97%, 50%, 37% and 11% of their respective areas facing the Sun. Data from these detectors are available between ~25 keV and 1.9 MeV, though we only utilize those data up to 400 keV. COMPTEL detected the event only in burst mode due to the event's

CP510, *The Fifth Compton Symposium*, edited by M. L. McConnell and J. M. Ryan

location 66° off COMPTEL's zenith. The spacecraft's orientation with respect to the event resulted in only 7 gm/cm^2 obstruction of the burst detectors, a relatively small mass. These COMPTEL data cover 0.6 - 10 MeV.

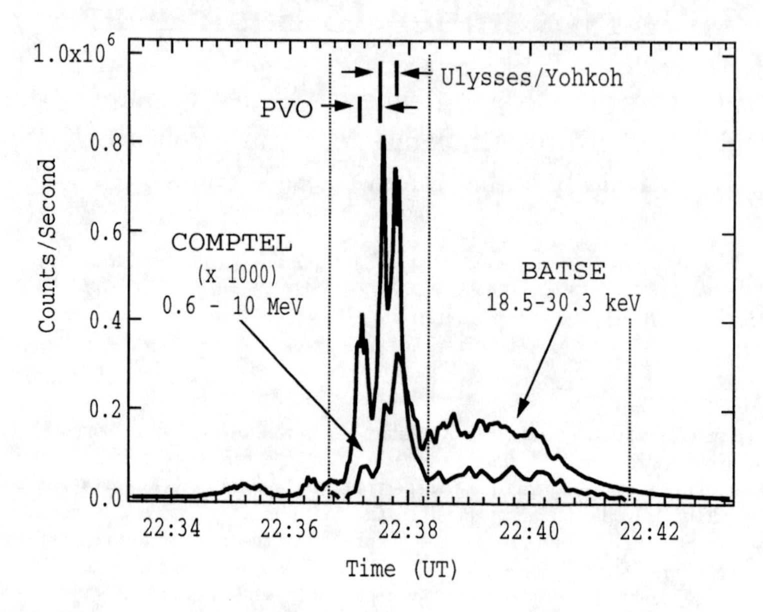

FIGURE 1. Light curve of BATSE (18.5 - 30.3 keV) and COMPTEL (0.6 - 10 MeV) data (magnified by 1000.) PVO and Ulysses/Yohkoh intervals are defined between arrows.

Figure 1 shows the light curve of the flare in the two extreme energy ranges: BATSE between 18.5 and 30.3 keV and COMPTEL integrated over the entire range of 0.6 to 10 MeV. The COMPTEL counts have been magnified by a factor of 1000. We define two time scales: the *whole event* and the *impulsive phase*. The *whole event* (2236:42 - 2241:37 UT) is defined by the interval during which COMPTEL detected significant emission from the flare. The *impulsive phase*, which ends at 2238:24 UT, is defined as the interval of intense emission detected by COMPTEL shortly after flare onset. PVO, Ulysses and Yohkoh data used in our composite spectrum were taken during two short intervals during the impulsive phase, PVO during the first interval (2237:10 - 2237:30 UT), Yohkoh and Ulysses during the second (2237:42 - 2241:37 UT.)

DATA ANALYSIS AND DISCUSSION

Composite Spectrum

A composite spectrum of X- and gamma rays provides information on the electron and nuclear emission from this event. Using BATSE CONT data [12] for the *whole event* between 40 and 400 keV, we determined the electron bremsstrahlung contribution to have a spectral index of -3.27 ± 0.01. This powerlaw is plotted in figure 2 along with powerlaws found from PVO, Ulysses and Yohkoh data with respective spectral indices of -3.37 ± 0.05 [8], -3.08 [6], and -3.82 [6]. The amplitudes of these three lines are greater than that from BATSE because the data were integrated over shorter time periods during the impulsive phase when the flux level was most intense. COMPTEL data from 400 keV to 2.5 MeV cover the energy range over which nuclear reactions dominate the electron brehmstrahlung. The COMPTEL spectra are decovolved using a Maximum Entropy Method [2]. The COMPTEL spectrum in figure 2 corresponds to the *impulsive phase*, and shows evidence for strong nuclear lines between 1 and 2.5 MeV.

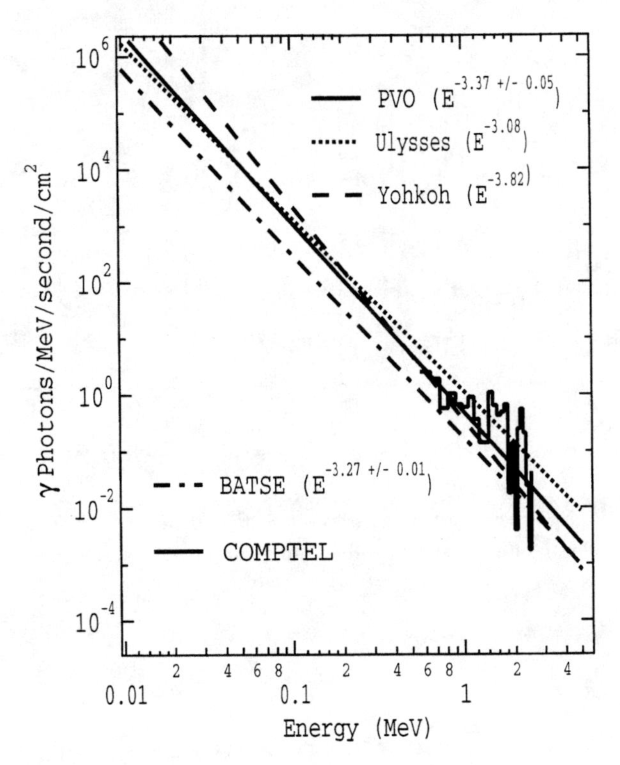

FIGURE 2. Composite spectrum with high energy data from BATSE, COMPTEL, PVO, Ulysses, and Yohkoh.

Spectral Fitting

Deconvolved spectra of the *whole event* and the *impulsive phase* are shown in figure 3. The error bars represent 1σ errors. The overlaid fits of the 25 data points were assumed to consist of a power law and either 3 or 5 gaussians. Peak locations were fixed where we expect to find common spectral lines: 0.85 MeV (^{56}Fe), 1.38 MeV (^{24}Mg), 1.63 MeV (^{20}Ne), 1.79 MeV (^{28}Si), and 2.223 MeV (^{2}H). The widths and amplitudes of the gaussians were free paramaters, as was the normalization factor for the power law. The 2.223 MeV neutron capture line is well centered in both spectra. The fit of the impulsive phase spectrum converged only after removing the gaussians centered on the ^{56}Fe and ^{28}Si lines. Comparison of the two fits (recall that the *impulsive phase* is a subset of the *whole event*) suggests spectral evolution with time, as evidenced by the relative heights of the ^{24}Mg and ^{20}Ne lines in each plot.

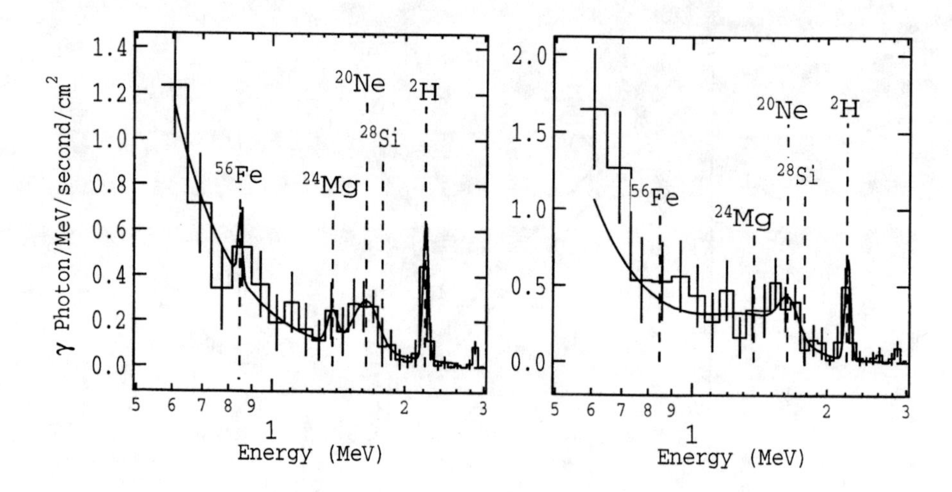

FIGURE 3. COMPTEL spectra of the *whole* event (left) and the *impulsive phase* (right). The solid lines are fits made up of a power law and either 3 or 5 guassians. Typical nuclear lines are marked.

Temporal Features

Analysis of the time evolution of the 2.223 MeV neutron capture line after the impulsive phase shows an expected exponential decay with a best-fit time constant of 155 ± 77 seconds. This result is in agreement (within error bars) with time constants of ~100 seconds found by Prince et. al. [10] and Hua and Lingenfelter [3]. The time profile of the 4 - 7 MeV data agrees with Yohkoh results presented by Kawabata et. al. [7]

4 - 7 Mev to 2.223 MeV Fluence Ratios

COMPTEL data in the 4 - 7 MeV range can be compared with the 2.223 MeV emission to estimate the energy spectrum of the parent ions. These ions interact with heavy nuclei in the ambient solar atmosphere, resulting in gamma ray emission. For this flare, the fluence ratio is $\frac{\Phi_{4-7}}{\Phi_{2.223}} = 0.45 \pm 0.11$. Combining this information with the heliocentric angle, we find a spectral index $-3.1 < S < -2.6$, suggesting a hard parent ion spectrum [9] .

CONCLUSIONS

Initial results from the analysis of the high energy emission from the 1991 November 15 solar flare show a strong nuclear line component above the bremstrahlung continuum, which has a power law of -3.27 $\pm$ 0.01. These nuclear lines exhibit spectral evolution with time, the 2.223 line with an expontial decay and typical time constant of 155 $\pm$ 77 seconds. Comparison of the 4 - 7 and 2.223 MeV line fluences suggests a hard parent ion spectrum, with a power law between -3.1 and -2.6. The data above 2.5 MeV need to be analyzed in more detail, in part to determine ambient solar atmosphere abundances. Data from other instruments, including interplanetary data, need to be added to this work to build a more comprehensive understanding of mechanisms at work within this flare.

REFERENCES

1. Fishman, G.J., et. al., *Proceedings of the Gamma Ray Observatory Science Workshop*, GSFC, MD (1989).
2. Gull, S., and Skilling, J., Quantified Maximum Entropy Users Manual, (MDEC: Meldreth) (1991).
3. Hua, X.-M., and Lingenfelter, R.E., *Ap. J.* **319**:555-566 (1987).
4. Hudson, H. et. al., *Publ. Astron. Soc. Japan* **44**:L77-L81 (1992).
5. Inda-Koide, M. et. al., *Publ. Astron. Soc. Japan* **47**:661-676 (1995).
6. Kane, S.R. et. al., *Ap. J.* **500**:1003-1008 (1998).
7. Kawabata, K. et. al., *Ap. J. Supp.* **90**:701-705 (1994).
8. McTiernan, J. et. al., Proc of Kofu Sympsoium, NRO Report No. 360 (1994).
9. Murphy. R., Ph.D. Dissertation (1985).
10. Prince, T.A. et. al., 18th ICRC, India **4**:79-82 (1983).
11. Schönfelder, V. et. al., *IEEE trans. Nucl. Sci.* NS-31, (1984).
12. Schwartz, R. A., *personal communcation*, (1999).

Detection of 6 November 1997 Ground Level Event by Milagrito

A.D. Falcone[1], for Milagro collaboration

[1] *Space Science Center, University of New Hampshire, Durham, NH 03824 USA*

Abstract. Solar Energetic Particles (SEPs) with energies exceeding 10 GeV associated with the 6 November 1997 solar flare/CME (coronal mass ejection) have been detected with Milagrito, a prototype of the Milagro Gamma Ray Observatory. While SEP acceleration beyond 1 GeV is well established, few data exist for protons or ions beyond 10 GeV. The Milagro observatory, a ground based water Cherenkov detector designed for observing very high energy gamma ray sources, can also be used to study the Sun. Milagrito, which operated for approximately one year in 1997/98, was sensitive to solar proton and neutron fluxes above ~4 GeV. In its scaler mode, Milagrito registered a rate increase coincident with the 6 November 1997 ground level event observed by Climax and other neutron monitors. A preliminary analysis suggests the presence of >10 GeV particles.

INTRODUCTION

Particle acceleration beyond 1 GeV due to solar processes is well established (1), but its intensity and energy still amazes researchers. However, few data exist demonstrating acceleration of protons or ions beyond 10 GeV (2, 3). The energy upper limit of solar particle acceleration is unknown but is an important parameter, since it relates not only to the nature of the acceleration process, itself not ascertained, but also to the environment at or near the Sun where the acceleration takes place. The Milagro instrument, a water Cherenkov detector near Los Alamos, NM, is at 2650 m elevation with a geomagnetic vertical cutoff rigidity of ~4 GV. It is sensitive to hadronic cosmic rays from approximately 5 GeV to beyond 1 TeV. These primary particles are detected via Cherenkov light, produced by secondary shower particles, as they traverse a large (80 × 60 × 8 m) water-filled pond containing 723 photomultiplier tubes (228 PMTs for the prototype, Milagrito). This energy range overlaps that of neutron monitors (in the region < 10 GeV) such that Milagro complements the worldwide network of these instruments. These ground-based instruments, in turn, complement spacecraft cosmic ray measurements at lower energies. This suite of instruments may then be capable of measuring the full energy range of solar hadronic

CP510, *The Fifth Compton Symposium,* edited by M. L. McConnell and J. M. Ryan

cosmic rays, with the goal of establishing a fundamental upper limit to the efficiency of the particle acceleration by the Sun.

Milagro's baseline mode (air shower telescope mode) of operation measures extensive air showers above 300 GeV from either hadrons or gamma rays. A description of Milagro's capabilities as a VHE gamma ray observatory is available elsewhere (4). Milagro measures not only the rate of these events but also the incident direction of each event, thereby localizing sources. While performing these measurements, the instrument records the rate of photomultiplier hits (the scaler mode), with an intrinsic energy threshold of about 4 GeV for the progenitor cosmic ray to produce at least one hit. The scaler mode provides data that are similar to those of a neutron monitor, while the telescope mode can significantly reduce background by pointing. With a proposed fast data acquisition system (DAQ) and modified algorithms for determining incident directions of muons, the energy threshold of Milagro's telescope mode will be reduced to ~4 GeV by detecting the (~300 kHz) single muons and mini muon showers. For now, this low energy threshold can only be achieved by using Milagro in the scaler mode, which is not capable of localizing sources. A description of the Milagro solar telescope mode can be found in another publication (5).

SOLAR MILAGRO/MILAGRITO SCALER MODE

In the scaler mode, a substantial portion of the rate recorded by Milagro (and Milagrito) is due to muons, and an integral measurement above threshold is performed. These data provide an excellent high energy complement to the network of neutron monitors, which has been, and continues to be, a major contributor to our understanding of solar energetic particle acceleration and cosmic rays. With Monte Carlo calculations, we estimated the effective areas of Milagrito to protons incident on the atmosphere isotropically, at zenith angles ranging from 0°-60° (Figure 1). The effective area curves for Milagro, which have been plotted for the sake of comparison, are for beamed protons. At 10 GeV, Milagro's scaler mode is at least an order of magnitude greater than the effective area of a sea level neutron monitor, with the effective area rising rapidly with energy, while Milagrito had approximately 4 times the effective area of a neutron monitor at 10 GeV. Pressure, temperature, and other diurnal corrections must be applied to the ground level scaler rate (6). We have begun to determine these correction factors for Milagro/Milagrito, and we find them to be reasonably consistent with past work with muon telescopes (7). However, these corrections are less important for transient (i.e. solar) events that rise above background quickly and have short durations.

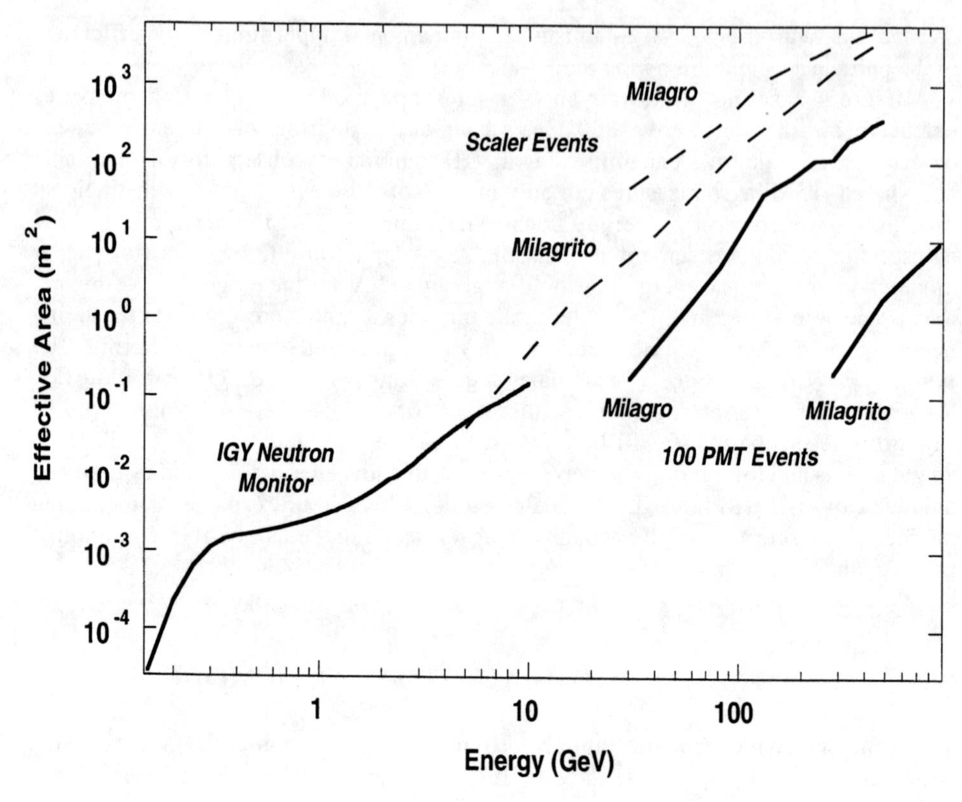

FIGURE 1. Preliminary effective area curves for Milagro and Milagrito, with a sea-level IGY neutron monitor for comparison. (The Milagro shower trigger is no longer set at 100 PMTs, but Milagrito shower data was recorded with a 100 PMT threshold.)

6 NOVEMBER 1997 GROUND LEVEL EVENT

On 6 November 1997 at approximately 12:00 UT, an X-class flare with an associated coronal mass ejection occurred on the Sun. This produced a nearly isotropic (8) ground level event registered by many neutron monitors. A preliminary analysis of neutron monitor data for this proton event yields a spectral index of 5.5 at event maximum in the 1-5 GV rigidity range, assuming a power law rigidity spectrum for protons (9). Climax, located in nearby central Colorado, is the closest of these neutron monitors to Milagro/Milagrito. Milagrito, a prototype version of Milagro with less effective area, registered a scaler rate increase coincident, within error, with that measured by Climax (Figure 2). If one accounts for meteorological fluctuations, the event duration and time of maximum intensity, as seen by Milagrito, are also consistent with that of Climax.

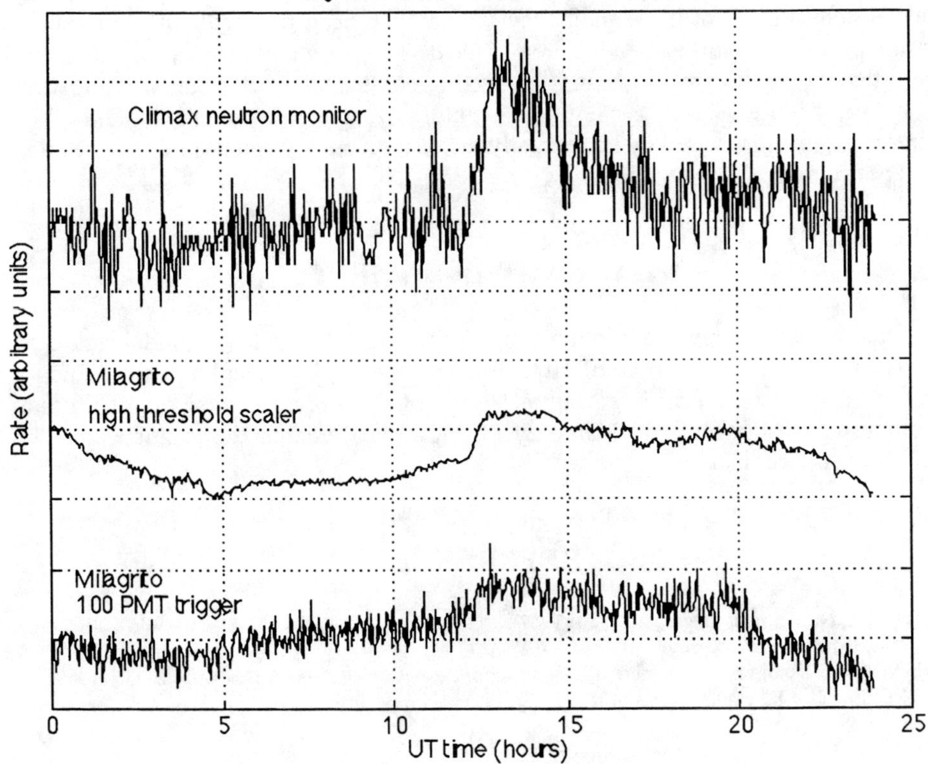

FIGURE 2. Milagrito registered a rate increase coincident with that of Climax during the GLE of Nov. 6, 1997. The y-axis units have been scaled and shifted for each plot to make comparison easier. (Climax data courtesy of C. Lopate, Univ. of Chicago)

The high threshold scaler rate increase can be used to derive characteristics of the primary proton spectrum. This is done by folding a trial power law spectrum of protons through the effective area of the detector. The parameters of the trial spectra are then varied until a good fit to the measured rate increase is achieved. When compared to the neutron monitor network's spectrum for protons < 4 GeV, the preliminary results of this analysis indicate the presence of protons in excess of 10 GeV. Further work is necessary to better determine the spectrum at these energies.

The 100 PMT shower trigger rate also experienced an increase, although the significance is not as great as that in scaler mode. It is not yet clear which of several possible mechanisms initiated the signal in the 100 PMT shower trigger, so the detector's sensitivity to various mechanisms is being investigated. This increase could have been caused by high energy primaries (> 100 GeV, see Figure 1) and/or secondary muons arriving from a nearly horizontal direction. If horizontal secondary muons contributed to this signal, they would have been the result of high energy proton primaries, but the effective area of the detector would be significantly different from that assumed here and cannot be used to constrain the spectrum without more extensive Monte Carlo calculations. Future work will address this issue by considering events caused by horizontally incident secondary muons and recalculating the spectrum.

ACKNOWLEDGMENTS

This work is supported in part by the National Science Foundation, U.S. Department of Energy Office of High Energy Physics, U.S. Department of Energy Office of Nuclear Physics, Los Alamos National Laboratory, University of California, Institute of Geophysics and Planetary Physics, the Research Corporation, and the California Space Institute.

REFERENCES

1. Parker, E.N., *Physical Review* **107**, 830 (1957).
2. Chiba, N., et al., *Astroparticle Physics* **1**(1), 27-32 (1992).
3. Lovell, J.L., Duldig, M.L., Humble, J.E., *Journal of Geophysical Research* **103**(A10), 23733 (1998).
4. McCullough, J.F., et al., "Status of the Milagro Gamma Ray Observatory," *Proc.26th Int. Cosmic Ray Conf, 1999.*
5. Falcone, A.D., et al., *Astroparticle Physics* **11**(1-2), 283-285 (1999).
6. Hayakawa, S., *Cosmic Ray Physics*, New York: John Wiley and Sons, 1969.
7. Fowler, G.N., Wolfendale, A.W., S.Flügge, eds., *Cosmic Rays I*, 1961.
8. Duldig, M.L. & Humble, J.E., "Preliminary Analysis of the 6 November 1997 Ground Level Enhancement," *Proc.26th Int. Cosmic Ray Conf, 1999, Vol. 6, pp. 403-406.*
9. Smart, D.F & Shea, M.A., "Preliminary Analysis of GLE of 6 November 1997A," *Proc. Spring American Geophysical Union Meeting, 1998.*

CATALOGS AND DATA ANALYSIS

The COMPTEL instrumental-line background

G. Weidenspointner*, M. Varendorff*, U. Oberlack‡, S. Plüschke*, D. Morris†, R. Diehl*, S.C. Kappadath||, M. McConnell†, J. Ryan†, V. Schönfelder*

* *Max-Planck-Institut für extraterrestrische Physik, Postfach 1603, 85740 Garching, Germany*
† *Space Science Center, University of New Hampshire, Durham, NH 03824, USA*
‡ *Astrophysics Laboratory, Columbia University, New York, USA*
|| *Louisiana State University, Baton Rouge, Louisiana, USA*

Abstract. The instrumental-line background of the Compton telescope COMPTEL onboard the Compton Gamma-Ray Observatory is due to the activation and/or decay of a number of different isotopes. The major components of the COMPTEL instrumental-line background can be attributed to eight individual isotopes, namely ^{2}D, ^{22}Na, ^{24}Na, ^{28}Al, ^{40}K, ^{52}Mn, ^{57}Ni, and ^{208}Tl. In addition, evidence for the presence of ^{27}Mg has been obtained in the search for gamma-ray lines from supernovae. The identification of the instrumental lines with specific isotopes is based on the line energies as well as on the variation of the activity with time, cosmic-ray intensity and deposited radiation dose during passages through the South-Atlantic Anomaly. The characteristic variation of the activity due to an individual isotope depends on its life-time, orbital parameters, and the solar cycle.

INTRODUCTION

Gamma-ray experiments in low-Earth orbit, such as the Compton telescope COMPTEL, are operated in an intense and variable radiation environment. The main constituents of the ambient radiation fields are primary cosmic-ray particles, geomagnetically trapped radiation-belt particles, and albedo neutrons as well as γ-ray photons emanating from the Earth's atmosphere. These particles interact with the spacecraft and detector materials through a multitude of processes, resulting in the emission of instrumental-background photons. A detailed qualitative and quantitative understanding of the intense and complex COMPTEL instrumental background is crucial for many astrophysical analyses.

The Compton telescope COMPTEL, sensitive to γ-ray photons from 0.8–30 MeV, consists of two planes of detector arrays, D1 and D2, separated by about 1.5 m [8]. The D1 detector consists of seven modules filled with organic liquid scintillator, the D2 detector consists of 14 NaI(Tl) crystals. Each of the two detector arrays

CP510, *The Fifth Compton Symposium,* edited by M. L. McConnell and J. M. Ryan

is completely surrounded by anti-coincidence domes to reject charged-particle triggers of the telescope. The COMPTEL instrument accepts and registers coincident triggers in a single D1 and D2 detector module in the absence of a veto signal as valid events. Among other parameters the time-of-flight (ToF) value and a so-called pulse-shape discriminator (PSD) value in D1 are recorded for each event. The ToF is a measure for the time difference between the triggers in the D1 and D2 detector. The PSD is a measure for the shape of the scintillation-light pulse in the D1 detector and is used to reject neutron-induced events. The total energy of the incident photon, E_{tot}, is measured by the summed energy deposits in the two detectors, E_1 and E_2.

The instrumental background experienced by COMPTEL can be subdivided into two major components according to their signature in energy space: first, a continuum background discussed in [7]; second, the instrumental-line background which is the focus of this paper.

IDENTIFIED ISOTOPES

The major components of the COMPTEL instrumental-line background can be attributed to eight individual isotopes, namely ^{2}D, ^{22}Na, ^{24}Na, ^{28}Al, ^{40}K, ^{52}Mn, ^{57}Ni, and ^{208}Tl (e.g. [5,12,14]). Identification of these isotopes was achieved in an iterative process, starting from the most prominent lines in E_{tot} and E_2 spectra. Viable isotope identifications were required to account self-consistently for spectral features in selected regions of the E_1-E_2 dataspace, as well as for their variation with time and/or incident cosmic-ray intensity. The telescope response to individual isotopes was modelled in Monte Carlo simulations. In addition, evidence for the presence of ^{27}Mg has been obtained from fits of E_{tot} spectra and excess counts in E_{tot} and E_2 following South-Atlantic Anomaly (SAA) transits [4]. The isotopes, their half-lifes, most important decay channels, and main production channels are summarized in Table 1.

The different radioactive isotopes result from activation of the instrument material (^{22}Na, ^{24}Na, ^{52}Mn, and ^{57}Ni), from thermal-neutron absorption (^{2}D and ^{28}Al), and from primordial radioactivity (^{40}K and ^{208}Tl). The majority of the instrumental-line background arises from the D1-detector material. The material composition of the D1-detector system therefore provides important clues as to which radioactive isotopes can effectively be produced.

LONG-TERM VARIATION

The activity of long-lived isotopes such as ^{22}Na, ^{24}Na, ^{52}Mn, and ^{57}Ni varies over long timescales due to the competing processes of activation/production and decay. In contrast, the activity of the primordial radio-nuclides ^{40}K and ^{208}Tl does not noticeably change with time over the duration of the COMPTEL mission of several years. Also, the activity of short-lived nuclei such as ^{28}Al, averaged over

TABLE 1. A summary of the isotopes identified in the COMPTEL instrumental-line background. For simplicity, only the photon energies of the most frequent decay modes are listed. If β-decays are involved, the β-particles have been included in the response simulations. The identification of ^{208}Tl has to be considered tentative. The evidence for the presence of ^{27}Mg in the instrumental background still is not unambiguous.

Isotope	Half-Life	Decay Modes and Photon Energies [MeV]	Main Production Channels
^{2}D	prompt	2.224	^{1}H(n$_{ther}$,γ)
^{22}Na	2.6 y	β^+ (91%): 0.511, 1.275 EC (9%) : 1.275	^{27}Al(p,3p3n), Si(p,4pxn)
^{24}Na	14.96 h	β^-: 1.37, 2.75	^{27}Al(n,α), ^{27}Al(p,3pn)
^{27}Mg ?	9.5 min	β^- (28%): 1.014 β^- (71%): 0.844	^{27}Al(n,p)
^{28}Al	2.2 min	β^-: 1.779	^{27}Al(n$_{ther}$,γ)
^{40}K	1.28×10^9 y	EC (10.7%): 1.461	primordial
^{52}Mn	5.6 d	EC (64%): 0.744, 0.935, 1.434 β^+ (27%): 0.511, 0.744, 0.935, 1.434	Fe(p,x), Cr(p,x), Ni(p,x)
^{57}Ni	35.6 h	β^+ (35%): 0.511, 1.377 EC (30%): 1.377	Ni(p,x), Cu(p,x)
^{208}Tl	1.4×10^{10} y (^{232}Th)	β^- (50%): 0.583, 2.614 β^- (25%): 0.511, 0.583, 2.614	primordial

time-scales long compared to the orbital period of about 90 min, is fairly constant because the isotopes' half-lifes are much shorter than the orbital period, which precludes a build-up of their radioactivity. Therefore, the study of the long-term variation of the activity of a background component provides valuable information concerning the half-life of the responsible decay.

It is expected that the long-term variation of the activity of long-lived radionuclides mostly arises from the combined effects of the isotopes' decay and the time history of the activation episodes during SAA passages, with activation outside the SAA by cosmic-ray particles being negligible (see e.g. [3]). Based on this assumption, a model for the activity of long-lived isotopes as a function of time (the "activity model") was developed [10]. The activity model describes the activity of a specific isotope as a function of six parameters: orbit altitude, geographic longitude and latitude, time since launch (to include variations due to the solar cycle), and the orientation (azimuth and zenith) of the satellite relative to its velocity vector.

A comparison of the measured activities of the long-lived isotopes ^{22}Na and ^{24}Na with the predictions of the normalized activity model over a time period of more than 6 years (May 1991 through July 1997) is depicted in Fig. 1. The activity

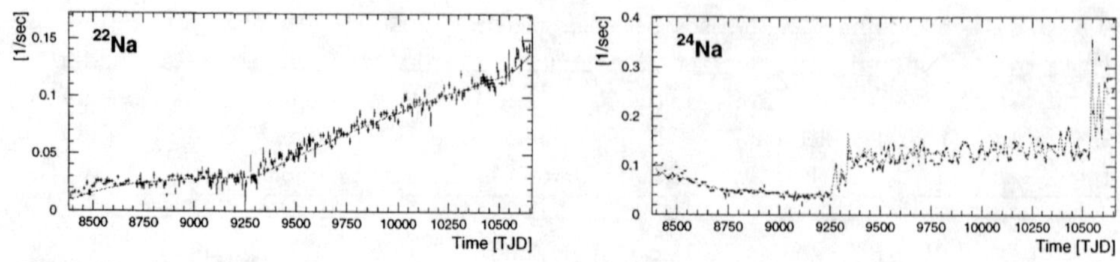

FIGURE 1. The long-term variation of the activity from ^{22}Na and ^{24}Na, compared to the activity model.

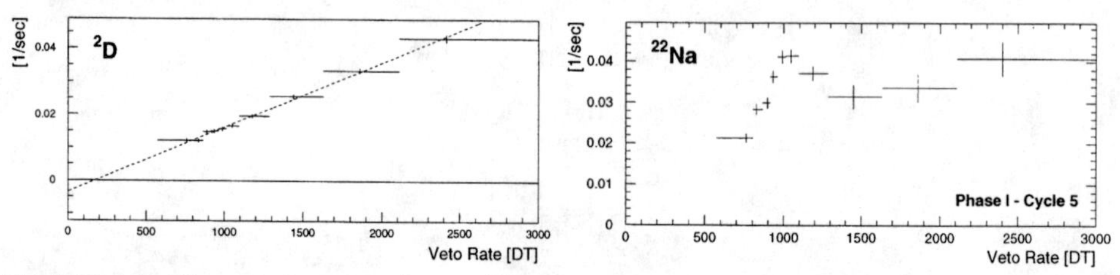

FIGURE 2. The variation of the ^{2}D and ^{22}Na activity with veto rate (DT ≡ veto deadtime-clock counts/2.048 sec). The dashed line in the left panel represents a linear fit to the ^{2}D count rate.

model can accurately reproduce the long-term activity-variation for these two isotopes, confirming the correctness of the basic model assumptions, in particular that activation predominantly occurs during transits through the SAA. In addition, the comparison of model and data provides a valuable verification of the isotope identifications, as the activity of an isotope depends, among other parameters, on its half-life.

VARIATION WITH COSMIC-RAY INTENSITY

The prompt (i.e. instantaneous) instrumental background closely follows the local cosmic-ray intensity, which can, e.g., be parameterized by a geomagnetic cut-off rigidity. Another way of parameterizing the incident cosmic-ray intensity is to use the count rate of the anti-coincidence domes of the COMPTEL instrument, referred to as "veto rate" in the following. To a good approximation, the prompt background varies linearly with the incident cosmic-ray intensity as monitored by the veto rate and can be eliminated from the data by an extrapolation technique in analyses of the cosmic diffuse gamma-ray background (CDG) [1]. As an example, the linear dependence of the event rate in the instrumental 2.2 MeV ^{2}D-line due to thermal-neutron capture on hydrogen in the organic scintillator of the D1 modules [11] is depicted in Fig. 2.

In contrast to prompt background components, the activity of the primordial isotopes ^{40}K and ^{208}Tl is independent of the incident cosmic-ray intensity and hence does not vary with cosmic-ray intensity or veto rate.

The variation of the activity of long-lived isotopes such as ^{22}Na with veto rate is complex and – in general – non-linear (as exemplified in Fig. 2), and depends on the isotopes' half-life as well as on the encountered geophysical environment and its time-variation (e.g. [12,14]). Similar to the study of the long-term variation, the study of the variation of the activity of a background component with cosmic-ray intensity can therefore provide valuable information concerning the half-life of the responsible decay.

SUMMARY AND DISCUSSION

The major components of the COMPTEL instrumental-line background can be attributed to eight different radioactive isotopes. Identification of the instrumental lines with specific isotopes is based on the line energies as well as on the long-term variation of the activity and its dependence on incident cosmic-ray intensity. The activity of long-lived isotopes can be described well by our activity model [10], implying that activation occurs predominantly during transits through the SAA, with the deposited radiation dose depending on orbital parameters such as the height of the satellite above Earth and the solar cycle. Considering the material composition of different γ-ray detectors, the isotopes identified in the COMPTEL line background were expected to be present based on experience from the SMM [9] and HEAO 3 instruments [15]. The different detection principles employed, however, result in significant differences in the count rates related to specific isotopes. Detailed understanding of the instrumental-line background and its variation is crucial for many astrophysical analyses, such as that of the CDG (e.g. [2,12,13]) or that of the galactic ^{26}Al 1.8 MeV line emission (e.g. [5,6]).

REFERENCES

1. Kappadath S.C., et al., *A&AS* **120**, C619 (1996).
2. Kappadath, S.C., Ph.D. Thesis, University of New Hampshire, USA (1998).
3. Kurfess J.D., et al., *High-energy rad. bgd. in space* (**AIP 186**), 250 (1989).
4. Morris D.J., et al., *1997 Conf. on the High-Energy Bgd. in Space (IEEE)*, **26** (1997).
5. Oberlack U., Dissertation, Technical University Munich, Germany (1997).
6. Plüschke S., et al., these proceedings.
7. Ryan J., et al., *1997 Conf. on the High-Energy Bgd. in Space (IEEE)*, 13 (1997).
8. Schönfelder V., et al., *ApJ Suppl. Ser.* **86**, 657 (1993).
9. Share G.H., et al., *High-energy rad. background in space* (**AIP 186**), 266 (1989).
10. Varendorff M., et al., *Proc. of the 4th Compton Symposium* (**AIP 410**), 1577 (1997).
11. Weidenspointner G., et al., *A&AS* **120**, C631 (1996).
12. Weidenspointner G., Dissertation, Technical University Munich, Germany (1999).
13. Weidenspointner G., et al., these proceedings.
14. Weidenspointner G., et al., *A&AS*, in preparation.
15. Wheaton W.A., et al., *High-energy rad. bgd. in space* (**AIP 186**), 304 (1989).

Improved COMPTEL maps of the Milky Way

H. Bloemen*, K. Bennett||, W. Collmar†, R. Diehl†, W. Hermsen*, A. Iyudin†, J. Knödlseder§, M. McConnell‡, J. Ryan‡, V. Schönfelder†, A. Strong†

*SRON-Utrecht, Sorbonnelaan 2, 3584 CA Utrecht, The Netherlands
†MPE, Postfach 1603, 85740 Garching, Germany
‡Space Science Center, UNH, Durham NH 03284, USA
||Astrophysics Division, ESTEC,2200 AG Noordwijk, The Netherlands
§CESR, CNRS/UPS, BP 4346, 31028 Toulouse Cedex, France

Abstract. In the course of the mission we have gradually developed an analysis method that separates in an iterative manner the celestial emission and the (*a priori* unknown) instrumental background. It has become our standard analysis tool for point sources. We illustrate here that this method is widely applicable now. It provides mutually consistent sets of model-fitting parameters (spectra) and sky maps, both for continuum and line studies. Because of the wide applicability, it has been possible to make various cross-checks while building up confidence in this procedure.

INTRODUCTION

COMPTEL has the capability of mapping the sky in the 1–30 MeV regime with an angular resolution of 1°–3° and an energy resolution of 5–10% FWHM [21]. The data analysis, however, is certainly not trivial. It is complicated by the dominant instrumental background and by the complex response of the instrument. The background exceeds the celestial (galactic) signal by a factor of roughly 50 and it has significantly changed during the mission. In mapping γ-ray line emission, the removal of celestial continuum radiation introduces an additional challenging problem. In order to study point sources along the Milky Way (apart from pulsars), the diffuse emission has to be accounted for carefully. We have gradually learned how to deal more accurately with this variety of important 'background' aspects.

The COMPTEL imaging data space is 3-dimensional, consisting of two spatial coordinates (the scatter direction χ, ψ) and the Compton scatter angle $\bar{\varphi}$ [21]. A sufficiently accurate *independent* estimate of the instrumental background in this data space is not available. We have therefore developed a filtering technique, of which the basic principle is described in [2]. This has become our standard method for point-source studies (through forward folding in maximum-likelihood analyses). It is based on the fact that the (χ, ψ) structure of the data cube is to first order

CP510, *The Fifth Compton Symposium,* edited by M. L. McConnell and J. M. Ryan

well described by known geometry characteristics, with deviations being largely independent of $\bar{\varphi}$. In order to enable accurate quantitative studies of extended emission (or to account for its presence in source studies), an iterative approach was added to this method, with background modeling and model fitting being performed simultaneously. Also, all viewing periods (VPs) are handled separately now to account accurately for changes in the background during the mission [10].

CONTINUUM EMISSION

Our prime goal here is to derive maps that can be used in a quantitative manner. Previous sky-mapping attempts have been presented by Strong et al. [22,25] and by Bloemen et al. [1,4], the latter [4] showing first results from the current approach. In all analyses addressed in this paper, our first step is the instrumental-background modeling and simultaneous (*all-sky*) model fitting mentioned above. The purpose of the model fitting is to estimate (and account for) the total sky flux in the map as well as its global distribution while generating the background model. Previously, background scaling factors had to be chosen and emission near intense sources was suppressed [25]. The resulting background model is then used to generate the actual map, for which we use here maximum-entropy imaging. In an iterative manner, through likelihood optimization, background models are first determined for each *individual* viewing period, while model intensity distributions are fitted simultaneously to the *combined* set of observations. For approximation of the expected continuum emission, we used a total gas column density map as a tracer of the bremsstrahlung emission (from HI and CO data), an inverse-Compton model [27], an isotropic component, and the strongest sources, Crab and Cyg X-1. Although the sky models used here provide a good global description of the data, contributions from ensembles of sources with similar large-scale distributions can of course not be excluded. But this is in principle irrelevant for the background modeling and subsequent imaging. We have verified that using other model intensity distributions (e.g. various combinations of 2 to 3 rather arbitrarily chosen smooth components) gives very similar results, provided, of course, that these models indeed enable a global description of the γ-ray Milky Way.

The model-fitting result is interesting in itself (spectra, hypothesis testing), but is not addressed here in further detail. The results of such large-scale model fitting procedures have already turned out to be strongly influenced by different handlings of the instrumental background (c.f. Strong et al. [22,23,26] & Bloemen et al. [3,4]).

Fig. 1a shows the 1–30 MeV map that we have extracted with this technique. Fig. 1b shows the corresponding map of candidate sources (see also [7]), which was obtained by adding the fitted sky models to the instrumental-background model. These maps are of the 'νF_ν' type; they are sums of 3 maps (1–3, 3–10, and 10–30 MeV), which we have added together with weight factors such that all 3 contribute equally under the assumption of an E^{-2} photon spectrum (so characteristics of each band are visible in these images). The significance of small-scale features has

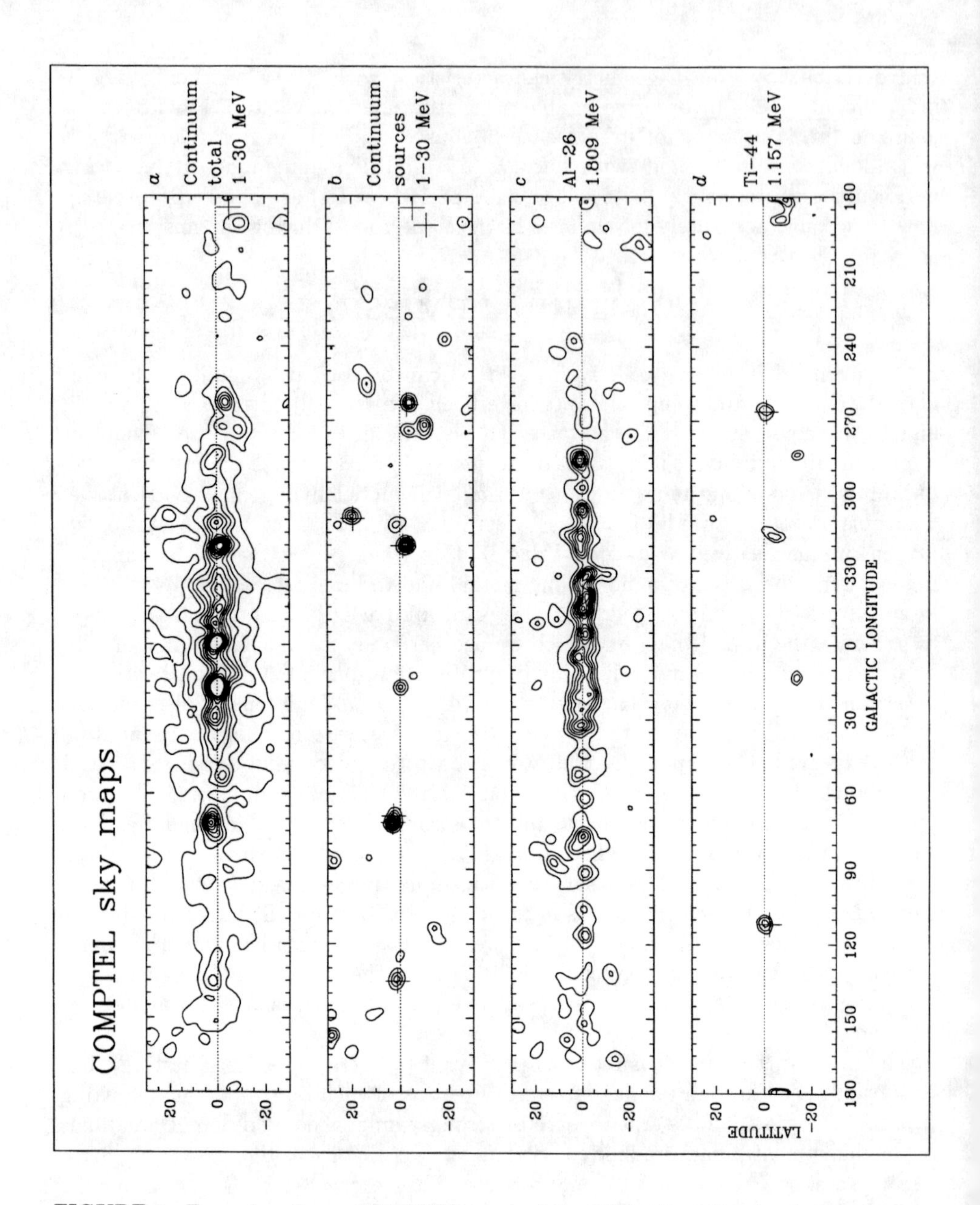

FIGURE 1. Examples of new COMPTEL maps of the Milky Way. Panels *abc* include data obtained during the first ~5 years of the mission (VPs 1–522.5); panel *d* includes data up to VP 617.1. The Crab is modeled out in the two continuum maps. Clearly visible in panel *b* are e.g. the Vela pulsar (263,–3), PSR 1509-58 (320,–1), Cyg X-1 (71,3), 2CG135+1 (135,1), Cen A (310,19), a strong unidentified source at (18,0), and a peculiar extended source centered at about (272,-8) [16]. In panel *d* the locations of Cas A and the new candidate ^{44}Ti source [11,12] are indicated.

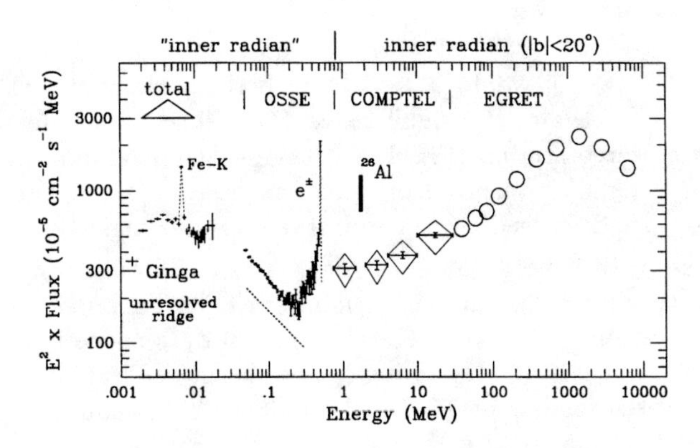

FIGURE 2. COMPTEL spectrum of the inner Galaxy (statistical errors and 25% errors, including systematics), together with results from *Ginga* [29], OSSE [19,20], and EGRET [24].

to be judged from dedicated studies (see [15] for a general discussion on maximum-entropy mapping of COMPTEL data). Figs. 2 shows the resulting spectrum for the inner galactic radian. Spectra for the individual model components are available as well [4]. These can be compared with e.g. bremsstrahlung and IC model spectra, but this is beyond the scope of this paper. Also a spectrum of the isotropic (cosmic) background emission is extracted in this analysis (see [6]), which turns out to be in good agreement with the result of a very different analysis technique [13,28].

LINE EMISSION

The background that needs to be accounted for in γ-ray line studies (main focus here is on the ^{26}Al 1.809 MeV line) consists of an instrumental component and the celestial continuum emission. We have previously treated these two components as one entity, both in our sky mapping [15,17,18] and in our sky-model fitting [8,9,14]. We determined in those studies the background model at 1.7–1.9 MeV (hereafter referred to as the 1.8 MeV band) by scaling of the total event distribution at adjacent energies ($\sim$ 1–10 MeV, excluding the 1.8 MeV band). This is an approximation, because the spectra of the instrumental and celestial components are different. As the instrumental component strongly dominates, the celestial continuum emission is not properly accounted for. We have now applied the same method as in our continuum study, but in a two-step approach, first extracting the celestial emission in narrow energy ranges adjacent to the 1.8 MeV band and then using the derived (interpolated) intensity distribution as a fixed ingredient in the analysis of the 1.8 MeV band itself. Further details can be found in [5].

Fig. 1c shows the resulting 1.809 MeV maximum-entropy map. The appearance of this map is similar to that of the map obtained with our previous background modeling method [17] (an update is presented in these proceedings [18]). In the present new analysis, however, we tend to extract a higher flux for the galactic disk at large (e.g $\sim 4\times10^{-4}$ γ cm^{-2} s^{-1} rad^{-1} for the inner galactic radian, $|b| < 20°$ — included in Fig. 2 — which would indeed be in better agreement with results from other experiments). Again, the significance of small-scale map features has to be judged from dedicated studies [15]. Fig. 1d presents a new map in the ^{44}Ti 1.157 MeV line. More detailed results from this analysis for Cas A and the new candidate ^{44}Ti source are discussed elsewhere in these proceedings [12].

In conclusion: We seem to have gradually learned how to account accurately for the time-variable intense instrumental background of COMPTEL, despite the fact that an independent estimate of it in the 3-dimensional data space is lacking. A particularly nice aspect of the developed analysis technique is that it can be applied to essentially all scientific topics covered by COMPTEL (sources, extended emission, lines) providing images as well as spectral information.

ACKNOWLEDGMENTS: The COMPTEL project is supported by the German government through DARA grant 50 QV 90968, by NASA under contract NAS5-26645, and by the Netherlands Organisation for Scientific Research (NWO).

REFERENCES

1. Bloemen H., et al., in *Compton Gamma-Ray Obs.*, eds. M. Friedlander et al., p. 30 (1993).
2. Bloemen H., et al., *ApJS* **92**, 419 (1994).
3. Bloemen H., et al., *Proc. 4th Compton Symp.*, eds. C.D. Dermer et al., p. 1074 (1997).
4. Bloemen H., et al., *Proc. 3rd Integral Symp.*, in press [*sky mapping*] (1999a).
5. Bloemen H., et al., *Proc. 3rd Integral Symp.*, in press [^{26}*Al study*] (1999b).
6. Bloemen H., et al., *Proc. 3rd Integral Symp.*, in press [*cosmic background*] (1999c).
7. Collmar W., et al., these proceedings
8. Diehl R., et al., *A&A* **298**, 445 (1995).
9. Diehl R., et al., *Proc. 4th Compton Symp.*, eds. C.D. Dermer et al., p. 1114 (1997).
10. van Dijk, R., Ph. D. thesis, University of Amsterdam (1996).
11. Iyudin, A., et al., *Nature* **396**, 142 (1998).
12. Iyudin, A., et al., these proceedings
13. Kappadath S.C., Ph. D. thesis, Univ. of New Hampshire (1998).
14. Knödlseder J., et al., *A&A*, **344**, 68 (1999a).
15. Knödlseder J., et al., *A&A*, **345**, 813 (1999b).
16. van der Meulen R.D., et al., *Proc. 3rd Integral Symp.*, in press (1999).
17. Oberlack U., et al., *A&AS* **120**, C311 (1996).
18. Plüschke S., et al., these proceedings
19. Purcell W.R., et al., *Proc. 24th Int. Cosmic Ray Conf.* **2**, 211 (1995).
20. Purcell W.R., et al., *A&AS* **120**, C389 (1996).
21. Schönfelder V., et al., *ApJS* **86**, 657 (1993).
22. Strong A.W., et al., *A&A* **292**, 82 (1994).
23. Strong A.W., et al., *A&AS* **120**, C381 (1996).
24. Strong A.W. & Mattox J., *A&A* **308**, L21 (1996).
25. Strong A.W., et al., in *The Transparent Universe*, eds. Winkler et al., p. 533 (1997a).
26. Strong A.W., et al., *Proc. 4th Compton Symp.*, eds. C.D. Dermer et al., p. 1198 (1997b).
27. Strong A.W., Moskalenko I. & Schönfelder V., *Proc. 25th ICRC*, OG 8.1.1 (1997c).
28. Weidenspointner G., et al., these proceedings
29. Yamasaki N.Y, et al., *A&AS* **120**, C393 (1996).

COMPTEL Time-Averaged All-Sky Point Source Analysis

W. Collmar*, V. Schönfelder*, A.W. Strong*, H. Bloemen+, W. Hermsen+, M. McConnell†, J. Ryan†, K. Bennett@

*Max-Planck-Institut für extraterrestrische Physik, Postfach 1603, 85740 Garching, Germany
+SRON-Utrecht, Sorbonnelaan 2, NL-3584 CA Utrecht, The Netherlands
†Universtity of New Hampshire, Durham NH 03824-3525, USA
@Astrophysics Division, ESTEC, NL-2200 AG Noordwijk, The Netherlands

Abstract. We use all COMPTEL data from the beginning of the CGRO mission (April '91) up to the end of CGRO Cycle 6 (November '97) to carry out all-sky point source analyses in the four standard COMPTEL energy bands for different time periods. We apply our standard maximum-likelihood method to generate all-sky significance and flux maps for point sources by subtracting off the diffuse emission components via model fitting. In addition, fluxes of known sources have been determined for individual CGRO Phases/Cycles to generate lightcurves with a time resolution of the order of one year. The goal of the analysis is to derive quantitative results – significances, fluxes, light curves – of our brightest and most significant sources such as 3C 273, and to search for additional new COMPTEL sources, showing up in time-averaged maps only.

INTRODUCTION

The imaging COMPTEL experiment aboard CGRO is the pioneering satellite experiment of the MeV-sky (~1 - 30 MeV). For a detailed description of COMPTEL see [1]. One of COMPTEL's prime goals is the generation of all-sky maps, which provide a summary on the MeV-sky in total. This goal has been achieved by e.g. [2], [3] who generated maximum-entropy all-sky images and by [4], who generated the first COMPTEL all-sky maximum-likelihood maps, which – compared to maximum-entropy ones – have the advantage of providing quantitative results like significances and fluxes of source features. Here we present all-sky maximum-likelihood maps from which models of the diffuse emission have been removed. Our emphasis is on AGN. For a discussion on the method see [5] in these proceedings.

The main analysis goals are 1) to derive a summary of known COMPTEL point sources, 2) to search for further point sources, 3) to derive time-averaged quantitative parameters ('first order') of our brightest point sources, i.e., significances, fluxes, MeV-spectra, and possible time variability, and 4) to further investigate our data and analysis methods.

CP510, *The Fifth Compton Symposium*, edited by M. L. McConnell and J. M. Ryan

1-56396-932-7/00/$17.00

DATA AND ANALYSIS METHOD

Using all data from the beginning of the CGRO mission (April '91) to the end of CGRO Cycle VI (Nov. '97), we generated a consistent database of relevant COMPTEL data sets (events, exposure, geometry) for individual CGRO viewing periods (VPs) in the 4 standard energy bands (0.75-1, 1-3, 3-10, 10-30 MeV) in galactic coordinates by applying consistent data selections. This database was supplemented by relevant data sets containing models describing the galactic diffuse γ-ray emission (HI, CO, and inverse-Compton components) and the isotropic extragalactic γ-ray background emission. To check for time variability of γ-ray sources these data sets were combined for different time periods: the six individual CGRO Phases/Cycles, the sum of all data (CGRO Phases I-VI; April '91 - Nov. '97) as well as the first (CGRO Phases I-III; April '91 - Oct. '94) and the second half (CGRO Phases IV-VI; Oct. '94 - Nov. '97). Each set of all-sky data is analysed by our standard maximum-likelihood method which simultaneously 'handles' individual VPs, generates, iteratively, a background model (see [6]), and finally generates significance and flux maps and/or significances and fluxes for individual sources. Because we are interested in point sources, the diffuse emission is always removed in the fitting procedure (e.g. Figure 1). For the derivation of the source fluxes (see Figure 2 as an example), the point sources of interest (e.g. 3C 273, Cyg X-1) have additionally been included in the fitting procedure. We like to mention however, that the results derived by such all-sky fits should be considered correct to first order only. To derive final/optimal results for a particular source, a dedicated analysis has to be carried out, which e.g. makes several cross checks by applying different background models and would take into account the presence of other source features in the region of interest. Also, along the galactic plane the results depend on the 'goodness' of the applied diffuse emission models for the MeV-band.

RESULTS

The significance maps in Figure 1, which contain all data of the first 6.5 years of the COMPTEL mission, are the first COMPTEL all-sky point source maps in the continuum bands. They provide a summary of the on-average brightest and most significant MeV-sources. Similar maps focussing on the Galactic plane only are given elsewhere in these proceedings ([5]). The Crab – for display reasons removed in all maps of Figure 1 – is by far the most significant COMPTEL point source. In the 1-3 MeV band for example it reaches a significance of $\sim$110σ (i.e. a likelihood ratio of $\sim$12000) for the CGRO Phase I-VI period. With significances of $\sim$11σ, $\sim$10σ, and $\sim$6σ in the 1-3, 3-10, and 10-30 MeV bands is the quasar 3C 273 found to be on average the second most significant point source. Its fluxes in these bands are between 10% and 15% of the Crab flux. Several other extragalactic (e.g. 3C 279, PKS 0528+134, Cen A) and galactic (e.g. Cyg X-1, PSR 1509-58, a known but unidentified source at l;b: 18°;0°) sources are visible as well. In addition there

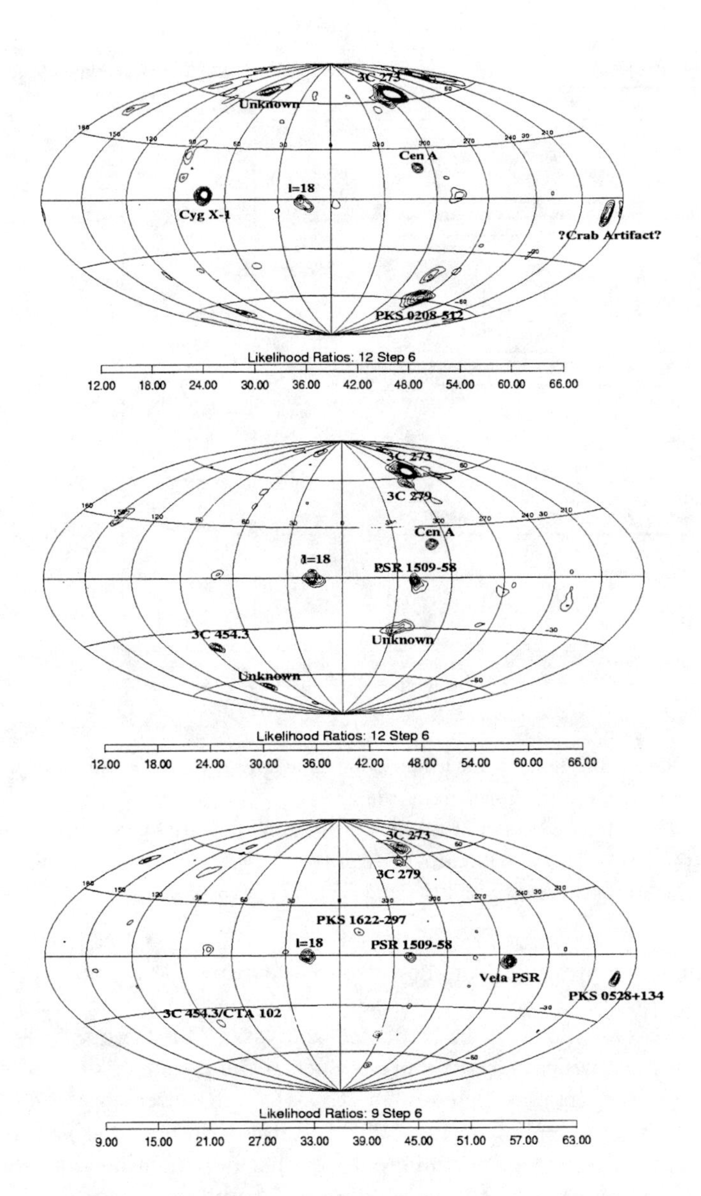

FIGURE 1. COMPTEL time-averaged maximum-likelihood point source all-sky maps in the 1-3 (upper), 3-10 (middle), and 10-30 MeV (lower panel) energy bands for the time period April '91 to November '97, i.e. CGRO phases I-VI. The galactic and extragalactic diffuse emission as well as the emission from the Crab have been subtracted off via model fitting. For the 1-3 MeV and the 3-10 MeV significance maps the contour lines start at a likelihood ratio value of 12 ($\sim 3.5\sigma$ for a known source; χ_1^2-statistics) and for the 10-30 MeV map at a likelihood ratio value of 9 (3.0σ for a known source) with steps of 6 for all maps. The most significant source features are labeled.

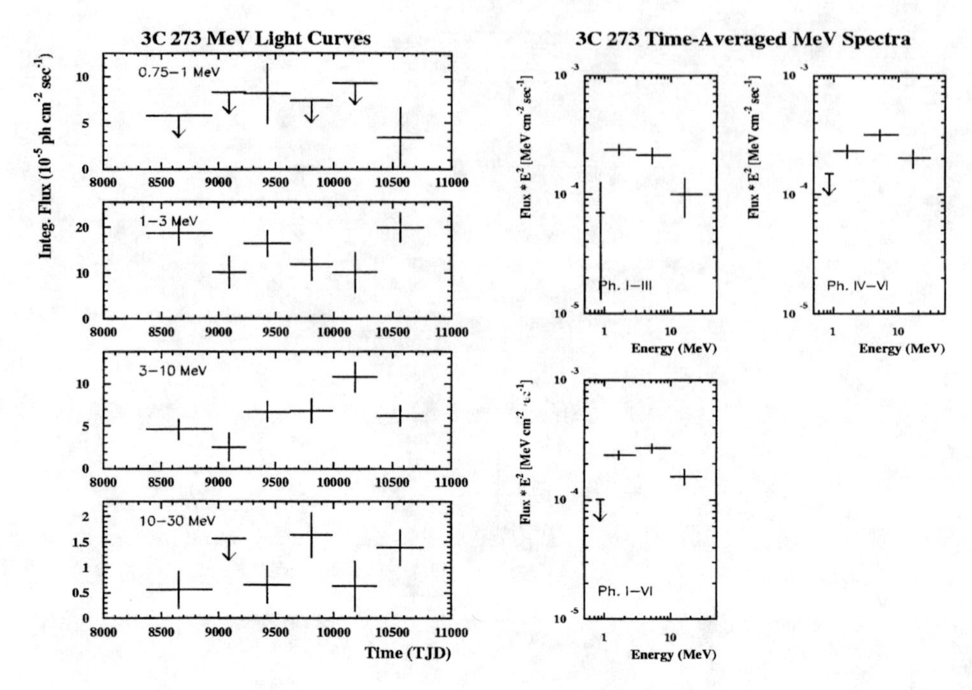

FIGURE 2. Left: COMPTEL light curves of the quasar 3C 273 in the 4 standard bands as derived from all-sky flux fitting. Each flux point is averaged over an individual CGRO phase. The error bars are 1σ and the upper limits are 2σ.
Right: COMPTEL MeV spectra of 3C 273 for the sum of all CGRO phase I-III (April '91 - Oct. '94), IV-VI (Oct. '94 - Nov. '97), and I-VI (April '91 - Nov. '97) data as derived from all-sky flux fitting. The error bars are 1σ and the upper limits are 2σ.

are indications for previously unknown source features like at l;b: 75°;$+65^o$ in the 1-3 MeV map and at l;b: 85°;-65° in the 3-10 MeV map for example. Such spots are promising candidates for further dedicated analyses. This time-averaged approach suppresses sources which flare up only in short time periods. Therefore the maps show fewer sources than are listed in the COMPTEL source catalog (see [7]).

For all bright and significant COMPTEL sources we have derived fluxes in our 4 standard energy bands for the different time periods mentioned above, and have combined them to MeV light curves and spectra. Some results for 3C 273 are shown as an example in Figure 2. In the 1-10 MeV energy band 3C 273 is detected in each CGRO Phase/Cycle i.e. in time periods of typically 1 year. The flux turns out to be rather stable and varies only within a factor of ~2 in the 1-3 MeV and within a factor of ~4 in the 3-10 MeV energy band. The spectra show the same trend. Whereas the flux below 3 MeV turns out to be same for both halves, there is an indication that at the upper COMPTEL energies (>3 MeV) the source was brighter during the second half. All three spectra clearly show the spectral turnover

TABLE 1. Detection significances (σ) in 3 different energy bands of some known COMPTEL AGN sources (plus Crab) as derived by this all-sky analysis for the sum of data of the first 6.5 years of the COMPTEL mission.

γ-ray Source	1-3 MeV	3-10 MeV	10-30 MeV	Source Type
Crab	~110	~76	~44	pulsar + nebula
3C 273	11.5	10.8	6.2	FSRQ[a]
3C 279	5.3	5.7	4.9	FSRQ
PKS 0528+134	—[b]	3.0	6.0	FSRQ
PKS 1622-297	<3	<3	4.1	FSRQ
3C 454.3	<3	5.3	<3	FSRQ
Cen A	5.7	5.3	3.0	Radio Galaxy

[a] Flat-Spectrum Radio Quasar

[b] Unknown due to uncertainties in removing the strong Crab signal

occuring at MeV-energies. However, we emphasize that for final conclusions a dedicated source analysis has to be carried out.

SUMMARY

We have applied the maximum-likelihood method to COMPTEL all-sky data of different time periods. By simultaneously fitting models for the different diffuse emission components, this analysis method provided quantitative all-sky results – significances and fluxes – on point sources. An all-sky summary on their time-averaged fluxes and significances is thereby provided. After the Crab – pulsar plus nebula – the quasar 3C 273 was found to be the most significant COMPTEL MeV-source, having time-averaged fluxes of the order of 10% to 15% of the Crab. Additional evidence for previously unknown source features has been found as well.

ACKNOWLEDGMENTS: The COMPTEL project is supported by the German government through DARA grant 50 QV 9096 8, by NASA under contract NAS5-26645, and by the Netherlands Organisation for Scientific Research (NWO).

REFERENCES

1. Schönfelder, V., et al., *ApJS* **86**, 657 (1993).
2. Strong, A.W., et al., *Proc. 2nd INTEGRAL Workshop, ESA SP*-**382**, 533 (1997).
3. Bloemen, H., et al., *Proc. 3rd INTEGRAL Workshop*, in press (1999).
4. Blom, J.J., et al., *COMPTEL High-Latitude Gamma-Ray Sources (Ph.D. Thesis)*, ISBN 90-9010945-5, (1997).
5. Bloemen, H., et al., these proceedings.
6. Bloemen, H., et al., *ApJS* **92**, 419 (1994).
7. Schönfelder, V., et al., *ApJS* submitted (1999).

Summary of the First COMPTEL Source Catalogue

V. Schönfelder*, K. Bennett||, J.J. Blom†, H. Bloemen†, W. Collmar*, A. Connors‡, R. Diehl*, W. Hermsen†, A. Iyudin*, R.M. Kippen‡, J. Knödlseder§, L. Kuiper†, G.G. Lichti*, M. McConnell‡, D. Morris‡, R. Much||, U. Oberlack*, J. Ryan‡, G. Stacy‡, H. Steinle*, A. Strong*, R. Suleiman‡, R. van Dijk||, M. Varendorff*, C. Winkler||, and O.R. Williams||

*Max-Planck-Institut für extraterrestrische Physik, D-85740 Garching, Germany
†SRON-Utrecht, Sorbonnelaan 2, NL-3584 CA Utrecht, The Netherlands
‡Space Sience Center, University of New Hampshire, Durham, NH 03824-3525, USA
||Astrophysics Division, ESTEC, NL-2200 AG Noordwijk, The Netherlands
§Centre d'Etude Spatiale des Rayonnements (CESR), BP 4336, F-31029 Toulouse Cedex, France

INTRODUCTION

The imaging Compton telescope COMPTEL aboard NASA's Compton Gamma-Ray Observatory has opened the MeV gamma-ray band as a new window to astronomy. COMPTEL provided the first complete all-sky survey in the energy range 0.75 to 30 MeV. The catalogue, presented here, is largely restricted to published results from the first five years of the mission (up to Phase IV / Cycle-5). In a few cases, more recent results have been added.

The catalogue contains firm as well as marginal detections of continuum and line emitting sources and presents upper limits for various types of objects. The numbers of the most significant detections are 32 for steady sources and 31 for gamma-ray bursters. Among the continuum sources, detected so far, are spin-down pulsars, stellar black-hole candidates, supernova remnants, interstellar clouds, nuclei of active galaxies, gamma-ray bursters, and the Sun during solar flares. Line detections have been made in the light of the 1.809 MeV ^{26}Al line, the 1.157 MeV ^{44}Ti line, the 847 and 1238 keV ^{56}Co lines, and the neutron capture line at 2.223 MeV. For the identification of galactic sources, a modelling of the diffuse galactic emission is essential. Such a modelling at this time does not yet exist at the reqired degree of accuracy. Therefore, a second COMPTEL source catalogue will be produced after a detailed and accurate modelling of the diffuse interstellar emission has become possible.

Here, only a summary of the First COMPTEL Source Catalogue is given. The

CP510, *The Fifth Compton Symposium,* edited by M. L. McConnell and J. M. Ryan

complete catalogue will appear in the Astronomy and Astrophysics Supplement Series [1].

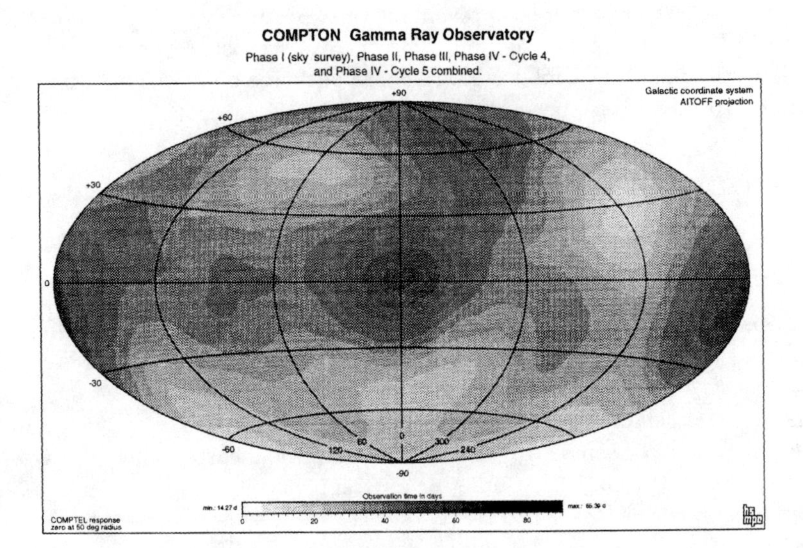

FIGURE 1. COMPTEL Exposure Maps. *Top:* Phase I to Phase IV / Cycle-5 (April 1991 to October 1996). *Bottom:* Phase I to Phase IV /Cycle-7 (April 1991 to December 1998).

OBSERVATIONS AND EXPOSURES

The effective COMPTEL exposure of the entire sky from the sum of all observations from the beginning of the mission to Phase IV /Cycle-5 is presented in Fig. 1 (top). For illustration, the improvement in exposure up to Phase IV / Cycle-7 is shown in Fig. 1 (bottom), as well. The deepest exposures were obtained in the Galactic Center and anticenter region, where effective observation times up to $6 \cdot 10^6$ (about 70 days) have been obtained.

COMPTEL SENSITIVITY

In its telescope mode COMPTEL has an unprecedented sensitivity. Within a 2-week observation period it can detect sources, which are about 10-times weaker than the Crab. By adding up all data from a certain source that were obtained over the entire duration of the mission, higher sensitivities can be obtained. Table 1 summarizes the achieved point-source sensitivities for a 2-week exposure in Phase I of the mission ($t_{eff} \sim 3.5 \cdot 10^5$ sec), and for the ideal cases, when all data from a certain source in the Galactic Center or anticenter (where the exposure is highest) are added from either Phase I to III ($t_{eff} \sim 2.5 \cdot 10^6$ sec) or Phase I to IV /Cycle-5 ($t_{eff} \sim 6 \cdot 10^6$ sec). (Phase I ended in November 1992, Phase III in October 1993, and Phase IV/Cycle-5 in October 1996.)

From this table rough upper limits can be derived for those objects, which are not contained in the catalogue by deriving the effective exposure from Figure 1.

TABLE 1. From this table rough upper limits can be derived for those objects, which are not contained in the source catalogue, by deriving the effective exposure from Fig. 1.

	3σ Flux Limits [10^{-5} photons cm^{-2} sec^{-1}]		
$E_\gamma [MeV]$	2 weeks in Phase 1	Phase 1+2+3	Phase 1+2+3+4 (Cycle-5)
0.75 – 1	20.1	7.4	3.7
1 – 3	16.8	5.5	3.8
3 – 10	7.3	2.8	1.7
10 – 30	2.8	1.0	0.8
1.157	6.2	2.0	1.6
1.809	6.6	2.2	1.6

RESULTS

Table 2 summarizes the most significant source detections ($\geq 3\sigma$) from the main source catalogue paper [1], and Fig. 2 gives an all-sky view of these sources. Six of the listed sources extend over larger areas. Their extent may actually be due to a larger number of - so far - unresolved point sources (see especially Cygnus region in 1.809 MeV [2], GRO J 1823-12 [3], and HVC complex C [3]).

TABLE 2. Summary of Most Significant COMPTEL Source Detections.

Type of Source	No. of Sources	Comments
Spin-Down Pulsars:	3	Crab, Vela, PSR B1509-58.
Stellar Black-Hole Candidates:	2	Cyg X-1, Nova Persei 1992 (GRO J0422+32).
Supernova Remnants: (Continuum Emission)	1	Crab nebula.
Active Galactic Nuclei:	10	CTA 102, 3C 454.3, PKS 0528+135, GRO J0516-609, PKS 0208-512, 3C 273, PKS 1222+216, 3C 279, Cen A, PKS 1622-297.
Unidentified Sources:		
• $\|b\| < 10°$	4	GRO J1823-12, GRO J2228+61 (2CG 106+1.5), GRO J0241+6119 (2CG 135+01), Carina/Vela region (extended).
• $\|b\| > 10°$	5	GRO J1753+57 (extended), GRO J1040+48, GRO J1214+06, HVC complexes M and A area (extended), HVC complex C (extended).
Gamma-Ray Line Sources:		
• 1.809 MeV (^{26}Al)	3	Cygnus region (extended), Vela region (extended, may include RX J0852-4621), Carina region.
• 1.157 MeV (^{44}Ti)	2	Cas A, RX J0852-4621 (GRO J0852-4642).
• 0847 and 1.238 MeV (^{56}Co)	1	SN 1991T.
• 2.223 MeV (n-capture)	1	GRO J0317-853.
Gamma-Ray Burst Sources: (within COMPTEL field-of- up to Phase IV/Cycle-5)	31	Location error radii vary from 0.34° to 2.79° (mean error radius: view 1.13°).

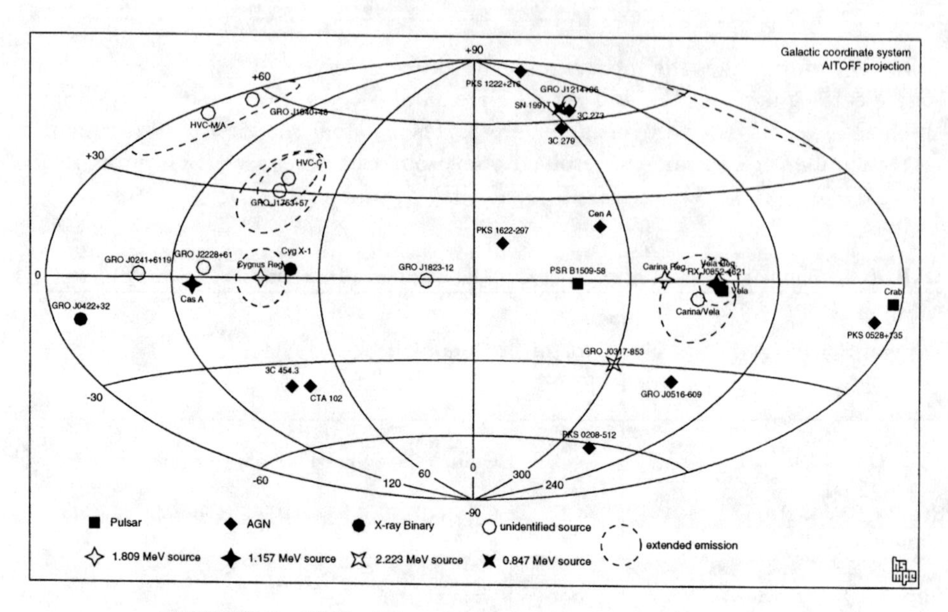

FIGURE 2. All-Sky View of the COMPTEL Source Catalogue.

ACKNOWLEDGEMENT

The COMPTEL project is supported by the German government through DLR grant 50 Q 9096 8, by NASA under contract NAS5-26645, and by the Netherlands Organization for Scientific Research NWO.

REFERENCES

1. Schönfelder, V. et al. *Astron. & Astrophys. Suppl. Series*, submitted for publication.
2. Plüschke, S. et al., these proceedings.
3. Williams, O.R. et al., these proceedings.

Unbinned Likelihood Analysis of EGRET Observations

Seth W. Digel

USRA/NASA Goddard Space Flight Center
Code 661
Greenbelt, Maryland 20771

Abstract. We present a newly-developed likelihood analysis method for EGRET data that defines the likelihood function without binning the photon data or averaging the instrumental response functions. The standard likelihood analysis applied to EGRET data requires the photons to be binned spatially and in energy, and the point-spread functions to be averaged over energy and inclination angle. The full-width half maximum of the point-spread function increases by about 40% from on-axis to 30° inclination, and depending on the binning in energy can vary by more than that in a single energy bin. The new unbinned method avoids the loss of information that binning and averaging cause and can properly analyze regions where EGRET viewing periods overlap and photons with different inclination angles would otherwise be combined in the same bin. In the poster, we describe the unbinned analysis method and compare its sensitivity with binned analysis for detecting point sources in EGRET data.

INTRODUCTION

The use of likelihood in gamma-ray astronomy dates at least to COS-B [1], and the likelihood function has been fundamental to EGRET data analysis [2]. In likelihood analysis, a model is fit to the observations using the likelihood function, which is literally the probability of the observation given the model, as the measure of goodness-of-fit. The likelihood test statistic, derived from the ratio of the maximum likelihoods for two models, can be used to quantitatively compare the significances of the two (see [2]).

The standard likelihood analysis compares models with observations in terms of the expected and observed numbers of photons in bins of solid angle, energy, and inclination angle (angle from the instrument axis). The likelihood of the data is then calculated as the joint Poisson probability for all of the bins.

Binning the data for analysis results in some loss of information, both from the binning itself and because the model must be defined in terms of (weighted) average instrumental response functions for the bins. In principle, if the bins are

CP510, *The Fifth Compton Symposium,* edited by M. L. McConnell and J. M. Ryan

small enough, the loss of information from binning and averaging can be made small.

Here we present the derivation of an unbinned likelihood function, one for which no binning of the data is required, and present the results from using this likelihood function to analyze a source from the 3EG catalog [3]. For this initial work, 3EG J0237+1635, an isolated, high-latitude source was selected for study and the emission was modelled as a point source observed against an isotropic background. Unbinned likelihood functions are also discussed by Tompkins [4].

DEFINITION OF THE MODEL: POINT SOURCE + FLAT BACKGROUND

The model intensity distribution for a point source of flux $F_s E^{-\alpha_s}$ against an isotropic background $I_b E^{-\alpha_b}$ is

$$I(x, y, E) = F_s E^{-\alpha_s} \delta(x - x_s, y - y_s) + I_b E^{-\alpha_b}, \tag{1}$$

where the coordinates of the source are (x_s, y_s). The corresponding model photon distribution is

$$C(x, y, E, \theta) = F_s E^{-\alpha_s} \epsilon(x, y, E, \theta) \delta(x - x_s, y - y_s) + I_b E^{-\alpha_b} \epsilon(x, y, E, \theta), \tag{2}$$

where $\epsilon(x, y, E, \theta)$ is the exposure as a function of direction on the sky, energy, and inclination angle from the instrument's axis.

The model observed photon distribution is the model photon distribution (Eqn. 2) convolved with the instrumental response functions (point-spread function, PSF, and energy dispersion, W):

$$C_m(x, y, E, \theta) = \iint C(x', y', E', \theta) PSF(x - x', y - y', E', \theta) W(E, E', \theta) \mathrm{d}E' \mathrm{d}\Omega'. \tag{3}$$

Both C_m and C have units of photons sr^{-1} MeV^{-1} deg^{-1}. Also note that no convolution is done over inclination angle. This is an approximation; the PSF and energy dispersion W do not vary strongly with inclination angle on degree scales. We make the further approximation here that the energy resolution is infinite. Equation (3) becomes

$$C_m(x, y, E, \theta) = F_s E^{-\alpha_s} \epsilon(x_s, y_s, E, \theta) PSF(x - x_s, y - y_s, E, \theta) + I_b E^{-\alpha_b} \epsilon_c(x, y, E, \theta), \tag{4}$$

where ϵ_c is the exposure convolved with the PSF.

The total number of photons predicted by the model for the region of the sky and energy and inclination angle ranges of interest is

$$C_{tot} = \iiint_{\Omega_{ROI}} C_m(x, y, E, \theta) \mathrm{d}\Omega \mathrm{d}E \mathrm{d}\theta, \tag{5}$$

with Ω_{ROI} denoting the solid angle of the region of interest.

DEFINITION OF THE LIKELIHOOD FUNCTION

In Mattox *et al.* [2] the likelihood function is defined in terms of the numbers of photons detected in each spatial (and energy and inclination angle) bin. With the bins enumerated with indicies ij for the spatial coordinates, and having fixed ranges of energy and inclination angle, Mattox *et al.* derive the likelihood function as

$$\ln L = \sum_{ij} n_{ij}(\text{obs}) \ln n_{ij}(\text{pred}) - \sum_{ij} n_{ij}(\text{pred}), \tag{6}$$

where $n_{ij}(\text{obs})$ is the number of photons observed in each bin and $n_{ij}(\text{pred})$ is the number of photons predicted by the model. Here, we wish to extend the derivation to the unbinned limit. If the data space is divided up into very small bins, of size $\Delta\Omega\Delta E\Delta\theta$, so that no bin has more than one photon, then the likelihood function may be written as

$$\ln L = \sum_{N} \ln(C_i \Delta\Omega\Delta E\Delta\theta) - \sum_{all} C_i \Delta\Omega\Delta E\Delta\theta. \tag{7}$$

where $C_i = C_m(x_i, y_i, E_i, \theta_i)$. The first summation is over just the N bins that contain photons, and the second is over all of the bins. Because only relative values of the likelihood function matter in practice (for comparing the significances of two models), the $\Delta\Omega\Delta E\Delta\theta$ factor may be removed from the logarithm in Equation (7). In the differential limit of tiny bins, the second summation may be replaced by its equivalent integral (Eqn. 5), so the likelihood function may be written as

$$\ln L = \sum_{N} \ln C_i - C_{tot}. \tag{8}$$

For the case of a point source observed against an isotropic background, and on the assumption of infinite energy resolution, Eqn. (8) may be written as

$$\ln L = \sum F_s E_i^{-\alpha_s} PSF(x_i - x_s, y_i - y_s, E_i, \theta_i)\epsilon(x_s, y_s, E_i, \theta_i) + \sum_{N} I_b E_i^{-\alpha_b} \epsilon_c(x_i, y_i, E_i, \theta_i) - C_{tot}. \tag{9}$$

On the assumptions that the source and background spectral indicies are known, the free parameters of the model are the source flux coefficient, F_s, the source position (x_s, y_s), the background intensity coefficient, I_b, and their corresponding spectral indicies, α_s and α_b. The model is fit to the observation by adjusting the parameters to maximize $\ln L$.

EGRET DATA

Photon data from EGRET viewing period 21.0 were used. Exposures were derived using the standard INTMAP program, for a range of inclination angles and the ten standard energy ranges. Exposure maps for different ranges of inclination angle were differenced to derive exposures for 5° bands of inclination.

The parameters of the PSF were derived from the EGRET calibration file psdfil07 (see [5]) by fitting the sum of two gaussians to the point-spread distributions.

RESULTS

To construct the localization map (Fig. 1), maximum likelihood fits were made at each point on a grid, with all photons within 20° of a grid point included in the analysis for that point. The background spectral index was assumed to be $\alpha_b = 2.1$.

Figure 1 compares the source localization contours found here with those of the source 3EG J0237+1635 in the Third EGRET Catalog [3]. Also indicated is the position of PKS 0235+164, the BL Lac object associated with the source [6].

Parameters derived for the source are compared in Table 1 with results from binned analyses. The relatively low detection significance (TS) in the 3EG catalog [3] may be due to the inclusion of another, fainter source in the analysis (3EG J0204+1458) close enough to decrease the significance of 3EG J0237+1635 (Hartman, priv. comm.). This additional source was not included here or in the analysis by Hunter *et al.* [6]. The source signifiance found here is greater and the confidence regions are smaller, and closer to the position of PKS 0235+164. The confidence

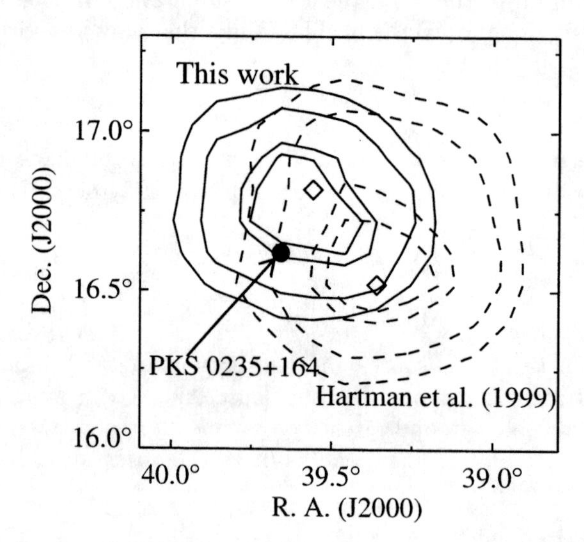

FIGURE 1. Localization of the EGRET source identified with PKS 0235+164. The solid contours are from the unbinned likelihood analysis described here. The dashed contours are for the source 3EG J0237+1635 from Hartman *et al.* Maximum likelihood positions are indicated with diamonds, and the 50, 68, 95, and 99% confidence regions are shown.

TABLE 1. Analyses of EGRET Data for 3EG J0237+1635

	$\sqrt{TS}$	α, δ [a]	r_{95}	r_{99}	F[b]	α_s
Hunter *et al.* 1993	13.0	39.7°, 16.7°	-	0.55°	7.0 ± 2.1	1.85 ± 0.19
Hartman *et al.* 1999	10.0	39.36, 16.59	0.37°	-	6.5 ± 0.9	1.85 ± 0.12
This work	16.0	39.56, 16.82	0.32	0.40	8.7 ± 0.2	1.61 ± 0.12

[a] J2000

[b] 10^{-7} cm^{-2} s^{-1}

region for the Hartman *et al.* analysis was derived from the sum of TS maps for 300–1000 MeV and $>$ 1 GeV. The fluxes are consistent across the analyses, although the unbinned analysis seems to indicate a harder spectrum.

CONCLUSIONS

The results from application of unbinned likelihood analysis to the EGRET data for 3EG J0237+1635 for one viewing period suggest that the method has promise for EGRET data analysis. By allowing each photon to be incorporated with the instrumental response functions that apply to it specifically, the source signficance and confidence region can be improved from the standard binned likelihood analysis.

Unbinned likelihood analysis may have greater advantages when data from multiple, partially overlapping viewing periods are analyzed together, where photons with different inclination angles would otherwise wind up in the same spatial bin.

For more general application of unbinned analysis to EGRET data, the model used here can be generalized to allow for multiple sources and structured diffuse emission.

REFERENCES

1. Pollock, A. M. T., *et al.*, *Astron. and Astrophys.* **94**, 116 (1981).
2. Mattox, J. R., *et al.*, *Astrophys. J.* **461**, 396 (1996).
3. Hartman, R. C., *et al.*, *Astrophys. J. Supp.* **123**, 79 (1999).
4. Tompkins, W., Ph. D. thesis, Stanford University (1999).
5. Thompson, D. J., *et al.*, *Astrophys. J. Supp.* **86**, 629 (1993).
6. Hunter, S. D., *et al.*, *Astron. and Astrophys.* **272**, 59 (1993).

Preliminary Results From A New Analysis Method For EGRET Data

D. J. Thompson, D. L. Bertsch, S. D. Hunter, P. Deines-Jones

NASA/Goddard Space Flight Center

B. L. Dingus

University of Utah

D. A. Kniffen

Hampden-Sydney/NASA HQ

P. Sreekumar

ISRO Satellite Center

Abstract. In order to extend the life of EGRET, the gas in the spark chamber was allowed to deteriorate more than was originally planned for the nominal two year Compton Observatory mission. Gamma ray events are lost because the pattern recognition analysis rules are not optimized for the poorer quality data. By changing the rules used by the data analysts, we can recover a significant fraction of the lost events, allowing improved statistics for detection and study of sources. Preliminary results from the Crab, Geminga, and BL Lacertae indicate the feasibility of this analysis.

CONCEPT

In EGRET, each trigger produces a picture in the spark chamber, with examples shown in Figure 1. Although the hardware trigger rejects the vast majority of the charged-particle background EGRET encounters, the useful pictures currently represent fewer than 10% of the total triggers. The useful pictures are those for which there is clear evidence of a pair-production event within the spark chamber. EGRET analysis uses a pattern recognition program (with manual verification using selection rules) to select the useful pair-production events. These methods were derived and optimized empirically (1), based on accelerator calibration data.

The performance of EGRET has diminished due to deterioration of the spark-chamber gas. The older gas produces fewer real signals and more spurious sparks, as can be seen in Figure 1. This was planned for by including a gas replenishment

CP510, *The Fifth Compton Symposium,* edited by M. L. McConnell and J. M. Ryan

system. The five refills, planned to allow a two-year lifetime, have been stretched out to more than eight years. As a result, the detection efficiency for EGRET is now less than 25% of what it was at the beginning of the mission. Correction for this loss of efficiency has been done by comparing overlapping regions of the sky, taking the diffuse emission as a steady reference source within any region.

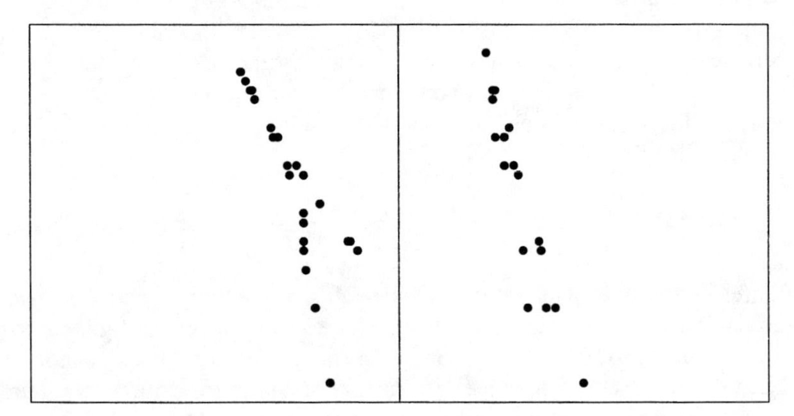

FIGURE 1. Left: gamma-ray pair event under good operating conditions. Right: gamma-ray pair event under poor operating conditions.

The principal motivation for this new analysis method is the fact that the hardware trigger for EGRET has not changed, only the ability of the pattern recognition program and the data analysts to accept events under the rules established when the instrument performance was much better. During periods of poorer EGRET performance there are nearly three times as many gamma rays in the EGRET data as appear in the final maps and event lists. Recovering "lost" events can substantially improve EGRET observations.

PROCEDURE

Under CGRO Cycle 7 and Cycle 8 proposals, EGRET data analysts have reviewed several sets of events that were rejected by the standard data processing system, using new techniques:

(1) We developed new selection criteria to screen sets of rejected events most likely to contain recoverable gamma rays.

(2) The analysts processed events with the characteristic pair structure, but with gaps in the tracks, spurious sparks, or other defects that had caused the events to be rejected.

(3) We processed these events for energy and direction, using software similar to that used for regularly-accepted events.

(4) Using a pointing toward the Galactic anticenter, we generated spectra for the bright Crab and Geminga pulsars using a combination of regular and "recovered" events. From these spectra, we derived sensitivity correction factors for each of the standard EGRET energy bands.

Even with the restricted selection criteria, a large number of events must be reviewed by the analysts; this is a labor-intensive process. With the limited budget, the number of experienced EGRET data analysts has decreased steadily, further slowing this review process. At present, we have only one full-time data analyst, who works on this recovery analysis part-time when not involved in regular data processing activities for current EGRET data.

PROOF OF CONCEPT - GALACTIC ANTICENTER, VP 5280

The figures below show analysis of spatial, timing, and energy spectral data with the recovered events. Both the intensity map (Figure 2) and the pulsar light curve (Figure 3) show the same basic features as seen in the standard data analysis, demonstrating that the recovered events can be used to enhance statistics for gamma-ray sources. By comparing the energy spectra derived from the standard analysis with that from the combined standard + recovered data, we find that more events are recovered at lower energies. The correction factors needed to recover absolute flux values (Figure 4) are derived from a combination of the two source spectra (the Crab has a softer spectrum than Geminga; therefore the combination should give a correction applicable to all but the most unusual spectra).

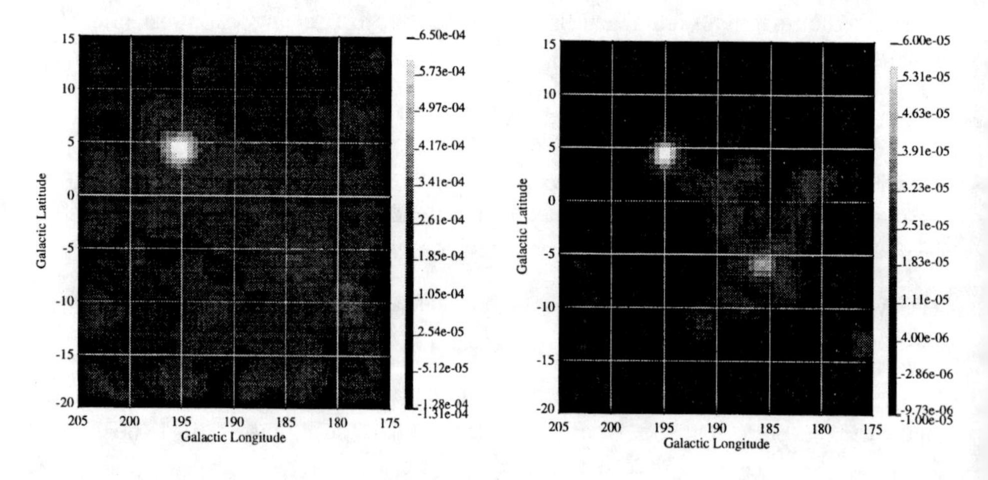

FIGURE 2. Left: Intensity map of the Galactic anticenter, VP5280, using standard analysis. Right: Intensity map of the Galactic anticenter, VP5280, using recovered event analysis. Note that the standard exposure has been used, so that the absolute intensity is not correct.

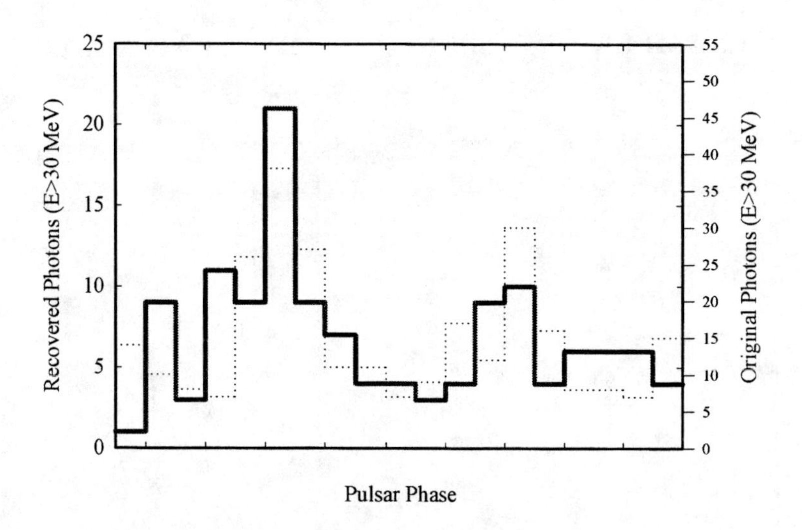

FIGURE 3. Light curve for the Crab pulsar. Solid line: recovered events only. Dotted line: standard events only. The two peaks of the typical Crab light curve are seen in both data sets, in phase.

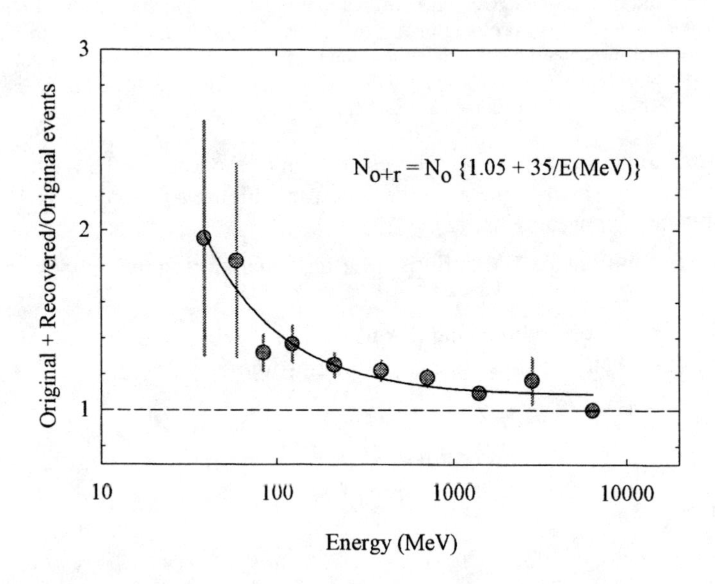

FIGURE 4. Energy-dependent correction factors for flux values that include recovered events compared to the original analysis. These factors are derived from the Crab and Geminga spectra measured in VP5280.

AN EXAMPLE - THE SPECTRUM OF BL LAC, VP6235

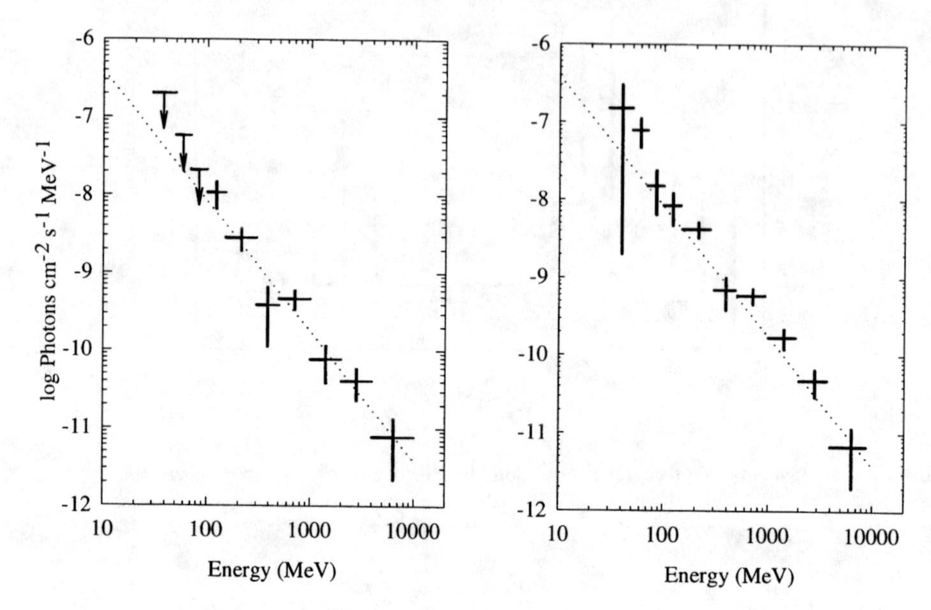

FIGURE 5. Left: BL Lac flare energy spectrum using original events only. Right: BL Lac flare spectrum using standard + recovered events. The flux in each energy bin has been scaled using the correction factors derived from the Crab/Geminga observation (Figure 4). The flux and power law index are consistent between the two spectra (1.68 ± 0.16 for the standard data, 1.73 ± 0.10 for the enhanced data), but better measured with the addition of the recovered events, because the error bars are smaller and upper limits have been converted into detections at low energies. The consistency indicates that the corrections derived from the Crab and Geminga are reasonable.

The recovery technique was applied to the observation of the flare of BL Lacertae in VP 6235. The improved measurement of the spectrum (Figure 5) strengthens previous conclusions about this flare (2):

- The spectrum is significantly flatter during the flare than during the only previous EGRET detection of BL Lac.

- The spectrum is consistent with a single power law, showing no evidence for a change of slope or curvature as might be expected in some models (3).

REFERENCES

1. Thompson, D.J. et al., *ApJS* **86**, 629-656 (1993).

2. Bloom, S.D. et al., *ApJ* **490**, L145-L148 (1997).

3. Madejski et al., *ApJ* **521**, 145-154 (1998).

A "Snapshot" Survey of the Gamma-Ray Sky at GeV Energies

W.T. Vestrand*, P. Sreekumar†, D. Bertsch‡, R. Hartman‡, D.J. Thompson‡, and G. Kanbach**

**NIS-2, Los Alamos National Laboratory, Los Alamos, NM 87545*
†*ISRO Satellite Center, Bangalore, India*
‡*NASA/Goddard Space Flight Center, Greenbelt, MD 20771*
***Max-Planck-Institut für Extraterrestrische Physik, Garching, Germany*

Abstract. The giant outbursts of GeV emission observed from some blazars pose a significant challenge to our understanding of blazar emission mechanisms. Unfortunately, the giant outbursts of GeV emission from blazars are not always predictable from observations in other energy bands. Here we describe a strategy that will be employed during Cycle 9 of the CGRO observing program which will allow us to search the sky for intense outbursts of GeV emission while having a minimal impact on the CGRO observing schedule and remaining spark chamber gas.

INTRODUCTION

Serendipity has played an important role in the remarkable success of the EGRET telescope. From the initial discovery of an intense outburst of GeV emission from the blazar 3C 279 during the first observations of the then only known extragalactic source 3C 273, through the all-sky survey, and into later phases of EGRET operation, the variability of the gamma-ray sky at GeV energies has led to many unexpected discoveries. Here we discuss an observational plan, that will be implemented during Cycle 9 of the CGRO observing program, which will maintain some capability to monitor the GeV gamma-sky sky for bright, serendipitous sources.

THE UNPREDICTABILITY OF GEV OUTBURSTS FROM BLAZARS

EGRET observations have shown that many gamma-ray blazars are highly variable. Most display outbursts, with uncertain duty cycle, during which some reach peak values more than 100 times their quiescent flux values (e.g. [1]). While this gamma-ray variability is often correlated with variations at lower energies, it is

CP510, *The Fifth Compton Symposium,* edited by M. L. McConnell and J. M. Ryan

important to understand that GeV outbursts are not always predictable from observations at other energies.

A good example of the unpredictability of GeV gamma-ray outbursts is provided by the well studied blazar 3C 279. In January 1996 an intense GeV outburst was detected from 3C279 which reached flux levels 3 times higher that those measured for any previous outburst and nearly 100 times those observed at gamma-ray quiescence [2]. During that intense outburst, the measurements at IR, optical, and UV wavelengths showed relatively modest flux increases and those fluxes never reached the levels measured during the smaller June 1991 gamma-ray outburst. Under the high thresholds now typical for triggering EGRET ToO observations, it is unlikely that the January 1996 observations would have been made. However, GeV outbursts like the January 1996 events arguably present the greatest challenge to our understanding because the measured peak gamma-ray flux significantly exceeds the predicted inverse Compton fluxes from standard formulations for either of the generally accepted External Compton or Synchrotron Self Compton models for blazars emission [2].

Another good example of the unpredictability of GeV outbursts is provided by the behavior of the TeV gamma-ray emitting blazar Mrk 501. Simultaneous multiwavelength observations of Mrk 501 have detected major outbursts at x-ray and TeV energies that are well correlated but not accompanied by significant activity at GeV energies. Recently Kataoka et al. [3] reported the first detection of GeV emission from Mrk 501 and Sreekumar et al. [4] reported the discovery of a significant outburst of GeV emission that was uncorrelated with the x-ray emission. Mrk 501 has displayed remarkable hard x-ray outbursts that are usually interpreted as events where the acceleration mechanism generates a population of synchrotron emitting electrons with an unusually high cutoff energy [5]. In that context, the GeV outbursts could be interpreted as major acceleration events where the cutoff appears at much lower energies. Understanding the differences between these types of outburst is likely to provide important clues about the nature of particle acceleration and the generation of high energy emission by blazars. Clearly, it is desirable to have a mechanism for detecting GeV outbursts that is not hampered by the selection effects imposed by triggering GeV observations through measurements in other energy bands.

While the sensitivity of the EGRET telescope has declined in recent years, EGRET still has the capability to discover new sources. A recent example is the discovery of a major outburst of GeV gamma-ray emission from the quasar PKS 2255-382 which was previously undetected at gamma-ray energies [6]. The peak >100 MeV flux measured during the outburst reached $\sim 4 \times 10^{-6}$ cm^{-2}s^{-1} and averaged $\sim 1.6 \times 10^{-6}$ cm^{-2}s^{-1} during the two week observation placing it among the brightest yet detected by EGRET. Identification and study of sources that can exhibit such extreme outbursts is likely to provide important clues about the underlying physical processes in blazars.

An important factor that has limited the ability of EGRET to make serendipitous discoveries in recent years has been the mode of telescope operation. After completion of the all-sky survey, the focus of EGRET research shifted to studies of selected sources or regions of the sky that were thought most likely to yield the greatest scientific return. The sky monitoring capability was further reduced during Cycle 5 when the standard operational mode of EGRET was altered from wide field-of-view (FOV) mode to a narrow FOV mode wherein only the vertical subtelescopes are triggered. That reduction in capability was compounded by the decision, prompted by the need to minimize degradation of the spark chamber gas, to switch the spark chamber off during the entire duration of many CGRO pointings. During the last few observing cycles, the EGRET spark chamber has been shut off for the entire duration of most CGRO pointings and has only been activated for repeat observations of a set of well known sources. In retrospect, we believe that limited operation of the spark chamber during high-latitude pointings would have been beneficial. For with data gathered in just a single day of operation, EGRET quicklook analysis can identify the outbursts of GeV emission from known or unknown sources that are best suited for follow up studies and most likely to have the greatest scientific impact.

THE "SNAPSHOT" SURVEY

A new strategy to monitor the sky for significant gamma-ray outbursts will be employed during Cycle 9 of the CGRO observing program. This so called "Snapshot Survey" approach is to activate the EGRET spark chamber in narrow FOV mode for one day during each CGRO viewing period when EGRET is pointed toward high Galactic latitudes ($|b| > 30°$) and is scheduled to be nominally turned off. High latitude pointings have been selected for two reasons. First, the GeV gamma-ray background is lower at high galactic latitudes. That leds to a greater sensitivity for detecting outbursts as well as minimizing the event rate and the associated degradation of the spark chamber gas. Second, outbursts detected at high latitudes are less subject to source confusion and are easier to associate with a counterpart observed at other energies.

Our feasibility studies, using actual EGRET data collected during the last few observing cycles, demonstrate that single-day snapshots have a surprising ability to detect gamma-ray flares from high-latitude sources. Using the current response properties of EGRET, we find that single day observations are able to achieve a source sensitivity which exceeds that achieved by COS-B during an entire month of observation. Specifically, our tests on archival data indicate that we can detect, with $> 3\sigma$ significance, outbursts within 6° of the pointing direction if they reach flux levels which exceed 1×10^{-6}photons cm^{-2}s^{-1}. Our survey of archival EGRET observations revealed approximately 30 blazar outbursts which exceeded this detection threshold in the data gathered through Cycle 4. Calibration studies have shown the instrument effective area decreases with increasing inclination angle, displaying a 50% drop at $\sim$9° in its narrow FOV mode. As a consequence, our detection threshold rises to $\sim 2 \times 10^{-6}$ photons cm^{-2}s^{-1} for sources at inclinations near 10°. EGRET one-day snapshots therefore have a significant capability to detect GeV outbursts within 10-15° of the CGRO pointing direction. Figure 1 shows maximum likelihood maps we derived for one-day snapshots constructed from actual Cycle 5 and 6 data. They clearly demonstrate that the current EGRET spark chamber performance is able to unambiguously detect bright outbursts from blazars located at high Galactic latitudes.

Encouraged by the remaining capability of EGRET to detect major GeV outbursts during a single day of observation, we will activate the EGRET spark Chamber in narrow FOV mode at the beginning of many scheduled high galactic pointing ($|b| > 30°$) during Cycle 9. The timing and pointing direction for these "snapshots" will be determined by the requirements of the other instruments aboard CGRO. A secondary benefit of this observing program is the information it will provide us about the spark chamber performance. The single-day snapshots scattered throughout the CGRO observing schedule will help us monitor variations in the spark chamber performance. This performance monitoring should be useful for determining the feasibility of other ToO observations during Cycle 9.

Multiwavelength observations are important for placing the EGRET measurements in their proper context. Of course it is impractical to implement a multiwavelength campaign that monitors the entire field-of-view for each EGRET snapshot. However, if the EGRET quicklook analysis discovers a major outburst, we plan to initiate a campaign of multiwavelength observations of the source via IAU telegrams and through notification of the established network of multiwavength observers that have collaborated with the EGRET team in the past. This technique for triggering a multiwavelength campaign has already had some notable successes. For example, in November 1997 our quicklook analysis detected a giant outburst of GeV emission from the BL Lac PKS 2155-304 [7] and we initiated a multiwavelength campaign that resulted in the detection of the largest x-ray [8,9] and TeV gamma-ray outbursts [10] yet seen from the source. The rapid variability detected at GeV energies together with the contemporaneous multiwavelength observations of PKS 2155-304 places interesting constraints on acceleration and particle energy loss in High frequency peaked BL Lacertids. To study variability in other sources, if we detect a major GeV outburst with our Snapshot Survey we will employ additional 1-day EGRET snapshots to follow the development of the outburst.

REFERENCES

1. Mattox, J.R. et al.,*Astrophysical Journal* **476**, 692 (1997).
2. Wehrle, A.E. et al.,*Astrophysical Journal* **497**, 178 (1998).
3. Kataoka, J. et al.,*Astrophysical Journal* **514**, 138 (1999).
4. Sreekumar, P. et al., these proceedings (2000).
5. Catanese, M. et al. *Astrophysical Journal* **487**, L143 (1997).
6. Macomb, D., Gehrels, N. & Shrader, C., *Astrophysical Journal* **513**, 652 (1999).
7. Sreekumar, P. and Vestrand, W.T., IAU Circ. 6774 (1997)

8. Chiappetti, L. et al.,*Astrophysical Journal* **521**,552 (1999).
9. Vestrand, W.T. & Sreekumar, P.,*Astroparticle Physics*,**11**, 197 (1999).
10. Chadwick, P. et al. *Astrophysical Journal*, **513**, 161 (1999).

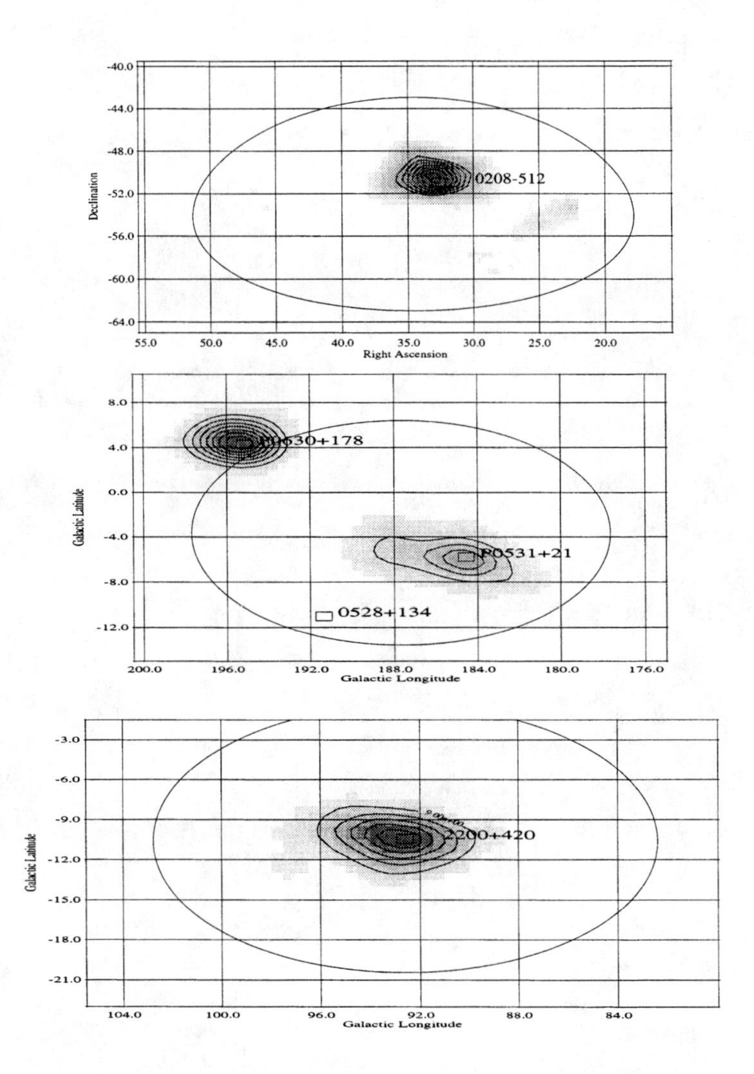

FIGURE 1. Maximum likelihood skymaps generated from 1-day samples of data from viewing periods 517.0 (PKS 0208-512), 526.0 (Geminga, Crab) and 623.5(BL Lac). The lowest contour level corresponds to a 3σ detection. The results show unambiguously, the ability of EGRET to study strong gamma-ray sources within 10° of the pointing axis.

GROUND-BASED GAMMA-RAY ASTRONOMY

Ground-Based Gamma-Ray Astronomy

Michael Catanese

Iowa State University[1], *Dept. of Physics and Astronomy, Ames, IA 50011*

Abstract. Ground-based γ-ray astronomy has become an active astrophysical discipline with four confirmed sources of TeV γ-rays, two plerionic supernova remnants (SNRs) and two BL Lac objects (BL Lacs). An additional nine objects (one plerion, three shell-type SNRs, one X-ray binary, and four BL Lacs) have been detected but have not been confirmed by independent detections. None of the galactic sources require the presence of hadronic cosmic rays, so definitive evidence of their origin remains elusive. Mrk 421 and Mrk 501 are weak EGRET sources but they exhibit extremely variable TeV emission with spectra that extend beyond 10 TeV. They also exhibit correlations with lower energy photons during multi-wavelength campaigns, providing tests of emission models. Next generation telescopes like VERITAS hold the promise of moving this field dramatically forward.

INTRODUCTION

Since the launch of the *Compton Gamma-Ray Observatory* (*CGRO*), ground-based γ-ray telescopes have come to play an important role in our understanding of the γ-ray sky. In many cases, it has required the results from both the ground and space to properly interpret the observations of a particular source. In this context, I review the status of ground-based γ-ray astronomy and consider the implications of these observations. I will concentrate on the results obtained with imaging atmospheric Cherenkov telescopes because they have produced most of the scientific results to date and because several papers in this proceedings address other ground-based telescope results. The interested reader is encouraged to seek out more complete reviews that have recently been published [1,2] for more information. To save space, I will not cite the original detection references for those objects that are included in the review articles.

---

[1]) Current address: Harvard-Smithsonian Center for Astrophysics, F.L. Whipple Observatory, P.O. Box 97, Amado, AZ 85645. This research is supported by the U.S. Department of Energy and by NASA

CP510, *The Fifth Compton Symposium*, edited by M. L. McConnell and J. M. Ryan
2000 American Institute of Physics 1-56396-932-7

TABLE 1. Galactic Sources of VHE γ-rays.

Source	Energy	Flux	Significance
Crab Nebula	>0.3 TeV	1.26×10^{-10}cm^{-2}s^{-1}	Conf.[a]
PSR 1706-44	>1.0 TeV	0.38 Crab	Conf.
Vela	>2.5 TeV	0.54 Crab	5.8 σ
SN 1006	>1.7 TeV	0.48 Crab	8.0 σ
RXJ 1713.7-3946	>2.0 TeV	0.40 Crab	5.0σ
Cassiopeia A	>0.5 TeV	???	4.7 σ
Centaurus X-3	>0.4 TeV	0.24 Crab	6.5 σ

[a] Significances are listed only for unconfirmed sources.

GALACTIC SOURCES

Seven sources of very high energy (VHE, E > 250 GeV) γ-ray emission associated with galactic objects have been detected at this time: three plerionic supernova remnants (SNRs) (Crab Nebula, PSR 1706-44, and Vela), three shell-type SNRs (SN 1006, RXJ 1713.7-3946 [3], and Cassiopeia A [4]), and the X-ray binary Centaurus X-3 [5]. A summary of the VHE properties of these objects is given in Table 1. The Crab Nebula and PSR 1706-44 have been confirmed as sources of VHE γ-rays by detections from independent groups. The Crab Nebula has the highest VHE γ-ray flux of these objects and this, along with its steady flux, has established it as the standard candle of ground-based γ-ray astronomy. Because of this and because ground-based γ-ray telescopes have a range of energy thresholds, I list source fluxes in units of the Crab flux to make comparisons of source strength easier.

All three of the detected plerions are EGRET sources [6], but the GeV emission is predominantly or entirely pulsed, while the TeV emission shows no evidence of pulsations. This is consistent with the VHE emission arising in the synchrotron nebulae of these objects. Several groups have measured accurately the spectrum of the Crab Nebula over an energy range spanning 250 GeV to 50 TeV. The spectrum is fit well by a simple power law with differential spectral index 2.5. The sub-GeV flux measurements, combined with the VHE measurements, are consistent with synchrotron self-Compton emission models for a magnetic field of $\sim$ 160 μG [7]. An interesting feature of the TeV emission detected from Vela is that the peak in the TeV emission is located $\sim 0.14°$ away from the pulsar position, coincident with the birthplace of the pulsar. An upper limit of 0.40 Crab above 300 GeV for Vela [8] implies that its spectrum must be harder than $E^{-2.3}$.

In addition to the plerions listed above, the pulsars detected by EGRET have been searched for VHE emission. None have been detected, and the pulsed flux from these objects must have a rapid decrease in power output between $\sim$ 1 GeV and 300 GeV. Evidence for such cut-offs is seen in the EGRET data for some of these objects [9], but for PSR 1951+32, the power output increases up to at least 10 GeV. The Whipple collaboration's upper limit on this object is $\sim$0.02 Crab [10], implying an extremely rapid fall off in the flux. Similarly, recent upper limits derived for

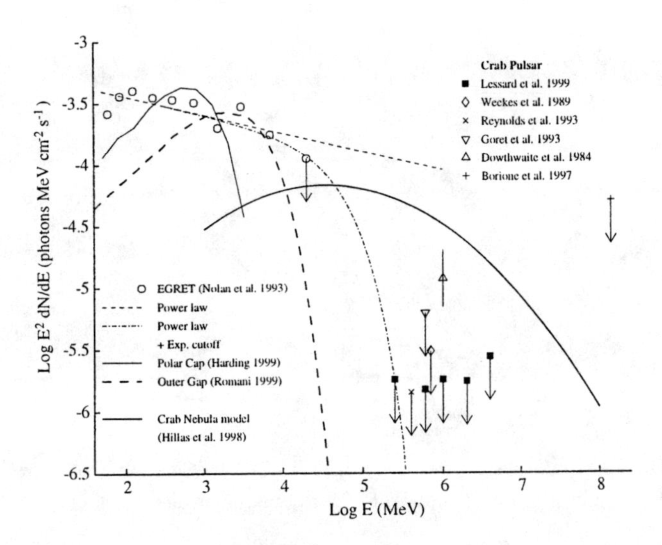

FIGURE 1. The pulsed γ-ray spectrum of the Crab Nebula. The open circles are EGRET flux points and the points with arrows are upper limits. The thick solid line is a model of the unpulsed inverse Compton spectrum. The thin solid and dotted lines are model fits and the dot-dashed line is an extension of the EGRET spectrum with a 60 GeV exponential cut-off. Figure adapted from [11].

pulsed emission from the Crab Nebula imply that if the cut-off is exponential, it must begin below 60 GeV, though this does not constrain the emission models (see Figure 1) [11].

Shell-type SNRs are believed to be sources of γ-rays produced by cosmic rays accelerated in the supernova shocks. In support of this, several EGRET sources are associated with shell-type SNRs [12]. However, the EGRET detections alone are not enough to claim definitively that the long-sought origins of cosmic rays have been found. Indeed, observations by the Whipple collaboration of several of the shell-type SNRs associated with the EGRET sources revealed no evidence of VHE γ-ray emission [13]. The limits derived from those observations imply that if the emission seen by EGRET is from the SNR shells and is produced by the interactions of cosmic rays, then the source cosmic-ray spectrum must be steeper than the $E^{-2.3}$ or that the spectrum cuts off below 10 TeV.

The three detected shell-type SNRs also do not require the presence of hadronic cosmic rays to produce the γ-rays. In all three objects, X-ray synchrotron emission has been detected, implying the presence of > 10 TeV electrons. Thus, the TeV detections can be explained as inverse Compton emission. This is supported by the EGRET's non-detection of these objects and by the positional coincidence of the TeV and X-ray synchrotron emission peaks in SN 1006 and RXJ 1713.7-3946 (see Figure 2, [3,14]). If the TeV emission from SN 1006 is produced from inverse Compton emission, the magnetic field in the shock region must be $6.5 \pm 2\mu$G [14].

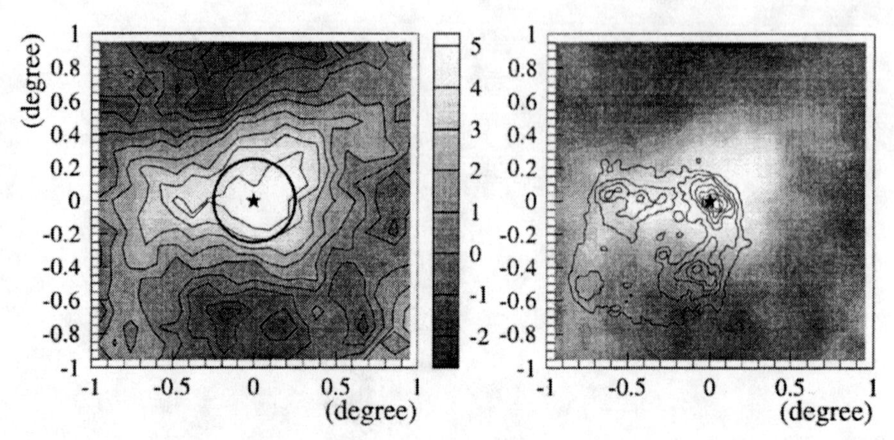

FIGURE 2. CANGAROO observations of RXJ 1713.7-3946. The left plot shows the excess in TeV γ-rays and the right plot superimposes the X-ray map of synchrotron emission over the TeV γ-ray map. Figure from [3].

TABLE 2. Extragalactic Sources of VHE γ-rays.

Source	z	Energy	Flux	Significance
Mrk 421	0.031	>0.50 TeV	0.3 Crab	Conf.[a]
Mrk 501	0.034	>0.30 TeV	0.08 Crab	Conf.
1ES 2344+514	0.044	>0.35 TeV	0.11 Crab	5.2 σ
PKS 2155-304	0.116	>0.30 TeV	0.48 Crab	6.8 σ
1ES 1959+650	0.048	>0.90 TeV	???	3.7 σ
3C 66A	0.444	>0.90 TeV	1.2 Crab	5.0 σ

[a] Significances are listed only for unconfirmed sources.

EXTRAGALACTIC SOURCES

Six BL Lacertae objects (BL Lacs) have been detected as sources of VHE γ-rays: Markarian 421 (Mrk 421), Mrk 501, 1ES 2344+514, PKS 2155-304, 1ES 1959+650 [15], and 3C 66A. A summary of their properties is given in Table 2. The results quoted in the table are values from the discovery papers. Only Mrk 421 and Mrk 501 have been confirmed as VHE sources. The other objects have been detected with high significance in limited time intervals, making confirmation difficult. Mrk 421, PKS 2155-304, and 3C 66A are sources in the third EGRET catalog [6]. Mrk 501 was first detected on the ground but it has recently been claimed as an EGRET source [16].

The most distinctive feature of the VHE emission from Mrk 421 and Mrk 501 is large amplitude, rapid variability. For Mrk 421, the average flux does not change much from year to year. Instead, flares develop and decay on day-scales or less and drop to a baseline emission level (if one exists at all) that is below the sensitivity of current telescopes [17]. Fluxes from 0.1 to 10 times the Crab flux have been

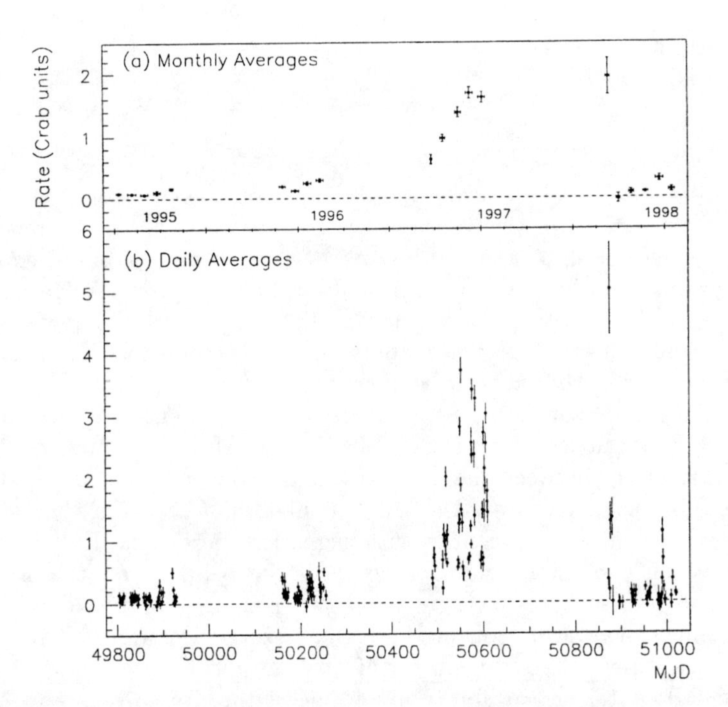

FIGURE 3. Whipple Observations of Mrk 501 between 1995 and 1998. The top plot shows monthly average fluxes and the bottom plot shows nightly average fluxes. Figure from [19].

detected and flares lasting as little as 30 minutes have been measured [18]. For Mrk 501, the flaring appears to be somewhat slower and of lower amplitude than that seen in Mrk 421 [19]. The most prominent features of the variability in Mrk 501 are large changes in its average flux and flaring activity, as shown in Figure 3. The yearly average flux has varied from 0.08 Crab in 1995 to 1.4 Crab in 1997 and the amount of day-scale flaring increases with increasing flux [19].

Perhaps the most important development in the TeV results on Mrk 421 and Mrk 501 since the 4th Compton Symposium has been accurate measurements of their spectra. Observations of several high (1 – 10 Crab) flux states between 1995 and 1996 from Mrk 421 by the Whipple collaboration show that the spectra are all consistent with a simple power law with photon index $-2.54 \pm 0.03_{\mathrm{stat}} \pm 0.10_{\mathrm{sys}}$ over the energy range from 0.25 – 10 TeV [20]. Observations by the HEGRA collaboration of Mrk 421 in a lower (<1 Crab) flux state in 1998 also indicate a power law spectrum, but the spectral index is $-3.09 \pm 0.07_{\mathrm{stat}} \pm 0.10_{\mathrm{sys}}$ over the energy range 0.5 – 7 TeV [21]. This difference could reflect a change in spectral index with flux, but neither Whipple nor HEGRA see evidence of spectral variability within their respective data sets. Further study may help resolve these differences.

Unlike Mrk 421, the spectrum of Mrk 501 during its high state in 1997 is not consistent with a simple power law. The Whipple [20] and HEGRA [22] collaborations derive spectra of the form:

$$\frac{dN}{dE} \propto E^{-2.22\pm0.04_{\mathrm{stat}}\pm0.05_{\mathrm{sys}}-(0.47\pm0.07_{\mathrm{stat}})\log_{10}(E)} \quad \text{(Whipple)}$$

$$\frac{dN}{dE} \propto E^{-1.92\pm0.03_{\mathrm{stat}}\pm0.20_{\mathrm{sys}}} \times e^{-E/6.2\pm0.4_{\mathrm{stat}}(^{+2.9}_{-1.5})_{\mathrm{sys}}} \quad \text{(HEGRA).}$$

The form of the curvature term in these two expressions reflects the preferences of the authors, since the data from both groups are indistinguishable when overlaid. The average spectrum for Mrk 501 measured by the CAT group is consistent with that observed by HEGRA and Whipple [23], but the CAT data show evidence of spectral hardening during high flux states while the Whipple and HEGRA data do not. Again, further study may resolve these issues.

Mrk 421 and Mrk 501 have been the target of several intensive multi-wavelength campaigns. Observations of Mrk 421 in 1995 [17] and Mrk 501 in 1997 [24] revealed day-scale correlations between the TeV γ-ray and X-ray emissions, suggesting that both sets of photons derive from the same population of particles. The variability of the synchrotron emission increases with increasing energy, and EGRET's lack of detected variability in these studies suggests similar behavior for the high energy emission. Thus, these flares seem to be caused primarily by impulsive increases in the efficiency for acceleration of the highest energy electrons. This is not to say that the multi-wavelength behavior of Mrk 421 and Mrk 501 is identical. For example, in Mrk 421, the variability amplitude of the TeV γ-rays and X-rays is comparable while for Mrk 501, the variability amplitude is larger in the TeV γ-rays. More spectacular, is the difference in the spectral energy distributions of these two objects, as shown in Figure 4. The spectrum of Mrk 421 is typical of high-frequency peaked BL Lacs: a synchrotron peak at $\sim$1 keV followed by a rapid drop-off. Mrk 501, on the other hand, appears to be an extreme version of a high-frequency peaked BL Lac, as its synchrotron spectrum peaked at 100 keV in 1997, the highest ever observed in a blazar. Also, the power output at X-ray and TeV energies in Mrk 421 is approximately equal but for Mrk 501, the TeV power can be much less than in X-rays.

Those observations do not resolve any of the hour-scale flares known to occur in Mrk 421 and Mrk 501. In 1998, a campaign involving the Whipple telescope and *BeppoSAX* had overlapping observations of an hour-scale flare from Mrk 421 as shown in Figure 5 [25]. The different energy bands exhibit a similar rise time, but the TeV γ-ray flux appears to fall-off much faster than the X-rays. Thus, at the same time that the first hour-scale correlations are seen in a TeV blazar, there is also evidence that the TeV γ-rays and X-rays in Mrk 421 may not be completely correlated on all time-scales.

CONCLUSIONS

From the previous paragraphs, it should be clear that ground-based γ-ray astronomy has become a vibrant branch of astrophysics. There are established sources

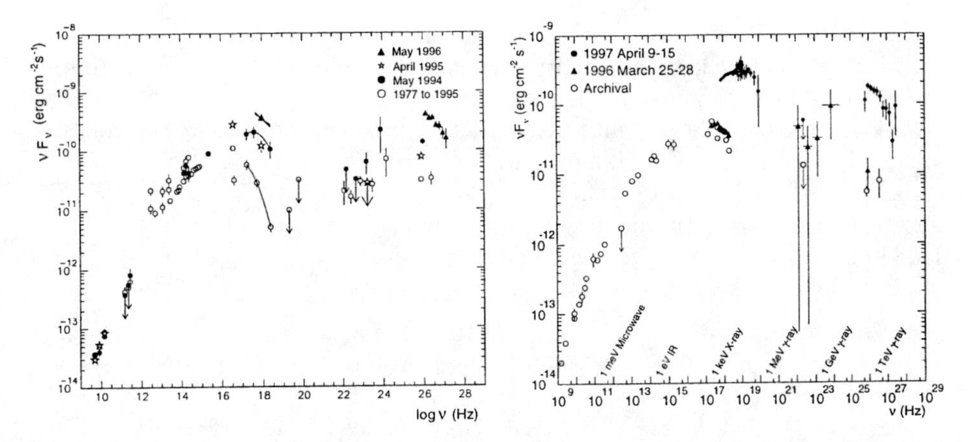

FIGURE 4. Spectral energy distributions of Mrk 421 and Mrk 501 from multi-wavelength campaigns and archival data. Figure adapted from [1].

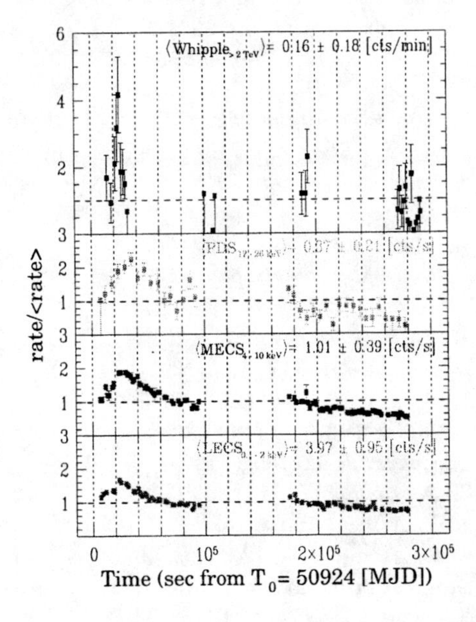

FIGURE 5. The light curve of Mrk 421 from 1998 April 21 to 24. The data are normalized to their mean during the observations (shown in each panel). The errors listed indicate the standard deviation of the data. The Whipple data are for $E > 2\,\mathrm{TeV}$. Figure adapted from [25].

with well-measured spectra and, in the case of the BL Lacs, variability light-curves. There are several unconfirmed sources which lead me to believe that more sources are to be found in this waveband. And, there are some controversies which need resolving (e.g., do the spectra in Mrk 421 and Mrk 501 vary or not?) - which I interpret as a healthy sign of a growing field.

However, it is also clear that many questions remain unanswered. For example, none of the sources show conclusive evidence of cosmic-ray acceleration. Also, we do not know the particle content in blazar jets, nor do we know where the emission spectra of most of the EGRET-detected blazars cut-off. Ground-based efforts, such as VERITAS [26], will dramatically improve the measurements in the VHE band. Combined with the next generation of space-based γ-ray telescopes (e.g., GLAST) and X-ray telescopes like Chandra and Astro-E, many of these questions will hopefully be answered.

REFERENCES

1. Catanese, M., and Weekes, T.C., *Publ. Astron. Soc. Pac.* **111**, 1193 (1999).
2. Ong, R.A., *Physics Reports* **305**, 93 (1999).
3. Muraishi, H., et al., in *Proc. of the 26th ICRC*, ed. D. Kieda, et al. **3**, 500 (1999).
4. Puehlhofer, G., et al., in *Proc. of the 26th ICRC*, ed. D. Kieda, et al. **3**, 492 (1999).
5. Chadwick, P.M., et al., *Astrophys. J.* **503**, 391 (1998).
6. Hartman, R.C., et al., *Astrophys. J. Suppl.* **123**, 79 (1999).
7. Hillas, A.M., et al., *Astrophys. J.* **503**, 744 (1998).
8. Chadwick, P.M., et al., in *Proc. of the 26th ICRC*, ed. D. Kieda, et al. **3**, 504 (1999).
9. Thompson, D.J., in *Neutron Stars and Pulsars*, ed. N. Shibazaki, et al., 273 (1997).
10. Hall, T.A., et al., in *Proc. of the 26th ICRC*, ed. D. Kieda, et al. **3**, 523 (1999).
11. Lessard, R.W., et al., *Astrophys. J.*, in press (1999).
12. Esposito, J.A., et al., *Astrophys. J.* **461**, 820 (1996).
13. Buckley, J.H., et al., *Astron. and Astrophys.* **329**, 639 (1998).
14. Tanimori, T., et al., *Astrophys. J.* **497**, L25 (1998).
15. Nishiyama, T., et al., in *Proc. of the 26th ICRC*, ed. D. Kieda, et al. **3**, 370 (1999).
16. Kataoka, J., et al., *Astrophys. J.* **514**, 138 (1999).
17. Buckley, J.H., et al., *Astrophys. J.* **472**, L9 (1996).
18. Gaidos, J.A., et al., *Nature* **383**, 319 (1996).
19. Quinn, J., et al., *Astrophys. J.* **518**. 693 (1999).
20. Krennrich, F., et al., *Astrophys. J.* **511**, 149 (1999).
21. Aharonian, F.A., et al., *Astron. and Astrophys.*, in press (astro-ph/9905032) (1999).
22. Aharonian, F.A., et al., *Astron. and Astrophys.*, in press (astro-ph/9903386) (1999).
23. Djannati-Atai, A., et al., *Astron. and Astrophys.*, in press (astro-ph/9906060) (1999).
24. Catanese, M., et al., *Astrophys. J.* **487**, L143 (1997).
25. Maraschi, L., et al., *Astrophys. J.*, submitted (1999).
26. Weekes, T.C., et al., *these proceedings* (1999).

The Solar Two 20-300 GeV Gamma-ray Observatory

J.A. Zweerink*, D. Bhattacharya*, G. Mohanty*, U. Mohideen*, A. Radu†, R. Rieben*, V. Souchkov*, H. Tom*, T.O. Tumer*

**IGPP, University of CA-Riverside, Riverside, California 92521*
†Institute of Space Sciences, Bucharest, Romania

Abstract. The Solar Two Gamma-Ray Observatory is designed to close the energy gap between 20-300 GeV that is inaccessible by current instruments, such as the satellite-borne EGRET detector and the ground-based air Cherenkov telescopes. Utilizing the facilities of the Solar Two Power Plant in Barstow, CA, the observatory will detect the Cherenkov light generated as high-energy gamma rays and charged cosmic-ray particles interact with the atmosphere. With over 2000 heliostats available, Solar Two has the largest heliostat mirror area in the world and, thus, the potential to be the most sensitive gamma-ray detector at these energies.

Construction of a secondary mirror system capable of imaging 32 heliostats is nearing completion with plans for the first observations of the Crab Nebula in late November. We report on the design, status and testing of this secondary mirror system including the optics, electronics, and heliostat field.

INTRODUCTION

Many of the questions in gamma-ray astronomy can only be answered by opening the unobserved 20-300 GeV energy window. Three different approaches are being pursued to open this window which is inaccessible by current ground-based and space-borne gamma-ray instruments.

On the space side, GLAST is an instrument in development intended to observe between 20 MeV-300 GeV and is planned for flight in 2005. The ground-based VERITAS experiment is an array of atmospheric Cherenkov telescopes (ACTs) with an estimated energy threshold of ~75 GeV and a planned completion in 2003. A third type of detector uses existing solar power plant facilities to observe 20-300 GeV gamma rays by collecting the light from individual heliostats with photomultiplier tubes (PMTs) placed on the central receiver (Figure 1). While all three types of detectors plan to observe the same energy region, the solar power plant detectors have two advantages over the ACT arrays and satellites: 1.) they will be ready for observations earlier [5,3] and 2.) because they use existing infrastructure, they can be built for a fraction of the ACT array and satellite costs.

CP510, *The Fifth Compton Symposium,* edited by M. L. McConnell and J. M. Ryan

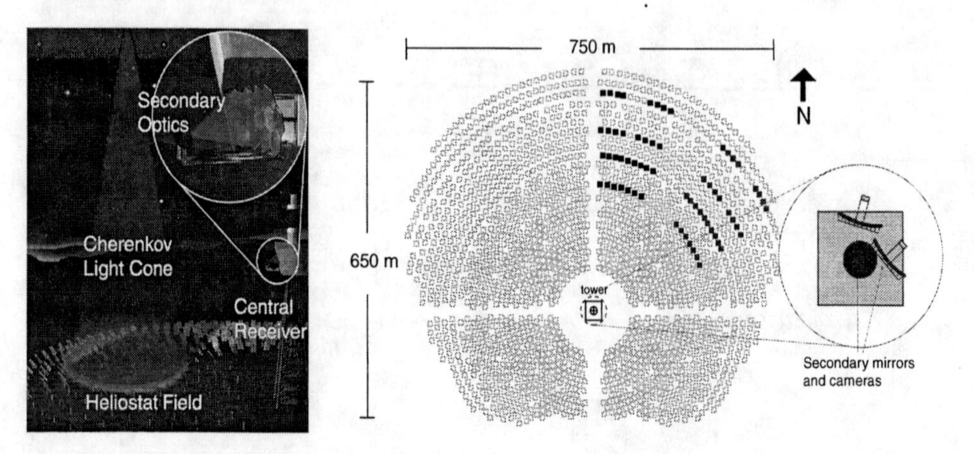

FIGURE 1. Left Conceptual drawing of the solar power plant gamma-ray telescope. The heliostat mirrors on the ground reflect the Cherenkov light to a secondary mirror located on the central receiver tower. The secondary mirror focuses the light from each heliostat onto a separate PMT. **Right** A schematic view of the heliostat field and central tower at Solar Two. The filled squares show the 64 heliostats that will be viewed–the first secondary will view the 32 heliostats on the left. Eventually, we hope to view more heliostats closer to the tower to cover the full 300 m diameter light pool of the Cherenkov shower.

Despite being proposed in 1982 [2], the solar power plant concept has only recently been utilized due to difficulties in overcoming the large background caused by overlapping heliostat images. Tümer [6] solved this background problem by using a secondary optic system on the central receiver to separate the individual heliostat images. Two such gamma-ray detectors under construction have already detected the Crab Nebula with thresholds below 100 GeV–STACEE [5,1] at the National Solar Thermal Test Facility, Sandia National Laboratories and CELESTE [3] at the THEMIS site in France. We are currently building a similar detector at the Solar Two Power Plant in Barstow, CA which is the largest solar power plant in the world with over 2000 heliostats (the next largest solar facility has ∼200 heliostats). By using 64 heliostats, Solar Two will have $> 2600m^2$ of mirror area giving it the largest light collection power of any Cherenkov telescope.

DETECTOR OVERVIEW

Initially, we plan to view 32 heliostats in the NE quadrant at Solar Two (dark filled squares in Fig 1) and expand to 64 heliostats within a year. Eventually, we hope to view ∼250-300 heliostats that uniformly cover a 350 m diameter circle in the NE quadrant so that the complete Cherenkov light pool of the shower will be sampled. Tests described by Ong, et al. [4] demonstrate the feasibility of using the Solar Two site to build the gamma-ray detector described here.

Optics

The secondary mirrors (see Fig 2) used at Solar Two have a 6 m radius of curvature and are made from thirteen 1 m hexagonal facets. The total secondary size is 4.5 m wide by 3.0 m high. Although measurements by ourselves and Ong, et al. [4] confirm that the heliostat spot sizes at the secondary are between 3-5 m, space limitations on the central tower at Solar Two require the secondary mirror to have a height <3 m. The mirror dimensions coupled with the heliostat field geometry insure that each heliostat is viewed with an off-axis angle less than 7°. Since the secondary will be located inside the tower for protection against the elements, a door has been installed in the tower wall that allows the secondary to view the heliostat field.

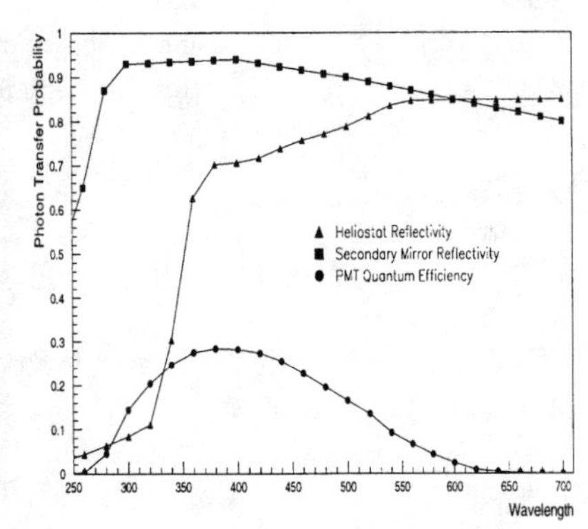

FIGURE 2. The secondary mirror has a radius of curvature R_c=6 meters and is 4.5 meters x 3 meters. Since the secondary is located inside the central receiver tower to protect it from the elements, a door in the tower wall must be opened in order to view the heliostat field. The figure on the right shows the reflectivity of the heliostat and secondary mirrors as well as the quantum efficiency of the Lanco XP2280B photomultiplier tubes.

Although the secondary reduces the heliostat image by a factor of ~100, the image is still larger than the PMT photocathodes. Thus, hollow Winston cones will be used to concentrate the light from the heliostat image onto the photocathode. We chose to use hollow cones over solid ones in the hope of eventually exploiting the UV content of the Cherenkov shower. Consequently, we are researching ways to modify the back-silvered heliostats since they are the only optical component that severely attenuates UV light (See Fig. 2).

Each Winston cone is designed to view a 3 m diameter section of the secondary, thus the system has an f-number$\simeq$1. To reduce the amount of albedo accepted by the cone the entrance apertures of the cones are smaller than the heliostat images as shown in Fig 3. By reducing the cone entrance aperture, the amount of albedo that each PMT receives is cut by a factor of 2, whereas the signal is only 5% less.

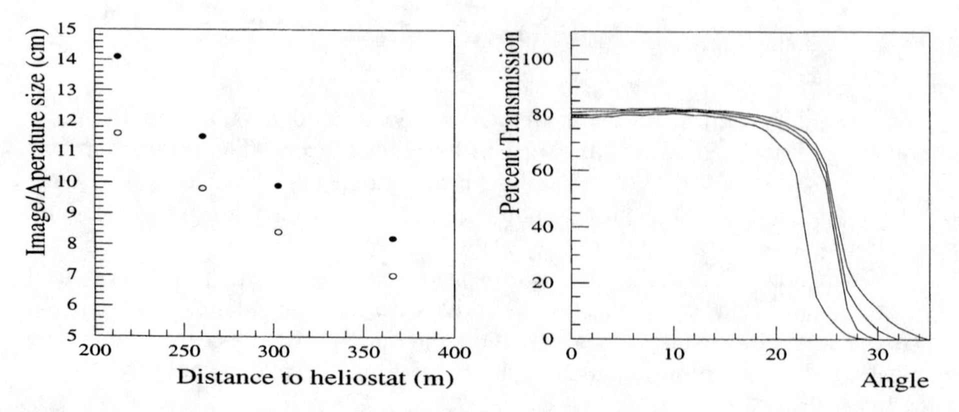

FIGURE 3. The left figure shows how the image size (filled circles) and entrance aperture (open circles) decrease with increasing heliostat distance from the tower. In order to decrease albedo collection, the entrance apertures are smaller than the image size. The right figure shows the transmission through the cones. The cone for the inner row has a smaller cutoff angle since its spot size on the tower can be made smaller than the secondary.

Electronics

Unlike the imaging ACTs, solar power plant detectors rely on accurate reconstruction of the Cherenkov wavefront to discriminate hadrons and gammas. At Solar Two, the signals from each PMT are RC-filtered and amplified 100 times since the gain of the PMTs is $\sim 10^5$. Each signal is individually discriminated and digitally delayed to account for the differing light travel times from the interaction point in the atmosphere to the PMT. Also, each discriminator pulse is time-stamped to a precision $<$1 ns using TDCs so that the wavefront may be accurately reconstructed in software.

After being delayed, the signals are again discriminated and combined in a two level process to form a detector trigger. In this two level process, the field is divided into clusters of 8 heliostats. In the current 32 heliostat configuration, discriminator signals from 5 heliostats in a cluster are required for a first level trigger and 3 of 4 clusters must fire to trigger the whole detector. Eventually, GHz digitizers (Acqiris model DC270) will be used to provide both precise timing and pulse height information from each PMT signal after a trigger is formed.

To have uniform sensitivity across the detector the gains of the different PMTs are needed. We are currently determining the absolute gains by measuring the single photoelectron peak for each PMT by placing the PMT under a low current and digitizing the signals using an oscilloscope which are then stored on disk. The integrated pulse from each trace is binned to find the single photoelectron peak. A method developed by STACEE using a blue LED to measure the relative gains *in situ* is being implemented at Solar Two so that it is possible to measure how the PMT gains change from month to month over the course of the observing season.

Data Acquisition

The digital delays used for this first system are only available in CAMAC, thus, the data acquisition software has been developed on a CAMAC system. A list processor from Hytec is used to reduce the large deadtime that would arise from the numerous CAMAC calls necessary to program the delays, read the scalers, TDCs, and ADCs if each call were done individually. Even so, the deadtime/event is on the order of 30 ms meaning that the detector cannot run at more than 3-4 Hz and have a reasonable deadtime. We have purchased 32 channels of the DC270 fast digitizers from Acqiris which use a compact PCI backplane and have a readout time of ~2.5 ms. We plan to develope a data acquisition system around these modules which would permit detector rates of ~40 Hz for the same deadtime.

CONCLUSIONS

Construction is nearly complete for the first secondary system capable of viewing 32 heliostats at the Solar Two site in Barstow, CA and the various subsystems of the detector are being aligned and calibrated. A CAMAC-based data acquisition system being tested with plans to move to compact PCI. We plan to begin observations of the Crab Nebula in December 1999 or January 2000.

ACKNOWLEDGEMENTS

We are thankful for the grant from the Keck Foundation, without which this work would not be possible. We are immensely grateful for the technical assistance and support provided by the STACEE collaboration. We acknowledge the support and assistance of Southern California Edison in coordinating the design and construction at the Solar Two Power Plant. We thank the advisory committee of our Keck Foundation grant for their guidance in developing the Solar Two detector. We also thank Prof. Steven Ahlen from Boston Univ., Physics Dept. for allowing us to use the ASAP software.

REFERENCES

1. Chantell, M., et al., *Nucl. Inst. Meth A*, **408**, 468 (1998).
2. Danaher, S., et al., *Solar Energy*, **28**, 335 (1982).
3. de Naurois, M., et al., Proc. 26th ICRC. **5**, 211 (1999).
4. Ong, R.A., et al., *Towards a Major Atmospheric Cherenkov Detector IV* (Padova), 295 (1995).
5. Oser A., et al., Proc. 26th ICRC. **3**, 464 (1999).
6. Tümer, T., et al., Proc. 22nd ICRC **2**, 635 (1991).

The Solar Tower Atmospheric Cherenkov Effect Experiment (STACEE): New Results at 100 GeV

C.E. Covault[1], D. Bhattacharya[4], L. Boone[3], M.C Chantell[1], Z. Conner[1], M. Dragovan[1], D. Gingrich[7] D. Gregorich[6] D.S. Hanna[2], G. Mohanty[4], R. Mukherjee[5], R.A. Ong[1], S. Oser[1], K. Ragan[2], R.A. Scalzo[1], C.G. Théoret[2], T.O. Tumer[4], D.A. Williams[3], J.A. Zweerink[4],

[1]*Enrico Fermi Institute, University of Chicago, Chicago, IL 60637, USA*
[2]*Department of Physics, McGill University, Montreal, Quebec H3A 2T8, Canada*
[3]*Santa Cruz Institute for Particle Physics, Univ. of California, Santa Cruz, CA 95064, USA*
[4]*Institute of Geophysics and Planetary Physics, Univ. of California, Riverside, CA 92521, USA*
[5]*Dept. of Physics & Astronomy, Barnard College & Columbia Univ., New York, 10027, USA*
[6]*Department of Physics and Astronomy, California State Univ., Los Angeles, CA 90032, USA*
[7]*Department of Physics, University of Alberta, Edmonton, Alberta, Canada*

Abstract. The Solar Tower Atmospheric Cherenkov Effect Experiment (STACEE) is a new ground-based instrument for observing astrophysical sources of gamma-rays in the energy range from 50 to 250 GeV. The first phase of STACEE, using 32 large heliostat mirrors, was completed in the fall of 1998. We describe the STACEE operations and observations during the 1998-1999 winter observing season. We have detected the Crab Nebula with high significance ($\sim 7\sigma$). This result demonstrates that the STACEE concept is sound and that we can expect to make sensitive measurements of gamma-ray sources at energies below 100 GeV. The full STACEE instrument, with 64 heliostats, will be completed during the year 2000. The first three years of observations with the complete instrument will include a range of sources with an emphasis on Active Galactic Nuclei (AGN) and supernova remnants.

INTRODUCTION

One of the last unexplored regions of the electromagentic spectrum is the energy range from 20 to 250 GeV corresponding to the energy region above the reach of the EGRET detector and below the energy threshold of current very high energy (VHE) atmospheric Cherenkov detectors. Astrophysical sources of interest in this energy range include galactic sources, such as gamma-ray plerions/pulsars, and extragalactic sources such as EGRET flat-spectrum blazars and BL Lac type active

CP510, *The Fifth Compton Symposium,* edited by M. L. McConnell and J. M. Ryan

galactic nuclei (AGN). Measurements of the energy spectra of AGN in this energy region are crucial in understanding the mechanism for production of gamma-rays, which are believed to result from relativistic jets coming from the central black hole. Expected spectral cutoffs in this energy region could indicate absorption of gamma-rays by the intergalactic infrared radiation field [1,2]. Gamma-rays are also expected from supernova remnants (SNR) and gamma-ray pulsars and the associated nebula. Measurements of gamma-ray from these sources should provide critical information on their particle acceleration and emission processes. In the most widely accepted models for gamma-ray emission from the Crab Nebula, the transition from a synchrotron-dominated flux to an inverse- Compton-dominated flux occurs in the few GeV energy range.

THE STACEE INSTRUMENT

The STACEE experiment has been described in detail elsewhere [3,4]. STACEE collects Cherenkov light from gamma-ray air showers using an array of "heliostat" solar-collector mirrors at the National Solar Thermal Test Facility (NSTTF), Sandia National Laboratories, Albuquerque, NM. The facility is used during daylight hours for solar power energy research. We use this facility at night for gamma-ray astronomy. Currently STACEE is using 32 of the 212 heliostats, each with a collection area of 37 m^2, with plans to expand to 64 heliostats in the near future (See Figure 1).

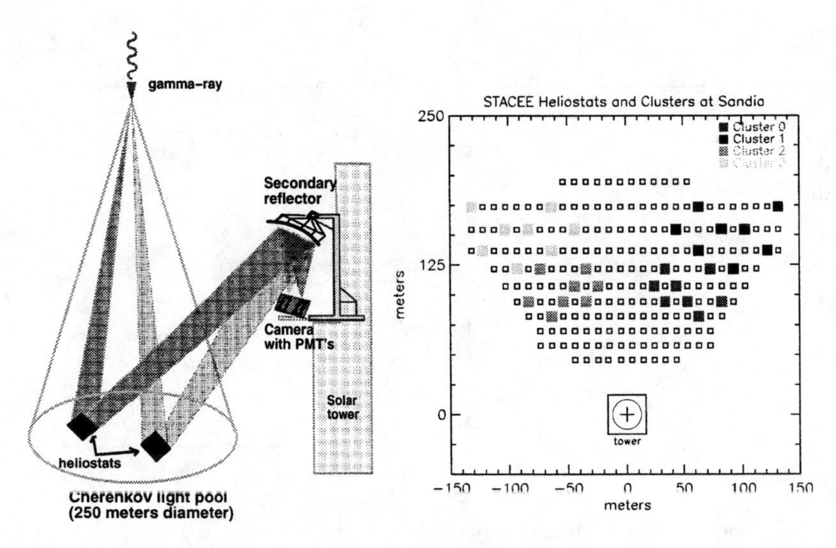

FIGURE 1. Left: STACEE concept. The large heliostat mirrors (37 m^2) collect Cherenkov light and direct it to a secondary mirror on the tower. **Right:** Plan view of NSTTF heliostat field showing 32 heliostats used for observations during 1998/1999.

The heliostats focus Cherenkov light onto secondary mirrors located at the top of a central tower; these in turn reflect the light onto an array of photomultiplier tubes (PMT) positioned within a camera box. Each PMT collects light from a single heliostat. A trigger is formed from a narrow time coincidence of several discriminated phototube signals after appropriate time-of-flight delays are applied. Accurate pulse timing is used to reconstruct the arrival direction of the primary, while pulse height measurements are used to estimate the primary's energy.

CALIBRATION AND INSTRUMENT PERFORMANCE

The arrival direction of each shower is determined from the measured times at which the Cherenkov shower front passes each heliostat, so careful calibration (< 1 ns) of STACEE's timing apparatus is important. This has been accomplished using a combination of LED and laser calibration systems, and geometrical surveying of the optical components. Variations in timing with pulse height ("time slewing") have also been measured via laser calibration. RMS timing residuals of about 1 ns lead us to expect an angular resolution of 0.25 degrees (Figure 2). An independent estimate of this accuracy, in which the STACEE heliostats are divided into two sub-arrays and the angular difference between the reconstructed direction from the two sub-arrays is calculated, confirms this expectation. After all these corrections are applied, the residual pointing bias is less than 0.1°..

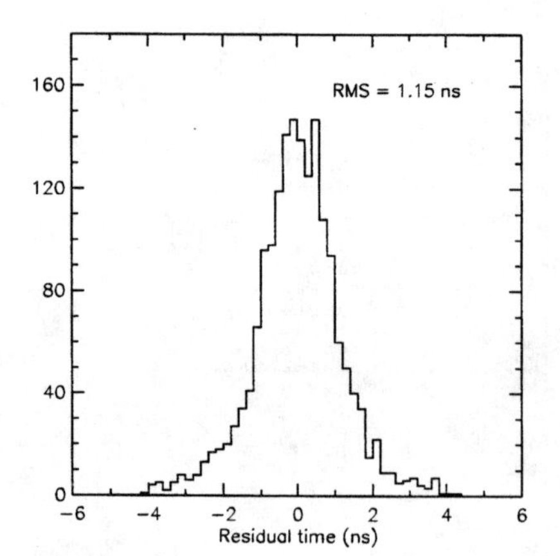

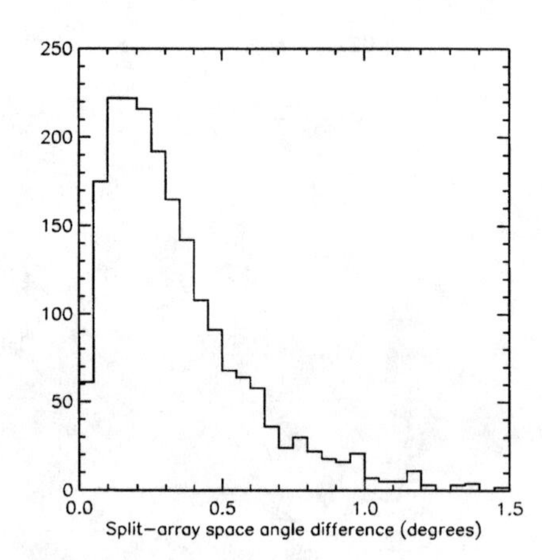

FIGURE 2. Left: Timing residuals for one STACEE channel relative to the best fit spherical wavefront. **Right:** Distribution of space-angle differences between overlapping sub-arrays for cosmic rays detected by STACEE. The median difference of about 0.2–0.3 degrees provides an independent estimate of STACEE's angular resolution.

DETECTION OF THE CRAB NEBULA AT 100 GEV

The Crab Nebula has been the primary astrophysical source observed with STACEE during the period November 15, 1998 through February 18, 1999. Data are collected during fixed intervals (typically 28 minute runs), alternating between observations tracking the Crab (on-source) and observations tracking a position 30 minutes to the east or west of the Crab (off-source). On-source and Off-source data are paired in such a way to obtain equal time on and off source. After run cuts which removed pairs affected by poor weather, poor atmospheric clarity or detector malfunctions, 113 "on-Crab/off-Crab" pairs of runs remained, with total on-Crab time of about 50 hours. Events were reconstructed by re-imposing the trigger in software using a short (12 ns) coincidence window, after which the arrival times were fit to a spherical wavefront with an assumed fixed core location to determin the primary's arrival direction. An excess in on-source events over off-source events is interpreted as a positive signal. The total significance of the excess before fitting the wavefront is approximately 7 sigma. The significances of the pairs within the data set are normally distributed with a standard deviation of 1, as is expected for a gamma-ray excess from a constant-flux source. We estimate the mean energy of detected gamma-rays from the Crab to be approximately 100 GeV. Further analyses to obtain a flux spectrum of the Crab are underway.

FUTURE PLANS

The STACEE instrument is still under construction. Data taken during 1998-1999 used a partially complete version of the experiment consisting of 32 heliostats in the field. During the year 2000, we will complete the construction of STACEE to the full experiment, upgrading to 64 heliostats (Figure 3) which will concentrate Cherenkov light onto five secondary mirrors.

We will also upgrade the electronics which will include a custom-designed trigger/delay and a new set of of high speed (1GHz) waveform digitizers. STACEE will use the Acqiris DC270 waveform digitizer, which is a 4-channel compact-PCI (cPCI) board. In comparison with STACEE-32, our simulations predict an increase in the trigger rate to $\sim 15\,\mathrm{Hz}$ for STACEE-64, and an improvement in gamma-ray sensitivity by a factor of five. We are also installing new IR monitors for detecting water vapor and an automated telescope/photometer for measuring sky clarity during observations.

By mid-2000, the construction of STACEE-64 will be completed and we will carry out several months of tests and calibrations to shake down the experiment. At this point, a three-year program of continuous astrophysical observation will begin. The majority of targets planned during this interval will be AGN which will be observed over a range of redshift values. We will also observe selected pulsars, SNR, and EGRET unidentified objects that are easily visible at the STACEE site.

ACKNOWLEGEMENTS

We are grateful to the staff at the NSTTF for their excellent support. This work was supported in part by the National Science Foundation, the Natural Sciences and Engineering Research Council, FCAR, and the California Space Institute. CEC and RM are Cottrell Scholars of Research Corporation.

REFERENCES

1. For a summary of the scientific motivations for the field of gamma-ray astronomy, see for example, Ong, R.A., *Physics Reports*, **305**, 93-202 (1998).
2. Mukherjee. R. et al. *Proc. 26th ICRC (Salt Lake City, 1999)* OG 2.1.19 (1999).
3. Chantell, M.C. et al. *Nucl. Instr. Meth. A* **408**, 468 (1998).
4. The STACEE Website contains all published papers and conference proceedings: http://hep.uchicago.edu/~stacee
5. Ong, R.A. *Towards a Major New Cherenkov Detector: A Workshop on GeV and TeV Gamma-Ray Astronomy*, Snowbird, Utah 1999 (in press) (2000). See also Oser, S. et al. *Proc. 26th ICRC (Salt Lake City, 1999)* OG 2.2.07 (1999).

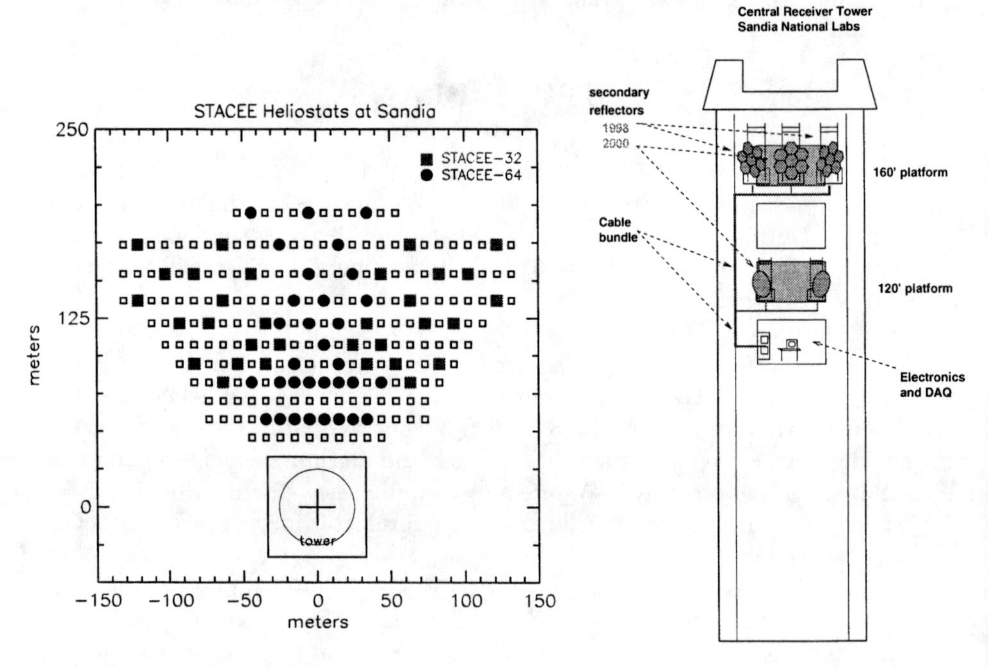

FIGURE 3. Left: Plan view of the Sandia heliostat field. The full experiment (STACEE-64) will use all shaded heliostats. (The coordinates are XY positions in meters.) **Right:** View of Sandia tower.

The Very Energetic Radiation Imaging Telescope Array System (VERITAS)

T.C. Weekes[1], S.M. Bradbury[2], I.H. Bond[2], A.C. Breslin[3], J.H. Buckley[4], D.A. Carter-Lewis[5], M. Catanese[1], B.L. Dingus[6], D.J. Fegan[3], J.P. Finley[7], J. Gaidos[7], J. Grindlay[1], A.M. Hillas[2], G. Hermann[8], P. Kaaret[1], D. Kieda[6], J. Knapp[2], F. Krennrich[5], S. LeBohec[5], R.W. Lessard[7], J. Lloyd-Evans[2], D. Müller[8], R. Ong[8], J. Quinn[3], H.J. Rose[2], M. Salamon[6], G.H. Sembroski[7], S. Swordy[8], V.V. Vassiliev[5]

[1] *Harvard-Smithsonian CfA, Amado, AZ 85645-0097, U.S.A.*
[2] *University of Leeds, Leeds, LS2 9JT, U.K.*
[3] *University College, Dublin, Ireland*
[4] *Washington University, St Louis, MO 63130, U.S.A.*
[5] *Iowa State University, Ames, IA 50011, U.S.A.*
[6] *University of Utah, Salt Lake City, UT 84112, U.S.A.*
[7] *Purdue University, West Lafayette, IN 47907, U.S.A.*
[8] *University of Chicago, Chicago, IL 60637, U.S.A.*

Abstract.
An overview of the current status and scientific goals of VERITAS, a proposed hexagonal array of seven 10 m aperture imaging Cherenkov telescopes, is provided. The selected site is close to Mt. Hopkins, the site of the Whipple Observatory, in Arizona. Each telescope, of 12 m focal length, will be equipped with a 499 element photomultiplier camera covering a 3.5 degree field of view. A central station will initiate the readout of 500 MHz FADCs upon receipt of multiple telescope triggers. The minimum detectable flux sensitivity will be 0.5% of the Crab Nebula flux at 200 GeV. VERITAS will operate primarily as a γ-ray observatory in the 50 GeV to 50 TeV range for the study of active galaxies, supernova remnants, pulsars and gamma-ray bursts.

I INTRODUCTION:

The present status of the atmospheric Cherenkov imaging technique has been reviewed by Ong (1998) and Catanese and Weekes (1999); its major contribution to ground-based VHE astronomy has led to plans for "next generation" instruments of increased collection area and complexity, one of the most ambitious of which is

CP510, *The Fifth Compton Symposium,* edited by M. L. McConnell and J. M. Ryan

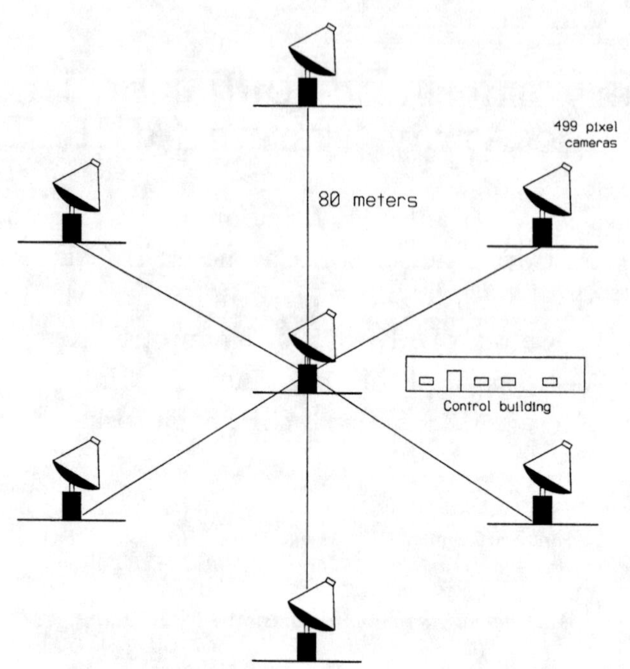

FIGURE 1. The seven telescopes of VERITAS, each of 10m aperture, will have the hexagonal distribution shown (Weekes et al. (1999).

the Very Energetic Radiation Imaging Telescope Array System (VERITAS), first proposed to the Smithsonian Institution in 1996. The design study has culminated in a detailed proposal by Weekes et al. (1999) to build an array of 10m aperture Cherenkov telescopes in southern Arizona. The chosen site is a topographically flat, dark site, at 1390 m a.s.l., close to the Whipple Observatory which will provide the necessary infrastructure.

VERITAS will consist of six telescopes located at the corners of a hexagon of side 80 m with a seventh at the center (Figure 1). The telescopes' structure will be similar to that of the Whipple 10m reflector, which has withstood mountain conditions for over thirty years. By employing largely existing technology in the first instance and stereoscopic imaging, the power of which has recently been demonstrated by HEGRA (Daum et al., 1997), we expect VERITAS to achieve the following:

- *Effective area*: $\gtrsim$0.1 km^2 at 1 TeV.
- *Effective energy threshold*: $\lesssim$100 GeV with significant sensitivity at 50 GeV.
- *Energy resolution*: 10% - 15% for events in the range 0.2 to 10 TeV.
- *Angular Resolution*: $\lesssim$0.05° for individual photons; source location to better than 0.005°.

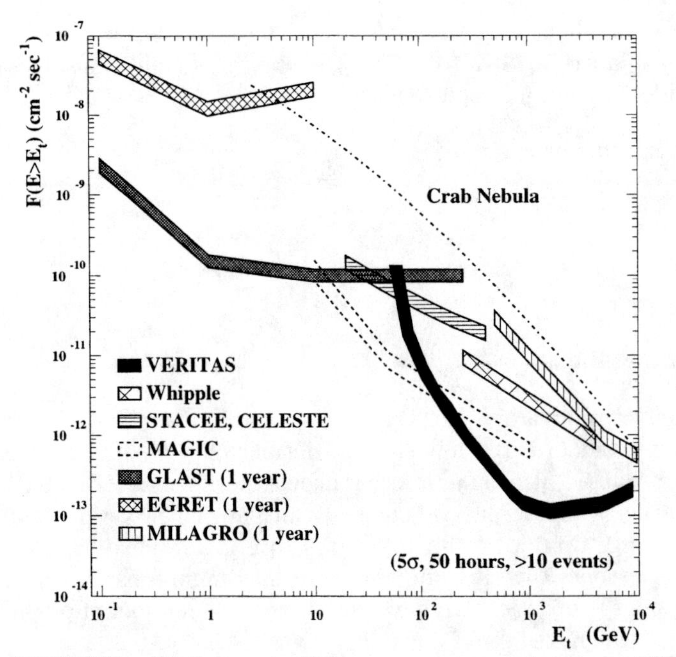

FIGURE 2. VERITAS sensitivity to point-like sources as compared to those of Whipple, MAGIC, CELESTE, STACEE, GLAST and MILAGRO (Weekes et al. (1999) and references therein)

The performance of VERITAS is perhaps best summarised by its flux sensitivity versus energy, shown in Figure 2 for an object of spectrum $dN/dE \propto E^{-2.5}$. Here we define a minimum detectable flux for VERITAS as that giving a 5σ excess of γ-rays above background (or 10 photons where the statistics become Poissonian). We expect to detect sources which emit at levels of 0.5% of the Crab Nebula at energies of 200 GeV in 50 hours of observation. VERITAS, together with the southern hemisphere Cherenkov telescope arrays HESS and CANGAROO-IV, will obtain high sensitivity in the 100 GeV to 10 TeV range between the ranges covered by space-borne instruments and air shower arrays. If sources of EHE cosmic rays are discovered by HiRes and the Auger experiments, Cherenkov telescopes such as VERITAS may further localise and identify them. Also, the MILAGRO wide-field water Cherenkov detector and the Tibet Air Shower Array will be sensitive to transient sources, which, once detected can be studied in more detail by VERITAS.

II SCIENTIFIC OBJECTIVES:

At a capital cost of ~$16 M, less than 10% of that of the Gamma-ray Large Area Space Telescope (GLAST), VERITAS will be an excellent investment in terms of scientific return. The scientific objectives of VERITAS and GLAST are remarkably similar and are complementary in many ways. The two experiments will have significant sensitivity in the 50 GeV to 200 GeV energy range and thus cross calibration will be possible. The major topics will be:

- Supernova remnants
- Pulsars
- AGNs
- Galactic Plane
- Gamma Ray Bursts

The large effective area of VERITAS ($\gtrsim 2.5 \times 10^4\,\mathrm{m}^2$ at 200 GeV) will allow accurate measurements of extremely short variations in γ-ray flux of strong sources. Unlike other proposed atmospheric Cherenkov experiments, VERITAS will be extremely versatile. With seven telescopes it can be used in a variety of configurations. These range from all seven telescopes covering seven different areas of the sky to seven telescopes covering one object for maximum sensitivity. Sub-arrays of 3 and 4 telescopes are particularly powerful. Thus, the number of possible scientific programmes is maximized.

III DESIGN:

The telescopes will be constructed following the Davies-Cotton reflector design with spherical and identical facet mirrors of float glass (slumped and polished, aluminised and anodised) to provide the optimum combination of optical quality and cost effectiveness. A study is underway to design a stress-free mounting scheme for the hexagonal (60 cm flat to flat) mirrors. The time spread in light across the proposed $f/1.2$ reflector is only 3-4 ns and 100% of the light from a point source is captured by a 0.125° pixel out to 1° from the optical axis, decreasing to 72% at 2°. It is then possible to match the inherent angular fluctuations in the shower images with a camera that has a reasonable number of pixels (499) and a field of view (3.5°) which is large enough to conduct surveys and observe extended ($\leq 1°$) sources efficiently. For the optical support structure of trussed steel on a commercially available pedestal, effects due to gravitational slumping during slewing will be less than 2.2 mm on the camera face. Slew speed will be as high as 1° per second on both axes.

At present, the need for a low noise, high gain ($> 10^6$), photon counting detector, with risetimes of a few ns is satisfied only by photomultipliers (PMTs). The Hamamatsu R7056, with a bialkali photocathode, UV glass window, and 25 mm active diameter meets our requirements. The collection efficiency will be increased by Winston cones. The spacing between the PMTs will correspond to a focal plane angular distance of 0.15°. A modular high voltage supply (e.g. LeCroy 1458) will be used where each PMT has a separately programmable high voltage, supplied from a system crate located at the base of the telescope.

Each PMT signal will be taken via a linear amplifier in the focus box to a custom CFD/scaler module. The targeted gain of the PMT plus amplifier (based on a standard 1GHz bandwidth integrated circuit chip, AD8009) provides a signal level of ~2 mV/photoelectron. Optical fiber signal lines may be an attractive alternative to RG58 coaxial cabling allowing the CFD and downstream electronics to be located in the central control building. A prototype multi-channel analog optical fiber system is now being installed on the Whipple 10 m telescope.

The CFD/scaler board, incorporating an adjustable channel by channel delay, will provide an analog fanout of the PMT signal to an FADC system. A prototype 500 MHz FADC system, each channel using a commercially available 8-bit FADC integrated circuit, has been developed and tested successfully on the Whipple telescope (Buckley et al. 1999).

VERITAS will operate at a minimal threshold by requiring a time coincidence of adjacent pixel signals to form a single telescope trigger and $>$ n coincident telescope triggers to initiate data recording. For example, at a threshold of 5 photoelectrons the CFD trigger rate of a single pixel will be ~300 kHz. A telescope trigger will then require a coincidence of ≥ 3 *neighboring* pixel signals. This topological trigger will be similar to that used on the Whipple telescope (e.g. Bradbury et al. 1999). Telescope triggers will be received at a central station where they can be used to immediately initiate a telescope readout, or delayed to account for orientation of the shower front (e.g. by a CAEN V486 digital delay) and combined in a more complex trigger requirement. For example, if an array trigger requires that 3 of 7 telescopes trigger within a 40 ns coincidence window then the *accidental* array trigger rate is $<$ 1Hz at the 5 photoelectron trigger threshold.

IV CONCLUSION:

Our aim is to commission the first VERITAS telescope in 2001, then two more per year until 2004 when the full array will come on-line. Hence VERITAS will be in a good position to complement AGILE in the interim before GLAST and to be in routine operation by the launch of GLAST in 2005.

V REFERENCES

Bradbury, S.M. et al. 1999, Proc. 26th ICRC (Salt Lake City, 1999), 4, 263
Buckley, J.H. et al. 1999, Proc. 26th ICRC (Salt Lake City, 1999), 4, 267
Catanese, M, Weekes, T.C., 1999, Publ. Ast. Soc. Pac. 111, 1193
Daum, A. et al. 1997, Astrop. Phys., 8, 1
Ong, R.A. 1998, Physics Reports, 305, 93
Weekes, T.C. et al. 1999, VERITAS, proposal submitted to the U.S. D.O.E.

Milagro: A TeV Gamma-Ray Monitor of the Northern Hemisphere Sky

B.L. Dingus[1], R. Atkins[1], W. Benbow[2], D. Berley[3,10], M.L. Chen[3,11], D.G. Coyne[2], D.E. Dorfan[2], R.W. Ellsworth[5], D. Evans[3], A. Falcone[6], L. Fleysher[7], R. Fleysher[7], G. Gisler[8], J.A. Goodman[3], T.J. Haines[8], C.M. Hoffman[8], S. Hugenberger[4], L.A. Kelley[2], I. Leonor[4], M. McConnell[6], J.F. McCullough[2], J.E. McEnery[1], R.S. Miller[8,6], A.I. Mincer[7], M.F. Morales[2], P. Nemethy[7], J.M. Ryan[6], B. Shen[9], A. Shoup[4], C. Sinnis[8], A.J. Smith[9], G.W. Sullivan[3], T. Tumer[9], K. Wang[9], M.O. Wascko[9], S. Westerhoff[2], D.A. Williams[2], T. Yang[2], G.B. Yodh[4]

(1) University of Utah, Salt Lake City, UT 84112, USA
(2) University of California, Santa Cruz, CA 95064, USA
(3) University of Maryland, College Park, MD 20742, USA
(4) University of California, Irvine, CA 92697, USA
(5) George Mason University, Fairfax, VA 22030, USA
(6) University of New Hampshire, Durham, NH 03824, USA
(7) New York University, New York, NY 10003, USA
(8) Los Alamos National Laboratory, Los Alamos, NM 87545, USA
(9) University of California, Riverside, CA 92521, USA
(10) Permanent Address: National Science Foundation, Arlington, VA ,22230, USA
(11) Now at Brookhaven National Laboratory, Upton, NY 11973, USA

Abstract. A new type of very high energy (> a few 100 GeV) gamma-ray observatory, Milagro, has been built with a large field of view of > 1 steradian and nearly 24 hours/day operation. Milagrito, a prototype for Milagro, was operated from February 1997 to May 1998. During the summer of 1998, Milagrito was dismantled and Milagro was built. Both detectors use a 80 m x 60 m x 8 m pond of water in which a 3 m x 3 m grid of photomultiplier tubes detects the Cherenkov light produced in the water by the relativistic particles in extensive air showers. Milagrito was smaller and had only one layer of photomultipliers, but allowed the technique to be tested. Milagrito observations of the Moon's shadow and Mrk 501 are consistent with the Monte Carlo prediction of the telescopes parameters, such as effective area and angular resolution. Milagro is larger and consists of two layers of photomultiplier tubes. The bottom layer detects penetrating particles that are used to reject the background of cosmic-ray initiated showers.

CP510, *The Fifth Compton Symposium,* edited by M. L. McConnell and J. M. Ryan
 1-56396-932-7/00/$17.00

I A NEW TYPE OF TEV γ-RAY OBSERVATORY

Several active galactic nuclei and supernova remnants have now been observed to emit TeV gamma-rays. [1], [2], [3] These TeV observations were made by detecting the atmospheric Cherenkov light produced by the extensive air shower of particles created when a high energy gamma ray interacts in the atmosphere. Due to the limitations of this detection technique, these observatories only operate on clear, moonless nights and must point at the source using a large mirror. Therefore, observations of unpredictable, short duration transients, such as gamma-ray bursts, and all sky surveys are difficult.

A new type of TeV γ-ray observatory with a large field of view and continuous operation has recently been built in the Jemez Mountains near Los Alamos, NM (106.7^o W, 35.9^o N, 2650 m above sea level). The observatory is called Milagro, and the prototype which operated at the same site from February 1997 to May 1998 is called Milagrito [4].

Both detectors used the large pond of water 80 m x 60 m x 8 m which can be seen in the photograph of Figure 1. The pond has a light-tight cover. Milagro contains 723 photomultiplier tubes (PMTs) which are placed on a 3 m x 3 m grid in 2 layers at 1.5 m and 7 m below the surface. The prototype Milagrito had only one layer of 228 PMTs on a 3 m x 3 m grid spread over the smaller area of 30 m x 50 m.

FIGURE 1. Aerial photograph of the 60 m x 80 m x 8 m pond and cover used by Milagro and Milagrito. The pond as instrumented for Milagro contains 5 million gallons of water.

An extensive air shower is detected when the relativistic particles radiate Cherenkov light in the water causing several tens of PMTs to observe the light within a few 100 nsec of each other. From the relative timing of the photomultiplier tube signals, the direction of the particle or gamma-ray initiating the shower can be determined to $\sim$1 degree depending on the number of PMTs hit. The field of view is such that 50% of the showers detected are within 20 degrees of zenith and

90% are within 50 degrees. Almost all of these triggers are due to the background of showers that are initiated by charged cosmic rays. Monte Carlo simulations correctly predict the observed rate and zenith angle distribution of cosmic-ray initiated showers. Simulations of γ-ray initiated showers show sensitivity to γ-rays as low as ~100 GeV with the effective area increasing as E $^{\sim 2}$, where E is the γ-ray energy, and flattening near 10 TeV.

II MILAGRITO OBSERVATIONS

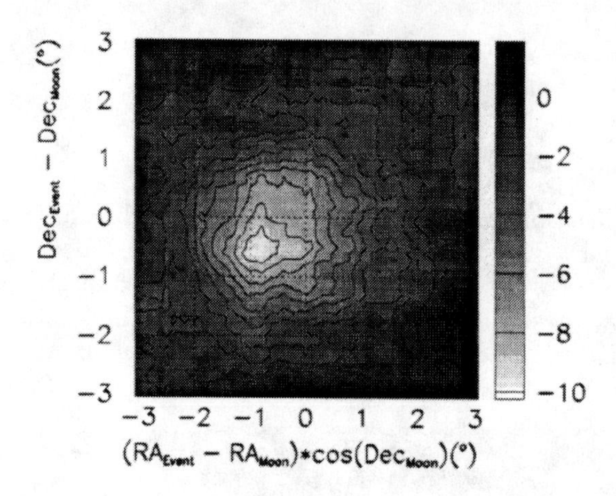

FIGURE 2. The shadow of the Moon due to the blockage of cosmic rays. The scale is in standard deviations of the Gaussian distributed background. The deficit is not located at the direction to the Moon, but is deflected in R.A. because the trajectories of the charged cosmic rays are bent by the Earth's magnetic field.

The performance of Milagrito has been confirmed by observations. The Moon blocks cosmic rays and a deficit of showers has been detected, which is a 10 sigma deviation from the background (Figure 2) [5]. The shape and size of the deficit is consistent with the Moon's angular size and Milagrito's angular resolution, and the deflection in Right Ascension is consistent with the Earth's magnetic field and the energy of the cosmic rays detected by Milagrito.

The simulations of gamma-ray initiated showers were verified by Milagrito observations of Mrk501, an x-ray selected BL Lac, which was a bright TeV source during 1997. The Milagrito detection of Mrk501 shown in Figure 3 was a 3.7 sigma deviation from the cosmic-ray background [6]. The flux and spectrum were well measured by several atmospheric Cherenkov telescopes and the significance of the Milagrito detection agrees with this spectrum folded with the Milagrito effective area. Simulations also indicate that the sensitivity of Milagrito was too low to detect the Crab nebula, the standard candle of TeV astronomy.

New observations have been performed by Milagrito that atmospheric Cherenkov telescopes have not been able to do because of their low duty cycle (5-10%) and small field of view (2-4 degrees in radius). Specifically, an all sky survey of the Northern Hemisphere was performed and no sources brighter than 5 times the Crab nebular flux were detected [7]. An all-sky search for 10 second duration transients was also done and none were found [8], which can place limits on the local density of evaporating primordial black holes. However, the most interesting results have

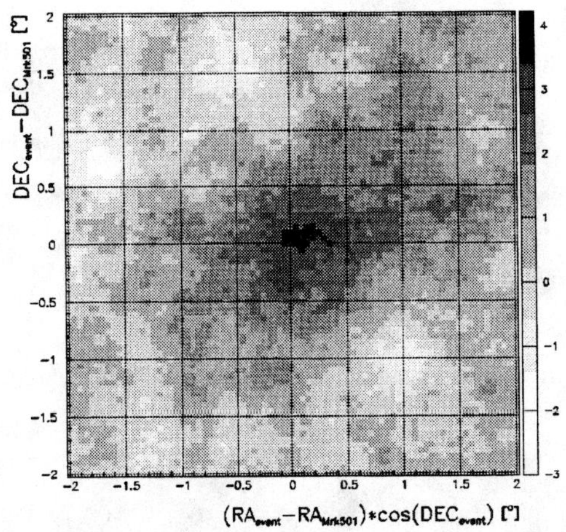

FIGURE 3. The number of showers in the vicinity of Mrk501 plotted in standard deviations of the background. For each position the number of showers is determined for directions within one degree of that position; thus neighboring bins are highly correlated. The position of Mrk 501 is at the center of the plot. The excess at Mrk 501 is 3624±990 events, or 3.7 σ.

been from the search for TeV emission correlated with BATSE detected bursts [9] and the detection of a coronal mass ejection from the Sun [10]. These analyses were both reported separately at this conference.

III MILAGRO EXPECTATIONS

Simulations and preliminary data indicate that Milagro will be more than 5 times more sensitive than Milagrito. The improvement comes from a combination of factors – larger area, improved angular resolution, and cosmic-ray background rejection. Milagro will be fully operational Fall 1999 with a data taking rate of 2000 showers per second resulting in more than 100 GBytes of data per day.

REFERENCES

1. M. Catanese & T.C. Weekes, *Astronomical Society of the Pacific* (1999) (in press)
2. C.M. Hoffman, C. Sinnis, P. Fleury, M. Punch *Rev. of Mod. Phys.* 71(4), 897 (1999)
3. R. Ong *Physics Reports* 305, 93 (1998)
4. R. Atkins et al., submitted to *Nucl. Inst. and Methods* (2000)
5. M. Wascko et al., *26th ICRC Proceedings* (Salt Lake City,UT) (1999)
6. R. Atkins et al., *Ap. J. Lett.* **525**, L25 (1999).
7. A.J. Smith et al. *AIP Proc. GeV-TeV Gamma-Ray Astrophysics– Toward a Major Atmospheric Cherenkov Telescope VI* (Snowbird, UT) (2000)
8. C. Sinnis et al., *26th ICRC Proceedings* (Salt Lake City,UT) (1999)
9. R. S. Miller et al., *AIP Proc. 5th Compton Symp.* (Portsmouth,NH) (2000)
10. A. Falcone et al., *AIP Proc. 5th Compton Symp.* (Portsmouth,NH) (2000)

PACHMARHI ARRAY OF ČERENKOV TELESCOPES

P. N. Bhat, B. S. Acharya, V. R. Chitnis, P. Mazumdar, M. A. Rahman, M. R. Krishnaswamy and P. R. Vishwanath

Tata Institute of Fundamental Research, Homi Bhabha Road, Mumbai 400 005, India

Abstract. Pachmarhi Array of Čerenkov Telescopes (PACT) has been designed to search for celestial TeV $\gamma-$ rays using the wavefront sampling technique. PACT consists of 25 telescopes deployed over an area of 80 $m \times 100$ m. A telescope consists of 7 parabolic reflectors, each viewed by a fast phototube behind a 3° mask at the focus. The density and arrival time of photons at the PMT are measured. The energy threshold and collection area of the array are estimated, from Monte Carlo simulations, to be ~ 1 TeV and 10^5 m^2 respectively. The accuracy in determination of arrival angle of a shower was estimated to be 0.1° in the near vertical direction. About 99% of the off-axis hadronic events could be rejected from directional information alone. Further, nearly 75% of the on-axis hadronic events could be rejected using species sensitive measurements like the photon density fluctuations. These cuts on data to reject background would retain $\sim$ 44% of the $\gamma-$ray signal. The sensitivity of the array for a 5σ detection of $\gamma-$ray signal has been estimated to be $\sim 4.1 \times 10^{-12}$ $photons$ cm^{-2} s^{-1}, for an on source exposure of 50 hrs.

INTRODUCTION

Atmospheric Čerenkov Technique is the only viable technique at present for the detection of celestial $\gamma-$rays of very high energy ($\sim$ TeV). It is based on the effective detection of Čerenkov light emitted by the secondary particles produced in the extensive air showers initiated by the primary $\gamma-$rays. Cosmic rays are a major source of background which are more abundant in number compared to $\gamma-$rays themselves. By reconstructing the various features of Čerenkov light pool the cosmic ray initiated showers, which are hadronic in nature, could be rejected. Second generation telescopes have achieved significant improvement in the sensitivity for the detection of $\gamma-$rays by rejecting cosmic ray background considerably [1–3]. Čerenkov imaging technique has been successfully demonstrated to be an efficient method of rejecting more than 99.9% of cosmic rays (see [4] for a review). Angular imaging and spatial sampling are two complimentary ways to examine the Čerenkov light pool. Simulation studies have shown that wavefront sampling technique could also distinguish between electromagnetic and hadronic cascades [5–7].

CP510, *The Fifth Compton Symposium,* edited by M. L. McConnell and J. M. Ryan

In the following we shall describe the status of an experiment based on the wavelength sampling technique for the detection of very high energy celestial $\gamma-$rays. Its performance is evaluated and compared with other similar experiments.

THE EXPERIMENTAL SET UP

Pachmarhi Array of Čerenkov Telescopes consists of a spaced array of 25 telescopes deployed in an area of about $80m \times 100m$ at the High Energy Gamma Ray Observatory (HEGRO) Pachmarhi (latitude = 22° 28′ N and longitude = 78° 28′ E and altitude = 1075 m). The separation between neighbouring telescopes in the N-S direction is 25 m while it is 20 m in the E-W direction. The layout of the telescopes is shown in Figure 1.

Each telescope consists of 7 paraxial parabolic mirrors ($f/d \sim 1$) of diameter 0.9 m mounted on a single equatorial mount. The reflectors are fabricated locally and their image size is < 1° FWHM. A fast phototube (EMI 9807B) is mounted at the focus of each mirror defining a field of view of $\sim \pm$ 1°.5. The movement of telescopes is remotely governed by a low-cost control system called Automatic Computerized Telescope Orientation System (ACTOS). The hardware consists of a semi-intelligent closed loop stepper motor system which senses the angular position using a gravity based transducer called clinometer with an accuracy of 1′. The two clinometers, one each in N-S and E-W direction, are accurately calibrated using the stars. The system can orient to the putative source with an accuracy of $\sim$ (0.003 $\pm$

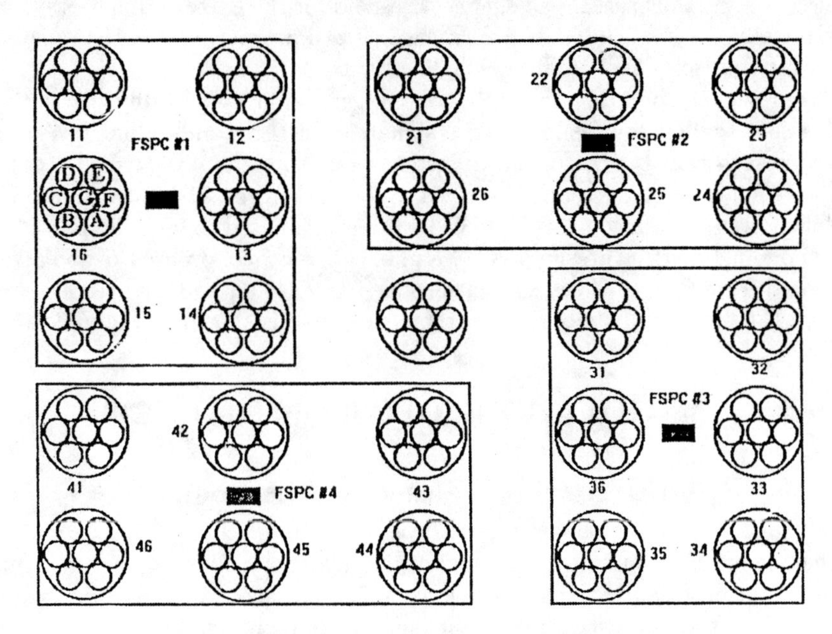

FIGURE 1. Layout of the telescopes at Pachmarhi

0.2)°. The source pointing is monitored with an accuracy of $\sim 0.05°$ and corrected in real time.

High voltages to PMT's are applied through a computerised control system and the volatge as well as the rate of individual PMT's are monitored through out. The array is sub-divided into 4 groups of 6 telescopes each for the sake of data acquisition. Data is acquired on-line using a network of computers for each group independently. The density and arrival time of photons are recorded for 6 of the 7 PMT's in each telescope for every event together with the event arrival time and other house-keeping informations. Relevent data for the whole array is also recorded in the central control room for every event. Data from the 4 groups of telescopes are collated offline using the event arrival times and the central control room data.

All 25 telescopes were fabricated and positioned at the respective locations. The data recording system is operational for 12 telescopes (2 of the 4 quassi-independent groups) while it is being commissioned for the rest.

ENERGY THRESHOLD

Čerenkov photons from air showers are to be detected in the presence of a night sky background (NSB) photons which is about 10^4 times the former. The intensity of NSB photons is a limiting factor in detecting Čerenkov photons due to primaries of lower energy, thus setting an energy threshold for the detection of $\gamma-$rays by the experimental set up. The NSB light measured at Pachmarhi, over the range of the spectral response of the PMT, is $\sim 3.3 \times 10^8\ ph\ cm^{-2}\ s^{-1}$.

A trigger corresponding to a telescope is generated by suitably discriminating analog sum (called royal sum) of the signals from the 7 individual PMTs of the telescope such that $\leq 10\%$ of the triggers are due to night sky noise. An optimum rate of $\sim$ 3-5 Hz per telescope is arrived at for the trigger. Using the above trigger rate the Čerenkov photon threshold of PACT is estimated to be $\sim 20\ ph\ m^{-2}$. The corresponding energy threshold (E_{th}) and effective collection area of the array are estimated from Monte Carlo simulations as $\sim 1\ TeV$ and $10^5\ m^2$ respectively for $\gamma-$rays. The energy distribution of triggered events is shown in Figure 2(a).

ANGULAR RESOLUTION OF PACT

Mirror and Telescope Alignments

The alignment of individual mirrors of the telescope is done using the drift scan method of a bright star across the telescope field of view. All the optic axes of the 7 mirrors are then adjusted, using this method, to be parallel to each other within an error of 0.2° to ensure that all mirrors view the same part of the sky.

Angular resolution of the system

In order to estimate the arrival direction of the shower one has to measure the relative arrival time of the Čerenkov shower front at different telescopes. This has been done using fast time to digital converters (TDC's). From the measurements carried out recently using a quarter of the array, the accuracy in time measurement has been estimated to be 1.3 ns. The angular resolution of the array is estimated using the split array method. Data were collected with the telescopes pointing to vertical direction. The arrival direction of Čerenkov front is obtained by fitting the relative time of arrival of pulses to a plane wavefront. The arrival direction is estimated using one of the 6 TDC informations from each telescope. The space angle between the two independently estimated arrival directions is shown in Figure 2(b). The distribution peaks at a space angle of $0^\circ.42$ yielding a value of $0^\circ.3$ for angular resolution. A group of 6 telescopes have 42 independent TDC informations and there are 4 independent groups of telescopes. Thus a conservative estimate for the angular resolution for the whole array is $0^\circ.1$ which could deteriorate to $\sim 0^\circ.12$ at zenith angles as large as 30°. The angular accuracy is expected to improve with increasing primary energy because of larger number of degrees of freedom available.

EXPECTED SENSITIVITY OF PACT

The signal to noise ratio, with $\gamma-$rays forming signal and cosmic rays as background, is $\propto \sqrt{\frac{A_p T}{\Omega}} E^{0.85-G}$, where G is the $\gamma-$ray spectral index; A_p, the effective

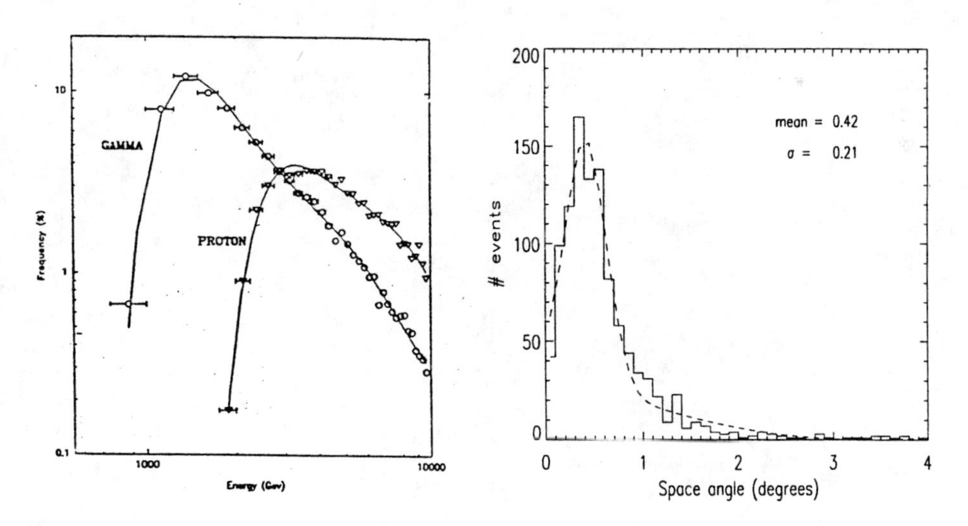

FIGURE 2. (a) Energy distribution of triggered events for γ and proton primaries, (b) The space angle between two independently estimated arrival directions

collection area; T, the observation duration and Ω, the solid angle of the telescope.

A 5σ sensitivity for an observation duration of 50 hours is estimated to be $\sim 4.1 \times 10^{-11}$ $ph\ cm^{-2}\ s^{-1}$ for no background rejection. The angular resolution of $\leq 0^{\circ}.1$ would enable a rejection of 99% off-axis background events using the arrival angle information alone. An additional rejection of $\sim$ 75% on-axis proton showers based on temporal and spatial distributions of Čerenkov photons could be achieved as suggested from Monte Carlo simulation studies [7]. Thus cosmic ray background could be rejected at a level of 99.75%. Accordingly, the sensitivity improves to $\sim 4.1 \times 10^{-12}$ $ph\ cm^{-2}\ s^{-1}$. The efficiency for retaining γ−ray showers while exercising cuts for rejecting hadronic showers is estimated to be $\sim$ 44%. Based on these conservative estimates, the minimum duration of observation required to detect Crab nebula at a significance level of 5σ is $\approx$ 8 hours (4 hrs ON source and 4 hrs OFF source). The sensitivity of PACT is compared with the present and future detectors in Figure 3.

REFERENCES

1. Vacanti G. *et al.*, *Ap. J.*, **377**,461 (1991).
2. Baillon P. *et al.*, *Astrop. Phys.*, **1**, 341 (1994).
3. Goret P. *et al.*, *A & A*, **270**, 401 (1993).
4. Fegan, D. J., *J. Phys. G: Nucl. Particle Phys.*, **23**, 1013 (1997).
5. Chitnis, V. R. and Bhat, P. N., *Astrop. Phys.*, **9**, 45 (1998);
6. Chitnis, V. R. nad Bhat, P. N. *Astrop. Phys.* **12**, 45 (1999).
7. Chitnis, V. R. and Bhat, P. N., *These proceedings*, (2000).

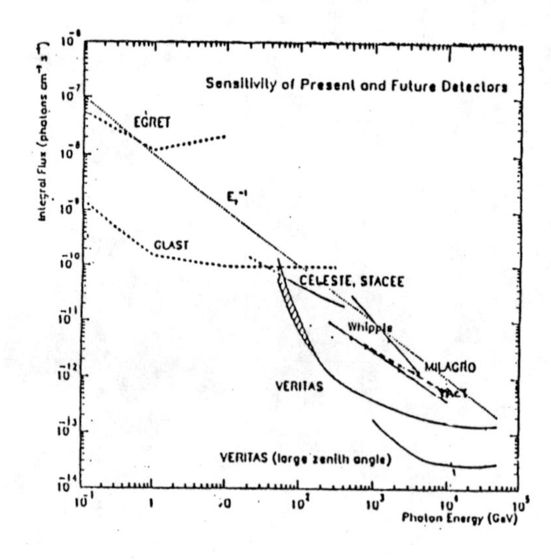

FIGURE 3. Comparison of sensitivity of PACT with other detectors

Gamma-hadron Separation Based on Čerenkov Photon Arrival Time Studies

V. R. Chitnis and P. N. Bhat

Tata Institute of Fundamental Research, Homi Bhabha Road, Mumbai 400 005, India

Abstract. We have carried out systematic Monte Carlo simulation studies of the timing information of Čerenkov photons produced by VHE gamma rays and hadronic primaries. We have investigated several measurable parameters such as radius of the shower front, pulse shape parameters (rise time, decay time and FWHM) and arrival time jitter of Čerenkov photons. The efficiency of these parameters to discriminate between gamma ray and hadronic showers has been studied in terms of quality factor. This study has been carried out for vertical as well as inclined showers and for various altitudes of observation. Among the parameters considered here, Čerenkov pulse decay time and relative arrival time jitter of Čerenkov photons are found to be particularly useful.

INTRODUCTION

It has been known for long that the spatial and temporal properties of Čerenkov photons from extensive air showers contain valuable information about the development and propagation of showers. There had been several attempts in the past to use Čerenkov pulse shape parameters to discriminate between photonic and more abundant hadronic showers at UHE range [1–3]. We have carried out systematic studies of the temporal and spatial profile of Čerenkov light generated by gamma-ray and hadron primaries of energies in the range 100 GeV - 2 TeV. We have studied the efficacy of following parameters derived from temporal properties of Čerenkov photons for discrimination between photonic and hadronic showers : 1. Radius of curvature of Čerenkov shower front, 2. Rise time, decay time and FWHM of Čerenkov pulse, and 3. Relative arrival time jitter of Čerenkov photons. Efficiency of the parameter to distinguish between showers generated by gamma-rays and hadrons is given in terms of quality factor, defined as:

$$q = \frac{N_a^\gamma}{N_T^\gamma}\left(\frac{N_a^h}{N_T^h}\right)^{-\frac{1}{2}} \tag{1}$$

where N_a^γ is the number of accepted gamma-ray showers, N_T^γ is the total number of gamma-ray showers, N_a^h is the number of accepted hadron showers and N_T^h is

CP510, *The Fifth Compton Symposium,* edited by M. L. McConnell and J. M. Ryan

TABLE 1. Quality of radius of the spherical wavefront as a discriminating parameter for vertical showers

Type of primary	Energy of primary (GeV)	Threshold radius of curvature (km)	Fraction of $\gamma-$ rays accepted (%)	Fraction of hadrons accepted (%)	Quality factor
$\gamma-$ rays and protons	100 250	12.0	100	78	1.13
$\gamma-$ rays and protons	500 1000	8.5	97	74.5	1.12
$\gamma-$ rays and protons	1000 2000	8.0	100	80	1.12
$\gamma-$ rays and Fe nuclei	1000 10000	6.0	80.6	16.7	2.0

the total number of hadron showers.

SIMULATIONS

CORSIKA version 5.6 [4] was used for simulation of air showers generated by gamma-rays and cosmic rays of various energies. Čerenkov radiation emitted by relativistic particles in the shower, over wavelength range 300-650 nm was propagated to ground level, which corresponds to Pachmarhi level. A rectangular array of 357 telescopes with area 2.11×2.11 m^2 each, with 17 telescopes with spacing 25 m in E-W direction and 21 telescopes with spacing 20 m in N-S direction is used in simulations. This configuration is similar to the Pachmarhi Array of Čerenkov Telescopes (PACT, [5]), but array size taken here is much larger, so that core distance dependence of various parameters can be studied. The position, arrival angle, time of arrival and production height of Čerenkov photons reaching the array are recorded. Wavelength dependent absorption of Čerenkov photons is not taken into account.

SHOWER FRONT PARAMETERS

The Čerenkov light from air showers generated by gamma-rays and cosmic rays reaching observation level can be approximated by a spherical shower front moving with the speed of light, so that all Čerenkov light originates from a fixed point on the shower axis. Fit with spherical shower front is carried out for gamma-ray and cosmic ray generated showers and the radius of curvature of the shower front is found to correspond to the height of shower maximum from observation level [6]. We have studied usefulness of radius of curvature of the shower front as a discriminant between gamma-ray and proton generated showers as well as showers initiated by iron nuclei. Results are given in Table 1.

TABLE 2. Quality of decay time as a discriminating parameter for vertical showers

Type of primary	Energy of primary (GeV)	Threshold value (ns)	Fraction of $\gamma-$ rays accepted (%)	Fraction of hadrons accepted (%)	Quality factor
$\gamma-$ rays and protons	250 500	3.5	50.1	11.5	1.48
$\gamma-$ rays and protons	500 1000	4.0	45.4	9.1	1.51
$\gamma-$ rays and protons	1000 2000	4.0	34.4	6.6	1.34
$\gamma-$ rays and Fe nuclei	1000 10000	4.5	46.2	8.6	1.58

TABLE 3. Quality of pulse shape parameter discrimination for showers of 500 GeV $\gamma-$ rays and 1 TeV protons, incident at 30° to the vertical

Type of primary	Pulse shape parameter	Threshold value (ns)	Fraction of $\gamma-$ rays accepted (%)	Fraction of protons accepted (%)	Quality factor
$\gamma-$ rays and protons	Rise time	1.5	95.5	84.7	1.04
$\gamma-$ rays and protons	Decay time	5.5	66.2	17.4	1.59
$\gamma-$ rays and protons	FWHM	3.6	82.0	44.5	1.23

PULSE SHAPE PARAMETERS

Even though pulse shape parameters contained information of shower development in the atmosphere, their use for identifying the primary species was always in doubt especially at lower primary energies. Here we have tried to investigate the possibility of using pulse shape parameters to discriminate between gamma-ray and hadron showers. Arrival time distributions of Čerenkov photons at each telescope are fitted with a lognormal distribution function (LDF) [6]. Using LDF, pulse shape parameters i.e. rise time, decay time and FWHM are estimated. Quality factors are calculated using derived pulse shape parameters for all 357 telescopes, for a sample of about 100 showers generated by gamma-rays, protons and iron nuclei. Decay time is found to be a particularly useful discriminant (see Table 2 and Figure 1). We have also investigated usefulness of pulse shape parameters as a discriminant for inclined showers. Quality of discrimination seems to improve for inclined showers compared to vertical showers (see Table 3). Also we have seen that sensitivity of pulse shape parameter discrimination is maximum around hump region. Table 4 gives quality factors for decay time for showers of 500 GeV gamma-rays and 1 TeV protons incident vertically at the top of the atmosphere at various altitudes of observation level, within pre-hump and hump region. Increase in quality factor at

TABLE 4. Quality of decay time as a discriminating parameter for 500 GeV $\gamma-$ rays and 1 TeV protons, incident vertically at the top of the atmosphere, measured at different altitudes of observation, in pre-hump and hump region

Type of primary	Altitude of observation	Threshold Value (ns)	Fraction of $\gamma-$ rays accepted (%)	Fraction of protons accepted (%)	Quality Factor
$\gamma-$ rays and protons	sea level	2.5	34.3	1.2	3.11
$\gamma-$ rays and protons	1.0 km	3.5	63.1	10.7	1.93
$\gamma-$ rays and protons	2.2 km	4.0	84.9	28.4	1.59

lower altitudes is due to increase in prominence of the hump in case of gamma-ray showers.

RELATIVE ARRIVAL TIME JITTER

Each telescope consists of 7 mirrors. Time jitter is the RMS of the mean arrival times of Čerenkov photons at 7 mirrors. Relative time jitter is the ratio of jitter to

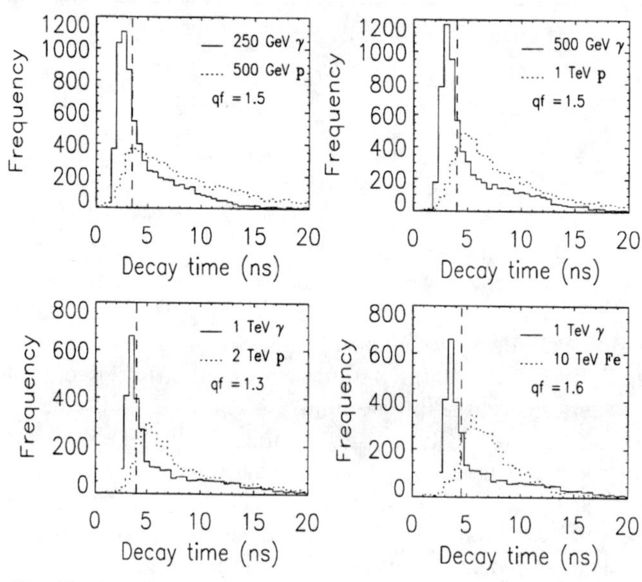

FIGURE 1. Distribution of decay times for $\gamma-$ rays and protons or iron nuclei of respective energies (a) 250 and 500 GeV, (b) 500 GeV and 1 TeV, (c) 1 TeV and 2 TeV and (d) 1 TeV and 10 TeV.

TABLE 5. Quality of relative arrival time jitter as a discriminating parameter.

Type of primary	Energy of primary (GeV)	Threshold value (ns)	Fraction of $\gamma-$ rays accepted (%)	Fraction of hadrons accepted (%)	Quality factor
$\gamma-$ rays and protons	100 250	0.24	53.5	5.6	2.26
$\gamma-$ rays and protons	1000 2000	0.07	61.9	10.1	1.95
$\gamma-$ rays and Fe nuclei	1000 10000	0.05	23.8	0.3	4.34
$\gamma-$ rays and protons at 30°	500 1000	0.0055	67.3	38.2	1.09

mean arrival time of Čerenkov photons and is found to be almost independent of core distance. We have checked for the usefulness of this parameter for identifying primary species. Results for vertical and inclined showers are given in Table 5. Sensitivity of the parameter decreases for inclined showers compared to vertical ones.

CONCLUSIONS

We have investigated usefulness of three types of measurable parameters based on Čerenkov photon arrival time at different core distances at the observation level, for discrimination between photonic and hadronic showers. It is found that radius of the shower front is particularly useful for discrimination against heavy primaries. Of all the three pulse shape parameters, decay time is the most useful discriminant. Quality of discrimination improves for inclined showers. Also discrimination is better in pre-hump and hump region and quality factor increases at lower altitudes of observation because of increase in the prominence of the hump. Relative jitter in arrival times of Čerenkov photons serves as a useful discriminant for near vertical showers.

REFERENCES

1. Tümer, O. T., *et al.*, *Proc. 21st ICRC (Adelaide)*, **2**, 155 (1990).
2. Patterson, J. R. and Hillas, A. M., *Nucl. Instr. Meth.*, **A 278**, 553 (1989).
3. Roberts, M. D., *et al.*, *J. Phys. G: Nucl. Particle Phys.*, **24**, 225 (1998).
4. Knapp, J. and Heck, D., *EAS Simulation with CORSIKA, V5.60: A User's Guide* (1998).
5. Bhat, P. N., *et al.*, *these proceedings* (2000).
6. Chitnis, V. R. and Bhat, P. N., *Astrop. Phys.*, **12**, 45 (1999).

FUTURE MISSIONS AND INSTRUMENTATION

Future Prospects for NASA's Gamma-Ray Astronomy Program

N. Gehrels & D. Macomb

Laboratory for High Energy Astrophysics, NASA Goddard Space Flight Center Greenbelt, MD 20771

Abstract. In 1997, NASA convened a working group to provide input on the future of gamma-ray astronomy. This group, the Gamma-Ray Program Working Group or GRAPWG, was reconvened in 1998 to provide further input. This paper describes the scope and recommendations of this group.

1. OVERVIEW

The GRAPWG is an ad-hoc NASA committee formed to provide recommendations about the future of gamma-ray astrophysics, especially to NASA's Structure and Evolution of the Universe (SEU) subcommittee. The specific charge of the GRAPWG is to evaluate scientific opportunities, survey technology opportunities, review existing mission concepts and develop a mission and technology roadmap. The GRAPWG originally convened in a set of four meetings from 1995-1997. Further deliberations took place in 1998-1999. Both sets of meetings were timed to give input to NASA's Office of Space Sciences (OSS) strategic plans.

The current GRAPWG committee membership is:

- Elena Aprile (Columbia)
- Jerry Fishman (NASA/MSFC)
- Neil Gehrels (chair, NASA/GSFC)
- Josh Grindlay (Harvard)
- Fiona Harrison (Caltech)
- Kevin Hurley (UC Berkeley)
- Neil Johnson (NRL)
- Steve Kahn (Columbia)
- Vicky Kaspi (MIT)
- Mark Leising (Clemson)
- Rich Lingenfelter (UC San Diego)
- Peter Michelson (Stanford)
- Mike Pelling (UC San Diego)
- Roger Romani (Stanford)
- Jim Ryan (UNH)
- Bonnard Teegarden (NASA/GSFC)
- Dave Thompson (NASA/GSFC)
- Trevor Weekes (Whipple/CfA)
- Stan Woosley (UC Santa Cruz)
- Daryl Macomb (Exec. Secretary, GSFC)
- Alan Bunner (ex-officio, NASA)
- Paul Hertz (ex-officio, NASA)
- Lou Kaluzienski (ex-officio, NASA)

NASA used input from the earlier GRAPWG meetings to help create a Strategic Plan in 1998. In it were two missions of particular interest to high-energy astrophysics and gamma-ray astronomy: The Gamma Ray Large Area Space Telescope (GLAST)

CP510, *The Fifth Compton Symposium,* edited by M. L. McConnell and J. M. Ryan

and a Hard X-ray Telescope (HXT) on Constellation-X. The most recent meetings of the GRAPWG concentrated on gamma-ray science opportunities beyond those two major missions and an assumed gamma-ray burst Explorer. The results of the most recent deliberations are discussed in this paper.

2. RECOMMENDATIONS

Current missions, including CGRO, RXTE and *Beppo*SAX, have stimulated tremendous interest and demonstrated the importance of hard X-ray and gamma-ray observations to our understanding of the universe. In the next decade, further discoveries are expected from these missions, from ESA's INTEGRAL mission (launch in 2001, http://astro.estec.esa.nl/SA-general/Projects/Integral/integral.html) and from HETE-II (launch in 2000; http://space.mit.edu/HETE/). However, there are currently no future approved major missions beyond INTEGRAL to advance the field. With NASA updating that plan, and the international community developing several missions that complement the NASA program, a new set of recommendations was developed. These recommendations fall in three categories; previously planned missions, science topics, and new missions.

2.1 MISSIONS FROM THE 1997 GRAPWG

In the previous GRAPWG report, the top-priority was given to the GLAST high-energy gamma-ray mission. Other high priority missions were a focusing hard X-ray telescope and a next-generation nuclear line and MeV continuum mission. For Explorer-class missions the top scientific opportunities were found to be for gamma-ray burst observations and a hard X-ray survey. In addition, the GRAPWG stated its support for both INTEGRAL and the development of new ground-based gamma-ray telescopes. NASA has now started some of these space missions.

GLAST is in the OSS strategic plan for new start in 2002 and launch in 2005. The continuation of the GLAST program remains the highest priority recommendation of the GRAPWG. GLAST will accomplish a broad range of science goals over the energy range from 30 MeV – 300 GeV (http://glast.gsfc.nasa.gov). GLAST will be complemented by the Italian Space Agencies AGILE (Astrorivelatore Gamma ad Immagini LEggero, http://www.ifctr.mi.cnr.it/Agile/) mission, a pair production telescope operating from 30 MeV – 50 GeV with smaller effective area than CGRO/EGRET but better angular resolution and larger field-of-view. AGILE will launch in 2002.

The Swift gamma-ray burst MIDEX has been selected for flight in 2003. Swift will detect about 300 gamma-ray bursts (GRBs) per year in the gamma-ray camera (coded aperture with CZT detectors) and make the preliminary arc-minute positions available in 10's of seconds (http://swift.gsfc.nasa.gov). By utilizing an X-ray and UV/optical telescope, Swift can localize around 100 bursts per year to arc-second resolutions and provide broadband spectra and lightcurves. In addition to determining the origin of GRBs, Swift will provide a hard X-ray survey.

Also, the OSS strategic plan contains the top mission of the X-ray community, Constellation X (launch in 2007, http://constellation.gsfc.nasa.gov). The mission includes a focusing hard X-ray telescope (HXT; 6-50 keV, 1500 cm^2 at 40 keV) onboard that achieves some of the objectives identified by the GRAPWG for a focusing hard X-ray mission. The GRAPWG finds that the scientific case for GLAST, Constellation-X HXT and Swift has grown since 1997.

2.2 SCIENCE PRIORITIES

Beyond the science covered by the aforementioned missions, the GRAPWG identified the following PRIORITIZED list to be the most compelling science that future gamma-ray missions can address. These are areas in which hard X-ray and gamma-ray astronomy offer unique capabilities for advancing our understanding of the universe. We anticipate that future missions will address each topic.

1) Nuclear Astrophysics and Sites of Gamma Ray Line Emission: Gamma-ray astronomy holds the promise of revolutionizing studies of nucleosynthesis in our galaxy and beyond. Through the detection of nuclear lines, sites of nucleosynthesis can be studied and elemental abundances can be measured. In addition, the configuration and dynamics of the emitting gas can be determined. Topics for future missions include:

- ♦ Abundance yields of explosive nucleosynthesis
- ♦ Mass cut between SN ejecta and core
- ♦ Supernova and nova explosion physics and dynamics
- ♦ Sites of nucleosynthesis in the Galaxy and universe
- ♦ Cosmic nucleosynthesis rate from redshifted SN Ia lines
- ♦ Supernova rate in the Galaxy
- ♦ Better understanding of SN Ia cosmological distance scale calibration
- ♦ Cosmic ray interactions with interstellar gas
- ♦ Positron diagnostics of compact objects

2) Gamma Ray Bursts: Appropriate to their nature, gamma-ray burst studies continues to change quickly and dramatically. The increasing number of counterparts at lower energies when coupled with the impressive BATSE database is leading to a new era in GRB studies. Aside from the intrinsic astrophysics of GRB's, bursts will become an important probe of the early universe. Topics for future missions include:

- ♦ Links to star formation
- ♦ Evolution and populations of massive stars
- ♦ Possible sites of black hole formation
- ♦ New GRB populations and mechanisms
- ♦ Probes of dusty matter in distant galaxies
- ♦ Probes of the intergalactic medium out to high redshift

3) Hard X-ray Emission from Accreting Black Holes and Neutron Stars: Hard X-ray and gamma-ray studies of accreting sources are becoming increasingly critical for full understanding of these objects. Detections of galactic and extragalactic black hole systems at high energies provide a laboratory for studying black holes across a wide range of masses. Topics for future missions include:

- First population study of absorbed Seyfert 2's
- Constraints on blazar spectra and diffuse IR back-ground
- Non-thermal components in galactic transients
- Jets associated with galactic BH's and AGN
- Black hole parameters (spin, mass)
- Accretion physics

4) Medium Energy (500 keV–30 MeV) Emissions: Distinct from nuclear lines, the continuum emission in the medium energy range has been shown to be important for understanding nonthermal emission from objects such as pulsars and AGN and sites of cosmic ray interaction with gas. This relatively unexplored band ties together studies at MeV and GeV energies. Topics for future missions include:

- Search for MeV blazars and spectral studies to understand emission
- Pulsar physics through broad-band spectral studies
- Components of diffuse galactic emission
- Extragalactic diffuse emission in poorly measured MeV band
- Nonthermal components from accretion-driven sources
- Cosmic ray interactions with the ISM

2.3 MISSION RECOMMENDATIONS

Figure 1 shows the mission roadmap that the GRAPWG has developed for hard X-ray and gamma-ray astronomy. In addition to the three legacy missions mentioned above, the GRAPWG found that three near-term missions and two long-term concepts to be the most exciting for addressing our top-priority science topics. These are:

Near-term Missions:

High-Resolution Spectroscopic Imager (HSI): HSI is a focusing optics telescope operating in the 10 to 170 keV range. With a factor of 100 improvement in sensitivity compared to RXTE, this mission will answer key questions on the nature of accretion onto neutron stars and black holes and will allow detailed studies of sites of nucleosynthesis in the universe. The development of new multi-layer mirror

TABLE 1. Main goals of the HSI mission concept

Parameter	Goal
Effective Area	700 cm^2 (68 keV)
	400 cm^2 (156 keV)
Angular resolution	10"
Energy range	2–170 keV
Spectral resolution (E/ΔE)	160 @ 158 keV
Line sensitivity(10^6 s)	6.5×10^{-8} cm^{-2} s^{-1} (44 Ti; 5σ)
	8×10^{-8} cm^{-2} s^{-1} (156 Ni ; 5σ)
Continuum sensitivity (10^6 s)	3×10^{-8} cm^{-2} s^{-1} keV^{-1}
	(@50 keV, E/ΔE = 0.5, 5σ)

technology will enable the upper energy bound of the mirrors to be as high as 200 keV. The mission addresses science areas (1) and (3).

Energetic X-ray Imaging Survey Telescope (EXIST): EXIST offers a factor of 100–1000 improvement in sensitivity compared to the only previous all-sky hard X-ray survey (HEAO-1). It will allow the discovery of the predicted, but so-far unobserved, class of absorbed Seyfert 2's that are thought to make up at least half of the total inventory of AGN's. A large area array of new-technology solid state detectors, used in conjunction with a wide field-of-view coded aperture, will cover the 5–600 keV region and address science areas (2) and (3) as well as significant portions of (1). The International Space Station is a possible platform for this instrument. Complementing these missions will be projects, which will extend and improve upon those already in the strategic plan.

TABLE 2. Deep hard X-ray survey mission (EXIST) goals

Parameter	Goal
Energy range:	–600 keV
Sensitivity (10^7 sec, 5σ)	0.05 mCrab (5–100 keV) 0.5mCrab (100–600 keV)
Field of view	40 deg×160 deg
Angular resolution	5arcmin
Source locations	<1arcmin
Energy resolution	2% (60 keV); 1% (500 keV)
Temporal resolution	100% sky each orbit;

Advanced Compton Telescope (ACT): ACT is a high-technology nuclear line and medium energy continuum Compton telescope mission operating in the 500 keV to 30 MeV range. A factor of 30 improvement in sensitivity compared to CGRO and INTEGRAL is envisioned, promising detailed studies of sites of nucleosynthesis in the universe and a deep survey of continuum sources. Extensive technology development is needed for ACT, currently focusing on large imaging detector arrays based on either semiconductor or high-density rare gases. The ACT mission addresses science areas (1), (2) and (4) in the above list.

TABLE 3. Main goals of the ACT mission using a high-resolution Compton telescope

Parameter	Goal
Energy Range	300 keV–20 MeV (Compton mode) 25–500 keV (Coded aperture mode)
Energy Resolution	<3 keV @ 2 MeV
Detector Area	~10,000 cm 2
Field-of-View	~60 degrees (Compton) ~10 degrees (Coded aperture)
Point Source localization	~5 arcmin
Line Sensitivity (@1 MeV)	$\sim 2\times10^{-7}$ cm^{-2} s^{-1} (Narrow Lines) $\sim 1\times10^{-6}$ (SN Ia broadened lines)
Continuum Sensitivity	$\sim 1\times10^{-5}$ cm^{-2} s^{-1} MeV^{-1}

Long-term Mission Concepts:

Next Generation Gamma-Ray Burst Mission (NGGRB): The GRAPWG believes that gamma-ray bursts will continue to be one of the most important and fascinating areas of astronomical research for tens of years to come. A mission will be needed in the post HETE-II and Swift era to further this field. Emphasis in that time frame may involve observations of nonelectromagnetic radiation such as gravitational waves and neutrinos and will certainly involve multiwavelength electromagnetic instrumentation. To correlate these data with known properties of bursts and to monitor the sky for infrequent special events, it will be essential to have a continuous gamma-ray burst monitor in space. The GRAPWG recommends that such a mission, NGGRB, be identified in NASA's program.

Next Generation High-Energy Gamma-Ray Mission (NGHEG): The discoveries of GLAST will produce strong interest in the astronomical community in high-energy gamma-ray phenomena and will undoubtedly raise new fundamental questions. The bandwidth of the high-energy range is huge, from 30 MeV to 300 GeV, and overlaps with the growing number of very high-energy (TeV and PeV) ground-based observatories. The GRAPWG recommends that a mission called NGHEG be identified in NASA's program to follow on GLAST.

2.4 GAMMA-RAY BURSTS

The gamma-ray burst problem and the promise that bursts offer for fundamental studies in astrophysics particularly intrigue the GRAPWG. There are many implications that bursts have on many future missions. Below are some topics and recommendations on gamma-ray burst astronomy. The GRAPWG recommends that:

- HETE-II and Swift should be flown on schedule.
- Support should continue for the Interplanetary Network as an effective means for deriving arc-minute GRB locations.
- Support should continue for BATSE and the Gamma-ray burst Coordinate Network (GCN).
- The GRB monitor currently planned for GLAST will greatly enhance its GRB capabilities.
- Synergism between space-borne GeV GRB observations and ground-base TeV observations should be recognized and exploited.
- A global network of small, dedicated GRB robotic telescopes be developed.
- It is highly desirable to establish a network of coordinated 1–3m telescopes to monitor light-curves and bright event spectra over the first few days.
- Time on major ground- and space-based observatories should continue to be provided for GRB follow-up observations.

2.5 OTHER RECOMMENDATIONS

Gamma-ray astrophysics is a broad enterprise covering many efforts. The GRAPWG recommends that the following items receive special consideration:

- **Technology Development**: Many exciting new technologies are arising in gamma-ray astronomy, including multilayer mirrors, Laue lenses (Bragg concentrators), complex coded masks, solid-state pixel and strip detectors, rare gas and liquid detectors, and VLSI electronics. These form the backbone and future of our field. The GRAPWG strongly recommends a vigorous program of technology development for hard X-ray and gamma-ray astronomy.

- **TeV Astronomy**: Aside from their independent successes, ground-based observatories will pro-vide an important complement to future high-energy gamma-ray missions such as GLAST. The GRAPWG endorses the continued development of TeV telescopes with low energy thresholds.
- **Balloon Program**: The ultra-long duration balloon (ULDB) program offers great potential for both instrument development and significant science in gamma-ray astronomy. The GRAPWG recommends strong NASA support for LDB's and ULDB's.
- **International Space Station**: The GRAPWG views the ISS as an opportunity for hard X-ray and gamma-ray research. It is particularly well suited for wide-field instruments and long-term monitors.
- **Optical Telescope Support**: Many areas of gamma-ray astronomy research, particularly GRB and AGN studies, benefit from a multiwavelength approach. In particular, optical telescopes can provide important monitoring capabilities that are difficult to achieve at other wavelengths. The development of a network of optical telescopes capable of near-continuous observation of gamma-ray transients is supported by the GRAPWG.
- **Data Analysis and Theory**: Making the most of the rich databases expected from future missions is an important concern of the GRAPWG. Adequate support for data analysis and theory is a cost-effective way of maximizing the return from current and future experiments.

3. STATUS

The goals and priorities described above represents the final conclusions of the GRAPWG committee in its current incarnation. A document that discusses these recommendations in more detail will be published in October 1999 and made available to the astronomy community. This publication will also be made available to NASA planning committees and the Decadal Survey of the National Academy of Sciences. The NAS SEU subcommittee is currently prioritizing radio, IR, UV, X-ray, gamma-ray and cosmic ray missions. Our short term new missions - HSI, EXIST and ACT – are all under consideration. The NASA Galveston meeting in November of 1999 will determine the NASA OSS priorities. An electronic version of this document is already available (as a PDF file) through the GRAPWG web site at http://universe.gsfc.nasa.gov/grapwg.html.

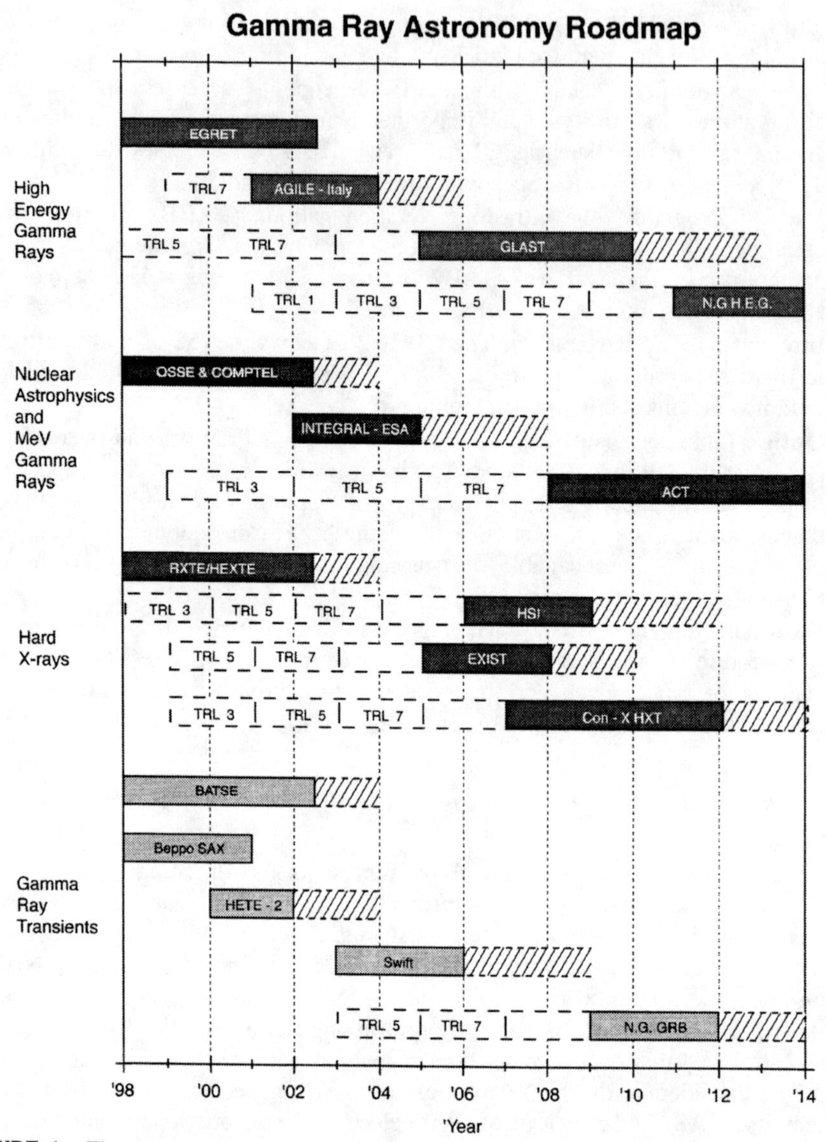

FIGURE 1. The gamma-ray astronomy timeline with current and proposed missions grouped by science topic.

Capability of the ASTRO-E Hard X-ray Detector for High-Energy Transients and γ-Ray Bursts

Y. Terada[1], K. Yamaoka[2], M. Kokubun[1], J. Kotoku[1], T. Mizuno[1], J. Kataoka[2], T. Takahasi[2], T. Murakami[2], K. Makishima[1], T. Kamae[1] and the HXD team

[1] *Department of Physics, University of Tokyo, Bunkyo-ku Tokyo, Japan*
[2] *The Institute of Space and Astronautical Science, Sagamihara, Saitama, Japan*

Abstract. The Hard X-Ray Detector (HXD) is one of the instruments on board ASTRO-E, scheduled for lanch in January–February 2000. The HXD consists of 16 Well-type phoswich counters, surrounded by 20 active shield counters (Anti Coincidence Counters: Anti–Counters). It covers the energy range 10–600 keV with a very low background. Because the Anti–Counters are made of thich high-Z material with a very large geometrical area, they retain a large effective area up to high energies. Therefore the Anti–Counters can be used for monitoring high-energy transient sources and γ-ray bursts. In this paper, the all sky monitoring function with the Anti–Counters and the result of their ground calibration tests are described.

I HARD X-RAY DETECTOR ON BOARD ASTRO-E

ASTRO-E is the fifth in the series of Japanese cosmic X-ray satellites after the successful Hakucho, Tenma, Ginga and ASCA satellites. Its launch is planned for January–February 2000, via the M-V launch vehicle. It features in excellent sensitivity with a broad-band energy range of 0.3 to 600 keV. There are three–type scientific instruments on board ASTRO-E, the X-ray calorimeter(X-ray Spectrometer, or XRS), four X-ray sensitive imaging CCD cameras (X-ray Imaging Spectrometers, or XISs), and the non-imaging, collimated Scintillator (Hard X-ray Detector, or HXD).

The HXD covers the energy range from 10 keV to 600 keV with a very low background, typically 1×10^{-5} c s^{-1} cm^{-2} keV^{-1} at 200 keV on ground, with a typical effective area of 330 cm^2 at 50 keV. Table 1 shows the basic parameters and the performance of HXD, based on pre-flight calibration tests in June 1999. Figure 1 shows the HXD sensor configuration. The X-ray detection part consists of an array of 4×4 counter units ("Wells"), surrounded on four sides by 20 active shield

CP510, *The Fifth Compton Symposium,* edited by M. L. McConnell and J. M. Ryan

counters ("Anti–Counters") working also as All Sky Monitor. The X-ray sensing material "inside the Well" is Gadolinium Silicate (GSO; $Gd_2SiO_5(Ce)$) crystal mainly sensitive above ~30 keV, and 2mm thich silicon PIN diodes sensitive over 10 ~40 keV. The Well are actively collimated by 3mm thick Bismuth Germanate (BGO; $Bi_4Ge_3O_{12}$) crystals, and passively by 50μm thick phosphor bronze plates.

FIGURE 1. HXD sensor

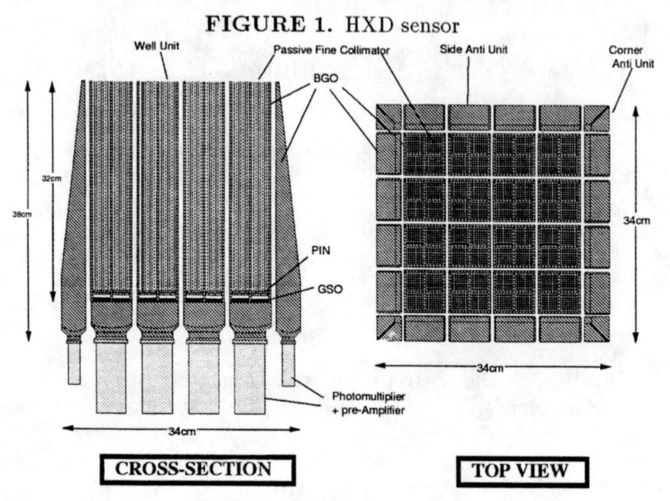

TABLE 1. Basic parameters of the HXD Well Units

Energy Range	10–600 keV (10–60 keV with PIN diodes, 30–600 keV with GSO Scintillators)
Effective Area (On Axis)	160 cm^2 for the 2 mm thick silicon PIN diode 350 cm^2 for the 5 mm thick GSO
Energy Resolution	3.5 keV (PIN diode), $2.4 \times \sqrt{E(\mathrm{keV})}$ keV (GSO)
Field of View	$0.^{\circ}57 \times 0.^{\circ}57$ ($\leq$ 100 keV), $4.^{\circ}57 \times 4.^{\circ}57$ ($\geq$ 100 keV)
Background Rate (†)	$\sim (1-8) \times 10^{-5}$ c sec^{-1} cm^{-2}keV^{-1}
Time Resolution	normally 61μsec (30.1μsec on condition)

† based on ground calibration test (June 1999)

II ALL SKY MONITOR WITH ANTI COUNTERS

The Anti–Counters are made of 4cm thick BGO crystals. They have a very large geometrical area of 800 cm^2 for each side, retaining the effective area of 400 cm^2 per one side at 1 MeV, as illustrated in figure 2 in comparison with the effective area of BATSE LAD. Thus the Anti-Counters are also used for monitoring high-energy transient sources and γ-ray bursts in an energy range of about 100 keV to 2 MeV.

The transient monitoring function is achieved by the Earth Occultation method. X-ray spectra from the Anti-counters are read out every 1 sec with an absolute timing accuracy of 30.5 μ sec. Each spectrum contains 54 energy bins covering

FIGURE 2. Effective area of the HXD Anti–Counters for one side. "corner Anti units"(see fig.1) are not included.

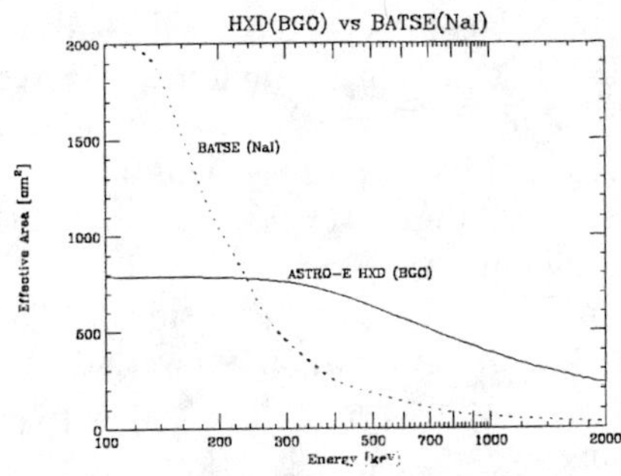

TABLE 2. Basic parameters of the HXD Anti–Counters

Energy Range	$\leq$100 keV – 2000 keV
Energy Resolution	$5.6 \times \sqrt{E}(\mathrm{keV})^{-1}$ (FWHM)
Effective Area	1200 cm^2 (one side)
Field of View	no collimation
Time Accuracy	31 μsec
Time Resolution	1 sec (0.5 sec on condition) for TRNSIENT MONITOR $\frac{1}{32}$sec ($\frac{1}{64}$sec on condition) before/after γ-Ray Burst Trigged

over 100 keV to 2 MeV. When the telemetry limitation is severe, the on-board CPU reduces these spectral energy bins.

The γ-ray burst detection is achieved by monitoring the Anti-Counter count rates with a time resolution of 1/4 sec and/or 1 sec. The automated burst detection algorithm utilizes both hardware circuits and the onboard software. The detail of burst detection algorithm of HXD is shown in figure 3. Once a γ-ray burst is detected, light curve data from the Anti-counters are acquired both before (16 sec) and after (112 sec) the burst.

FIGURE 3. Summary of the HXD Burst Detection

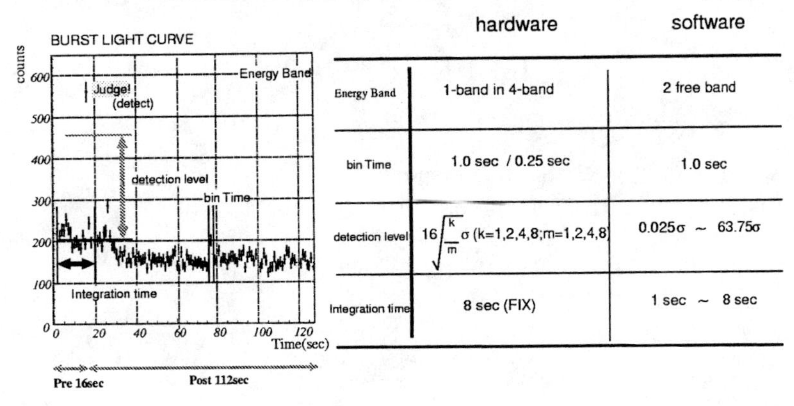

III CALIBRATION OF ANTI COUNTER

The location of γ-ray bursts can be determined coarsely by comparing signal intersities from the four sides of the Anti–Counters. This requires accurate knowledge of the angular response of the Anti–Counter. Accordingly, we calibrated the angular response on ground in June 1999. The sensor and data processing components for HXD are all in flight configuration. We irradiated the Anti–counters with radio isotopes from various angles, and accumulated spectra. Figure 4 shows the measured longitudinal response of one particular side of the Anti–Counters (SIDE-0), to the ^{88}Y isotope irradiated from various azimuthal angles ($\theta = 90$). Figure 5 shows the measured latitudinal response to the ^{88}Y isotope irradiated from various elevation in the $\phi = 0$ plane.

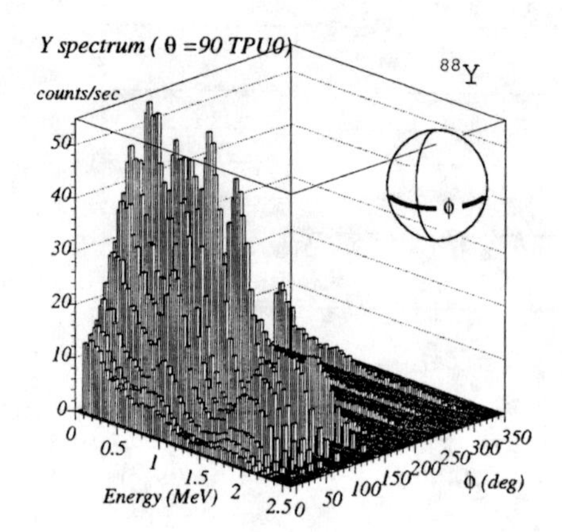

FIGURE 4. Longitudinal response for ^{88}Y

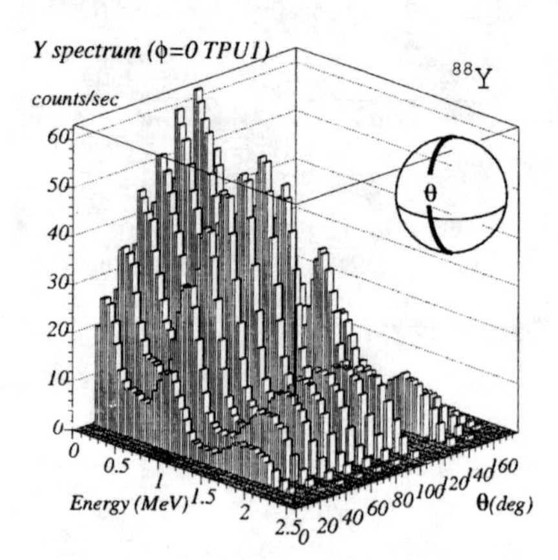

FIGURE 5. Latitudinal response for ^{88}Y

We are now constructing the detector response matrix for Anti–Counters, comparing the measured spectrum with Monte Carlo simulated spectrum. With the latter, we can calculate the response for continuous incident angles at any specified energy.

REFERENCES

1. Kamae, T., et al. *SPIE,* 2806, 314 (1996)
2. Nakazawa, K., et al. *SPIE,* in press (1999)
3. Tanihata, C., et al. *SPIE,* in press (1999)
4. Takahashi, T., et al. *SPIE,* 3445, 155 (1998)
5. Kokubun, M., et al. *IEEE Trans. Nucl. Sci.*, in press (1999)
6. Terada, Y., Master Thesis ,The University of Tokyo (1999)
7. Terada, Y., et al. Astron. Nachr. 320 (1999)

[1] http://www.astro.isas.ac.jp/xray/mission/astroe/astroeE.html ASTRO-E HOME PAGE
[2] http://heasarc.gsfc.nasa.gov/docs/astroe/astroegof.html ASTRO-E Guest Observer Facility
[3] http://www-utheal.phys.s.u-tokyo.ac.jp/hxd/index.html ASTRO-E HXD home page

Extra-Solar Astrophysics with the High Energy Solar Spectroscopic Imager (HESSI)

D. M. Smith, R. P. Lin, J. McTiernan, A. S. Slassi-Sennou, and K. Hurley

Space Sciences Laboratory, University of California, Berkeley, Berkeley, CA 94720-7450

Abstract.
The High Energy Solar Spectroscopic Imager (HESSI) is a NASA Small Explorer mission being built at the University of California at Berkeley (Prof. Robert P. Lin, Principal Investigator), the NASA Goddard Space Flight Center, the Paul Scherrer Institut in Switzerland, and Spectrum Astro, Inc., with participation by a number of other institutions. It is scheduled for launch into low-Earth orbit in July 2000. Here we describe some of the non-solar astrophysical research which will be performed with HESSI by our group, and which can also be performed by the community at large (all of HESSI's data will be immediately public).

DESCRIPTION OF THE INSTRUMENT

HESSI's primary science goals are imaging spectroscopy (3 keV to several MeV), and high-resolution nuclear spectroscopy of solar flares during the next solar maximum. The instrument consists of 9 large germanium detectors (cooled to 75 K by a mechanical cooler) which cover the full energy range (see Figure 1). The detectors sit below a Rotation Modulation Collimator (RMC) system for high resolution imaging capability (2 arcsec at hard x-ray energies). The rotation is provided by spinning the whole spacecraft at about 15 rpm.

The spectrometer sits at the bottom of the spacecraft, below an imager tube carrying two trays of grids, one near the spectrometer and one about 1.5 m away. The top of the spacecraft always points toward the Sun. Each of the 9 germanium detectors is a closed-end coaxial cylinder with a volume of over 300 cm^3, and is electronically segmented into a thin front segment and thick rear segment. The rear segments view nearly half the sky through side walls of only 4mm of aluminum, giving them an effective energy range of 20 keV to 15 MeV. The front segments shield the rear segments from solar photons below 100 keV and view the sun through beryllium windows and a small amount of thermal blanket material, giving them a

CP510, *The Fifth Compton Symposium,* edited by M. L. McConnell and J. M. Ryan

useful energy range down to about 3 keV.

Every photon is tagged with time and energy and stored in spacecraft memory for telemetry to the ground during several daily contacts from a ground station at U. C. Berkeley.

HESSI ASTROPHYSICS

Although HESSI is primarily a solar mission, the HESSI team is committed to making sure its capabilities for extra-solar astrophysical observations are fully exploited. All HESSI data and analysis software will be public, with no proprietary period.

The astrophysics program the HESSI team is planning to pursue combines aspects of what has been done with the CGRO/BATSE and Wind/TGRS instruments, as well as techniques unique to HESSI. The scientific goals include Galactic gamma-ray line spectroscopy by Earth occultation, RMC imaging of the Crab Nebula, x-ray pulsar spectroscopy and period-drift measurements by folding on the pulsar periods, detection of bright Galactic transients by folding on the spacecraft spin period, and high-resolution spectroscopy of gamma-ray bursts and soft gamma repeater (SGR) events. The last will be accomplished without the need for a burst trigger, since every photon event is telemetered to the ground.

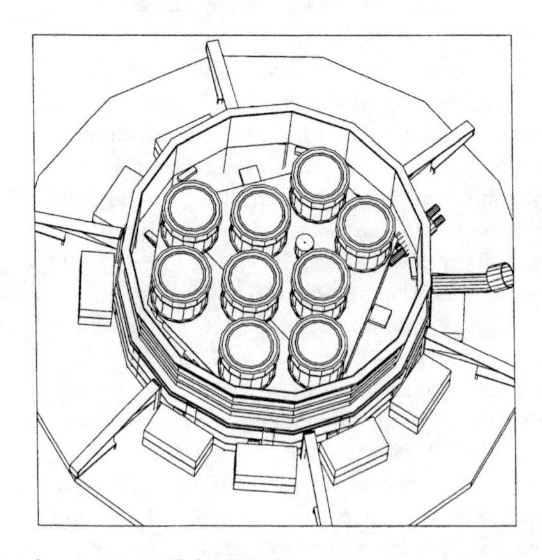

FIGURE 1. Arrangement of the HESSI germanium detectors in their cryostat

RMC IMAGING OF THE CRAB NEBULA

Two bright Galactic sources will drift into the $\sim 2°$ field of view of the RMC imaging system as the spacecraft follows the Sun across the sky: the Crab nebula and pulsar (Ecliptic latitude -1.3°), and the "Z" source (low mass x-ray binary) GX 5-1 (Ecliptic latitude -1.6°). Modulation by the imager will allow us to cleanly separate source and background during these pointings (the average count rate from the sources will be on the order of 10% of the background rate in the front segments), resulting in very clean high-resolution spectra.

More importantly, we should be able to image the Crab nebula in the hard x-ray range. Only one image above a few keV has ever been produced [11], with a resolution of about 15 arcsec, as compared to HESSI's 2 arcsec. The ROSAT soft x-ray image [7] shows shows features at this scale, as do the radio wisps, so the hard x-ray image should be very informative. Simulated HESSI images with appropriate photon statistics can reproduce the same level of structure. The Crab radio wisps near the pulsar are known to evolve rapidly, so it will be interesting to compare hard x-ray images of that region from one year to the next.

GALACTIC GAMMA-RAY LINES

By using the Earth as an occulter, we can produce background-subtracted spectra of the Galactic Center region. In this analysis, the whole HESSI array will be treated as a single detector. Spectra will be summed over a time on the order of a minute (several revolutions), and background will be constructed from data taken during other orbits when the Galactic Center was blocked by the Earth. A similar technique has been used to measure the Galactic 511 keV line with BATSE to the highest precision of any experiment [12].

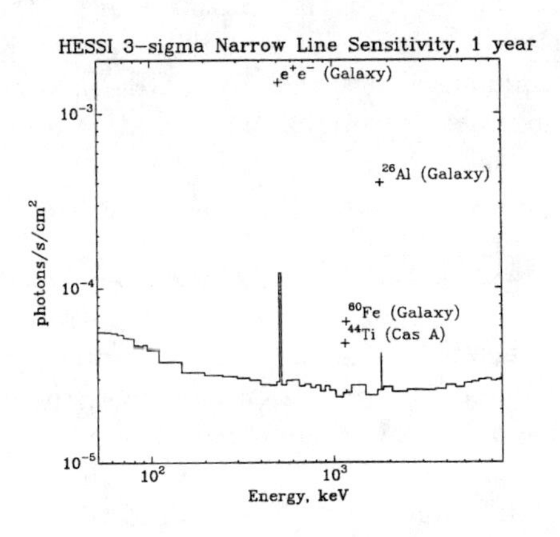

FIGURE 2. HESSI narrow-line sensitivity

Figure 2 shows HESSI's 3σ sensitivity to narrow lines in one year of observations. The sensitivity is not as good as the INTEGRAL SPI, since HESSI is unshielded, but HESSI also observes a much larger portion of the sky at once, and will therefore receive a larger (albeit unimaged) signal in the diffuse Galactic lines. This will make HESSI a good complement to INTEGRAL; subtracting the fluxes in INTEGRAL line maps from HESSI fluxes will allow us to find large scale, low-surface-brightness components in the 511 keV and 1809 keV lines.

As can be seen from the figure, HESSI will make very sensitive measurements of the total flux and lineshape of the Galactic 511 keV and 1809 keV lines. Important results awaiting confirmation include:

The small (or zero) amount of Galactic 511 keV flux which is in the relatively broad, 6.4 keV FWHM component expected from annihilation in flight after charge exchange with neutral hydrogen. This result [4] implies that Galactic positrons are mostly magnetically excluded from cold cloud cores.

The large width (5.4 (+1.4,-1.3) keV FWHM) measured for the integrated Galactic 1809 keV line by the GRIS balloon instrument [10]. This unexpectedly high width means that ^{26}Al ejected in supernovae maintains high velocities long after it would be expected to slow down in the ISM.

The low upper limit on ^{60}Fe, also from GRIS, constraining models of supernova nucleosynthesis when compared to ^{26}Al (and assuming most of the ^{26}Al is produced in supernovae).

Classical novae are expected to produce annihilation-line radiation in the first day after outburst, with most of the positrons produced by ^{18}F and ^{13}N [2,6]. Only upper limits have been observed to date, with Wind/TGRS [5] and BATSE [6]. Because the line is blueshifted due to the velocity of the ejecta, high-resolution instruments like TGRS and HESSI benefit from having the line offset from the strong annihilation background line. Our predicted 3σ sensitivity for a 12-hour observation of a slightly blueshifted annihilation line is 8×10^{-4} ph/cm^2/s, which is about a factor of three better than the existing limits. The existing limits are just about at the predicted flux levels [6], so HESSI gives us a good chance at the first real detection of annihilation in novae. INTEGRAL, although more sensitive, cannot point to a nova in time: the annihilation radiation occurs before the optical detection.

PULSAR PERIOD AND SPIN PERIOD FOLDING

By folding the rear-segment data on the period of known accreting pulsars, we will produce some of the best high-resolution spectra of the pulsed emission from these objects. Figure 6 shows our expected spectrum from the pulsed emission of Her X-1 in one month of observation. The upper curve was generated under the assumption that the cyclotron absorption line is of the same width as the resolution of the scintillators which have generally observed it. The lower curve, divided by 10 for clarity, shows the spectrum we would observe if the absorption line were narrower than the resolution of HESSI's germanium detectors.

In addition, pulsar period folding will allow us to follow the period evolution of the sources, a project which has been extremely fruitful for BATSE [1]. Although HESSI will have less than 10% of BATSE's effective area for these observations, there are still a number of known sources which will be bright enough to follow. Furthermore, since every photon will be recovered with a time tag, HESSI will not have the limitation of BATSE's normal operating mode, which samples rates every 2 seconds. We will therefore be able to do a long-term survey of the undersampled range of periods < 4 sec.

Finally, HESSI will perform by far the most sensitive search for redshifted annihilation lines from the Crab and other pulsars. Of the many reports of such lines, the most challenging to test is a phase-dependent line seen at 440 keV by the FIGARO II balloon instrument at 8×10^{-5} ph/cm^2/s [9]. HESSI should be able to observe this flux at 3σ in roughly 2 months if the line is narrow.

By folding the rear-segment data on the spin period of the spacecraft, we will observe bright Galactic point sources by analyzing the repeated occultation of one detector by another with respect to the source. BATSE's success with occultation by the Earth is well known [3,13,8]. Although HESSI is much smaller than BATSE, we have the advantage of gaining many more occultations per orbit: about 750 detector/detector occultations due to spin per orbit in addition to the 2 Earth occultations. We will therefore be able to monitor transients and persistent sources above a few hundred mCrab.

Although the detectors are not completely opaque at 511 keV, we may be able to obtain some spatial information on the 511 keV line by detector/detector occultation, in a way similar to the analysis done for Wind/TGRS [4], but taking advantage of HESSI's extra order of magnitude of germanium volume.

REFERENCES

1. Bildsten, L. et al. 1997, ApJ 113, 367
2. Gomez-Gomar, J. et al. 1998, MNRAS, 296, 913
3. Harmon, B. A. et al. 1992, Proc. Compton Observatory Workshop, p. 69
4. Harris, M. J. et al. 1998, ApJ, 501, 55
5. Harris, M. J. et al. 1999, ApJ 522, 424
6. Hernanz, M. et al. 1999, Poster #A100, this meeting
7. Hester, J. J. et al. 1995, ApJ, 448, 240
8. Ling, J. C. et al. 1996, A&AS 120, 677
9. Massaro, E. et al. 1991, ApJL, 376, L11
10. Naya, J. E. et al. 1996, Nature, 384, 44
11. Pelling, R. M. et al. 1987, ApJ 319, 416
12. Smith, D. M. et al. 1998, Proc. of the 4th Compton Symposium, AIP Conf. Proc. #410, p. 1012
13. Zhang, S. N. et al. 1993, Nature 366, 245

Science with the International Gamma-Ray Astrophysics Laboratory INTEGRAL

Christoph Winkler* and Wim Hermsen[†1]

*ESA-ESTEC, Space Science Department, Astrophysics Division, Keplerlaan 1, NL-2201 AZ Noordwijk, The Netherlands
†SRON, Sorbonnelaan 2, NL-3584 CA Utrecht, The Netherlands

Abstract.
The International Gamma-Ray Astrophysics Laboratory (INTEGRAL), to be launched in 2001, is dedicated to the fine spectroscopy (ΔE: 2 keV FWHM @ 1.3 MeV) and fine imaging (angular resolution: 12′ FWHM) of celestial gamma-ray sources in the energy range 15 keV to 10 MeV with concurrent source monitoring in the X-ray (3 - 35 keV) and optical (V, 550 nm) range. The mission is conceived as an observatory led by ESA with contributions from Russia and NASA. The INTEGRAL observatory will provide to the science community at large an unprecedented combination of imaging and spectroscopy over a wide range of gamma-ray energies. This paper summarises the key scientific goals of the mission, the current development status of the payload and spacecraft and it will give an overview of the science ground segment including the science data centre, science operations and key elements of the observing programme.

INTRODUCTION

The International Gamma-Ray Astrophysics Laboratory (INTEGRAL) is dedicated to the fine spectroscopy (ΔE: 2 keV FWHM @ 1.3 MeV) and fine imaging (angular resolution: 12′ FWHM) of celestial gamma-ray sources in the energy range 15 keV to 10 MeV. The INTEGRAL observatory will provide to the science community at large an unprecedented combination of imaging and spectroscopy over a wide range of X-ray and gamma-ray energies including optical monitoring. The mission is conceived as an observatory led by ESA with contributions from Russia and NASA and will be launched in 2001. ESA is responsible for the overall spacecraft and mission design, instrument integration into the payload module, spacecraft integrations and testing, spacecraft operations including one ground station, science operations, and distribution of scientific data. Russia will provide a PROTON launcher and launch facilities, and NASA will provide ground station support

[1] on behalf of the INTEGRAL Science Working Team.

CP510, *The Fifth Compton Symposium,* edited by M. L. McConnell and J. M. Ryan

through the Deep Space Network. The scientific instruments and the INTEGRAL Science Data Centre will be provided by large collaborations from many scientific institutes of ESA member states, USA, Russia, Czech Republic and Poland, nationally funded, and led by Principal Investigators (PI's).

SCIENTIFIC OBJECTIVES

INTEGRAL is a 15 keV - 10 MeV gamma-ray mission with concurrent source monitoring at X-rays (3 - 35 keV) and in the optical range (V, 500 - 600 nm). All instruments - co-aligned with large FOV's - cover simultaneously a very broad energy range of high energy sources (Tables 1, 2).

The scientific goals of INTEGRAL will be attained by fine spectroscopy with fine imaging and accurate positioning of celestial sources of gamma-ray emission. Fine spectroscopy over the entire energy range will permit spectral features to be uniquely identified and line profiles to be determined for physical studies of the source region. The fine imaging capability of INTEGRAL within a large field of view will permit the accurate location and hence identification of the gamma-ray emitting objects with counterparts at other wavelengths, enable extended regions to be distinguished from point sources and provide considerable serendipitous science which is very important for an observatory-class mission. In summary the scientific topics will address: (i) compact objects: *white dwarfs, neutron stars, black hole candidates, high energy transients and GRB's*; (ii) extragalactic astronomy: *galaxies and clusters, AGN, Seyferts, Blazar, cosmic diffuse background*; (iii) stellar nucleosynthesis: *hydrostatic nucleosynthesis (AGB and WR stars), explosive nucleosynthesis (Supernovae and novae)*; (iv) Galactic structure and the Galactic Centre: *cloud complex regions, mapping of the continuum and line emission, ISM, cosmic-ray distribution*; (v) particle processes and acceleration: *transrelativisic pair plasms, beams, jets*, (vi) identification of high energy sources: *unidentified gamma-ray objects as a class*; PLUS: (vii) unexpected discoveries.

Recent new results from high-energy observations of millisecond radio pulsars increase substantially the expectation that INTEGRAL can also contribute significantly to the study of rotation powered neutron stars. To date, three millisecond pulsars have been detected at energies above 2 keV with very hard non-thermal pulsed emission, namely PSR B1821-24 (ASCA, [5]; RXTE, [4]), PSR B1937+21 (ASCA, [6]) and PSR J0218+4232 (ROSAT, [2]; BeppoSAX, [3]). The pulsed emissions have power-law spectral shapes, with indexes between $\sim$ -0.6 and $\sim$ -1.2, and exhibit X-ray pulses with intrinsic widths $\leq$ 100 μsec. Surprisingly, for PSR J0218+4232 convincing evidence has been reported in this meeting [1] for detection in the EGRET data of pulsed high-energy γ-ray emission between 100 MeV and 1 GeV. Extrapolations of the pulsed spectra measured between 1 and 10 keV into the INTEGRAL energy range, as well as interpolation between the spectra measured for PSR J0218+4232 below 10 keV and above 100 MeV, demonstrate that INTEGRAL will be capable to bridge the intermediate gap, most importantly the energy interval between 50 keV and 1 MeV. These studies will provide important parameters for modelling the production of pulsed high-energy emission in the magnetosphere of (millisecond) pulsars. In addition, we cannot exclude that part of the unidentified high-energy gamma-ray sources are millisecond pulsars. INTEGRAL might also shed some light on this problem.

SCIENTIFIC PAYLOAD

TABLE 1. INTEGRAL science and payload complementarity

Instrument	Energy range	Main purpose
Spectrometer SPI	20 keV - 8 MeV	Fine spectroscopy of narrow lines Study diffuse emission on >deg scale
Imager IBIS	15 keV - 10 MeV	Accurate point source imaging Broad line spectroscopy and continuum
X-ray Monitor JEM-X	3 - 35 keV	Source identification Monitoring @ X-rays
Optical Monitor OMC	500 - 600 nm	Optical monitoring of high energy sources

The INTEGRAL payload consists of two main gamma-ray instruments: Spectrometer SPI and Imager IBIS, and of two monitor instruments, the X-ray Monitor JEM-X and the Optical Monitoring Camera OMC. The design of the INTEGRAL instruments is largely driven by the scientific requirement to establish a payload of scientific complementarity. As shown in Table 1, the payload does meet this goal.

Each of the main gamma-ray instruments, SPI and IBIS, has both spectral and angular resolution, but they are differently optimised in order to complement each other and to achieve overall excellent performance. The two monitor instruments (JEM-X and OMC) will provide complementary observations of high energy sources at X-ray and optical energy bands. An overview of the INTEGRAL payload is given below, detailed descriptions can be found in the various instrument papers presented at this symposium. Also part of the payload is a small particle radiation monitor, which continuously measures the particle environment of the spacecraft. Therefore it is possible to provide essential information to the payload in case high particle background (radiation belts, solar flares) is being encountered. This information is used to decide on switch - off and switch - on of instrument high voltages and to provide actual background information for sensitivity estimates.

Spectrometer SPI

The Spectrometer SPI (Table 2) will perform spectral analysis of gamma-ray point sources and extended regions with an unprecedented energy resolution of 2 keV (FWHM) at 1.3 MeV. This will be accomplished using an array of 19 hexagonal high purity Germanium detectors cooled by two pairs of Stirling Coolers to 85 K. The total detection area is 500 cm^2. A hexagonal coded aperture mask is located 1.7 m above the detection plane in order to image large regions of the sky (fully coded field of view = 16°) with an angular resolution of 2°. In order to reduce background radiation, the detector assembly is shielded by an active BGO veto system which extends around the bottom and side of the detector almost completely up to the coded mask. A plastic veto between mask and upper veto shield ring further reduces background events.

Imager IBIS

The Imager IBIS (Table 2) provides powerful diagnostic capabilities of fine imaging (12′ FWHM), source identification and spectral sensitivity to both continuum and broad lines over a broad (15 keV - 10 MeV) energy range. The energy resolution is 7 keV @ 0.1 MeV and 60 keV @ 1 MeV. A tungsten coded aperture mask (located at 3.2 m above the detection plane) is optimised for high angular resolution imaging. Sources ($> 10\sigma$) can be located to $< 60''$. As diffraction is negligible at gamma-ray wavelengths, the angular resolution obtainable with a coded mask telescope is limited by the spatial resolution of the detector array. The IBIS design takes advantage of this by utilising a detector with a large number of spatially resolved pixels, implemented as physically distinct elements. The detector uses two planes, a front layer (2600 cm^2) of CdTe pixels, each (4x4x2) mm, and a second one (3100 cm^2) of CsI pixels, each (9x9x30) mm. The division into two layers allows the paths of the photons to be tracked in 3D, as they scatter and interact with more than one element. The aperture is restricted by a thin passive shield. The detector array is shielded from the sides and below by an active BGO veto.

X-Ray Monitor JEM-X

The Joint European X-Ray Monitor JEM-X (Table 2) supplements the main INTEGRAL instruments (Spectrometer SPI and Imager IBIS) and plays a crucial role in the detection and identification of the gamma-ray sources and in the analysis and scientific interpretation of INTEGRAL gamma-ray data. JEM-X will make observations simultaneously with the main gamma-ray instruments and provides images with 3′ angular resolution in the 3 - 35 keV prime energy band. The photon detection system consists of two identical high pressure imaging microstrip gas chambers (Xenon at 5 bar) each viewing the sky through a coded aperture mask (4.8° fully coded FOV), located at a distance of 3.2 m above the detection plane. The total detection area is 1000 cm^2.

Optical Monitoring Camera OMC

The Optical Monitoring Camera OMC (Table 2) consists of a passively cooled CCD in the focal plane of a 50 mm lens. The CCD (1024 x 2048 pixels) uses one section (1024 x 1024 pixels) for imaging, the other one for frame transfer before readout. The FOV is $5° \times 5°$ with a pixel size of 17.6″. The OMC will observe the optical emission from the prime targets of the INTEGRAL main gamma-ray instruments with the support of the X-Ray Monitor JEM-X. Variability patterns on timescales of 1 s and longer, up to months and years will be monitored. The limiting magnitude of 19.7^{m_v} (3σ, 10^3 s), corresponds to ~40 photons $cm^{-2}s^{-1}keV^{-1}$ (@ 2.2 eV) in the V-band. Multi-wavelength observations are particularly important in high-energy astrophysics where variability is typically rapid. The wide band observing opportunity offered by INTEGRAL is of unique importance in providing for the first

TABLE 2. Key parameters of the INTEGRAL scientific payload.

	SPI	IBIS	JEM-X	OMC
Energy range	20 keV - 8 MeV	15 keV - 10 MeV	3 keV - 35 keV	(500 - 600) nm
Detector area (cm^2)	500	2600 (CdTe) 3100 (CsI)	1000 (2 units each 500)	CCD (2k×1k pxl)
Spectral resolution (FWHM, keV)	2 @ 1.3 MeV	7 @ 100 keV 60 @ 1 MeV	1.5 @ 10 keV	–
Field of view (fully coded)	16°	9° x 9°	4.8°	5.0° x 5.0°
Angular resolution	2° FWHM	12′ FWHM	3′ FWHM	17.6″/pixel
Typical source location	20′	$< 1'$	$< 20''$	$< 8''$
Continuum sensitivity[a]	7×10^{-8} @ 1 MeV	4×10^{-7} @ 100 keV	1×10^{-5} @ 6 keV	19.7^{m_v} (3σ, 10^3 s)
Line sensitivity[b]	5×10^{-6} @ 1 MeV	1×10^{-5} @ 100 keV	2×10^{-5} @ 6 keV	–
Timing (3σ)	100 μs	67μs – 1000 s	128 μs	$>$ 1s
Mass (kg)	1309	628	65	17
Power (W)	373	275	55	18
Data rate (kbps)	20 (avge)	57 (avge)	7	2

[a] Units are (ph cm^{-2} s^{-1} keV^{-1}) for 3σ detection in 10^6 s.
[b] Units are (ph cm^{-2} s^{-1}) for 3σ detection in 10^6 s.

time simultaneous observations over seven orders of magnitude in photon energy for some of the most energetic objects in the Universe.

MISSION SCENARIO

The INTEGRAL spacecraft consists of a service module (commonly designed with the service module of the ESA XMM mission) containing all spacecraft subsystems, and a payload module containing the scientific instruments. During summer 1998, the service module and the payload have succesfully completed the structural and thermal test (STM) programme and the electrical test (EM) programme has been completed in August 1999.

INTEGRAL (with a payload mass of 2019 kg and a total launch mass of ~4000 kg) will be launched in 2001 by a Russian PROTON launcher into a highly eccentric orbit with high perigee in order to provide long periods of uninterrupted observation with nearly constant background and away from trapped radiation. The parameters for the orbit are: period 72 hours, inclination 51.6°, initial perigee height 10 000 km, initial apogee height 153 000 km. The particle background radiation affects the performance of high-energy detectors, and scientific observations will therefore be carried out while the spacecraft is above an altitude of nominally 40 000 km. The particle background of the local spacecraft environment will be continuously measured by the on-board radiation monitor: this device allows the optimisation

FIGURE 1. The INTEGRAL spacecraft. The cylindrically shaped SPI is next to the larger rectangular payload module (PLM) structure housing the IBIS and JEM-X detectors inside. The top of the PLM carries the coded mask for IBIS (squared) and two coded masks for the two redundant JEM-X detectors. The OMC and two star trackers are located at the top of the PLM. The size of the spacecraft (w/o solar arrays) is ~ 4x4x6 m (lxwxh).

of the observing time before or after radiation belt passages and solar flare events, and provides essential information about the actual background. Data from the onboard radiation monitor will be routinely checked to verify and possibly update the nominal altitude above which scientific observations will be performed. A nominal altitude of 40 000 km implies that 90% of the time spent on the orbit can be used for scientific observations. However, a number of in-orbit activities have an influence on the net amount of orbit time (e.g. slews, eclipses, resctrictive spacecraft operations, instrument calibrations) such that the average observation efficency becomes ~85% per year. The real-time scientific data rate (including instrument housekeeping) is 86 kbps.

The spacecraft employs fixed solar arrays: this means, that the target pointing of the spacecraft (at any point in time) will remain outside an avoidance cone around the sun and anti-sun. This leads to a minimum angle between any celestial source and the sun/anti-sun of 50° during the nominal mission life (2 years) outside eclipse seasons and 60° during extended mission life (year 3+). During eclipse seasons of the nominal mission (few weeks per year) 60° will be applied.

Because of imaging deconvolution requirements by SPI, the spacecraft will routinely, during nominal operations, perform a series of off-source pointing manouevres, known as "dithering". These dithering patterns consist of sets of different pointings at sky positions around the nominal target position (at the centre). The dithering points are separated by 2°. The exposure time per point is 30 minutes. Two dither patterns will be employed: a 7 point hexagone and a 5×5 point raster, both centred on the target position. If required by observers, dithering can be disabled.

GROUND SEGMENT AND SCIENCE OPERATIONS

The ground segment (Figure 2) consists of three major elements, ESA's Mission Operations Centre (MOC), the INTEGRAL Science Operations Centre (ISOC), and the INTEGRAL Science Data Centre (ISDC) plus two ground stations (ESA, NASA). MOC will implement the observation plan received from the ISOC within the spacecraft system constraints into an operational command sequence.

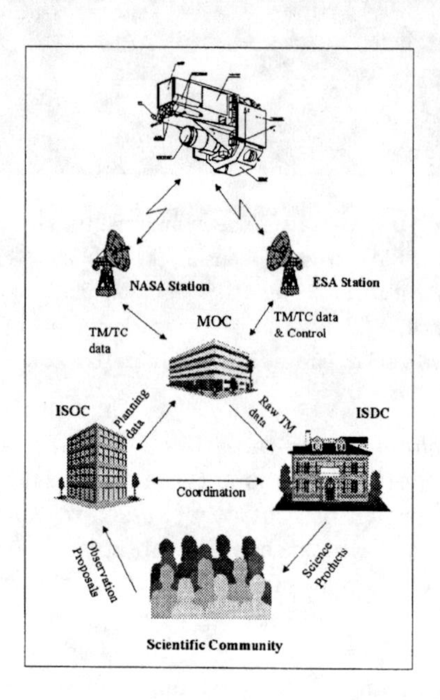

FIGURE 2. The INTEGRAL ground segment.

In addition, MOC will perform all classical spacecraft operations, real-time contacts with spacecraft and payload, maintenance tasks and anomaly checks (including payload critical health and safety). MOC will determine the spacecraft attitude and orbit, and will provide raw science data to the ISDC.

The ISOC, provided by ESA and located at ESTEC, will issue the AO for observing time and will handle the incoming proposals which will be processed into an optimised observation plan which consists of a timeline of target pointings plus the corresponding instrument configuration. This observation plan will then be forwarded to MOC to be uplinked to the spacecraft. Furthermore, the ISOC will validate any changes made to parameters describing the on-board instrument configuration and it will keep a copy of the scientific archive produced at the ISDC. Finally, the ESA Project Scientist at the ISOC will decide on the generation of TOO alerts (Target of Opportunity) in order to update and reschedule the observing programme.

The ISDC, located in Versoix, Switzerland, will receive the complete raw science telemetry plus the relevant ancillary spacecraft data from the MOC. Science data will be processed, taking into account the instrument characteristics, and raw data will be converted into physical units. Using incoming science and housekeeping information, the ISDC will routinely monitor the instrument science performance and conduct a quick-look science analysis. Most of the Targets of Opportunity (TOO) showing up during the lifetime of INTEGRAL will be detected at the ISDC during the routine scrutiny of the data and will be reported to ISOC. Scientific data products obtained by standard analysis tools will be distributed to the observer and archived for later use by the science community. INTEGRAL will be an observatory-type mission with a nominal lifetime of 2 years, an extension up to 5 years is technically possible. Most of the observing time (65% during year 1, 70% (year 2), 75% (year 3+)) will be awarded to the scientific community at large as the General Programme. Typical observations will last from 10's of minutes up to two weeks. Proposals, following a standard AO process, will be selected on their scientific merit only by a single Time Allocation Committee. These selected observations are the base of the General Programme. The first call for observation proposals is scheduled for release during first half of 2000. In principle, observers will receive data from all co-aligned and simultaneously operating instruments on-board INTEGRAL. The remaining fraction of the observing time will be reserved, as guaranteed time, for the INTEGRAL Science Working Team for its contributions to the programme. This fraction, the Core Programme, will be devoted to: (i) a Galactic Plane Survey, (ii) a deep exposure of the central radian of the Galaxy, and (iii) pointed observations of selected regions/targets including TOO follow up observations. The current status of the Core Programme is described in detail by [7]. The full details of the Core Programme will be made available at the issue of the first AO. In accordance with ESA's policy on data rights, all scientific data will be made available to the scientific community at large one year after they have been released to the observer. This guarantees the use of the scientific data for different investigations beyond the aim of a single proposal.

REFERENCES

1. Hermsen W., et al., *these proceedings* (1999).
2. Kuiper L., et al., *A&A* **335**, 545 (1999).
3. Mineo T. et al., *A&A*, submitted (1999).
4. Rots A.H. et al., *ApJ* **501**, 749 (1998).
5. Saito Y. et al., *ApJ* **477**, L37 (1997).
6. Takahashi M. et al., *Proc. IAU Coll.* **177**, Bonn (1999)
7. Winkler C., et al., *Proc. 3rd INTEGRAL workshop*, in press (1999)

The IBIS Gamma-Ray telescope on INTEGRAL

P. Ubertini[1], F. Lebrun[2], G. Di Cocco[3], L. Bassani[3], A. Bazzano[1], A. J. Bird[4], K. Broenstad[5], E. Caroli[3], M. Cocchi[1], G. De Cesare[1], M. Denis[6], S. Di Cosimo[1], A. Di Lellis[1], F. Giannotti[3], P. Goldoni[2], A. Goldwurm[2], G. La Rosa[7], C. Labanti[3], P. Laurent[2], O. Limousin[2], G. Malaguti[3], I. F. Mirabel[2], L. Natalucci[1], P. Orleansky[6], M. J. Poulsen[1,*], M. Quadrini[8], B. Ramsey[9], V. Reglero[10], L. Sabau[11], B. Sacco[7], A. Santangelo[7], A. Segreto[7], R. Staubert[12], J. Stephen[3], M. Trifoglio[3], L. Vigroux[2], R. Volkmer[12], M. C. Weisskopf[12], A. Zdziarski[13], A. Zehnder[14]

*[1]IAS, CNR, Rome, Italy - [2]CEA - Saclay, France - [3]ITESRE, CNR, Bologna, Italy - [4]Southampton University, Southampton, U.K - [5]University of Bergen, Bergen, Norway - [6]SRC, PAC, Warsaw, Poland - [7]IFCAI, CNR, Palermo, Italy - [8]IFCTR, CNR, Milano, Italy - [9]MSFC, NASA, Huntsville, Alabama, USA - [10]Univ. of Valencia, Burjasot, Spain - [11]INTA, LAEF, Villafranca de Castillo, Spain - [12]AIT, Tubingen, Germany - [13]NCAC, Warsaw, Poland - [14]PSI, Villigen, Switzerland - *Also: Laben Spa, Vimodrone, Italy*

Abstract. The IBIS Telescope is the high angular resolution Gamma-Ray imager on-Board the INTEGRAL Satellite. IBIS features a coded aperture mask and a novel large area (~3,000 cm^2) multilayer detector which utilises both Cadmium Telluride (16,384 detectors) and Caesium Iodide elements (4,096 detectors) to provide the fine angular resolution ~12 arcmin, wide spectral response (20 keV to 10 MeV), high resolution timing (61 μs) and spectroscopy (6% at 100 keV) required to satisfy the mission's imaging objectives.
This paper will focus on the IBIS hardware characteristics while the Scientific Performance of the telescope have been recently addressed elsewhere [1].

IBIS HARDWARE CONFIGURATION

IBIS has a tungsten coded-mask imaging system in order to achieve a high angular resolution over a wide field of view. The instrument consists of the following main items: Mask and Passive Shielding System, Detector Unit, Data Handling System and Calibration Unit. The detector includes the low energy CdTe detector Layer (ISGRI) and the high energy CsI detector Layer (PICSIT). The separation between the crystal upper surfaces of the two detectors is about 94 mm. The double-layer-discrete-element design allows the paths of interacting photons to be tracked in 3D if the event involves detection units of both ISGRI and PICsIT. The application of Compton reconstruction algorithms to this type of events (between few hundred keV and few MeV) allows an increase in signal to noise ratio attainable by rejecting those events

CP510, *The Fifth Compton Symposium*, edited by M. L. McConnell and J. M. Ryan

likely to correspond to source photons outside the FOV. The detector aperture is restricted, in the soft gamma-ray part of the spectrum, by a thin lead passive shield, covering the distance between mask and detection plane. The active BGO scintillator veto system shields the detector bottom as well as the four sides up to the bottom of ISGRI. This configuration ensures a good broad line sensitivity and outstanding angular resolution on a large field of view over the wide IBIS energy range.

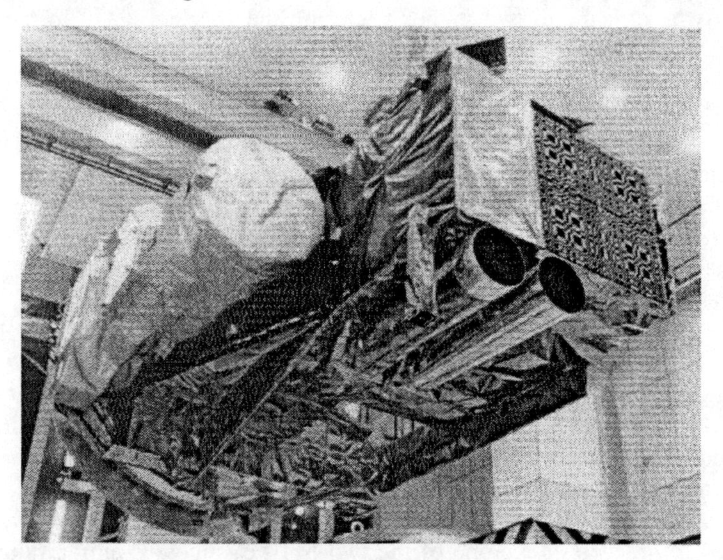

FIGURE 1. The INTEGRAL STM Model during test campaign. The IBIS Coded Mask is clearly visible, due to the absence of the thermal blanket.

CdTe Layer: The Cadmium Telluride is a II-VI semiconductor operating at ambient temperature. The CdTe detectors with their small area are ideal to build up a pixellated imager with good spatial resolution. On the other hand, their small thickness restricts their use to the low energy domain (50% efficiency at 150 keV). Providing spectral performances intermediate between that attained by the cooled germanium spectrometers and those of the scintillators, the CdTe can be used way down in the X-Ray domain (E >15 keV).

CsI Layer: The CsI(Tl) bars are optically bonded to custom made low leakage silicon PIN photodiodes, designed for optimal performance at 511 keV. The design provides a high degree of modularity. The CsI layer is divided in eight rectangular modules of 512 detector elements, each module is integrated into a stand-alone testable sub-system. Modules may be assembled into the layer. The CsI modules have the same cross-sectional shape as those of the CdTe ones.

Active Anticoincidence System: The two detector planes are surrounded by an active BGO shield. Like the detector array, the Veto shield will be modular with 8 lateral shields, i.e. 2 modules per side, and 8 bottom modules. Each Veto Detector Module includes: the BGO crystal and related housing, two PMT's optically coupled to the BGO and assembled with the dedicated FEE and HV divider, one HV Power Supply and one Veto Module Electronics for Module control.

Collimating system and passive shields: The collimating system of IBIS is made of two subsystems mechanically independent: (a) a tungsten hopper structure on the top of the ISGRI layer that is a truncated (at 550 mm height) pyramid with a profile that ideally joins the detector perimeter to the active mask sides; (b) a lead tube made of four walls (two vertical and two inclined) closing the aperture down to the hopper level. The passive shielding is designed to substantially reduce the celestial diffuse background component up to about 200 keV.

The Coded Mask: The coded mask assembly (1.064 x 1.064 m^2) lays at 3.4 m above the detector bench and is composed of 11.2 mm squared pixels, 16 mm thick, interconnected by 2 mm ribs. The coded mask projects a shadowgram onto the detector plane and images of the sky will be reconstructed by decoding the detector shadowgram with the mask pattern.

Half of the cells are "opaque" to photons offering a ~70% opacity at 3 MeV, while the other are "open", i.e. with an on-axis transparency of 60% at 20 keV.

The Mask/detector set up will permit to achieve ~10 arcsec point source location accuracy.

On Board Calibration Unit: On-board calibrations are necessary in order to monitor and control the instrument inflight performances and to reach the limit sensitivity on weak astronomical sources to be studied.

The Calibration Unit consists of 0.4 μCi ^{22}Na radioactive source placed near to the centre of the largest face of a BGO Detector Module. A tagging system is implemented in order to detect one of the two 511 keV photons emitted during the ^{22}Na decays and additionally can detect the 1275 keV photon which is emitted toward the tagging detector in 50% of the cases. In this way the gain and the thresholds of the detector will be calibrated to better than 1% accuracy in 3-4 hours @511 keV and 8 hours @1275 keV.

Data Handling System: The Data Handling System is based on the Data Processing Electronics and Hardware Event Pre-processor for IBIS (HEPI) that is necessary to cope with the overall processing requirements due to the high detector event rates that would correspond to a throughput up to 300kbit/s if directly transmitted to ground. All scientific data and housekeeping from the two imaging detectors as well as from the veto and the calibration system are collected, handled and finally channelled to the On Board Data Handling system for telemetry generation.

IBIS Calibration Plan. In order to meet the scientific objectives of IBIS, a comprehensive series of on ground and inflight calibrations of the overall telescope are required. The calibrations will enable determination of the instrument response function(s) allowing observations to deduce location, spectral and temporal behaviour of X and Gamma-ray sources.

Ground Calibration will be performed at Component/Module, Detector, Instrument and Payload Level. The IBIS calibration will proceed throughout a number of phases from the individual components to the final integrated payload, under the control of the IBIS calibration implementation team. The key products of this process are the *calibrated model* and the *response functions* that it will generate. Since the response of IBIS is extremely complicated, some parts of the response can only be calculated (not measured directly) and so a highly accurate computer model of the instrument is essential. This model will therefore encapsulate both a comprehensive instrument description and the results of all the calibration measurements.

The First EM Scientific Tests. The IBIS requirement to provide a moderately high energy resolution over the entire telescope energy range is achieved by the use of innovative technologies. At low energy the instrument employs a novel type of solid state CdTe detector (4x4x2 mm^3) working at room temperature, to achieve a large area (~2600 cm^2) matrix of 128x128 elements that provides good spectroscopic capabilities and outstanding angular resolution.

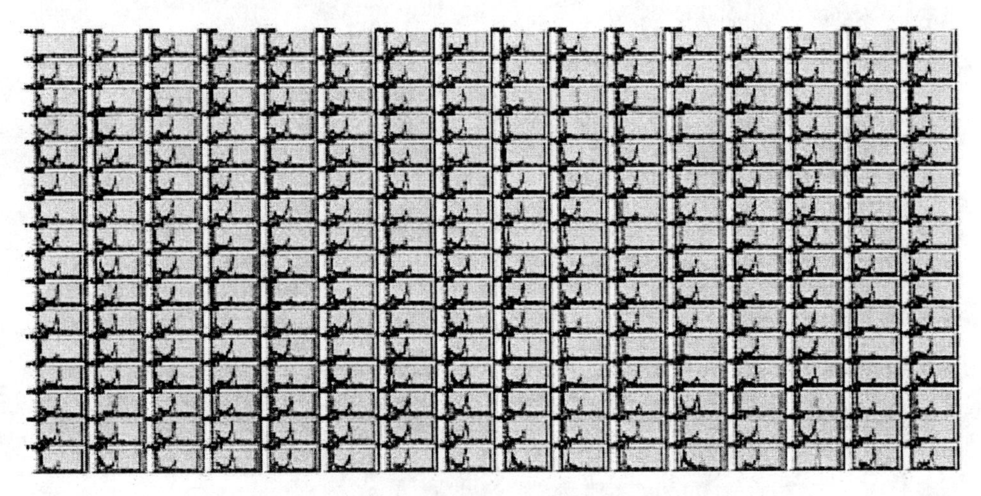

FIGURE 2. Collection of all ISGRI EM detector spectra (^{241}Am source).

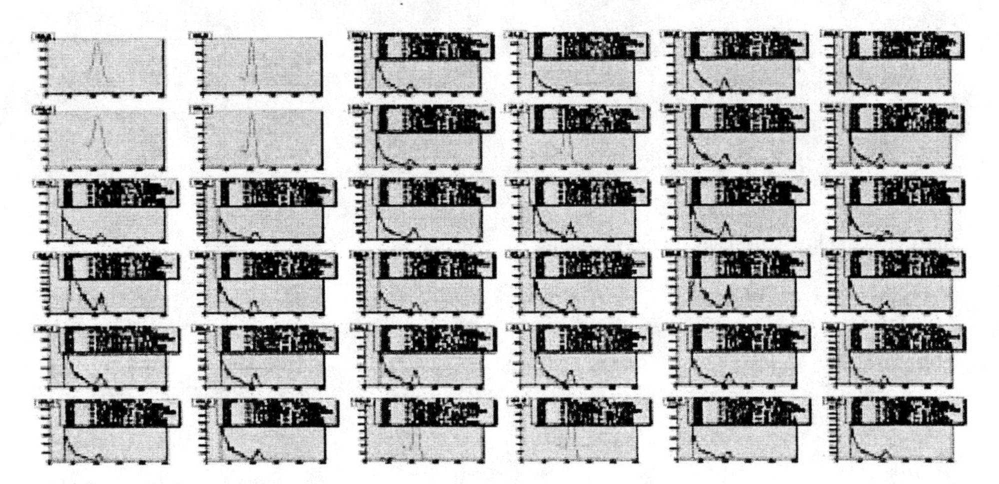

FIGURE 3. Collection of all PICsIT EM detector spectra (^{137}Cs source).

The figure 2. shows a collection of all ISGRI spectra obtained with the EM Model exposed to a ^{241}Am radioactive source. The EM contains 256 detectors out of the 16,384 of the FM for end-to end tests.

The high energy detection system is based on an array of small (8.55x8.55x30 mm^3) CsI detectors treated with a diffusing coating read out by a high efficiency photodiode coupled with a low noise asic and the large area matrix (~3000 cm^2) of 64x64 elements that provides good spectroscopic capability and outstanding angular resolution up to 10 MeV.

The figure 3. shows a collection of all PICsIT spectra obtained with the EM Model exposed to a ^{137}Cs radioactive source. The EM contains 32 detectors out of the 4,096 of the FM for end-to end tests.

IBIS SENSITIVITY TO CONTINUUM AND LINES

The IBIS continuum and line sensitivities has been evaluated using a Monte Carlo (MC) simulation. In this way the total and photo-peak detection efficiencies have been obtained via the MC simulation of the transport of the x and γ-ray photons and secondary particles inside the instrument materials. The evaluation of the hadronic background is a key parameter in the sensitivity estimation. In the following figures the predicted values have been independently obtained from simulations and extrapolation from the γ-ray SIGMA/GRANAT satellite.

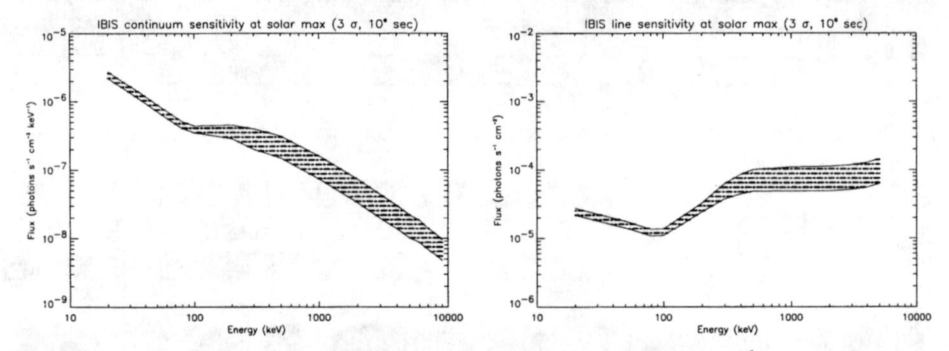

FIGURE 4. IBIS continuum (left) and line sensitivities (3σ, observing time 10^6 s). The grey area is due to uncertainty in the background evaluation.

REFERENCE

1. Ubertini P. et al., Proceedings of "The Extreme Universe, 3rd Integral Workshop", 14-18 Sept. 1998, Taormina, Italy, 1999, in press.

The PICsIT high-energy detector of IBIS: Calibration of the Engineering Model

G. Di Cocco[*], G. Malaguti[*], O. Pinazza[*], F. Schiavone[*], J.B. Stephen[*], C. Labanti[*], A. Spizzichino[*], M. Trifoglio[*], F. Gianotti[*], G. Ferro[§]

[*]*ITESRE / CNR, Bologna (Italy)*
[§]*ENEA, Bologna (Italy)*

Abstract. IBIS is the gamma-ray imaging telescope onboard the ESA satellite INTEGRAL, which will be launched in 2001. PICsIT, the high-energy (140 keV – 10 MeV) detector of IBIS, consists of a 64×64 units array. Each detection unit is a ≅0.8 cm^2, 3 cm thick CsI(Tl) crystal coupled with a photodiode. The engineering model (EM) of PICsIT has now been tested, calibrated, and integrated in IBIS for delivery to ESA. The calibration of PICsIT EM has allowed for the first time its scientific qualification in terms of: energy threshold, linearity, energy resolution, and photopeak efficiency (for events of various multiplicity).

INTEGRAL AND IBIS

INTEGRAL (1) (INTErnational Gamma-Ray Astrophysics Laboratory) is the ESA scientific satellite for γ-ray astrophysics to be launched in 2001.

Scientific observations onboard the satellite will be performed by two main instruments: a spectrometer (SPI, SPectrometer on INTEGRAL) and an imager (IBIS, Imager onBoard INTEGRAL Satellite). The IBIS imager is designed to produce images of the γ-ray sky in the 15 keV to 10 MeV energy range using two pixellated detectors layers: ISGRI (2) (INTEGRAL Soft Gamma-Ray Imager, the top layer, operating in the low energy range) and PICsIT (3) (Pixellated Imager CsI Telescope, the bottom layer, operating in the high energy range).

PICSIT DESIGN AND FUNCTIONALITY

PICsIT consists of 4096 physically and optically independent 0.85×0.85 cm^2 in cross-section and 3 cm thick CsI(Tl) pixels, arranged in a 64×64 array, which is subdivided into 8 identical modules. The detector matrix is coupled with a coded mask placed 3.25 m above ISGRI. Large field of view (9°×9° is the Fully Coded Field of View, FCFV), fine resolution (12 arcmin) images of the gamma-ray sky are obtained covolving the detected shadowgram with the transmission function of the mask.

The discrete pixel design of PICsIT allows the introduction of the concept of multiplicity of events, which is defined as the number of pixels triggered. Due to

CP510, *The Fifth Compton Symposium,* edited by M. L. McConnell and J. M. Ryan

downlink telemetry budget, events with multiplicity greater than one are transformed onboard into single events with energy equal to the sum of all energy deposits. Multiple events incidence pixels are as well reconstructed on-board with an algorithm based on Monte-Carlo simulations.

Telemetry budget limitations have also imposed the use of on-board data handling and compression. In particular, 256 channels energy spectra will be integrated for each pixel, and for single (multiplicity order 1), and multiple events (multiplicity order 2, and 3) over a ~1000÷2000 s time period (spectral imaging mode). In parallel, timing data based on 8 channels spectra, and with 1÷512 ms resolution are sent to ground after on-board compression integrated over the full detection plane (spectral timing mode). In the case of very low count rate, on-board data compression is not needed, and events are down linked with full information.

PICSIT SCIENTIFIC REQUIREMENTS

The scientific requirements of PICsIT are: angular resolution (FWHM) $\omega = 12$ *arcmin*, lower energy threshold $E_{TH} \cong 140$ keV (the upper energy threshold is limited by the thickness of the crystal that must allow a $\cong 8\%$ detection efficiency at 10 MeV, and by the dynamic range of the front-end electronics), and energy resolution $\Delta E/E_{\mathrm{FWHM}}$ better than 12% at 662 keV (for single events).

Expected PICsIT low energy threshold is given by: $E_{TH} = \dfrac{5N}{n_e}\ keV$, where N is the electronic noise in number of e^- rms, and n_e is the light output in e^-/keV. Extensive tests on single pixels have shown that $n_e \geq 33$ e^-/keV, and $N \leq 950$, which translate into an energy threshold $E_{TH} \cong 144$ keV.

The energy resolution is given by: $\left(\dfrac{\Delta E}{E}\right)_{FWHM} = 2.35\dfrac{\sqrt{n_e E + N^2}}{n_e E} + k$, where E is the energy deposit in keV, and k accounts for light production and collection in-homogeneity in the crystal. From the value of k (PICsIT pixels qualification lot measurements indicate that $k \leq 2\%$), n_e, and energy resolution, it is possible to calculate N, and therefore the threshold.

PICSIT ENGINEERING MODEL

PICsIT Engineering Model (EM) was finalized in spring 1999. PICsIT EM consisted of 32 active pixels positioned in an "L"-shaped area on one corner of the detector plane. Its main goals were the verification of all electrical and software interfaces, and of operational procedures. Bearing in mind the limitations implied by the low number of pixels, the EM has given also the occasion for measuring PICsIT scientific performances, and for a first scientific calibration of PICsIT. This has implied, for each pixel, the measurement of noise (i.e. lower energy threshold), energy resolution, linearity (integral and differential), and count rate (for background only, and in the presence of a radioactive source). Detector level analysis included the distribution of single vs. multiple events (for different energies), the spread of gain, energy resolution, count rate, and the study of edge effects (4).

PICsIT EM calibration campaign

The measurements were performed at LABEN laboratories in Milan. Table 1 shows the measurements performed at PICsIT level.

TABLE 1. PICsIT EM performance evaluation and detector calibration: columns 1 to 3 report the source, its intensity and integration time; columns 4 and 5, show the detected count rate in counts/s/pixel.

Source	Activity	Integration time	Count rate (cts/s/pixel)	
			Whole energy range	E > 300 keV
^{203}Hg	10 μCi	1600 s	48.28	28.80
Background		1800 s	0.83	0.69
^{22}Na	100 μCi	1800 s	28.10	24.40
Background		3600 s	0.84	0.70
^{137}Cs	10 μCi	3600 s	15.90	13.60
Background		3600 s	0.87	0.72

Table 1 shows that PICsIT EM has been tested at four energies: 279 keV (^{203}Hg), 511 and 1275 keV (^{22}Na), and 662 keV (^{137}Cs). The integration time has been chosen to acquire $\geq 10^4$ counts in the photopeak of single pixel spectra. Data acquisition, control and archive were done with a dedicated Test Equipment (TE), which also included two workstations: one for real time Quick-Look verification and analysis, and the other for detailed off-line analysis (5).

Energy Resolution And Gain Spread

Single pixel photopeak have been fitted with a gaussian to measure the peak position and energy resolution. The results are shown in Figure 1. With the exception of one pixel, which shows a very low gain, PICsIT EM gain spread is comprised within ±5%. Again with the exception of one pixel, the energy resolution at 662 keV is, on the other hand, comprised between 11.5% and 13.5%. PICsIT EM pixels and ASICs have not been selected and coupled for scientific performance optimization, and not all pixels were FM representative (e.g. some have $n_e > 33$ e^-/keV). Notwithstanding this limitation, Figure 1 shows that 19 pixels (out of 32) show an energy resolution at 662 keV better than 12.5%. This means that for these pixels $N \leq 950$, and, since $k=2$, the lower energy threshold meets the requirement.

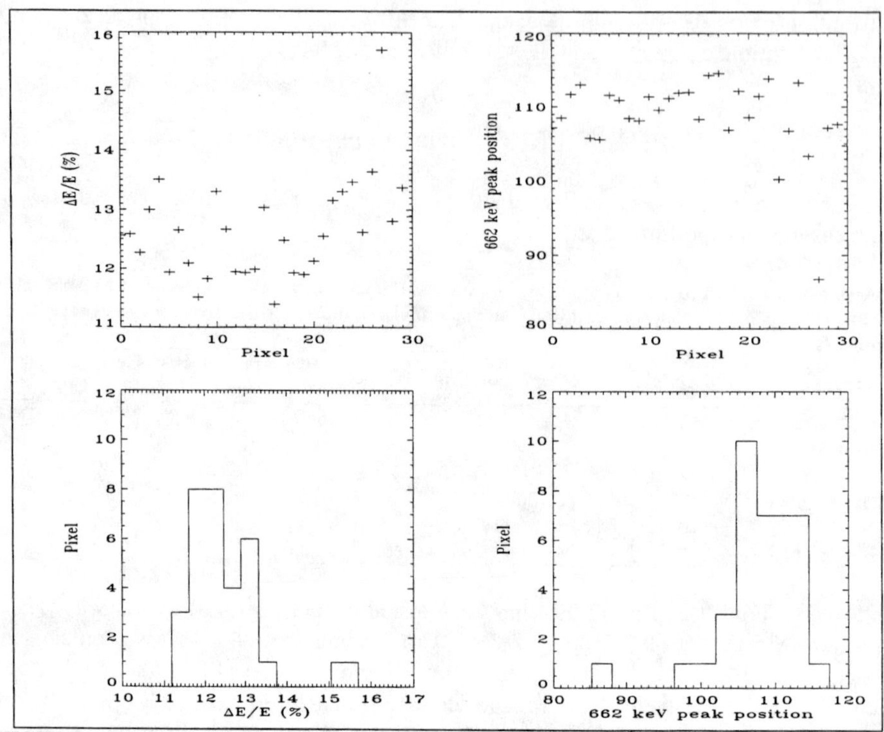

FIGURE 1. Energy resolution (FWHM, left panels) and photopeak centroid position (channel, right panels) at 662 keV for PICsIT EM single pixel spectra.

ACKNOWLEDGEMENTS

Italian participation to the IBIS project is financed by the Italian Space Agency (ASI). Fruitful collaboration with LABEN S.p.A. is kindly acknowledged.

REFERENCES

1. Winkler, C., "INTEGRAL: the current status", Proc of 3rd INTEGRAL Workshop The Exterme Universe, Taormina, Sept. 1998, in press.
2. Lebrun, F., et al., "Coded aperture imaging in gamma-ray astronomy", *Spa. Sci. Rev.,* **A380**, 414 (1996)
3. Labanti, C., et al., " PICsIT: the high-energy detection plane of the IBIS instrument onboard INTEGRAL", Proc. of 1996 SPIE Conference, San Diego.
4. Di Cocco, G. et al., "The scientific calibration of the PICsIT Engineering Model detector of the IBIS telescope", Proc. of 1999 SPIE Conference, Denver.
5. Trifoglio, M. et al., "The science Test Equipment for the INTEGRAL-PICsIT Instrument", Proc. of 1999 SPIE Conference, Denver.

The EGSE Science Software of the IBIS instrument on-board INTEGRAL satellite

Giovanni La Rosa[1], Fulvio Gianotti[2], Giacomo Fazio[1], Alberto Segreto[1], John Stephen[2], Massimo Trifoglio[2]

1. IFCAI, Consiglio Nazionale delle Ricerche, via Ugo La Malfa 153, Palermo, Italy
2. ITESRE, Consiglio Nazionale delle Ricerche, via Gobetti 101, Bologna, Italy

Abstract. IBIS (Imager on Board INTEGRAL Satellite) is one of the key instrument on-board the INTEGRAL satellite, the follow up mission of the high energy missions CGRO and Granat. The EGSE of IBIS is composed by a Satellite Interface Simulator, a Control Station and a Science Station. Here are described the solutions adopted for the architectural design of the software running on the Science Station. Some preliminary results are used to show the science functionality, that allowed to understand the instrument behaviour, all along the test and calibration campaigns of the Engineering Model of IBIS.

INTRODUCTION

IBIS (Imager on Board INTEGRAL Satellite) is one of the main instruments on-board INTEGRAL satellite, the second Medium Sized Mission of the ESA programme Horizon 2000. IBIS is a coded mask telescope designed to produce sky images in the 20KeV÷10MeV band, with a FOV of 9x9 degrees (fully coded) and an angular resolution of 12 arcmin. It also exhibits a good spectroscopic capability.
IBIS consists of a coded mask and a double layer imaging detector. The front layer is a plane of 128x128 pixels of CdTe named ISGRI, the lower one is composed by 64x64 pixels of CsI, named PICSIT.
The Electrical Ground Support Equipment (EGSE) of IBIS is composed by:
- The Spacecraft Interface Simulator (SIS). It provides the correct interface to the instrument by the emulation of the On Board Data Handling (OBDH) bus for telemetry (TM) and telecommands (TC) exchange. It provides, also, interfaces to the engineering housekeepings (HK) as well as the appropriate power lines.
- The Central Check-Out Equipment (CCOE). It manages a whole test session, sending TCs to the instrument and receiving the TM. The CCOE controls the HKs parameters. It distributes, in real time, to the science station, the TM packets and the TCs sent.
- The Experiment Check-Out Equipment (ECOE). It is the component that provides the on-line science capabilities required to support a whole test session. In particular, it is able to archive, accumulate, analyse and display HKs and science data, contained in the telemetry packets provided by the CCOE (or by the INTEGRAL Core EGSE, for satellite level tests). The following sections present the architecture and functionality of the software running on the ECOE, together with some example of its specific tools.

CP510, *The Fifth Compton Symposium,* edited by M. L. McConnell and J. M. Ryan

DESIGN CONCEPT

The ECOE software is designed to perform some basic operation required during a test session. In order to provide the capabilities required, the following functions are implemented: TM acquisition, verifying and archiving, TM unpacking, Archiving of the reduced data (in FITS format), Quick-Look, Analysis Tools, TM distribution to other users, Man/Machine Interface, TM dump and some generic system functions like hard-copy of the screen, data backup and other functions.

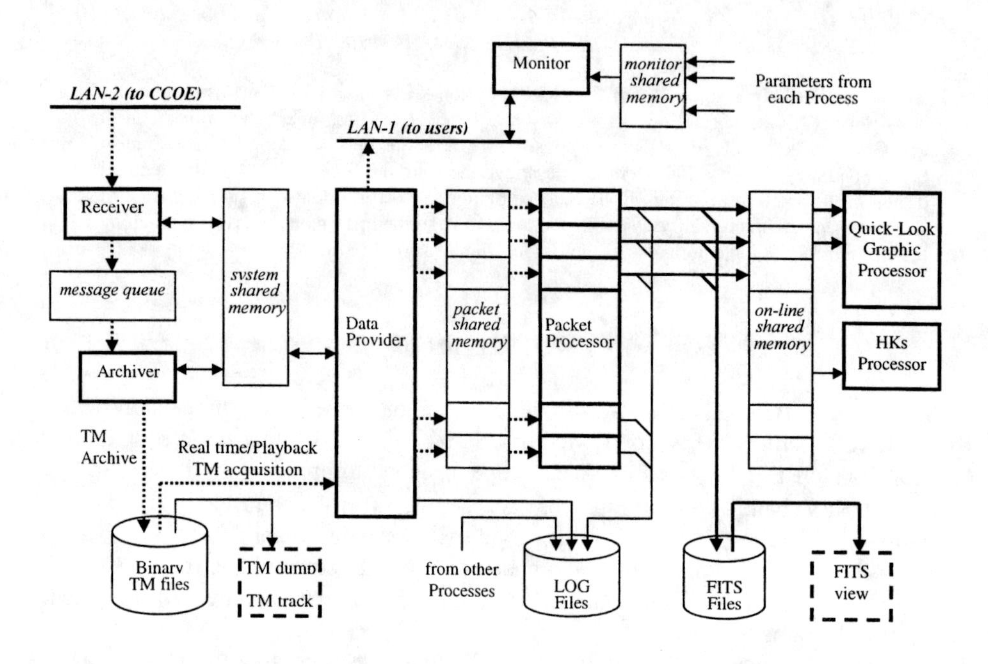

FIGURE 1. ECOE software architectural design. Dotted lines indicate raw TM data flow, bold lines indicate reduced data flow. Bold boxes indicate the processes running during TM acquisition (for TM Playback, the *Receiver* and the *Archiver* are not required).

To guarantee a better efficiency and robustness, each of the above tasks is realised as a separate process. The better efficiency is gained by the fact that many processes allow the dead time associated with I/O operations to be used, thereby obtaining a kind of "parallelism" even for a mono-processor architecture. The robustness is guaranteed by the modularity, in particular by concentrating the most complex calculation after the acquisition one obtains a system which is fault tolerant and, in case of problems in the packet data processing, is still able to independently archive the raw telemetry data.
The processes required to run for real-time TM acquisition are described below.
The *Receiver* is devoted to the activation of the TM/TC link on a LAN. It initialise the shared memories and the message queue. Then waits for a connection request from the CCOE, and, before the first packet is read, forks itself and generates the *Archiver*

process. The packets arriving from the LAN are written into the message queue, which provides to the *Archiver* the mechanism to synchronise the data reading.
The *Archiver* reads the packets from the message queue, writing them in a file. There is a file naming convention that univocally associates a file to a given measurement. As before, a signal is sent to the *Provider* to notify that a new packet is available.
The *Provider* reads a packet from the current acquisition file (or from any previous acquired file, when in playback mode) and writes it in a channel of the shared memory on the basis of his APID. Then the *Provider* generates a signal directed to the relevant *Packet Processor* to notify that a new packet is available.
The *Monitor* process allows the operator to monitor in real time the status of all the processes running in the chain. The *Monitor* updates the screen every second.
The *Packet Processor* reads a packet from the relevant channel of the shared memory. First of all, it verifies the correctness of the TM format, then it unpacks the science data contained in the packet, writing the reduced data in a file using the widely used FITS format. The *Packet Processor* manages any packet type/subtype of IBIS, so that, the execution of the *Packet Processor* have to be replicated *n* times for *n* different APIDs. The *Packet Processor* produces a log file, where there are written all the errors detected in the packets together with some statistic parameter. It also writes the unpacked data in a shared memory channel, available to the *Quick-Look* process.
The *Quick-Look* reads a block of data (only for the photon-by-photon operational mode) from the relevant shared memory and after their accumulation, it shows, in a graphic window, images and spectra of the two detector layers. The *Quick-Look* follows, in real time, the data evolution; in particular, the zoom function allows to the operator a better look of some interesting part of the spectra or of the images.

REAL-TIME ACQUISITION

The real-time acquisition of TM packets (or for TM playback) is controlled by the *Control Panel,* an user friendly console from where the ECOE software can be commanded by simply clicking with the mouse on the function buttons. The *Monitor* is the first process to run, then the *Quick-Look* and, in the end, all the other processes.
While IBIS is stand-by, only HK packets are produced (1pck/8sec). The *Housekeeping* window displays the most important science parameters, for each IBIS subsystem, contained in the HK packets. In particular, for each detection layer, status, ratemeters and executed/rejected commands are displayed. The window is updated in real time as soon as a new HK packets is received.
When the operational status change to science mode, the ECOE software closes the current acquisition file and creates a new one in which the science packets are saved.
The *Quick-Look, now,* shows the images and the histograms growing up meanwhile new science packets are received. This allows the ECOE operator to control the instrument behavior, looking if the data received are as expected. At the end of the acquisition run, the instrument goes back to stand-by. At this point, all the files produced are automatically closed and a new raw TM file is created to save the HK packets received during the stand-by. Now, a new acquisition run can be restarted.

SPECIFIC TOOLS

To complete the ECOE software, there are two kind of on-line tools, one for the analysis of the raw contents of the TM packets (*TM_track* and *TM_dump*) and the other (*FITSview*) for the analysis of the science data written in the FITS files.

TM_track is able to inspect the TM stream. It displays the most important information contained in the Packet Header and in the Data Field Header for each incoming packet. In a single line is displayed the APID, the Packet Sequence Counter, the Packet Length, the packet type/subtype and the value of the time field. It is also possible to show the previous information only for packets belonging to a selected APID.

TM_dump allows to inspect the contents of a single packet, showing in a separate fields: Packet Header, Data Field Header and Data Field. The contents of the packet is displayed in raw hexadecimal format. Anyway, it also possible to display the contents of the Data Field in clear decimal format in a event by event table.

FITSview allows to easily analyse the science data of a selected acquisition run, showing the accumulated spectra and images in a graphic screen window. *FITSview* automatically detect the contents of a specific FITS file, thus allowing the selection of a given data set to be shown in graphical format. Typically, is possible to see the Time Profile of the events countrate, the PHA spectrum, the Rise Time spectrum (where applicable), the counts profile along the Y and the Z axis. *FITSview* is able to show also images in 2D/3D. It is possible to filter the data on the basis of a window selection in PHA, Rise Time, Y, Z and time.

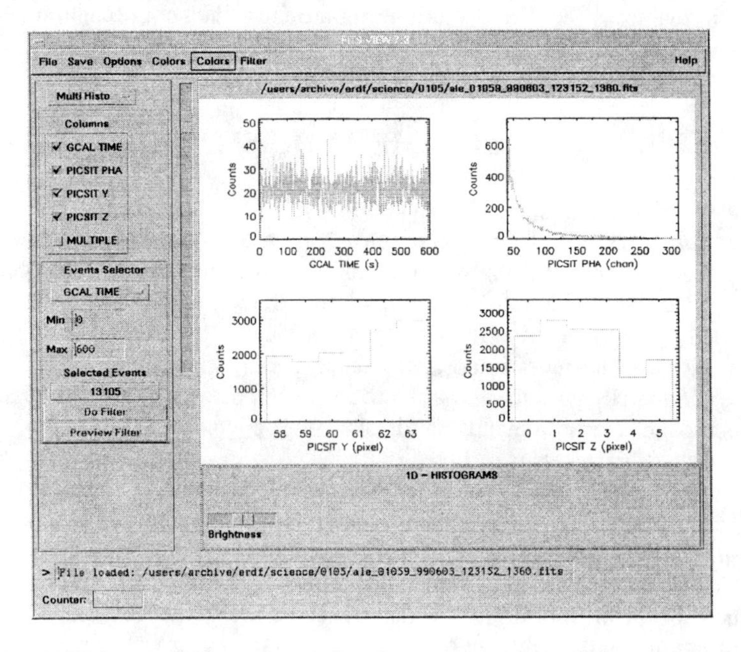

FIGURE 2. The FITSview multi-histograms window. In a unique graphic window are composed four spectra: Time Profile, PHA, Y Profile, Z Profile (data from a typical PICSIT background acquisition).

FITSview is able to display, also, the histogram of the interarrival times. Infact, the data distribution of the interarrival times for photons leaving a radioactive source, commonly used to test radiation detectors, are expected to follow a decaying exponential, due to the poissonian statistics underlying the physics of the event. Any deviation from this shape, indicates the presence of some noise in thc detector.
FITSview is designed to work also during an acquisition run, reading the data from the current FITS file, thus allowing a pseudo real time analysis mode. *FITSview* is able to save any of the images/histograms in a file, in any of the most popular graphic formats.

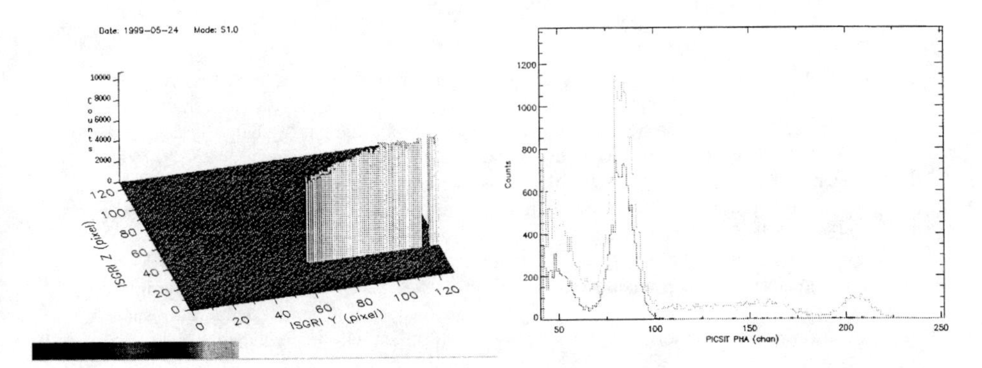

FIGURE 3. Left: A typical 3D display of the ISGRI layer illuminated by a ^{22}Na uncollimated source. The image shows the counts for the pixels (4x64) assembled in the ISGRI for the Engineering Model of IBIS. **Right:** The PHA spectrum of one of the pixels of PICSIT illuminated by a strongly collimated and tagged ^{22}Na source. The figure shows the PHA spectrum of all events (shaded line) against the PHA spectrum of the tagged events only (gray line), in this case, since only the 511 KeV events are tagged, the 1.27 MeV line is completely removed, showing the cleaned spectrum of the 511 KeV photons.

CONCLUSION

The ECOE software has been used in a massive way during all the functional and science tests performed on the Engineering Model of IBIS, at instrument level and at satellite level. It showed very high stability against crashes. A lot of graphic hard-copies were produced, thus accompanying adequately the test reports in the various test campaigns. Nevertheless, most of the functions have to be ameliorated and upgraded in view of the severe tests foreseen for the Flight Model of IBIS.

REFERENCES

1. INTEGRAL - *Experiment Interface Document*, part A, ESA Document (1998)
2. La Rosa G., Stephen J.B., Trifoglio M., *IBIS ECOE S/W Requirements*, IN-IB-IFP-SD-0005, Issue 1, IBIS Document (1998)
3. La Rosa G., Gianotti F., Trifoglio M., Fazio G., Segreto A., *IBIS ECOE S/W Architectural Design*, IN-IB-IFP-SD-0006, Issue 2.0, IBIS Document (1998)
4. Research Systems Inc., *Interactive Data Language*, ver. 5.2, Boulder, CO, USA

IBIS detectability of the Hard X-Ray Tailed sample of Bursters

A. Bazzano, G. De Cesare, M. Cocchi, L. Natalucci and P. Ubertini

Istituto di Astrofisica Spaziale, C.N.R., Via del Fosso del Cavaliere 00133 Roma- Italy

ABSTRACT. During the 2.5 years of Galactic Bulge Monitoring with the WFC on board the Beppo-SAX satellite new X-Ray Burst sources have been discovered. Some of these sources have also been promptly observed on a wide energy band by using the Narrow Field Instruments of the Satellite and a high energy component up to 200 keV has been detected. Moreover bursts behaviour has been discovered with the WFC for 4 more sources previously observed with other high energy instruments enlarging the sample of studied objects.
These results support the existence of such a class of previously suggested hard X-ray emitters that are very promising for the future IBIS/INTEGRAL CORE Programme observations (Galactic Plane Survey and the Galactic Centre Deep Exposure). Simulation results on the IBIS view on this class of object will be presented.

INTRODUCTION

X-ray Busters have been proven to be a new class of Hard X-ray emitters [1, 2]. In fact, the long term monitoring of the Galactic Bulge region performed by the SIGMA experiment, the first Gamma-Ray Imager ever flown on a Satellite, revealed for the first time a high energy component emission in the spectra of such a sources demonstrating the hard X-ray emission is not an exhaustive proof of the black hole candidacy. Later on a long term program of monitoring has been performed using BATSE experiment on board CGRO to investigate a sample of known X-Ray Busters using earth occultation and occultation transform imaging technique. At least 11 objects have been studied and the hard X-ray tail characteristic has been confirmed by the BATSE result on this sample of objects [1]. This issue has been then very challenging to stimulate the first simultaneous observations covering the X and Gamma ray domain, from 2 to at least 150 keV, with both the RossiXTE and BeppoSAX Telescope in order to discern successful criteria distinguishing Black Hole and Neutron Star on the basis of their broad band spectral properties. As a result the number of known Bursters is increased and detailed spectral parameters have been obtained in most cases [3, 4, 5, 6, 7, 8, 9, 10, 11, 12, 13, 14, 15]. So far the total number of hard tailed X-ray Bursters is at least 20 with a weaker indication for the existence of a high energy component in 2 other cases (XTE 1709-267 and XTE 1723-376).

CP510, *The Fifth Compton Symposium,* edited by M. L. McConnell and J. M. Ryan

IBIS ON INTEGRAL

IBIS [16] is an essential element of the INTEGRAL programme because of its powerful diagnostic capabilities of fine imaging source identification and spectral sensitivity to both continuum and broadened lines. It will observe, simultaneously with the other instruments on board INTEGRAL, celestial objects of all classes ranging from the most compact galactic systems to extragalactic objects.

IBIS (see [16]) features a coded aperture mask and a novel large area (~3,000 cm^2) multilayer detector which utilises both Cadmium Telluride (16,384 detectors) and Caesium Iodide elements (4,096 detectors) to provide ~12 arcmin fine angular resolution, wide spectral response (20 keV to 10 MeV), high resolution timing (60 μs) and spectroscopy (7% at 100 keV) required to satisfy the mission's imaging objectives. One of the major requirements for the new generation of Gamma-Ray telescopes is an outstanding angular resolution over a wide field of view (FOV). The current situation for IBIS is described in Table 1.

TABLE 1. IBIS field of view

Field of view	Extension(squared degrees)	Type of coding
Squared	9 x 9	Fully coded
Squared	19 x 19	FWHM
Squared	29 x 29	Zero response

The choice of a large IBIS field of view, similar to the one of the other main instrument SPI, will allow joint observations of any region of the sky with complementary capability ensuring the main goal of the INTEGRAL mission. Such an extension of the IBIS FOV will also enable a better sky coverage and hence to increase the number of detectable sources, and monitoring capability which is mandatory due to the high variability of the X and Gamma-Ray sky.

The IBIS angular resolution is 12 arcmin, whereas the point source location accuracy (for a relatively strong source) would be around 1 arcmin. Assuming no error in pointing axis reconstruction or other systematic effects ISGRI will locate a 30 sigma point like source within an error box smaller than 30 arcsec radius.

IBIS continuum sensitivity to different source fluxes during an exposure time of 9000 s is shown in Figure 1 while in Figure 2 the IBIS simulated count rate spectrum for a Crab-like source (50 milliCrab intensity) is shown.

OBSERVING STRATEGY

The CORE Programme (CP), i.e. the guaranteed time for the INTEGRAL Science Working Team as return for the contribution of both hardware teams and other participants team to execution of the INTEGRAL programme, will consist of 3 elements:

- Galactic Plane Regular Scans

- Deep Exposure of the Galactic Central radian
- Pointed Observation of selected sources

The amount of time devoted to CORE programme is 35% during the first year of observations, than decreasing to 25% after the second year. During both the Regular scans of the Galactic Plane and the Central Radian Deep Exposure comparison of spectral and time characteristic of compact objects, Neutron Stars and Black Hole Candidates, will be performed.

In particular 2.3 x 10^6 s and 4.8 x 10^6 s have been allocated respectively for this 2 key elements of the CP in the hypothesis of a PROTON launcher for the first year operative life of the Satellite.

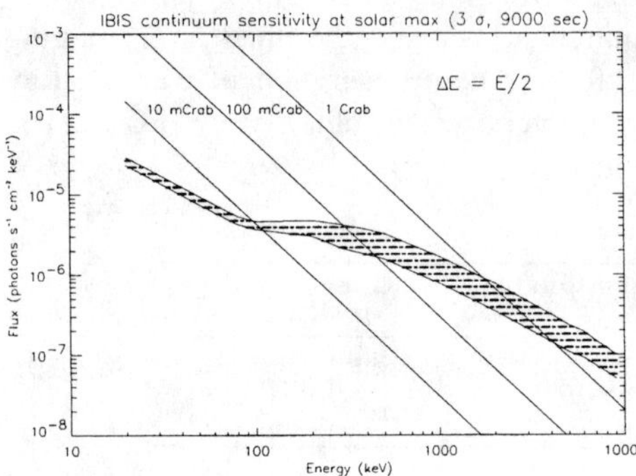

FIGURE 1. IBIS continuum sensitivity for the GCDE assuming 9000 s observing time. At low energy (less than 200 keV) the sensitivity depends on ISGRI detector only. This curve has been evaluated using Monte Carlo simulation in order to compute the ISGRI efficiency.

FIGURE 2. Count rate spectrum for X1724-30.1 in Terzan 2 corresponding to a 50 milliCrab flux as will be detected by ISGRI. This spectrum has been generated by a physical Monte Carlo simulation of the transport of the electro-photonic shower inside the instrument using a simplified Mass Model of IBIS. The CERN GEANT3 package has been used to implement the model.

The central radian of the Galaxy will be observed using a rectangular pointing grid of 21x 31 pointings centred on the Galactic Centre with a pitch angle of 2 degree. This grid will cover a celestial region between l ≤|30 °| and b l≤|20 °| in order to optimise the Spectrometer imaging response. The preliminary plan is to survey inner part of the region with a double exposure time and the nominal dither exposure will be 30 minutes [17].

As a result of an optimisation process taking into account scientific objectives, different instruments characteristics, Spacecraft and Ground Segment elements, observing strategy for the Galactic Plane survey is as follows:

- regular scans will be performed once a week by slew and stare manoeuvre of the spacecraft across the accessible part of the plane within ± 10 ° latitude with a 6° scan path (separation between 2 different exposures). Scan will be executed as a sawtooth with inclination of 21° with respect to the Galactic Plane and with consecutive scans shifted by a 27.5°.

During both the GPS and GCDE IBIS will be able to constrain the position and the detailed spectral and timing behaviour for X-ray burst sources as can be seen in the simulated images of fully coded region around the Galactic Centre. In the fully coded IBIS FOV the known high energy Busters detected so far are X1724-30.1, A1742-294, SAX J 1747-285 and SLX 1735-269, as it is shown in figure 3.

In the simulated images besides these objects, the persistent, though variable, source 1E1740.7-2942 has been taken into account. In fact, 1E1740.7-2942 is the only persistent high energy source reported by the SIGMA experiment along all the lifetime of operation in the FOV under study [18].

In Table 2 we report both flux variability ranges and spectral indices, assuming a power law model, as found in the literature for the 5 objects in the FCFOV.

TABLE 2.

Source	Flux (mCrab)	Spectral index
X1724-30.1	20-50	2.1
A1742-294	5-200	3.1
SAX J1747-285	20	2.05
SLX 1735-269	20	2.05
1E1740.7-294	13-140	2.03

As is shown in Figure 4a, b and c, even simply adding 5 GCDE pointings of 1800s each, all the 5 sources are clearly detectable up to at least 50 keV, while for the higher energy images (50-300 keV) only A1742-294 does not appear because of its very soft spectral index.

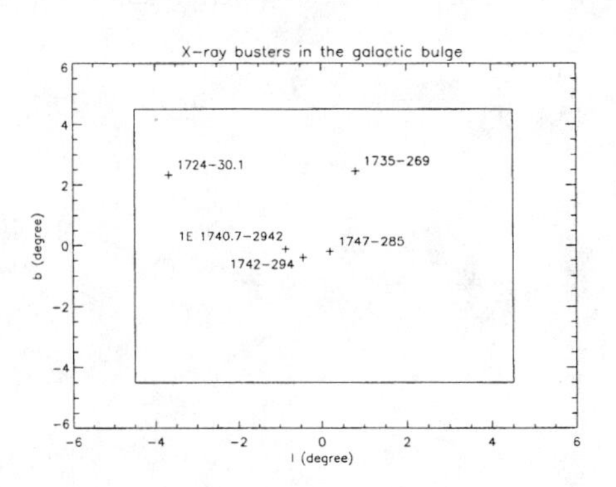

FIGURE 3. Known X-ray Busters in the IBIS fully coded field of view (9 x 9 degree) amenable for observation during the INTEGRAL Galactic Centre Deep Exposure.

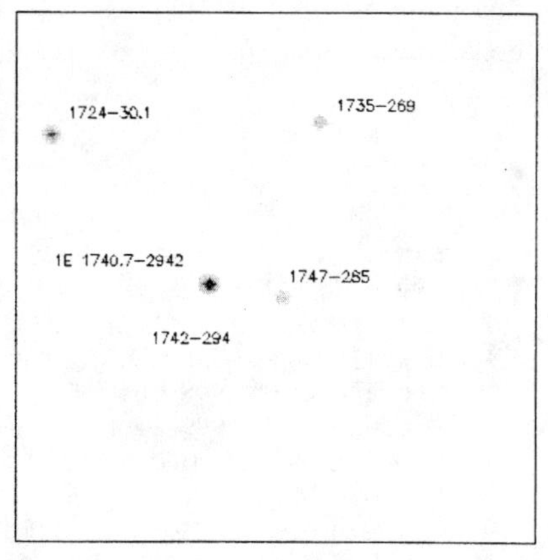

FIGURE 4a. 15 - 50 keV ISGRI image generated by Monte Carlo simulation. The observing time correspond to GCDE 5 pointings (9000,i.e. 1800x 5). Fluxes values for the X-ray busters has been assumed as the maximum of the variability range.

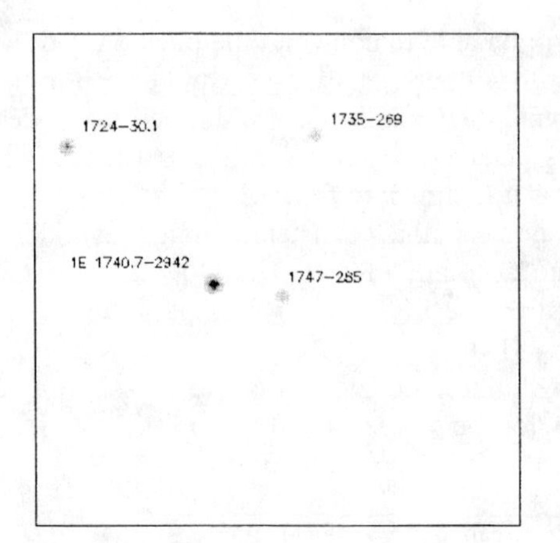

FIGURE 4b. 50 - 300 keV ISGRI image generated by Monte Carlo Simulation. Observing time and flux for the sources as for Figure 4a.

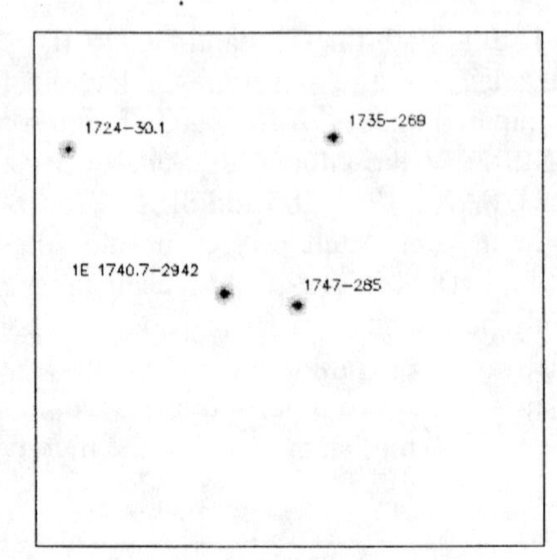

FIGURE 4c. 15 - 50 keV ISGRI image generated by Monte Carlo simulation. Observing time as for 4a and 4b while the flux of the sources has been assumed at the average level

Aim of these observations will be to: *a*) Finally establish whether the hard X-ray emission is associated (anti-correlated) with low state, i.e. low accretion rate, as firstly suggested by Zhang et al. [19]. *b)* Establish the existences and values for cut-off energy that has been suggested as possible parameter to distinguish between Black Holes and Neutron Stars for as many as possible sources. Recent results are in fact indicating cut off or break at ~50 keV for NS and above 100 keV for BHC; *c)* detect new transient sources and discern their nature by their spectral and timing behaviour.

REFERENCES

1. Barret D., et al., 1996, *A. & AS* **120**,121
2. Tavani M and Barret D., 1997, *AIP*, **410**, 75
3. in 't Zand J. et al., 1998, *A&A*, **331**, L 25
4. Heindl W. A. and Smith D. M., et al., 1998 *ApJL*, **506**, L35
5. Cocchi M. et al., 1998, *ApJL*, **508**, L163
6. Cocchi M. et al., 1999, *A&A* , **346**, L 48
7. Guainazzi M. et al., 1998 *A&A* **339**, 802
8. Guainazzi M., et al., 1999 submitted to *A&A*,
9. in 't Zand J. et al., 1999 *A&A*, **345**,100
10. in 't Zand J. et al., 1999b *A&A* submitted
11. Olive et al., 1999 in press, *proceedings* of 32nd Cospar meeting
12. Bazzano A. et al., 1998 IAU circular 6873
13. Bazzano A. et al., 1997 IAU circular 6668
14. Ubertini P. et al., 1998, IAU circular 6843
15. Natalucci L. et al., 1999,in preparation
16. Ubertini et al., 1996, *SPI*, Vol. **2806**, 246
17. Winkler C. et al., 1999, INT-TN 22523
18. Vargas et al., 1996 ESA SP, **382**, 129
19. Zhang S.N. et al., 1996, *A&AS*, **120**, 279

Observation of X-ray Novae with INTEGRAL

A. Goldwurm*, P. Goldoni*, P. Laurent* and F. Lebrun*

*Service d'Astrophysique/DAPNIA, CEA/Saclay, F-91191 Gif-Sur-Yvette, France

Abstract. The INTEGRAL satellite, with the hard X-ray/soft gamma ray telescope IBIS onboard, will be launched in 2001. The core program observations include the Galactic Plane Survey (GPS) and the Galactic Plane Deep Exposure (GCDE). They will respectively provide a weekly coverage of a significant part of the Galactic plane and a deep exposure of the central region of the Galaxy. Both will lead to the discovery of new transient sources, X-ray Novae being the best example. We discuss the possible strategies and results of IBIS/INTEGRAL detection and follow-up observations of X-ray Novae in the framework of the Galactic Plane Survey.

INTRODUCTION

The Imager on Board Integral Satellite (IBIS) [8] is the imaging instrument of the INTEGRAL satellite, the hard-X/soft-gamma ray ESA mission to be launched in 2001. The INTEGRAL observing program is divided in a core program reserved to instrument teams and collaborators and an open observing program. The core program is mainly formed by two kind of galactic surveys: the Galactic Plane Deep Exposure and the Galactic Plane Survey (GPS) [11]. The first one will consist of deep pointings of the central region of the Galaxy. The GPS (for a total of 2.3 10^6 s per year) instead is composed of a series of separate short pointings arranged to perform a scan of a section of the galactic plane every week.
This portion of the core program will monitor the galactic plane to search for galactic transient sources at the beginning of their outbursts. The discovery and arcminute positioning of a new transient source will trigger INTEGRAL TOO observations, and will generate an alert to the astronomical community enabling ground and space observatories to start observations at different wavelengths. If a TOO observation is not triggered, the GPS will anyway follow the transient during its outburst with an observation every week, if the pointing constraints allow it. We present here (but see also [2]) simulations of IBIS Galactic plane survey observations. They show the capabilities of this instrument in discriminating between different sources while at the same time monitoring a huge FOV. Prime target of the INTEGRAL GPS program in particular with the IBIS telescope are

CP510, *The Fifth Compton Symposium,* edited by M. L. McConnell and J. M. Ryan

X-Ray Novae [7,1] a class of hard X-ray transients known to contain black holes accreting matter from a low mass companion star.

THE IBIS TELESCOPE

IBIS provides diagnostic capabilities of fine imaging, source identification and spectral sensitivity to both continuum and broad lines over a broad (15 keV–10 MeV) energy range. It has a continuum sensitivity of 4 10^{-7} ph cm^{-2} s^{-1} (1 mCrab) at 100 keV and 2 10^{-7} ph cm^{-2} s^{-1} (25 mCrab) at 1 MeV for a 10^6 seconds observation (ΔE=E) and a spectral resolution better than 7 % at 100 keV and of 6 % at 1 MeV. The imaging capabilities of the IBIS are characterized by a 12′ FWHM angular resolution, a 3′ to <1′ point-source location accuracy and a very wide field of view (FOV), namely 9° × 9° fully coded and 29° × 29° partially coded at 0 sensitivity.

The IBIS detection system is composed of two planes, an upper layer made of 16384 squared CdTe pixels (ISGRI) and a lower layer made of 4096 CsI scintillation bars (PiCsIT). This system enables high sensitivity continuum spectroscopy and a wide spectral range (15 keV – 10 MeV).

The simulation we performed are for the moment limited to the ISGRI upper layer [6]. The ISGRI pixels are 4× 4 mm^2, 2 mm thick crystals of Cadmium Telluride, a semiconductor operating at ambient temperature, providing a spectral resolution of about 8% at room temperature. The 128×128 pixels are arranged in 8 modules separated by dead zones 2 pixel wide needed by the mechanical structures which sustain the detector plane. The sensitivity loss caused by dead zones is not large, however the absence of sensitive elements in the detector plane must be properly taken into account during deconvolution procedures [5].

GPS PARAMETERS AND SIMULATIONS

One GPS scan will involve 18 to 57 (depending on visibility and satellite constraints) 1050 sec pointings spaced by 6° on a saw-tooth pattern inclined by 21° with respect to the galactic plane [11]. This effectively results in an coverage of regions extending from 100° up to 320° in Galactic Longitude and 40° in Galactic Latitude. Every week the scan is repeated with a 27° shift in longitude. Taking into account the scan pointings and the sensitivity over the FOV, the effective IBIS exposure for each sky point along the path is about 3250 seconds, which translates in a 3σ sensitivity of ~20 mCrab in the 50-150 keV band. This is vastly better than the sensitivity level of the BATSE occultation analysis on similar timescales, assuring the monitoring of hard X-ray sources activity at a level never reached by past all-sky monitors.

We simulated a GPS scan in the 50-150 KeV band containing 2 galactic regions from -30° to 60° and from 90° to 180° in longitude. We selected a sample of hard X-ray galactic sources extracted from the Heasarc X-ray binaries catalog [9] and containing the sources monitored by BATSE in its occultation program

TABLE 1. Sources, their simulated 50-150 keV flux and measured signal to noise level at source position (level of detection for Novae).

GRS 1716-249	100 mCrab	16 σ	Bright Nova	1 Crab	170 σ
SLX 1735-269	90 mCrab	16 σ	4U 1630-472	100 mCrab	10 σ
GX 1+4	40 mCrab	7 σ	GX 339-4	200 mCrab	22σ
GRS 1758-258	60 mCrab	13 σ	OAO 1657-415	100 mCrab	13 σ
GS 1826-38	30 mCrab	2.6 σ	GRO J1655-40	200 mCrab	27 σ
GS 1845-03	100 mCrab	21 σ	4U 1700-377	250 mCrab	38 σ
A 1845-024	40 mCrab	8 σ	GX 354-00	50 mCrab	8 σ
GRS 1915+105	300 mCrab	51 σ	Faint Nova	30 mCrab	6 σ
QSO 0241+622	100 mCrab	20 σ	1E1740.7-2942	100 mCrab	18 σ
LSI+61 303	20 mCrab	2.5 σ	A 1742-294	20 mCrab	2.5 σ
4U0115+63	50 mCrab	8 σ	Terzan 2	30 mCrab	5 σ

(http://cossc.gsfc.nasa.gov/cossc/batse/ hilev/occ.html) and transient and persistent sources detected by the SIGMA hard X-ray/soft γ-ray telescope in the Galactic plane [4]. The sample was selected in order to give a representative example of GPS results and possibilities but we consider that several other sources could be detected in an average scan, especially at lower energies. The simulated fluxes are representative of the average behaviour of these sources. To this sample we added two serendipitous transient events to check the sensitivity of our procedures to these occurrences. A bright (1 Crab) one to test the localization capability of our system and a faint (25 mCrab) one in a crowded region to test the source discrimination capability. We simulated the bright one in the second scan (l between 90° and 180°) while the faint transient in the Galactic Bulge, where most of the weak X-ray novae like GRS 1739-278 [10] were observed.

The simulated images are deconvolved, analysed, cleaned and summed [5]. In this process point-sources are detected and localized by fitting with the proper Point Spread Function. A section of the resulting image of the first scan (around the Galactic Center) is displayed in Fig. 1. The simulation parameters and the deconvolution results are shown in Table 1. Note that the signal to noise level depends both on source strength and on its effective exposure. The GPS sensitivity is maximal (RMS noise 6 mCrab) on the galactic plane and decreases towards the image limits. From the results reported in Table 1 it is clear that hard X-ray sources will be easily detected if they have a flux $\geq$ 30 mCrab.

We here focus on IBIS capability of positioning a new source in the two cases we simulated. The 1 Crab transient can be positioned by IBIS in principle with a 90% error radius of <10 arcseconds. In reality some systematic effects (e.g due to thermal distorsions) are expected to limit the accuracy to $\sim$20″ (but proper in-flight calibrations and modelling should allow to correct part of these effects

after the first months of the mission). This gives reasonable possibility of promptly identifiying the optical/radio counterpart from the ground directly from the IBIS derived position.
The 30 mCrab transient is easily detected in a crowded region at a 6 σ level (see detailed image in [2]) with a $2'$ error radius, a value comparable to the one obtained by XTE/PCA for stronger ($\sim$200 mCrab) transients like XTEJ1755-324. On the other hand this error box is too big to allow a direct identification from the ground. An optical/radio search of the counterpart may however be pursued from the ground trying to identify highly variable objects in the error box. Follow up observations with more sensitive X-ray instruments will in principle allow to identify the origin of the emission. If the object is an X-ray Nova at its maximum [10], then it is likely located on the other side of the Galaxy allowing us to probe this largely unknown region.

CONCLUSIONS

We have presented simulations of IBIS/INTEGRAL GPS observations, the main result are the following:
1) Several known sources will be clearly ($\geq$ 10 σ) detected allowing to determine their spectral state.
2) Bright transient sources will be readily localised with a precision previously unavailable in this energy range. The high level of detection will allow to search for high energy features [3] and other spectral signatures of BHs in X-ray Novae.
3) Fainter, more distant transients will be detected and localised with lower precision allowing to determine the number of these sources in the Galaxy.

REFERENCES

1. Chen W., Shrader C.R. & Livio M., *ApJ* **491**, 312, (1997).
2. Goldoni P.et al., *X-Ray Astronomy 1999 Conf. Proc.*, Bologna 6-10 Sep 1999, in press (1999).
3. Goldwurm A. et al., *ApJ* **389**, L79, (1992).
4. Goldwurm A. et al., *Nat* **371**, 589 (1994).
5. Goldwurm A. et al., *25th ICRC Conf. Proc.* **OG 10**, 309 (1997).
6. Lebrun F. et al., *SPIE* **2806**, 258 (1996).
7. Tanaka Y. and Shibazaki N., *ARAA* **34**, 607 (1996).
8. Ubertini P. et al., *Proc. of 3rd INTEGRAL Workshop, Astrophysical Letters & Communications* , in press (1999).
9. Van Paradijs J., *X-ray Binaries*, eds. W.H.G. Lewin, J.v.Paradijs, and E.P.J. van den Heuvel, Cambridge University Press, 1995, P536.
10. Vargas M. et al., *ApJ* **476**, L23, (1997).
11. Winkler C. et al., *Proc. of 3rd INTEGRAL Workshop, Astrophysical Letters & Communications* , in press (1999).

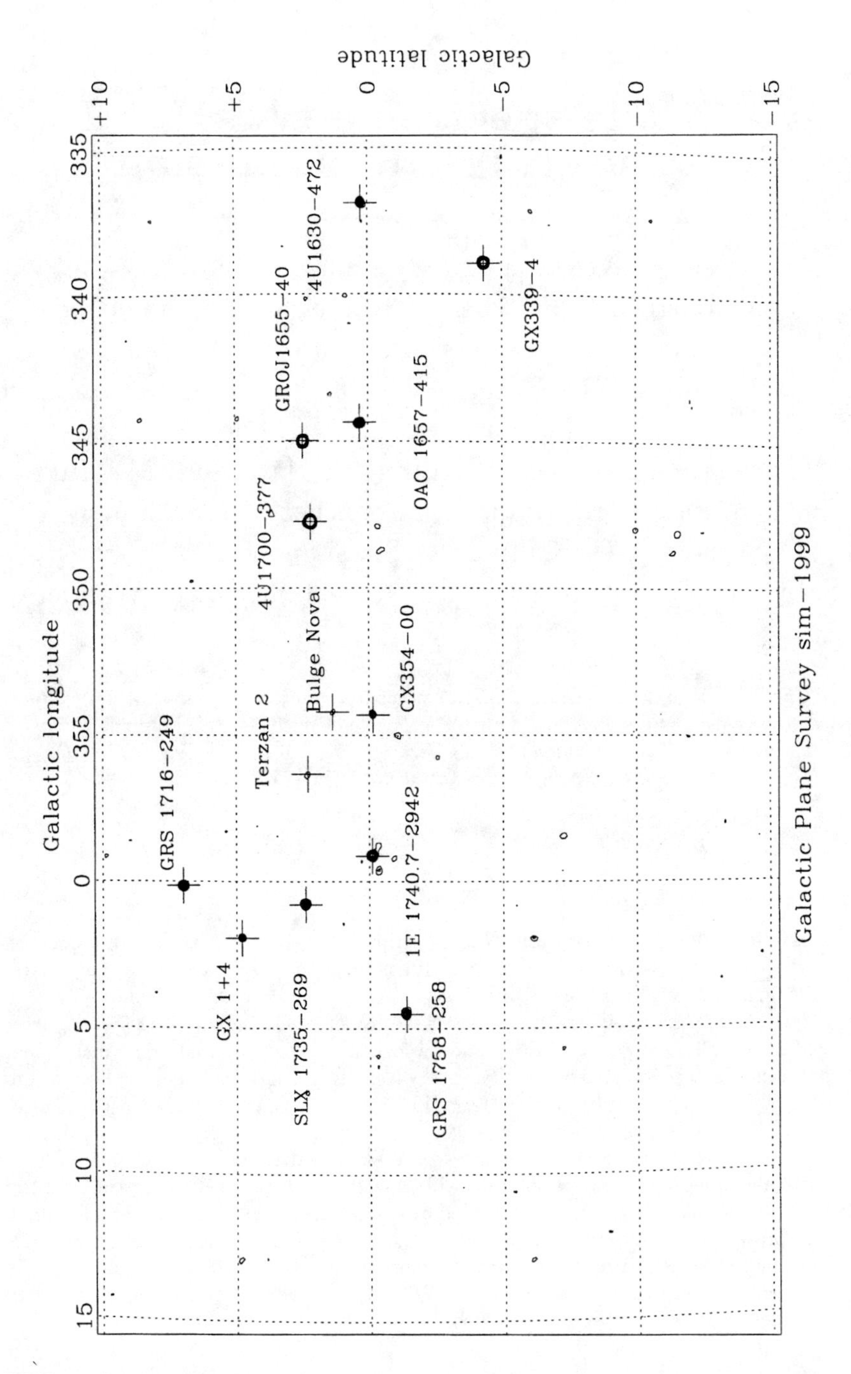

FIGURE 1. Contour level image of part of the 50-150 keV reconstructed sky region of a simulated GPS scan with the IBIS telescope. Levels start from 2.5 σ, spaced by 1 σ. All marked sources are detected at more then 4.5σ

The spectrometer SPI of the INTEGRAL mission

P. Jean[1], G. Vedrenne[1], V. Schönfelder[2], F. Albernhe[1], V. Borrel[1], L. Bouchet[1], P. Caraveo[3], P. Connell[4], B. Cordier[5], M. Denis[6], R. Coszach[7], R. Diehl[2], Ph. Durouchoux[5], R. Georgii[2], J. Juchniewicz[6], A. von Kienlin[2], J. Knödlseder[1], Th. Larque[5], J.M. Lavigne[1], P. Leleux[7], G. Lichti[2], R. Lin[8], P. Mandrou[1], J. Matteson[9], M. Mur[5], Ph. Paul[1], J.P. Roques[1], F. Sanchez[11], S. Schanne[5], C. Shrader[10], G. Skinner[4], S. Slassi-Sennou[8], A. Strong[2], S. Sturner[10], B. Teegarden[10], P. von Ballmoos[1], C. Wunderer[2]

1 CESR, Toulouse, France; 2 MPE, Garching, Germany; 3 IFC/CNR, Milano, Italy; 4 University of Birmingham, UK; 5 CEA, Saclay, France;6 CBK, Warsaw, Poland; 7 University of Louvain, Belgium; 8 SSL, Berkeley, USA; 9 UCSD/CASS, La Jolla, USA; 10 GSFC, Greenbelt, USA;11 University of Valencia, Spain

Abstract. The spectrometer on INTEGRAL (SPI) is one of the two main telescopes of the future INTEGRAL observatory. SPI is made of a compact hexagonal matrix of 19 high-purity germanium detectors shielded by a massive anticoincidence system. A HURA type coded aperture modulates the astrophysical signal. The spectrometer system, its physical characteristics and performances are presented. The instrument properties such as imaging capability, energy resolution and sensitivity have been evaluated by means of extensive Monte-Carlo simulations. With the expected performances of SPI, it will be possible to explore the γ-ray sky in greater depth and detail than it was possible with previous γ-ray telescopes like SIGMA, OSSE and COMPTEL. In particular, the high-energy resolution will allow for the first time the measurement of γ-ray line profiles. Such lines are emitted by the debris of nucleosynthesis and annihilation processes in our Galaxy. Lines from these processes have already been measured, but due to the relatively poor energy resolution, details of the emission processes in the source regions could not be studied. With the high-resolution spectroscopy of SPI such detailed investigations will be possible.

CP510, *The Fifth Compton Symposium,* edited by M. L. McConnell and J. M. Ryan

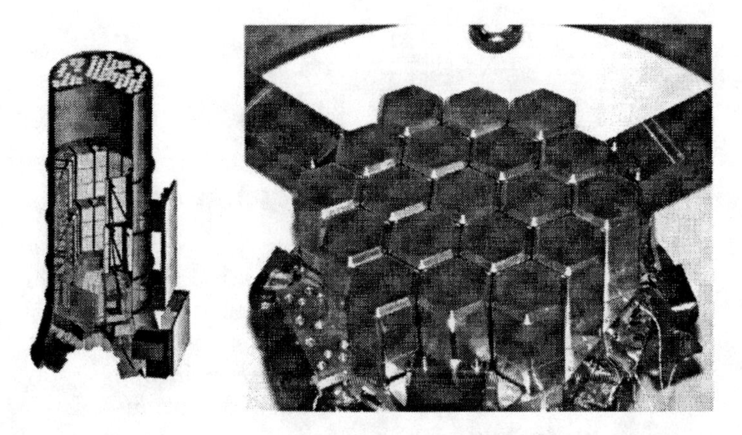

FIGURE 1. View of the spectrometer SPI (left) and its flight model Ge detector array (right).

DESIGN OF SPI

SPI will be aboard the INTEGRAL observatory which is an ESA's high-energy astrophysics mission to be launched on a Proton vehicle into a high eccentric orbit end 2001. Figure 1 (left) shows a view of the spectrometer. The detection plane is made of an array of 19 hexagonal cooled (85 K) high-purity Ge detectors providing a total area of ≈500 cm^2. The diameter and the thickness of detectors are 5.6 cm (flat to flat) and 7 cm respectively. Figure 1 (right) shows a view of the SPI flight-model Ge-detector array. It will be mounted on a beryllium plate and housed in a beryllium cryostat. A Pulse Shape Discrimination (PSD [1]) electronic system has been implemented in order to reduce the β decay background in Ge. The anticoincidence system consists of: (i) 91 bismuth germanate scintillator blocks that define the instrument field-of-view (FOV) and reduce the background, and (ii) a plastic scintillator underneath the coded-aperture in order to reduce the background originated in the mask. The coded aperture is 171 cm from the detection plane. It is made of 63 hexagonal 3 cm thick tungsten alloy elements. The design of the different parts of the spectrometer (shield thickness, detector nature...) has been optimised to reduce the instrumental background.

PERFORMANCES OF SPI

The anticipated performance for narrow-line spectroscopy is characterized by an energy resolution in the parts per thousand in the energy range relevant for nuclear astrophysics. The table below shows the SPI performance parameters.

Energy range	20 keV - 8 MeV
Energy resolution	2 keV at 1 MeV
Fully coded FOV	16 °
Partially coded FOV	35 °
Angular resolution	3. °
Point source location	<2 °

The sensitivity of the spectrometer is limited by the instrumental background induced by primary and secondary cosmic-ray particles and cosmic diffuse emission. The instrumental continuum background has been estimated using mainly the GEANT code and semi-empirical and empirical nuclear cross-sections. The physical effects inducing such background events are: the cosmic diffuse γ-ray emission, the radioactive decays in Ge, the spacecraft γ-ray and neutron emissions and the mask γ-ray emission. Figure 2 presents the point source narrow-line sensitivity that has been estimated using predicted background [2] and calculated efficiency which agrees with measurements [3]. Background reduction techniques (PSD, rejection of multiple events with a 511 keV signature) have been applied. Since positrons are induced by interaction of cosmic-rays in the instrument, a strong 511 keV background line limits the sensitivity at 2.8 10^{-5} photons cm^{-2} s^{-1} at this energy of astrophysical interest.

SCIENTIFIC OBJECTIVES

The energy range, spatial and spectral resolution of SPI are optimised for the measurement of astrophysical γ-ray lines. Such lines are emitted by radioactive nuclei produced in supernovae (^{56}Co, ^{57}Fe, ^{44}Ti, ^{26}Al, ^{60}Fe), in novae (^{22}Na, ^{7}Be) and WR stars (^{26}Al). The measured characteristics (intensity, shape, location) of these

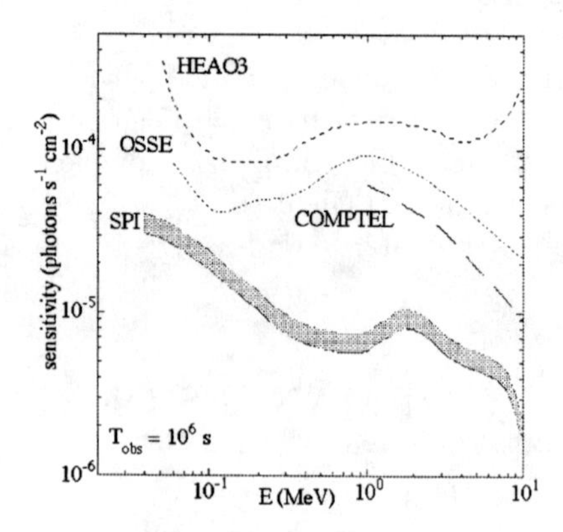

FIGURE 2. SPI on-axis point source narrow-line sensivity (3σ) for an observation time of 10^6 s ($\approx$11 days). Sensitivities of existing experiments are also shown.

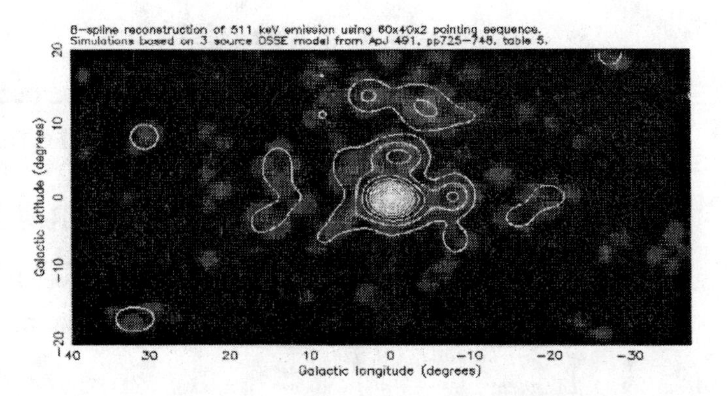

FIGURE 3. Simulated images of a possible 511 keV galactic emission model (see text) for the Galactic Center Deep Exposure ($\approx$ 56 days) of INTEGRAL.

γ-ray lines yield valuable information on the physics (e.g. SNIa explosion mechanism [4], mass-cut in core-collapse supernovae...) involved in the nucleosynthesis sites and their distribution in our Galaxy. Gamma-ray lines are also produced in the interstellar medium (ISM) such as the 511 keV emission due to annihilation of galactic e^+, that has been observed since the early 1970s. Figure 3 shows a simulated SPI observation of this emission based on a particular model [5]. Spectroscopic analysis of the e^+e^- line will provide insight into the physical conditions (temperature, density) of the annihilation regions, and its spatial distribution will provide clues for the e^+ origin. Nuclear de-excitation lines from energetic particle interactions with the ISM nuclei are also expected but remain to be discovered.

Although designed for γ-ray line observation, SPI will also be able to provide constraints on models of continuum emission from hard X-ray and γ-ray sources such as galactic center sources and active galactic nuclei. Other continuum emissions of interest are the galactic continuum emission and cosmic γ-ray background (CGB). In the 300 keV to 3 MeV range, the CGB may be due to the cumulative emission of cosmological SNIa. Its spectrum would contain edge features with amplitude depending on the cosmological parameter values. With its expected performances, the scientific objectives of SPI will cover a large field of astrophysical investigation such as stellar nucleosynthesis, physic of compact objects and of ISM, Galaxy structure and cosmology.

REFERENCES

1. Skelton J., et al., these proceedings (1999).
2. Jean P., Ph. D. Dissertation, Université Paul Sabatier, Toulouse (1996).
3. Kandel B., Ph. D. Dissertation, Université Paul Sabatier, Toulouse (1998).
4. Gomez-Gomar J., Isern J., Jean P., MNRAS 295, 1 (1998).
5. Purcell W.R., et al., ApJ 491, 725 (1997).

Pulse Shape Discrimination on the INTEGRAL Imaging Spectrometer

R. T. Skelton*, J. L. Matteson*, S. A. Slassi-Sennou†, R. P. Lin†, N. W. Madden‡, P. von Ballmoos§, J. Knoedlseder§

*Center for Astrophysics and Space Science, University of California San Diego La Jolla, CA 92093-0424
†Space Sciences Laboratory, University of California, Berkeley CA 94720-7450
‡Lawrence Berkeley National Laboratory, University of California, Berkeley, CA 94720
§Centre d'Etude Spatiale des Rayonnements, 31029 Toulouse, France

Abstract. NASA is providing a Pulse Shape Discrimination (PSD) system for the Ge detectors of the Imaging Spectrometer (SPI) on ESA's INTEGRAL mission. This will reduce the background and improve the sensitivity between approximately 400 keV and 2 MeV. The dominant background contributor for the detectors in this energy range will be β^- decays from Ge isotopes activated by cosmic rays. This induced activity has mostly single-site energy losses, while photons in this energy range usually deposit energy in multiple sites. The differences in these types of events' pulse shapes can be exploited to reject single-site events. The effectiveness of this technique is enhanced by using Ge detectors with small central bores and a high-bandpass preamplifier output (separate from the spectroscopy output); pulses from single-site events are strongly peaked, and pulse duration is around 400 ns. This signal is sampled at 100 MHz, digitized to 9 bits, and analyzed on the basis of its resemblance to a single-site event versus a linear combination.

Recent progress includes an analysis algorithm that runs fast enough in the flight processor yet still approaches the theoretical increase in sensitivity, rejecting more than 95% of the single-site events while accepting about 60% of the multiple-site events. For the background expected in the interplanetary INTEGRAL orbit, sensitivity should be improved by a factor of 2.1 at 845 keV.

INTRODUCTION

From its orbit in interplanetary space, SPI on INTEGRAL will conduct gamma-ray line observations with a resolving power, $E/\Delta E$, around 600 at 1 MeV and a sensitivity to narrow lines of $(2 \text{ to } 5)\times10^{-6}$ γ/cm^2 s for 4×10^6 s observation time in its nominal energy range from $\approx$ 20 keV to $\approx$ 8 MeV.

Its observations will be background limited, with the background rate $\sim$ 90% of the total count rate. The most significant background component will be delayed

CP510, *The Fifth Compton Symposium,* edited by M. L. McConnell and J. M. Ryan

β^- decays over most of the PSD energy range (400-2000 keV), by nearly an order of magnitude [1]. The largest single contributor to these β^- decays is ^{75}Ge, which decays directly to the ground state 87% of the time (no γ). Since the range of a 1-MeV electron in Ge is less than 1 mm, the energy deposition for these events will be essentially at a single site in the detector. Above several hundred keV, photons depositing their full energy will generally Compton-scatter at least once, with the recoil photon traversing $\sim$ 1 cm before being fully absorbed. Hence, photons normally deposit their energy at multiple sites.

PSD rejects most single-site events and accepts most multiple-site events by analyzing the shape of the current pulse from the Ge detector, yielding a sensitivity improvement between 400 keV and 2 MeV, which peaks around 800 keV at $\approx$ 2.1. PSD will further improve SPI performance by recognizing and marking piled up pulses. It will also improve energy resolution by correcting for ballistic deficit and mitigate radiation damage [2,3] by providing information on interaction location.

SYSTEM AND DISCRIMINATION ALGORITHM

SPI contains 19 closely packed hexagonal Ge detectors which view the sky through a coded mask to achieve $\approx 2°$ position resolution. The overall SPI instrument is described more fully elsewhere [4,5]. Each of the 19 Ge detectors has a pre-amplifier with two separate output channels. The 19 spectroscopy output channels feed the energy analyzers. The 19 buffered fast output channels feed the PSD, which has pulse recognition logic, a multiplexer, and an Analog-to-Digital Converter (ADC) system which samples to 9-bit accuracy at 100 MHz. The digitized data is analyzed by the PSD Digital Signal Processor, which determines whether the pulse shape is valid and in the PSD energy range; if so, the shape is analyzed to determine whether the event was a single or a multiple, and the results are formatted for the downlink telemetry, with a throughput of 800 events/s.

The discrimination algorithm is based on that of Philhour et al. [6], which involves several steps: (a) A template library of single shapes, sorted by time to peak (TTP), is assembled and stored in memory (Fig. 1a); (b) the actual pulse, ψ_{test}, is fit as $\psi_{\text{fit}} = \alpha\psi_i + (1-\alpha)\psi_j$, where $\psi_{i,j}$ are library templates and α is a number between

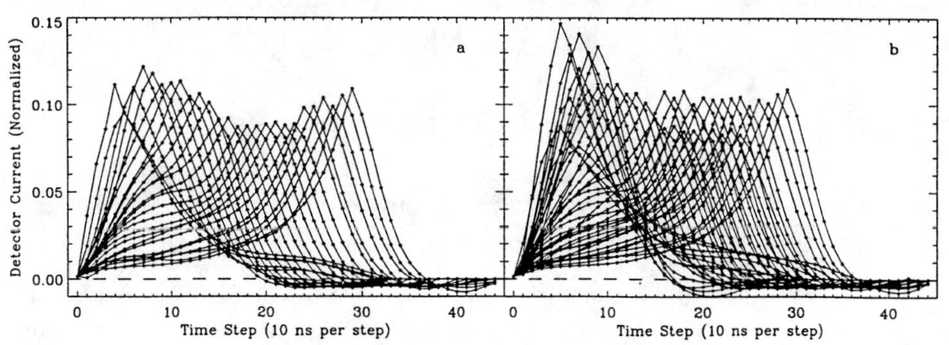

FIGURE 1. (a) Basic library, 1 template per TTP. (b) Expanded library.

0.5 and 1.0; and the template indices i and j and the value of α are determined to minimize the χ^2 between ψ_{test} and ψ_{fit}; and (c) the event is deemed a multiple if, e.g., α is less than 0.85 *and* ΔTTP between ψ_i and ψ_j exceeds 20 ns; exact values depend on energy and are to be optimized.

The variety of pulse shapes in Fig. 1 stems from differences in interaction location in the detector, since hole and electron mobilities, electric fields, weighting potentials, and electronic circuitry time constants all influence the shape [7]. In general, events close to the central bore give rise to pulses with shorter TTP, and events in the closed end give rise to pulses with shorter fall times.

We have found that the ψ_i always has a TTP within 20 ns of the pulse TTP, reducing the search space. We have also found that using a three-element fit does not improve the discrimination results. Several promising enhancements are being considered: (a) a library with more than one template per TTP (Fig. 1b); (b) more sophisticated discrimination, including use of the absolute χ^2 and dynamically adjusting criteria; and (c) using the χ^2 to "quick-ID" an event – if χ^2 for an *individual* template is low enough, the event can rapidly be deemed a single.

VALIDATION OF COMPUTER SIMULATION

One key element in the PSD development is a computer program due to Slassi [7,8] to generate simulated pulses, based on energy deposition at locations in the detector. The modeling includes the hexagonal shape of the detectors. Locations can be generated manually, randomly, or via Monte Carlo simulation of photon and electron transport. Noise modeling includes frequency dependent effects from the electronic circuitry, and noise parameters are adjustable. Fig. 2 shows comparisons of this model to actual laboratory data. For the simulations, all events were singles, with a random interaction location in the detector, and noise

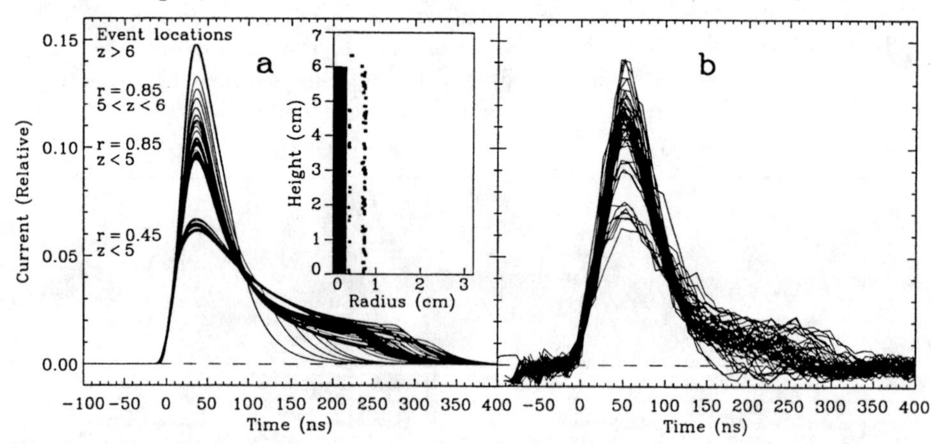

FIGURE 2. (a) Simulated pulses with TTP between 34 and 36 ns and their locations in the detector (inset). Locations are annotated adjacent to the peaks of the pulse families. (b) Laboratory pulse shapes with TTP from 44 to 52 ns.

set to zero for ideal pulses. Laboratory data were generated by Compton scattering gamma rays in one Ge detector with a second Ge detector collimated to capture recoil photons at 90°. By demanding coincidence and appropriate energy depositions, nearly pure singles are recorded.

Fig. 2a shows simulated pulses with TTP between 34 and 36 ns, slightly above the shortest TTP of all simulated pulses; the diagram in the right shows a cross-section of the detector, with the central bore in black, and the interaction location for each pulse. Selecting this TTP range results in selecting events from specific detector regions. Interactive analysis provided the correlation between interaction location and pulse shape indicated on the figure. Fig. 2b shows comparable laboratory pulses, i.e., with TTP slightly above the shortest TTP observed in all laboratory data. These laboratory shape families agree qualitatively with the computer simulations. Similar comparisons were made at all TTP, with comparable agreement, validating the simulations as a tool for exploring the space of pulse shapes.

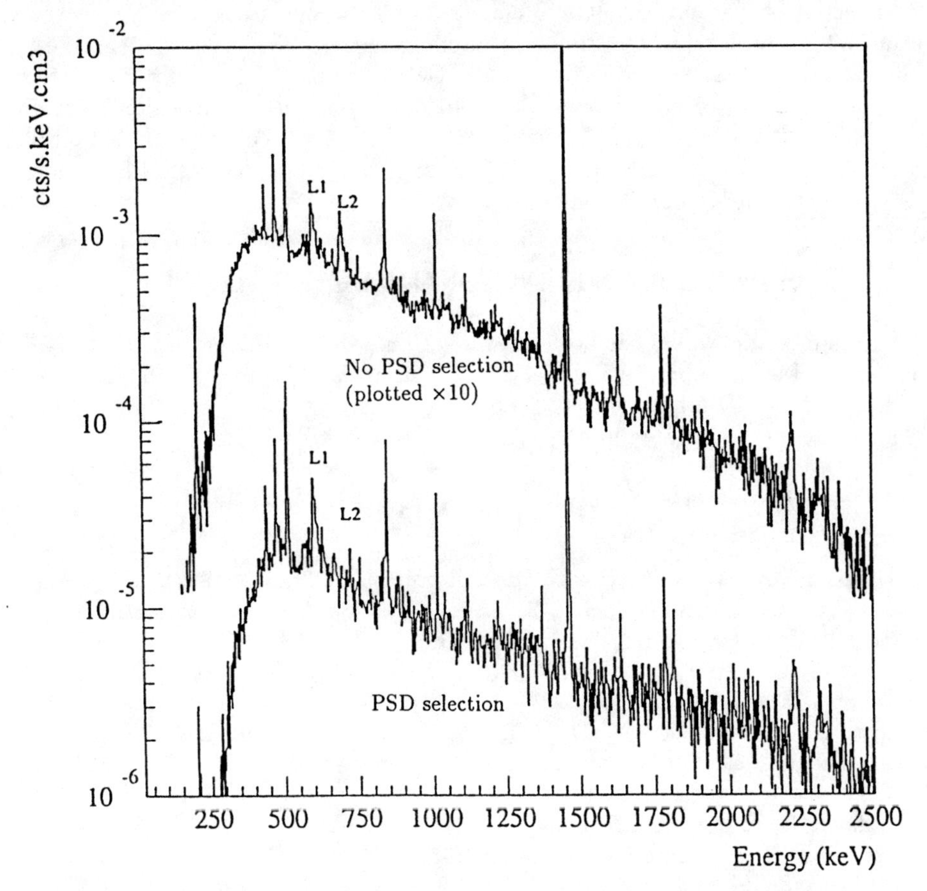

FIGURE 3. Use of PSD on balloon flight data, showing a continuum reduction by a factor of 4, photon acceptance of ≈ 40%, and virtual elimination of the electron feature L2. See text.

TEST RESULTS

Tests of PSD have been done with simulated pulses, laboratory data, and balloon flight data. Using simulated pulses at 800 keV with the basic template set and basic algorithm, discrimination of 97.5% of the electrons while accepting 56% of the photons has been demonstrated. With the expanded template set and more sophisticated discrimination, photon acceptance can be improved to 61% with the same discrimination of electrons. With laboratory data, the basic template set and basic algorithm have achieved discrimination of 97% of electrons while accepting 53% of the photons; use of the expanded set for laboratory data remains in progress. PSD results from a balloon flight are shown in Fig. 3. The shield on the balloon instrument was comparable to SPI's, but the radiation environment leads to a background wherein singles and multiples are roughly equal. Hence a given technical performance of PSD will lead to a lesser sensitivity improvement than in SPI, where the expected background is 90-95% singles. Inelastic neutron scattering features L1 and L2 are particularly pertinent. Feature L1 arises $^{74}\mathrm{Ge}(\mathrm{n},\mathrm{n}')^{74}\mathrm{Ge}^*$(596 keV), while L2 arises similarly from $^{72}\mathrm{Ge}(\mathrm{n},\mathrm{n}')^{72}\mathrm{Ge}^*$(694 keV). The difference is that $^{74}\mathrm{Ge}^*$ de-excites by γ emission, while $^{72}\mathrm{Ge}^*$ de-excites via a conversion electron. Events in L1 are thus photons, and PSD accepts about 40% of them. Events in L2 are electrons, and PSD rejects at least 92% of them. Hence, a basic PSD system clearly works on a balloon instrument.

STATUS AND ACKNOWLEDGMENT

The Engineering Model has been delivered to CESR and integrated into the SPI Engineering Model, where it is functioning nominally. The Flight Model is being assembled at UCSD for delivery to CESR in late 1999.

This work was performed under NASA Contract NAS-5-97121.

REFERENCES

1. Gehrels, N., *Nucl. Inst. Meth. in Physics Research* **A 239**, 513 (1992).
2. Ho, W. C. G., Boggs, S. E., Lin, R. P., Slassi-Sennou, S., et al., *Proc. IEEE Nucl. Science Symp.*, Del Guerra, A. ed., 894 (1996)
3. Ho, W. C. G., Boggs, S. E., Lin, R. P., Slassi-Sennou, S., et al., *Nucl. Inst. Meth. in Physics Research* **A 412**, 507 (1998).
4. Lichti, G. G., Schönfelder,V., Diehl, R., et al., *Proc. SPIE* **2806**, 217 (1996).
5. Vedrenne, G., Jean, P., Mandrou, P., et al, "The Spectrometer SPI of the INTEGRAL Mission," these proceedings (1999).
6. Philhour, B., Boggs, S. E. Primbsch, J. H., Slassi-Sennou, S. A., Lin, R. P., et al., *Nucl. Inst. Meth. in Physics Research* **A 403**, 136 (1998).
7. Slassi-Sennou, S. A., Boggs, S. E., Feffer, P. T., and Lin, R. P., *Proc. 2nd INTEGRAL Workshop (ESA SP-382)*, 627 (1997).
8. Slassi-Sennou, S. A., Boggs, S. E., Philhour, B., et al., *Proc. SPIE* **2806**, 483 (1996).

INTEGRAL/SPI Spectral Deconvolution

C.R. Shrader*, S.J. Sturner* & B.J. Teegarden

Laboratory for High-Energy Astrophysics, NASA Goddard Space Flight Center
** also with Universities Space Research Association*

Abstract. The INTEGRAL Spectrometer or "SPI", is a cooled-germanium instrument covering the 20 keV to ~10 MeV energy range with a resolution of about 500. The unique nature of the instrument, 19 separate detector elements viewing the sky through a coded mask aperture, poses a number of difficulties in analyzing the data. For example, a typical observation consists of multiple pointing directions (dithering) with an instrument response that is highly directional. Multiple sources within the nominal 16° FoV are likely to be common, and most high-energy point sources are variable over observable time scales. Thus a typical deconvolution of the detector count-rate data involves a complex global minimization problem over large data and parameter spaces. Strategies for dealing with these difficulties, are discussed and some preliminary results based on simulated data are presented.

INTRODUCTION

The spectrometer SPI (SPectrometer on INTEGRAL) will study gamma-ray point sources and diffuse emission in the 20 keV - 8 MeV energy range with an energy resolution of 2 keV (FWHM) at 1 MeV[1,2]. This will be accomplished using an array of 19 hexagonal high purity, cooled Germanium detectors. A hexagonal coded aperture mask is located *1.7 m* above the detection plane providing a fully coded field of view $\approx 16^{\circ}$, with an angular resolution of $\approx 2^{\circ}$. The complex nature of the instrument makes the analysis of its data a challenge. In this contribution, we discuss a specific strategy for SPI spectral deconvolution under study at the NASA Goddard Space Flight Center.

SPI response matrices are expected to have substantial off-diagonal terms, thus a forward-folding approach should provide the most accurate possible spectroscopy. However, conventional approaches are problematic with a coded-mask, multi-detector instrument and "dithered" observation strategies. The instrumental response is highly directional and owing to the wide FoV, multiple sources are typically observed simultaneously. Large data and parameter spaces are thus involved and computational difficulties, such as memory limitations or distingushing local from global minima are likely.

We have been studying the SPI spectral deconvolution problem and incorporation of

CP510, *The Fifth Compton Symposium,* edited by M. L. McConnell and J. M. Ryan

the required functionality into the XSPEC package. Following its introduction during the EXOSAT era, XSPEC has become the defacto standard analysis tool for X-ray spectroscopy. It offers a rich database of astrophysical models, robust and efficient computational performance and multiple platform support. Additionally, it facilitates combined analyses of multiple instrument or multiple observation datasets. This in turn facilitates instrumental cross-calibrations and/or broad-band spectral analyses. Recently, the use of XSPEC has been extended into the domain of gamma-ray spectroscopy, having been applied for example to data obtained with the BATSE, OSSE and EGRET experiments on CGRO, as well as the hard-X-ray instruments on RXTE and SAX.

SPI-SPECIFIC REQUIREMENTS

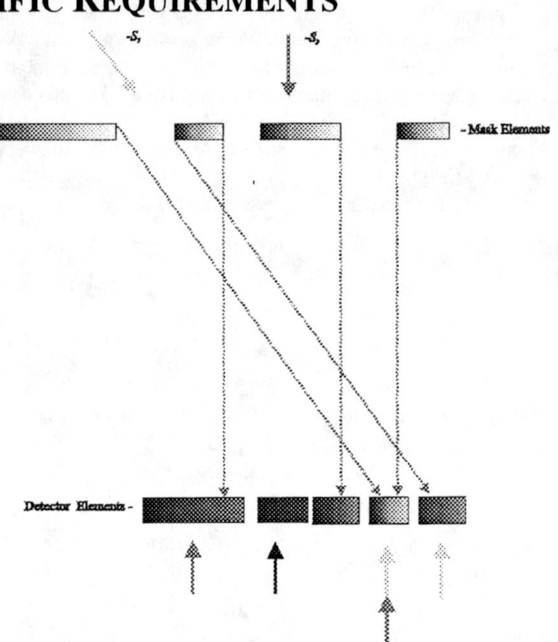

Figure 1 A rigorous spectral deconvolution using a coded-mask instrument must take into account the distinct types of possible photon events. Here for example, a detector elements is illuminated by S1 only, S2 only S1+S2, or it may be completely shadowed. This information, which is stored in the response matrix, can be exploited in the deconvolution procedure.

Conventional spectral deconvolution procedures, involve a single source model (and background), detector response model, and a best-guess at a physical source model. One then performs a chi-square minimization varying model parameters until the "forward-folded" count spectrum nearly matches the observed net count rates. The situation will be more complicated here. SPI consists of 19 separate Ge detector elements, illuminated through a coded mask aperture. It is sensitive from approximately *20* keV - 10 MeV, and the FoV is *~16°*. A diverse range of astrophysical targets (and thus spectral models) such as nucleosynthesis sites and black holes are anticipated. A typical observation is expected to consist of a series of N_p~*25* "dither" positions about a central line of sight. The $N_p \times N_{det}$ resulting spectra will be stored as rows of a table extension in the FITS "PHA-II" file format, utilized by XSPEC and adopted by the Integral Science Data Center (ISDC) as the standard SPI binned-event file format.

We will assume in the following discussion that an image reconstruction has been performed, and the directions of individual point sources are known. This type of image reconstruction has already been accomplished using simulated data. We further note that simultaneous spatial plus spectral deconvolutions are in principle possible, but there are complications yet to be addressed.

Our basic data space is thus detector count-rate arrays of the form: C=C(*d, p, ch*); where the indices are *d*≡detector id, *p*≡pointing, and *ch*≡channel number. There can in general be *n* sources in the total FoV for a given observation, denoted here by $\{s_1,...,s_n\}$. For any given individual dither position, a subset of the sources, may be visible to a given detector. A simplistic example is illustrated in Figure 1 for which two distinct point sources are contained within the FoV. Different detector elements are illuminated by either source individually, by both sources, or by neither source. This information, which is contained in the response matrices, can in principle be exploited in deconvolving the count spectra into photon space. Enumerating over individual detectors and pointings there are then $d \times p$ coupled Chi-Square minimizations to perform. The general form of these for a given detector and pointing is:

$$\vec{C} = \sum_{s \in \psi} [R] \vec{F} + \vec{B} \quad \text{,where} \quad F = \sum_{s} F_k \quad \text{,and} \quad R = R(d,p,ch,E,\phi)$$

The incident spectrum F has N_E elements for each source *s* illuminating a detector, and [R] is the $ch \times N_e$ response matrix, and *B* is the background count rate. Information regarding *B* can in principle be derived empirically if there are dither positions for which the relevant detector elements are not illuminated, but in general we will assume it is deconvolved along with the source model. Thus, the Chi-Square minimization is of the

$$\chi^2 = \sum_{d} \sum_{p} \sum_{ch} \frac{[C(d,p,ch) - \sum_{s_k \in \psi} \sum_{E} R_k(d,p,ch,E,\theta,\varphi) F_k(E,\lambda_m) - B(d,p,ch,\lambda_m)]^2}{\sigma(d,p,ch)^2}$$

form:

For $p \times d$ ~10^2 or even ~10^3 and perhaps ~10^2 pulse-height channels, the scope of the computational of problem, and the associated difficulties are evident. Furthermore, in general, each source model is dependent on some parameter space; $F \equiv F(E, \lambda_m)$; where the number of parameters is typically *m* ranges from 2-4. XSPEC currently uses by default the Levenburg-Marquardt algorithm to minimize chi-square. However, alternative minimization algorithms — a "simulated annealing" algorithm and a "genetic" algorithm — already installed, and additional ones can be added easily. Such algorithms may be suitable for global minimization problems, where many local-minima are likely to be encountered.

Aside from the scope of the computational problem, it is evident that some additional functionality is needed to accommodate multiple sources, source models, and detectorresponses. XSPEC currently supports multiple "data groups", which might for example involve multiple detectors and responses. However, a common model and parameter set is applied to each group, although normalizations can vary individually. Work to add a capability for the incorporation of multiple source models with associated response matrices is ongoing. Testing at the prototype level is already proceeding as described below.

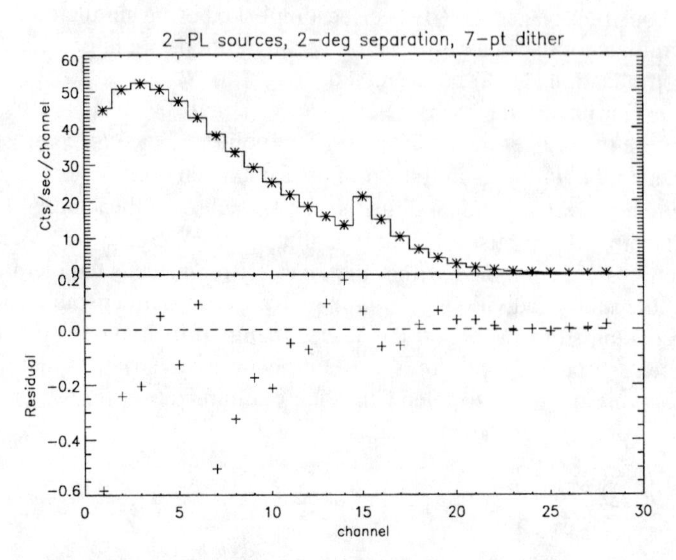

Figure 2 Simulated data was generated for 2 point sources with distinct power-law spectra, separated by 2° which is close to the nominal SPI PSF. The input parameters were recovered to within statistics following the procedure described in the text. The reduced chi-square was 1.5 for 3722 DoF if 1% systematic error level was included. The apparent discontinuity at channel 15 is a result of (larger) binning.

SOME TEST RESULTS

We have carried out a number of tests on simulated data created using our current SPI mass model and our MGEANT software[3]. The instrument matrices used for these test were also created using MGEANT simulations. A coarse energy binning – 28 logarithmically spaced energy bins from 20 keV to 8 MeV – was employed, since we are currently storage and CPU limited. A 7-point, hexagonal dither pattern, which may be the standard observing mode for isolated point sources, was used. Spectral deconvolutions tested thus far include a single Crab-like power-law spectrum, 2 power-law sources separated by 2°, 3 power-law sources, at 2°, right triangular vertices, and a power-law source plus a cutoff power law with 2° separation. Each of these was carried out initially with no background. An example is illustrated in Figure 2. More realistic analyses, however, must contend with the major nemesis of mid-energy gamma-ray astrophysics, namely the instrumental background.

An instrumental background has been modeled semi-empirically based on GRIS flight data. The line contribution has been smoothed over in the current analysis, since our current response matrices lack the required resolution for a rigorous treatment. This limitation, as noted, is a result of the large amount of storage and computation involved

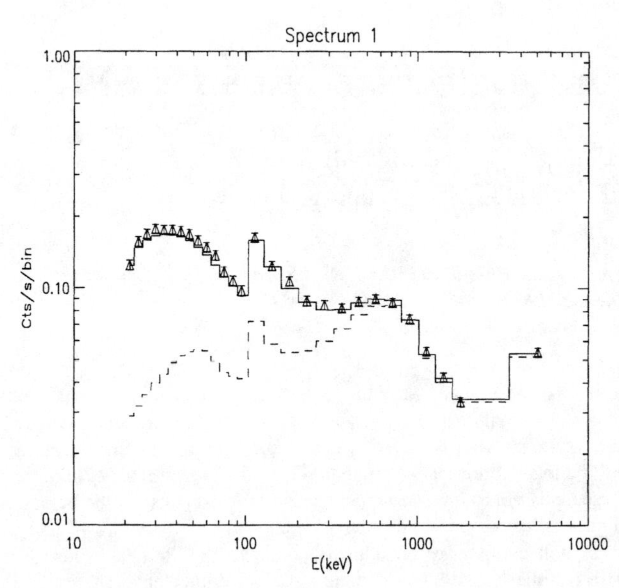

Figure 3 Fit of Crab-like power-law point source plus a 28-parameter (one per energy bin) background model for a detector which is nearly fully illuminated. Note that the source is dominant at low energies but the background becomes dominant at the energies above a few hundred keV.

in creating them. We note however, that techniques have been developed to overcome this problem. Several functional forms have been used to model the background for the purpose of our source plus background deconvolution. We tried both piecewise smooth polynomial model and a 1-parameter per bin piecewise-constant model, the latter is the preferred approach as it will likely be used later with flight data once a realistic background model is developed. It is further noted that we have thus far assumed that the background is the same for all of the detectors. Since the background model is developed in detector count space, a unit response is applied (see the χ^2 minimization equation above). The number of degrees of freedom is then $DoF = N_{dith} \times N_{det} \times N_{Ebin} - N_{modpar} - N_{Ebin}$ For the tests performed thus far, $DoF \sim 3\text{-}4 \times 10^3$.

Following this procedure, again for a hexagonal 7-point dither pattern and a Crab-like point source we were able to obtain a reasonable fit and successfully recover our source and background model parameters - provided, the background parameters could be constrained to within about $\pm 20\%$. An example is shown in Figure 3 for the case of a (mostly) unshadowed detector. Note that Figure 3 represents the model fitted to one particular detector/dither point. The fit parameters are derived for all 133 (i.e. 7 pointings times 19 detectors).

The results obtained thus far are encouraging and give us some confidence that the method can be successfully applied to at least some realistic SPI observation programs. How far the method can be pushed – for example some observing strategies may entail ~10^3 or even 10^4respones – is a question we are planning to investigate in the near future.

REFERENCES

1. Winkler, C. et al., (these proceedings)
2. Jean, P., et al,. (these proceedings).
3. Sturner, S., et al., (these proceedings)

The γ-Ray Burst-Detection System of SPI*

G. G. Lichti[1], R. Georgii[1], A. von Kienlin[1], V. Schönfelder[1], C. Wunderer[1], H.-J. Jung[2], and K. Hurley[3]

1) *Max-Planck-Institut für extraterrestrische Physik, PO Box 1603, 85740 Garching, Germany*
2) *Dornier Satellitensysteme GmbH, 88039 Friedrichshafen, Germany*
3) *University of California, Space-Sciences Laboratory, Berkeley, CA 94708, USA*

Abstract. The determination of precise locations of γ-ray bursts is a crucial task of γ-ray astronomy. Although γ-ray burst locations can be obtained nowadays from single experiments (BATSE, COMPTEL, BeppoSax) the location of bursts via triangulation using the interplanetary network is still important because not all bursts will be located precisely enough by these single instruments. In order to get location accuracies down to arcseconds via triangulation one needs long baselines. At the beginning of the next decade several spacecrafts which explore the outer planetary system (the Mars-Surveyor-2001 Orbiter and probably Ulysses) will carry γ-ray burst instruments. INTEGRAL as a near-earth spacecraft is the ideal counterpart for these satellites. The massive anticoincidence shield of the INTEGRAL-spectrometer SPI allows the measurement of γ-ray bursts with a high sensitivity. Estimations have shown that with SPI some hundred γ-ray bursts per year on the 5σ level can be measured. This is equivalent to the BATSE sensitivity. We describe the γ-ray burst-detection system of SPI, present its technical features and assess the scientific capabilities.

INTRODUCTION AND DESCRIPTION OF THE ACS OF SPI

INTEGRAL will be launched into an eccentric earth orbit at the end of 2001. The anticoincidence subsystem (ACS) of the INTEGRAL-spectrometer SPI will have the capability to measure the arrival times and light curves of γ-ray bursts with a high accuracy. It will therefore allow in conjunction with other distant spacecrafts the determination of precise γ-ray burst locations using the 4th interplanetary network (IPN) [1].

The ACS surrounds, with the exception of the field of view, the central Ge-detector array from all sides. It consists of 512 kg of BGO crystals. Each crystal is viewed by two photomultiplier tubes whose signals are measured. These signals are ORed in the Veto-Control Unit (VCU) yielding the overall-veto count rate R which is measured and for the purpose of γ-ray burst detection transmitted via the spectrometer's Digital Processing Electronics (DPE) to ground. A more detailed description of SPI can be found in [2] and [3].

* A detailed version of the paper can be found under http://ww.gamma.mpe-garching.mpg.de/integral/SPI/DOCUMENTATION/doclist.html under the document number SPI-MPE-TN-1-29

CP510, *The Fifth Compton Symposium,* edited by M. L. McConnell and J. M. Ryan

SENSITIVITY ESTIMATION

The sensitivity for measurements of γ-ray bursts is calculated using the total veto count rate R. This count rate is measured during a time interval Δt. A γ-ray burst manifests itself as a sudden and rapid increase of the veto rate. Assuming Poisson statistics the counts/bin fluctuate around the average bin count rate whereby the 1σ-statistical fluctuation is given by the square root of the bin count rate. To detect a γ-ray burst with a high certainty a significant positive deviation from this average count rate is required. In the following a deviation by at least 5σ is required. It is assumed that a burst has occurred when the number of counts per bin N exceeds a value of $5 \cdot \sqrt{R \cdot \Delta t}$.

The number of counts from a burst can be calculated for an average burst spectrum:

$$N = A \cdot \Delta t \cdot \int_{E_s}^{\infty} F(E)dE \tag{1}$$

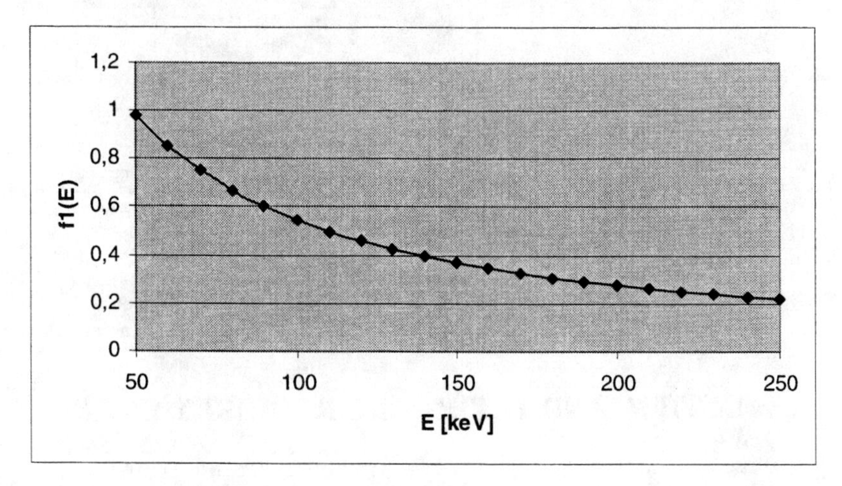

FIGURE 1: The dependence of $f_1(E_s)$ on the threshold energy E_s for the canonical value of E_o (= 150 keV).

Here A is the effective area of the ACS whose energy dependence is neglected and F(E) is the differential energy spectrum of a γ-ray burst approximated by the Band function [4] for which we use the canonical parameter E_o = 150 keV. E_s is the threshold of the anticoincidence subsystem.

Solving (1) for the trigger condition given above one can find the weakest γ-ray burst which can be detected with the ACS of SPI. Knowing the weakest burst one is able to derive the number of bursts/year which can be detected by SPI by calculating the energy flux P of a γ-ray burst.

To avoid a divergence in the calculation an upper cut-off energy E_m (= 10 MeV) has been introduced. One gets finally the minimal-detectable energy flux P as a function of all relevant parameters (the dimensionless function $f(E_s)$ is given in Figure 1):

$$P = \frac{5 \cdot \sqrt{R} \cdot E_o}{A \cdot \sqrt{\Delta t} \cdot f_1(E_s)} \cdot \left[e^{-\frac{E_s}{E_o}} + e^{-1} \cdot \left(\ln \frac{E_m}{E_o} - 1 \right) \right] \tag{2}$$

In Figure 2 the dependence of P on the normal (i. e. non-burst) rate R is shown. The total rate of the ACS varies depending on the solar activity between ~80 000 cts/s and ~160 000 cts/s. According to Figure 2 this corresponds to minimal detectable burst energy fluxes between $4 \bullet 10^{-7}$ erg/(cm^2 s) and $7 \bullet 10^{-7}$ erg/(cm^2 s) for $\Delta t = 1$ s. The difference between the two fluxes is small and within the uncertainty of the estimation. Using the corresponding figure of [5] one finds that with SPI within one year ~280 bursts for $\Delta t = 1$ s can be observed.

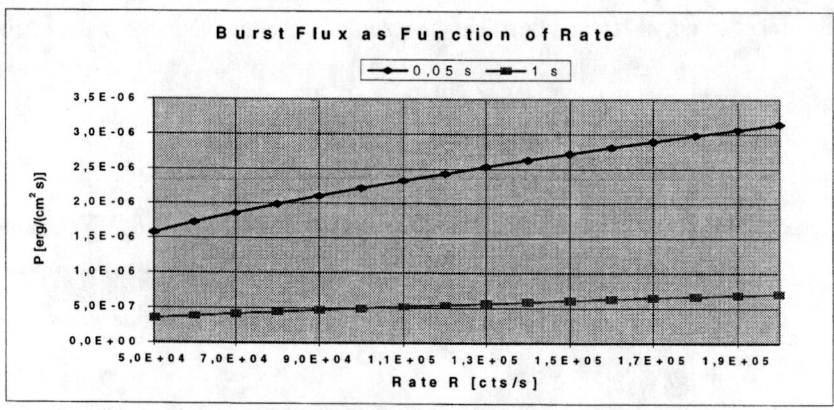

FIGURE 2: The minimal detectable energy flux P as a function of the overall-count rate R for two time intervals Δt, for a threshold of 80 keV, an effective area of 3000 cm^2 and for a cut-off energy of 10 MeV.

SCIENTIFIC AND TECHNICAL REQUIREMENTS

The width of the arrival-direction annulus of a γ-ray burst depends on the accuracy of the arrival-time measurement Δt and the distance d between the spacecrafts of the IPN:

$$\Delta\alpha = \frac{c \cdot \Delta t}{d \cdot \sin\alpha} \tag{4}$$

c is the speed of light and α the angle of the arrival cone. If one takes d = 4.3 AU (= $6.45 \bullet 10^8$ km, the distance to the Ulysses spacecraft), $\alpha = 45^\circ$ and requires that $\Delta\alpha < 10''$ one finds $\Delta t < 74$ ms. Therefore a time resolution of 50 ms was chosen for the SPI-ACS γ-ray burst-detection subsystem.

Since the arrival time of a γ-ray burst has to be known in universal time (UT) this time has to be allocated to the start or end of the integration time of the count-rate measurement. The accuracy of this time shall be much less than 50 ms. A safe requirement is to know it with a precision of ~5%. Therefore the acquisition of the veto count rate shall be performed with an accuracy of ~2.5 ms with respect to UT.

REALISATION OF THE TECHNICAL REQUIREMENTS

The realisation of this requirement is achieved with the overall-veto-counter of the ACS, the 8 Hz clock pulse and the the on-board-time (OBT) of the DPE.

The overall-veto counter is a start/stop counter of the VCU which collects the ORed vetos from the 91 FEEs and is set to zero every 50 ms. The 8 Hz clock is a continuous rectangular pulse coming from the DPE which is directly linked to the VCU micro-controller. The OBT is a relative mission time with known relation to the UT. Inside the VCU a 50 ms VCU-timer interrupt is generated that is synchronized with the OBT every 8 s.

The overall-counter readings are stored in a "toggle buffer" with 10 records each inside the VCU-RAM. Each record contains the 8-bit record count s and the 16 overall-counter values. So in total one buffer contains 160 overall-counter values allowing the storage of count rates within a time interval of 8 s. The two buffers A and B will be filled and read out alternately. In addition the toggle buffer contains one header which holds the toggle-buffer start time t_B. This time is the VCU time derived from the OBT which is updated every 125 ms (see Figure 3).

Whenever SPI has been switched on by command the DPE 8 Hz clock will update the VCU internal time every 125 ms. At this moment it has the same accuracy as the OBT time of the DPE. Every 640 s the DPE sends the value of the OBT to the ACS. This value is the time of the last falling edge of the 8 Hz pulse prior to the reception of the OBT (see Figure 3).

The first interrupt from the 8 Hz pulse after the first start-time reading from the DPE will also start the acquisition sequence. The current value of the overall-counter register will be transferred to the first sample of buffer A and the overall-counter value is set to zero. Parallel to the writing of the first counter value to buffer A the actual time of the VCU (as derived from the OBT) is written into the header of the buffer. This time, the toggle-buffer start time t_B, lies, as is shown in Figure 3, between the start and end time of the current 50 ms interval. The time intervals τ_1 and τ_2 from this time to the previous or succeeding edges of the corresponding 50 ms (= τ_o) interval have been chosen to be 40 ms (= τ_1) and 10 ms (= τ_2). Thus the toggle-buffer start time t_B refers to the 50 ms time interval that started 40 ms (= τ_1) before.

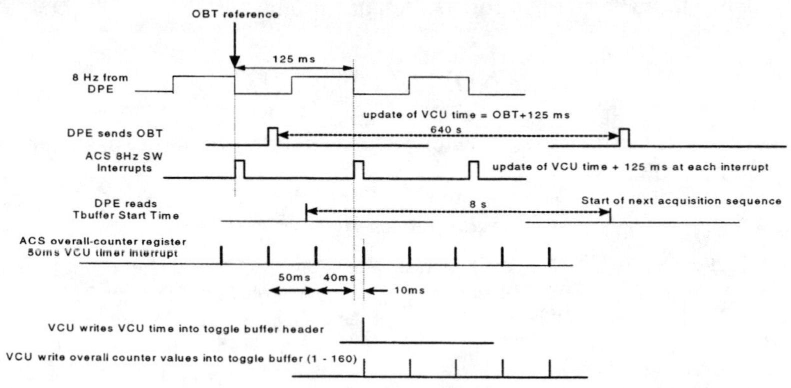

FIGURE 3: The time synchronization and γ-ray burst-acquisition procedure.

The start and end times t_s and t_e of a 50 ms interval are given by $t_s = t_B + s \cdot \tau_o - \tau_1$ and $t_e = t_B + s \cdot \tau_o + \tau_2$. Here s is the record counter and t_B the toggle-buffer start time, which is equivalent to the OBT in the moment of the first count writing into the buffer. The uncertainty of the time allocated to the count value is $\pm$ 1ms.

The total timing error depends on the accuracy of the falling edge (with a fall time of < 20 ns) of the 8 Hz interrupt signal with a time jitter < 0.1 μs, the numerical resolution of the OBT of ± 1.907 μs and the uncertainty of the falling edge of the 50 ms pulse which could be shifted by ± 1 ms due to the DPE interrupts.

Since every 125 ms the internal VCU clock is updated the error of τ_o is not given by $s \cdot \Delta\tau_o$ but by $2.5 \cdot \Delta\tau_o$ (2.5 = 125 ms/50 ms). The timing-error budget can therefore – under the assumption that the errors are of statistical nature - calculated according to the following formula:

$$\Delta t_s = \sqrt{\Delta t_B{}^2 + (2.5 \cdot \Delta\tau_o)^2 + \Delta\tau_1{}^2} \qquad (5)$$

The errors of the three parameters are $\Delta\tau_B = 1.91$ μs, $\Delta\tau_o = 1$ ms and $\Delta\tau_1 = 1$ ms, yielding a total error of $\Delta\tau_s = 2.69$ ms.

DATA-ANALYSIS SYSTEM AND CONCLUSIONS

A software which will be installed at the ISDC in Geneva will check automatically the stream of count-rate data for the occurrence of a γ-ray burst. Once this automatic system will recognize a burst an alert will be issued to the institution which is responsible for the 4th IPN. From the measured arrival times at the different spacecraft locations the arrival direction of the γ-ray burst will be calculated. This information will then be distributed worldwide to those observatories which participate in the search for counterparts of γ-ray bursts at other wavelengths.

For an assumed absolute timing error of the different arrival times of 100 ms an error region in the sub-arcminute range can be achieved due to the long baselines of the 4th IPN as was shown by [1]. For a γ-ray burst with a fluence of $2 \cdot 10^{-6}$ erg/cm^2 an error box of 0.5 square-arcminutes was obtained. With the γ-ray burst-detection subsystem of SPI as described above even smaller error regions can be achieved. This is small enough to make a search for counterparts promising.

ACKNOWLEDGMENTS

The development of the spectrometer SPI is supported by DLR under the contract number 50.OG.9503.0.

REFERENCES

1. Hurley, K., Proc. of the 2nd INTEGRAL workshop St. Malo **SP-382**, 1997, pp. 491-493.
2. Lichti, G. G. et al., Proc. of the SPIE conference in Denver, Vol. **2806**, 1996, pp. 217-233.
3. Mandrou, P., G. Vedrenne, P. Jean et al., Proc. of the 2nd INTEGRAL workshop in St. Malo, ESA **SP-382**, 1997, pp. 591-598.
4. Band, D., J. Matteson, L. Ford, B. Schaefer, D. Palmer et al., Ap. J. **413**, 1993, pp. 281-292.
5. Fenimore, E. E., R. I. Epstein, C. Ho, R. W. Klebesadel et al., Nature **366**, 1993, pp. 40-42.

The Scientific Role of JEM–X: the X–ray Monitor on INTEGRAL

Niels Lund, Niels J. Westergaard, Søren Brandt, Allan Hornstrup, and Carl Budtz-Jørgensen

Danish Space Research Institute
Juliane Maries Vej 30
DK 2100 Copenhagen, Denmark

Abstract.
INTEGRAL, ESA's next venture in gamma–ray astronomy is due for launch in 2001. In addition to the two large gamma–ray instruments INTEGRAL will carry an X–ray monitor, JEM–X, and an optical monitor, OMC. The primary role of JEM–X will be to provide data on the X–ray flux and its variations from the targets observed by the gamma–ray instruments. JEM–X will also provide accurate positions for transient X– and gamma–ray sources observed during the mission. Together with the optical monitor, JEM–X can carry out a special programme on stellar activity, independent of the two high energy instruments on INTEGRAL.

INTRODUCTION

JEM–X consists of two co–aligned, coded mask telescopes with position sensitive detectors. The primary energy range is 3 to 35 keV; between 35 and 60 keV the instrument will still work, but with reduced efficiency.

JEM–X is a monitoring instrument that will support the gamma–ray observations of INTEGRAL with IBIS, the imager, and SPI, the spectrometer. The major observations will be determined by the gamma–ray instruments. One important aspect of this is that most targets will be observed with long exposure times since the gamma–ray photon flux is low. In the X–ray energy range, however, photons are much more ubiquitous and, usually, shorter exposure times would suffice.

As a baseline the INTEGRAL pointing will not remain on a single source for an extended period of time, since the optimal SPI background subtraction requires pointings in a pattern of directions around the source in question, the "dithering" mode. This pattern will have pointings offset from the source with 2 or 4° implying a lower JEM–X efficiency [1]. On the other hand, it gives JEM–X an opportunity to observe other sources in the vicinity of the primary source.

CP510, *The Fifth Compton Symposium,* edited by M. L. McConnell and J. M. Ryan

JEM–X can pinpoint the source positions with a better precision than IBIS, and it will extend the spectral energy range of INTEGRAL down to 3 keV. Thus JEM–X is the only instrument on INTEGRAL capable of observing the important iron–line complex around 7 keV.

JEM–X can make a contribution to the understanding of the various types of sources by filling the gap in the energy range where most focussing telescopes have only limited throughput.

INSTRUMENT SPECIFICATIONS

Detector. The JEM–X detectors are Xe–filled microstrip gas detectors operating at 5 bar. Each detector has a sensitive area of 420 cm^2 including the effect of the collimator/strongback supporting the beryllium window. The position accuracy is better than 1 mm in the range from 6 to 10 keV.

Coded Mask. A mask with 25% open area has been selected for JEM–X. The low transmission value is expected to improve the instrument sensitivity in the crowded fields during the galactic plane scans and the observations of the galactic center region. The low transmission value also reduces the telemetry load. This is important as all instruments on INTEGRAL are telemetry limited. The masks are manufactured from 0.5 mm tungsten by wire spark machining. The mask pixels are hexagons, 3.3 mm between faces.

SCIENCE DATA

The telemetry rate from JEM–X (both units together) is restricted to 6.7 kbit s^{-1} (10% of the total INTEGRAL telemetry rate). In the full imaging mode, where all the important parameters for each recorded photon are transmitted to the ground the count rate cannot exceed 105 cts s^{-1} per JEM–X unit. Of this, the DXB takes ~7 cts s^{-1} and the particle induced background ~8 cts s^{-1} leaving ~90 cts s^{-1} for distinct sources in the range from 3 to 60 keV — the nominal energy–band for JEM–X. The Crab count rate is 180 cts s^{-1} in each JEM–X unit implying that when the intensity af sources in the FOV (with efficiency correction) add up to 500 mCrab the telemetry will be saturated. Rather than leaving the decision about which events to lose to the telemetry system, a grey filter mechanism has been introduced in the front–end computer that will, in a stochastic manner, reject a fraction of the events so that the data buffer will not overflow.

Five distinct data formats have been set up in order to optimize the information content of the transmitted data when the data rate exceeds the full imaging limit. For strong sources the imaging may be discarded emphasizing spectral data or timing data, or in crowded fields the timing information can be minimized.

SCIENTIFIC PERFORMANCE

A table of instrument parameters and more references can be found in [2].

Field of view. The collimator defines the transmission at off axis angles. At large off axis angles, the partial coverage of the mask on the detector further reduces the transmission. The Zero Range FOV (diameter) is 13.2°.

Background. The DXB is a major source of the background counts in the detector. The high energy end of the DXB as well as energetic particles will interact in the detector gas and in the materials around the detector producing X–ray photons and ionization tracks in the detector. This background has been estimated by [3] and [4].

Sensitivity. Figure 1, left panel, shows the on–axis source detection sensitivity as a function of exposure time. A power law spectrum with photon index of 1.7 and column density of 10^{21} cm^{-2} has been used as a typical source. All energies above 20 keV have been discarded since the source intensity for this range is so low that the background removes the advantage of including source photons. See also [5]. Figure 1, right panel, shows the sensitivity for detection of ~6 keV lines.

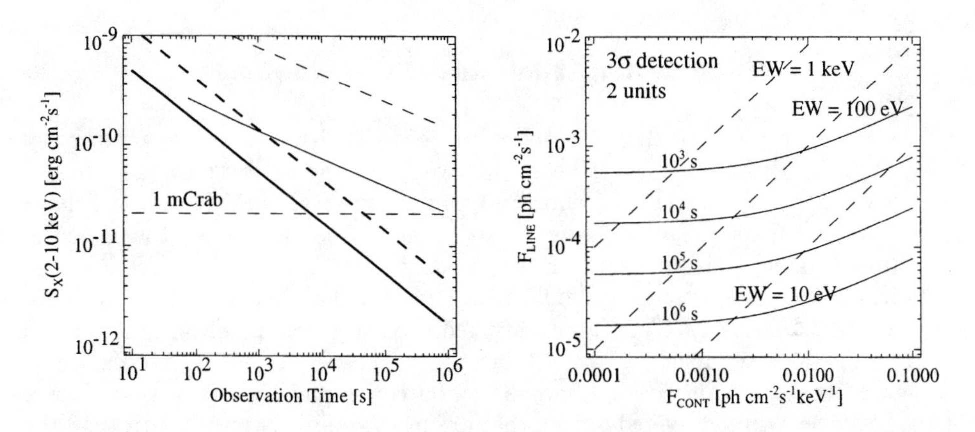

FIGURE 1. Left Panel: 5σ source detection limit as a function of observation time. The thick curve has been calculated assuming no other significant sources present in the FOV. The thick dashed curve is for the case with a total of 1 Crab of other sources in the FOV. The detection of a single source in the reconstructed image gives a limit as indicated by the thin line. With 1 Crab of other sources in the FOV the thin dashed curve applies. The energy range from 3 to 20 keV has been used for the detection. **Right Panel:** The line detection sensitivity at 6 keV as a function of continuum strength. The energy interval is 1.15 keV, the FWHM of the detector energy resolution.

Source Localization. For sources with S/N ratio of 10σ the source localization will be around 20 arcsec.

Time Resolution. The event time tagging is done with a precision of 122 μs. This can only be fully exploited for pulsar studies due to the limited telemetry allocation.

Angular Resolution. JEM–X is designed to be able to separate two equally strong sources at a distance of 3 arcmin. This capability has been verified by raytracing simulations.

THE INTEGRAL CORE PROGRAMME

The INTEGRAL core programme consists of three parts:

- Galactic Plane Scans (GPS) once per week. The galactic plane will be scanned in steps of 6° which gives some overlap between contiguous JEM–X fields.
- Galactic Center Deep Exposure (GCDE) once per month.[1] The grid pitch in the 2–dimensional scan will be ~2°.
- Observations of selected sources. These have not been selected yet but 1 Ms of observation time have been assigned for this purpose.

JEM–X in the Core Programme

The low energy end of the spectrum of the sources observed by the gamma ray instruments, SPI and IBIS, can be studied with very good accuracy since observation times will be quite long. Investigations of the ~6.5 keV Fe line will have high priority. In the Galactic Plane Scan the JEM–X sensitivity limit is around 6 mCrab but in the Galactic Center Deep Exposure the exposure time is longer and fainter sources will appear except right in the center, where the background is high.

Galactic Compact Objects. Continued (several years) monitoring of sources in the galactic plane (each source will be visited between 2 and 4 times per year) in search for state transitions. Figure 2 demonstrates that the majority of known binary sources will be covered during the core programme. Together with IBIS the broad spectral coverage can be used for observing spectral changes. For pulsars phase dependent spectral shifts will be observed. INTEGRAL will discover many transients in the galactic center of plane and JEM–X will be able to make a spectral characterisation in a short time.

AGN. The study of AGNs over a broad energy range is important for the understanding of the emission mechanism. The time variability in the various energy regimes i.e. the time lag between high energy photons and low energy photons can tell about the geometry of the high energy X/gamma–ray source and the absorbing gas columns. This will be possible for a few AGNs.

---

1) provided the galactic center fulfills the Sun–angle constraints

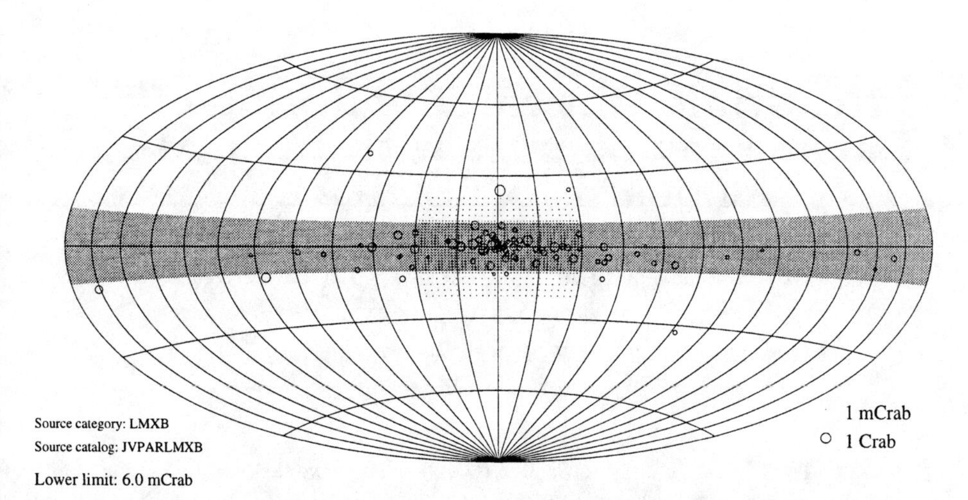

FIGURE 2. The Low Mass X-ray Binary sources from the J. van Paradijs catalogue [6] in galactic coordinates. The intensity has been shown as a geometric mean between maximum and minimum recorded X-ray brightness. The region of the galactic scan is shown as a grey area and the galactic center exposures are shown as a grid of points.

CONCLUSIONS

JEM–X will expand the scientific return of the INTEGRAL mission by contributing the exact position of the gamma–ray sources and the low–energy part of the spectrum including the time variability. In addition to that a class of sources will be too faint in gamma–rays but observable in X–rays and optical light by OMC where simultaneous lightcurves can be obtained.

REFERENCES

1. N. Lund and N.J. Westergaard, Physica Scripta **T77**, 39 (1998).
2. Budtz–Jørgensen, C. *et al.*, Proceedings 3rd INTEGRAL Workshop (Taormina, Italy) (1999) *in press.*
3. Feroci, M., Costa. E., Budtz–Jørgensen, C., Schnopper, H.W., Westergaard, N.J., Rapisarda, M., SPIE Proceedings **2806**, p.494 (1996).
4. Feroci, M. *et al.*, Proceedings 3rd INTEGRAL Workshop (Taormina, Italy) (1999) *in press.*
5. Westergaard, N.J. and Lund, N., Proceedings 3rd INTEGRAL Workshop (Taormina, Italy) (1999) *in press.*
6. van Paradijs, J., in "X-Ray Binaries", Cambridge Astrophysics Series, page 536 (1995).

Real-time Optical Monitoring of Gamma-ray Sources with the OMC onboard INTEGRAL

Alvaro Giménez*, J. Miguel Mas-Hesse* and the OMC team

*LAEFF - INTA, POB 50727, E-28080 Madrid

Abstract. The Optical Monitoring Camera (OMC) will observe the optical emission from the prime targets of the gamma-ray instruments onboard the ESA mission INTEGRAL with the support of the JEM-X monitor in the X-ray domain. This capability will provide invaluable diagnostic information on the nature and the physics of the sources over a broad wavelength range.

The main scientific objectives are: (1) to monitor the optical emission from the sources observed by the gamma- and X-ray instruments, measuring the time and intensity structure of the optical emission for comparison with variability at high energies, and (2) to provide the brightness and position of the optical counterpart of any gamma- or X-ray transient taking place within its field of view.

The OMC is based on a refractive optics with an aperture of 50 mm focused onto a large format CCD (1024 x 2048 pixels) working in frame transfer mode (1024 x 1024 pixels imaging area). With a field of view of 5°x5° it will be able to monitor sources down to V = 19 magnitudes. Typical exposures will consist of 10 integrations of 100 seconds each allowing photometric uncertainties below 0.005 magnitudes for objects with V = 14.

INTRODUCTION

The International Gamma-ray Astrophysics Laboratory (INTEGRAL) will be dedicated to the fine spectroscopy and imaging of sources in the energy range between 15 keV and 10 MeV, but it also includes two monitors providing data of the measured objects in the X-ray and optical domains. The Optical Monitoring Camera (OMC) offers the first opportunity to make observations of long duration in the optical band simultaneously with those at X- and gamma-rays, since it has the same field of view (FOV) than the fully coded FOV of the X-ray Monitor and is coaligned with the central region of the Spectrometer and Imager. Descriptions of the Spectrometer (SPI), Imager (IBIS) and X-Ray Monitor (JEM-X) can be found in this volume. Variability patterns ranging from minutes or hours, up to months and years will be monitored. For bright sources, fast optical monitoring at intervals down to 1 s will be possible.

CP510, *The Fifth Compton Symposium*, edited by M. L. McConnell and J. M. Ryan

Multiwavelength observations are particularly important in high-energy astrophysics, where variability is typically rapid, unpredictable, and of large amplitude. In particular, transient events are associated with many kinds of astrophysical phenomena and are of paramount importance in the X- and gamma-ray Universe. On the other hand, arranging multifrequency observations that are simultaneous for both ground-based and space-borne instruments is extremely difficult due to weather conditions, scheduling constraints, or technical problems. Therefore, having onboard INTEGRAL an optical monitor like the OMC, adapted to the spatial resolution and field of view of the high-energy instruments is a powerful additional tool for the understanding of high-energy astrophysical processes. The limiting magnitude will be 19.2 magnitude in the V filter (at 3 σ level in 10 x 100 s).

The main scientific objectives of the OMC are:

- To monitor during extended periods of time the optical emission of all high-energy targets within its field of view, simultaneously with the gamma- and X-ray instruments. This will allow the determination of the optical lightcurves for comparison of variability patterns with the hard X-ray and gamma-ray measurements.
- To provide simultaneous and calibrated standard V filter photometry of the high-energy sources. This will allow the comparison of their behaviour with previous or future ground-based optical observations.
- To monitor optically variable sources within its field of view, in a serendipitous way, which may require long periods of continuous photometry for their physical interpretation, allowing to deliver at the end of the mission a catalogue of thousands of variable sources with a well calibrated optical monitoring, covering periods of minutes to weeks and months.

In addition, as a byproduct, the OMC will provide the precise pointing of the platform, with an accuracy of few arcseconds, every 10 to 100 seconds. This information will be available on ground to improve the image reconstruction of the high-energy instruments.

CAMERA DESIGN

The Optical Monitoring Camera (OMC) consists of an optical system focused onto a large format CCD detector working in frame transfer mode. The optics is based on a refractive system with entrance pupil of 50 mm, focal length of 154 mm, and a field of view of 5°x5°. A once-only deployable cover will protect the optics from contamination during ground preparations and early operations in orbit.

The CCD (1024 x 2048 pixels) uses one section (1024 x 1024 pixels) for imaging and the other one for frame transfer before readout. The frame transfer time will be around 0.2 ms so that the need for a mechanical shutter is avoided. The selected chip for the OMC is an EEV CCD 47-20, with an imaging area of 13.3 x 13.3

TABLE 1. OMC Scientific Performances.

Parameter	Baseline value
Field of view	5°x5°
Aperture	50 mm
Focal length	154 mm (f/3.1)
Optical throughput	> 60% at 550 nm
Straylight reduction factor (within UFOV)	$<10^{-4}$ (for diffuse background)
Point spread function	>70% of energy within 1 pixel
CCD pixels	1024 x 2048, 1024 x 1024 image area (13 mm x 13 mm per pixel)
Image area	13.3 x 13.3 mm
Angular pixel size	17".6 x 17".6
CCD Quantum efficiency	85% at 550 nm
Full well capacity	100k electrons per pixel
Frame transfer time	<0.2 ms
Time resolution	> 1 s
Typical integration times	10 - 100 s
Wavelength range	V filter (centered at 550 nm)
Limit magnitude (10x100 s, 3σ)	19.2 (V)
Sensitivity to variations (10x100 s, 3σ)	$\delta < 0.1$ mag for $V < 16$
Average number of stars per pixel	0.6 (full sky)
($m_V < 19.5$)	2.0 (b = 0°); < 0.1 (b > 40°)

mm and 1024 x 1024 pixels. The CCD head will be cooled by means of a passive radiator to an operational temperature of -90 C degrees. The OMC camera includes also the CCD readout electronics, the necessary power conditioning electronics and the corresponding interface with the standard dedicated DPE (Data Processing Electronics) of the spacecraft. The camera will restrict the wavelength range of the CCD to around 550 nm by means of a V filter in the optical system to allow for a straightforward photometric calibration. An LED light source within the optical cavity will provide "flat-field" illumination of the CCD for on-board calibration. We show in Figure 1 the layout of the OMC camera unit. The scientific performances and additional parameters of the OMC are summarized in Table 1.

PHOTOMETRIC PERFORMANCES

With the current OMC design, we will be able to measure variations in V smaller than 0.1 mag for objects brighter than magnitude 18 and smaller than 0.02 mag for objects brighter than magnitude 16. The limiting magnitude of the OMC depends mainly on the amount of background light produced by the stellar contribution and zodiacal sunlight. At 3 σ level, for a minimum level of expected background (outside the galactic plane), the limiting magnitude will be m(V) = 19.2 for 10 integrations of 100 s each. This will be the standard observing mode of the camera. For maximum expected background level, within the galactic plane, the limiting magnitude will not go further than m(V) = 18.4.

The mentioned limiting magnitude of 19.2 implies that the limiting sensitivity of the OMC is 7 x 10^{-17} erg cm^{-2} s^{-1} $Å^{-1}$ for integrations of 10 x 100 s. With the typical integrations required for the gamma-ray instruments of 10^5 s (around one day), the OMC will then provide in total close to 100 individual points for the optical light curve, at the limiting magnitude of 19.2. Concerning the bright limit, the full-well capacity of the CCD provides the value of the brightest stars that can be measured without pixel saturation for a given integration time. An observing strategy combining consecutive integrations of different duration will permit to extend the effective dynamic range in a given field. For example, we can take a 10 s integration immediately after every 100 s exposure during the period of time when the spacecraft has the same pointing (before dithering).

In Table 2, we show the expected error expressed in magnitudes of a measurement for given integration times and magnitudes. These values have been derived from the signal-to-noise ratio obtained by comparing the expected count rate from the target source and the uncertainties in the total counts, dark counts and read-out noise. Concerning the levels of the background, minimum and maximum values were considered in Table 2 for the stellar and zodiacal light contributions. What we learn from this table is that very good V photometry can be performed with the OMC for objects of different brightness. For the faintest optical sources to be observed by INTEGRAL, around magnitude V = 19, measurements with enough accuracy can be obtained in a relatively short span of time compared to that devoted to the high-energy observations. Several individual points can be produced during the gamma-ray measurements in the optical and thus light variations as small as 0.3 magnitudes can be detected in those objects. In brighter sources, like many of the objects to be observed by INTEGRAL, fast photometry can be obtained with the OMC as well as very accurate photometry of few mmag.

The values presented for the photometric accuracy and the limiting magnitudes also include the effect of source confusion, since the stellar contribution to the background was used to determine the expected number of counts per pixel and the noise of the measurements. The amount of sources which are expected to be within one pixel for the OMC field of view (5°x5°) and the adopted CCD of 1024 x 1024 pixels imaging area are given approximately in Table 1.

SCIENCE WITH THE OMC

In addition to the monitoring of high-energy sources, the operation of INTEGRAL from a high orbit means that particular areas of the sky can be observed virtually continuously for periods of several weeks. This will provide a unique photometric capability for the investigation of complex, perhaps multiperiodic, light curves or those with periods that cannot be addressed by ground based observers.

For this purpose, an input catalogue is being compiled containing: all known optical counterparts of gamma-ray sources, X-ray sources, quasars and AGNs within the photometric limits of the OMC, all known late-type active stars, all known

TABLE 2. Photometric accuracy in V magnitudes of the OMC for different V values. Top: minimum background level; bottom: maximum background. First column indicates the integration time, assuming single images or the combination of several (5, 10 and 100) images of 100 s integration each.

Integration time (s)	8	10	12	14	16	18	20
1	0.011	0.027	0.075	0.24	-	-	-
10	0.003	0.008	0.022	0.058	0.20	-	-
50	-	0.004	0.010	0.025	0.08	-	-
5 x 100	-	-	0.003	0.008	0.024	0.10	-
10 x 100	-	-	0.002	0.006	0.017	0.07	-
50 x 100	-	-	0.001	0.003	0.008	0.03	0.19
1	0.011	0.027	0.075	0.27	-	-	-
10	0.003	0.009	0.022	0.070	0.32	-	-
50	0.001	0.004	0.010	0.031	0.13	-	-
5 x 100	-	0.001	0.003	0.010	0.042	0.24	-
10 x 100	-	-	0.002	0.007	0.030	0.17	-
50 x 100	-	-	0.001	0.003	0.013	0.08	0.48

erupting variable stars (including novae and cataclysmics), and additional variable objects which might require optical monitoring. The catalogue currently contains about 300 gamma-ray sources, 60,000 X-ray objects and 105,000 optically variable sources of different types. On the other hand, HIPPARCOS reference stars for auto-centering algorithms (100,000 selected stars) and photometric calibration (150,000 selected stars) are also included. During the mission, additional sources of interest will be included in the catalogue, namely, newly discovered optical counterparts of high-energy sources (specially sources discovered during the Galactic Plane Survey), regions of special interest for INTEGRAL science, new supernovae and transient sources, or any other Target of Opportunity.

During normal operations of INTEGRAL, the OMC will monitor around 100 targets within its field of view, extracted from this catalogue, in addition to the main high-energy sources. Taking into account the way in which INTEGRAL will be operated, changing its pointing by few degrees every 30 minutes, the OMC will have covered a very large area at the end of its lifetime. The OMC final catalogue will therefore provide at the end of the mission a complete set of optical light curves for thousands of objects. An estimation has been made of around 200,000 objects monitored with an average of around 800 points per object. These light curves will have been observed with the same instrument and under similar circumstances, covering both large and short time periods, and will be furthermore simultaneous to the X-ray and gamma-ray light curves.

The INTEGRAL Science Data Center

P. Dubath, A. Aubord, P. Bartholdi, M. Beck, J. Borkowski, T. Contessi, T. Courvoisier, D. Cremonesi, D. Jennings, P. Kretschmar, L. Lerusse, T. Lock, M.T. Meharga, N. Morisset, B. O'Neel, S. Paltani, R. Rohlfs, J. Soldan, R. Walter, R. Zondag

INTEGRAL Science Data Center, 16 ch. d'Ecogia, CH-1290 Versoix, Switzerland

Abstract. The INTEGRAL Science Data Center (ISDC) is designed as an interface between the INTEGRAL data and the user community. Its main task is to receive and process all the INTEGRAL data in order to (1) detect GRBs and other bright transient sources, for alerting the astronomical community and for initiating possible INTEGRAL Target Of Opportunity (TOO) observation, and (2) produce and archive a set of standard data products, and to distribute them to the observers. It is also foreseen to provide help and support to the observers in analyzing INTEGRAL data. The ISDC is funded by a consortium of European and US institutes. It is located in Versoix near Geneva.

THE INTEGRAL GROUND SEGMENT

Figure 1 illustrates the position of the ISDC within the different elements of the INTEGRAL ground segment. The INTEGRAL Science Operations Center (ISOC) is in charge of handling observation proposals from the users community, and of transforming the accepted proposals into an observation plan. The Mission Operation Center (MOC) main task is to perform all spacecraft and instrument commanding. It is also to provide information such as orbit and attitude data, and the spacecraft telemetry including science data and housekeeping data to the ISDC.

ISDC OPERATIONAL TASKS

Data Receipt and pre-processing consist of the reception of the INTEGRAL data, and the transformation of the telemetry stream into numerous FITS-formatted **raw data** products.

The **GRB Detection System** screens the telemetry stream to detect gamma-ray bursts occurring in the field of view of the imager or of the spectrometer. In

CP510, *The Fifth Compton Symposium,* edited by M. L. McConnell and J. M. Ryan

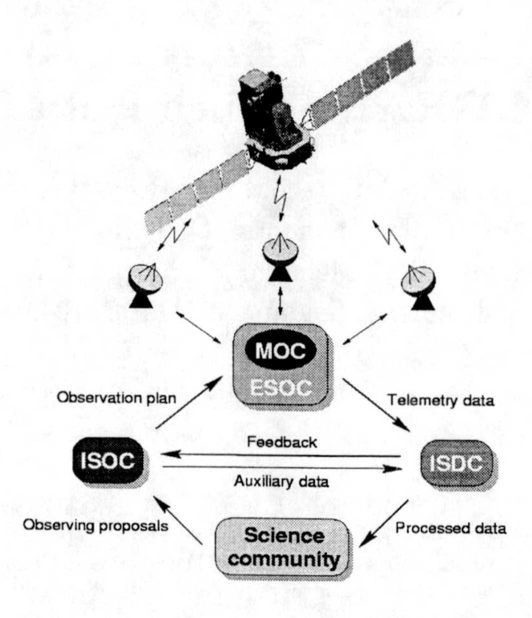

FIGURE 1. The INTEGRAL ground segment – from the spacecraft to the science community.

case of burst detection, an automatic process localizes the burst and, within few minutes, can send alerts to interested scientists for possible follow up observations from ground based and other space borne facilities. Bursts detected in the field of view of the imager can be located with a precision of about 1 arc minute.

Operation status and performance monitoring is concerned with the monitoring of selected operation and instrument parameters (background, sensitivity) to build a description of their temporal evolution behavior and to verify that the operations and the instruments are progressing as expected.

Quick-look analysis is a semi-automatic processing carried out to obtain sky images, source positions, fluxes and spectra in a relatively short processing time. In the case of observation of bright transient sources or any large discrepancy in the expected properties of the field the ISDC will alert the INTEGRAL Science Operation Center which can modify the observation program when appropriate.

Calibration data analysis is the regular analysis of calibration data to derive an updated description of the instrumental calibrations and responses.

The **standard analysis** of the data generates a set of standard data products, such as images of the sky, detected source positions and fluxes, spectra and light curves. These data, together with the raw data, will be made available to the guest observers with a one year proprietary period, and archived for later use by the community. Figure 2 shows a break down of one high-energy instrument standard analysis. The standard results and data products are in most cases derived to serve as a preview, and as a starting point for deeper off-line analyses.

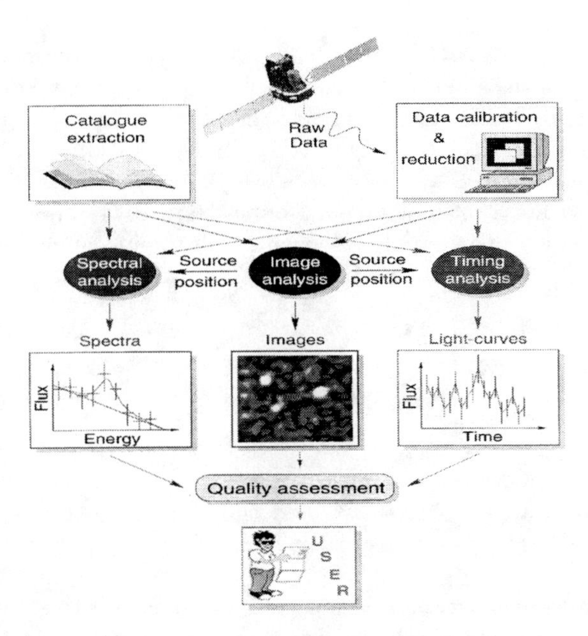

FIGURE 2. Illustration of the scientific analysis of an INTEGRAL instrument data

The **off-line processing** groups all the analyses carried out at the ISDC which are not part of the automatic operational processing. All tools available at the ISDC can be used in off-line analyses, and a sub-set of these tools will be provided to the users for performing their scientific analysis.

The **archive and distribution** system is to provide the long term storage and management of INTEGRAL data products, as well as an efficient and easy data distribution system for the user community. The archive and related database are to be remotely accessible through a computer network. It will become available during the first year of the mission, and will be maintained and regularly backed-up at least until 2 years after the end of the mission.

PRE-LAUNCH ACTIVITIES

The **software development** of all the components required to support the above listed operational tasks is the major pre-launch ISDC activity. This development is achieved in close cooperation with the teams developing and building the instruments. They provides the so-called Instrument Specific SoftWare (ISSW) which is integrated into the ISDC software system.

Some **support for calibration activities** is proposed to the instrument teams, for the processing and the archiving of the ground calibration data. The ground calibration data is also being used for the ISDC software testing.

The **operations preparation** consists of participating to the concept definition of the overall operations and of defining and preparing the ISDC operations in line with this concept. ISDC also participates to the program level tests, the simulation campaigns and organizes the training of the operational staff.

An ISDC **observation simulator** provides simulation tools to determine the expected signal to noise ratio of various source features, taking into account the observation strategy, the expected background and the instrument efficiency. This simulator, or its results, can be used by scientists in the preparation of their proposals.

STATUS

The ISDC has started its activities since its selection by ESA in 1995. It is hosted by the University of Geneva, in an institute attached to the Geneva Observatory. The current staff of about 22 scientists and engineers from several European and American institutes is increasing regularly, and is to reach approximately 30 at the time of launch.

The ISDC software has been specified, and the architecture design defines a number of ISDC sub-systems. Most of these sub-systems are in the coding and testing phase, and ISDC in the process of formalizing Interface Control Documents (ICDs) with the instrument teams. Overall tests involving the full ground segment will be conducted during the year 2000. One major goal is to have available in due time a first complete system for processing the final ground calibration data, from the telemetry stream to the high-level data products, such as images, source positions and fluxes, and spectra.

Among all activities performed by the ISDC staff, scientific research holds an important place with the aim to be well aware of the questions to be studied with INTEGRAL data.

CONCLUSION

The ISDC aims at being a service to the astronomical community at large and an element that will help astronomers active in other regions of the electro-magnetic spectrum to gain access to gamma-ray data when this can further the understanding of astrophysical problems.

We are happy to receive comments and/or contributions to these efforts. This can be done through many different channels, e.g., visits of long or short durations. We can also be contacted through the web (http://isdc.unige.ch) where a more detailed description of our work is available.

Modelling the effects of a solar flare on INTEGRAL

C.L. Perfect, C. Ferguson, A.J. Bird, F. Lei, J.J. Lockley and A.J. Dean

University of Southampton, Southampton, England, SO17 1BJ.

Abstract. The delayed effects of a large solar flare proton flux on the instruments on-board INTEGRAL have been modelled. We simulated exposing INTEGRAL to a flux of 1.5×10^8 protons/cm^2 over a period of five days. The induced count rates due to this proton flux over an energy range of 30 MeV - 2 GeV one minute after the end of the flare are 4170 $\pm$ 30 c/s for IBIS (the imager) and 190 $\pm$ 2 c/s for SPI (the spectrometer). We show that lowering the minimum incident proton energy range in the simulations below 30 MeV has no effect on the delayed count rate. Energy spectra indicate that there is little evolution of the spectral shape of the induced background up to 24 hours after the flare.

INTRODUCTION

INTEGRAL (INTErnational Gamma Ray Astrophysics Laboratory) is dedicated to fine imaging and high resolution spectroscopy at γ-ray energies. The two gamma ray instruments on-board INTEGRAL are SPI, the spectrometer and IBIS, the imager. There are also two other instruments: JEM-X, the X-ray monitor and the OMC, the Optical Monitoring Camera which we do not simulate here. The GGOD software suite [1] has been used to construct a model of INTEGRAL, called The Integral Mass Model, and to simulate the background noise in SPI and IBIS shortly after solar flare activity. Solar flares vary greatly in intensity and duration and it is therefore difficult to select a typical flare. The event chosen for our Monte-Carlo simulations occurred in March 1991 [2] and represents a five day flare with a steep power law spectrum.

THE MODEL

The SPI detector plane has 19 hexagonal germanium detectors and IBIS features two detector planes, a front layer of 16384 CdTe pixels (ISGRI) and a back layer of 4096 CsI pixels (PICsIT). Further information on The Integral Mass Model (TIMM) can be found in [1].

CP510, *The Fifth Compton Symposium,* edited by M. L. McConnell and J. M. Ryan

The input proton spectrum used for the simulations is 2.51×10^5 $E^{-3.4}$protons/cm^2/s/sr/MeV. The observed count rate for the 26 March 1991 [2] solar flare is approximately constant for a day before declining over a period of five to six days. This has been simplified in our model by using a constant flux over a period of five days with the same total integrated flux as the observed flare (1.5×10^{14} protons in the energy range 30 MeV - 2 GeV.). Further work will include a variable input proton flux over the duration of the flare.

THE SIMULATIONS

The decay profiles (Figures 1, 2 and 3) show how the count rates vary with time after the end of the flare for each instrument. The delayed count rates at various times after the flare indicate the input proton energy threshold below which there is no appreciable increase in the delayed count rate. The induced spectra for each instrument are also compiled in order to give an idea of the shape of the background noise spectra and the cause of the γ-ray lines produced.

The Decay Profiles

The input proton energy range has been varied and the decay profile (count rate versus time) for each instrument plotted (from 1 min to 24 hours after the flare). The quiescent background has not been included and the count rates have been integrated over the energy ranges 20 keV - 8 MeV for SPI, 10 keV - 2 MeV for ISGRI and 100 keV - 10 MeV for PICsIT.

Figures 1, 2 and 3 show the decay profiles for SPI, PICsIT and ISGRI respectively for input energy ranges 10 MeV - 2 GeV, 30 MeV - 2 GeV and 100 MeV - 2 GeV. Lowering the input proton energy threshold from 100 MeV to 30 MeV increases the count rate, one minute after the end of the flare, from ~165 c/s to ~190 c/s for SPI, from ~2700 c/s to ~3600 c/s for PICsIT and from ~250 c/s to 550 c/s for ISGRI. These increases in count rates are directly due to an increase in the input proton flux. Lowering the threshold further, however, to 10 MeV has no effect on the delayed count rates since these lower energy protons are absorbed by materials surrounding the detectors. The decay profiles show that the count rates 24 hours after the flare are about 25% of the count rates one minute after the flare (not including the quiescent background count rate). This indicates decaying isotopes with half-lives of the order of, or greater than, one day.

Energy Spectra

Figure 4a shows that, apart from a reduction in count rates and a number of specified lines, there is little evolution of the underlying spectral distribution for ISGRI for up to 24 hours after the flare. Figures 4b and 5a and b show the energy

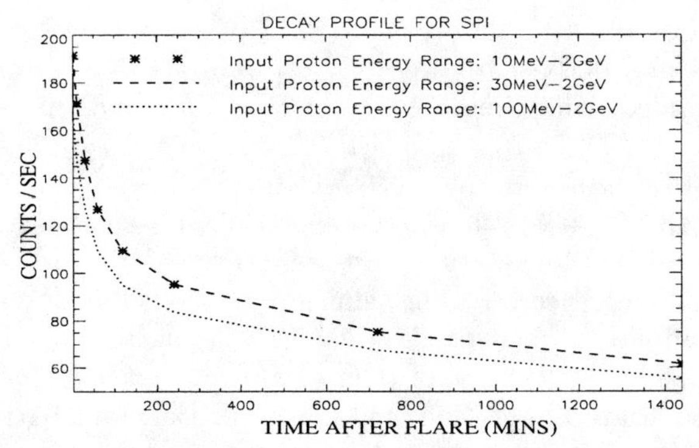

FIGURE 1. Decay profiles for SPI

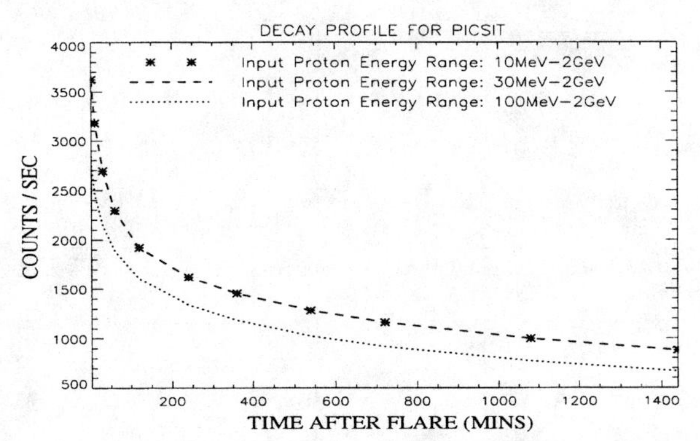

FIGURE 2. Decay profiles for PICsIT

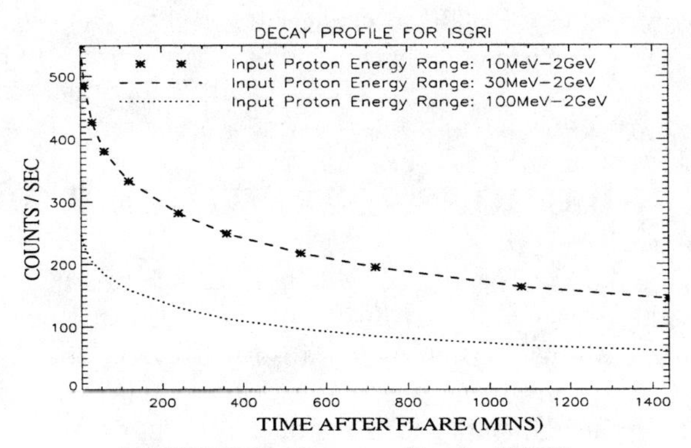

FIGURE 3. Decay profiles for ISGRI

spectra one minute after the flare for ISGRI, PICsIT and SPI respectively. The quiescent background derived from cosmic ray protons and cosmic diffuse γ-rays has not been included. The integrated count rate one minute after the flare is 550 $\pm$ 10 c/s for ISGRI, 3620 $\pm$ 20 c/s for PICsIT and 190 $\pm$ 2 c/s for SPI.

Some of the strong lines in the SPI spectrum (93.31 keV, 101.95 keV, 101.93 keV, 184.57 keV, 393.53 keV) are from the decay of $^{67}_{31}$Ga created in the germanium detectors. The strongest line features in the ISGRI spectrum (27.47 keV and 27.0 keV) are due to fluorescence from the tellurium in the CdTe. We are developing a fully automated approach where the Monte-Carlo simulation itself identifies the key isotopes decaying and the location of their origin within the INTEGRAL payload. A more complete study of the line emissions will therefore be presented in the future.

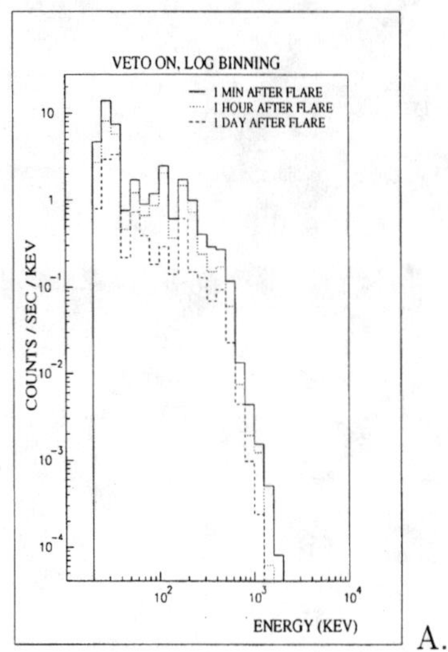

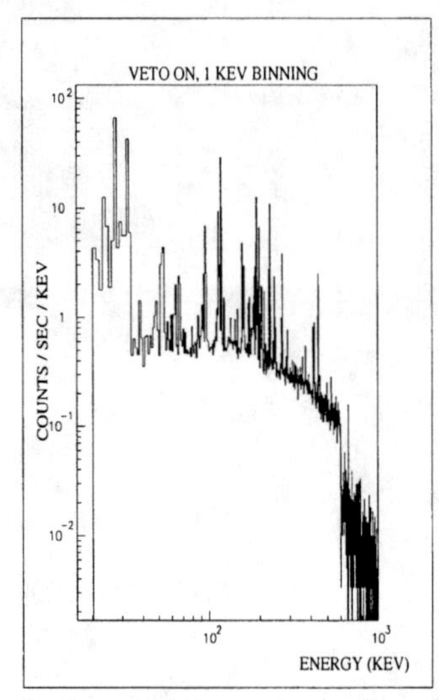

FIGURE 4. Energy spectra for ISGRI a) one minute, one hour and one day after the flare showing the spectral evolution and b) one minute after the flare showing the line features.

CONCLUSIONS AND FUTURE WORK

Monte-Carlo simulations for the 26 March 1991 solar flare indicate that solar flare protons in the energy range 10 - 30 MeV do not contribute to the delayed count rate. These low energy protons are absorbed by the materials surrounding the detectors. Protons in the energy range 30 MeV - 2 GeV give rise to activation of the spacecraft and instruments which is still significant days after the irradiation. The complex nature of the spectra and decay profiles tends to suggest that many isotopes are involved.

Preliminary results show that, apart from the reduction in count rates, there is

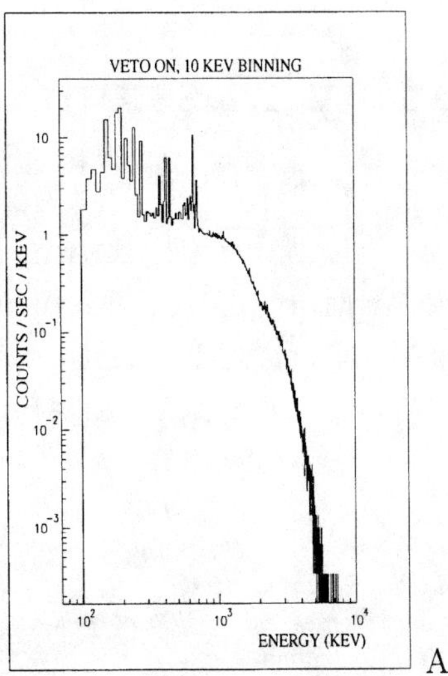

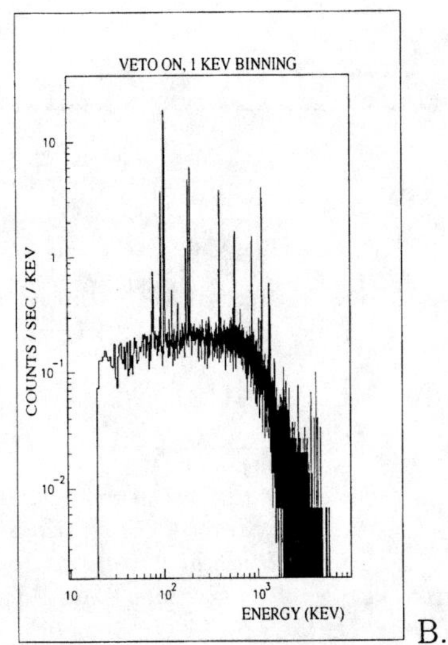

FIGURE 5. Energy spectra one minute after the flare with veto on showing line features for a) PICsIT and b) SPI.

little evidence of spectral evolution for SPI and IBIS for up to a day after the flare. Some line features for SPI and IBIS have been identified though a more detailed study is currently under way in order to identify further line features and their corresponding parent isotopes.

ACKNOWLEDGEMENTS

This work is partly funded by an EPSRC studentship and the author would like to thank Dr J Reeve of the Electronics and Computer Science Department, University of Southampton, for his support.

REFERENCES

1. Lei, F., Green, A.R., Bird, A.J., Ferguson, C. and Dean A.J., *Conference on the High Energy Radiation Background in Space 1997*, IEEE press:97TH8346, 1999, p 66.
2. Sanderson, T.R. Marsden, R.G., Heras, A.M., Wenzel, K.-P., Anglin, J.D., Balogh, A. and Forsyth, R., *Geophys. Res. Let.*, **19**, 12, (1992), p1263.

AGILE: a Gamma-Ray Mission

M. Tavani[1], G. Barbiellini[2], G. Budini[2], P. Caraveo[1], V. Cocco[5], E. Costa[3], G. Di Cocco[4], M. Feroci,[3] C. Labanti[4], I. Lapshov[3], F. Longo[2], S. Mereghetti[1], E. Morelli[3], A. Morselli[5], A. Pellizzoni[6], F. Perotti[1], P. Picozza[5], C. Pittori[5], M. Prest[2], M. Rapisarda[7], A. Rubini[3] P. Soffitta[3], M. Trifoglio[4], E. Vallazza[2], S. Vercellone[1].

[1]*Istituto di Fisica Cosmica del CNR "G. Occhialini", Milano, Italy*
[2]*Dipartimento di Fisica, Università di Trieste and INFN, Italy*
[3]*Istituto di Astrofisica Spaziale del CNR, Roma, Italy*
[4]*Istituto di Tecnologie e Studio della Radiazione Extraterrestre, CNR, Bologna, Italy*
[5]*Dipartimento di Fisica, Università di Roma II,"Tor Vergata" and INFN, Italy*
[6]*Agenzia Spaziale Italiana,*
[7]*ENEA, Italy*

Abstract. AGILE is an innovative, cost-effective gamma-ray mission selected by the Italian Space Agency for a Program of Small Scientific Missions. The AGILE gamma-ray imaging detector (GRID, made of a Silicon tracker and CsI Mini-Calorimeter) is designed to detect and image photons in the 30 MeV–50 GeV energy band with good sensitivity and very large field of view (FOV $\sim$ 3 sr). The X-ray detector, Super-AGILE, sensitive in the 10–40 keV band and integrated on top of the GRID gamma-ray tracker will provide imaging (1–3 arcmin) and moderate spectroscopy.

For selected sky areas, AGILE might achieve a flux sensitivity (above 100 MeV) better than 5×10^{-8} ph cm^2 s^{-1} at the completion of its scientific program. AGILE will operate as an Observatory open to the international community and is planned to be operational during the year 2002 for a nominal 2-year mission. It will be an ideal 'bridge' between EGRET and GLAST, and the only mission entirely dedicated to high-energy astrophysics above 30 MeV during that period.

INTRODUCTION

AGILE (*Astro-rivelatore Gamma a Immagini LEggero*) is a gamma-ray mission [7] selected by the Italian Space Agency (ASI) as the first project for the Program of Small Scientific Missions. AGILE ideally conforms to the *faster, cheaper, better* philosophy for a scientific mission. Gamma-ray detection by AGILE is based on silicon tracking detectors developed for space missions by INFN and Italian University laboratories during the past ten years [1]. AGILE is both very light ($\sim$ 70 kg) and highly efficient in detecting and monitoring gamma-ray sources in the energy range 30 MeV–50 GeV. The accessible field of view is unprecedently large ($\sim$ 1/4

CP510, *The Fifth Compton Symposium,* edited by M. L. McConnell and J. M. Ryan

of the whole sky) because of state-of-the-art readout electronics and a segmented anticoincidence system. The goal is to achieve an on-axis sensitivity comparable to that of EGRET on board of CGRO (a smaller background resulting from an improved angular resolution more than compensates the loss due to a smaller effective area) and a better sensitivity for large off-axis angles (up to $\sim 60°$). We refer to a companion paper for more details on the instrument (Barbiellini et al., these Proceedings). Table 1 shows the expected scientific performance of the baseline AGILE GRID and Super-AGILE detectors. The AGILE scientific program is optimized for three main goals.

(1) Optimal imaging capabilities, reaching $1' - 3'$ resolution by the combined gamma-ray/hard X-ray detection.

(2) Excellent timing capabilities (absolute time resolution $\lesssim 10\,\mu$s, deadtime $\sim 10\,\mu$s for the Mini-Calorimeter and Super-AGILE detectors and $\lesssim 100\,\mu$s for the GRID).

(3) Fast reaction to transients, with dissemination of consolidated quicklook results to allow multiwavelength observations.

SCIENTIFIC OBJECTIVES

Because of the large field of view ($\sim$ 3 sr) AGILE will discover a large number of gamma-ray transients, monitor known sources, and allow rapid multiwavelength follow-up observations because of a dedicated data analysis and alert program. We summarize here some of AGILE's scientific objectives (listed without any meaning to the ordering).

• **Active Galactic Nuclei**. For the first time, simultaneous monitoring of a large number of AGNs per pointing will be possible. Several outstanding issues concerning the mechanism of AGN gamma-ray production and activity can be addressed by AGILE including: (1) the study of transient vs. low-level gamma-ray emission and duty-cycles; (2) the relationship between the gamma-ray variability and the radio-optical-X-ray-TeV emission; (3) the correlation between relativistic radio plasmoid ejections and gamma-ray flares. A program for joint AGILE and ground-based monitoring observations is being planned. On the average, AGILE will achieve deep exposures of AGNs and substantially improve our knowledge on the low-level emission as well as detecting flares. We conservatively estimate that for a 3-year program AGILE will detect a number of AGNs 2–3 times larger than that of EGRET. Super-AGILE will monitor, for the first time, simultaneous AGN emission in the gamma-ray and hard X-ray ranges.

• **Gamma-ray bursts**. About ten GRBs have been detected by EGRET spark chambers during $\sim$ 7 years of operations [4]. This number is limited by the EGRET FOV and sensitivity and apparently not by the GRB emission mechanism. GRB detection rate by the AGILE-GRID is expected to be at least a factor of ~ 5 larger than that of EGRET, i.e., $\gtrsim$ 5–10 events/year). The small AGILE deadtime ($\leq$ 1000 times smaller than that of EGRET) allows a better study of the initial phase of GRB pulses (for which EGRET response was in many cases inadequate). The remarkable discovery of 'delayed' gamma-ray emission up to $\sim$ 20 GeV from GRB 940217 [3] is of great importance to model burst acceleration processes. AGILE is expected to be highly efficient in detecting photons above 10 GeV because of limited backscattering. Super-AGILE will be able to locate GRBs within a few arcminutes, and will systematically study the interplay between hard X-ray and gamma-ray emissions above 30 MeV. AGILE fast timing allows the study of ultra-short GRB pulses of duration $\sim 100\,\mu$s [2].

• **Diffuse Galactic and extragalactic emission**. The AGILE good angular resolution and large average exposure will further improve our knowledge of cosmic ray origin, propagation, interaction and emission processes. We also note that a joint study of gamma-ray emission from MeV to TeV energies is possible by special programs involving AGILE and new-generation TeV observatories of improved angular resolution.

• **Gamma-ray pulsars**. AGILE will contribute to the study of gamma-ray pulsars in several ways: (1) improving photon statistics for gamma-ray period searches by dedicated observing programs with long observation times of 1-2 months per source; (2) detecting possible secular fluctuations of the gamma-ray emission from neutron star magnetospheres; (3) studying unpulsed gamma-ray emission from plerions in supernova remnants and searching for time variability of pulsar wind/nebula interactions, e.g., as in the Crab nebula.

• **Galactic sources, new transients**. A large number of gamma-ray sources near the Galactic plane are unidentified, and sources such as 2CG 135+1 or transients suggesting the existence of a new class of gamma-ray sources (e.g., GRO J1838-04) [6] can be monitored on timescales of months/years. Also Galactic X-ray jet sources (such as Cyg X-3, GRS 1915+10, GRO J1655-40 and others) can produce detectable gamma-ray emission for favorable jet geometries, and a TOO program is planned to follow-up new discoveries of *micro-quasars*.

• **Solar flares**. During the last solar maximum, solar flares were discovered to produce prolonged high-intensity gamma-ray outbursts [5]. AGILE will be operational during part of the next solar maximum and several solar flares may be detected by the Mini-Calorimeter and by the Si-Tracker for favorable pointings. Particularly important for analysis will be the flares simultaneously detected by AGILE and HESSI (sensitive in the band 20 keV–20 MeV).

• **Fundamental Physics: Quantum Gravity**. AGILE detectors are suited for Quantum Gravity studies [8]. The existence of GRB micro-pulses lasting hundreds of microseconds [2] opens the way to study QG delay propagation effects with AGILE detectors. Particularly important is AGILE's Mini-Calorimeter with its independent readout for each of the 32 CsI bars with small deadtime and absolute timing accuracy ($10-20\,\mu$s). Energy dependent time delays near $\sim 100\,\mu$s for GRB micro-pulses in the energy range 0.3–3 MeV can be detected (requiring the detection of a minimum of 5-10 photons). If GRB micro-pulses originate at cosmological distances, sensitivity to the Planck's mass can be reached [8].

AGILE Scientific Performances

	Gamma-Ray Imaging Detector	
Energy Range	30 MeV – 50 GeV	
Field of View	3 sr	
Sensitivity at 100 MeV	6×10^{-9} ph cm^{-2} s^{-1} MeV^{-1}	(5σ in 10^6 s)
Sensitivity at 1 GeV	4×10^{-11} ph cm^{-2} s^{-1} MeV^{-1}	(5σ in 10^6 s)
Angular Resolution at 1 GeV	36 arcmin	(68% contain. radius)
Source Location Accuracy	~5-20 arcmin	for S/N $\gtrsim$ 10
Energy Resolution	$\Delta E/E\sim1$	at 400 MeV
Absolute Timing Accuracy	$\lesssim 10\,\mu$s	
	Super-AGILE	
Energy Range	10-40 keV	
Field of View	$107°\times68°$	Full Width Zero Sens.
Sensitivity	5–10 milliCrab	(5σ in 1 day)
Angular Resolution (Pixel Size)	6 arcmin	
Source Location Accuracy	~1-3 arcmin	for S/N~10
Energy Resolution	$\Delta E<4$ keV	
Absolute Timing Accuracy	$\lesssim 10\,\mu$s	

MISSION

The ideal orbit for AGILE is a low-background orbit of 550-650 km. The AGILE satellite weight is 200-220 kg, and its pointing will be obtained by a three-axis stabilization system with an accuracy near 0.5–1 degree. Pointing reconstruction reaching an accuracy of $\lesssim$ 1 arcmin will be obtained by two star sensors. The total downlink telemetry rate of science data is 500 kbit s^{-1}. The AGILE mission is being planned as an Observatory open to the Italian and international scientific community. The AGILE mission emphasizes a rapid response to the detection of gamma-ray transients. AGILE will ideally 'fill the vacuum' between the upcoming end of EGRET operations and the beginning of the GLAST mission.

REFERENCES

1. Barbiellini G. et al., *Nucl. Instrum. & Methods*, **354**, 547 (1995).
2. Bhat, C.L., et al., *Nature*, **359**, 217 (1992).
3. Hurley K. et al., *Nature*, **372**, 652 (1994).
4. Schneid E.J. et al., in *AIP Conf. Proc.*, **384**, p.253 (1996a).
5. Schneid E.J. et al., *Astron. Astrophys. Suppl. Ser.*, **120**, 299 (1996b)
6. Tavani, M., et al., *Astroph. Journal*, **479**, L109 (1997).
7. Tavani, M., et al., *AGILE Phase A Report* (1998) (see also the *Science with AGILE*, http://www.ifctr.mi.cnr.it/Agile).
8. Tavani, M., et al., to be submitted (2000),
9. Thompson D.J. et al., Astrophys. J. Suppl., **86**, 657 (1993).

AGILE: the Scientific Instrument

G. Barbiellini[2], M. Tavani[1], G. Budini[2], P. Caraveo[1], V. Cocco[5], E. Costa[3], G. Di Cocco[4], M. Feroci,[3] C. Labanti[4], I. Lapshov[3], F. Longo[2], S. Mereghetti[1], E. Morelli[3], A. Morselli[5], A. Pellizzoni[6], F. Perotti[1], P. Picozza[5], C. Pittori[5], M. Prest[2], M. Rapisarda[7], A. Rubini[3], P. Soffitta[3], M. Trifoglio[4], E. Vallazza[2], S. Vercellone[1].

[1]*Istituto di Fisica Cosmica del CNR "G. Occhialini", Milano, Italy*
[2]*Dipartimento di Fisica, Università di Trieste and INFN, Italy*
[3]*Istituto di Astrofisica Spaziale del CNR, Roma, Italy*
[4]*Istituto di Tecnologie e Studio della Radiazione Extraterrestre, CNR, Bologna, Italy*
[5]*Dipartimento di Fisica, Università di Roma II,"Tor Vergata" and INFN, Italy*
[6]*Agenzia Spaziale Italiana,*
[7]*ENEA, Italy*

Abstract.
The scientific instrument of the AGILE mission is innovative in many ways. It is an integrated instrument based on three detecting systems: (1) a Silicon Tracker, (2) a Mini-Calorimeter, and (3) an ultralight coded mask system with Si-detectors (Super-AGILE). For a relatively low mass ($\sim$ 70 kg) and large ratio of expected performance over cost, AGILE is planned to provide

(i) Optimal imaging in the energy bands 30 MeV–50 GeV (5–10 arcmin for intense sources) and 10-40 keV (1–3 arcmin).

(ii) Optimal timing capabilities, with independent readout systems and minimal dead-times for the Silicon Tracker, Super-AGILE and Mini-Calorimeter.

(iii) A very large field of view for the gamma-ray imaging detector (3 sr) and Super-AGILE (1 sr).

INTRODUCTION

We present a brief outline of the baseline design for the AGILE scientific instrument. Its main goal is to provide an integrated system made of a Gamma-Ray Imaging Detector (GRID, Si-Tracker and Mini-Calorimeter) and Super-AGILE, with imaging capabilities in the energy ranges 30 MeV–50 GeV and 10–40 keV, respectively. The instrument is designed to be light ($\sim$ 70 kg) and with an optimal ratio of expected performance over cost because of the crucial detector development carried out by our group in previous years [1,2]. Timing capabilities of the whole instrument are optimized reaching $\lesssim 10\,\mu$s absolute time tagging (using an

CP510, *The Fifth Compton Symposium,* edited by M. L. McConnell and J. M. Ryan

on-board GPS), and unprecedently small deadtimes ($\sim 100\ \mu s$ for the GRID, and $10 - 20\ \mu s$ for each detecting unit of the Mini-Calorimeter and Super-AGILE).

INSTRUMENT OVERVIEW

Fig. 1 shows schematically the AGILE configuration (with a height of ~ 50 cm, a Si-plane geometric area of 38×38 cm^2, and a weight of ~ 70 kg including Si-Tracker, Mini-Calorimeter, Anticoincidence system and electronic unit). The baseline instrument is made of the following elements.

- **Silicon-Tracker**, the gamma-ray converter/tracker is made of 14 Si-planes with microstrip pitch equal to 121 μm (readout pitch equal to 242 μm). The fundamental unit is a module of area 9.5×9.5 cm^2 and thickness 410 μm for a total of 384 readable microstrips. The first 12 planes are made of three layers: a first layer of tungsten ($0.07\ X_0$) for gamma-ray conversion, and two Si-layers with microstrips orthogonally placed to obtain the plane coordinates of charged particles produced by gamma-ray pair creation interactions. For each Si-plane there are then $2 \times 1,536$ readable microstrips. Since the GRID trigger requires at least three Si-planes to be activated, two more Si-planes are inserted at the bottom of the tracker without a tungsten layer. The total readable microstrip number for the GRID tracker is then $\sim 43,000$. The Front End Electronics (FEE) is based on commercially available TA1 chips. We emphasize that the use of TA1 chips makes available **both digital and analog** information for track analysis. The distance between mid-planes equals 1.6 cm as optimized by Montecarlo simulations. The AGILE photon tracking system has an *on-axis* total radiation length larger than $0.8\ X_0$ for an interaction probability above 400 MeV above 35%.

- **Mini-Calorimeter**, made of two planes of Cesium Iodide (CsI) bars, for a total (on-axis) radiation length of $1.5\ X_0$. The signal from each CsI bar is collected by two photodiodes placed at both ends. The aims of the Mini-Calorimeter (MCAL) are: *(i)* obtaining information on the energy deposited in the CsI bars by particles produced in the tracker, and therefore contributing to the determination of the total photon energy; *(ii)* detecting GRBs and other impulsive events with spectral and intensity information in the energy band $\sim 0.3 - 100$ MeV with optimal deadtime ($\sim 10 - 20\ \mu s$) for a readout system treating each CsI bar independently. We note that the problem of particle backscattering for this configuration is much less severe than in the case of EGRET. AGILE allows an efficient photon detection above 1 GeV.

- **Super-AGILE**, made of 4 Silicon detectors (19×19 cm^2 each) and associated FEE placed on the first GRID tray plus an ultra-light coded mask system supporting an Au mask placed at a distance of 14 cm from the silicon detectors. The Super-AGILE goals are: *(i)* photon-by-photon detection and imaging of

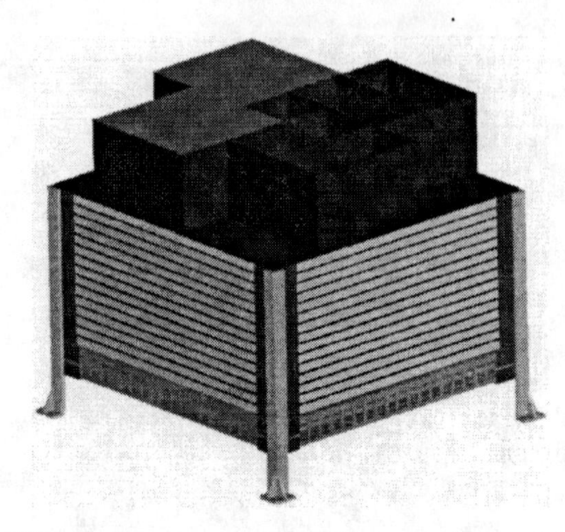

FIGURE 1. The AGILE baseline instrument (AC system and electronic boxes not shown). The GRID is made of 14 Si planes and a Mini-Calorimeter at the bottom of the istrument. Super-AGILE has its 4 Si-detectors on the first GRID tray, and the ultra-light coded mask system positioned on top. The baseline payload size is $\sim 53 \times 53 \times 50\,\mathrm{cm}^3$, including Super-AGILE.

sources in the energy range 10–40 keV, with a large field-of-view (FOV) of $\sim$ 1 sr, good angular resolution (1-3 arcmin, depending on source intensity and geometry), and good sensitivity ($\lesssim$ 10 mCrab for 50 ksec integration, and $\lesssim$ 1 Crab for a few seconds integration); *(ii)* simultaneous X/γ spectral studies of high-energy sources; *(iii)* excellent timing (1-10 μs); *(iv)* burst trigger for the GRID; *(v)* burst alert (and on-board quick positioning) capability.

- **Anticoincidence (AC) system**, aimed at both charged particle background rejection and preliminary direction reconstruction for triggered photon events. The AC system completely surrounds all AGILE detectors (Super-AGILE and GRID). Each lateral face is segmented with three overlapping plastic scintillator layers (0.6 cm thick) connected with photomultipliers placed at the bottom. A single plastic scintillator layer (0.5 cm thick) constitutes the top-AC whose signal is read by four light photomultipliers placed externally to the AC system and supported by the four corners of the structure frame.

DATA HANDLING

The GRID trigger logic for the acquisition of gamma-ray photon data and background rejection is structured in Level-1 and Level-2 trigger phases. The Level-1 phase is fast (1-2 μs) and requires a signal in at least three out of four contiguous tracker planes and a proper combination of fired FEE chip number and AC signals. An intermediate Level-1.5 stage is also envisioned (lasting $\sim 20\ \mu s$), with the acquisition of the event topology based on the identification of fired FEE chips. Both Level-1 and Level-1.5 have a hardware-oriented veto logic providing a first cut of background events. Level-2 data processing includes the complete FEE readout and pre-processing, "cluster data acquisition" (both the analog and digital information provided by the TA1 chips), and processing by a dedicated CPU. The Level-2 processing is asynchronous (estimated duration $\sim$ a few ms). The GRID deadtime is therefore $\sim 100\ \mu s$ and is dominated by the tracker FEE readout.

In order to maximize the GRID FOV and detection efficiency for large-angle incident gamma-rays (and minimize the effect of particle backscattering from the mini-calorimeter), the data acquisition logic uses proper combinations of top and lateral AC signals and a coarse direction reconstruction in the Si-Tracker. For events depositing more than 200 MeV in the Mini-Calorimeter, the AC veto can be disabled to allow the acquisition of photon events with energy larger than 1 GeV.

Appropriate data buffers and burst search algorithms are envisioned to maximize data acquisition for impulsive gamma-ray events for all AGILE detectors (Si-Tracker, Mini-Calorimeter and Super-AGILE) independently.

Table 1: AGILE vs. EGRET

	EGRET	AGILE
Mass	1830 kg	70 kg
Energy band	30 MeV – 30 GeV	30 MeV–50 GeV; 10–40 keV
Field of view (FOV)	~ 0.8 sr	3 sr (GRID); 1 sr (Super-A)
PSF (68% containment radius)	5.5° 1.3° 0.5°	4.7° (@ 0.1 GeV) 0.6° (@ 1 GeV) 0.2° (@ 10 GeV)
$\Delta E/E$ at 400 MeV	~ 0.2	~ 1
Deadtime	$\gtrsim 100$ ms	$\lesssim 100\ \mu s$ (GRID) $\lesssim 20\ \mu s$ (MCAL, Super-A)
Sensitivity for pointlike γ-ray sources[†] ($\mathrm{ph\,cm^{-2}\,s^{-1}\,MeV^{-1}}$)	8×10^{-9} 1×10^{-10} 1×10^{-11}	6×10^{-9} (@ 0.1 GeV) 4×10^{-11} (@ 1 GeV) 3×10^{-12} (@ 10 GeV)
Super-A sensitivity (1 day)		$\lesssim 10$ mCrab (10-20 keV)
Pointing reconstruction	$\sim$10 arcmin	$\sim$1 arcmin

(†) Obtained for a typical exposure time near 2 weeks for both AGILE and EGRET.

REFERENCES

1. Barbiellini G. et al., *Nucl. Instrum. & Methods*, **354**, 547 (1995).
2. Golden R.L., Morselli A., Picozza P. et al., *Il Nuovo Cimento B*, **105**, 191 (1990)
3. Tavani, M., et al., *AGILE Phase A Report* (1998) (see also *Science with AGILE*, http://www.ifctr.mi.cnr.it/Agile).

The GLAST Silicon-Strip Tracking System

Robert P. Johnson

Santa Cruz Institute for Particle Physics
University of California at Santa Cruz

Abstract. The GLAST instrument concept is a gamma-ray pair conversion telescope that uses silicon microstrip detector technology to track the electron-positron pairs resulting from gamma-ray conversions in thin lead foils. A cesium iodide calorimeter following the tracker is used to measure the gamma-ray energy. Silicon strip technology is mature and robust, with an excellent heritage in space science and particle physics. It has many characteristics important for optimal performance of a pair conversion telescope, including high efficiency in thin detector planes, low noise, and excellent resolution and two-track separation. The large size of GLAST and high channel count in the tracker puts demands on the readout technology to operate at very low power, yet with sufficiently low noise occupancy to allow self triggering. A prototype system employing custom-designed ASIC's has been built and tested that meets the design goal of approximately 200 _W per channel power consumption with a noise occupancy of less than one hit per trigger per 10,000 channels. Detailed design of the full-scale tracker is well advanced, with non-flight prototypes built for all components, and a complete 50,000 channel engineering demonstration tower module is currently under construction and will be tested in particle beams in late 1999. The flight-instrument conceptual design is for a 4×4 array of tower modules with an aperture of 2.9 m^2 and an effective area of greater than 8000 cm^2.

OVERVIEW

This paper describes the tracker-converter section of a proposed instrument (1) for the Gamma Large Area Space Telescope mission (2), currently in its formulation phase. GLAST is a gamma-ray pair conversion telescope that operates in much the same way as the EGRET experiment on the Compton Gamma Ray Observatory (3). As a successor to EGRET, however, GLAST is expected to improve upon EGRET's sensitivity to astronomical point sources by factors of 10 to 100. That is accomplished in this design primarily by taking advantage of silicon-strip detector technology developed during the past

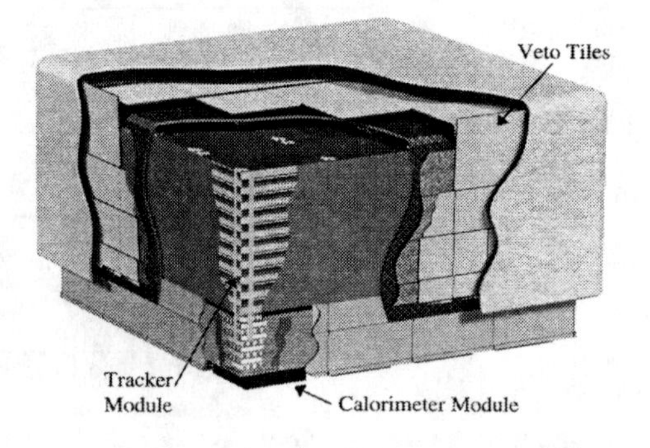

FIGURE 1. Cutaway view of the GLAST instrument, composed of a 4×4 array of tower modules surrounded by a veto shield and thermal blanket.

CP510, *The Fifth Compton Symposium,* edited by M. L. McConnell and J. M. Ryan

decade for applications in elementary particle and space physics experiments.

The GLAST detector consists of a square 4×4 array of nearly identical tower modules, as indicated in Fig. 1. Each tower has a scintillator veto counter on the top (and on the sides for the edge towers), followed by a multilayer silicon-strip tracker and, finally, a cesium iodide calorimeter. Each of the tracking layers has two planes of single-sided silicon-strip detectors with strips oriented at 90 degrees with respect to each other. All but the bottom few layers have a thin lead foil preceding the detector planes. The foils convert the incident gamma-ray photons into electron-positron pairs, which are subsequently tracked by the remaining detector layers to determine the photon direction. Finally, the calorimeter absorbs the electrons and thereby measures the photon energy.

Besides providing optimal angular resolution for this type of device, the silicon-strip technology is fast, yielding a system with very little dead time, provides excellent multi-track separation, which is important for pattern recognition and background rejection, and can be made self triggering. The latter two points eliminate the need for a time-of-flight system, such as that used by EGRET for triggering, and thereby result in a very compact instrument with a wide, 2.3 sr fwhm, field of view. The silicon-strip technology is by now well developed, is known to be robust, requires no consumables, such as gas, and operates at a relatively low voltage, compared with spark or drift chambers. It therefore appears to be ideally suited for space applications and, in fact, already has a substantial heritage in space applications. The AMS experiment is a notable example of a large silicon-strip system used in space (4).

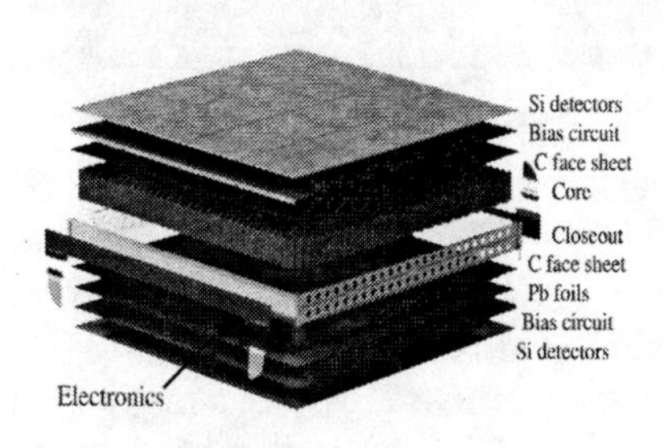

FIGURE 2. Exploded view of a tray.

TRACKER TOWER MECHANICAL STRUCTURE

The support structure for the silicon-strip detectors must prevent damage to the detectors, electronics, and wire bonds during 10 g static accelerations and random vibrations in excess of 14 g rms. A tracker tower is made up of 19 carbon-composite sandwich structures, called "trays," each of which supports silicon-strip sensors on both sides and readout electronics on two edges. Figure 2 shows an exploded view of a tray.

FIGURE 3. Cutaway view of a single tracker tower.

Figure 3 shows a view of a single tower, with the

trays stacked up and held in alignment by pins in the four corners. Vectran cables under tension also pass through the corners to hold the stack in compression. Two thin flex-circuit cables on each of the four sides connect the readout electronics to the data acquisition system (DAQ). All four sides are covered by 1.5 mm carbon-fiber walls, which act as shear panels and conduct heat from the electronics to the base of the tower. A complete engineering model of this structure has been fabricated with aluminum tray panel closeouts and subjected to vibration testing in excess of the Delta-II launch vehicle qualification levels without damage or excessive displacement (5). A fully instrumented tray has also been subjected to the same qualification levels with no resulting damage.

SILICON-STRIP READOUT ELECTRONICS

The GLAST instrument design has more than a million silicon-strip channels. Two clear limitations on operating such a system in space are the availability of power for the electronics and the difficulty of dissipating the resulting heat. Previous silicon-strip systems, designed for operation in ground-based experiments or in space with a significantly smaller number of readout channels, have not needed to contend with such severe power limitations. For those reasons, a major goal of the research and development effort within the GLAST collaboration has been to design and test readout electronics that can meet the signal-to-noise requirements with minimal power dissipation.

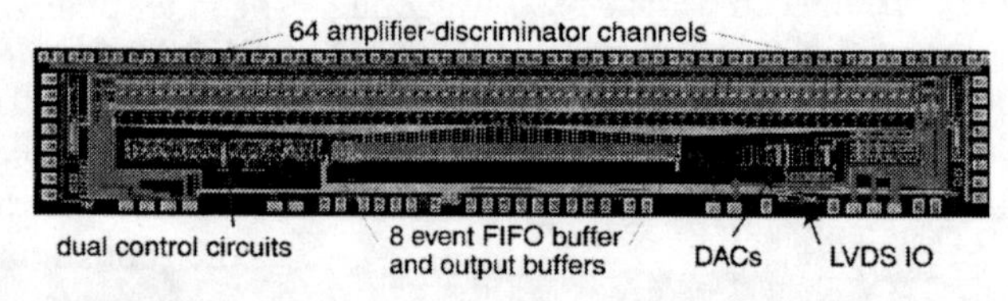

FIGURE 4. Layout of the front-end readout chip.

The readout is based upon the 64-channel CMOS ASIC (GTFE64) illustrated in Fig. 4. The amplifier-discriminator chain, described in (6), operates at a power level of only 140 μW per channel, with about 1600 *e* rms noise for a 30-cm strip (≈38 pF), and has been well tested in a silicon-strip system operated in particle beams (7). In the GLAST design 28 chips, plus two digital ASICs, are arranged in a readout section attached to the side of a tray (Fig. 5). The data and control flow for a set of trays is illustrated in Fig. 7. This novel design provides two redundant data and control paths for each chip, protecting the system from catastrophic single-point failures. The digital ASIC (GTRC) acts as an interface to the DAQ and also formats and buffers the data. The data are delivered to the DAQ in the form of a zero-suppressed hit list. This entire system has been proto-

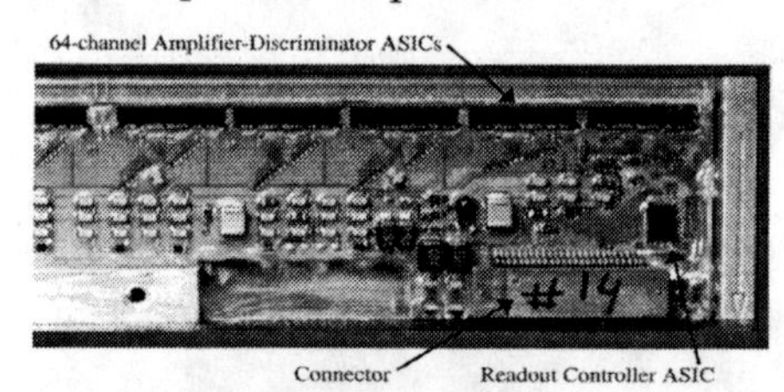

FIGURE 5. Photograph of one end of a completed prototype readout section, mounted on a prototype tray (aluminum closeout).

typed in an engineering model with 32 readout sections and operates with a power consumption of only 203 μW per channel at a 12.5 kHz trigger rate.

The principal trigger of the GLAST instrument is provided by the tracker itself. A logical OR is formed of all channels in each readout section and sent to the DAQ where coincidences between x and y layers in the same tracking plane are detected. If 3 consecutive x,y planes in a tower are in coincidence, then a trigger signal is sent to all 16 towers. The trigger signal latches all discriminator outputs into an 8-event deep FIFO buffer, where the data await a readout command from the DAQ.

SILICON STRIP DETECTORS

The tracker design is based upon single-sided, AC-coupled silicon-strip detectors, with p-type strip implants on n-intrinsic silicon. For good operational stability, polysilicon resistors are used to bias the implants. The baseline design calls for four 9.5-cm square detectors, cut from 6-inch wafers, to be wire bonded into 38-cm "ladders," with 4 ladders on each face of each tray. The strip pitch is 208 microns, making a total of 448 strips per ladder. More than 200 detectors in this technology have already been procured from Hamamatsu Photonics and tested, with excellent results. The fraction of bad strips is well below 10^{-3}, and the leakage current is typically less than 10 nA/cm^2.

CONCLUSIONS

The design of the silicon-strip tracker for the GLAST instrument is already well advanced. Completely functional engineering models have been built of all components to verify the technological approach. Recently, a complete tracker tower has been constructed, to be operated in test beams in December 1999. Figure 6 shows a photograph of one of 17 trays of that tower. This development effort has verified that the GLAST detector and electronics requirements can be readily achieved by existing technology and has already solved many engineering and fabrication issues, thus minimizing risks for the flight-instrument development.

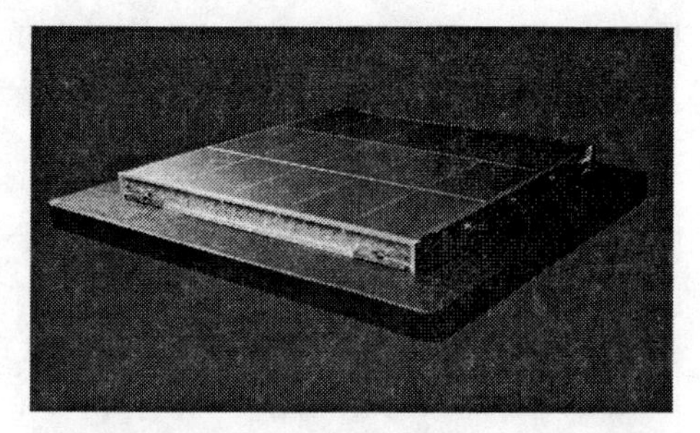

FIGURE 6. A completed prototype tracker tray, with detectors on both top and bottom (not visible) faces. The bottom detectors are read out by electronics on the far side. The heavy base plate that the tray is resting upon is not part of the tower.

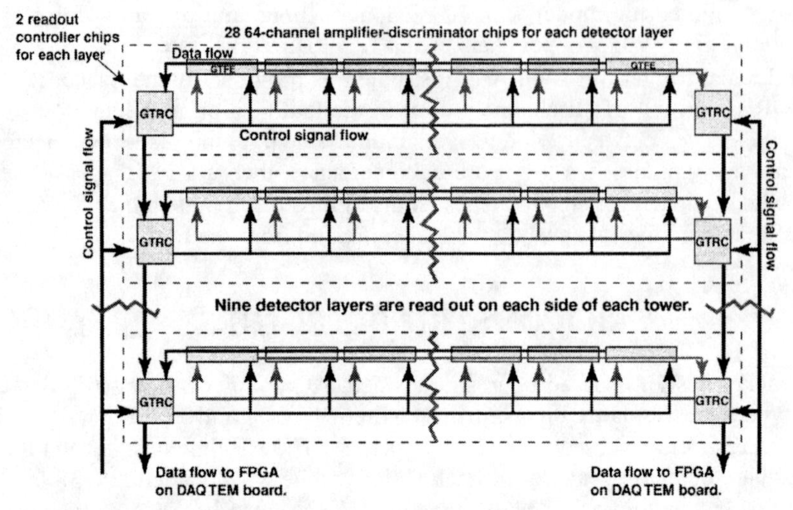

FIGURE 7. Block diagram of the tracker readout electronics. Each pair of redundant cables connecting to the DAQ handles nine readout modules, or one side of one tower.

ACKNOWLEDGMENTS

This work is supported by NASA contract NAS5-98039 and DoE grant DEFG03-92ER40689. I would like to acknowledge the work of all of my colleagues in the GLAST collaboration, in particular those at U.C. Santa Cruz, Stanford Linear Accelerator Center, Tokyo University, Hiroshima University, and Hytec Inc. who have contributed to the silicon-strip tracker hardware design and prototyping work described here.

REFERENCES

1. P. Michelson, *et al., Proc. SPIE* **2806**, 31 (1996).
2. NASA, Office of Space Science, AO 99-OSS-03, *Gamma-Ray Large Area Space Telescope (GLAST), Flight Investigations* (1999).
3. Thompson, D.J., *et al., ApJ Suppl.* **86**, 629 (1993).
4. Pauluzzi, M., *Nucl. Instrum. Meth. A* **383,** 35–43 (1996).
5. Ponslet, E., Ney, S.A., Miller, W.O., *Innovative, Low-Mass, Passively Cooled, All Composite Material Tower Structure for High Resolution Charged Particle Tracking in a Gamma-Ray Space Telescope,* NASA SBIR 97-1 17.01-5179, Phase I Final Report, HYTEC Inc., Los Alamos, NM (1998).
6. Johnson, R.P., *et al., IEEE Trans. Nucl. Sci.* **45**, 927 (1998).
7. Atwood, W.B., *et al.,* to be published in *Nucl. Instrum. Meth. A*, SLAC-PUB-8166 (1999).

Detecting the Attenuation of Blazar Gamma-ray Emission by Extragalactic Background Light with GLAST

Andrew Chen, Steven Ritz

NASA/Goddard Space Flight Center
Code 661, Greenbelt, MD 20771

Abstract. Gamma rays with energy above 10 GeV interact with optical-UV photons resulting in pair production. Therefore, a large sample of high redshift sources of these gamma rays can be used to probe the extragalactic background starlight (EBL) by examining the redshift dependence of the attenuation of the flux above 10 GeV. GLAST, the next generation high-energy gamma-ray telescope, will for the first time have the unique capability to detect thousands of gamma-ray blazars up to redshifts of at least $z = 4$, with enough angular resolution to allow identification of a large fraction of their optical counterparts. By combining recent determinations of the gamma-ray blazar luminosity function, recent calculations of the high energy gamma-ray opacity due to EBL absorption, and the expected GLAST instrument performance to produce simulated samples of blazars that GLAST would detect, including their redshifts and fluxes, we demonstrate that these blazars have the potential to be a highly effective probe of the EBL.

LUMINOSITY FUNCTION OF GAMMA-RAY BLAZARS

The Energetic Gamma Ray Experiment Telescope (EGRET) has detected more than 60 blazar-type quasars (Mukherjee et al. 1997) emitting gamma rays with $E > 100$ MeV. These sources are flat-spectrum radio-loud quasars (FSRQs) and BL Lac objects, often exhibiting non-thermal continuum spectra, violent optical variability, and/or high optical polarization. They are also highly variable gamma-ray sources. The EGRET blazars whose optical redshifts have been measured range from $z = 0.03$ to 2.28. The redshift distribution is consistent with the distribution of FSRQs, which extends up to $z = 3.8$.

Chiang & Mukherjee (1998) modeled the evolution and luminosity function of the parent gamma-ray blazar distribution, taking careful account of selection biases, without assuming a correlation between luminosities at gamma-ray and other wavelengths. Parameterizing the luminosity function by

$$\frac{dN}{dVdL_0} = \rho(z)N_0L_0^{-\gamma} \text{ and } L_0 = L/(1+z)^{\beta} \quad (1)$$

with a maximum cutoff redshift of $z_{max} = 5$, they found $\rho(z)$ consistent with a constant (no density evolution) and $\beta = 2.7$ provided the best fit. However, they found that a simple power law in L failed to model the dearth of blazars below $z = 1$ adequately.

CP510, *The Fifth Compton Symposium,* edited by M. L. McConnell and J. M. Ryan

The best fit was a simple power law with $\gamma = 2.2$ at high luminosities and a luminosity cutoff of $L_B = 1.1 \times 10^{46}$ erg s^{-1}.

EXTRAGALACTIC BACKGROUND LIGHT

Gamma-rays traveling through intergalactic space will interact through pair production with the extragalactic background starlight (EBL) emitted by galaxies. The cross section depends on the energies of the target and incident photons. Gamma rays with energy $E > 10$ GeV are significantly attenuated. Thus, the apparent spectra of gamma-ray emitting objects will be modified at those energies, increasing with increasing redshift. Salamon & Stecker (1998) calculated the opacity out to $z = 3$. To estimate the stellar emissivity and spectral energy distributions vs. redshift they adapted the analysis of Fall, Charlot, & Pei (1996), consistent with the Canada-France Redshift Survey. They included corrections for metallicity evolution. They found that the stellar emissivity peaks between $z = 1$ and 2, leading to a significant redshift-dependent absorption.

GLAST

The Gamma-ray Large Area Space Telescope (GLAST) is under development with a planned launch in 2005. It is intended to be the successor to EGRET, with a much larger effective area, especially at higher energies (≥ 8000 cm^2 at > 1 GeV), larger field of view and better angular resolution than EGRET. GLAST should be able to reach a flux limit of 4×10^{-9} photons cm^{-2} s^{-1} after one year, resulting in the detection of thousands of blazars. The improved angular resolution should, in theory, allow a high percentage of optical identifications and possible redshift measurements, depending on the available ground-based resources. The improved high-energy performance should yield accurate flux determinations above 10 GeV for many of these sources.

PROCEDURE

Using the de-evolved luminosity function proportional to $L^{-2.2}$ with a minimum cutoff at 10^{46} erg s^{-1} according to Chiang & Mukherjee, we generated 10^6 blazars. We assigned each one a random redshift z between 0 and 5 distributed according to the following relations:

$$\frac{dN}{dz} = \frac{dN}{dV}\frac{dV}{dz}, \quad \frac{dN}{dV} = \text{constant},$$

$$\frac{dV}{dz} = \frac{4\pi c}{H_0}\frac{d_l^{\,2}(z)(1+z)}{\sqrt{1+2q_0 z}}, \tag{2}$$

$$d_l = \frac{c}{H_0 q_0^{\,2}}\left[1 - q_0 + q_0 z + (q_0 - 1)(2q_0 z + 1)^{1/2}\right],$$

where $q_0 = 0.5$ and $H_0 = 75$ km s^{-1} Mpc^{-1}. The flux of each blazar was then calculated according to

$$F = L_0 \frac{(1+z)^{1-\alpha+\beta}}{4\pi d_l^{\,2}(z)}, \qquad (3)$$

where $\beta = 2.2$ and $\alpha = 2.15$, the average spectral index of the EGRET blazars. This yielded ~ 5000 blazars with observed flux greater then 4×10^{-9} photons cm^{-2} s^{-1} above $E > 100$ MeV.

We calculated the $E > 10$ GeV flux of each by adding two effects. First, each blazar got a random, normally distributed spectral index, -2.15 ± 0.04. An index of -2.15 yields a ratio of ~ 0.07 between the two fluxes. We also included the redshift-dependent absorption above 10 GeV. The form of the dependence was parameterized from the graph in Salamon and Stecker with metallicity corrections. We set the absorption for $z > 3$ equal to the absorption at $z = 3$, both because it is a conservative assumption and because it is physically realistic (little stellar emissivity for $z > 3$). To produce observed fluxes from these intrinsic fluxes, we assigned each blazar a random position on the sky and added isotropic and Galactic backgrounds appropriate to each flux. The isotropic backgrounds were set to 4×10^{-6} photons cm^{-2} s^{-1} for $E > 100$ MeV with a power law index of –2.15, under the assumption that GLAST may resolve as much as two thirds of the EGRET isotropic background. The Galactic backgrounds were derived from the diffuse model used in EGRET analysis (Hunter et al. 1997).

The fluxes of each blazar at $E > 1$ GeV and 10 GeV were used to generate observed fluxes using the appropriate Poisson distributions. We removed from the sample any blazar within 10° of the Galactic plane and any blazar whose observed flux was less than 3 σ above the background flux at $E > 1$ GeV. We calculated the ratio between these fluxes and the error in that ratio.

We tested the hypothesis that each blazar had the same ratio between the intrinsic fluxes in each energy range (which would result in different measured ratios because of the backgrounds). Figure 1 shows the mean ratio in each redshift bin vs. the ratio predicted by the model of Salamon & Stecker. For comparison, the dashed lines show the same results with the intrinsic spectral variation left intact, but with no extragalactic absorption. Figure 2 shows the same results as Figure 1 when the intergalactic absorption is removed from the observed blazar fluxes.

We repeated the analysis with the mean blazar spectral index changed from -2.15 to –2.7. Although more blazars have no detected flux above 10 GeV, the effects of absorption are still apparent.

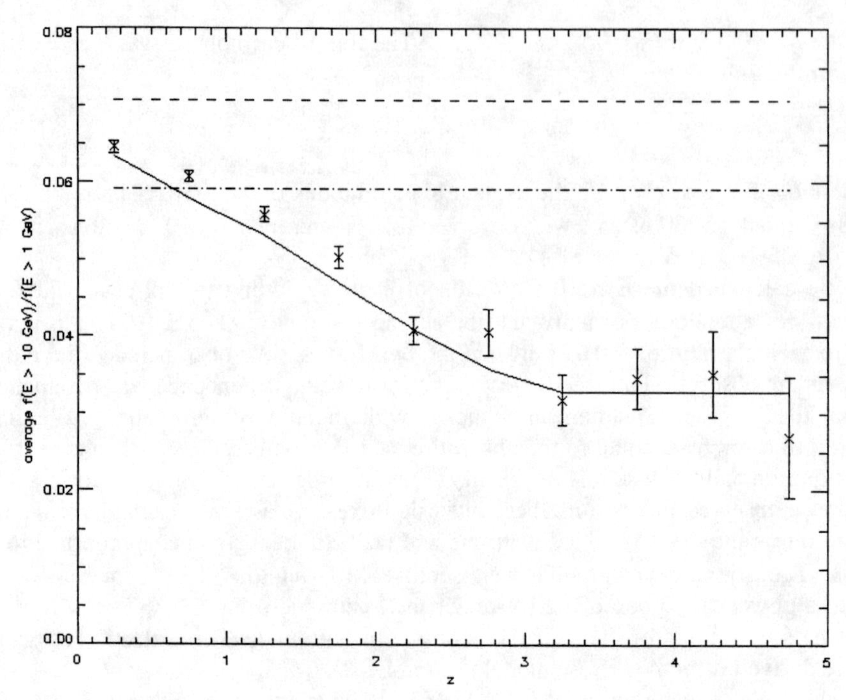

FIGURE 1. Mean flux ratio vs. redshift with extragalactic attenuation.

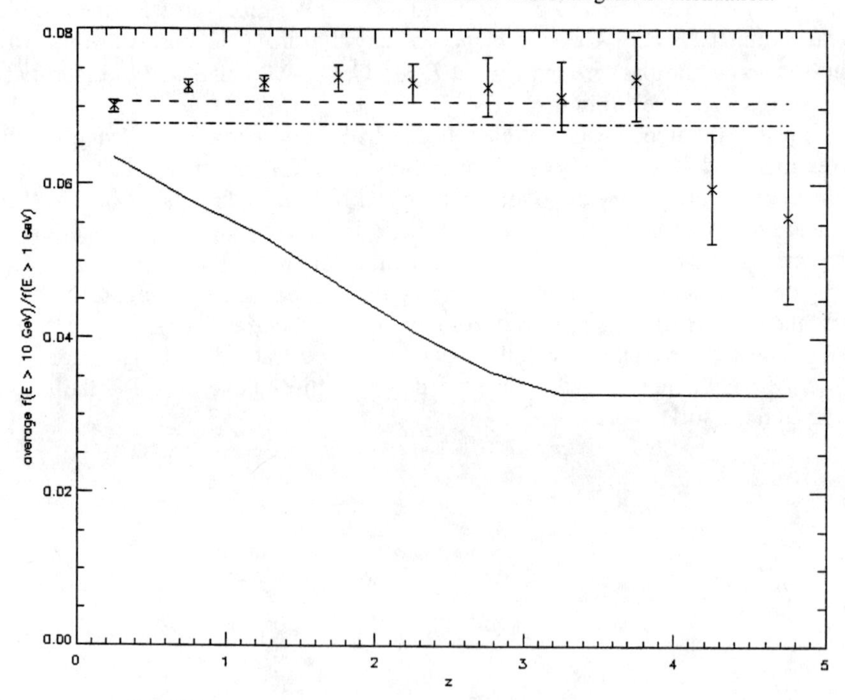

FIGURE 2. Mean flux ratio vs. redshift with no extragalactic attenuation.

DISCUSSION

Extragalactic attenuation of gamma-rays by low-energy background photons produces a distortion in the spectra of gamma-ray blazars that increases with increasing redshift. Because we cannot distinguish the difference between extragalactic attenuation, internal attenuation, or an intrinsic rolloff in individual blazar spectra, statistical analysis of a large sample of blazars is useful to study EBL absorption. GLAST will be the first instrument capable of observing this large sample at these energies. Our results indicate that the redshift dependence of the attenuation should be easily detectable by GLAST even when the diffuse background is taken into account and possible high energy intrinsic rolloffs are considered.

Selection effects, both from GLAST itself and from optical coverage of redshift determinations, will primarily affect sources with low flux. These sources will have poorly measured flux ratios, and will suffer from optical selection effects due to their more poorly determined positions. Other biases include the location of telescopes, source clustering, and other effects. It will be important to catalog these effects explicitly; in particular, insuring adequate optical coverage may require active preparation and participation. Also, we are repeating this analysis with other EBL models and other blazar luminosity functions that are consistent with EGRET data.

Finally, even after observation of a redshift-dependent effect, the possibility would remain that the spectral evolution of gamma-ray blazars happened to mimic redshift-dependent EBL absorption. Note that blazars are variable, and there are some indications that blazar spectra can become harder when they flare. Evolution in flaring probability could produce the same effect as actual spectral evolution from a statistical standpoint if, for example, a higher percentage of high-redshift blazars were observed in a quiescent phase; however, one would expect the GLAST flux limit to produce a selection effect in the opposite direction. In any case, observation of a redshift-dependent spectral softening will provide an important constraint. Theorists will have to decide the likelihood of an evolutionary conspiracy.

ACKNOWLEDGEMENTS

We acknowledge useful conversations with Bill Atwood, who first suggested using the large statistics of GLAST AGNs to look for systematic effects of extragalactic background light attenuation with redshift.

REFERENCES

1. Bloom, E. D., Space Sci. Rev. 75, 109 (1996).
2. Chiang, J., and Mukherjee, R., ApJ 496, 752 (1998).
3. Chiang, J., et al., ApJ 452, 156 (1995).
4. Hunter, S. D., et al., ApJ 481, 205 (1997).
5. Mighell, K. J., ApJ 518, 380 (1999).
6. Mukherjee, R., et al., ApJ 490, 116 (1997).
7. Salamon, M. H., and Stecker, F. W., ApJ 493, 547 (1998).

High Altitude Balloon Flights of Position Sensitive CdZnTe Detectors for High Energy X-Ray Astronomy

Kimberly R. Slavis*, Paul Dowkontt*, Fred Duttweiler†, John Epstein*, Paul L. Hink*, George L. Huszar†, Philippe C. Leblanc†, James L. Matteson†, Robert T. Skelton†, Ed Stephan†

**Dept. of Physics and McDonnell Center for the Space Sciences, Washington U., St. Louis, MO*
†Center for Astrophysics and Space Sciences, Univ. of California - San Diego, La Jolla, CA

Abstract. Cadmium Zinc Telluride (CZT) is a semiconductor detector well suited for high energy X-ray astronomy. The High-Energy X-ray Imaging Spectrometer (HEXIS) program is developing this technology for use in a hard X-ray all-sky survey and as a focal plane imager for missions such as FAR_XITE and Constellation X. We have designed a novel electrode geometry that improves interaction localization and depth of interaction determination. The HEXIS program has flown two high altitude balloon payloads from Ft. Sumner, NM to investigate background properties and shielding effects on a position sensitive CZT detector in the energy range of 20–350 keV.

INTRODUCTION

We flew both a cross-strip CZT detector and a standard planar CZT detector [5,6] on two high altitude balloon flights from Ft. Sumner, NM. The first flight was at float for 4 hours and utilized passive shielding; the second flight was at float for 22 hours and utilized both active and hybrid passive-active shielding. The cross-strip detector is 1.2cm×1.2cm×2mm with orthogonal readout electrodes. The 450 μm cathodes are on the front face of the detector, with a 500 μm pitch. The 100 μm anodes are on the back face also with a 500 μm pitch. Interleaved between the anodes is a 100 μm wide steering electrode. The electrode geometry is tailored to the characteristics of electrons and holes in CZT. The holes have poor mobility and short trapping lengths. Therefore we chose wide cathode electrodes which yielded a weighting potential that reaches deep into the detector, thereby improving charge collection efficiency [3,2]. In contrast, electrons have good mobility and long trapping lengths [3]; hence we chose narrow anode electrodes, whose weighting potential leads to most of the charge development occurring near the

CP510, *The Fifth Compton Symposium,* edited by M. L. McConnell and J. M. Ryan

anode. Consequently, the anode signal is comparatively insensitive to depth. The steering electrode improves localization of charge collection on the anode and provides very valuable depth of interaction (DOI) information.

DETECTOR PERFORMANCE

In-flight ^{241}Am calibrations were conducted periodically during both balloon flights and have been discussed previously [5,6]. The full width at half maximum (FWHM) at 60 keV during float was 6.4 keV which is dominated by electronic noise (pedestal FWHM is 5.5 keV). We also performed a laboratory calibration of the flight detector using ^{57}Co and have developed three energy-deposition correction methods (Fig. 1a). Adding nearest neighbors with significant energy depositions ($\geq$ 10 keV) to the spectrum of the maximum anode signal (dashed, 7.7 keV FWHM) improves the photopeak efficiency (dash-dot-dot-dot, 8.6 keV FWHM). Corrections based on the DOI make further improvements on the photopeak because the anode shows a small charge deficit with deeper interactions due to hole trapping. There are two indicating DOI parameters: the ratio of the cathode signal to the anode signal (C/A) and the signal on the Steering electrode (ST). Figure 1b is a contour plot of the C/A vs. the anode signal. The C/A decreases with decreasing anode signal. As the interaction site occurs deeper in the detector the charge collection efficiency of the cathode decreases, whereas the anode detection efficiency is comparatively insensitive to the depth of interaction. A linear least absolute deviation fit results in a correction factor based on the C/A that can be applied to the anode signal to recover full energy deposition (dotted, 6.5 keV FWHM). The photopeak was isolated to perform this fit (two solid lines in Fig. 1b). The dotted line is the resulting linear fit. A similar

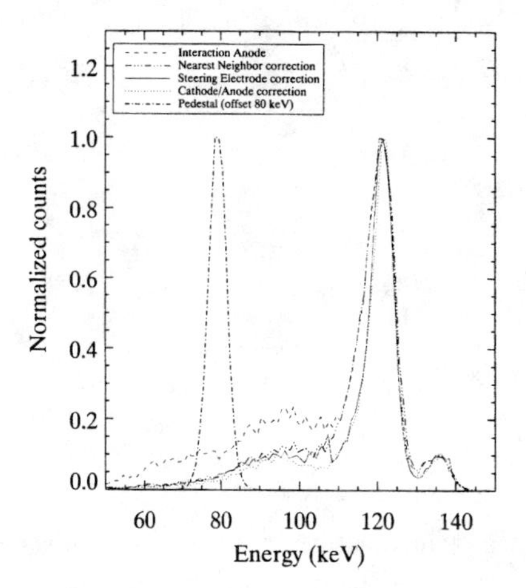

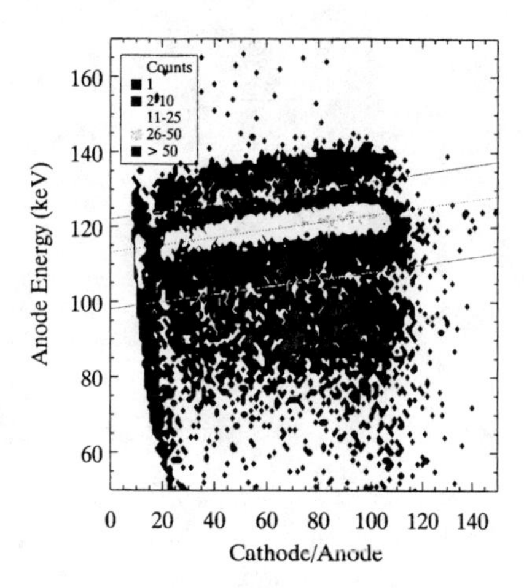

FIGURE 1. (a) Laboratory ^{57}Co calibration utilizing three energy corrections that greatly improve the photopeak efficiency. (b) Determination of the DOI correction factor to the anode energy.

linear correlation exists between the anode signal and the steering electrode signal [6] and has been discussed previously (solid, 6.3 keV FWHM). The DOI corrections yield a photopeak with similar features of the pedestal (dash-dot, 5.3 keV FWHM) and hence the current energy resolution is limited by electronic noise.

BACKGROUND LEVELS

The background levels at float for the seven shield configurations are shown in Fig. 2 for the central 30 pixels. Passive shielding configurations with depth of interaction cuts, as described previously [5], are in Fig. 2a. The 2 mm shielding results in the lowest passively-shielded background levels, as expected. All 1998 shielding configurations had a base configuration which consisted of side and rear CsI anticoincidence shields. The four 1998 configurations are: (1) "No Collimator"— the base configuration, (2) "Pb Collimator"— the base plus a 2 mm thick graded lead collimator, (3) "NaI Collimator"— the base plus a NaI collimator with the same field of view as (2) and (4) "NaI Collimator with Pb Shutter"— configuration (3) plus a 2 mm graded lead shutter over the NaI collimator's aperture. Figure 2b shows the 1998 flight data with 10 MeV shield anticoincidence cuts, for the rejection of most cosmic ray events. We find that side and rear active shielding are very beneficial, and a passive collimator yields results comparable to an active collimator. The resulting background level is 3.7×10^{-4}cts/cm^2-sec-keV in 50–90 keV (NaI Collimator with Pb Shutter). These levels will be comparable to those using charge particle anticoincidence shields based on results from GEANT models performed by Bloser *et al.* [1] and are suitable for moderate to wide field of view instruments.

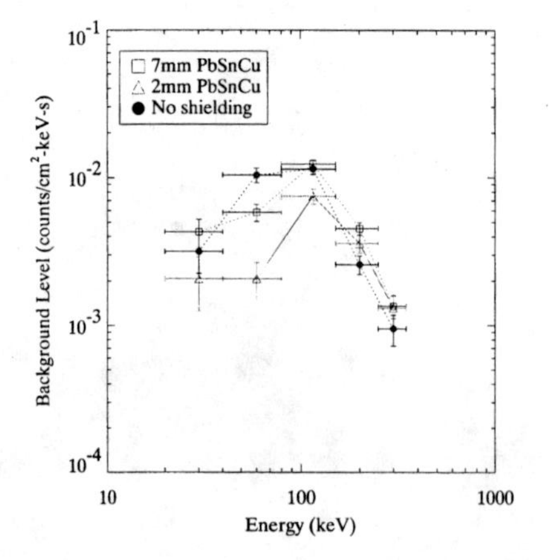

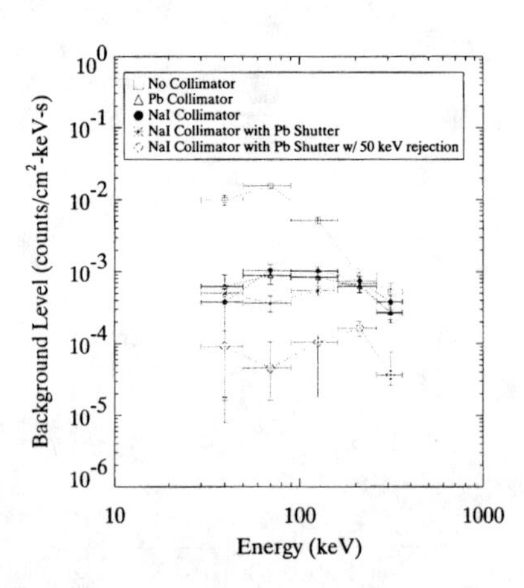

FIGURE 2. In flight background levels for the central 30 pixels utilizing seven different shielding configurations. (a) October 1997 flight data with passive shielding. (b) May 1998 flight data with active and hybrid active-passive shielding with 10 MeV shield anticoincidence cuts on all four configurations plus a 50 keV cut on configuration (4). Note configurations (2) and (3) are comparable indicating an active collimator is not required over a passive collimator.

Applying a veto at the sensitivity limit of the shields ($\sim$ 50 keV) yields a background of 5×10^{-5}cts/cm^2-sec-keV in 50–90 keV (NaI Collimator with Pb Shutter). This level is suitable for narrow field of view instruments.

ACTIVATION

The long duration at float in 1998 and the maneuverability of the shielding system permitted us to cycle through all the shield configurations at least twice. Background levels for two data sets accumulated 12 hours apart for the cross-strip CZT detector in the "NaI Collimator, Pb shutter" shield configuration are shown in Fig. 3. The two data sets were nominally at the same altitude and latitude. Data are presented with and without

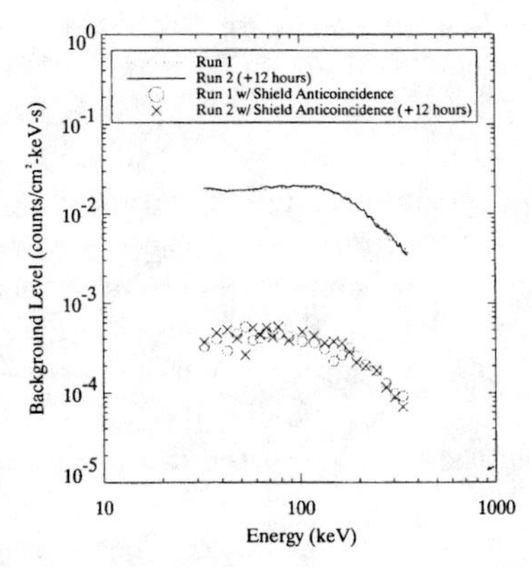

FIGURE 3. Study for evidence of activation in full cross-strip detector during 12 hours at float. No statistically significant activation was found.

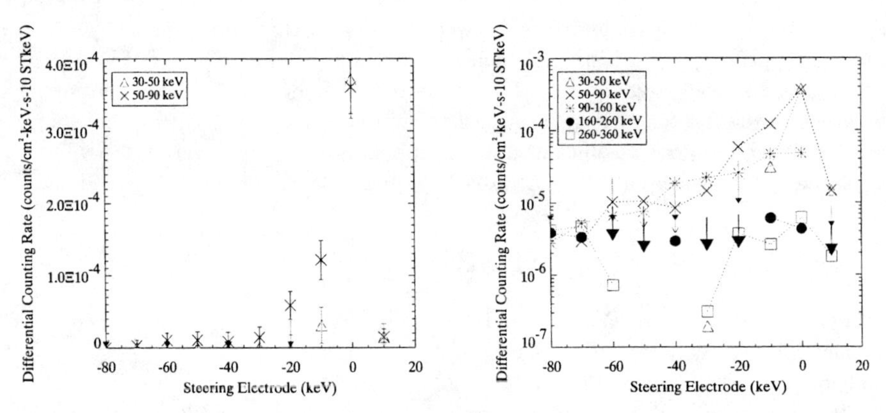

FIGURE 4. Aperture Component of the Background as a function of depth of interaction (DOI). (a) Linear plot of 30-50 keV and 50-90 keV. (b) Log plot of all energies. Note error bars are 1σ confidence level, whereas upper limits are at the 2σ confidence level.

maximum shield anticoincidence rejection for the two data sets. We find no statistically significant activation after 12 hours at a 3σ upper limit of 4×10^{-5}cts/cm^2-sec-keV in 30–350 keV. The same analysis was performed on the planar detector [6] and a 3σ upper limit of 7×10^{-5}cts/cm^2-sec-keV in 30–350 keV was determined.

DOI STUDY OF THE APERTURE COMPONENT

We measured the aperture component of the background by removing the NaI collimator's shutter. The "background subtracted aperture flux" shown in Fig. 4 has been energy binned and plotted as a function of DOI in the full cross-strip detector. The error bars represent a 1σ confidence level, whereas upper limits are at the 2σ confidence level. Figure 4a is a linear plot of two energy intervals as a function of the steering electrode signal (ST). Figure 4b is a log plot of all energy intervals, without error bars for clarity, as a function of ST. The predominant aperture flux is low energy photons interacting in the top portion of the detector (ST $\sim$ 0 keV). One expects this because low energy X-rays have a higher flux than high energy X-rays at float [4] and the mean free path of these photons is only a few hundred microns in CZT. As incident energy increases, we see an increasingly uniform interaction-site distribution within the detector due to the increasing photon mean free path. These results are consistent with our DOI models [2]. The steering electrode signal decreases (increases in absolute value) with deeper interactions and is induced by both the electrons and the holes. We have also studied the aperture component as a function of C/A [6], and the trend of the energy dependence on the DOI agrees with the trend presented here. The C/A is more sensitive to depth in the top portion of the detector than the ST, due to the difference in their weighting potentials [2].

CONCLUSIONS

The tailored electrode geometry is a successful design that improves the photopeak efficiency and provides valuable supplementary DOI information. Studies of the aperture component of the background agree with our detector models [2]. We are continuing to make progress toward the next generation hard X-ray instrumentation by achieving low background, good energy resolution, and an upper limit on the level of activation has been placed. Our future plans include a balloon flight incorporating ASIC readout.

REFERENCES

1. Bloser P., *et al.*, *Proc. SPIE* **3765**, (1999) *in press.*
2. Kalemci E., *et al.*, *Proc. SPIE* **3768**, (1999) *in press.*
3. Matteson J., *et al.*, *Proc. SPIE* **3446**, 192 (1998).
4. Matteson J., *et al.*, *Space Science Instrumentation* **3**, 491, (1977).
5. Slavis K., *et al.*, *Proc. SPIE* **3445**, 169 (1998).
6. Slavis K., *et al.*, *Proc. SPIE* **3765**, (1999) *in press.*

Modelling of CZT Strip Detectors

E. Kalemci*, J.L. Matteson*, R.T. Skelton*
P. Hink†,K. Slavis†

*Center for Astrophysics and Space Sciences, University of California, San Diego
La Jolla, CA, 92093-0424
†Department of Physics and McDonnell Center for the Space Sciences
Washington University, St. Louis, MO 63130

Abstract. Position-sensitive CZT detectors for astrophysical research in the five – several hundred keV range are being developed at UCSD and WU. These can be used for large area detector arrays in coded mask imagers and small-area focal plane detectors for focusing X-ray telescopes. The detectors have crossed-strip readout and optimized strip widths and gaps to improve energy resolution. A model of charge drift in the detectors and charge induction on the electrodes has been developed to allow for a better understanding of these types of detectors and to improve their design. The model is described and its predictions are compared with laboratory measurements. In general, there is good agreement between the model and the measurements.

INTRODUCTION

CdZnTe is a very promising detector material for X-rays and gamma-rays. This is due to its large bandgap allowing room temperature operation, and high Z making photoelectric effect dominant. At CZT's energy range, there are three very important objectives of high energy astronomy; (1) a full sky survey at a sensitivity level that will yield thousands of new sources, (2) localizing gamma ray bursts with arcminute spatial resolution, and (3) deep exposures of individual sources by using a focusing hard X-ray telescope.

Position-sensitive CZT X-ray detectors, used with coded masks and focusing optics, have been proposed for space and balloon missions to achieve these objectives. Some of the examples are HEXIS [1], FAR-XITE [2], MARGIE [3](strip readout), CONSTELLATION X [4](pixel readout) and SWIFT [5](array of discrete detectors).

Basic operation principles of CZT strip detectors can be summarized as follows: After the photoelectric interaction, a photoelectron is emitted. This photoelectron creates electron-hole pairs. Those electron-hole pairs drift under the applied field and induce charges on electrodes. While drifting, they can be trapped and detrapped. At each position, they induce $Q_{\rm ind}(x) = -Q(x) \times W(x)$ of charge on

CP510, *The Fifth Compton Symposium,* edited by M. L. McConnell and J. M. Ryan

each electrode, where W is the weighting potential which is calculated separately for each electrode [6].

MODELLING

By using the principles explained above, a computer simulation of a CZT detector was developed that predicts induced charge on each electrode for various geometries. It consists of two parts: (1) calculation of electric fields and weighting potentials, and (2) charge drift, trapping and calculations of induced charge using the fields and potentials. The former is done with a commercial program, MAXWELL [7], which, for a given geometry, calculates the potentials and fields. Then another program developed by us (EK) drifts charges on a rectangular grid and calculates induced charge on each electrode. Cathodes are at the top and the electrons move down on grid points following the electric field lines. The model uses point charge approach and does not include detrapping. Details can be found in Kalemci et al [6].

EXPERIMENTAL SET UP

The detector modelled is a laboratory prototype [1,6] which was developed to study techniques for X-ray imaging instruments. There are 22 anode and 22 orthogonal cathode strips on the $12 \times 12 \times 2$ mm^3 detector. In addition, there are steering electrodes in between anodes to enhance the charge collection which are connected to each other. Anodes and steering electrodes are 100 μm wide, cathodes are 450 μm wide, and all have 500 μm pitch. The anodes are biased at 200 V and steering electrodes at 180 V. The electrodes are coupled to Amptek A250 charge sensitive preamplifiers, whose signals then are shaped, amplified and digitized by ADC's to give the signal observed.

An additional setup was used to measure pulse shapes from various electrodes. The detector was illuminated by X-ray sources with two Amptek A250 preamplifiers connected. Preamp outputs were recorded using a digital oscilloscope.

RESULTS

Results of electric field calculations (See Fig. 1) showed that most of the electron trajectories are directed onto the anodes by the steering electrode. For interactions at random positons near the cathodes, 85% go to the anodes, 10% to the steering electrode, and 5% to the gaps between them.

Figure 2a shows weighting potential distributions of all three types of electrodes. The cathode and steering electrode weighting potentials extend deep into the detector. If the weighting potential of an electrode is extended, signals from that electrode are affected by hole trapping. On the other hand, the weighting potential

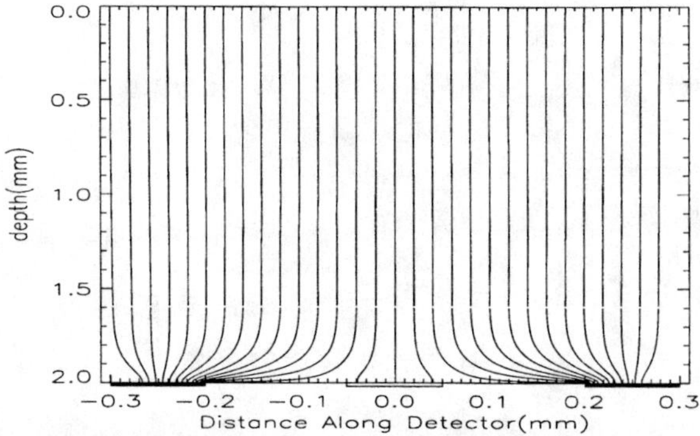

FIGURE 1. Model calculation of charge drift trajectorics. Filled boxes are anodes and empty boxes are steering electrodes here and in Fig. 2 . Reproduced with permission from [6], © 1999 SPIE.

of the anodes is localized to the immediate vicinity of the anodes. Thus, the anodes are insensitive to hole trapping and most of the time, full signal is collected. This is called "the small pixel effect" [8]. It is well established that addition of steering electrodes to the detector improves energy resolution [9]. Figure 2b. shows the main reason for this. Without the steering electrode, weighting potential of the anode extends further, reducing the small pixel effect.

Weighting potential curves along with hole trapping indicate that the amplitude of the signals is a function of depth of interaction. This effect was studied by calculating the total signal for interactions at various depths. Fig. 3 shows the results. For the anodes, the response is uniform to $\pm 2\%$ for depths less than 1.7 mm. Typically, useful signals are produced on several neighboring cathode strips.

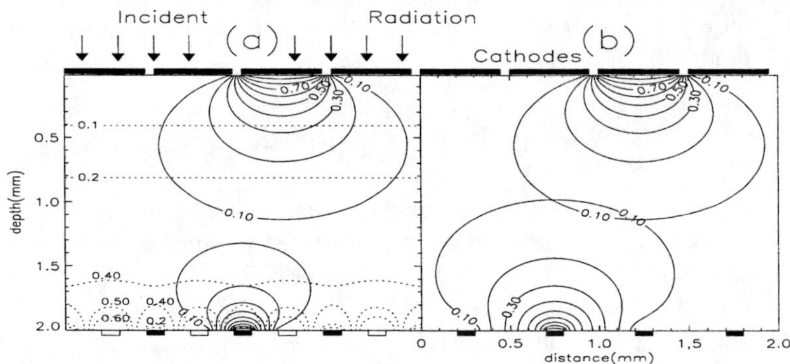

FIGURE 2. (a) Model calculation of weighting potential of the prototype detector. The dashed lines are steering electrodes weighting potential. (b) Model calculation of weighting potential without the steering electrode. Reproduced with permission from [6], © 1999 SPIE.

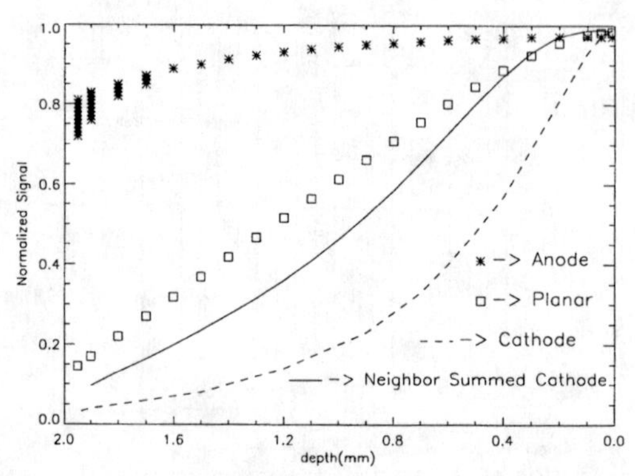

FIGURE 3. Model calculations of total signals at electrodes of prototype strip detector and planar detector. The spread in anode signal for small depths is due to its lateral position dependence. . Reproduced with permission from [6], © 1999 SPIE.

For the single cathode with the largest signal, the signal decreases rapidly with depth. When the two neighboring cathode signals are added, the depth dependence is much less severe. The model results can be used to correct for the dependence of signal on depth of interaction and thus improve energy resolution. An example is discussed in Kalemci et al [6].

Depending on the interaction position, there are various types of signal. An example is shown in Fig. 4. In this case, the depth of interaction is within 100 μm from the cathode. The anode signal stays constant until the electrons reach close to the anodes, then it rises sharply. On the other hand, the cathode signal starts to build up instantly. After the electrons reach the anodes at around 200 ns, the cathode signal continues to build due to holes slowly moving towards the cathodes. Due to trapped holes, full cathode signal cannot be obtained. Typical parameters are used for this calculation i.e. electron and hole mobilities of 1000 and 40 cm^2 / V s respectively, and electron and hole trapping lengths (average distance travelled by carriers before being trapped) of 6 cm and 0.3 mm respectively.

CONCLUSIONS

The model is able to predict correctly the fundamentals of how our detector works. In terms of predicting total signals, it agrees within 1% with the measurements [6]. In terms of temporal behavior, the model is within 15%. (See Fig. 3)

It is shown that the small pixel effect is a result of not only the size of the electrode, but also the distance between electrodes. (See Fig. 2.)

Both cathode and steering electrode signals are strongly depth dependent. Depth of interaction information, which can be obtained using the model, can be used to reject background [10] and to improve the energy resolution, as demonstrated in Kalemci et al [6].

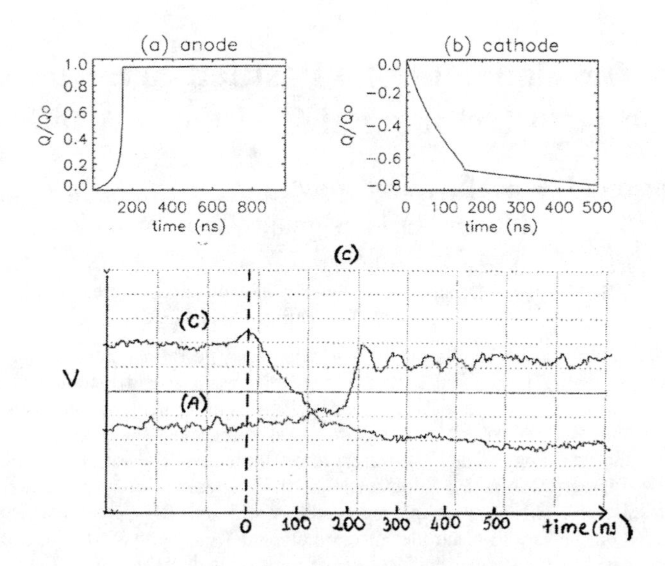

FIGURE 4. Comparison of model calculation (Figs. 4a-b) of charge induction on anode and cathode strips with laboratory measurement(Fig. 4c). All the features can be explained by model parameters (See text). The dashed line in Fig. 4c indicates the time of interaction. (A) denotes the anode signal and (C) denotes the cathode signal. Reproduced with permission from [6],

ACKNOWLEDGMENTS

This work was supported by NASA SR&T Grant NAG5-5111 and at WU by NASA SR&T Grant NAG5-5114 and NASA GSRP Grant NGT-50170. We would also like to thank to Fred Duttweiler, Sandy O'Brien, Rick Rothschild and George Huszar from UCSD.

REFERENCES

1. Matteson et al., *Proc. SPIE* **3445**, 445 (1998)
2. Ulmer et al., *Proc. SPIE* **3768**, (1999) (in press)
3. Cherry et al., *Proc. SPIE* **3768**, (1999) (in press)
4. Harrison et al., *Proc. SPIE* **3768**, (1999) (in press)
5. swift.gsfc.nasa.gov (1999)
6. Kalemci et al., *Proc. SPIE* **3768**, (1999) (in press)
7. MAXWELL 3D Field Simulator, Ansoft Corporation, Four Station Square, Suite 200, Pittsburgh, PA 1521-1119, www.Ansoft.com
8. Luke P. N., *Proc. 9th Int. Workshop on Room Temp. Sem. X- and Gamma-Ray Det.* (1995)
9. Matteson et al., *Proc. SPIE* **3446**, 192 (1998)
10. Slavis et al., *Proc. SPIE* **3445**, 169 (1998)

The Development of a Position-Sensitive CZT Detector with Orthogonal Co-Planar Anode Strips

K.A.Larson[a], L.Hamel[b], V.Jordanov[c], J.R.Macri[a], M.L.McConnell[a], J.M.Ryan[a], O.Tousignant[b], A.Vincent[b]

[a]Space Science Center, University of New Hampshire, Durham, NH, USA
[b]GCM, Physics Department, University of Montreal, Montreal, CA
[c]Yantra, 12 Cutts Rd., Durham, NH, USA

Abstract. We report on the simulation, construction, and performance of prototype CdZnTe imaging detectors with orthogonal coplanar anode strips. These detectors employ a novel electrode geometry with non-collecting anode strips in one dimension and collecting anode pixels, interconnected in rows, in the orthogonal direction. These detectors retain the spectroscopic and detection efficiency advantages of single carrier (electron) sensing devices as well as the principal advantage of conventional strip detectors with orthogonal anode and cathode strips, i.e. an N x N array of imaging pixels are with only 2N electronic channels. Charge signals induced on the various electrodes of a prototype detector with 8 x 8 unit cells (1 x 1 x 5 mm^3) are compared to the simulations. Results of position and energy resolution measurements are presented and discussed.

INTRODUCTION

One of the limiting technologies of hard X-ray and gamma-ray imaging is that of simultaneous spectroscopy and imaging. The requirements for high resolution images with good spectral resolution in an efficient and compact design drives research in monolithic or arrays of position-sensitive, spectroscopic detectors. A leading candidate is CsZnTe strip detectors. They offer high atomic number and density stopping power, good spectral resolution and good spatial resolution in a strip detector format without the need for cryogenic cooling. We describe here the advances made with a new design that satisfies many of these criteria.

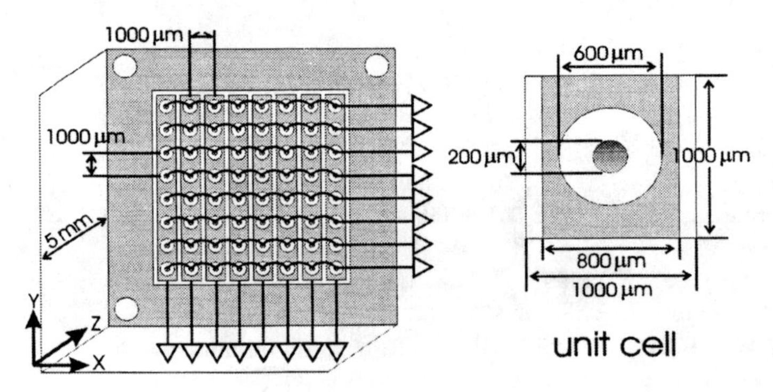

FIGURE 1. Schematic layout showing the prototype coplanar anode detector design and a close up of a unit cell.

CP510, *The Fifth Compton Symposium,* edited by M. L. McConnell and J. M. Ryan

DETECTOR DESIGN

CdZnTe (CZT) detectors have faced two major limitations. Pixellated CZT detectors have demonstrated good energy and spatial resolution, but require n^2 electronic channels, where n is the number of pixels in one dimension. Traditional strip detectors reduce the number of electronic channels to 2n, but the poor hole transport properties limit the effective energy range of the device. The detector presented here (figure 1) overcomes both problems by employing a new coplanar anode geometry. An array of anodes (pixels) collects the electron current in this design. Rows of pixels parallel to the x axis are interconnected providing the y coordinate of a γ-ray interaction. Strips parallel to the y-axis circumscribe the pixels as shown, but collect no real charge. They are biased at a level between that of the cathode and the pixels, and see only the induced signal generated by electrons drifting towards the collecting pixels. The pixels therefore provide the energy and one coordinate of the position, while the strips provide the other coordinate. This eliminates any dependence on the signal generated by the holes, and allows all electrical connections to be made on one side of the detector.

A low temperature polymer flip chip bonding process was used for the first time to form the electrical and mechanical connection for the patterned CZT and ceramic substrates of our prototype (1). The result is a rugged assembly with no wire bonds to the CZT anode surface. We demonstrated the applicability of our prototype for balloon or space flight in thermal cycle and vibration tests. The results presented were obtained using inexpensive, multi-crystal, counter grade CZT material from eV Products.

LABORATORY MEASUREMENTS

For these measurements, the pixels were set at zero potential, while the cathode, strips, and guard ring all have adjustable potentials. All output signals are AC-coupled to Ev-5093 charge sensitive preamplifiers with a sensitivity of 3.6 mV/fC. For pulse-height measurements, the signals are processed with standard CAMAC lab equipment. For signal shape analysis, the signals were sampled with digital oscilloscopes (Tektronix TDS360) and then stored for later analysis.

Measurements with ^{57}Co and ^{137}Cs showed a non-uniform response across all pixels. We down-selected a 3x3 region of cells that produced good and uniform signals. We used the cell in the middle of this region for all measurements.With this cell, the bias levels were varied for the best energy resolution. This meant having full collection on the pixels without significant pixel to strip leakage currents. A cathode bias of –800 Volts with a strip and guard ring bias of –30 to –70 Volts produced the best results.

Energy resolution measurements using a 0.5 μs gaussian shaping time were made at room temperature with events triggering only from the cell under test. Figure 2 shows the spectra from two γ-ray sources, along with the pulser for a noise estimate.

For ^{137}Cs, the FWHM is 5.9 keV, or 0.9%, while the pulser FWHM is 3.3 keV. For ^{241}Am, the resolution is 3.4 keV FWHM, or 5.7 %. If the electronic noise from the pulser is quadratically subtracted from the resolution, an intrinsic resolution of 0.7% has been achieved at 662 keV.

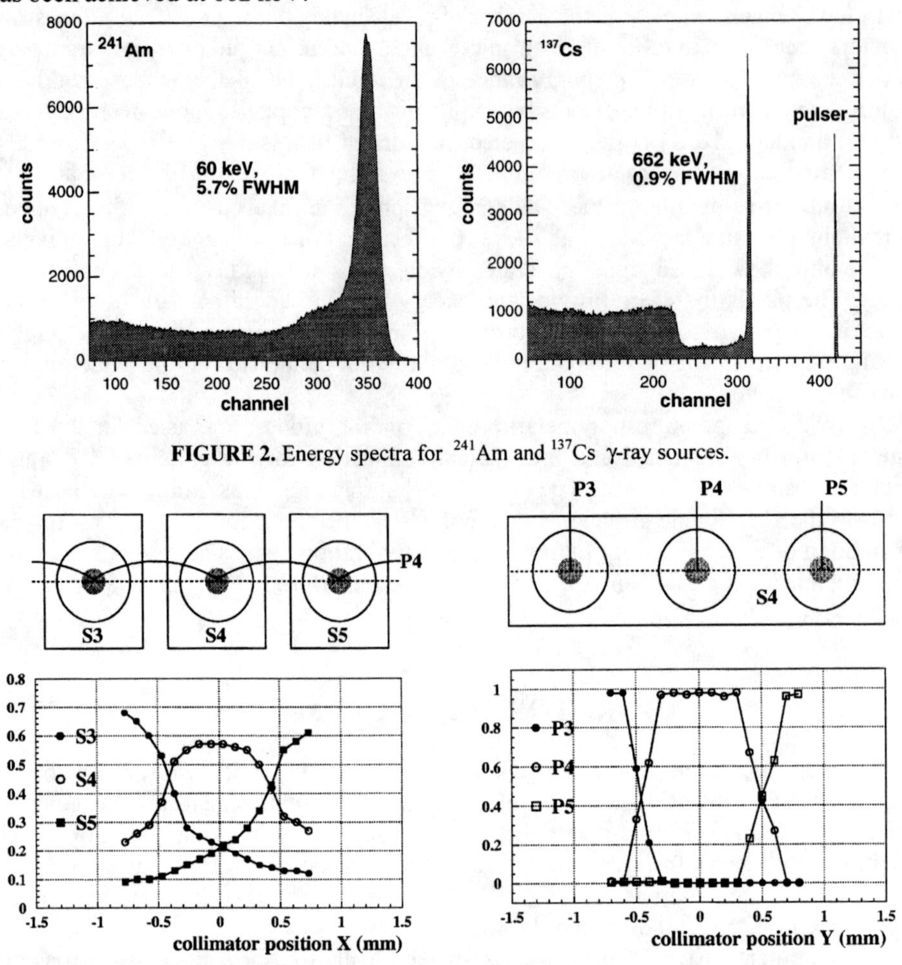

FIGURE 2. Energy spectra for ^{241}Am and ^{137}Cs γ-ray sources.

FIGURE 3. Collimated α-ray scans. For both plots, the relative amplitudes on three adjacent channels (normalized to their sum) are plotted as a function of source position. The X-scan plot (left) shows very little charge sharing between pixels, while the Y-scan plot (right) shows greater sharing.

For position resolution measurements, a collimated ^{244}Cm α source on a 3-axis micro-manipulator illuminated a spot approximately 200 μm in diameter. The source was scanned across the 3x3 cell area and events for each pixel and strip were recorded on the digital oscilloscopes. The center cell was scanned to determine the location of the maximum signal with the signal of its neighboring cells being roughly equal. Since we could not see the location of the spot illuminated with α's, we took this location to be the center of the cell. The 3x3 region was then scanned in x and y in 100 μm steps, and the pulse heights were recorded. The results of the scan, using mean amplitudes normalized for each event, are shown in figure 3.

While significant charge sharing occurs between strips, little charge sharing occurs between pixels. What there is is limited to a small region directly between the pixels (figure 3). As a result, the analysis of relative pixel pulse heights yields little information. Consequently, the resolution in y is on the order of 1 mm. For strips, however, position information is shared among as many as three strips. The analysis of the relative pulse heights of those strips yields a resolution on the order of 300 μm.

The measured signals have been compared to simulations (2). Figure 4 shows simulation results compared to scope traces for ^{137}Cs γ-rays interacting at three different depths. Three events were selected for which signals on neighboring strips were equal, ensuring that all three occurred at the center of the cell. Measured signals were matched with simulation signals to determine the approximate depth of interaction.

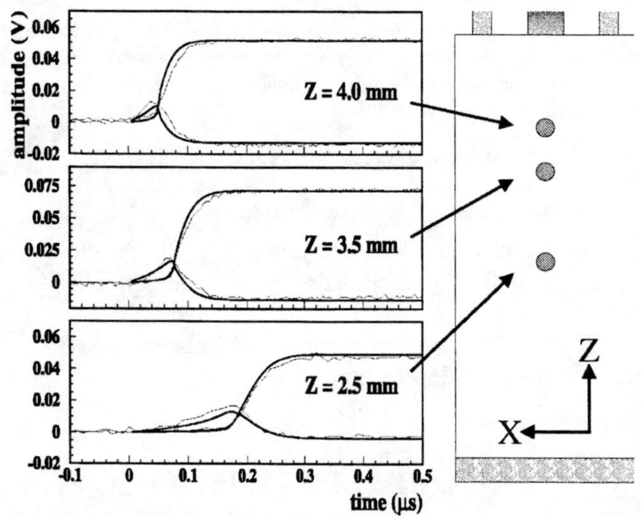

FIGURE 4. Simulated and measured signals from ^{137}Cs γ-rays interacting at three different depths. The depth position is inferred from the simulation.

It is clear from the simulation results that the strip signal shape carries information about the depth of interaction. Both the rise time and residual level vary with depth. Figure 5 shows these parameters plotted against one another for both simulated and measured events. The scatter in the points indicates a z position resolution of approximately 1 mm. Figure 5 also shows a plot of the strip residual versus depth for ^{137}Cs γ-rays for both simulated and measured events using the normalized large amplitude cathode signal as a measure of depth. Depth information can also be gained strictly from examination of the cathode signal, but due to poor hole mobility the main signal component is induced charge from the movement of electrons. Small events far away from the cathode may therefore not produce signals above the noise.

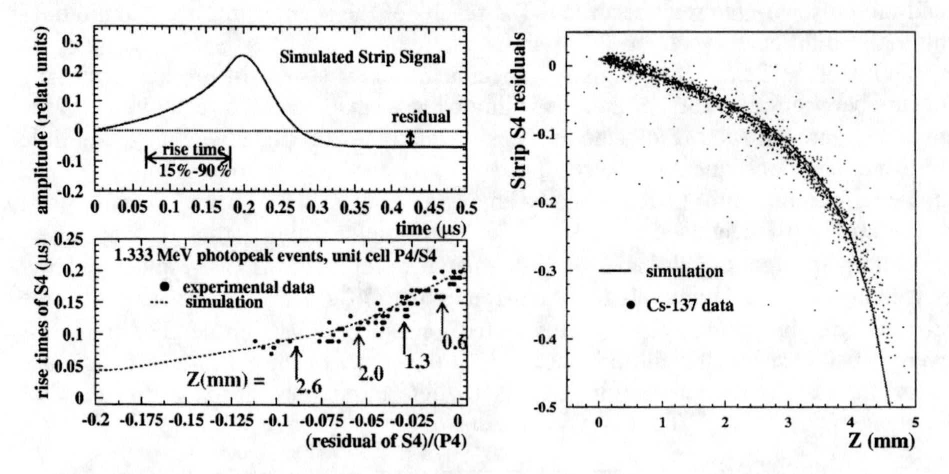

FIGURE 5. Depth of interaction derived from measurement of strip signal risetime and residual, both measured and simulated, for 1.333 MeV (left). Depth of interaction derived solely from strip signal residual for ^{137}Cs γ-rays, both measured and simulated (right).

CONCLUSIONS

A prototype of a CZT imaging detector with orthogonal coplanar anodes has been fabricated. Tests have confirmed that this detector could provide excellent energy resolution and imaging performanc. Signals observed on both pixels and strips are in good agreement with the simulations. Future work will include complete characterization of the detector response, as well as imaging with a 16 channel digital acquisition system currently in developement.

ACKNOWLEDGEMENTS

This work is supported by NASA's High Energy Astrophysics SR&T program and by the Natural Sciences and Engineering Research Council (NSERC) of Canada.

REFERENCES

1. Jordanov, V.T., et al., "Multi-Electrode CZT Detector Packaging Using Polymer Flip Chip Bonding," presented at 11th International Workshop on Room Temperature Semiconductor X- and Gamma-Ray Detectors and Associated Electronics, Vienna, Austria, October 1999.

2. Hamel, L.A., et al., "An Imaging CdZnTe detector with coplanar orthogonal strips", Proceedings of the 1997 Fall Meeting of the Materials Research Society, Boston, 1-5 December, 1997.

Minute-of-Arc Resolution Gamma Ray Imaging Experiment – MARGIE

S.C. Kappadath*, P.P. Altice*, M.L. Cherry*, T.G. Guzik*, J.G. Stacy*,‡, J. Macri†, M.L. McConnell†, J.M. Ryan†, D.L. Band§,$, J.L. Matteson§, T.J. O'Neill‖, A.D. Zych‖, J. Buckley#, P.L. Hink#

* *Dept. of Physics & Astronomy, Louisiana State Univ., Baton Rouge, LA 70803*
† *Univ. New Hampshire, Durham, NH 03824;* § *UC San Diego, La Jolla, CA 92093*
‖ *UC Riverside, Riverside, CA 92521;* # *Washington Univ., St. Louis, MO 63120*
‡ *Southern Univ., Baton Rouge, LA 70813;* $ *Los Alamos Natl. Lab., Los Alamos, NM 87545*

Abstract. MARGIE (Minute-of-Arc Resolution Gamma-ray Imaging Experiment) is a large area ($\sim 10^4$ cm^2), wide field-of-view ($\sim$1 sr), hard X-ray/gamma-ray ($\sim$20–600 keV) coded-mask imaging telescope capable of performing a sensitive survey of both steady and transient cosmic sources. MARGIE has been selected for a NASA mission-concept study for an Ultra Long Duration (100 day) Balloon flight. We describe our program to develop the instrument based on new detector technology of either cadmium zinc telluride (CZT) semiconductors or pixellated cesium iodide (CsI) scintillators viewed by fast-timing bi-directional charge-coupled devices (CCDs). The primary scientific objectives are to image faint Gamma-Ray Bursts (GRBs) in near-real-time at the low intensity (high-redshift) end of the logN–logS distribution, thereby extending the sensitivity of present observations, and to perform a wide field survey of the Galactic plane.

INTRODUCTION

Gamma-Ray Bursts (GRBs) are intense bursts of γ radiation, lasting from fractions of a second to minutes, which emit the bulk of their energy above 0.1 MeV (see e.g., Fishman and Meegan 1995; Band 1998). The origin and emission mechanism of GRBs are still quite uncertain. It has been long recognized that the key to unraveling the GRB mystery is the identification of burst counterparts at other wavelengths. Over the past two years, the BeppoSAX X-ray mission has localized over a dozen bursts to sufficiently small spatial regions (a few arc-minutes), on short enough timescales (a few hours) so that X-ray, optical and radio telescopes have detected the fading GRB afterglows. The recent multiwavelength observations which now include over four redshift measurements (Kulkarni et al. 1999) suggest that GRBs are cosmological in origin.

CP510, *The Fifth Compton Symposium,* edited by M. L. McConnell and J. M. Ryan

A relativistic fireball model (see e.g., Piran 1998; Sari et al. 1998) has been reasonably successful in explaining the observed X-ray and optical afterglows. The γ-ray burst itself appears to be the result of internal shocks in the relativistic expanding ejecta and the forward shock moving through the interstellar medium produces the afterglows. Multiwavelength observations triggered by rapid GRB localization and notification provide a wealth of information about the physical conditions in the GRB environment. The richness and complexity of the fireball phenomenon and the observational differences between GRB events require a large statistical sample of observed bursts in order to fully unravel the physics of these sources.

The most recent report of NASA's Gamma Ray Working Group outlined a plan for the next years that included a hard X-ray (<200 keV) survey as a high priority item. The basis for promoting a hard X-ray instrument was that the most recent survey was conducted with the HEAO-1 mission, more than 20 years ago. Since then most of the information we have collected about the hard X-ray sky comes from balloon instruments on day-long flights, from occultation monitoring of strong point sources with BATSE on GCRO, from investigations of point sources with the OSSE instrument on CGRO, and from partial sky surveys with moderate spatial resolution ($\sim 13'$) by the SIGMA instrument on GRANAT.

MARGIE (Minute-of-Arc Resolution Gamma-ray Imaging Experiment) is a large area ($\sim 10^4$ cm^2), wide field-of-view ($\sim$1 sr), hard X-ray/γ-ray ($\sim$20–600 keV) coded-mask imaging telescope capable of performing a sensitive survey of both steady and transient cosmic sources. MARGIE has been selected for a NASA mission-concept study for an Ultra Long Duration (100 day) Balloon flight. The instrument is designed to observe 30–40 GRBs in a 100 day balloon flight with sufficient S/N to permit GRB localizations to within $\sim$2 arc-minutes. MARGIE is also ideal to conduct a hard X-ray mapping of the diffuse emission from the Galactic plane and to survey the hard X-ray sky.

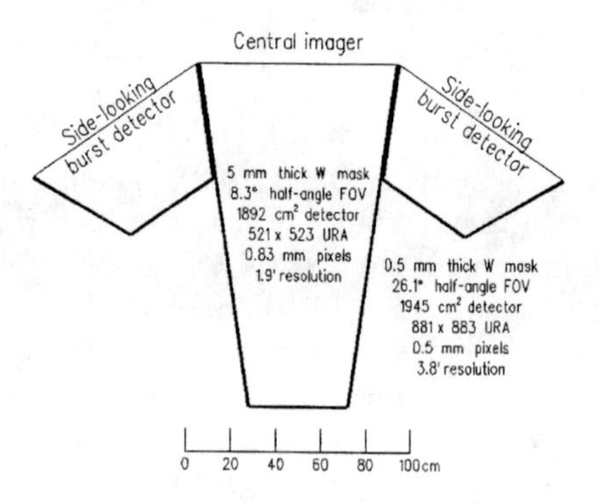

FIGURE 1. Schematic of the MARGIE γ-ray telescope.

MARGIE INSTRUMENT CONCEPT

The essential components of the MARGIE experiment (Cherry et al. 1999; Stacy et al. 1999; McConnell et al. 1996) are the coded aperture mask and the CsI/CCD or CZT central detector. As shown in Figure 1, the instrument consists of five separate telescopes. The central telescope has an $87 \times 87 \text{cm}^2 \times 5$ mm thick tungsten mask viewed by an 1892 cm^2 CsI/CCD array with $(0.8 \text{ mm})^2$ pixels at a mask-detector separation of 150 cm. The central telescope is designed for high resolution ($1.9'$), narrow ($8.3°$ half angle) field-of-view (FOV) measurements of point sources in addition to GRBs. The four side telescopes, each with $88 \times 88 \text{cm}^2 \times 0.5$ mm thick masks, a 1945 cm^2 detector array, 0.5 mm pixels, and 45 cm mask–detector separation, provide excellent sky coverage and sensitivity ($26.1°$ half angle FOV, each with $3.8'$ resolution) for GRBs and point sources. Plastic scintillators covering the masks provide an anticoincidence veto for charged cosmic rays interactions in the detector.

The coded aperture (or "multi-pinhole") technique works by allowing an absorbing mask to cast a shadow pattern on a position-sensitive detection plane (e.g., Caroli et al. 1987; Skinner et al. 1987). With a proper choice of mask pattern to minimize artifacts from the imaging process, the encoded pattern can then be processed to reproduce an image of the sky. The mask element geometry is defined by the mask thickness and the mask element size (or width). The mask thickness must be sufficient to attenuate photons (hence, modulate the incident flux) in the desired energy range. On the other hand, the thickness of the mask must be limited so as to maintain uniformity of mask transmission for off-axis sources. The telescope angular resolution corresponds to the angular size of a mask element as seen from the detection plane, and so is dictated by the mask element size and the mask-detector separation. Therefore the detector must be able to resolve the individual mask elements in the projected pattern; i.e., it must be able to locate events with an accuracy no larger than the mask element size. Any technology which improves the detector plane spatial resolution can therefore lead to an improvement in telescope angular resolution and sensitivity.

DETECTOR PLANE TECHNOLOGIES

The key enabling technology for MARGIE is the central detector. In order to obtain the fine-grained position resolution required in the plane of the central detector, two alternate position-sensitive γ-ray detector technologies are under development for use in MARGIE. The central detector will consist of five two-dimensional arrays of either 0.3 mm pitch cadmium zinc telluride (CZT) strip detectors or 0.5–0.9 mm pitch segmented cesium iodide (CsI) scintillator. The CsI scintillators will be viewed by a Bi-Directional Charge Coupled Device (CCD) array designed for low-noise spectroscopy and fast timing (10 μs) applications.

CZT Detectors

Prototype cross-strip and planar CZT detectors have been flown and flight tested. The measured background in flight is low compared to the atmospheric flux entering the telescope aperture. The latest results on the overall CZT performance and background levels at balloon altitudes are presented by Slavis et al. in these proceedings. CZT detectors with orthogonal coplanar anode strips have also been developed. Sub-millimeter position resolutions and excellent energy resolution have been demonstrated. These results are presented by Ryan et al. elsewhere in these proceedings.

Segmented-CsI arrays coupled to Bi-Directional CCDs

Segmented CsI: CsI(Tl) is an efficient X-ray/γ-ray scintillator due to its high density (~ 4.51 gm/cm^3) and large Z. Since the light is emitted essentially isotropically, the spot size diameter is comparable to the detector thickness. Therefore, one has competing requirements: a thicker detector for higher detection efficiency and a thinner detector for better position resolution. For MARGIE, the position resolution requirements demand a spot size <0.5 mm.

Pixellated CsI arrays offer a possible solution. Large area (50×50 cm^2) sub-millimeter pixel arrays up to 4 cm thick are commercially available (Krus et al. 1999) in several different scintillator materials (e.g., CsI, BGO, $CdWO_4$) and produced with white paint, white epoxy or metal reflectors between the pixels. The pixel sizes are dictated by the mechanical properties of the crystal (e.g., hardness, cleavage plane). Evaluation of these arrays is currently underway.

Bi-Directional CCD: The segmented CsI scintillator output will be detected with a CCD. Standard CCDs are integrating devices operating at video rates (typically ~30 Hz). A balloon-borne γ-ray telescope, however, demands a faster time

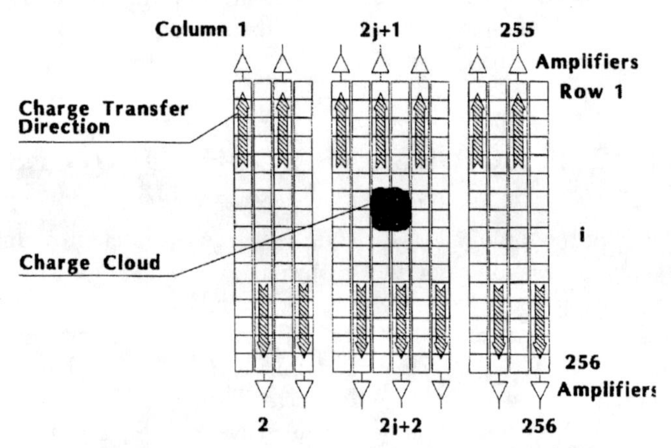

FIGURE 2. Schematic view of the Bi-Directional CCD.

resolution to veto cosmic rays hitting the detector (~10 kHz) and isolate individual photons to measure their energies. We have therefore developed a Fast Timing Bi-Directional CCD with 10 μs timing resolution and 50 μm position resolution (Cherry et al. 1996).

The Bi-Directional CCD employs a continuous readout scheme in which charge collected in alternate pixel columns is clocked separately to the top and bottom of the CCD chip where it is amplified and read out (see Figure 2). The basic operation and performance characteristics of this device have been demonstrated with a set of prototype Bi-Directional CCDs fabricated for us at the Orbit Semiconductors foundry, based on a design by Suni Imaging Microsystems, Inc. (Cherry et al. 1999; Stacy et al. 1999). We are currently implementing the data readout circuitry for the Bi-Directional CCD in a 1.2 μm CMOS ASIC design. Ultimately, both the CCD and readout electronics will be incorporated into a single monolithic CMOS chip. These further improvements will lead to an imaging device with exceptional low-noise performance and fast timing ideally suited for our astrophysical objectives.

SUMMARY

The MARGIE instrument will be a large-area, wide field-of-view, hard X-ray/γ-ray imaging telescope. It will be capable of providing accurate positions, and of characterizing the temporal and spectral behavior of faint transient sources (e.g., GRBs) in near-real-time, for rapid counterpart searches. It will also carry out sensitive surveys for both steady and transient cosmic sources over the course of a 100-day Ultra Long Duration Balloon flight.

REFERENCES

1. G.J. Fishman and C.A. Meegan, Ann. Rev. Astron. Astrophys. 33, 415 (1995)
2. D. Band, in 8th Marcel Grossmann Mtg. on General Relativity, ed. T. Piran, in press (astro-ph/9712193, 1998)
3. S. R. Kulkarni et al., Nature, submitted (astro-ph/9902272 v2, 1999).
4. T. Piran, preprint, astro-ph/9810256 (1998)
5. R. Sari, T. Piran and R. Narayan, ApJ 497 L17 (1998)
6. M.L. Cherry et al., Proc. SPIE 3765, in press (1999)
7. J.G. Stacy et al., Proc. 19th Texas Symposium (Paris), in press (1999)
8. M. McConnell et al., Proc. SPIE 2806, 349 (1996)
9. E. Caroli et al., Space Sci. Rev. 45, 349 (1987)
10. G.K. Skinner et al., Astrophys.Space Sci. 136, 337 (1987)
11. K. Slavis et al., these proceedings
12. J. Ryan et al., these proceedings
13. D. Krus et al., Proc. SPIE Conf. 3768, in press (1996)
14. M.L. Cherry et al., Proc. SPIE 2806, 551 (1996)

EXIST: A High Sensitivity Hard X-ray Imaging Sky Survey Mission for ISS

J. Grindlay[1], L. Bildsten[2], D. Chakrabarty[3], M. Elvis[1], A. Fabian[4], F. Fiore[5], N. Gehrels[6], C. Hailey[7], F. Harrison[8], D. Hartmann[9], T. Prince[8], B. Ramsey[10], R. Rothschild[11], G. Skinner[12], S. Woosley[13]

[1]*CfA,* [2]*ITP/UCSB,* [3]*MIT,* [4]*IOA/Cambridge,* [5]*Rome Obs./BeppoSAX,* [6]*NASA/GSFC,* [7]*Columbia Univ.,* [8]*Caltech,* [9]*Clemson Univ.,* [10]*NASA/MSFC,* [11]*UCSD,* [12]*Birmingham Univ./UK,* [13]*UC Santa Cruz*

Abstract. A deep all-sky imaging hard x-ray survey and wide-field monitor is needed to extend soft (ROSAT) and medium (ABRIXAS2) x-ray surveys into the 10-100 keV band at comparable sensitivity (~0.05 mCrab). This would enable discovery and study of $\gtrsim$ 3000 obscured AGN, which probably dominate the hard x-ray background; detailed study of spectra and variability of accreting black holes and a census of BHs in the Galaxy; Gamma-ray bursts and associated massive star formation (PopIII) at very high redshift and Soft Gamma-ray Repeaters throughout the Local Group; and a full galactic survey for obscured supernova remnants. The Energetic X-ray Imaging Survey Telescope (EXIST) is a proposed array of 8 $\times$ 1m^2 coded aperture telescopes fixed on the International Space Station (ISS) with 160° $\times$ 40° field of view which images the full sky each 90 min orbit. EXIST has been included in the most recent NASA Strategic Plan as a candidate mission for the next decade. An overview of the science goals and mission concept is presented.

I INTRODUCTION

The full sky has not been surveyed in space (imaging) and time (variability) at hard x-ray energies. Yet the hard x-ray (HX) band, defined here as 10-600 keV, is key to some of the most fundamental phenomena and objects in astrophysics: the nature and ubiquity of active galactic nuclei (AGN), most of which are likely to be heavily obscured; the nature and number of black holes; the central engines in gamma-ray bursts (GRBs) and the study of GRBs as probes of massive star formation in the early universe; and the temporal measurement of extremes: from kHz QPOs to SGRs for neutron stars, and microquasars to Blazars for black holes.

A concept study was conducted for the Energetic X-ray Imaging Survey Telescope (EXIST) as one of the New Mission Concepts selected in 1994 (Grindlay et al 1995). However the rapid pace of discovery in the HX domain in the past ~2 years,

CP510, *The Fifth Compton Symposium,* edited by M. L. McConnell and J. M. Ryan

coupled with the promise of a likely 2-10 keV imaging sky survey ABRIXAS2 (see http://www.aip.de/cgi-bin/w3-msql/groups/xray/abrixas/index.html) in c.2002-2004 and the recent selection of Swift (see http://swift.gsfc.nasa.gov/) which will include a ~10-100 keV partial sky survey (to ~1 mCrab) in c.2003-2006, have prompted a much more ambitious plan. A dedicated HX survey mission is needed with full sky coverage each orbit and ~0.05 mCrab all-sky sensitivity in the 10-100 keV band (comparable to ABRIXAS2) and extending into the 100-600 keV band with ~0.5 mCrab sensitivity. Such a mission would require very large total detector area and large telescope field of view. These needs could be met very effectively by a very large coded aperture telescope array fixed (zenith pointing) on the International Space Station (ISS), and so EXIST-ISS was recommended by the NASA Gamma-Ray Program Working Group (GRAPWG) as a high priority mission for the coming decade. This mission concept has now been included in the NASA Strategic Plan formulated in Galveston as a post-2007 candidate mission. In this paper we summarize the Science Goals and briefly present the Mission Concept of EXIST-ISS. Details will be presented in forthcoming papers, and are partially available on the EXIST website (http://hea-www.harvard.edu/EXIST/EXIST.html).

II SCIENCE GOALS AND OBJECTIVES

EXIST would pursue two key scientific goals: a very deep HX imaging and spectral survey, and a very sensitive HX all-sky variability survey and GRB spectroscopy mission. These Survey (**S**) and Variability (**V**) goals can be achieved by carrying out several primary objectives:

S1: *Sky survey for obscured AGN & accretion history of universe*
It is becoming increasingly clear that most of the accretion luminosity of the universe is due to obscured AGN, and that these objects are very likely the dominant sources for the cosmic x-ray (and HX) diffuse background (e.g. Fabian 1999). No sky survey has yet been carried out to measure the distribution of these objects in luminosity, redshift, and broad-band spectra in the HX band where, as is becoming increasingly clear from BeppoSAX (e.g. Vignati et al 1999), they are brightest. EXIST would detect at least 3000 Seyfert 2s and conduct a sensitive search for Type 2 QSOs. Spectra and variability would be measured, and detailed followup could then be carried out with the narrow-field focussing HX telescope, HXT, on Constellation-X (Harrison et al 1999) as well as IR studies.

S2: *Black hole accretion on all scales*
The study of black holes, from x-ray binaries to AGN, in the HX band allows their ubiquitous Comptonizing coronae to be measured. The relative contributions of non-thermal jets at high $\dot{m}$ requires broad band coverage to $\gtrsim$ 511 keV, as does the transition to ADAFs at lower $\dot{m}$ values. HX spectral variations vs. broad-band flux can test the underlying similarities in accretion onto BHs in binaries vs. AGN.

S3: *Stellar black hole content of Galaxy*
X-ray novae (XN) appear to be predominantly BH systems, so their unbiased detection and sub-arcmin locations, which allow optical/IR identifications, can provide a direct measure of the BH binary content (and XN recurrence time) of the Galaxy. XN containing neutron stars can be isolated by their usual bursting activity (thermonuclear flashes), and since they may solve the birth rate problem for millisecond pulsars (Yi and Grindlay 1998), their statistics must be established. A deep HX survey of the galactic plane can also measure the population of galactic BHs not in binaries, since they could be detected as highly cutoff hard sources projected onto giant molecular clouds. Compared to ISM accretion onto isolated NSs, for which a few candidates have been found, BHs should be much more readily detectable due to their intrinsically harder spectra and (much) lower expected space velocities, V, and larger mass M (Bondi accretion depending on M^2/V^3).

S4: *Galactic survey for obscured SNR: SN rate in Galaxy*
Type II SNe are expected to disperse $\sim 10^{-4}\ M_{\odot}$of ^{44}Ti, with the total a sensitive probe of the mass cut and NS formation. With a ~87y mean-life for decay into ^{44}Sc which produces narrow lines at 68 and 78 keV, obscured SNe can be detected throughout the entire Galaxy for $\gtrsim$ 300y given the $\sim 10^{-6}$ photons cm^{-2} s^{-1} line sensitivity and $\lesssim$ 2 keV energy resolution (at 70 keV) possible for EXIST. Thus the likely detection of Cas A (Iyudin et al 1994) can be extended to more distant but similarly (or greater) obscured SN to constrain the SN rate in the Galaxy. The all-sky imaging of EXIST would extend the central-radian galactic survey planned for INTEGRAL to the entire galaxy.

V1: *Gamma-ray bursts at the limit: SFR at z $\gtrsim$ 10*
Since at least the "long" GRBs located with BeppoSAX are at cosmological redshifts, and have apparent luminosities spanning at least a factor of 100, it is clear that even the apparently lower luminosity GRBs currently detected by BeppoSAX could be detected with BATSE out to z ~4 and that the factor ~5 increase in sensitivity with Swift will push this back to z ~5-15 (Lamb and Reichart 1999). The additional factor of ~4 increase in sensitivity for EXIST would allow GRB detection and sub-arcmin locations for z $\gtrsim$ 15-20 and thus allow the likely epoch of Pop III star formation to be probed if indeed GRBs are associated with collapsars (e.g. Woosley 1993) produced by the collapse of massive stars. The high throughput and spectral resolution for EXIST would enable high time resolution spectra which can test internal shock models for GRBs.

V2: *Soft Gamma-ray Repeaters: population in Galaxy and Local Group*
Only 3 SGR sources are known in the Galaxy and 1 in the LMC. Since a typical ~0.1sec SGR burst spike can be imaged (5σ) by EXIST for a peak flux of ~200 mCrab in the 10-30 keV band, the typical bursts from the newly discovered SGR1627-41 (Woods et al 1999) with peak flux $\sim 2 \times 10^{-6}$ erg cm^{-2} s^{-1} would be

detected out to $\sim$200 kpc. Hence the brightest "normal" SGR bursts are detectable out to $\sim$3 Mpc and the rare giant outbursts (e.g. March 5, 1979 event) out to $\sim$40 Mpc. Thus the population and physics of SGRs, and thus their association with magnetars and young SNR, can be studied throughout the Local Group and the rare super-outbursts beyond Virgo.

V3: *HX blazar alert and spectra: measuring diffuse IR background*
The cosmic IR background (CIRB) over $\sim$1-100μ is poorly measured (if at all) and yet can constrain galaxy formation and the luminosity evolution of the universe (complementing **S1** above). As reviewed by Catanese and Weekes (1999), observing spectral breaks (from $\gamma - \gamma$ absorption) for blazars in the band $\sim$0.01-100 TeV can measure the CIRB out to z $\sim$1 *if* the intrinsic spectrum is known. Since the γ-ray spectra of the detected (low z) blazars are well described by synchrotron-self Compton (SSC) models, for which the hard x-ray ($\sim$100keV) synchrotron peak is scattered to the TeV range, the HX spectra can provide both the required underlying spectra and time-dependent light curves for all objects (variable!) to be observed with GLAST and high-sensitivity ground-based TeV telescopes (e.g. VERITAS).

V4: *Accretion torques and X-ray pulsars*
The success of BATSE as a HX monitor of bright accreting pulsars in the Galaxy (cf. Bildsten et al 1997), in which spin histories and accretion torques were derived for a significant sample, can be greatly extended with EXIST: the very much larger reservoir of Be systems can be explored, and wind vs. disk-fed accretion studied in detail. The wide-field HX imaging and monitoring capability will also allow a new survey for pulsars and AXPs in highly obscured regions of the disk, complementing **S4** above.

V5: *QPOs and accretion disk coronae*
The rms variability generally and QPO phenomena appear more pronounced above $\sim$10keV for x-ray binaries containing both BH and NS accretors, suggesting the Comptonizing corona is directly involved. Thus QPOs and HX spectral variations can allow study of the poorly-understood accretion disk coronae, with extension to the AGN case. Although the wide-field increases backgrounds, and thus effective modulation, the very large area ($\gtrsim$ 1m^2) of HX imaging area on any given source means that multiple $\sim$100 mCrab LMXBs could be simultaneously measured for QPOs with 10% rms amplitude in the poorly explored 10-30 keV band.

III EXIST-ISS MISSION CONCEPT

To achieve the desired $\sim$0.05 mCrab sensitivity full sky up to 100 keV (and beyond) requires a very large area array of wide-field coded aperture (or other modulation) telescopes. The very small field of view ($\sim$10$'$) of true focussing (e.g. multi-layer) HX telescopes precludes their use for all sky imaging and monitoring surveys. EXIST-ISS would take the coded aperture concept to a practical limit, with 8 telescopes each with $1m^2$ in effective detector area and $40° \times 40°$ in field of view (FOV). The individual FOVs are offset by 20° for a combined FOV of $160° \times 40°$, or ~2sr. By orienting the 160° axis perpendicular to the orbit vector, the full sky can be imaged each orbit if the telescope array is fixed-pointed at the local zenith. This gravity-gradient type orientation, and the large spatial area of the telescope array, are ideally matched for the ISS, which provides a long mounting structure (main truss) conveniently oriented perpendicular to the motion, as depicted on the EXIST website.

The sensitivity would yield $\sim 10^4$ AGNs full sky, thus setting a confusion limit resolution requirement (~1/40 "beam") of $\sim 5'$. With this coded mask pixel size, high energy occulting masks (5mm, W) can be constructed with 2.5mm pixel size for minimal collimation. The mask shadow is then recorded by tiled arrays of CdZnTe (CZT) detectors with effective pixel sizes of ~1.3mm, yielding a compact (1.3m) mask-detector spacing. The CZT detectors would likely be 20mm square × 5mm thick (for $\gtrsim$ 20% efficiency at 500 keV) and read out by flip-chip bonded ASICs (e.g. Bloser et al 1999, Harrison et al 1999).

The 8-telescope array is continuously scanning (sources on the orbital plane drift across the 40° FOV in 10min; correspondingly longer exposures/orbit near the poles), with each photon time-tagged and aspect corrected ($\sim 10''$) so that ISS pointing errors or flexure are inconsequential over the large FOV. Source positions are centroided to $\lesssim 1'$ for $\gtrsim 5\sigma$ detections. The resulting sky coverage is remarkably uniform with $\lesssim$ 25% variation in exposure full sky over the ~2mo precession period of the ISS orbit. More details of the current mission concept are given in Grindlay et al (2000), and will be further developed in the implementation study being conducted by the EXIST Science Working Group (EXSWG).

REFERENCES

Bildsten, L. et al, *ApJS*, **113**,367 (1997).
Bloser, P., Grindlay, J., Narita, T. and Jenkins, J., *Proc. SPIE*, **3765**, 388 (1999).
Catanese, M. and Weekes, T., *PASP*, **111**, 1193 (1999).
Fabian, A. *MNRAS*, **308**, L39 (1999).
Grindlay, J.E. et al, *Proc. SPIE*, **2518**, 202 (1995).
Grindlay, J.E. et al, *Proc. STAIF-2000*, in preparation (2000).
Harrison, F.A. et al, *Proc. SPIE*, **3765**, 104 (1999).
Iyudin, A.F. et al, *A&A*, **284**, L1 (1994).
Lamb, D.Q. and Reichart, D.E., *ApJ*, submitted (astro-ph/9909002) (1999).
Vignati, P. et al, *A&A*, **349**, 57L (1999).
Woods, P.M. et al, *ApJ*, **519**, L139 (1999).
Woosley, S.E. *ApJ*, **405**, 273 (1993).
Yi, I. and Grindlay, J.E., *ApJ*, **505**, 828 (1998).

Considerations for the Next Compton Telescope Mission

J.D. Kurfess[1], W.N. Johnson[1], R.A. Kroeger[1], and B.F. Phlips[2]

[1]*Naval Research Laboratory, Washington, DC 20375-5352*
[2]*George Mason University, Fairfax, VA 22030-4444*

Abstract. A high resolution Compton telescope has been identified by the Gamma Ray Astronomy Program Working Group (GRAPWG) as the highest priority major mission in gamma ray astrophysics following GLAST. This mission should provide 25-100 times improved sensitivity, relative to CGRO and INTEGRAL, for MeV gamma ray lines. It must have good performance for narrow and broad lines and for discrete and diffuse emissions. Several instrumental approaches are being pursued to achieve these goals. We discuss issues relating to this mission including alternative detector concepts, instrumental configurations, and background reduction techniques. We have pursued the development of position-sensitive solid-state detectors (Ge, Si) for a high spectral resolution Compton telescope mission. A ~1 m^2 germanium Compton telescope of position-sensitive germanium detectors was the basis for one of the GRAPWG concepts. Preliminary Monte Carlo estimates for the sensitivities of this instrument are encouraging. However, there are technical challenges of cooling large volumes of Ge and providing the large number of spectroscopy channels. We also show that with only two Compton scatter interactions followed by a third interaction, the incident gamma ray energy and direction cone can be precisely determined in detectors with excellent energy and position resolution. Full energy deposition is not required. We present a promising concept for a high efficiency Compton instrument for which thick silicon strip detectors might be preferred.

INTRODUCTION

From the first balloon-borne gamma ray observations in the 1960s to the *COMPTON* Observatory, instrumental sensitivities have improved by about a factor of 100. The *CGRO* instruments achieve these improved sensitivities through a combination of larger areas and longer observation times. ESA's *INTEGRAL* mission will provide modestly improved sensitivities for narrow gamma ray lines from discrete sources. However, significant improvement in sensitivity is required to achieve the desired advances in gamma ray astrophysics. There is a consensus that a high resolution Compton telescope is the best way to meet the broad scientific objectives. The GRAPWG [1] has endorsed such a mission as the highest priority major mission following *GLAST*.

There is a broad range of scientific objectives for the next mission. These include studies of supernovae, novae, compact galactic objects, diffuse galactic emissions, active galactic nuclei, gamma ray bursts, the cosmic diffuse background, and solar activity. Many of the compelling objectives involve gamma ray lines. A target

CP510, *The Fifth Compton Symposium,* edited by M. L. McConnell and J. M. Ryan

Table 1. Lines of Astrophysical Interest	
Science Objective	**Isotopes and Lines (MeV)**
Understand Type Ia SN explosion mechanism and dynamics	^{56}Ni (0.158, ***0.812***, ...) ^{56}Co (***0.847, 1.238***, ...) ^{57}Co (0.122)
Map the Galaxy in nucleosynthetic radioactivity	^{26}Al (***1.809, 0.511***) ^{60}Fe, ^{60}Co (***1.173, 1.332***) ^{44}Ti (0.068, 0.078, ***1.16***)
Map Galactic positron annihilation radiation	e^+–e^- annihilation (***0.511***, 3 photon continuum) SN Ia ^{56}Co positrons (***0.511***) ^{26}Al and ^{44}Ti positrons (***0.511***)
Understand the dynamics of Galactic Novae	^{13}N, ^{14,15}O, ^{18}F positrons (***0.511***) ^{7}Be (***0.478***), ^{22}Na (***1.275, 0.511***)
Cosmic Ray Interactions with the ISM	^{12}C (4.4), ^{16}O (6.1), ^{20}Ne(***1.634***), ^{24}Mg(***1.369***,2.754),^{28}Si(***1.779***), ^{56}Fe(***0.847,1.238***)
Neutron Star Mass-Radius	p-n (***2.223***)

sensitivity of 10^{-7} γ/cm^2-s (10^6 s observation) has been established as a goal. This is about two orders of magnitude better than *CGRO* or *INTEGRAL*!

Table 1 lists the lines of astrophysical interest. Note that many are in the ~0.5-2 MeV region (bolded in table), an important consideration in the instrumental design.

INSTRUMENTAL CONSIDERATIONS

A high resolution Compton telescope is the preferred instrument for several reasons. It has a very large field-of-view and associated multiplex advantage. Relative to coded aperture or collimated instruments the efficiency, and hence sensitivity, can improve substantially with instrumental configuration. Note that the efficiency of COMPTEL is typically 1% or less. The sensitivities of coded aperture or collimated instruments scale approximately with the square root of the size, making significant advances prohibitive in terms of instrument size and cost. In Compton telescopes, however, the use of position-sensitive detectors with excellent spectral resolution reduces the error in the width of the Compton scatter angle dramatically, thereby providing direct improvement of about a factor of 10 relative to COMPTEL for a similar size instrument.

Finally, a key to improved sensitivity is rejection of internal background. COMPTEL uses time-of-flight to provide good rejection of instrumental background, but it is still the limiting factor in sensitivity. Time-of-flight is not possible with the higher efficiency designs under development or investigation. However, other techniques, using the electron direction of motion and/or background re-construction of events have potential for even better background rejection. The latter depends critically on the energy and position resolution achieved in the detectors.

Candidate Instruments

Several groups are developing instruments or instrument concepts and these are addressed in more detail in other paper in these proceedings. UCR and MPE are developing the TIGRE [2] and MEGA [3] instruments that utilize arrays of thin, position-sensitive silicon detectors as the Compton scatterer. CsI is used as an absorber (the UCR concept also may employ arrays of Ge or CZT pixel detectors for part of the absorber to improve performance). The key advantages are the ability to track the electron through several layers of Si detector, thereby getting the scattered electron's direction and restricting the direction of the incoming gamma ray to a segment of a cone (background reduction). A second advantage is the improved efficiency achieved with these concepts.

Aprile et al. [4] have developed, LXeGRIT, a liquid xenon time projection chamber, which has been successfully flown on a balloon flight. The advantages are excellent 3-D position resolution, moderate efficiency, and the ability to reconstruct the events to reduce internal background. The latter capability is compromised by the relatively poor energy resolution of liquid xenon, but may be improved in gas detectors. [5]

NRL is pursuing an High Resolution Compton Telescope (HRCT) [6] using germanium strip detectors. This would provide the best energy resolution performance but with rather low efficiency. Germanium also requires cooling to liquid nitrogen temperatures, a significant technical challenge for the large volume detector arrays proposed.

The advantages and disadvantages of the candidate approaches are summarized the Table 2.

TABLE 2. Comparison of Proposed Advanced Compton Telescope Concepts

	Instrument				
	TIGRE	MEGA	LXeGRIT	HRCT	New Concept
Efficiency	moderate	moderate	moderate	low	high
Energy resolution	Moderate (with Ge or CZT)	poor	poor	excellent	excellent
Position Resolution	good	good	good	good	good
Background rejection	good	good	good	good	excellent
Electron tracking	> 1 MeV	> 1 MeV	> 1 MeV	none	possible
Event reconstruction	no	no	good	excellent	excellent
Line sensitivity	moderate	moderate	moderate	excellent	excellent
Continuum sensitivity	excellent	excellent	excellent	good	excellent

A NEW CONCEPT

Consider a gamma ray, E_1, incident on a detector or detector array which has good position resolution and good energy resolution. Consider two successive Compton scatter interactions followed by a third interaction as shown in Figure 1, where E_1, E_2, and E_3 are the incident photon energies for each interaction. The energy losses (to the scattered electrons) are L_1, L_2, and L_3, and the Compton scatter angles are φ_1, φ_2.

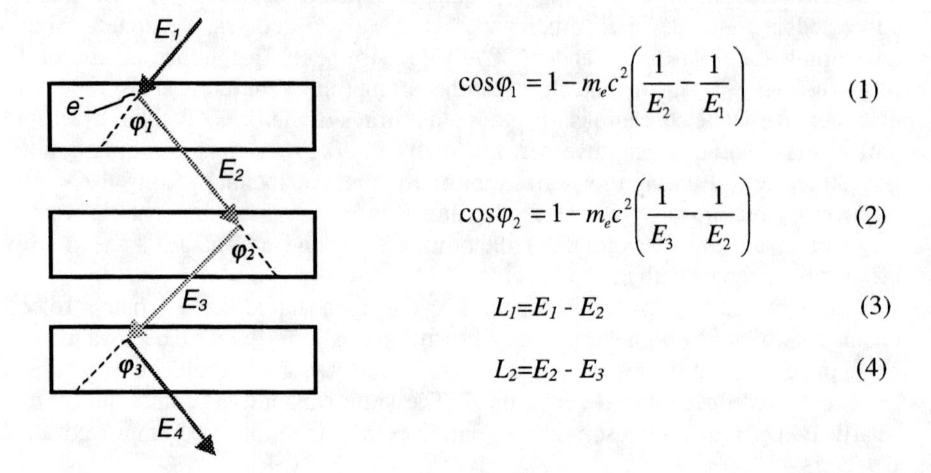

$$\cos\varphi_1 = 1 - m_e c^2 \left(\frac{1}{E_2} - \frac{1}{E_1} \right) \quad (1)$$

$$\cos\varphi_2 = 1 - m_e c^2 \left(\frac{1}{E_3} - \frac{1}{E_2} \right) \quad (2)$$

$$L_1 = E_1 - E_2 \quad (3)$$

$$L_2 = E_2 - E_3 \quad (4)$$

We solve eq. (4) for E_3 and substitute into (2). This yields and equation with E_2 as the only unknown, since φ_2 is determined from the locations of the three interactions. Thus the energy E_2 is known, and therefore the incident gamma ray energy, E_1 is also determined from (3), and is given by:

$$E_1 = L_1 + \frac{L_2}{2} + \frac{1}{2}\left[L_2^2 + \frac{4 m_e c^2 L_2}{1 - \cos\varphi_2} \right]^{\frac{1}{2}} \quad (5)$$

where $m_e c^2$ is the rest mass of the electron. The direction cone of the incoming gamma ray can then be determined from eq (1) just as for a conventional Compton telescope. The uncertainties in E_1 and φ_1 can also be determined, and setting $X = 4m_e c^2/(1-\cos\varphi_2)$, are:

$$dE_1 = \left[dL_1^2 + \left(\frac{1}{2} + \frac{1}{4}\left[L_2^2 + XL_2\right]^{-\frac{1}{2}}\left[2L_2 + X\right] \right)^2 dL_2^2 + \left(\frac{\sin\varphi_2}{4}\left[L_2^2 + XL_2\right]^{-\frac{1}{2}}\left[\frac{XL_2}{(1-\cos\varphi_2)} \right] \right)^2 d\phi_2^2 \right]^{\frac{1}{2}} \quad (6)$$

$$d\varphi_1 = \frac{m_e c^2}{\sin\varphi_1}\left[\left(\frac{1}{(E_1 - L_1)^2} - \frac{1}{E_1^2} \right)^2 dE_1^2 + \frac{dL_1^2}{(E_1 - L_1)^4} \right]^{\frac{1}{2}} \quad (7)$$

where dL_1 and dL_2 are the errors in the energy loss at the Compton scatter sites and $d\varphi_2$ is the error in the scatter angle φ_2. The uncertainty in the width of the scattering angle must also be combined with an uncertainty in the axis of the direction cone. There is also an additional uncertainty in both E_1 and $d\varphi_1$ associated with the atomic velocity of the scattered electron [7]. With excellent energy and position resolution in the individual array elements, it will be possible to determine the incoming gamma ray energy and scatter angle to typically a few keV and 1° or less, respectively.

This concept offers several new possibilities. The requirement for only two Compton scatter interactions plus a third interaction means that high-Z materials are not required. Therefore, high resolution detectors (e.g. Ge, Si, CZT, strip or pixel detectors or possibility high purity gas detectors) are good candidates. Si is an attractive choice [8] since Si detectors can be operated at or near room temperature, thereby avoiding the cryogenic requirements of Ge. The performance of an instrument using this approach is very dependent on the energy resolution and position resolution achieved.

One of the major advantages of an instrument using this concept is a very high efficiency. Efficiencies as high as 25-50% in the MeV region are possible. Achieving high efficiencies will require a large fraction of active detector volume relative to passive materials (structure, housings, and electronics). The instrument will inherently have a very large field-of-view. Achieving the desired 10^{-7} γ/cm^2-s sensitivity will also require good background reduction. This should be achieved through the use of event re-construction for all events, both external and internal. The internal background events, such as radioactive spallation products and neutron capture cascades, should be efficiently rejected because the re-construction of these events do not lead to valid Compton scattering sequences. The efficiency of such background rejection will be very dependent on the energy resolution and position resolution of the detectors [9], again placing very high priority on use of the best position-sensitive detectors available (e.g. Ge or Si strip detectors).

ACKNOWLEDGEMENTS

We acknowledge support from the Office of Naval Research and from NASA under Grant W19390.

REFERENCES

1. Recommended Priorities for NASA's Gamma Ray Astrophysics Program, NP-1999-04-072-GSFC
2 O'Neill, T., et al., these proceedings (2000)
3. Schopper, F., et al., these proceedings (2000)
4. Aprile, E., et al., these proceedings (2000)
5. Aprile, E., et al., SPIE **3446**, 88 (1998)
6. Kroeger, R.A., et al., these proceedings (2000)
7. Kamae, T., et al., Nucl. Instr. Meth. **35** 352 (1988)
8. Momayesi, M., Warburton, W.K., and Kroeger, R.A., SPIE Vol. **3768** 530 (1999)
9. van der Marel, J. and Cederwall, B., preprint submitted to Elsevier(1999)

Position Sensitive Germanium Detectors for the Advanced Compton Telescope

R.A. Kroeger[1], W.N. Johnson[1], J.D. Kurfess[1], B.F. Phlips[2], P.N. Luke[3], M. Momayezi[4], W.K. Warburton[4]

[1]*Naval Research Laboratory, Washington, DC*
[2]*George Mason University, Fairfax, VA*
[3]*Lawrence Berkeley National Laboratory, Berkeley, CA*
[4]*X-Ray Instrumentation Associates, Mountain View, CA*

Abstract. The nuclear line region of the gamma ray spectrum remains one of the most challenging and elusive goals of high energy astrophysics. The scientific objectives are well defined, but require well over a factor of 10 increase in sensitivity compared to present day instruments to be achieved. The most promising approach to achieve this sensitivity and a broad range of scientific objectives is offered by an Advanced Compton Telescope (ACT) that would function in the 0.5 to 30 MeV energy range. The ACT builds on the successful COMPTEL instrument on NASA's Gamma Ray Observatory by substituting modern detectors with over an order of magnitude better energy and spatial resolution. Germanium detectors are a natural choice, as they are available in large volumes, provide the best possible energy resolution, and are capable of fine spatial resolution. These improvements alone provide the required gain in sensitivity. Further optimization and the use of more sophisticated techniques promise even greater improvements. We discuss the current status of the germanium detector technology. New results from the characterization of a crossed strip detector using amorphous-germanium contacts and a demonstration of three-dimensional position resolution are presented.

INTRODUCTION

Instruments such as OSSE and COMPTEL on the Compton Gamma Ray Observatory (CGRO) have made many hundreds of observations in the eight years since launch. In all this time, however, many important objectives have yet to be achieved, such as the firm detection of ^{56}Co from Type Ia supernovae. The goal goes beyond simple detection, but to measure the ^{56}Co with an accuracy of better than 10%, thus tightly constraining the models, and the relationship between brightness and width of the light curve. This is important in the application of supernovae as cosmic distance indicators. The upcoming INTEGRAL mission will provide only comparable sensitivity as OSSE and COMPTEL for broad supernova lines. For practical reasons, INTEGRAL and OSSE represents a sensitivity limit for calorimeter types of instruments. Since sensitivity scales roughly as the square root of size and therefore cost, a different detection technology is therefore required.

COMPTEL on the Compton Gamma Ray Observatory (CGRO) represents the first space-based Compton telescope. It has produced impressive results, such as mapping the ^{26}Al distribution in the galaxy. The performance of a Compton telescope depend

CP510, *The Fifth Compton Symposium,* edited by M. L. McConnell and J. M. Ryan

critically on the energy and spatial resolution of its detectors. COMPTEL uses a liquid scintillator to scatter incoming gamma rays and a NaI scintillator to absorb the scattered energy. This combination permits an instrumental angular resolution on the order of 2-4 degrees rms, and an energy resolution on the order of 5-8% FWHM. Dramatic improvement in both of these measurements are possible using state-of-the art germanium detectors. Our baseline concept for the High-Resolution Compton Telescope (HRCT) will achieve ~1 degree angular resolution and 0.3% FWHM energy resolution at 1 MeV. This should provide a factor of at least 20-30 improvement over COMPTEL for broad line (30 keV) point sources such as Type Ia supernovae, using similar geometry.

The key advance that makes HRCT possible is a new generation of imaging radiation detectors with 10^4 to 10^6 channels of low noise spectroscopy electronics. Germanium detectors are a natural choice, as they: (1) are available in large volumes (up to ~700 cm^3 for a coaxial detector); (2) provide the best possible energy resolution among the semiconducting, gas, and liquid detectors (as good as 1.3 keV FWHM at 1 MeV, and 410 eV FWHM at 100 keV without any electronics contribution); and (3) are capable of fine spatial resolution (<1 mm).

SCIENTIFIC CAPABILITIES

The HRCT will address a broad range of scientific topics [1]. One of the chief objectives of the HRCT is to study line emission from nucleosynthetic radioactivity from supernovae and novae. The HRCT sensitivity should permit the detection of ^{56}Co (847, 1238 keV) to distances ~30 Mpc. The HRCT should then measure the total mass of ^{56}Co to a precision better than 10% in several supernovae per year, thus discriminating between Type Ia models and validating brightness estimates in the application as a cosmological "standard candle." In a zenith pointed mode, HRCT will provide a survey of >70% of the sky on an orbital time scale. Thus, the high sensitivity and sky coverage should provide both an effective monitor for weak transient phenomena and a discovery potential not possible with narrow field-of-view instruments.

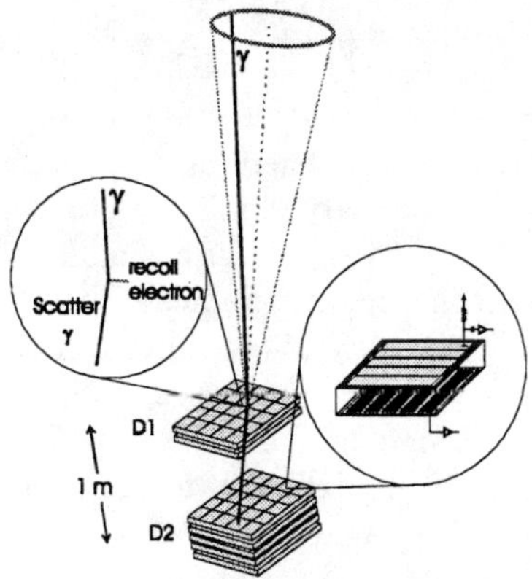

FIGURE 1. Classic Compton telescope in traditional D1/D2 configuration.

PRINCIPLE OF OPERATION

The classic Compton telescope (Figure 1) such as COMPTEL depends on those gamma rays that Compton scatter once in an upper detector, then go on to be totally absorbed in a lower detector. The restricted geometry of a "good event" and the demand for total absorption result in a relatively low efficiency compared with simple calorimeters. This is easily offset by the improved background rejection possible in a Compton instrument. The origin of each individual gamma-ray event is constrained by the measurement to have originated from a cone as shown in Figure 1. The superposition of many such cones forms an image. A detailed discussion of the germanium high-resolution Compton telescope is presented elsewhere [2].

Most good events undergo multiple interactions before they are totally absorbed in D2. It is very important in a high resolution Compton telescope to correctly identify the first two interactions (*i.e.* a single interaction in D1 and the first interaction in D2). These two positions are used to determine the direction of the scattered gamma ray, and thus the axis of the cone shown in Figure 1. This is achieved by accurately measuring the position and energy of each interaction, then sorting out their order based on simple consistency and probability tests [3].

There are several competing concepts for next generation Compton telescopes, generically called the Advanced Compton Telescope (ACT). Some of the primary candidates include highly segmented semiconductor detectors such as germanium, silicon, and CdZnTe detectors. High-pressure gas detectors also have potential advantages, particularly in making large, fully-active volumes. Good spatial and energy resolution are essential in unraveling the interaction sequence and reconstructing the energy of the incoming gamma ray in all of these concepts.

The generalized ACT concept can make full use of events that interact 3 or more times in any order within the detectors, even if the total energy is not totally absorbed by the active detector volumes [4]. The distinction between detectors D1 and D2 is no longer important. Utilization of partially absorbed events dramatically increases the telescope efficiency, but requires sophisticated detectors. The energy of the incident gamma ray, E_i, is fully determined from the first three interactions by,

$$E_i = \Delta E_1 + \frac{\Delta E_2 + \sqrt{\Delta E_2^2 + 4\Delta E_2 m_e c^2 / (1 - \cos\theta_2)}}{2}, \quad (1)$$

where ΔE_1 and ΔE_2 are the energy losses of the first two interactions, and θ_2 is the scatter angle of the second interaction, determined by measuring the positions of the first three interactions. It is not necessary to fully absorb the energy of the gamma ray in order to measure its energy. Monte Carlo simulations of the ACT concept implemented using only silicon detectors indicate that very high efficiencies are achievable. Roughly 30-40% of 1 MeV photons incident on a detector are usable (37 g-cm^{-2} total active thickness), depending on assumptions about passive material and practical considerations in the design.

We have constructed an efficient algorithm to determine the interaction order of events within ACT. Essentially, all permutations of possible events orders are evaluated. Those orders that are inconsistent with the pattern of energy loss are rejected. We have found that most 1 MeV events with four interactions are uniquely reconstructed in the proper order. Events of three interactions typically have one, two, or three solutions for event order. Future work will apply a test to determine which is the most probable of these possible solutions. The simple algorithm becomes computationally intensive for more than about 6 interactions, where pattern recognition or other techniques will be required to simplify the search.

Further work remains to be done to study the background rejection efficiency of these event-order reconstruction techniques. Using all of the information provided by the detectors has the promise of significantly reducing background, and potentially detecting Type Ia supernovae to >100 Mpc for a 1 m^2 sized instrument.

3-D POSITION OF INTERACTION

Accurate measurement of the 3-dimensional positions of each interaction is essential in order to reconstruct the events properly. We achieved this in germanium strip detectors [5]. The germanium strip detector can be made in a variety of sizes, ranging in thickness from a few mm to about 2 cm, and in area from very small to 8×8 cm (possibly 8×16 soon). The position resolution is determined by the segmentation of the electrodes and timing of the signal development on each face. The segmentation, or strip pitch, can be fabricated as fine as required, approaching a practical limit ~100 to 200 microns pitch [6]. The segmentation measures position in two dimensions in a crossed strip detector where strips on one face of the device are perpendicular to the other. The 2-dimensional position of an interaction is easily determined in a crossed strip detector by identifying the strips on which matching signals are measured.

The depth of interaction is measured by the timing of the induced charge on each face of the detector. The bulk of the induced charge develops when the primary ionization is near the strip that collects it. A time difference between the electron and hole collection is observed when the strip pitch is small compared to the detector size, and is approximately ±100 ns in a 1 cm thick germanium detector. The pulse shape can be computed by considering weighting potentials of the electrodes within the detector, and folding these with the motion of the drifting charge carriers [5].

The depth of interaction is measured rather simply: the time that the preamplifier output rises to 50% of the full signal is recorded. The difference between the times of crossing for the signals on each face is proportional to depth.

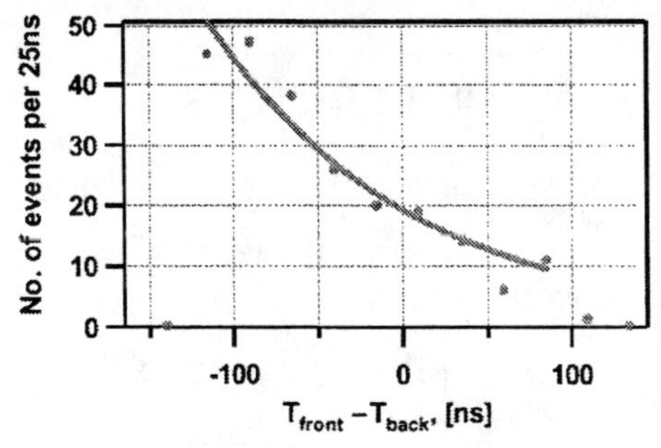

FIGURE 2. Depth of interaction for 122 keV gamma rays in 1 cm thick germanium strip detector.

A test was performed using a 122 keV gamma ray source. The time to rise to 50% full scale was computed from the preamplifier trace for each event. A histogram of the time differences between front and back are plotted in the Figure 2. The full-scale range of time differences corresponds to 10 mm depth in the detector. The exponential fit has a decay length of 5.6 ± 0.6 mm, in good agreement with the 5.75 mm attenuation length of a 122 keV gamma ray in germanium.

AMORPHOUS GERMANIUM CONTACTS

Until recently, the n^+ contacts on germanium strip detectors were fabricated with a lithium diffusion process. While lithium diffused contacts have excellent electrical properties, the diffusion depth is >300 microns, thus limiting the finest pitch that can be reliably produced to >1 mm, and creating a significant dead layer around the contact. An alternative to lithium diffusion has been developed at Lawrence Berkeley National Laboratories using a thin layer of amorphous germanium to form the contact [7]. It is expected that these devices will require less labor and fewer critical steps to produce than the lithium-diffused contact devices that are commercially available today. By removing the diffusion depth restriction, amorphous germanium contacts allow much finer contact structures to be fabricated.

We have tested a crossed strip detector that was produced using this technology [8]. This detector has a negligible surface dead layer, and has been demonstrated to be stable, even after annealing at 100°C for 24 hours (annealing may be required on orbit to recover from prolonged exposure to radiation over a period of years). The device was 1 cm thick with an active area of 1.5×1.5 cm^2, and 5 strips on each face. The strips were spaced on a 3 mm pitch with 0.25mm gaps separating them. The device performed well as expected, with sharp strip boundaries and good energy resolution (<2 keV FWHM limited by the electronics used). We are in the process of fabricating a larger device with 2 mm pitch strips and 5×5 cm^2 active area.

ACKNOWLEDGMENTS

This work was supported, in part, by ONR under funding document N001499WX40025, DOE under DE-A107-97ER6250 and NASA under W19390.

REFERENCES

1. Recommended Priorities for NASA's Gamma Ray Astrophysics Program, NP-1999-04-072-GSFC.
2. W.N. Johnson et al., 1995, SPIE Vol 2518, pp. 74.
2. J. van der Marel and B. Cederwall, 1999, submitted to Elsevier
4. J.D. Kurfess et al., 2000, this conference.
5. M. Momayezi,et al., 1999, SPIE Vol. 3768, pp 530.
6. R.A. Kroeger, et al., 1999, NIM A 422, pp 206.
7. P.N. Luke et al., 1992, IEEE Trans. Nucl. Sci., vol 39, no 4, pp. 590.
8. P.N. Luke et al., 1999, IEEE Conference Record, Seattle WA.

LXeGRIT: The Liquid Xenon Gamma-Ray Imaging Telescope

E. Aprile*, A. Curioni*, V. Egorov*, K. L. Giboni*, T. Kozu*, U. Oberlack*, S. Ventura*‡, T. Doke§, J. Kikuchi§, K. Takizawa§, E. L. Chupp†, P. P. Dunphy†

*Columbia Astrophysics Laboratory, Columbia University, New York, NY, USA
§Waseda University, Tokyo, Japan
†University of New Hampshire, Durham, NH, USA
‡INFN - University of Padua, Italy

Abstract. The feasibility of a large-volume liquid xenon Compton telescope based on full event imaging in a time projection chamber has been demonstrated with the development of the balloon-borne instrument LXeGRIT. With its 400 cm^2 sensitive area and 7 cm drift gap, the liquid xenon detector images γ-rays in the energy range from 200 keV to 25 MeV. The precise 3-dimensional localization of γ-ray interactions within the sensitive volume provides excellent background reduction capabilities. Together with the large efficiency of a homogeneous detector volume, LXeGRIT addresses the primary instrumental limitations encountered in this energy band. Following engineering tests at balloon altitude in 1997, LXeGRIT has been upgraded with a new trigger and data acquisition system, integrated with the existing readout electronics. Enhanced data transfer capability and onboard data storage were also implemented. LXeGRIT was successfully flown from Ft. Sumner, NM on May 7, 1999. The instrument worked as expected at balloon altitude. The large amount of data gathered during the 9.5 h flight are currently being analyzed. We describe the instrument and present some results from calibration data and preliminary results from the flight.

INTRODUCTION

γ-ray astronomy in the energy regime of nuclear lines has long been recognized for its high potential in the field of nuclear astrophysics (e.g., [3]). Yet, progress has been slow because of the difficulty of imaging, high background, and low source fluxes at energies ranging from about 0.05 – 10 MeV. Indeed, background remains the key limiting factor to the sensitivity of γ-ray telescopes. The necessary escape from the opaque atmosphere exposes instruments to the intense radiation environment in space, and causes detector and structural materials to glow just in the observed energy range. The need for a very significant improvement in sensitivity over current and planned telescopes (by a factor of at least 30) has recently been formulated by NASA's Gamma-Ray Astronomy Working Group (GRAPWG). While

CP510, *The Fifth Compton Symposium*, edited by M. L. McConnell and J. M. Ryan

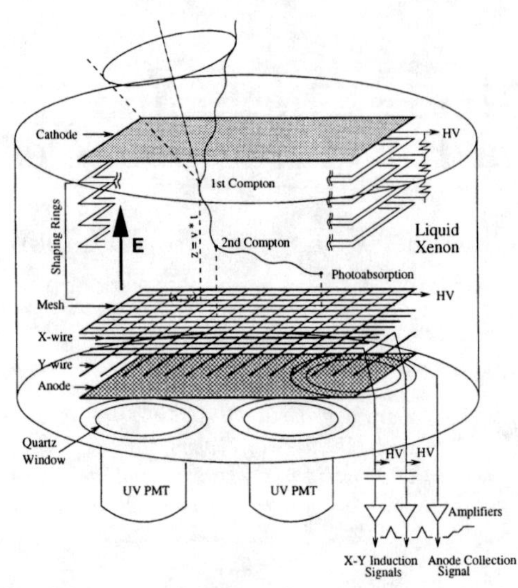

FIGURE 1. Schematic of the liquid xenon time projection chamber.

imaging with coded apertures has proven succesful at the lower energy end, where focussing optics promise significant progress in the future, Compton-telescope imaging remains the most promising approach at energies greater than a few 100 keV.

LXeGRIT is a Compton telescope based on a high-purity liquid xenon time projection chamber (LXeTPC) with $\sim$ 400 cm^2 active area and a 7 cm drift region. It is sensitive to γ-rays and charged particles between 200 keV and 25 MeV. Fig. 1 shows a schematic of the detector and its principle of operation with an incoming γ-ray, which typically loses its energy in the liquid via multiple Compton scatters and a final photoabsorption. Each of the released electrons produces an ionization cloud and UV scintillation photons. The UV photons are registered by photomultiplier tubes, which trigger the data acquisition and define the time of the event. The charges are drifted in a uniform electric field and, after passing a Frisch grid, induce signals on orthogonal sensing wires (62 by 62 wires, 3 mm pitch) which measure the (x, y) interaction locations and, via the drift time, also the z coordinates. Finally, all charges are collected on one or several of four anodes, measuring the energy depositions in each γ-interaction. This event-imaging capability translates directly into an effective background discrimination, based solely on event pattern recognition, independent of the spatial or spectral extent of a celestial source. γ-rays, which lose energy in successive Compton scatterings, are well distinguished from single-interaction events, such as point-like charge depositions from β^- background, and from charged-particle tracks. The incoming direction of multiple-interaction γ-rays is reconstructed as in a standard Compton telescope. Yet, the single homogeneous detector volume makes the LXeTPC much more efficient than classical Compton telescopes, which consist of separate converters and calorimeters and suffer from efficiency loss due to solid angle effects and spacing between sensitive modules.

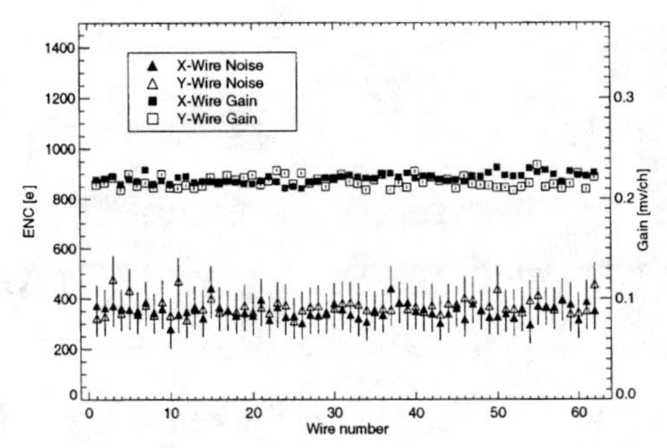

FIGURE 2. Readout noise and gain for the LXeTPC wires.

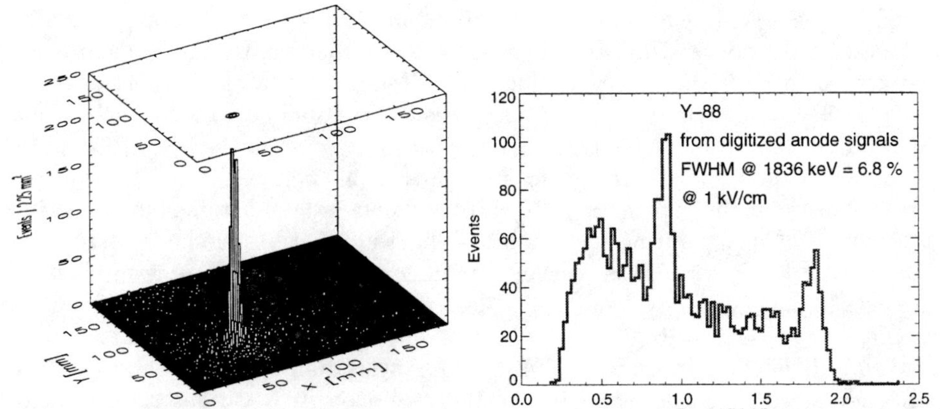

FIGURE 3. Left panel: x/y-positions of single-interaction events from a collimated ^{137}Cs beam.

FIGURE 4. Right panel: Energy spectrum from single-interaction events of a collimated ^{88}Y source.

CALIBRATION RESULTS

Three-dimensional imaging of MeV γ-rays, interacting in multiple points within the detector, requires low noise conditions, which was an important goal for the development of the LXeTPC readout electronics [2]. LXeGRIT achieves an electronics noise level of $\sim$ 350 e$^-$ rms on the x/y induction wires (in units of equivalent noise charge, ENC) and 800 e$^-$ rms on the anodes (of higher capacitance). This permits a minimum energy threshold of $\sim$ 100 keV. Fig. 2 shows the ENC and the gain of the LXeTPC readout chain for the wires.

LXeGRIT has been calibrated with γ-ray lines from radioactive sources. Fig. 3 shows the 2D image of a (2 mm) collimated ^{137}Cs source, consistent with a spatial resolution of $\sigma_x = \sigma_y \approx 1$ mm. Even higher precision, $\sigma_z = 0.14$ mm, is achieved from the drift time measurement. Fig. 4 displays the energy spectrum of ^{88}Y γ-rays with lines at 898 keV and 1836 keV, reconstructed from fully digitized anode signals. Our calibration yields an energy resolution of $\sim$ 9.8% at 1 MeV at a field of 1 kV/cm, scaling with $E^{-1/2}$. (This value includes electronics noise, grid inefficiency and other systematic effects.) Our result is consistent with previous measurements in small liquid xenon chambers [1,4], when operated at the same field.

PRELIMINARY RESULTS FROM THE MAY/7/1999 BALLOON FLIGHT

The goal of the LXeGRIT project has been to demonstrate the feasibility of the liquid xenon technology for astrophysical γ-ray imaging. The 10 liter LXeTPC, developed for laboratory studies, was turned into a balloon-borne instrument and tested for the first time in near space environment during a 1997 balloon campaign in Palestine, TX. Several shortcomings identified in that campaign were addressed and overcome with the successful flight of LXeGRIT on May 7, 1999 from Ft. Sumner, NM. The main goal of this flight was the measurement of the background rate of LXeGRIT at balloon altitude. The entire system, from the LXeTPC and its electronics readout, to the cryogenics and data acquisition performed as expected throughout the flight. Telemetry at 1 Mbps and onboard data storage provided full data taking efficiency during the 7 hours of flight at an atmospheric depth of 3.7 – 5.2 g cm^{-2}. The payload was recovered in good condition, with some mechanical damages to parts of the NaI side shields and minor structural elements. The Crab nebula was well within the large FOV for a few hours. While these data will be used to test the TPC imaging response, a longer duration balloon flight is planned for the year 2000 to allow imaging of MeV emission from various sources, with increased statistics and optimized onboard data reduction.

The top panels of Fig. 5 show the power of the TPC 3D event imaging to distinguish between photons that undergo multiple Compton interactions, which can be used for source imaging, from those that are absorbed in a single interaction or within a small volume of the detector. This provides very significant background rejection. Only photons that produce multiple scattering in a volume on the order of the spatial resolution of the detector are confused with single-interaction events and therefore lost for imaging. Charged cosmic-ray particles are also easily recognized, both from their large energy loss and long track length. Other event topologies, easily distinguishable from Compton γ-events, result from fast neutrons generating nuclear reactions within the detector volume (bottom panels of Fig. 5).

The analysis of the LXeGRIT data from the May 7, 1999 balloon flight is underway. We are confident that we will fulfill our primary goal of this flight, namely the measurement and understanding of the background spectrum in our detector in the near-space environment. At the current stage of analysis, we show the count spectrum of γ-rays with multiple interactions in Fig. 6 as a *preliminary* first result.

We would like to thank the NSBF launching team for excellent support, a successful flight, and the recovery of a healthy payload. Support for the LXeGRIT project is provided by the NASA High-Energy Astrophysics SR&T program under grant # NAG5-5108.

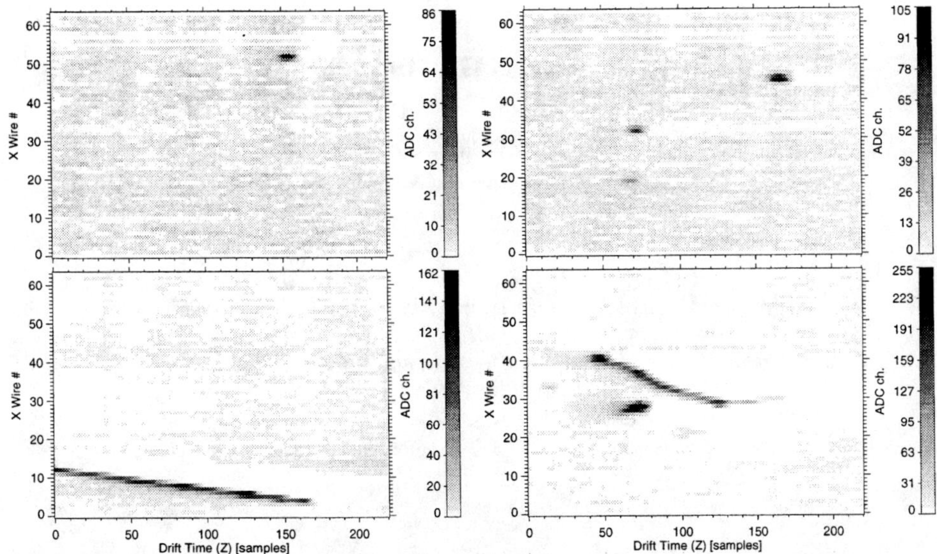

FIGURE 5. Different event types are easily recognized by their signature on the sensing wires, as shown in these images of (x, z)-coordinates. Single- (top left panel) and multiple-interaction (top right panel) γ-events, a cosmic-ray track (bottom left), and a more complicated event topology involving some nuclear reaction (bottom right). Events are from the May 1999 balloon flight.

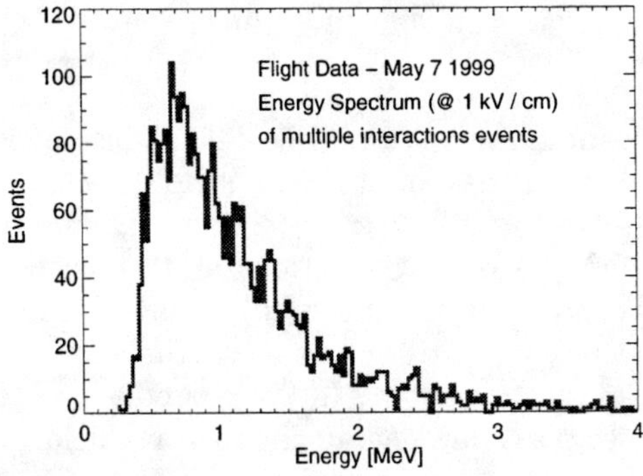

FIGURE 6. Preliminary count spectrum of multiple-interaction γ-events from a sample of the May 1999 balloon flight data.

REFERENCES

1. Aprile E., Mukherjee R., Suzuki M., 1991, NIM A **302**, 177
2. Aprile E., Egorov V., Giboni K.L., et al., 1998, NIM A **412**, 425
3. Clayton D.D., Craddock W.L., 1965, ApJ **142**, 189
4. Xu F., 1998, Ph.D. thesis, Columbia University, New York, USA

The TIGRE Gamma-Ray Telescope

T.J. O'Neill[1], D. Bhattacharya[1], D.D. Dixon[1], M. Polson[1], R.S. White[1], A.D. Zych[1], J. Ryan[2], M. McConnell[2], J. Macri[2], J. Samimi[3], A. Akyuz[4], W.M. Mahoney[5], L. Varnell[5]

[1]*Institute of Geophysics and Planetary Physics, University of California, Riverside, CA 92521, USA*
[2]*Space Science Center, University of New Hampshire, Durham, NH 03824, USA*
[3]*Dept. of Physics, Sharif University of Technology, Tehran, Iran*
[4]*Dept. of Physics, Cukurova University, Adana, Turkey*
[5]*Jet Propulsion Laboratory*

Abstract. TIGRE is an advanced telescope for gamma-ray astronomy with a few arcmin resolution. From 0.3 to 10 MeV it is a Compton telescope. Above 1 MeV, its multi-layers of double sided silicon strip detectors allow for Compton recoil electron tracking and the unique determination for incident photon direction. From 10 to 100 MeV the tracking feature is utilized for gamma-ray pair event reconstruction. Here we present TIGRE energy resolutions, background simulations and the development of the electronics readout system.

TIGRE Instrument Description: The TIGRE instrument [1] [2] [3], shown in Figure 1, features multilayers of silicon strip detectors (SSD) as both the Compton converter and recoil electron tracker. Double-sided SSDs provide submillimeter x and y spatial resolutions as the recoil electron is tracked through successive layers until it is fully absorbed. Position sensitive Ge and/or CdZnTe detectors are used as a calorimeter for the scattered photons and electron-positron pairs. CsI(Tl)-Photodiode detector arrays on the bottom and sides serve as a low energy gamma-ray shield. This is important at energies below 1 MeV where up-down discrimination of gamma-rays cannot be determined. The shield also serves as a Compton suppressor for gamma rays leaving the Ge and/or CdZnTe. The CsI(Tl) arrays serve the dual purpose of a calorimeter for large scatter angle events as well as of a shield. This is particularly important for polarization measurements below 2 MeV. A particle anticoincidence plastic scintillator surrounds the entire sensitive material.

Gamma ray pair events are also detected in the traditional manner by tracking the electron and positron individually through successive layers of silicon strip detectors until at least one particle or annihilation photon exits and

CP510, *The Fifth Compton Symposium,* edited by M. L. McConnell and J. M. Ryan

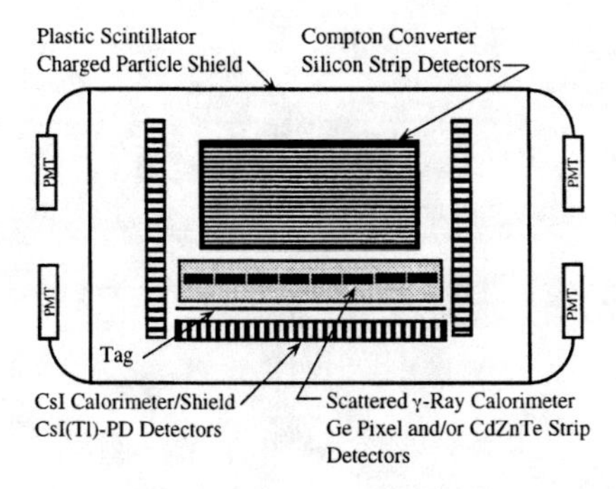

FIGURE 1. Baseline TIGRE instrument.

interacts in the Ge/CdZnTe calorimeter. Both the energy losses and positions of the pair particles are measured in each Si layer as these particles are tracked through the array. For Compton events, where the electron direction is not apparent, we use the energy loss and position in each Si layer traversed to determine a Direction-of-Motion (DOM) parameter for each track. This parameter is similar to Time-of-Flight (TOF) that is used for up-down discrimination in conventional Compton telescopes.

Silicon Strip Detector Measurements: A mini TIGRE instrument was assembled using seven 4 cm × 4 cm double-sided Si strip detectors and a single coaxial Ge detector. Figure 2 shows the laboratory setup for calibrating the

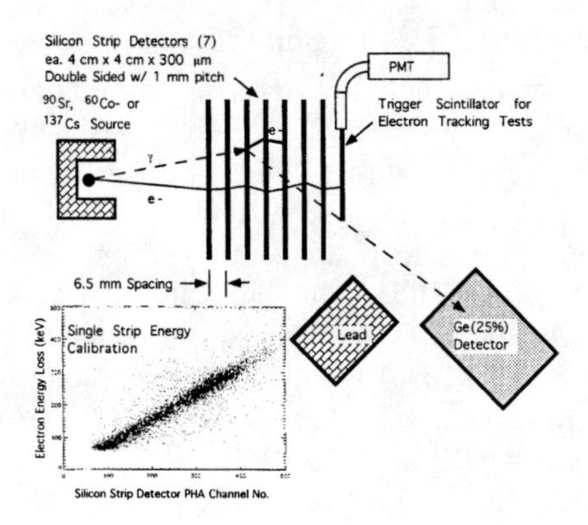

FIGURE 2. The laboratory setup used for the silicon germanium double scatter experiment.

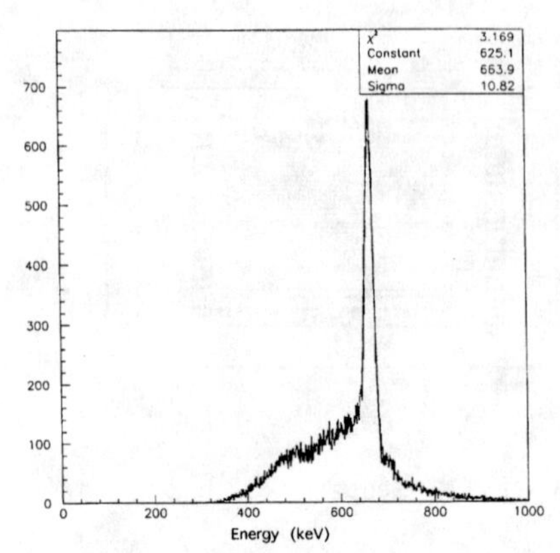

FIGURE 3. The spectrum of ^{137}Cs from the Si-Ge double scatter experiment.

Si detectors while the insert in the figure shows the calibration results for a single strip of a single detector. In a double scatter (Compton telescope) configuration, the wide range of scatter angles provides an equally wide range of energy losses in the Si. The true Si energy loss is determined from the measured Ge energy of the scattered gamma-ray and the source gamma-ray energy. This is compared to the measured Si energy thereby producing a calibration curve for each strip. A ^{137}Cs energy spectrum obtained by this method is shown in Figure 3. The measured energy resolution is 10.8 keV (1σ) or 3.8%. This resolution will improve with the new 10 cm $\times$ 10 cm Si detectors and low noise IDE-AS TA-1 chips with their trigger capability.

Figure 4 shows a nearly completed Si detector board containing a 10 cm $\times$ 10 cm double-sided 300μ thick Si detector with one TA-1 hybrid wire-bonded to the 128 ohmic strips and one wire-bonded to the 128 junction strips. The hybrid consists of a custom ceramic carrier and a TA-1 chip. Figure 5 shows the shaper output signal (top curve) and the trigger for a test calibration pulse on a single channel. Figure 6 shows the gain, pedestal and noise for all 256 channels of the two TA-1 chips. The noise specification for this chip should allow us to achieve 1.1 keV (1σ, 318 ENC) energy resolution. For 0.5 MeV electrons, where the total charge collected is 140,000 e in three layers, the energy resolution should be 1.9 keV (1σ).

Simulation: We have modeled the TIGRE instrument described above with the general-purpose MCNP code developed at LANL. Energy resolutions for each individual Si and Ge detector element are taken as a conervative 3 keV (1σ) and 1 keV (1σ), respectively. No CdZnTe was used in the simulation. The spatial resolutions are 0.75 mm and 2.0 mm. Thresholds of 30 keV for the Si and Ge detector elements are used. For the CsI(Tl) shield and plastic scintillators 100 keV thresholds are used. For pair events, either the pair

FIGURE 4. The double sided 10cm x 10 cm silicon strip detector with 2 TA-1 hybrid readout chips connected.

particle or an annihilation photon needs to be detected in the calorimeter. Event reconstruction is used to identify gamma-ray with directions within the instrument's FOV. The TIGRE angular resolutions and sensitivities are presented in [4].

TIGRE has a broad maximum efficiency of 5% and a high effective area of 80 cm^2 in the Compton regime above 0.5 MeV. All Compton events are either "tracked" or "non-tracked" depending on the number of silicon layers the recoil electron traverses. The percentage of tracked events increase from 50% at 0.5 MeV to >90% above 2 MeV. In the pair regime the efficiency remains constant with a value of 5% up to 100 MeV. The effective combination of shielding and kinematics reduces the background contribution from outside the FOV ($\sim \pi$ sr). Figure 7 shows the relative efficiencies at 1.8 MeV for an isotropic background as a function of the incident zenith angles for cases of no shielding, shielding and shielding with the tracking cuts.

REFERENCES

1. O'Neill, T. J. et al. 1995, *IEEE Trans. Nuc. Sci.*, 42, 933.
2. Akyuz A. et al. 1995, *Experimental Astronomy*, 6, 274.
3. Tumer et al. 1995, *IEEE Trans. Nuc. Sci.*, 42, No. 4, 933.
4. Bhattacharya, D. et al. 1999, *26th ICRC Conf. Proc.*, 5, 72.

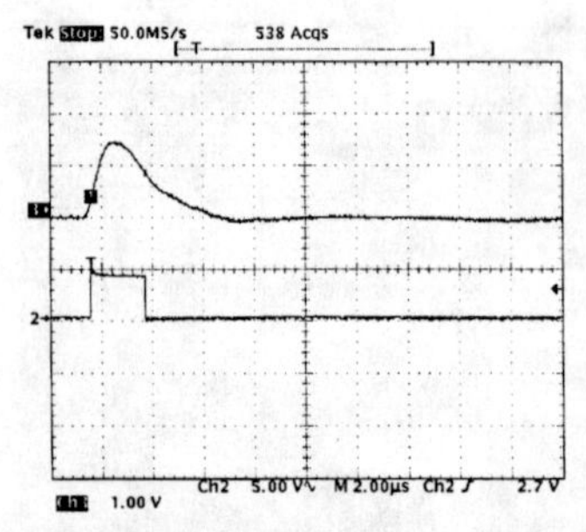

FIGURE 5. The TA-1 shaper output signal (top curve) and the trigger for a test calibration pulse. The trigger is necessary to eliminate accidental events.

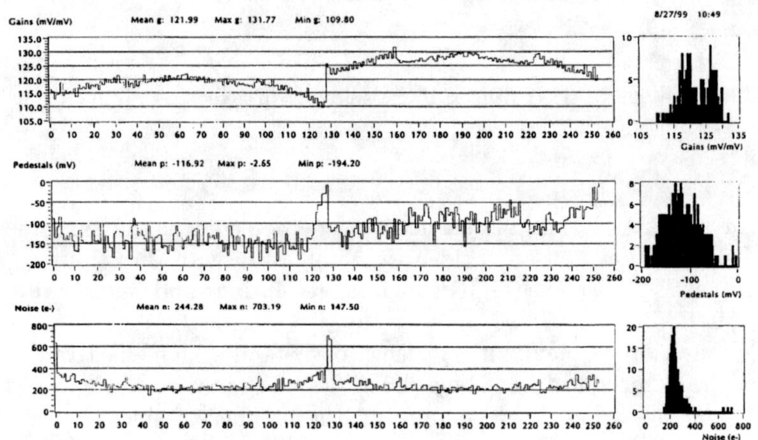

FIGURE 6. The preliminary gain, pedestal and noise performance for two 128 channel TA-1 chips serially connected.

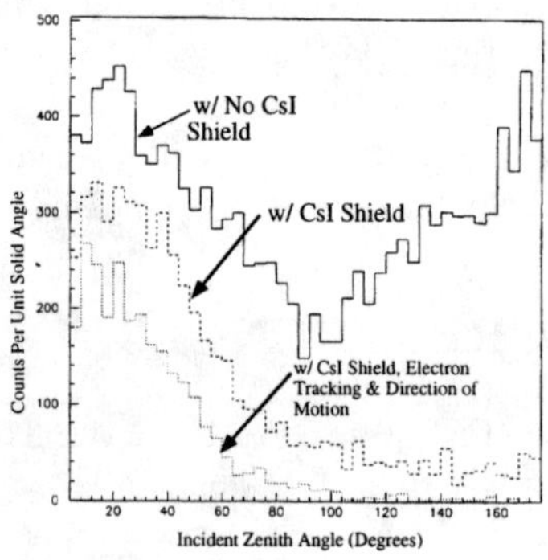

FIGURE 7. Relative efficiencies at 1.8 MeV for an isotropic background as a function of the incident zenith angles for cases of no shielding, shielding and shielding with the tracking cuts.

The CIPHER Telescope for Hard X and Soft γ-Ray Polarimetry

E. Caroli[a], J.B. Stephen[a], W. Dusi[a], A. J. Bird[b], A. J. Dean[b], G. Bertuccio[c], M. Sampietro[c], W. Yu[d], C. Zhang[d], R. M. Curado da Silva[e], P. Siffert[e], V. Reglero[f]

[a]*TeSRE/CNR, Bologna, Italy*
[b]*Physics Dept., University of Southampton, UK*
[c]*Dip. di Elettronica ed Informazione, Politecnico of Milano, Italy*
[d]*Institute of High Energy Physics, Academia Sinica, Beijing, China*
[e]*Laboratoire PHASE/CNRS, Strasbourg, France*
[f]*Dpto. de Astronomia y Astrofisica, University of Valencia, Spain*

Abstract. CIPHER (Coded Imager and Polarimeter for High Energy Radiation) is a hard X and soft γ-ray spectroscopic and polarimetric coded mask telescope based on an array of Cadmium Telluride microspectrometers. The position sensitive detector (PSD) is arranged in 4 modules of 32×32 crystals, each of 2×2 mm^2 cross section and 10 mm thickness giving a total active area of about 160 cm^2. The PSD is actively shielded by CsI crystals on the bottom and sides in order to reduce background and operates over a wide energy range (~10 keV to 1 MeV). The mask, based on a modified uniformly redundant array (MURA) pattern, is about four times the area of the PSD and, being situated at about 100 cm from the CdTe array top surface, provides a wide field of view. The CIPHER instrument is proposed for a balloon experiment, both in order to assess the performance of such an instrumental concept for a small/medium size satellite (or an external ISS-alpha payload) survey mission and to perform an innovative measurement of the Crab polarisation level. CIPHER's wide field of view allows it to keep a single source within the field of view for a long observation period without requiring a precise pointing system. Herein we describe the instrument design, together with results obtained in our development studies on CdTe detectors. Furthermore we present the expected performance in terms of image and spectral quality (angular and energy resolution) and polarimetric capabilities for an observation of the Crab pulsar from balloon altitudes

INTRODUCTION

To date the study of compact X and γ-ray sources has been limited to their spectral characteristics and timing variability, however further observational parameters are required to determine the precise emission mechanisms and physical conditions of the objects. Polarisation is one of these additional parameters because almost all mechanisms that generate high energy emission from astrophysical sources involve strong magnetic fields and lead to the production of polarised photons. Even if polarisation measurements in lower energy bands have been extremely useful (1), and although telescopes such as the COMPTEL instrument on the Compton Gamma Ray

CP510, *The Fifth Compton Symposium,* edited by M. L. McConnell and J. M. Ryan

Observatory are theoretically able to perform polarisation measurements, the sensitivity is such that no actual measurements have been performed at energies greater than about 10 keV (2).

For several years our group has studied the development of a compact telescope suitable for hard X- and soft γ-ray sky surveys, based on the use of thick Cadmium Telluride position sensitive spectrometers (3). The design concept of these instruments allows their operation also in polarimetric mode with promising performance. In the following section we describe a small payload (CIPHER: Coded Imager and Polarimeter for High Energy Radiation) suitable for use as a stratospheric balloon borne imager and polarimeter experiment. In particular we give a first evaluation of CIPHER's ability to perform a polarisation measurement of the Crab pulsar emission in the 10-1000 keV energy range.

THE CIPHER PAYLOAD

The CIPHER instrument is a coded mask telescope based on a CdTe position sensitive detector operating from 10keV to 1000 keV. The CIPHER design was conceived as a balloon borne payload, with a first flight from China, primarily intended to perform measurements of the polarisation level of strong hard X- and soft γ-ray astrophysical sources (the target of the first flight will be the Crab pulsar) and secondarily to assess and verify the performance of such an instrumental concept for a small size satellite or an ISS external payload (4). CIPHER will be operated in photon by photon mode, requiring a star sensor in order to have real time reconstruction of the telescope pointing direction, while the field of view allows the use of a low cost and weight platform with only moderate pointing stability (~10') and coarse pointing accuracy (~1°). Table 1 gives a summary of the CIPHER operational characteristics and required resources.

TABLE 1. CIPHER Operational Summary.

Operational Characteristics	
Energy Range	10-1000 keV
Field of view (FCFV)	8°×10°
Angular resolution	27'×22'
Point source location accuracy	2' for a 5σ detection
Timing accuracy	50 μs
Required Floating altitude	1-3 mbar
Required Resources	
Pointing Accuracy - Pointing Stabilisation	~1° - 10'
Telemetry - On board data storage	<32 kbits/s - 250 MB
Length - Footprint	140 cm - 60×70 cm
Power - Mass	60 W - 70 kg
Chinese Balloon Facility	
Balloon characteristics	Volumes: 1.2-6×10^5 m^3: internally produced
Guarantee flight duration at float	6 hours at 40 km altitude
Flight window limitation	Depend on the observation goal
Available telemetry	32 kb/s

In order to reduce the background rate from the CIPHER detector, a shield system is foreseen. This system comprises 2 elements: (a) an active veto shield of CsI(Tl) scintillating crystals in a well-shaped geometry around the detector and extending above the detector top surface in order to restrict the FOV to the telescope aperture; (b) a passive shield with a stopping power equivalent to 1 mm tungsten and shaped as a truncated pyramid that encloses the volume between the detector and the coded-mask and is mainly devoted to reducing the cosmic ray diffuse background that is the dominant component up to ~200 keV.

The imaging capability of CIPHER is obtained through the use of a coded aperture mask. The mask is positioned 100 cm from the PSD surface, and the design is based on a Modified Uniformly Redundant Array (MURA) basic pattern of 23x23 elements (5). The mask elements are made of 0.5 mm thick Tungsten and so are almost totally opaque (95%) up to 150 keV. Above this energy the mask becomes increasingly transparent so exposing a larger detector area for polarisation measurements. In the pure imaging regime the mask element size dictates an angular resolution of 27'×22' within a fully coded field of view (FCFV) of ~8°×10°. The mask pattern is almost self-supporting because each element is joined to those adjacent, and so requires only a very light structure (e.g. carbon fibre honeycomb) to support the elements and stiffen the mask, without affecting the transparency of the open elements at low energy.

The position sensitive detector (PSD) comprises four identical modules of 32×32 Cadmium Telluride micro-spectrometers. The basic sensitive unit is a crystal of 2×2× 10 mm^3. These units are used in the configuration in which the optical axis is orthogonal to the charge collecting field, and are assembled on thin (300 μm) ceramic plates in linear modules that contain 32 units together with their integrated analogue readout (low noise and low power consumption) electronics and bias circuits (4).

The linear modules are packed together (2.6 mm spacing is technically achievable) to form a 32×32 matrix module inside an Aluminium case. Below the linear module, two layers are foreseen containing the hybrid front-end electronics (FEE) with multiplexer and ADCs for the 1024 channels. The matrix module is therefore a complete and independent detector that can be tested and calibrated separately. Finally, the matrix modules are integrated (for a total sensitive area of 164 cm^2 with 4096 pixels) and supported by a metallic (Al) or carbon fibre grid that also provides the mechanical interface for the active veto shield.

THE CIPHER DETECTOR PERFORMANCE

In the traditional use of CdTe detectors the photons are incident on one of the electrodes, thus limiting the absorption thickness and the operative range up to 200-300 keV. Several years ago, the TeSRE group proposed a geometry in which the charge collection field is orthogonal to the optical axis of the detector (PTF: planar transverse field) (6). In this configuration the charge collection distance is independent from the photon absorption thickness that can be increased up to 1-2 cm, i.e. suitable for higher energy applications (up to the MeV band). The CdTe configuration proposed for the realisation of the CIPHER PSD has been tested for a long time in the

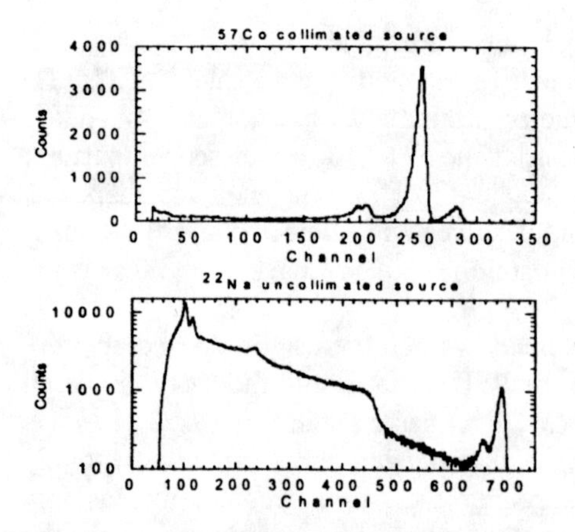

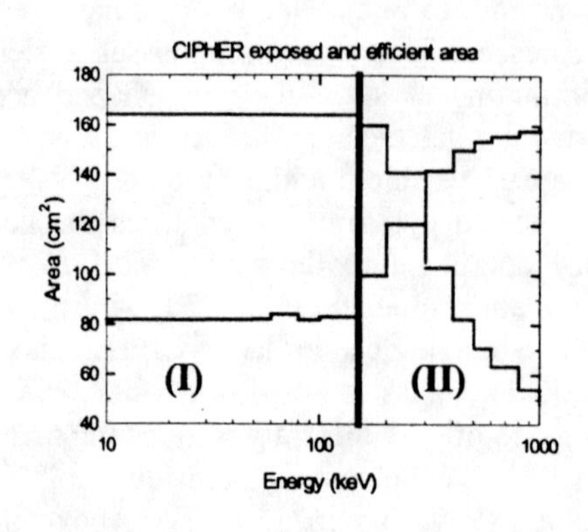

FIGURE 1. **(a)** Energy resolution measured in laboratory tests for 2×2×10 mm PTF CdTe spectrometers. These results was obtained with a selection of short rise time signals (we expect similar results by compensating signals with double shaping technique, see ELBA FEE panel); **(b)** The CIPHER exposed and efficient area: the exposed area is the sensitive area exposed through the mask; the efficient area is the sensitive area times the total PSD efficiency. The blue line indicate the limit between CIPHER pure imaging (I) and polarimetry (II) mode .

laboratory (7,8): the achievable energy resolution range from 9% at 60 keV to ~2% at 0.5 MeV. A pair of spectra obtained with 2×2×10 mm^3 PTF micro-spectrometers are shown in Figure 1.a.

In order to evaluate the sensitivity to polarisation we have performed Montecarlo simulations of a CIPHER module to estimate the total detection efficiency and the efficiency for scattered, and in particular double, events. The simulation has shown that both total and double event efficiency does not vary appreciably (~1%) with photon incidence angle within the field of view (up to 10°). Using this data and the mask transparency as function of energy we have evaluated both the exposed detector area through the coded aperture and detector efficient area (Figure 1.b). From this figure it is possible to see that the CIPHER design is optimised to be a pure spectroscopic imager below 150 keV, while above this energy the instrument becomes suitable for polarimetric measurements.

THE CIPHER POLARIMETRIC CAPABILITIES

In order to investigate the efficiency of this telescope design for polarimetric measurements we have calculated the response of the instrument to a 100% polarised beam of photons in order to obtain the Q polarimetric modulation factor. This is obtained by integrating the formula for the Compton polarimetric differential cross section over the solid angles defined by the physical geometry of the detection plane. It is then defined as:

$$Q = \frac{N_{\perp} - N_{\parallel}}{N_{\perp} + N_{\parallel}} \qquad (1)$$

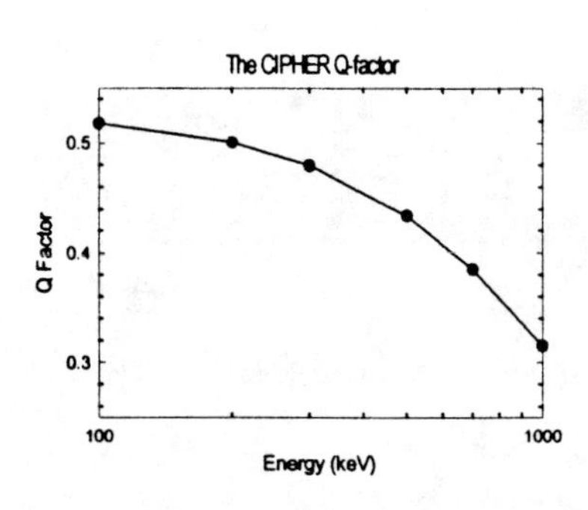

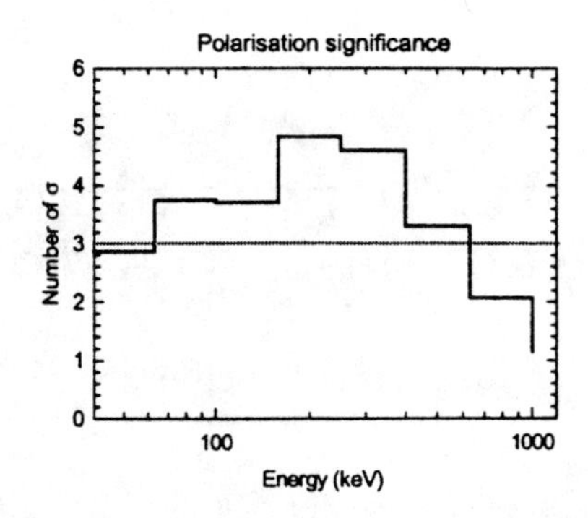

FIGURE 2. (a) The calculated polarimetric Q factor for the CIPHER PSD as a function of energy. The Q factor is the detector's ability to measure the asymmetry in the scattering angle distribution that is expected for polarised radiation; (b) The significance of a CIPHER detection of the Crab Pulsar (10% polarised) for a typical balloon observation time of 10^4 s.

where $N_\perp$ and $N_\parallel$ are the count rates in orthogonal detectors in the X/Y plane. In Figure 2.a we show the Q-factor as a function of photon energy from 100 keV to 5 MeV, showing that this design is very competitive with other proposed instrumentation (2). From the Q-factor it is possible to assess the significance of an observation of a polarised source. By means of error propagation relation, we can evaluate the uncertainty in Q and therefore the significance for polarisation detection that is defined by:

$$n\sigma = \frac{Q}{\sigma(Q)} = \frac{Q\sqrt{C}}{\sqrt{(1+Q^2)}} \qquad (2)$$

where $C=(N_\parallel + N_\perp)$ is the total number of recorded counts. In particular this number is related to double-hit event efficiency that for the CIPHER PSD typically ranges from ~8% to 16% in the 0.2-1. MeV energy band (9).

We have simulated a polarimetric observation of the CRAB pulsar, taking a typical balloon observation time of 10^4 seconds, a standard CRAB spectrum consisting of a power law of slope -2 and assuming, conservatively, that the pulsar emission is 1/5 of the total emission and that it will be 10% polarised. It can be seen from Figure 2.b that CIPHER will be able to obtain highly significant results between 200 keV and 1 MeV.

REFERENCES

1. T. Velusamy, *Monthly Notices Roy. Astron. Soc.*, **212**, p. 359 (1985)
2. F. Lei, A.J. Dean, and G.L. Hills, *Space Sci. Rev.*, **82**, p. 309 (1997)
3. E. Caroli, et al., *Photon Detectors for Space Instrumentation*, ESA SP-356, p. 27 (December 1992).
4. E. Caroli et al., *EUV, X-ray, and Gamma-Ray Instr. for Astronomy,*. SPIE Proceeding, **3765** (1999)
5. S.R. Gottesman and E.E. Fenimore, *Applied Optics*, **28** 4344 (1989)
6. F. Casali et al., *IEEE Trans. Nucl. Sci.*, **NS-39**, p. 598 (1992)
7. E. Caroli et al., *Il Nuovo Cimento*, **16C**, p. 727 (1993)
8. W. Dusi, et al, *Hard X-Ray, Gamma-Ray, Neutron Detector Physics,* SPIE Proceeding, **3768** (1999)
9. E. Caroli, et al., *Nuclear Instr. and Meth,* **A322**, p. 639 (1992)

MGEANT - A GEANT-Based Multi-Purpose Simulation Package for Gamma-Ray Astronomy Missions

S. J. Sturner[1,2], H. Seifert[1,2], C. Shrader[1,2], B. J. Teegarden[2]

[1] *Universities Space Research Association, Seabrook, MD*
[2] *NASA/Goddard Space Flight Center, Greenbelt, MD*

Abstract. MGEANT is a multi-purpose simulation package based on the CERN GEANT package and Program Library. This package allows for the rapid prototyping of a wide variety of gamma-ray detector systems. The two main advantages of this package over standard GEANT simulations are 1) the detector geometry and material data are supplied via input files instead of being hard coded and 2) MGEANT has several built in spectral and beam options including the ability to simulate observations with multiple pointings and diffuse sources. We are currently using MGEANT to produce detailed observation simulations and response matrices for INTEGRAL/SPI. We present a sample of the type of work that has been done at GSFC using MGEANT simulations. This software package is available from our website http://lheawww.gsfc.nasa.gov/docs/gamcosray/legr/integral/ .

INTRODUCTION

MGEANT, developed by the Low Energy Gamma Ray (LEGR) Group at NASA/GSFC, is based on the GEANT program which is supported by the Application Software Group, Computing and Networks Division, CERN Geneva, Switzerland. GEANT is an extensive system of general purpose, detector design and simulation tools which are used in the development, optimization, and testing of complex high energy physics experiments, as well as in the interpretation of experimental data. The program is well suited for this purpose since it is able to simulate the transport of elementary particles through an experimental setup, and to represent the setup and the particle trajectories graphically. Although originally designed for high energy physics experiments, it has to date found applications ranging from the medical and biological sciences to radiation protection and space science. GEANT has been used extensively at NASA/GSFC, in particular by the LEGR Group to model the Transient Gamma Ray Spectrometer (TGRS) on the WIND spacecraft, the Gamma-Ray Imaging Spectrometer (GRIS), as well as to prototype new detectors.

CP510, *The Fifth Compton Symposium,* edited by M. L. McConnell and J. M. Ryan

GEANT offers the framework and the raw tools for experimental setup, event simulation, and visualization. But it is the responsibility of the user to code all the relevant subroutines describing the experimental environment, to assemble the appropriate program segments and utilities into an executable program, and to compose the appropriate input parameters or "data cards" which control the execution of the program. The process of setting up a simulation of a particular instrument and/or experiment requires sometimes difficult and/or nonintuitive programming steps. Often, even simple modifications to an existing setup require changes in the program code itself, which may be difficult to perform by a user who is not the author of the program.

To allow the rapid prototyping of a wide variety of detector systems, the LEGR group has developed this GEANT based generic multi-purpose program appropriate for gamma ray astronomy using a modular, "object oriented" approach. Rather than hardcoding instrument specific geometries and materials, these data are supplied via input files. Also, the user can select from a standard set of event generators and beam options, which are implemented as "plug-ins" and are included at the compilation level.

MGEANT OVERVIEW

The NASA/GSFC MGEANT program has been developed using the same tools which are used for the releases of CERNLIB 96a and later, i.e. using CVS, Imake, Makedepend, gmake, and cpp (the C Preprocessor) [1]. The MGEANT releases at our website are distributed as a set of source code, header files, and libraries together with an Imakefile and associated configuration files for installation.

The MGEANT source code contains several beam and spectral generators that allow the user to customize their executable. These options are chosen at compile time. MGEANT can also be customized to run in either a batch or interactive mode and the format of the output event list file can be selected to be FITS, ASCII, or HBOOK format. These formats as well as the detailed use of MGEANT are fully described in the MGEANT User Guide available on our website.

Current Beam Models:

astrophysical point source
diffuse emission (requires input FITS image)
calibration source
isotropic beam
plane wave beam
homogeneous and isotropic emission within a volume
models for response matrix calculation

Current Spectral Models:

power-law
broken power-law

exponential
power-law with exponential cut-off
single line emission
multiple line emission
cosmic ray spectra [2]
gamma-ray burst [3]
user defined model using ASCII file input

Once the user-specified version of MGEANT is built, all the required parameters to describe a simulation are given in four ASCII input files:

input.dat: This file contains all the information that is required to fully specify the beam geometry and the spectrum. It can also contain additional information such as the shield veto threshold, the observation length, the observing pattern (dithering), the level of debug output, etc.

setup.geo: This ASCII file contains all the information needed to define and position the volumes of the instrument to be studied.

materials.mat: This file contains a list of the materials needed to define the volumes in the instrument. We have included a simple mechanism for defining a mixture which contains elements in their naturally occuring isotopic abundances.

media.med: This file contains the information needed by MGEANT to define tracking media using the GEANT GSTMED subroutine or to modify a medium parameter using GSTPAR.

The use of ASCII input files to give the specifics of a simulation allows the user to perform a suite of simulations without editing source code. This is an important advantage over standard GEANT. For instance, variations in detector geometry can be quickly tested to determine their effect on instrument sensitivity. Alternatively, a suite of simulations in which the detector is the same but the source location and/or source spectrum change can be accomplished quickly with no re-coding.

APPLICATIONS

To illustrate the power and versitility of MGEANT, we now present a sample of the research carrried out at GSFC where we are developing software in support of the upcoming INTEGRAL mission. Specifically, we are tasked with providing simulated data sets, analysis software, and response matrices for the SPI spectrometer.

Simulated Astrophysical Observations

The SPI imaging spectrometer on INTEGRAL consists of 19 Ge detectors located within a massive BGO active shield and a tungsten coded mask. To utilize the coded mask to make imaging observations, a typical observation will consist of many different pointings [4]. These pointings will generally be in a rectangular or hexagonal grid with $\sim 2°$ spacing. To make simulations of these observations easier, MGEANT has a built in subroutine to generate the pointing or "dither" pattern. A simple rectangular grid with uniform spacing can be implemented within the input.dat file or an ASCII input file which contains a more complex pattern can be specified.

The number of events to be generated for an observation can be specified in several ways within the input.dat file. It can be given explicitly, in which case MGEANT will use the specified spectrum to calculate the observation time. Alternatively, the observation time can be specified and MGEANT will calculate the number of events using the specified spectrum. The user can also input a source flux for a given energy band. MGEANT will then recalculate the spectrum normalization and use this new spectrum to either calculate the observation time or the number events to generated. Shrader et al. (2000) [5] show several simulated spectra generated with MGEANT.

Response Matrix Generation

The generation of response matrices for SPI is a sizable undertaking. The fine spectral resolution of the Ge detectors coupled with the spatial variation of the response that results from the coded mask requires approximations when calculating the response, otherwise the amount of CPU time and disk space to calculate and store them would be immense. We have thus implemented a plan in which the detector response is broken into 2 parts. One which describes the attenuation due to the mask and shield and is calculated using ray tracing techniques. The second part is generated with Monte Carlo techniques and describes interactions within the cryostat and detectors. We have two routines within MGEANT that perform these calculations in a manner that allows them to be easily combined to generate the full instrumental response. The first routine tracks geantinos (imaginary non-interacting particles) outward from a rectangular grid of points while recording the pathlengths through each material as function of grid point and direction. The second routine tracks photons which are generated on the same grid of points but are fired inward toward the detector plane. The output of the first routine is then used to calculate the attenuation of a photon beam incident from a given direction on a given grid point. The output of the second routine contains the attenuation of this beam by cryostat materials and the energy redistribution of the incoming beam. We have found that this method significantly reduces the computation and storage requirements while retaining the desired accuracy [5].

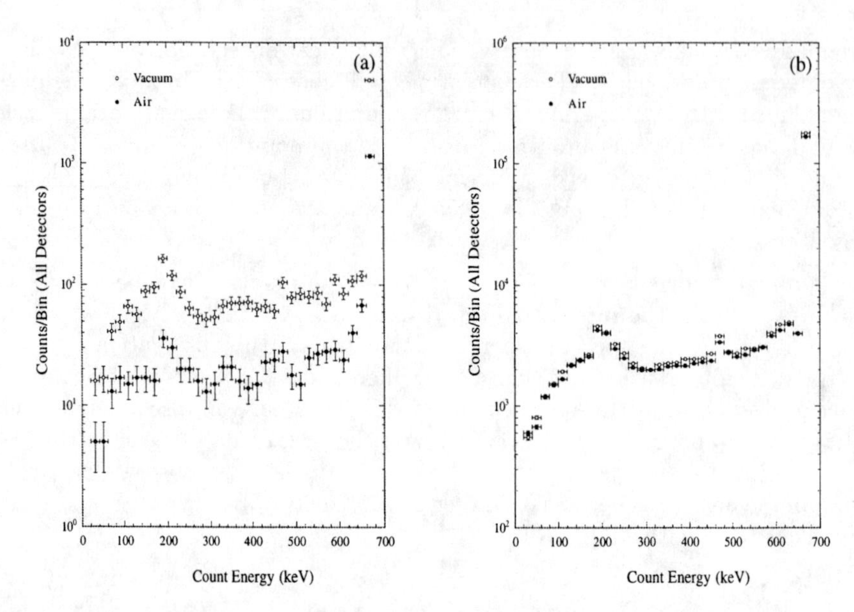

FIGURE 1. Simulated spectra of a ^{137}Cs calibration source located at 150 m (a) and 8 m (b)

Calibration Studies

We have used MGEANT to perform studies of detector calibrations in the laboratory. These studies are necessary to determine what corrections need to be added to response matrices produced using simplified/idealized computer models. In Figure 1 we show simulated spectra for two different calibration runs using ^{137}Cs which emits one line at 661.6 keV. In Figure 1a we show a spectrum for the source located at 150 m from the detector. This large distance makes the incoming beam almost a plane wave. In Figure 1b we show the results when the source is located at 8 m. In both figures the filled circles are the result when the setup is in air. The open circles show what happens when the air is removed. Note how the air in the 150 m run produces significant attenuation of the beam and how a low energy wing to the photo-peak is caused by small-angle scattering in the air. Because MGEANT uses ASCII input files to read in the detector geometry, the user can perform all four runs shown above without altering any code.

REFERENCES

1. see website at http://wwwinfo.cern.ch/asd/index.html
2. Webber, W.R., & Lezniak, J.A. *Astrophys Space Sci.* 30, 361 (1974).
3. Band, D. *ApJ*, 413, 282 (1993).
4. Jean, P.. et al. , this proceeedings.
5. Shrader, C. et al. , this proceeedings.

ONLINE INFORMATION

The Dedicated Centaurus A Web Pages

Helmut Steinle

Max-Planck-Institut für extraterrestrische Physik
Garching, Germany

Abstract. Centaurus A is the nearest active radio galaxy and thus it is the object of many detailed studies. A set of dedicated web based Centaurus A pages has been established in an effort to collect and present all available information and make it available "by a mouse click" at the URL :
"http://www.gamma.mpe-garching.mpg.de/~hcs/Cen-A/".

INTRODUCTION

Centaurus A is the nearest active radio galaxy and thus it is the object of many detailed studies. A wealth of information is available in literature and in the World Wide Web.

In an effort to collect and present all available information, a set of dedicated web based Centaurus A pages has been established at the Max-Planck-Institut für extraterrestrische Physik. The aim of this pages is to provide the user with all (electronically) available information "by a mouse click".

DESCRIPTION OF THE PAGES

To date (October 1999), there are eight pages:

- the **main** page with latest news and pictures.
- a **facts** page with a summary of basic parameters of Centaurus A.
- all **observations** which have been carried out during a campaign to measure the multiwavelength spectrum from radio to gamma-ray wavelengths in 1995 and other observations.
- the newest results of **monitoring** the intensity variation of Cen A in the 20 - 100 keV band with BATSE. (The most recent data are usually not older than one week.)

CP510, *The Fifth Compton Symposium,* edited by M. L. McConnell and J. M. Ryan

- a complete, up-to-date active **literature data base** brings almost all available literature on Cen A on-line "by a mouse click" through links to the ADS data base or other originating sites. At present there are already more than 200 papers available.
- a collection of **pictures** obtained in all wavelength bands.
- the observational **history** of Cen A.
- and finally, **links** to other web sites with information on Centaurus A.

The Dedicated Centaurus A Web Pages are available at:

http://www.gamma.mpe-garching.mpg.de/~hcs/Cen-A/

The Web-Based COMPTEL Bibliography

Helmut Steinle

Max-Planck-Institut für extraterrestrische Physik
Garching, Germany

Abstract. A complete bibliography of all CGRO-COMPTEL related[1] publications has been established and is maintained on WWW accessible pages at the URL :
"http://www.gamma.mpe-garching.mpg.de/~hcs/CBIB/".

INTRODUCTION

More than 500 publications which refer to the COMPTEL instrument on-board the *Compton* Gamma Ray Observatory (CGRO; launched in April 1991) or to results obtained by it have been published since 1982.

A complete bibliography[1] has been established on WWW accessible pages and is kept up-to-date at the Max-Planck-Institut für extraterrestrische Physik, the site of the COMPTEL principal investigator. All publications are accessible using the WWW. Wherever possible, active links to publications have been included. Preference has been given to links to either ADS (NASA Astrophysics Data System) or to electronic versions of the papers at the publisher site. In case neither of those official links exist, links to the author's verions (PostScript files etc.) or to preprint servers are established.

DESCRIPTION OF THE PAGES

To date (October 1999), there are six pages:

- the **Main Page** with a description of the bibliography and all its options.
- the **Full Bibliography** page, which lists all publications with
  - full title,
  - full author list,

[1] This bibliography contains all publications **about** COMPTEL, or containing **results** from COMPTEL, or making **reference** to COMPTEL

CP510, *The Fifth Compton Symposium*, edited by M. L. McConnell and J. M. Ryan

– the complete reference.

In addition

– there are attributed key-words,

– the publications are categorized,

– the actual publication status as well as

– links to electronic versions of the paper

are given. This list is ordered by the year of publication

- the **Short Bibliography** page, which is ordered by first author name. This list only lists the first author, the full title and the full reference. A link to the reference in the Full Bibliography is available.

- The **Input Form** consists of two pages and its use is restricted to members of the COMPTEL team. This shall be the main tool to enter new papers as they are published.

  The first page requires the identification of the user. Once submitted, the validated user is presented with a form to input a new publication or to change an existing entry. (The latter function is not yet implemented.)

  A new entry is entered simultaneously in the Full- and in the Short Bibliography lists.

- the **Request Hardcopies** page opens up a new window into which the user can copy a reference from the Full- or Short Bibliography pages by "cut and paste". The address the hardcopy should be sent to has to be given as well.

It is planned to include a **search** function which should enable the user to search for authors, keywords, etc..

The bibliography is based on frames, however the **option to use a frameless version is offered**.

The COMPTEL Bibliography is available at:

http://www.gamma.mpe-garching.mpg.de/~hcs/CBIB/

LIST OF PARTICIPANTS

Carl Akerlof
University of Michigan
akerlof@umalp1.physics.lsa.umich.edu

Elena Aprile
Columbia University
age@astro.columbia.edu

Martina Arndt
University of New Hampshire
martina.b.arndt@unh.edu

Sandy Barnes
USRA/NASA/GSFC
barnes@grossc.gsfc.nasa.gov

Kevin Bennett
ESTEC-ESA
kbennett@astro.estec.esa.nl

Dave Bertsch
NASA/GSFC
dlb@mozart.gsfc.nasa.gov

P.N. Bhat
Tata Institute of Fundamental Research
pnbhat@tifr.res.in

Dipen Bhattacharya
UC Riverside
dipen@tigre.ucr.edu

Hans Bloemen
SRON - Utrecht
h.bloemen@sron.nl

Markus Böttcher
Rice University
mboett@spacsun.rice.edu

Konstantin Borozdin
Los Alamos National Laboratory
kbor@lanl.gov

J. Braga
INPE-Brazil
braga@das2.inpe.br

Alberto Carramiñana
INAOE
alberto@inaoep.mx

Mike Catanese
Iowa State University
mcatanese@cfa.harvard.edu

Andrew Chen
NASA GSFC
awc@egret.gsfc.nasa.gov

Mike Cherry
Louisiana State University
cherry@phunds.phys.lsu.edu

Chul-Sung Choi
Korea Astronomy Observatory
cschoi@hanul.issa.re.kr

Ed Chupp
University of New Hampshire
edward.chupp@unh.edu

Thomas Cline
NASA/GSFC
cline@apache.gsfc.nasa.gov

Wayne Coburn
UCSD
wcoburn@ucsd.edu

W. Collmar
MPE
wec@mpe.mpg.de

Jorge Combi
IAR
combi@venus.fisica.unlp.edu.ar

Jim Connell
University of Chicago
connell@odysseus.uchicago.edu

Alanna Connors
Eureka Scientific
aconnors@maria.wellesley.edu

Jim Cravens
Southwest Research Institute
jcravens@swri.edu

Giancarlo Cusumano
IFCAI/CNR
cusumano@ifcai.pa.cnr.it

Hansford Cutlip
University of New Hampshire
hansford.cutlip@unh.edu

Bozena Czerny
Copernicus Astronomical Center
bez@camk.edu.pl

Daniele Dal Fiume
ITESRE/CNR
daniele@tesre.bo.cnr.it

A. J. Dean
University of Southampton
ajd@astro.soton.ac.uk

Stefano Del Sordo
IFCAI-CNR
desordo@ifcai.pa.cnr.it

Chuck Dermer
Naval Research Laboratory
dermer@gamma.nrl.navy.mil

Monty Di Biasi
Southwest Research Institute
ldibias@ibm.net

Cathy Dicks
RITSS/NASA/GSFC
cathy@milkyway.gsfc.nasa.gov

G. Di Cocco
ITESRE/CNR
dicocco@tesre.bo.cnr.it

Roland Diehl
MPE
rod@mpe.mpg.de

Seth Digel
USRA/NASA/GSFC
digel@gsfc.nasa.gov

Brenda Dingus
University of Utah
dingus@physics.utah.edu

A. Djannati-Atai
PCC-College de France-Paris
djannati@in2p3.fr

Eduardo do Couto e Silva
SLAC
eduardo@slac.stanford.edu

Vladimir Dogiel
P.N. Lebedev Institute
dogiel@lpi.ru

Pierre Dubath
ISDC
pierre.dubath@obs.unige.ch

Phil Dunphy
University of New Hampshire
phil.dunphy@unh.edu

Philip Edwards
ISAS
pge@vsop.isas.ac.jp

Don Ellison
North Carolina State University
don_ellison@ncsu.edu

Gordon Emslie
Univ. of Alabama in Huntsville
emslieg@email.uah.edu

Abe Falcone
University of New Hampshire
afalcone@comptel.sr.unh.edu

Jerry Fishman
NASA/MSFC
fishman@msfc.nasa.gov

Giovanni Fossati
UCSD
gfossati@ucsd.edu

Neil Gehrels
NASA/GSFC
gehrels@gsfc.nasa.gov

Robert Georgii
MPE
rog@mpe.mpg.de

Alvaro Giménez
LAEFF/CSIC/INTA
ag@laeff.esa.es

A. Goldwurm
SAp/CEA Saclay
agoldwurm@cea.fr

Paola Grandi
IAS/CNR
grandi@alphasax2.ias.rm.cnr.it

Isabelle Grenier
CEA Saclay
isabelle.grenier@cea.fr

Jonathan Grindlay
SAO
josh@cfa.harvard.edu

J. Eric Grove
Naval Research Laboratory
grove@gamma.nrl.navy.mil

Rodolfo Gurriaran
University of Southampton
rg@astro.soton.ac.uk

Takanobu Handa
SLAC
handa@slac.stanford.edu

Diana Hannikainen
University of Helsinki
diana@astro.helsinki.fi

Mike Harris
USRA/NASA/GSFC
harris@tgrs2.gsfc.nasa.gov

Biff Heindl
UCSD
wheindl@ucsd.edu

Wim Hermsen
SRON - Utrecht
w.hermsen@sron.nl

Margarita Hernanz
IEEC/CSIC
hernanz@ieec.fcr.es

Bill Howard
USRA
whoward.usra.edu

Kevin Hurley
UC Berkeley
khurley@sunspot.ssl.berkeley.edu

Giampiero Iannone
University of Bologna
iannone@iras.ucalgary.ca

Anatoli Iyudin
MPE
ani@mpe.mpg.de

Pierre Jean
CESR
jean@cesr.fr

Robert Johnson
UC Santa Cruz
johnson@scipp.ucsc.edu

Phil Kaaret
Harvard Smithsonian CfA
pkaaret@cfa.harvard.edu

Emrah Kalemci
UCSD/CASS
emrahk@mamacass.ucsd.edu

Gottfried Kanbach
MPE
gok@mpe.mpg.de

Cheenu Kappadath
Louisiana State University
cheenu@rouge.phys.lsu.edu

Vicky Kaspi
MIT
vicky@space.mit.edu

Don Kniffen
NASA / HQ
dkniffen@hq.nasa.gov

Jurgen Knödlseder
CESR
knodlseder@cesr.fr

Henric Krawczynski
MPI Kernphysik
henric.krawczynski@mpi-hd.mpg.de

Peter Kretschmar
ISDC/IAAT
peter.kretschmar@obs.unige.ch

Dick Kroeger
Naval Research Laboratory
kroeger@gamma.nrl.navy.mil

Sasha Kudryashov
INASAN
akud@inasan.rssi.ru

Jim Kurfess
Naval Research Laboratory
kurfess@gamma.nrl.navy.mil

Masaaki Kusunose
Kwansei Gakuin University
kusunose@kwansei.ac.jp

Anne Lähteenmäki
Metsahovi Radio Observatory
alien@kurp.hut.fi

G. LaRosa
IFCAI-CNR
larosa@ifcai.pa.cnr.it

Mark Leising
Clemson University
lmark@clemson.edu

Edison Liang
Rice University
liang@spacsun.rice.edu

Y. C. Lin
Stanford University
lin@egret0.stanford.edu

Niels Lund
Danish Space Research Institute
nl@dsri.dk

Svetlana Marchenko
Boston University
jorstad@rjet.bu.edu

Alan Marscher
Boston University
marscher@bu.edu

John Mattox
Boston University
mattox@bu.edu

Lowry McComb
University of Durham
t.j.l.mccomb@dur.ac.uk

Mark McConnell
University of New Hampshire
mark.mcconnell@unh.edu

Maura McLaughlin
Cornell University
mam36@cornell.edu

Bernie McNamara
New Mexico State University
bmcnamar@nmsu.edu

Dan Messina
SFA/NRL
messina@gamma.nrl.navy.mil

Peter Michelson
Stanford University
peterm@leland.stanford.edu

Rich Miller
University of New Hampshire
richard.miller@unh.edu

Peter Milne
NRC/NRL
milne@gamma.nrl.navy.mil

G. Mohanty
UC RIverside
mohanty@in2p3.fr

Daniel Morris
University of New Hampshire
dmorris@comptel.sr.unh.edu

Reshmi Mukherjee
Barnard College
muk@astro.columbia.edu

Ronald Murphy
Naval Research Laboratory
murphy@gamma.nrl.navy.mil

Stephen Murray
Smithsonian Astrophysical Obs.
ssm@cfa.harvard.edu

Lorenzo Natalucci
IAS/CNR
lorenzo@ias.rm.cnr.it

Sergei Nayakshin
NASA/GSFC
serg@milkyway.gsfc.nasa.gov

Uwe Oberlack
Columbia University
oberlack@astro.columbia.edu

A. P. Martinez
IAS-CNR
andrea@ias.rm.cnr.it

Bill Paciesas
University of Alabama in Huntsville
paciesasw@cspar.uah.edu

Giorgio Palumbo
University of Bologna
ggcpalumbo@astbo3.bo.astro.it

Robbin Pendexter
University of New Hampshire
robbin.pendexter@unh.edu

Geoff Pendleton
University of Alabama in Huntsville
pendletong@cspar.uah.edu

Liz Pentecost
USRA
lpenteco@usra.edu

Charlotte Perfect
University of Southampton
clp@astro.soton.ac.uk

Bernard Phlips
Naval Research Laboratory
phlips@gamma.nrl.navy.mil

Glenn Piner
JPL
b.g.piner@jpl.nasa.gov

Stefan Plüschke
MPE
stp@mpe.mpg.de

Craig Pollock
Southwest Research Institute
cpollock@swri.edu

Mark Polsen
UC Riverside
mpolsen@citrus.ucr.edu

Katja Pottschmidt
IAA Tuebingen
katja@astro.uni-tuebingen.de

Bill Purcell
Ball Aerospace & Technologies Corp.
bpurcell@ball.com

O. Reimer
MPE
olr@mpe.mpg.de

Natale Robba
Palermo University
robba@gifco.fisica.unipa.it

Brian Rodriquez
New Mexico State University
bmcnamar@nmsu.edu

Gustavo Romero
IAR
romero@venus.fisica.unlp.edu.ar

Jim Ryan
University of New Hampshire
james.ryan@unh.edu

Andrea Santangelo
IFCAI
andrea@ifcai.pa.cnr.it

Re'em Sari
Caltech
sari@tapir.caltech.edu

V. Schönfelder
MPE
vos@mpe.mpg.de

Chris Shrader
USRA/NASA/GSFC
shrader@grossc.gsfc.nasa.gov

Frank Six
NASA/MSFC
frank.six@msfc.nasa.gov

Tom Skelton
UCSD/CASS
rskelton@ucsd.edu

Kim Slavis
Washington University
slavis@hbar.wustl.edu

P. Sreekumar
ISRO Satellite Center
pskumar@isac.esnet.in

Greg Stacy
LSU
gstacy@phys.lsu.edu

Ruediger Staubert
IAA Tuebingen
staubert@ait.physik.uni-tuebingen.de

Helmut Steinle
MPE
hcs@mpe.mpg.de

Mark Strickman
Naval Research Laboratory
strickman@gamma.nrl.navy.mil

A. Strong
MPE
aws@mpe.mpg.de

Steve Sturner
USRA/NASA/GSFC
sturner@swati.gsfc.nasa.gov

Jean-Pierre Swings
Astrophys. Inst. Liege
swings@astro.ulg.ac.be

Michel Tagger
CEA Saclay
tagger@cea.fr

F. Takahara
Osaka University
takahara@vega.ess.sci.osaka-u.ac.jp

Tadayuki Takahashi
ISAS
takahasi@astro.isas.ac.jp

Marco Tavani
Istituto di Fisica Cosmica
tavani@ifctr.mi.cnr.it

Bonnard Teegarden
NASA/GSFC
bonnard@lheamail.gsfc.nasa.gov

Yukikatsu Terada
University of Tokyo
terada@amalthea.phys.s.u-tokyo.ac.jp

Harri Teräsranta
Metsahovi Radio Observatory
hte@kurp.hut.fi

Lih-Sin The
Clemson University
tlihsin@hubcap.clemson.edu

Dave Thompson
NASA/GSFC
djt@egret.gsfc.nasa.gov

Merja Tornikoski
Metsahovi Radio Observatory
merja.tornikoski@hut.fi

Norman Trams
ESA-ESTEC
ntrams@astro.estec.esa.nl

Pietro Ubertini
IAS-CNR
ubertini@ias.rm.cnr.it

Tom Vestrand
LANL
vestrand@lanl.gov

Osmi Vilhu
Helsinki University Observatory
osmi.vilhu@helsinki.fi

Paul Wallace
Berry College
wallace@blazar.gsfc.nasa.gov

Christopher Wanjek
NASA/GSFC
wanjek@gsfc.nasa.gov

Ken Watanabe
USRA/NASA/GSFC
watanabe@grossc.gsfc.nasa.gov

Trevor Weekes
SAO/Whipple Observatory
weekes@cfa.harvard.edu

G. Weidenspointner
NASA / GSFC
ggw@milkyway.gsfc.nasa.gov

Mathew Westmore
University of Southampton
mjw@astro.soton.ac.uk

Rees Williams
ESA-ESTEC
owilliam@astro.estec.esa.nl

Jorn Wilms
IAA Tuebingen
wilms@astro.uni-tuebingen.de

Colleen Wilson-Hodge
NASA/MSFC
colleen.wilson@msfc.nasa.gov

Kazutaka Yamaoka
ISAS
yamaoka@astro.isas.ac.jp

T. Yamasaki
Osaka University
yamasaki@vega.ess.sci.osaka-u.ac.jp

Insu Yi
Korea Inst. for Advanced Study
iyi@kias.re.kr

Alex Young
University of New Hampshire
c.alex.young@unh.edu

Bob Zavala
New Mexico State University
rzavala@nmsu.edu

Ming Zhang
University of Chicago
mzhang@odysseus.uchicago.edu

Bing Zhang
NASA /GSFC
bzhang@twinkie.gsfc.nasa.gov

Jeff Zweerink
UC Riverside
zweerink@solar2.ucr.edu

Allen Zych
UC Riverside
allen.zych@ucr.edu

Author Index

D

E

F

G

H

I

J

K

L

M

N

O

P

Q

R

S